# Plant Molecular Biology 2

# NATO ASI Series

## Advanced Science Institutes Series

*A series presenting the results of activities sponsored by the NATO Science Committee, which aims at the dissemination of advanced scientific and technological knowledge, with a view to strengthening links between scientific communities.*

The series is published by an international board of publishers in conjunction with the NATO Scientific Affairs Division

| | | |
|---|---|---|
| A | **Life Sciences** | Plenum Publishing Corporation |
| B | **Physics** | New York and London |
| C | **Mathematical and Physical Sciences** | Kluwer Academic Publishers |
| D | **Behavioral and Social Sciences** | Dordrecht, Boston, and London |
| E | **Applied Sciences** | |
| F | **Computer and Systems Sciences** | Springer-Verlag |
| G | **Ecological Sciences** | Berlin, Heidelberg, New York, London, |
| H | **Cell Biology** | Paris, Tokyo, Hong Kong, and Barcelona |
| I | **Global Environmental Change** | |

***Recent Volumes in this Series***

*Volume 207*—Bioorganic Chemistry in Healthcare and Technology
edited by Upendra K. Pandit and Frank C. Alderweireldt

*Volume 208*—Vascular Endothelium: Physiological Basis of Clinical Problems
edited by John D. Catravas, Allan D. Callow, C. Norman Gillis, and Una S. Ryan

*Volume 209*—Molecular Basis of Human Cancer
edited by Claudio Nicolini

*Volume 210*—Woody Plant Biotechnology
edited by M. R. Ahuja

*Volume 211*—Biophysics of Photoreceptors and Photomovements in Microorganisms
edited by F. Lenci, F. Ghetti, G. Colombetti, D.-P. Häder, and Pill-Soon Song

*Volume 212*—Plant Molecular Biology 2
edited by R. G. Herrmann and B. A. Larkins

*Volume 213*—The Midbrain Periaqueductal Gray Matter: Functional, Anatomical, and Neurochemical Organization
edited by Antoine Depaulis and Richard Bandler

*Series A: Life Sciences*

# Plant Molecular Biology 2

Edited by

R. G. Herrmann

Ludwig-Maximilians-Universität
München, Germany

and

B. A. Larkins

University of Arizona
Tucson, Arizona



Plenum Press
New York and London
Published in cooperation with NATO Scientific Affairs Division

Proceedings of a NATO Advanced Study Institute
on Plant Molecular Biology,
held May 14-23, 1990,
in Elmau, Bavaria, Germany

Library of Congress Cataloging-in-Publication Data

NATO Advanced Study Institute on Plant Molecular Biology (6th : 1990 :
Schloss Elmau)
Plant molecular biology 2 / edited by R.G. Herrmann and B. A.
Larkins.
p. cm.
Proceedings of the Sixth NATO Advanced Study Institute on Plant
Molecular Biology held May 14-23, 1990, in Elmau, Bavaria, Germany.
Includes bibliographical references and index.
ISBN 0-306-44024-5
1. Plant molecular biology--Congresses. I. Herrmann, R. G.
(Reinhold G.) II. Larkins, B. A. (Brian A.) III. Title.
QK728.N38 1990
581.8--dc20 91-26944
CIP

ISBN 0-306-44024-5


A Division of Plenum Publishing Corporation
233 Spring Street, New York, N.Y. 10013



Printed in the United States of America

# PREFACE

The VI NATO Advanced Study Institute on Plant Molecular Biology, held in Elmau, Bavaria, Germany, from 14 to 23 May, 1990, brought together representative scientific leaders from all over the world to review their lastest results. They presented lectures or posters, participated in lively discussions, educated students, and exchanged views and plans for future research in this highly exciting field of science.

The experiments, data and questions were naturally varied, but all of them illustrate that the modern techniques of molecular biology, complemented by developments in immunology, genetics, and ultrastructural research, have pervaded nearly every branch of biology. The presentations show that these approaches have tremendously increased our potential both for fundamental research, our understanding of life, and by analogy to the precedents of physics and chemistry, have led and will continue to lead to "engineering sciences" and implicitly, to new industrial processes.

Some of these applications are a matter of debate in the public domain today and many feel that the development of industrial gene technology requires the attention of the whole scientific community. Nevertheless, the implications of this research for the genetic improvement of agricultural plants are profound. Some of the near term technologies being developed provide novel approaches for improving the utility of food crops. They can also result in reduced dependence on the use of pesticides for food production. Scientists have the knowledge to evaluate these developments and we hope that this ASI and volume have served, and will continue to serve this purpose as well.

The contributions in this book summarize lectures of invited speakers, as well as poster presentations. They have not been subjected to any major cutting or editing, in order to preserve the individual nature of each contribution.

We appreciate greatly the financial support of the following organizations and institutions: NATO, FEBS, ISPMB, Ministerium für Forschung und Technologie der Bundesrepublik Deutschland, Deutsche Forschungsgemeinschaft, Fonds der Chemischen Industrie, Bayerisches Staatsministerium für Wissenschaft und Kunst, Schering AG, CIBA-GEIGY, Bundesverband Deutscher Pflanzenzüchter, Monsanto Agricultural Products, Boehringer Mannheim GmbH, Amersham Buchler, Kleinwanzlebener Saatzucht, Pharmacia as well as the Stiftung Volkswagenwerk for an associated workshop. Without their support this meeting could not have taken place. We are also indebted to our colleagues in the staff of the Institute of Botany, University München, for their enthusiastic handling of the logistic and administrative aspects of the conference.

Reinhold G. Herrmann
München, Germany

Brian Larkins
Tucson, USA

CONTENTS

## PLANT VIRUSES, SUBVIRAL RNAs and VIROIDS



## NITROGEN FIXATION, NITROGEN METABOLISM

PHYTOPATHOLOGY



AGROBACTERIUM, TRANSFORMATION

## GENOME ANALYSIS, RFLP



## TRANSPOSONS



## MITOCHONDRIA, CHLOROPLASTS, PHOTOSYNTHESIS

GENE EXPRESSION, PHOTORECEPTORS

PLANT DEVELOPMENT



PROTEIN TRANSPORT



STORAGE PROTEINS

STRESS REACTIONS



INTERESTING PATHWAYS

# CONTRIBUTION OF PLANT AND VIRUS GENES TO CAULIFLOWER MOSAIC VIRUS PATHOGENICITY

Simon N. Covey, David S. Turner, Rebecca Stratford,
Keith Saunders, Andrew Lucy,
Sarah Riseborough and Pierre Ray

Department of Virus Research, John Innes Institute
John Innes Centre for Plant Science
Research, Colney Lane, Norwich NR4 7UH, UK

## INTRODUCTION

Viruses are useful entities for the molecular biologist since they can be exploited as simple genetic systems to study molecular processes of plant cells, and because they are pathogens and can be used to understand and potentially control plant virus diseases. The development of a plant virus disease typically involves the interaction of virus and host gene products such that the virus exploits host processes to ensure its multiplication and dissemination within the plant. Progeny virus must also become available for transmission to other plants. Methods of studying these processes have centred on the analysis of viral gene products, on changes in host gene expression, and more recently by genetic manipulation techniques. Genetic variability in both virus and host can be used to pinpoint possible sites of interaction and potential targets for ameliorating the outcome of the disease.

We have been exploiting cauliflower mosaic virus (CaMV) to study host/virus interactions because it has been extensively characterised at the molecular level and its complex multiplication cycle has many potential sites of interaction with the host (for reviews, see Covey 1985; Gronenborn, 1987; Covey *et al.*, 1990). CaMV encapsidates a DNA genome which is replicated through an RNA intermediate by a process of reverse transcription. The replication cycle in the cell can be divided into two principal phases which take place in different cellular compartments. In the nuclear phase, a supercoiled form of the CaMV 8kbp DNA genome associates with host proteins in the conformation of a minichromosome which is transcribed under the direction of the host transcription machinery (Olszewski *et al.*, 1982). Viral polyadenylated transcripts move to the cytoplasm where translation takes place. Here, reverse transcription of the RNA form of the CaMV genome (35S RNA) also occurs apparently in replication complexes (Pfeiffer and Hohn, 1983; Thomas *et al.*, 1985) possibly peripherally associated with virus-generated inclusion bodies (Stratford *et al.*, 1988).

In addition to the processes of viral gene expression and multiplication in which host gene products are specifically involved, there are likely to be several other aspects of virus activity in which the host is exploited. For example, movement of virus between cells, presumably via plasmodesmata (Linstead *et al.*, 1988), is probably a highly specific process involving the interaction of the CaMV putative transport protein (gene I product) and a possible host receptor. Long distance systemic movement of virus between the site of inoculation and other plant organs in the vascular tissue might also require specific interactions.

*Plant Molecular Biology 2*, Edited by R.G. Herrmann and
B. Larkins, Plenum Press, New York, 1991

Once inside the cell, virions must uncoat. For CaMV this process could involve a receptor, perhaps in a membrane complex near to the nucleus, since the high chemical stability of CaMV virions would require an active process to liberate the viral DNA. It is possible that uncoating and targeting of viral DNA to the nucleus are linked events although there is, as yet, no experimental evidence to support this speculation. Moreover, very little is yet known about targeting and compartmentalisation of specific viral functions in the cytoplasm. For example, the mechanisms which target translational and replicative forms of viral transcripts are not known. Similarly, progeny virions must be differentially targeted to electron dense cytoplasmic inclusion bodies derived from the CaMV gene VI protein, or to electron lucent inclusion bodies containing the gene II protein involved in aphid acquisition and transmission of virus from plant to plant (Espinoza, 1989). The role of progeny virion DNA in maintaining the level of the viral minichromosome is a further regulatory feature of the multiplication cycle.

## CaMV SYMPTOM DETERMINANTS

We have been studying how subcelluar events are reflected at the whole plant level in terms of symptom expression and disease severity, and host susceptibility, by investigating the genetic contribution of both virus and host in specific functions determined at both molecular and whole plant levels. To this end we have exploited the genetic variability found in naturally-occurring strains of CaMV and differences in susceptibility exhibited by plant species that are hosts to CaMV. Our first approach has been to locate genetic determinants on the CaMV genome specifying particular symptomatic effects observed in host turnip plants (*Brassica campestris* cv. "Just Right"). The logic behind this approach was to attempt to link specific viral genome domains controlling symptoms with the known or suspected role of that domain in the virus multiplication cycle and hence suggest possible sites of molecular interaction. Clearly, some symptomatic effects are likely to have resulted from non-specific or secondary influences on host metabolism and these must be distinguished from more specific effects.

The strategy to locate symptom determinants was to construct hybrid viral genomes from strains of CaMV exhibiting contrasting symptom characters. CaMV strain Cabb B-JI produces fairly typical severe symptoms in turnip plants which include rapid appearance and spread of systemic vein clearing symptoms, severe plant stunting, and generalised leaf chlorosis accompanied by a reduction in chlorophyll content. Plants infected with the mild CaMV strain, Bari 1, are slow to develop systemic vein clearing, plants are only slightly stunted and leaves do not show chlorosis but produce dark green islands sometimes reflected in an increase in chlorophyll content compared with healthy controls. We have created forty or more hybrid viruses between the two CaMV strains either by recombinational rescue *in vivo* following co-inoculation of plants with defective complementary genome fragments, or by restriction enzyme fragment exchange *in vitro*. Recombinants fall into one of three groups: (i) those with complete symptom characteristics of one or other of the parental strains; (ii) those with a mixture of characters of both parental strains; (iii) those producing novel symptoms not observed in infections with either parental strain. These latter examples include effects such as atypical stunting, novel coloration effects and patterns on leaves, and presence or absence of leaf crinkling. Symptomatic effects of the severe Cabb B-JI and mild Bari 1 strain and some hybrids are shown in Fig. 1.

By comparing the induction of particular symptomatic effects with the presence of a specific fragment of DNA from one or other parental strain determined by restriction mapping experiments, it has been possible to build up a symptom map of the viral genome (Stratford and Covey, 1989; Fig. 2). Thus, differences between the two strains which influence the degree of leaf chlorosis were mapped to a domain containing part of gene VI and the 35S RNA promoter region. Timing of symptom appearance was located between nucleotides 2150 and 4438 although we believe that this determinant is probably located in gene V (reverse transcriptase) between nucleotides 3956 and 4438. This domain also contains determinants influencing plant stunting.

**Figure 1.** Turnip leaves (A) and plants (B) infected with the severe Cabb B-JI CaMV strain (A2, B2), the mild strain Bari 1 (A5, B3), and virus hybrids with exchanged DNA fragments in the gene VI region producing intermediate levels of leaf chlorosis (A3, A4). Plant B4 contains a CaMV hybrid with the stunting characteristics of the parental strain Bari 1 but the leaf chlorosis characteristics of Cabb B-JI.

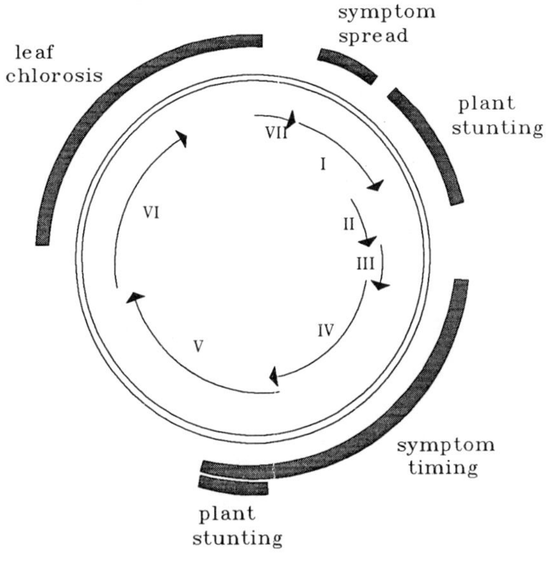


**Figure 2.** Map of the CaMV genome showing open reading frames (I-VII) and regions responsible for differences in symptom characteristics produced in turnip plants infected with the CaMV strains Cabb B-JI and Bari 1. Gene functions are: I, putative transport protein; II, aphid acquisition factor; III, virion protein; IV, coat protein; V, reverse transcriptase; VI, inclusion body protein.

A second domain between nucleotides 905 and 1785 also influenced plant stunting. In those CaMV hybrids where a crossover occurred in this region, a hypersevere stunting effect was observed. The rate of spread of systemic vein clearing symptoms was controlled by a domain within gene I between nucleotides 331 and 780. From a comparison of the nucleotide sequences of Cabb B-JI and Bari 1 in this region of gene I containing the N-terminal 140 amino acids, eight changes were observed six of which were to familial amino acids (Fig. 3).

| AMINO ACID POSITION: | 18 | 30 | 33 | 44 | 56 | 60 | 114 | 124 |
|---|---|---|---|---|---|---|---|---|
| Cabb B-JI | E | D | S | K | E | K | S | M |
| Bari 1 | D | *A* | T | R | D | R | *Q* | I |

**Figure 3.** Differences in amino acid sequence in the first 140 residues of gene I responsible for the rapid rate of symptom spread in turnip leaves characteristic of CaMV strain Cabb B-JI and the slow rate of Bari 1. Non-familial amino acid changes are shown in italics.

We are presently subjecting this region of gene I to site directed mutagenesis to determine which of the eight changed amino acids is responsible for the slower rate of symptom spread in the Bari 1 infections. There is accumulating evidence that the polypeptide product of gene I is responsible for mediating movement of CaMV from cell-to-cell and this view is consistent with our observation since rate of spread of systemic symptoms must be related to virus movement around the plant. Using this kind of approach we hope to delimit further those domains of viral proteins which interact with the host to mediate symptom development. This should lead us to develop strategies to interrupt these interactions and attenuate symptom development and viral pathogenesis.

## HOST INVOLVEMENT IN CaMV PATHOGENESIS

Our studies of host involvement in CaMV pathogenesis have centred upon an analysis of different phases of the virus multiplication cycle in different plant species showing a range of responses to infection, and in plant tissues which can or cannot support CaMV replication. We use a method of 2-dimensional agarose gel electrophoresis coupled with Southern blotting to characterise the complex population of unencapsidated CaMV DNA intermediates which accumulate in infected plants (Covey and Turner, 1986). This has already permitted us to show (Covey *et al.*, 1990; Saunders *et al.*, 1990) that the relative levels of the nuclear (transcriptional) phase of the CaMV multiplication cycle and the cytoplasmic (reverse transcriptional) phase are not the same under all circumstances. For example, in highly susceptible systemically-infected turnip leaves which are actively producing progeny virus, a relatively small amount of supercoiled (SC) DNA in the nucleus is transcribed to produce copious RNA which in turn in responsible for generating reverse transcription replication DNA intermediates. These DNA forms are readily resolved following 2-D gel electrophoresis of leaf unencapsidated DNA (Fig. 4A). Various hairpin (hp) and linear (L) DNA forms characteristic of reverse transcription, are shown.

In contrast, other tissues in infected turnips were found not to be able to support all phases of the CaMV multiplication cycle. For instance, we have found that turnip roots contain much less of the reverse transcription intermediates indicating only a very low level of virus replication. Unexpectedly, these organs were found to contain elevated levels of the SC DNA component of the viral minichromosome. However, analysis of viral transcripts by Northern blotting suggested that the minichromosome was not active in maintaining levels of RNA sufficient to produce DNA replication intermediates by reverse transcription. A similar but lesser effect was observed in stems; the greatest levels of SC DNA were found in callus cells derived from infected turnip leaves but with virtually no detectable RNA transcripts (Rollo and Covey, 1985; Covey *et al.*, 1990). Taken together these observations suggest that not all organs or tissues in turnip plants are capable of supporting the entire CaMV multiplication cycle. Leaves seem to be most productive and callus tissue the least. Moreover, this phenomenon seems likely to be related to specific regulation of the CaMV multiplication cycle at the stage of minichromosome transcription. Thus, transcriptional regulation of this complex appears to be organ or cell-type specific, a phenomenon presumably related to the presence of specific host trans-acting factors.

We used a similar approach to analyze the CaMV multiplication cycle in various *Brassica* species which showed differential susceptibility or tolerance to infection. From our studies of a broad range of plants representative of the brassicas, we found broadly that *Brassica campestris* variants were especially susceptible and developed the most severe symptoms whilst *Brassica oleracea* subspecies were most tolerant and developed mild, or sometimes no symptoms of infection. *Brassica nigra* exhibited intermediate characteristics. Various allotetraploid species were also analyzed and these, in general, also responded to CaMV infection by developing symptoms with severity somewhere between that of the parental diploid species.

The unencapsidated DNA replication products isolated from representative brassicas are shown in Fig. 4. Leaves from CaMV-infected pak choy (*B. campestris*) contain abundant reverse transcription intermediates but little SC DNA (Fig. 4A) whereas the relatively tolerant cauliflower (*B. oleracea*) contained abundant SC DNA with little detectable reverse transcription products (Fig. 4C). Quantitation of the RNA transcripts in the various species showed a direct correlation between the amount of CaMV RNA, abundance of reverse transcription products and

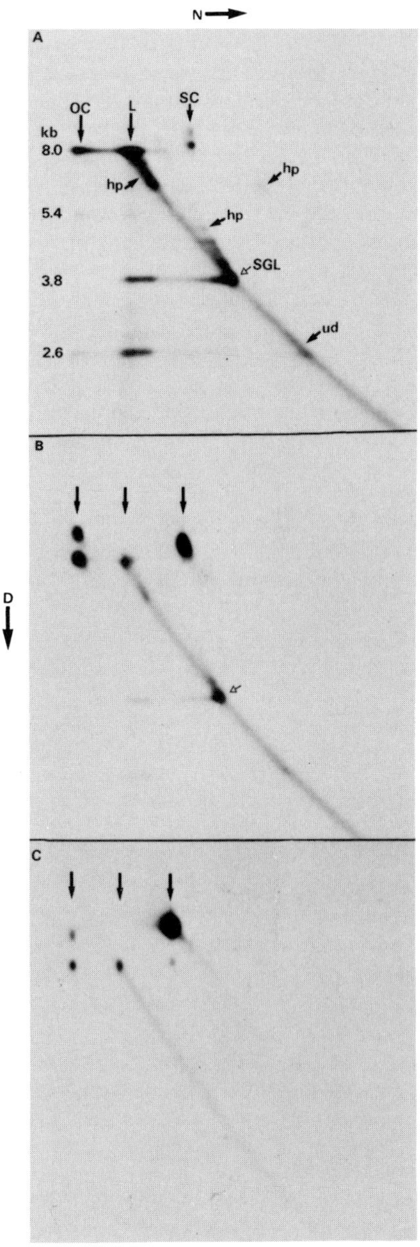


**Figure 4.** (Right Panel) Analysis by 2-D gel electrophoresis of unencapsidated CaMV DNAs isolated from leaves of (A) *B. campestris* (pak choy), (B) *B. nigra* (mustard) and (C) *B. oleracea* (cauliflower) plants infected with CaMV. Unencapsidated DNAs were electrophoresed first in neutral (N) buffer and then at 90$^{0}$ orientation relative to the first dimension in denaturing (D) alkaline medium. The major DNA forms resolved in the first dimension are open circles (OC), linears (L), and supercoiled (SC) forms. The unit diagonal (ud) represents fragmented linear dsDNAs. Hairpin (hp) molecules generated by reverse transcription *in vivo*, and a subgenomic linear (SGL) are also shown. Note the abundance of SC DNA in the tolerant cauliflower (C) and the low level of SC DNA in the susceptible pak choy (A) sample.

symptom severity, and an inverse correlation with the level of SC DNA observed. This was similar to the situation seen in certain organs/tissues of susceptible turnip plants.

We wondered how CaMV DNAs, especially the SC form, might accumulate in plants or organs in which little or no reverse transcription was occurring. We checked for accumulation of a CaMV clone containing a lethal mutation in the reverse transcriptase gene but found none when plants were analyzed about three weeks post inoculation. Infected tissue from leaves of different species was also examined by dissection to determine whether the viral DNA was limited to certain tissues less active in transcription of the minichromosome. However, we found no evidence for this since the same DNA forms were found in leaf vein tissue, the vein border tissues and leaf tissue containing predominantly mesophyll cells.

These findings suggest that CaMV DNA replication is occurring in parts of the plant or at a time not analyzed in the above experiments. However, *B. oleracea* roots and stems showed the same pattern of DNAs as those in leaves (Fig. 4C). We have undertaken time-course analysis of *B. oleracea* leaves and our early findings suggest that young leaves pass through a short phase of development during which time they are capable of supporting CaMV replication. In fact, old leaves of susceptible species like turnip also show accumulation of SC DNA reminiscent of less active tissues. Apparently, susceptibility of a particular species is related to the length of that period in leaf development during which virus multiplication can occur and that this period appears much longer in more susceptible species. Moreover, this putative susceptible period seems to be specifically related to an ability of the host to regulate expression of the viral minichromosome.

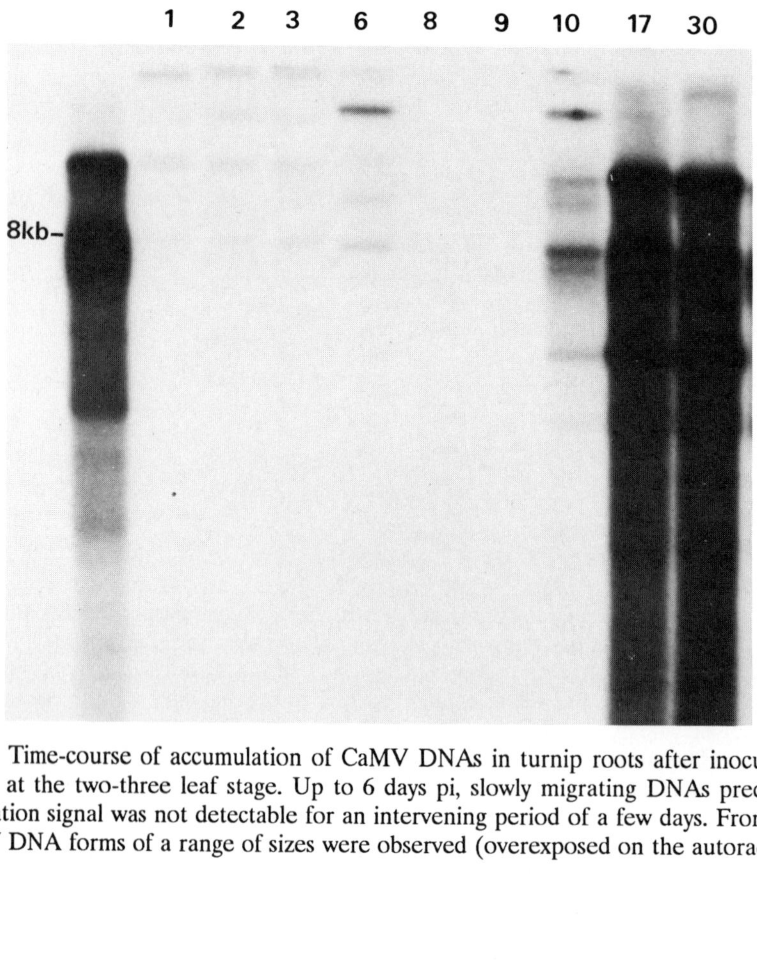


**Figure 5.** Time-course of accumulation of CaMV DNAs in turnip roots after inoculation of seedlings at the two-three leaf stage. Up to 6 days pi, slowly migrating DNAs predominate. Hybridisation signal was not detectable for an intervening period of a few days. From 10 days pi CaMV DNA forms of a range of sizes were observed (overexposed on the autoradiogram).

During a time-course analysis of the early events of the CaMV multiplication cycle, we observed viral DNA forms in leaves and roots of turnip plants, following systemic spread, very soon after leaf inoculation. These forms were unlike other DNAs previously observed in that they were predominantly high molecular weight. A time-course of the accumulation of unencapsidated DNA forms arising from systemic spread to the roots following inoculation of a single leaf, is shown in Fig. 5. After only 24 hours post inoculation, CaMV DNA was found in the roots. In different experiments these forms have been observed up to a period of about a week post inoculation and they then disappear in both roots and leaves before a second wave of DNA forms are detectable.

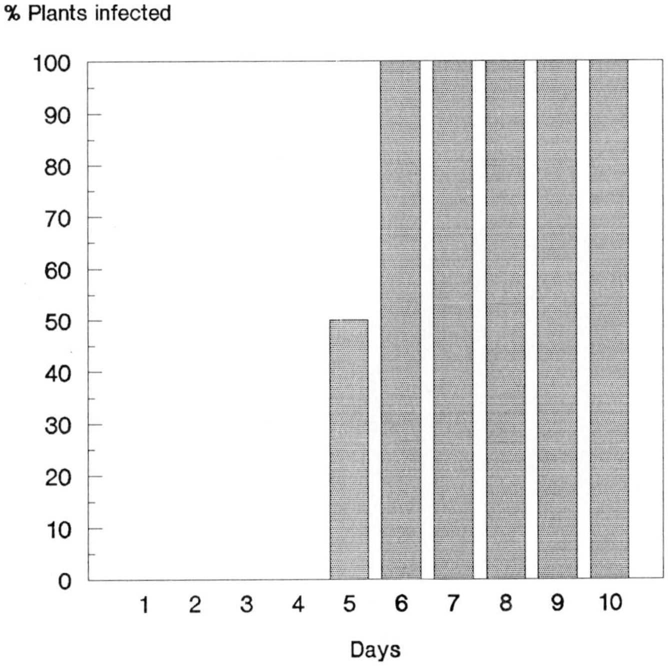


**Figure 6.** Effect of removal of the inoculated leaf on development of a systemic CaMV infection in turnip seedlings.

We were surprised by the rapid systemic movement of viral DNA from the inoculated leaf both to roots and to other leaves. This suggested that movement might occur before the onset of reverse transcription. To test this, we determined the minimum amount of time required between inoculation and establishment of a full systemic infection by observation of visible leaf systemic symptoms. A number of plants were inoculated and, from different plants, the inoculated leaf was removed at various times post inoculation. Figure 6 shows that up to the first four days after inoculation, removal of the inoculated leaf prevents establishment of a full systemic infection. From 6 days or more, systemic infection was established. Half of the plants inoculated showed symptoms following removal of the inoculated leaf five days post inoculation. Systemic leaf symptoms were observed 12-14 days post inoculation.

A comparison of the results in Figures 5 and 6 shows that viral DNA becomes distributed systemically around the plant before establishment of a full infection takes place. One inference from this is that the very early systemic DNAs result from movement of the inoculum before the onset of reverse transcription. For this to occur, it would be necessary to postulate that inoculation occurred directly into the phloem of the inoculated leaf. The fact that novel DNA forms were observed indicates the presence of the DNA in active cells. We have analyzed some of these DNAs by 2-D gel electrophoresis and found they comprise multiple SC forms of unit-length, half unit-length and others of greater-than-unit-length. Some of these novel forms are shown in Fig. 7 in which a root sample was taken at the start of the second wave of DNA accumulation (see Fig. 5). We have not yet analyzed the RNA populations of the tissues containing the 'early' SC DNAs to determine whether transcription is taking place.

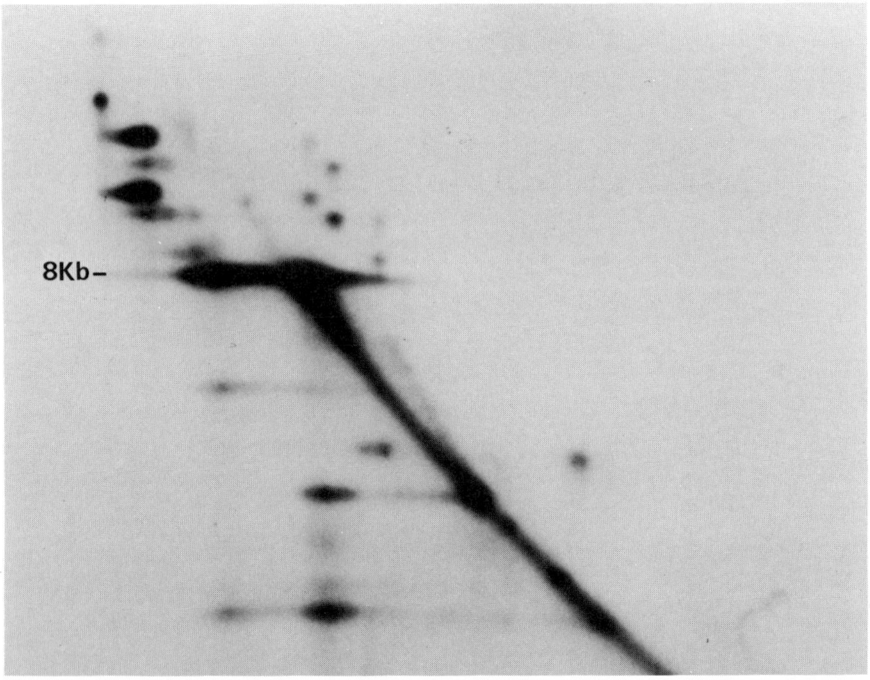


**Figure 7.** 2-D gel electrophoresis of CaMV unencapsidated DNAs isolated from young turnip roots at the onset of reverse transcription. Compare the pattern of DNAs with that observed in leaves when viral replication is fully underway (see Fig. 4A). In particular young roots accumulate a range of higher molecular weight forms of SC and open circular conformation.

A picture is beginning to emerge of three phases during the CaMV multiplication cycle in which differential patterns of SC DNA accumulation and activity occur (Fig. 8). The phases appear to be influenced by host genetic factors such as tissue/organ specificity and host specificity. In an apparent 'pre-replicative' phase, both roots and leaves accumulate viral DNA which has moved systemically from the site of inoculation. This DNA is predominantly of multimeric SC forms and different from that in the inoculum and has therefore undergone re-arrangement presumably mediated by the host. This process occurs during the first week post inoculation. Following a short lag phase, a second wave of CaMV DNA accumulation occurs in which the major viral products and progeny virions are synthesised. The length of this period appears to depend upon the developmental stage of the tissue, or on the particular host species infected. The genome-length SC DNA, although active, is present as a single species in relatively small amounts. In older tissues, as the replicative phase declines, SC DNA once more accumulates. In some tissues, such as callus, it will undergo re-arrangements typically to produce subgenomic forms. This suggests that a viral gene product might be involved in regulating the level of unit-length SC DNA available for the minichromosome. The absence of this putative regulatory function in the pre- and post-replicative phases might account for the accumulation and re-arrangement of the SC DNA. It remains possible, however, that the pre-replicative re-arrangements are not an essential part of the replication cycle although they might play a role in recombinational events observed by many workers in CaMV infections.

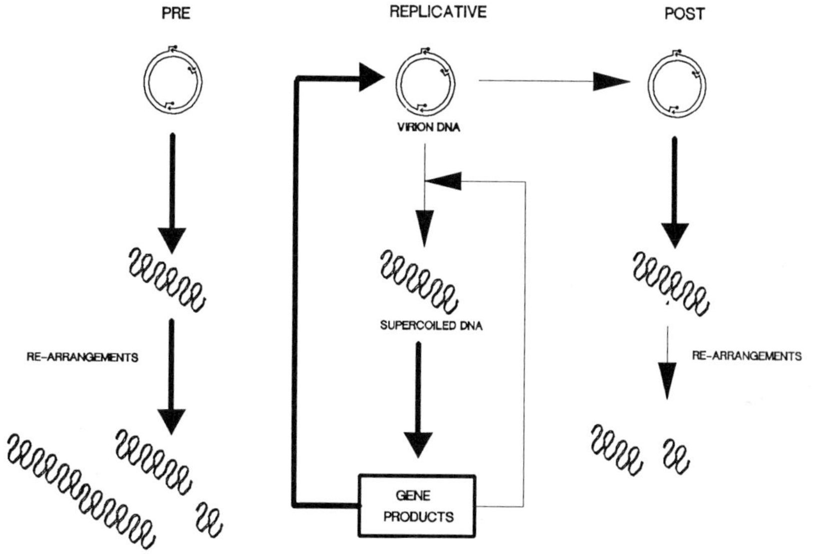


**Figure 8**. Scheme showing three phases of the CaMV multiplication cycle during which SC DNA is generated. The size of the arrows indicates the importance of a particular pathway. The substrate in each case is virion DNA which contains site-specific discontinuities. These are repaired, presumably in the nucleus by a host DNA gyrase system, to generate SC DNA. In the proposed pre-replicative phase, the SC DNA undergoes re-arrangements and multimerisation. During the replicative phase, SC DNA is maintained at a relatively low level with the sequestration of progeny virions in inclusion bodies a possible mechanism to prevent recycling back to the nucleus. It is during this major stage that symptom development occurs although in tolerant host species this phase might be of relatively short duration. In the post replicative phase, SC DNA once more accumulates and becomes re-arranged possibly because regulatory factors produced during the replicative phase are no longer present.

## CONCLUSIONS

Two aspects of CaMV pathogenesis have been investigated. In the first, we have attempted to locate the CaMV genetic determinants controlling specific symptom characters. By linking symptomatic effects to the functions of particular viral gene products or domains thereof, it should lead to a greater understanding of molecular interactions of host and virus and the development of novel strategies of disease control. The second aspect addresses the role of the host in maintaining viral functions. This has highlighted complex interactions in the CaMV/*Brassica* system. A key stage is the way in which the nuclear transcriptional and cytoplasmic reverse transcriptional phases are regulated. It appears that the virus imposes most regulation in the cytoplasm, whilst the host has greater influence in the nucleus.

## REFERENCES

Covey, S. N., In *Molecular Plant Virology*, vol. 2 (ed. J. W. Davies), CRC Press, Boca Raton, Florida, pp121-159 (1985).

Covey, S. N. & Turner, D. S., *EMBO J.* **5**, 2763-2768 (1986).

Covey, S. N., Stratford, R., Saunders, K., Turner, D. S. & Lucy, A. P., In *Genetic Engineering of Crop Plants* (eds. G. Lysett and D. Grierson), Butterworths, UK. pp33-49 (1990).

Covey, S. N., Turner, D. S., Lucy, A. P. & Saunders, K., *Proc. Natl. Acad. Sci. (USA)* **87**, 1633-1637 (1990).

Espinoza, A. M., Ph.D. Thesis, University of East Anglia, UK (1989).

Gronenborn, B. In *Plant DNA Infectious Agents* (eds. T. Hohn & J. Schell), Springer-Verlag, Vienna, pp1-29 (1987).

Linstead, P. J., Hills, G., Plaskitt, K. A., Wilson, I. G., Harker, C. L. & Maule, A. J., *J. Gen. Virol.* **69,** 1809-1818 (1988).

Olszewski, N., Hagen, G. & Guilfoyle, T. J., *Cell* **29**,395-402 (1982).

Pfeiffer, P. & Hohn, T., *Cell* **33,** 781-789 (1983).

Stratford, R., Plaskitt, K. A., Turner, D. S., Markham, P. G. & Covey, S. N., *J. Gen. Virol.* **69,** 2375-2386 (1988).

Stratford, R. & Covey, S. N., *Virology* **172,** 451-459 (1989).

Saunders, K., Lucy, A. P. & Covey, S. N., *J. Gen. Virol.* In press.

Thomas, C. M., Hull, R., Bryant, J. A. & Maule, A. J., *Nucleic Acids Res.* **13,** 4557-4576 (1985).

# BROMOVIRUS RNA REPLICATION AND HOST SPECIFICITY

Paul Ahlquist, Richard Allison[1], Walter DeJong, Michael Janda, Philip Kroner[2], Radiya Pacha, and Patricia Traynor[1]

Institute for Molecular Virology and Department of Plant Pathology
University of Wisconsin-Madison
Madison, WI 53706 U.S.A.

## INTRODUCTION

Positive strand RNA viruses, i.e., viruses that encapsidate single stranded messenger-sense RNA, are a numerically and practically important class of viruses in eukaryotic hosts. Such viruses are particularly significant and well diversified in plants, comprising by far the largest single class of known plant viruses (Matthews, 1982). Research interest in positive strand RNA viruses derives from many perspectives, including their role as highly successful pathogens, their facility at high level gene transmission and expression, and the novelty of their RNA-dependent replication, gene expression, and recombination pathways. The value of understanding RNA-based genetic processes is underscored by the growing suggestion that RNA genomes may have preceded DNA genomes as one stage in the early evolution of life (Joyce, 1989). Thus, as well as illuminating their own unusual biochemistry and genetics, RNA virus studies may provide general insights into replication, survival, and evolution in the putative primordial RNA world. In addition, since many of its features are widely conserved among otherwise diverse viruses of both plants and animals (see below), the viral RNA replication apparatus is a potential target for the development of broad spectrum antiviral agents, which would have great practical benefits. The genetic pathways of RNA viruses have also provided and should continue to provide further new ways to amplify and regulate gene expression in applied genetic engineering (French et al., 1986; Xiong et al., 1989). Finally, viruses possess a number of attractive features as general models for plant-microbe interactions, including uniquely small genomes, ready manipulation by recombinant DNA techniques, and a growing understanding of their life cycle. This chapter discusses some aspects of replication, recombination, and host specificity in one representative group of positive strand RNA plant viruses, the bromoviruses.

---

[1] Present address: Department of Botany and Plant Pathology, Michigan State University, East Lansing, MI 48824 U.S.A.

[2] Present address: The Blood Center of Southeastern Wisconsin, 1701 West Wisconsin Ave., Milwaukee, WI 53233 U.S.A.

*Plant Molecular Biology* 2, Edited by R.G. Herrmann and
B. Larkins, Plenum Press, New York, 1991

## BROMOVIRUS BACKGROUND

The bromoviruses are a relatively well-characterized group of positive strand RNA plant viruses (Lane, 1981) that have long been used as model systems to study viral replication and gene expression. Bromovirus are characterized by a genome divided among three separate RNAs, designated RNA1 (ca. 3.2 kb), RNA2 (ca. 2.8-2.9 kb) and RNA3 (ca. 2.1-2.2 kb) (Fig. 1). RNA3 also gives rise during infection to a subgenomic coat protein mRNA, RNA4 (0.8-0.9 kb). Each of the three genomic RNAs is encapsidated in a separate 28 nm icosahedral virion whose outer capsid is composed of 180 copies of the 20 kD coat protein. RNA3 is co-encapsidated with RNA4, giving each virion a total RNA content of approximately 3 kb.

RNAs 1-4 each bear a 5′ $m^7$GpppG cap and a conserved 3′ terminal sequence of approximately 200 bases. The conserved 3′ sequence is 95-99% identical among the genomic RNAs of each bromovirus and forms an extensive secondary structure that is highly similar among all bromoviruses (Ahlquist et al., 1981a). This region directs (-) strand RNA synthesis and is also responsible for interaction of the RNA with a number of tRNA-specific cellular factors, including tRNA nucleotidyl transferase, aminoacyl tRNA synthetase, and a translation elongation factor (Bujarski et al., 1986). Aminoacylatable 3′ ends are not unique to bromoviruses, but are also found on a number of other plant viral RNAs (Haenni et al., 1982).

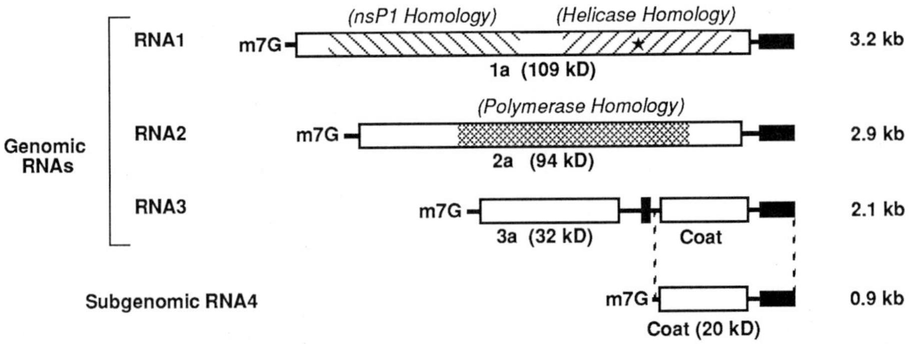


Fig. 1 - Schematic diagram of bromovirus genome organization, showing the structure of the four encapsidated RNAs. The open reading frames for proteins 1a, 2a, 3a and coat are boxed and labelled. The three variously shaded regions define the segments of the 1a and 2a frames encoding amino acid similarity to other viral replication proteins (see text). The star marks the position of a nucleotide binding consensus in the 1a open reading frame. Filled boxes at the ends of the RNAs show the extent of 3′-terminal sequence conservation. The smaller filled box in the central intercistronic region of RNA3 maps the site of an oligo(A) tract important for subgenomic promoter activity (French and Ahlquist, 1988).

In addition to the coat protein, bromoviruses encode three noncapsid proteins, which are each translated directly from one of the genomic RNAs (Fig. 1). RNAs 1 and 2 encode the 109 kD 1a and 94 kD 2a proteins, respectively. These proteins function as trans-acting factors in viral RNA replication and are discussed further below. RNA3 serves as mRNA for the 32 kD 3a protein. Neither 3a protein, coat protein, nor RNA3 itself is required for bromovirus RNA replication (Kiberstis et al., 1981; French et al., 1986). However, both 3a protein and coat protein are required for systemic infection of whole plant hosts (Sacher and Ahlquist, 1989; Allison et al., 1990).

The type and best studied member of the bromovirus group is brome mosaic virus (BMV), which infects a number of grasses including cereal crops. The closest known relative of BMV is cowpea chlorotic mottle virus (CCMV), which infects legumes including cowpeas and soybeans. The BMV and CCMV genomes have been completely sequenced (Ahlquist et al., 1981b; Ahlquist et al., 1984b; Allison et al., 1989; Jozef Bujarski, personal communication). The various genes of BMV and CCMV diverge from each other by 30-50% at both the RNA and protein sequence level. The noncoding regions of the two viruses are further differentiated by certain sequence rearrangements (see below).

For both BMV and CCMV, biologically active cDNA clones have been constructed from which infectious *in vitro* transcripts can be readily synthesized (Ahlquist et al., 1984c; Janda et al., 1987; Allison et al., 1988). These clones provide a means to engineer and test any desired change in these virus genomes, and have facilitated most of the studies described below. Suitable transcripts from such clones are infectious not only to the normal host plants of each virus, but also to protoplasts from a variety of different plants. Barley protoplasts, e.g., can be used for *in vivo* replication studies of BMV, CCMV, and hybrids between the two viruses (French and Ahlquist, 1987; Allison et al., 1988; Traynor and Ahlquist, 1990).

## VIRUS-ENCODED TRANS-ACTING RNA REPLICATION FACTORS

RNAs 1 and 2 are both necessary and together sufficient to direct bromovirus RNA replication in protoplasts (Kiberstis et al., 1981; French et al., 1986). Furthermore, any of a variety of frameshift mutations in the 1a and 2a genes block RNA replication (Traynor and Ahlquist, 1990; unpublished results), and point substitutions or two- to three-codon insertions at positions throughout either gene produce a wide array of aberrant RNA replication phenotypes (Kroner et al., 1989; P. Kroner et al., unpublished results). Together these results show that RNAs 1 and 2 encode trans-acting functions required for RNA replication, and suggest that these functions are mediated by the 1a and 2a proteins rather than being solely due to possible enzymatic properties of RNAs 1 and 2 themselves.

Domains within the 1a and 2a proteins show marked amino acid sequence similarity with noncapsid proteins from many other plant and animal RNA viruses, such as the rod-shaped tobacco mosaic virus (TMV) and the membrane-enveloped Sindbis virus, the type member of the animal alphavirus group (Fig. 1). 1a shares similarity with the TMV 126 kD protein and Sindbis proteins nsP1 and nsP2, while the central domain of 2a shares similarity with the readthrough domain of the TMV 183 kD protein and Sindbis nsP4 (Haseloff et al., 1984; Cornelissen and Bol, 1984; Ahlquist et al., 1985). In all three of these viruses most of the genome encodes RNA replication functions, while additional genes for noncapsid and capsid proteins complete the ability of the virus to move from cell to cell and host to host.

Further sequence comparisons have suggested possible functions for the conserved domains in proteins 1a and 2a (Fig. 1). The central core of the 2a gene is homologous with a number of known polymerases, including the RNA polymerase subunits of bacteriophage Qβ and poliovirus (Kamer and Argos, 1984; Argos, 1988). The phenotypes of many mutations in this 2a region are also consistent with a possible role in general elongation of viral RNA synthesis (Kroner et al., 1989). The C-terminal domain of 1a is homologous with a number of known bacterial helicases (i.e., enzymes that separate the two strands of a double helix) and with herpesvirus proteins previously suggested to be helicases (Hodgman, 1988; Gorbalenya et al., 1989). Helicase activity may be required at several stages of RNA replication, including unwinding the strongly base paired ends of the RNA to allow initiation of synthesis, and opening up internal secondary structure in the template to facilitate elongation. Hybrid arrest experiments show that an *in vitro* BMV polymerase extract can elongate through RNA template regions made double stranded by annealing to defined cDNA fragments, indicating that a substantial strand separating ability is associated with elongation in BMV RNA synthesis (Ahlquist et al., 1984a). Theoretical considerations of RNA replication in the putative primordial "RNA world" have also emphasized the need for strand separating activitie(s), as in releasing the product and template RNA strands after synthesis (Joyce, 1989). Inasmuch as the C-terminal 1a domain contains a nucleotide binding site consensus (a subset of the helicase similarity), it has also been suggested that this region might function as a guanylyl transferase in viral RNA capping (Dreher and Hall, 1988b). Complementing this suggestion, the N-terminal conserved domain in protein 1a is related to Sindbis virus protein nsP1, to which has recently been mapped a mutation affecting a methyltransferase activity implicated in viral RNA capping (Mi et al., 1989). Recent experiments have also implicated the 1a gene in some aspects of template specificity in replication (Traynor and Ahlquist, 1990).

Protoplast replication studies show that heterologous combinations of BMV and CCMV RNAs 1 and 2 fail to support RNA replication (Allison et al., 1988). One possible explanation is that the 1a and 2a proteins may need to interact for one or more stages of replication, and that the heterologous protein combinations are unable to support effective interaction. 1a-2a interaction could produce a complex analogous to the covalent linkage of the corresponding domains in the TMV 183 kD readthrough protein (Haseloff et al., 1984). Such interaction would associate the 1a helicase-like domain with the 2a polymerase-like domain; precedents for helicase-polymerase interactions are of course available from both DNA replication and DNA transcription. It is interesting to note that the polymerase-like domain in protein 2a is flanked by N- and C-terminal extensions that lack analogs in TMV, where translational readthrough fuses the polymerase-like domain directly to the helicase-like domain. Artificially engineered hybrids exchanging RNA2 segments between BMV and CCMV show that effective function with BMV 1a segregates with an N-terminal segment of the BMV 2a gene, and that partial activity with CCMV 1a segregates with a C-terminal segment of the CCMV 2a gene (Traynor and Ahlquist, 1990). Thus, the extensions flanking the putative polymerase domain in 2a might contribute to noncovalent interaction of the separately translated 1a and 2a proteins.

## CIS-ACTING SIGNALS DIRECTING VIRAL RNA SYNTHESIS

In addition to trans-acting factors such as the 1a and 2a proteins, viral RNA replication depends on cis-acting RNA signals that direct the specific selection of viral RNAs as appropriate templates for copying. When compared to some early a priori considerations of the minimal signals necessary to define a viral replication template and to the small size of well known regulatory elements such as the high efficiency promoters for phage T7- and SP6-encoded RNA polymerases, bromovirus replication signals appear relatively large and elaborate. These attributes might be a consequence of the differential regulation that the virus maintains with regard to (+), (-) and subgenomic RNA synthesis, and the need to coordinately regulate the participation of viral RNA templates in the competing processes of translation, replication and encapsidation.

*In vitro*, the last 134 bases from the aminoacylatable 3′ ends of BMV RNAs are sufficient to direct (-) strand RNA initiation by a BMV polymerase extract (Dreher and Hall, 1988a). *In vivo* studies confirm the involvement of this segment and show that even farther upstream sequences in the 3′ noncoding region are also necessary for efficient replication (French and Ahlquist, 1987; Pacha et al., 1990). By corollary, it was expected that 5′-proximal sequences would be involved in initiation of (+) strand synthesis, and indeed 5′ segments of 60-90 bases are found to be required for replication (French and Ahlquist, 1987; Pacha et al., 1990). Unexpectedly, it was also found that replication of BMV RNA3 was strikingly reduced (though not abolished) when a segment was deleted from the central intercistronic noncoding region, more than 1 kb from either RNA end. Inspection revealed that this segment contained a conserved motif, GGUUCAAnnCCCU, common to the 5′ ends of most RNAs from BMV, CCMV, and cucumber mosaic virus (French and Ahlquist, 1987; Allison et al., 1989). This motif was later recognized to correspond to the box B motif of RNA polymerase III promoters (Marsh and Hall, 1987), which is equivalent to the conserved residues of the TψC loop of tRNAs. Thus, not only the aminoacylatable 3′ ends but most of the regions found to contribute in cis to bromovirus RNA replication contain features related to tRNAs or their genes. One possible implication of this repeated association is that replication by these viruses may involve or be modulated by cellular as well as viral factors.

In contrast to BMV RNA3, the nucleotide sequence of CCMV RNA3 revealed that its intercistronic region did not contain the TψC loop motif (Allison et al., 1989). In keeping with the suspected significance of this element, later deletion analysis showed that removing the intercistronic region from CCMV RNA3 did not inhibit its replication, as had been found for BMV RNA3 (Pacha et al., 1990). Replication signals can thus be organized in significantly different ways on even closely related bromovirus RNAs. This apparent flexibility in replicon design has presumably contributed to virus evolution by facilitating successful gene rearrangements as seen in the relationships between bromoviruses, TMV, Sindbis, and other RNA viruses.

Subgenomic RNA4 is produced by internal initiation of RNA synthesis on (-) strand RNA3 templates (Miller et al., 1985). The sequences responsible for directing this event have been studied both *in vitro* (Marsh et al., 1988) and *in vivo* (French and Ahlquist, 1988). Rearrangement and deletion experiments show that the full subgenomic promoter active *in vivo* covers about 100 bases. This RNA-dependent RNA promoter can be functionally subdivided into a core promoter element and upstream domains which stimulate activity. The entire promoter can be moved as a cassette to new sites in the viral genome and there directs the production of novel subgenomic RNAs (French and Ahlquist, 1988).

## RECOMBINATION AS A FORCE IN RNA VIRUS EVOLUTION, VARIABILITY, AND SURVIVAL

A combination of diverse observations reveals that, whether considering long-range evolutionary time scales or short term laboratory experiments, recombination is an important factor in the genetics of bromoviruses and many other RNA viruses. Just as for DNA genomes, recombination allows RNA virus genomes to alter genome structure by sudden, quantum leaps. This potential for discontinuity not only accelerates virus evolution but also opens it to fundamentally new possibilities and directions, since if RNA virus evolution were limited to more gradual, continuous mechanisms, necessary evolutionary intermediates between many of the virus types seen today would likely not be viable. Although a number of alternatives have been considered, the actual mechanism(s) of recombination in RNA viruses have not been conclusively established (Haseloff et al., 1984). One widely favored model for RNA virus recombination is template switching during RNA synthesis. This model would require only known enzyme activities and RNA substrates, and experimental evidence favoring such a mechanism has been presented for poliovirus (Kirkegaard and Baltimore, 1986). Recombination may thus be a byproduct of the normal RNA virus replication process.

The importance of recombination in early virus evolution is well illustrated in the relation of single and multi-component genomes. For example, in TMV the helicase-like and polymerase-like domains are encoded by contiguous segments, while in Sindbis they are separated by coding sequences for the alphavirus-specific nsP3 protein, and in BMV they are encoded on separate RNAs (Ahlquist et al., 1985). Under any divergent evolutionary model, then, recombination must have played a major role in the redistribution of the relevant gene segments and the emergence of these distinct RNA virus groups. Recombination also offers a ready explanation for the initial generation of long conserved sequences at the 3′ ends of the separate genomic RNAs of the bromoviruses and certain other multicomponent viruses. Since the highly conserved regions at the 3′ ends of bromovirus RNAs are approximately 200 bases long and satisfy a number of complex functional constraints, it is difficult to imagine how a viable divided genome dependent on such 3′ functions could emerge by more gradual, convergent processes rather than by recombinational events. Though initially proposed on such theoretical grounds (Ahlquist et al., 1984b), the role of recombination in establishing and subsequently maintaining common 3′ termini on bromovirus RNAs has more recently been substantiated by direct evidence (Bujarski and Kaesberg, 1986).

As well as playing an important role in establishing some distinct virus groups, recombination seems to have contributed to the finer scale differentiation of related members within some groups. Comparison of the BMV and CCMV genomic sequences, e.g., reveals insertion/deletion differences in the interior of several otherwise similar regions. These sites appear to represent the signatures of recombination events that occured at some point in the divergence of the two viruses (Allison et al., 1989).

The role of recombination as a sufficiently dynamic process to be measurable in individual RNA virus infections, and not just on evolutionary time scales, was first recognized for animal viruses. Important contributions in this area were made by the pioneering work of Cooper with poliovirus (Cooper, 1969). The study of defective interfering (DI) RNAs, which are generally deletion/rearrangement mutants, also made the role of recombination apparent in viruses such as Sindbis (Schlesinger, 1988). Nevertheless, the extent to which recombination occurred in many RNA viruses remained uncertain for some time. For example, though DI RNA studies made it clear that nonhomologous recombination occurred, early attempts to demonstrate homologous recombination in Sindbis infections were unsuccessful (King, 1988). The recognized absence of observed DI RNAs in most plant virus infections was also taken to suggest that recombination might be relatively inactive in these systems.

The first direct observation of recombination in a plant RNA virus was reported by Bujarski and Kaesberg (1986), who demonstrated that recombination with RNA1 or RNA2 led to repair of an RNA3 deletion mutation in the 3′ noncoding region of BMV RNA3. Several additional examples have now been provided, including the observation of plant RNA virus DI RNAs (Li et al., 1989; Burgyan et al., 1989) and the loss of an inserted foreign gene (Dawson et al., 1989). Recently it was also observed that intermolecular recombination rescued a wild type genome when plants were co-inoculated with CCMV RNAs 1 and 2 and two RNA3 mutants bearing independent deletions, each of which prevented systemic infection (Allison et al., 1990). Because the number of cells infected by the input genomes was severely limited, this suggests that co-infecting plant RNA viruses might be able to exchange genetic information even when one or both are not completely adapted for efficient spread in the relevant host plant. Since achieving adaptation for systemic infection would constitute a dramatic selection pressure, such situations might provide one avenue for extension or alteration of host range. Another implication of the CCMV genome rescue results is that recombination in RNA viruses may act as a useful genome repair mechanism to counter the frequent substitution errors of RNA-based RNA replication. In this way viruses would reap the benefit of higher genetic variability at a reduced cost.

A currently unresolved issue is the extent to which RNA virus recombination and evolution involve the recruitment of cellular genes and sequences. Precedents for viral incorporation of cellular sequences are best established for retroviruses and oncogenes, but are also known for other kinds of RNA viruses: Among positive strand RNA viruses, some Sindbis DI RNAs are known to incorporate cellular tRNA sequences (Schlesinger, 1988). A recent report also described the presence of cellular 28S rRNA sequences in the hemagglutinin gene of a novel isolate of influenza, a (-) strand RNA virus (Khatchikian et al., 1989). Thus, recombination may give RNA viruses access to the genetic pool of their hosts, a possibility with dramatic implications for virus diversification.

## SYSTEMIC INFECTION AND HOST SPECIFICITY

To address certain fundamental aspects of virus-host interaction, some studies were recently begun to investigate the features responsible for the adaptation of BMV and CCMV to infect certain monocots and dicots, respectively. This system was chosen to exploit contrasting host/virus genetic relationships: The wide genetic gulf between monocots and dicots makes it likely that BMV and CCMV necessarily differ in many or most of the virus-host interactions that commonly affect the success of infection. At the same time, our previous work has shown that BMV and CCMV are sufficiently related for informative genetic exchanges (Traynor and Ahlquist, 1990).

In most cases, the practical host range of plant viruses is not determined by limitations on viral RNA replication or packaging in individual cells, but by the ability of the virus to initiate and sustain sufficient spread of infection between cells to achieve an effective systemic infection. As discussed above, e.g., both BMV and CCMV replicate and package their RNAs to high levels in barley protoplasts, but only BMV is able to systemically infect whole barley plants. The process of infection movement and the constitutive and induced constraints imposed by the host are still poorly understood, but are attracting growing research attention (Hull, 1989).

As described above, bromovirus RNA3 is dispensable for RNA replication but required for systemic infection. Targeted mutations show that both RNA3-encoded genes, the coat gene and the 3a noncapsid gene, are required for systemic infection (Allison et al., 1990; Sacher and Ahlquist, 1989). The first 25 amino acids of BMV coat protein contain a highly basic region implicated in RNA binding. Engineered deletion of this region simultaneously blocks both encapsidation and systemic infection, even though the remainder of the 188 amino acid coat protein is expressed to nearly normal levels (Sacher and Ahlquist, 1989). In the absence of coat protein expression, there is evidence for slight spread of bromovirus infection within inoculated leaves (Allison et al., 1990). However, no bromovirus mutant that fails to express a functional coat protein has yet been observed to move into an uninoculated leaf, even though coat protein deletion mutants of some single component RNA viruses are capable of moving between leaves, albeit at impaired efficiency. As with the required role of the basic region at the coat protein N-terminus, the possibly greater dependence of bromoviruses on coat protein for long range transport of infection is consistent with a possible role of encapsidation, as opposed to some additional hypothetical coat protein function, in long range transport. If relevant to infection movement, the instability of unencapsidated RNA should be a greater burden on bromoviruses than single component viruses because of the greater difficulty of independently transporting three separate genomic RNAs intact to a new infection site.

In the absence of 3a gene expression, no evidence for bromovirus infection spread has yet been seen even in inoculated leaves (Allison et al., 1990). In this and in several other respects, the 32 kD 3a protein appears similar to the 30 kD protein of TMV. 3a and the 30 kD protein are the only noncapsid proteins of their respective viruses that do not detectably influence RNA replication, both are required for spread of infection, and both occupy relatable positions in their respective genomes (Haseloff et al., 1984; Allison et al., 1989). Further supporting the possible analogy between these genes is weak sequence similarity between CCMV 3a and the 30 kD gene of the cowpea strain of TMV (Allison et al., 1989) and the clear one-to-one correspondence between all remaining bromovirus and TMV gene products. Recent studies showing that the TMV 30 kD protein influences plasmodesmatal exclusion limits (Wolf et al., 1989) and functions as a single strand nucleic acid binding protein (Citovsky et al., 1990) appear to provide important insights for future investigations. Further studies of infection movement and the role of such proteins may well reveal new general insights on intercellular connections and selective transport between plant cells.

Since both of its genes are necessary for infection spread, the level at which host specificity is most frequently seen, it is not surprising that reassortment experiments show that RNA3 carries host specificity determinants (Allison et al., 1988). These same experiments also show that RNA3 is not the sole determinant of bromovirus host specificity. I.e., although the only known functions encoded by RNAs 1 and 2 are in RNA replication, which shows no evident host specificity when assayed in protoplasts, RNA1 and/or RNA2 also encode host specificity determinant(s) (Allison et al., 1988). Host adaptation in bromoviruses thus involves characteristics encoded on two or more separate genomic RNAs. Tests of more defined exchanges between BMV and CCMV are in progress and promise to be one informative approach to identify and characterize the particular virus functions that determine host specificity.

## ACKNOWLEDGMENTS

We thank Ben Young and Craig Thompson for excellent technical assistance in a variety of experiments. The research discussed here was supported by the National Institutes of Health under Public Health Service Grant GM35072 and by the National Science Foundation under Presidential Young Investigator Award DMB-8451884.

## REFERENCES

Ahlquist, P., Bujarski, J., Kaesberg, P., and Hall, T. C., 1984a, Localization of the replicase recognition site within brome mosaic virus RNA by hybrid-arrested RNA synthesis, Plant Mol. Biol., 3:37.

Ahlquist, P., Dasgupta, R., and Kaesberg, P., 1981a, Near identity of 3' RNA secondary structure in bromoviruses and cucumber mosaic virus, Cell, 23:183.

Ahlquist, P., Dasgupta, R., and Kaesberg, P., 1984b, Nucleotide sequence of the brome mosaic virus genome and its implications for viral replication, J. Mol. Biol., 172:369.

Ahlquist, P., French, R., Janda, M., and Loesch-Fries, L. S., 1984c, Multicomponent RNA plant virus infection derived from cloned viral cDNA, Proc. Natl. Acad. Sci. USA, 81:7066.

Ahlquist, P., Luckow, V., and Kaesberg, P., 1981b, Complete nucleotide sequence of brome mosaic virus RNA3, J. Mol. Biol., 153:23.

Ahlquist, P., Strauss, E. G., Rice, C. M., Strauss, J. H., Haseloff, J., and Zimmern, D., 1985, Sindbis virus proteins nsP1 and nsP2 contain homology to nonstructural proteins from several RNA plant viruses, J. Virol., 53:536.

Allison, R., Janda, M., and Ahlquist, P., 1988, Infectious *in vitro* transcripts from cowpea chlorotic mottle virus cDNA clones and exchange of individual RNA components with brome mosaic virus, J. Virol., 62:3581.

Allison, R., Janda, M., and Ahlquist, P., 1989, Sequence of cowpea chlorotic mottle virus RNAs 2 and 3 and evidence of a recombination event during bromovirus evolution, Virology, 172:321.

Allison, R., Thompson, C., and Ahlquist, P., 1990, Regeneration of a functional RNA virus genome by recombination between deletion mutants and requirement for cowpea cholorotic mottle virus 3a and coat genes for systemic infection, Proc. Natl. Acad. Sci. USA, 87:1820.

Argos, P., 1988, A sequence motif in many polymerases, Nucl. Acids Res., 16:9909.

Bujarski, J. J., Ahlquist, P., Hall, T. C., Dreher, T. W., and Kaesberg, P., 1986, Modulation of replication, aminoacylation and adenylation *in vitro* and infectivity *in vivo* of BMV RNAs containing deletions within the multifunctional 3′ end, EMBO J., 5:1769.

Bujarski, J., and Kaesberg, P., 1986, Genetic recombination between RNA components of a multipartite plant virus, Nature, 321:528.

Burgyan, J., Grieco, F., and Russo, M., 1989, A defective interfering RNA molecule in cymbidium ringspot virus infections, J. gen. Virol., 70:235.

Citovsky, V., Knorr, D., Schuster, G., and Zambryski, P., 1990, The P30 movement protein of tobacco mosaic virus is a single-strand nucleic acid binding protein, Cell, 60:637.

Cooper, P. D., 1969, The genetic analysis of poliovirus, pp. 177-218, in: "The Biochemistry of Viruses," H. B. Levy, ed., Marcel Dekker, New York.

Cornelissen, B. and Bol, J., 1984, Homology between the proteins encoded by tobacco mosaic virus and two tricornaviruses, Plant Mol. Biol., 3:379.

Dawson, W. O., Lewandowski, D. J., Hilf, M. E., Bubrick, P., Raffo, A. J., Shaw, J. J., Grantham, G. L., and Desjardins, P. R., 1989, A tobacco mosaic virus-hybrid expresses and loses an added gene, Virology, 172:285.

Dreher, T., and Hall, T., 1988a, Mutational analysis of the sequence and structural requirements in brome mosaic virus RNA for minus strand promoter activity, J. Mol. Biol., 201:31.

Dreher, T., and Hall, T., 1988b, RNA replication in brome mosaic virus and related viruses, pp. 91-113, in: "RNA Genetics, Vol. I, RNA-Directed Virus Replication," E. Domingo, J. Holland and P. Ahlquist, eds., CRC Press, New York.

French, R., and Ahlquist, P., 1987, Intercistronic as well as terminal sequences are required for efficient amplification of brome mosaic virus RNA3, J. Virol., 61:1457.

French, R., and Ahlquist, P., 1988, Characterization and engineering of sequences controlling *in vivo* synthesis of brome mosaic virus subgenomic RNA, J. Virol., 62:2411.

French, R., Janda, M., and Ahlquist, P., 1986, Bacterial gene inserted in an engineered RNA virus: efficient expression in monocotyledonous plant cells, Science 231:1294.

Gorbalenya, A., Koonin, E., Donchenko, A., and Blinov, V., 1989, Two related superfamilies of putative helicases involved in replication, recombination, repair and expression of DNA and RNA genomes, Nucl. Acids Res. 17:4713.

Haenni, A.-L., Joshi, S., and Chapeville, F., 1982, tRNA-like structures in the genomes of RNA viruses, pp. 85-104, in: "Progress in Nucleic Acid Research and Molecular Biology," Vol. 27, Academic Press, Inc., New York.

Haseloff, J., Goelet, P., Zimmern, D., Ahlquist, P., Dasgupta, R., and Kaesberg, P., 1984, Striking similarities in amino acid sequence among nonstructural proteins encoded by RNA viruses that have dissimilar genomic organization, Proc. Natl. Acad. Sci. USA. 81:4358.

Hodgman, T., 1988, A new superfamily of replicative proteins, Nature 333:22-23 and 333:578.

Hull, R., 1989, The movement of viruses in plants, Ann. Rev. Phytopathol. 27:213.

Janda, M., French, R., and Ahlquist, P., 1987, High efficiency T7 polymerase synthesis of infectious RNA from cloned brome mosaic virus cDNA and effects of 5′ extensions on transcript infectivity, Virology 158:259.

Joyce, G., 1989, RNA evolution and the origins of life, Nature 338:217.

Kamer, G., and Argos, P., 1984, Primary structural comparison of RNA-dependent polymerases from plant, animal and bacterial viruses, Nucl. Acids Res., 12:7269.

Khatchikian, D., Orlich, M., and Rott, R., 1989, Increased viral pathogenicity after insertion of a 28S ribosomal RNA sequence into the haemagglutinin gene of an influenza virus, Nature, 340:156.

Kiberstis, P., Loesch-Fries, L. S., and Hall, T., 1981, Viral protein synthesis in barley protoplasts inoculated with native and fractionated brome mosaic virus RNA, Virology, 112:804.

King, A., 1988, Genetic recombination in positive strand RNA viruses, pp. 149-165, in: "RNA Genetics, Vol. II: Retroviruses, Viroids, and RNA Recombination," E. Domingo, J. Holland and P. Ahlquist, eds., CRC Press, New York.

Kirkegaard, K., and Baltimore, D., 1986, The mechanism of RNA recombination in poliovirus, Cell, 47:433.

Kroner, P., Richards D., Traynor, P., and Ahlquist, P., 1989, Defined mutations in a small region of the brome mosaic virus 2a gene cause diverse temperature-sensitive RNA replication phenotypes, J. Virol., 63:5302.

Lane, L., 1981, Bromoviruses, pp. 333-376, in: "Handbook of Plant Virus Infections and Comparative Diagnosis," E. Kurstak, ed., Elsevier/North-Holland Biomedical Press, Amsterdam.

Li, X. H., Heaton, L. A., Morris, T. J., and Simon, A. E., 1989, Turnip crinkle virus defective interfering RNAs intensity viral symptoms and are generated *de novo*, Proc. Natl. Acad. Sci. USA, 86:9173.

Marsh, L., and Hall, T. C., 1987, Evidence implicating a tRNA heritage for the promoters of positive-strand RNA synthesis in brome mosaic and related viruses, Cold Spring Harbor Symp. Quant. Biol. 52:331.

Marsh, L., Dreher, T., and Hall, T. C., 1988, Mutational analysis of the core and modulator sequences of the BMV RNA3 subgenomic promoter, Nucl. Acids Res., 16:981.

Matthews, R. E. F., 1982, Classification and nomenclature of viruses, Intervirology, 17:no.1-3.

Mi, S., Durbin, R., Huang, H., Rice, C., and Stollar, V., 1989, Association of the Sindbis virus methyltransferase activity with the nonstructural protein nsP1, Virology, 170:385.

Miller, W. A., Dreher, T., and Hall, T., 1985, Synthesis of brome mosaic virus subgenomic RNA *in vitro* by internal initiation on (-)-sense genomic RNA, Nature, 313:68.

Pacha, R., Allison, R., and Ahlquist, P., 1990, Cis-acting sequences required for *in vivo* amplification of genomic RNA3 are organized differently in related bromoviruses, Virology, 174:436.

Sacher, R., and Ahlquist, P., 1989, Effects of deletions in the N-terminal basic arm of brome mosaic virus coat protein on RNA packaging and systemic infection, J. Virol., 63:4545.

Schlesinger, S., 1988, The generation and amplification of defective interfering RNAs, pp. 167-185, in: "RNA Genetics, Vol. II: Retroviruses, Viroids, and RNA Recombination," E. Domingo, J. Holland and P. Ahlquist, eds., CRC Press, New York.

Traynor, P., and Ahlquist, P., 1990, Use of bromovirus RNA2 hybrids to map cis- and trans-acting functions in a conserved RNA replication gene, J. Virol., 64:69.

Wolf, S., Deom, C. M., Beachy, R. N., and Lucas, W. J., 1989, Movement protein of tobacco mosaic virus modifies plasmodesmatal size exclusion limit, Science, 246:377.

Xiong, C., Levis, R., Shen, P., Schlesinger, S., Rice, C., and Huang, H., 1989, Sindbis virus: an efficient, broad host range vector for gene expression in animal cells, Science, 243:1188.

# ANALYSIS OF TOBACCO MOSAIC VIRUS-HOST INTERACTIONS BY DIRECTED GENOME MODIFICATION

James N. Culver, Alwyn G. C. Lindbeck, Paul R. Desjardins, and William O. Dawson

Department of Plant Pathology
University of California
Riverside, CA 92521 USA

## INTRODUCTION

Small single stranded RNA viruses cause disease and reduce yields of many crop plants. Because these viruses carry limited amounts of genetic information into plants, they are dependent upon their hosts to provide the cellular machinery needed for replication. This dependency requires that virus and host components interact at the molecular level. It is at this level that host plants recognize and respond to infecting viruses. Our understanding of these interactions has been extremely limited. Only recently have advances in technology allowed a systematic study of virus-host interactions. Understanding the mechanisms behind these interactions should provide insight for the development of new techniques for the control of plant virus diseases.

Tobacco mosaic virus (TMV) is typical of this type of viral pathogen. TMV is a positive strand RNA virus having a genome length of 6395 nucleotides that codes for at least four proteins (Goelet *et al.*, 1982). Of these four proteins the coat protein is by far the most highly expressed TMV product, accounting for over 10% of total host cellular protein (Fraser, 1987). Although the main function of the coat protein is structural, it has been shown to play a role in a number of different host responses (Knorr and Dawson, 1988; Dawson *et al.*, 1988). The ability to manipulate the coat protein of TMV via a full-length infectious cDNA clone (Dawson *et al.* 1986) has enabled us to study the effects of numerous coat protein mutations on a number of different host responses. In this chapter we describe the involvement of the TMV coat protein in the development of a specific host resistance response as well as its association with chloroplasts during a disease response.

## INVOLVEMENT OF TMV COAT PROTEIN IN A RESISTANCE RESPONSE

Expression of resistance by the plant can occur in a number of different ways. A resistant plant may lack some component required by the virus for pathogenesis (passive resistance) or the plant may specifically recognize the virus and produce some component that inhibits the infection process (active resistance). One type of active resistance response is expressed by the plant as the production of necrotic lesions at the sites of virus

*Plant Molecular Biology 2*, Edited by R.G. Herrmann and B. Larkins, Plenum Press, New York, 1991

infection, limiting the further spread of the virus. Lesion development is generally associated with a single dominant host gene that confers the ability to recognize a specific virus gene (product) resulting in the induction of host active defense mechanisms (Flor, 1971; Keen and Staskawicz, 1988). This "gene for gene" response is termed the hypersensitive response (HR). We are examining the molecular mechanisms responsible for the HR in *Nicotiana sylvestris* Speg. & Comes.

Induction of N' gene hypersensitive reaction

*N. sylvestris* carries the N' gene for resistance to most strains of TMV. TMV 204, an infectious cDNA clone of the U1 strain of TMV (Dawson *et al.*, 1986) does not induce the HR in *N. sylvestris* but instead systemically invades the plant producing mosaic symptoms. However, it is easy to obtain mutants of TMV 204 either spontaneously or through chemical mutagenesis that will induce the development of the HR in *N. sylvestris* (Aldaoud, 1987). We have utilized such mutants to identify five independent single point mutations, occurring in the coat protein open reading frame, that were responsible for the induction of the N' gene HR (Knorr and Dawson, 1988; Culver and Dawson, 1989a). Additionally, all of the identified point mutations resulted in amino acid substitutions in the coat protein. Thus, mutations in either the RNA or protein could have induce the N' gene HR.

Coat protein: an elicitor of HR

To determine if mutated RNA or protein was directly responsible for inducing the HR in *N. sylvestris*, a virus mutant was created from one of the previously identified HR inducing point mutants, TMV 25. This new mutant, TMV [-25], was made using site-directed mutagenesis and resulted in an alteration of the coat protein translational start from AUG to AGA (Culver and Dawson, 1989b). This alteration left the other nucleotides of the coat protein open reading frame intact, including the original point mutation associated with the induction of HR, but prevented the production of the altered coat protein.

Inoculation of mutant TMV [-25] onto expanded leaves of *N. sylvestris* failed to induce the development of local lesions. Western immunoblot analysis of proteins extracted from TMV [-25] inoculated leaves revealed the presence of TMV replicase protein, demonstrating that TMV [-25] replicated in the inoculated leaves, but an absence of detectable TMV coat protein (Culver and Dawson, 1989b). This demonstrates that the coat protein of TMV 25 and not the RNA was responsible for the induction of the N' gene HR.

To further examine the role of the coat protein in the induction of N' gene HR, transgenic plants expressing either of two different HR-inducing TMV coat proteins or wild type TMV coat protein were examined (Culver and Dawson in preparation). The two HR-inducing coat protein genes used to transform *N. sylvestris* originated from mutant TMV 11, having an amino acid substitution of Val to Met at position 11 in the coat protein, and mutant 25, having an amino acid substitution of Asn to Ser at position 25 in the coat protein (Culver and Dawson, 1989a). The two coat protein mutants produce distinctly different local lesion phenotypes in *N. sylvestris*. Mutant TMV 11 was characteristic of a "weak" elicitor of the HR, with local lesions developing 5 to 7 days after inoculation and continually spreading from the original necrotic area eventually collapsing the entire leaf. In contrast, mutant TMV 25 was characteristic of a "strong" elicitor of the HR, with local lesions developing 2 to 3 days after inoculation and not spreading from the original point of necrosis. These two different HR-inducing coat protein genes were used to transform *N. sylvestris* to examine whether their distinctly different lesion phenotypes would be expressed in the transgenic plants.

Each coat protein open reading frame, TMV 11, TMV 25, and wild type TMV 204, was individually placed behind the 35S promoter of CaMV in the *Agrobacterium* binary vector pMON530 (Rogers, *et al*., 1987). Transformed plants testing positive for the production of coat protein were subsequently observed for phenotypic differences. *N. sylvestris* plants expressing the wild type coat protein developed normally being only slightly stunted when compared to healthy non-transformed *N. sylvestris* plants. However, transformed *N. sylvestris* plants expressing either of the two HR inducing coat proteins developed necrotic symptoms characteristic of the HR. Symptoms included mild to severe stunting and the development of small necrotic spots occurring across the entire leaf surface. These necrotic spots slowly enlarged over the course of several weeks eventually coalescing and collapsing the entire leaf. The development of necrosis in *N. sylvestris* plants expressing HR-inducing coat proteins demonstrates that these coat proteins are singly responsible for eliciting the N' gene HR.

A comparison of transformed plants expressing either coat protein 11, the weak elicitor, or coat protein 25, the strong elicitor, showed that necrosis developed faster and more severely in plants expressing the strong elicitor than in those expressing the weak elicitor. Coat protein levels produced by transformants expressing these two different elicitor coat proteins showed that lower amounts of strong elicitor coat protein resulted in stronger plant necrotic responses then did higher amounts of weak elicitor coat protein. This demonstrates that the transformed plants specifically recognized the two different coat protein elicitors, resulting in phenotypic differences in the transgenic plants producing them. Thus, no virus component or replication process, other than the presence of specific elicitor coat proteins, was necessary for the induction of the HR in *N. sylvestris*.

Structural analysis of elicitor coat proteins

The identification of a number of single amino acid substitutions in the coat protein of TMV that lead to the induction of the N' gene HR suggests that the coat protein directly interacts with some host component which then triggers the development of necrotic local lesions. It appears unlikely that the wild type coat protein acts only to suppress the HR since mutants that do not express coat protein do not induce HR in *N. sylvestris* (Culver and Dawson, 1989b) and transgenic plants that express wild type coat protein will develop necrotic local lesions when inoculated with an HR inducing strain of TMV. Therefore, it seems likely that the amino acid substitutions are being recognized by plants carrying the N' gene.

To understand this coat protein-host recognition, the smallest active coat protein unit that will induce HR needs to be known. Recent studies by Saito *et al*., (1989) with the HR inducing TMV-L strain suggest that nearly the entire coat protein may be necessary for the induction of the N' gene HR. This was shown by constructing various deletions and frameshifts in the coat protein gene of TMV-L as well as chimeric coat proteins between TMV-L and a non-HR inducing strain TMV-OM. Substantial changes in several regions of the TMV-L coat protein were found to result in a loss of ability to induce the N' gene HR.

It now appears that some aspect of the three dimensional coat protein structure is responsible for eliciting HR. Locating the positions of the initial HR eliciting amino acid substitutions on the structural configuration of a single coat protein subunit revealed that these substitutions occurred at different locations throughout the subunit. However, in a collaboration with Dr. G. J. Stubbs, Vanderbilt University, we have been able to define a possible site on the surface of the coat protein that may be directly responsible for triggering the HR. All of the HR inducing amino acid substitutions occur on or directly interfere with the surface structure of the coat protein subunit. When two subunits are positioned together as they naturally would occur in a protohelix or virion, all of the identified amino acid substitutions line up in a "footprint" located between the two subunits (Fig. 1). A number of additional amino acid substitutions have supported this footprint region as the active site for the induction of the N' gene HR (Culver *et al.* in preparation). All changes made inside the predicted footprint and interfering with the surface of the structure were found to elicit the HR, while changes made outside of this region did not.

Currently, we are refining the structure of this "footprint" by the creation of specific amino acid substitutions throughout this region. We are also examining whether coat protein subunits must form multimers to induce the N' gene HR. Information obtained from transgenic plants expressing elicitor coat proteins indicates that virion formation is not required for the induction of the HR. However, under physiological conditions, individual coat protein subunits assemble into protohelices in the absence of viral RNA. It should be possible to interfere with subunit assembly by altering those amino acids that interact between subunits and thus, identify the type of coat protein subunit configuration that is recognized by the plant.

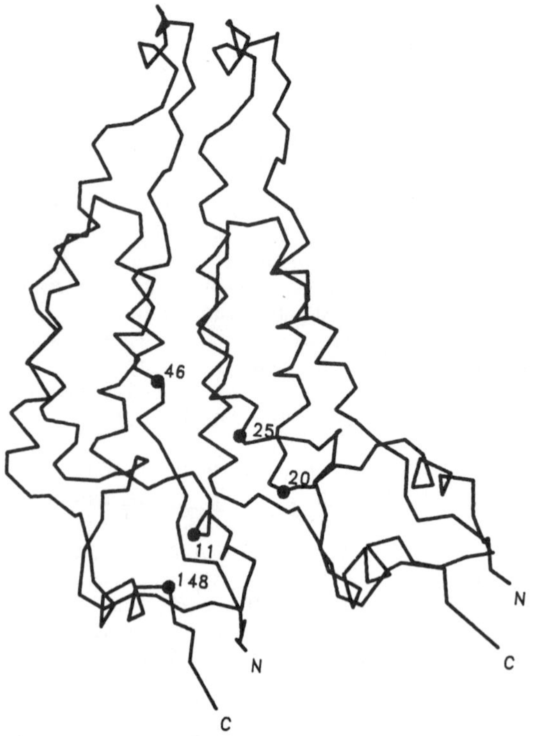


Fig. 1. Diagrammatic representation of the structural configuration of two tobacco mosaic virus coat protein subunits (Bloomer *et al.*, 1978), showing the approximate location of five independent amino acid substitutions responsible for the induction of the hypersensitive reaction.

## ASSOCIATION OF COAT PROTEIN WITH A DISEASE RESPONSE

A major response of plants to viruses that results in disease is the loss of chloroplast function, often visualized as chlorotic or mosaic leaves. This symptom usually results from improperly developed or degraded chloroplasts.

### Association of TMV with chloroplasts

TMV replicates in the cytoplasm where most of the virions are found, although there have been reports of virus particles within chloroplasts. Different strains of TMV differ greatly in the amount of virions found in chloroplasts. Chloroplasts from tissue infected with TMV strain U5 often contain many virions while chloroplasts from tissue infected with TMV strain U1 contain few, if any, virions. However, there is little difference in the symptoms caused by these strains.

There is evidence that suggests an association of TMV with chloroplasts. In infections with some TMV strains, pseudovirions (Shalla *et al.*, 1975) consisting of viral coat protein encapsidating chloroplast RNA (Rochon and Siegel, 1984) have been seen in chloroplasts. More recently both TMV coat protein (Reinero and Beachy, 1986, 1988) and TMV RNA (Schoelz and Zaitlin, 1989) have been isolated from Percoll-purified chloroplasts. In chloroplasts isolated from spinach infected with TMV (Hodgson *et al.*, 1989), it has been demonstrated that coat protein is associated with photosystem II complexes in the thylakoids. Further, a ubiquitinated coat protein has been found accumulated in chloroplasts (Dunigan *et al.*, 1988). Also TMV coat protein and the large subunit of ribulose-1,5-bisphosphate carboxylase have an immunological cross-reactivity and some amino acid sequence homology (Dietzgen and Zaitlin, 1986). On the basis of the above observations it has been suggested (Beachy *et al.*, 1987; Grumet *et al.*, 1987; Sherwood, 1987) that coat protein plays a role in the development of disease symptoms in chloroplasts.

It is not clear how viral RNA and proteins enter the chloroplast. It was recently reported (Schoelz and Zaitlin, 1989) that full-length genomic RNA can be isolated from intact chloroplasts from directly inoculated and systemic leaves from U1 infected plants. It has been demonstrated that chloroplast ribosomes *in vitro* (Camerino *et al.*, 1982; Sela and Kaesberg, 1969) and 70S ribosomes of *E. coli* (Glover and Wilson, 1982) can translate TMV RNA to yield coat protein. However, attempts to synthesize coat protein "in organello" in chloroplasts from infected leaves have been unsuccessful (M. Zaitlin, unpublished; cited by Reinero and Beachy, 1989). There are reports (Hirai and Wildman, 1969; Mohamed and Randles, 1972) of a repression of chloroplast ribosomal RNA and chloroplast protein synthesis as a result of viral infection while cytoplasmic synthesis is largely unaffected. Fraser and Gerwitz (1980) report that host protein synthesis is reduced by up to 75% during virus replication but then recovers. Their evidence suggests that the controls are at the translational level, possibly due to competition between viral and host mRNA's. Further, the principal site of coat protein synthesis appears to be the cytoplasm rather than the chloroplast and the native protein does not contain any of the published signal sequences (see Keegstra *et al.*, 1989 for review) that would target it to the chloroplast. Electron microscope immunocytochemistry of U1 infected systemic leaves (Hills *et al.*, 1987) suggests that the level of coat protein in chloroplasts of these leaves is either very low or zero. Our own data (Lindbeck *et al.*, in preparation) suggests that this is also true in chloroplasts in yellow tissue from chlorotic areas on leaves inoculated with coat protein deletion mutants.

Degradation of developed chloroplasts

Some strains of TMV and other viruses cause chlorosis of mature leaves. This suggests that these viruses are able to degrade the structure and/or function of developed chloroplasts. Dawson *et al*. (1988) have produced a series of TMV mutants with insertions and/or deletions at the unique AccI site in the coat protein gene that produce a variety of symptoms in inoculated leaves. Three types of symptoms were observed in mutant infected Xanthi tobacco: no symptoms, yellowing and necrosis. The effects of the coat protein mutants is summarized in the table 1.

Table 1. Summary of coat protein mutants.

| Strain | Deletion 5' ⟵ | Deletion 3' ⟶ | Symptoms on Xanthi Local | Symptoms on Xanthi Systemic |
|---|---|---|---|---|
| cp 5 | +1 | +1 | none | none |
| cp +2 | -1 | -1 | yellow spots | yellowing |
| cp 10 | -14 | -26 | necrosis | --- |
| cp 4 | -37 | -97 | yellowing | yellowing |
| cp 25 | -38 | -15 | yellow spots | yellowing |
| cp 27 | -108 | -23 | necrosis | --- |
| cp 28 | -164 | -135 | yellow spots | none |
| cp 35 | -216 | -163 | none | none |
| cp 35-5 | -216 | +1 | yellow areas | yellowing |
| cp 5-35 | +1 | -163 | none | none |
| cp S3-28 | -345 | -135 | none | none |
| cp S3-CAT-28 | -345 | -135 | none | none |
| TMV 204 | 0 | 0 | none | mosaic |

Most of the TMV coat protein mutants produced coat protein-related polypeptides *in vivo* although the size and amount of protein produced in inoculated leaves varied among the different mutants. Most mutants which produced no or low levels of coat protein did not induce symptoms. The mutants which retained the carboxyl terminus of the coat protein tended to produce more coat protein and induced yellowing symptoms. Two of the mutants induced necrosis. These results suggest that the TMV coat protein gene has a multi functional role during infection.

Ultrastructure of altered chloroplasts

We have examined the effects of the coat protein deletion mutants on mature chloroplasts. A survey of tissue from chlorotic and necrotic areas on inoculated leaves indicated that the coat protein mutants cause significant alterations in chloroplast structure in these areas. Chloroplasts from wild-type infected leaves were ultrastructurally similar to chloroplasts from healthy leaves.

Chloroplasts from chlorotic areas of leaves inoculated with mutants which cause mild to strong yellowing (cp 4 and cp 35-5) contained vesicle-like structures (fig. 2). In some chloroplasts these structures appeared to be transverse sections of tubules extending through the stroma of the chloroplasts. Stereo-pair electron micrographs of affected tissue confirmed that these structures are tubules. The tubules appear to be inter-connected. Similar structures have also been observed in chloroplasts in tissue from the thin chlorotic band that surrounds the necrotic lesion caused by cp 10 and in the chlorotic area that precedes the necrotic lesion caused by cp 27. In a number of the plastids examined the tubules appeared to run in a circular fashion at 90 degrees to the longitudinal axis of the plastids. In most of the plastids examined structurally normal photosynthetic membranes were also present.

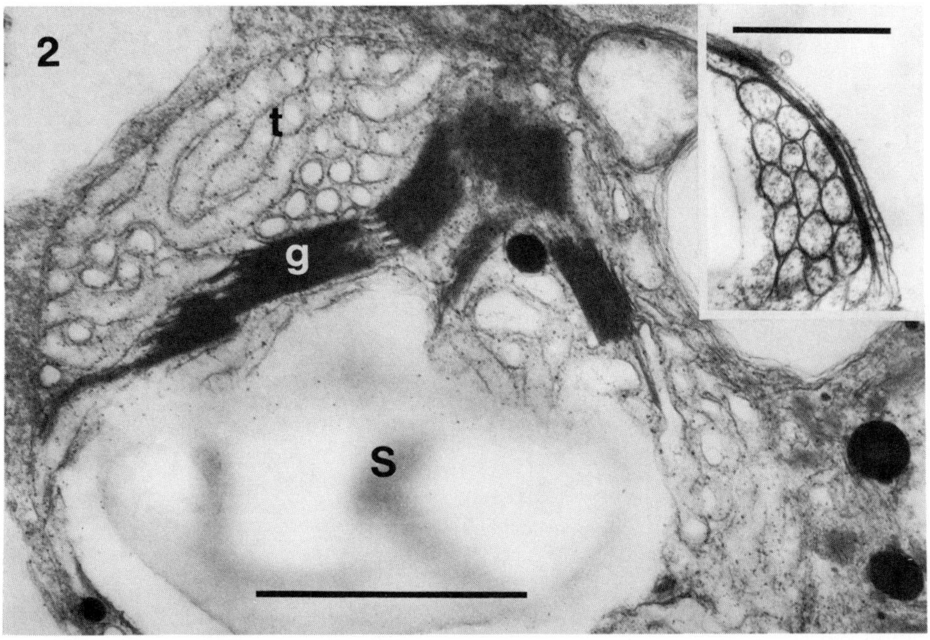


Figure 2. Chloroplast from the center of a cp 35-5-induced chlorotic area. Note that an extensive tubule network (t) as well as normal photosynthetic membranes occur in the same chloroplast. (g = granum; S = starch granule; bar = 1 um). Inset: Section of a chloroplast from the center of a cp 4-induced chlorotic area showing the circular organization of the photosynthetic membranes observed 4 days post-inoculation. (bar = 0.5 um).

The tubules appear to be derived from rearrangements of the photosynthetic membranes in affected chloroplasts. Chloroplasts from tissue inoculated with cp 4 contain grana which appear to be unstacking, forming convex membrane structures, four days post-inoculation. In some chloroplasts the photosynthetic membranes appear to be completely unstacked and arranged in circular structures (inset fig. 2). Seven days after inoculation tubules had formed in these chloroplasts.

Chloroplasts from chlorotic areas of leaves inoculated with mutants that cause weak yellowing (cp +2 and cp 28) appear to be structurally normal. Chloroplasts from necrotic tissue from leaves inoculated with the mutants that cause necrosis (cp 10 and cp 27) are degraded to the point where no internal structure is visible. Chloroplasts from tissue outside the necrotic areas appear to be structurally normal.

Localization of mutant coat proteins

We have attempted to localize mutant coat protein in inoculated leaves infected with coat protein deletion mutants. We were particularly interested in whether the coat protein of the mutants that degrade chloroplasts enters the chloroplasts.

Coat protein from wild type TMV was localized to virus inclusions in the cytoplasm and vacuoles of infected cells. Little, if any, coat protein was found in the chloroplasts even when the chloroplasts were adjacent to large virus inclusions (fig. 3). These results are consistent with those reported previously (Hills *et al.*, 1987).

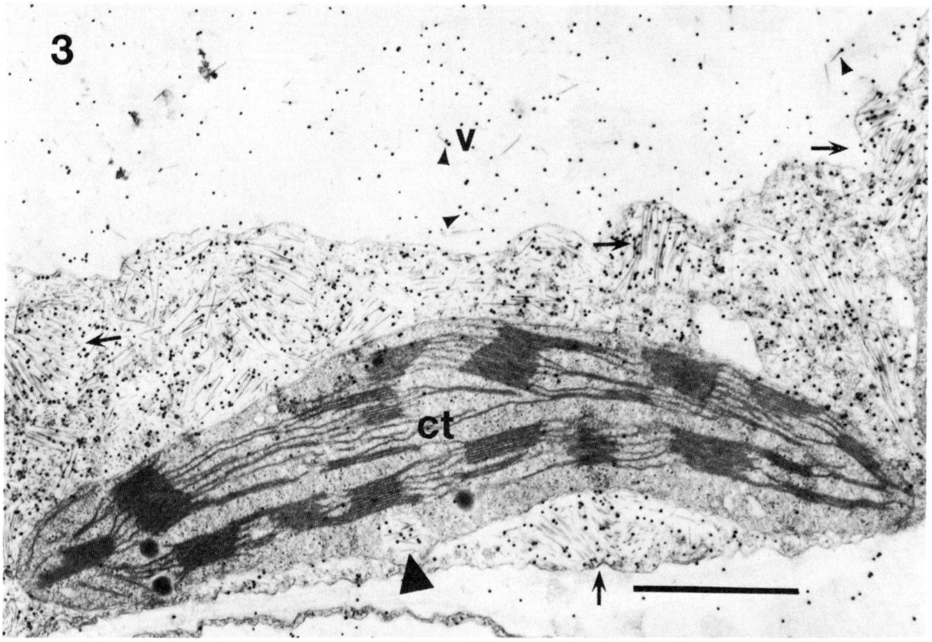


Figure 3. Anti-coat protein antibody labeling of tissue from a leaf infected with the wildtype strain TMV 204. The bulk of the label is associated with virus particles in the cytoplasm ( → ). Some of the label is associated with individual virions (►) in the vacuole (v). Very little label can be found in the chloroplast (ct). A virus inclusion can be seen in the chloroplast with label associated with the virions ( ▶ ). (bar = 1 um).

Cells from chlorotic tissue from expanded leaves inoculated with coat protein mutants that cause yellowing on directly inoculated leaves contained dark staining bodies in the cytoplasm (fig. 4). The mutant coat protein was localized to these structures. These 'coat protein bodies' were often associated with X-material in the cytoplasm of infected cells. In some mutants, where X-material did not appear to be present, the coat protein bodies were in the cytoplasm of the cell.

Gold particles were also observed in the vacuoles of infected cells. In the case of wildtype infected cells this label was often associated with single virions in the vacuole (see fig. 3). However, some of the label in the vacuoles of these cells and all the label in the vacuoles of cells infected with the coat protein mutants (which cannot form virions) is presumably associated with free coat protein.

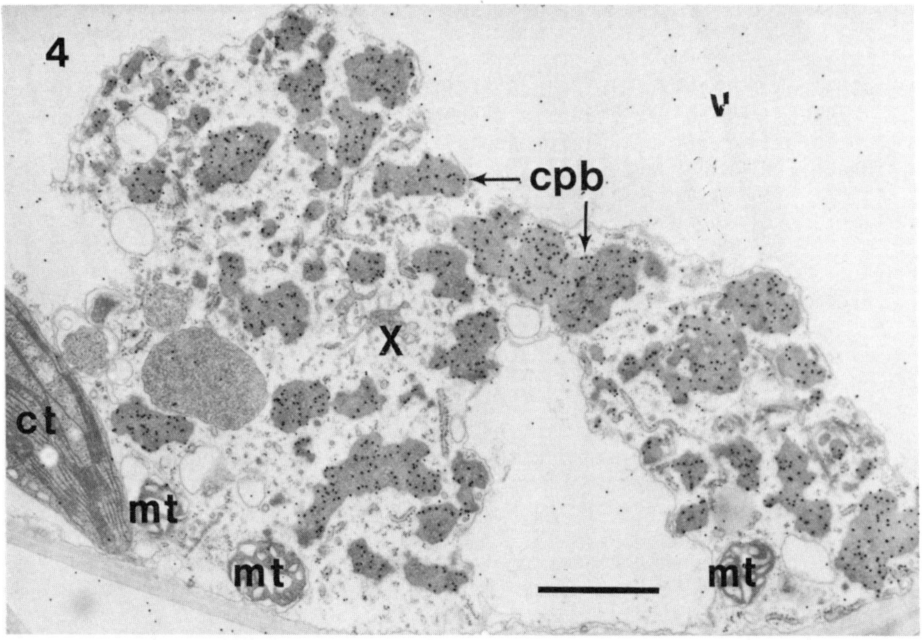


Figure 4. Anti-coat protein antibody labeling of tissue from a leaf inoculated with the coat protein mutant cp 10. The gold label is associated with dark staining bodies in the X-material (X). We have termed these structures coat protein bodies (cpb). Virtually no label occurs outside these structures. (ct = chloroplast; mt = mitochondrion; bar = 1 um).

These results suggest that coat protein does not directly cause the degradation of chloroplasts by entering the chloroplasts. Coat protein probably subverts the chloroplast indirectly. A number of important chloroplast proteins including enzymes involved in chlorophyll synthesis, components of both photosystems and the ATP synthetase, and the regulatory subunit of ribulose-1,5-bisphosphate carboxylase are synthesized in the cytoplasm and transported into the chloroplast (see Dyer, 1984 for review). It is possible that TMV exerts its influence by interfering with the synthesis and transport of some or all of these proteins.

<u>Degradation of developing chloroplasts</u>

One of our observations suggests that the coat protein may not be involved in the induction of mosaic symptoms in developing leaves as has been suggested (Beachy *et al*., 1987; Grumet *et al*., 1987; Sherwood, 1987). Mutants unable to produce coat protein because of an alteration to the translational start codon of the coat protein gene, TMV [-25] and TMV [-CP], were found to induce the development of mosaic symptoms in *N. sylvestris* (Culver and Dawson, 1989b). Although these mutants moved systemically much slower than wild type TMV, when they did move into the apical area the mosaics they produced were identical to the mosaics produced by wild type TMV. Since these mutants did not produce any coat protein, some other viral component must initiate the development of mosaic symptoms in *N. sylvestris*.

## REFERENCES


Aldaoud, R., 1987, Biological characteristics, variations, and factors affecting variation in cloned tobacco mosaic virus, Ph.D. Thesis, University of California, Riverside.

Beachy, R. N., Powell-Abel, P., Nelson, R. S., Rogers, S. G., and Fraley, R. T., 1987, Transgenic plants that express the coat protein gene of TMV are resistant to infection by TMV, in: "Molecular strategies for crop protection", C. J. Arntzen and C. Ryan, eds, A.R. Liss, New York. pp 205-213.

Bloomer, A. C., Champness, J. N., Bricogne, G., Staden, R., and Klug, A., 1978, Protein disks of tobacco mosaic virus at 2.8 A resolution showing the interactions within and between subunits, Nature, 276:362-368.

Camerino, G., Savi, A., Cifferi, A.O., 1982, A chloroplast system capable of translating heterologous mRNAs, FEBS Lett., 150:94-98.

Culver, J. N. and Dawson, W. O., 1989a, Point mutations in the coat protein of tobacco mosaic virus induce hypersensitivity in *Nicotiana sylvestris*, Mol. Plant Microbe Interact., 2:209-213

Culver, J. N. and Dawson, W. O., 1989b, Tobacco mosaic virus coat protein: an elicitor of the hypersensitive reaction but not required for the development of mosaic symptoms in *Nicotiana sylvestris*, Virology, 173:755-758.

Dawson, W. O., Beck, D. L., Knorr, D. A., and Grantham, G. L., 1986, cDNA cloning of the complete genome of tobacco mosaic virus and the production of infectious transcripts, Proc. Natl. Acad. Sci. USA, 83:1832-1836.

Dawson, W. O., Bubrick, P., and Grantham, G. L., 1988, Modifications of the tobacco mosaic virus coat protein gene affect replication, movement, and symptomatology, Phytopathology, 78:783-789.

Dietzgen, R.G., Zaitlin, M., 1986, Tobacco mosaic virus coat protein and the large subunit of ribulose - 1,5 - bisphosphate carboxylase share a common antigenic determinant, Virology, 155:262-266.

Dunigan, D.D., Dietzgen, R.G., Schoelz, J.E., Zaitlin, M., 1988, Tobacco mosaic virus particles contain ubiquitinated coat protein subunits, Virology, 165:310-312.

Dyer, T.A. 1984, The chloroplast genome: Its nature and role in development, in:" Chloroplast Biogenesis" (N.R. Baker, J. Barber, eds), Elsevier Publishers BV, Amsterdam pp23-69.

Flor, H. H., 1971, Current status of the gene for gene concept, Ann. Rev. Phytopathol., 9:275-296.

Fraser, R. S. S., 1987, "Biochemistry of virus-infected plants", Research Studies Press LTD, Letchworth, England, pp1-7.

Fraser, R.S.S., Gerwitz, A., 1980, Tobacco mosaic virus infection does not alter the polyadenylated messenger RNA content of tobacco leaves, J. Gen. Virol., 46:139-149.

Glover, J.F., Wilson, T.M.A., 1982, Efficient translation of the coat protein cistron of tobacco mosaic virus in a cell - free system from *Escherichia coli*, Eur. J. Biochem., 122:485-492.

Grumet, R., Sanford, J.C., Johnston, S.A., 1987, Pathogen - derived resistance to viral infection using a negative regulatory molecule, Virology, 161:561-569.

Goelet, P., Lomonossoff, G. P., Butler, P. J. G., Akam, M. E., Gait, M. J., and Karn, J., 1982, Nucleotide sequence of tobacco mosaic virus RNA, Proc. Natl. Acad. Sci. USA, 79:5818-5822.

Hills, G.J., Plaskitt, K.A., Young, N.D., Dunigan, D.D., Watts, J.W., Wilson, T.M.A., Zaitlin, M., 1987, Immunogold localization of the intracellular sites of structural and nonstructural tobacco mosaic virus proteins, Virology, 161:488-496.

Hirai, A., Wildman, S.G., 1969, Effect of TMV multiplication on RNA and protein synthesis in tobacco chloroplasts, Virology, 38:73-82.

Hodgson, R.A.J., Beachy, R.N., Pakrasi, H.B., 1989, Selective inhibition of photosystem II in spinach by tobacco mosaic virus: an effect of the viral coat protein, FEBS Lett., 245:267-270.

Keegstra, K., Olsen, L.J., Theg, S.M., 1989, Chloroplastic precursors and their transport across the envelope membranes, Ann. Rev. Plant Physiol. Plant Mol. Biol., 40:471-501.

Keen, N. T. and Staskawicz, B., 1988, Host range determinants in plant pathogens and symbionts, Ann. Rev. Microbiol., 42:421-440.

Knorr, D. A. and Dawson, W. O., 1988, A point mutation in the tobacco mosaic virus capsid protein gene induces hypersensitivity in *Nicotiana sylvestris*, Proc. Natl. Acad. Sci. USA, 85:170-174.

Mohamed, N.A., Randles, J.W., 1972, Effect of tomato spotted wilt virus on ribosomes, ribonucleic acid and Fraction 1 protein *Nicotiana tabacum* leaves, Physiol. Plant Pathol., 2:235-245.

Reinero, A., Beachy, R.N., 1986, Association of TMV coat protein with chloroplast membranes in virus - infected leaves, Plant Mol. Biol., 6:291-301.

Reinero, A., Beachy, R.N., 1989, Reduced photosystem II activity and accumulation of viral coat protein in chloroplasts of leaves infected with tobacco mosaic virus, Plant Physiol., 89:111-116.

Rochon, D., Siegel, A., 1984, Chloroplast DNA transcripts are encapsidated by tobacco mosaic virus coat protein, Proc. Natl. Acad. Sci. USA, 81:1719-1723.

Rogers, S. G., Klee, H. J., Horsch, R. B., and Fraley, R. T., 1987, Improved vectors for plant transformation: expression cassette vectors and new selectable markers, Methods Enz., 153:253-277.

Saito, T., Yamanaka, K., Watanabe, Y., Takamatsu, N., Meshi, T., and Okada, Y., 1989, Mutational analysis of the coat protein gene of tobacco mosaic virus in relation to hypersensitive response in tobacco plants with the N' gene, Virology, 173:11-20.

Schoelz, J.E, Zaitlin, M., 1989, Tobacco mosaic virus RNA enters chloroplasts *in vivo*, Proc. Natl. Acad. Sci. USA, 86:4496-4500.

Sela, I., Kaesberg, P., 1969, Cell - free synthesis of tobacco mosaic virus coat protein and its combination with ribonucleic acid to yield tobacco mosaic virus, J. Virol., 3:89-91.

Shalla, T.A., Petersen, L.J., Guichedi, L., 1975, Partial characterization of virus - like particles in chloroplasts infected with the U5 strain of TMV, Virology, 66:94-105.

Sherwood, J.L., 1987, Demonstration of the specific involvement of coat protein in tobacco mosaic virus (TMV) cross protection using a TMV coat protein mutant, J. Phytopathol., 118:358-362.

# CELL-TO-CELL MOVEMENT OF PLANT VIRUSES

T. Godefroy-Colburn, C. Erny, F. Schoumacher,
A. Berna, M.-J. Gagey and C. Stussi-Garaud

Institut de Biologie Moléculaire des Plantes
du Centre National de la Recherche Scientifique
12, rue de Général Zimmer
67084 Strasbourg, France

## INTRODUCTION

Cell-to-cell movement of plant viruses has attracted a great deal of attention in the last few years. Since the subject has been reviewed extensively (Atabekov and Dorokhov, 1984; Zaitlin and Hull, 1987; Hull, 1989; Godefroy-Colburn et al., 1990; Atabekov and Taliansky, submitted to Adv. Virus Res.), we will focus on newer data and systematically omit the references which can be found in Hull's review.

Plant viruses are usually inoculated through wounds caused by an animal or fungal vector, or by abrasion of the epidermis. Thus a few cells become infected and the virus replicates there. In order to have any macroscopic effect on the plant, however, the infection has to spread from these primarily-infected cells to their healthy neighbours. This is what we refer to as cell-to-cell movement, cell-to-cell spread or short-distance transport. If the infection reaches the phloem, the virus may be taken up by the sap and transported throughout the plant, mainly to the apex and the roots. This second type of movement, which occurs only in systemic infections, is referred to as long-distance transport.

To move over short distances, viruses make use of the natural connections between cells, the plasmodesmata. The problem here is one of size. Plasmodesmata are the plant equivalent of the animal gap junctions. In spite of their different shapes and structures, both gap junctions and plasmodesmata are permeable to solutes but exclude large molecules, and both have a molecular exclusion limit in the order of 1000 Da, as measured by microinjection of calibrated fluorescent probes. This figure is to be compared with the molecular weight of an average viral RNA, 1000 kDa. Obviously, plant viruses must drastically increase the exclusion limit of

*Plant Molecular Biology 2*, Edited by R.G. Herrmann and
B. Larkins, Plenum Press, New York, 1991

plasmodesmata if they are to use them for their movement. This often goes with spectacular ultrastructural changes (Gibbs, 1976; Francki et al., 1985). On the other hand, some viruses such as tobacco mosaic virus (TMV) have never been reported to affect the ultrastructure of plasmodesmata.

It has been recognized over the last decade that virus movement, like the other viral functions, is dependent on one or several viral gene products. These are the so-called movement (or transport, or spread) proteins. Many viruses (for example TMV) can spread from cell to cell as unencapsidated RNA. In other cases, e. g. cowpea mosaic virus (CPMV; Wellink and van Kammen, 1989), the coat protein is required. In contrast, all viruses tested so far require the coat protein for long distance transport (in addition to the movement protein which is required for movement in and out of the phloem), except bipartite gemini-viruses. The current working hypothesis is that the movement proteins make plasmodesmata competent for the transit of infectious particles. These proteins may be responsible for the spectacular modifications which were mentioned earlier, or may cause more subtle ones such as in TMV infections. A distinction must be made between movements through the parenchymal tissue and through the conductive tissue. Some viruses (the luteo and some geminiviruses) are strictly phloem-limited. It has been proposed that they do not have a movement protein or that their movement protein does not allow them to cross the barrier between phloem and mesophyll (Hull, 1989; Harrison et al., 1990; Atabekov and Taliansky, submitted).

We will try to summarize recent findings concerning the identification of movement proteins, their subcellular localization and their mechanism of action.

## IDENTIFICATION AND LOCALIZATION OF THE MOVEMENT PROTEINS

### Genetic Evidence

Few movement proteins have been identified thus far on the basis of mutations: those of TMV, of CPMV and of tobacco rattle virus (TRV). Mutational data also exist for geminiviruses, both bipartite (cassava latent and tomato golden mosaic) and monopartite (beet curly top: Briddon et al., 1989; maize streak: Lazarowitz et al., 1989; Boulton et al., 1989), but the lack of distinction between short- and long-distance movement makes them difficult to interpret. They will not be discussed here.

Identification of the TMV movement protein rests mainly on the isolation of a mutant of the L strain, Ls1, which is temperature-sensitive for short-distance spread. This mutant gives a subliminal infection in tobacco at high temperature, i.e. the virus replicates in the primarily-infected cells but does not infect the surrounding cells. Ls1 carries a non-conservative point mutation in the cistron coding for the 30K protein and multiplies normally in transgenic plants expressing wild-type 30K. Conversely the same mutation, introduced into the wild-type strain by reverse genetics, renders the virus temperature-sensitive for cell-to-cell movement. This proves beyond doubt that the 30K is the movement protein of TMV. Data on CPMV and TRV are based on the construction of deletion and frameshift mutants. Thus CPMV needs the 48K/58K protein and the coat protein (both encoded by RNA-M) for cell-to-cell spread (Wellink and van Kammen, 1989), whereas TRV (strain SYM) needs the 29K protein encoded by RNA-1 Ziegler-Graff et al., submitted) but has been known for decades not to require the capsid protein.

By extension, the movement proteins of other tobamo, tobra and comoviruses are believed to be viral gene products homologous to the 30K, 29K and 48/58K respectively.

For monopartite viruses of other families, the arguments leading to identification of the movement protein(s) are mainly based on amino-acid sequence homology and on subcellular localization (see further below). In the case of multipartite viruses, it often happens that one of the genome segments is not required for replication in protoplasts but only in whole plants. This segment is thus identified as carrying the movement protein gene. For example, RNA-2 of red clover necrotic mosaic dianthovirus (RCNMV) and RNA-3 of the tricornaviruses (alfalfa mosaic, AlMV; brome mosaic, BMV; cucumber mosaic virus, CMV) have this characteristic. RNA-2 of RCNMV has only one cistron which codes for a 35K non-structural protein but RNA-3 of tricornaviruses codes for two proteins, a 30-32K non-structural protein (P3 or P3a) and the coat protein. Identification of P3/P3a as a movement protein rests on subcellular localization and on limited sequence homology between P3a of BMV and the 35K of RCNMV.

## Sequence Homology

The tobamo and tobravirus movement proteins have significant primary sequence homologies between each other and so have the P3 and P3a proteins of tricornaviruses (CMV, BMV, AlMV and tobacco streak ilarvirus), in addition to the already-mentioned homology to the 35K of RCNMV. However, there is no obvious similarity between these two groups. Localized homologies have been found between the movement protein of TMV and the gene 1 product of caulimoviruses (P1). This, combined with subcellular location and with data showing that gene 1 is involved in symptom spread (Stratford and Covey, 1989), identifies P1 as a movement protein. Similarly, there is some homology between the 48K/58K protein of CPMV and a domain of the polyprotein encoded by RNA-2 of tomato black ring nepovirus (TBRV). The significance of this finding is strengthened by two facts: (i) the homologous domains are encoded by equivalent parts of the genome and (ii) the TBRV domain also has some homology with the 30K of TMV.

Occasional homologies have been reported between the TMV 30K protein and non-structural proteins of poty, potex, geminiviruses, etc. Their significance is uncertain, given the different genetic organizations of the viruses.

In short, unlike core RNA polymerases, movement proteins cannot be identified on the basis of amino-acid sequence only. There is, apparently, no such thing as a "movement protein consensus sequence".

## Subcellular Localization

A few movement proteins have been detected in plants with the help of specific antisera, raised against chemically-synthesized peptides or against proteins obtained by recombinant DNA techniques. Their subcellular location was studied in some detail to discriminate between two possible mechanisms of action. The movement proteins could either activate (presumptive) cellular genes responsible for opening the plasmodesmata or act directly on the plasmodesmata. In the former case they should associate with the nucleus and in the latter case with the plasmodesmata or cell walls. The consensus to date is for the second hypothesis. Indeed the movement proteins of five viruses have been detected in the cell-wall/plasmodesmatal compartment of their host plant, both by subcellular fractionation and by immmuno-electron microscopy. A slight doubt remains however, because the movement protein of CMV was observed in the nucleoli of infected plants by immuno-electron microscopy (without confirmation by subcellular fractionation).

The movement proteins of AlMV and TMV were the first ones to be detected in virus-infected plants. Their kinetics of appearance in mechanically-inoculated leaves (non-synchronous, systemic infection) were found to be very similar. These proteins were detected in the cytoplasm early after inoculation but after a peak at 2 or 3 days (at optimum temperature) they disappeared from this compartment. In parallel they accumulated in a cell-wall fraction (washed free of cytoplasm with a buffer containing Triton X100) where they resided for several days after the cytoplasmic pool had been exhausted. A recent study (Lehto et al., 1990a) refined these data by using the "differential-temperature inoculation" method to achieve nearly-synchronous infection of tobacco leaves by TMV. As expected, accumulation of the 30K was faster than in the non-synchronous situation. Most of the 30K was synthesized between 12 and 24 h post-inoculation (p.i.) whereas replicase and coat proteins were synthesized until at least 72 h p.i. It is thus clear that the synthesis of 30K shuts down long before the bulk of viral particles is made. Although synchronicity of infection was not achieved in whole plants with AlMV, the kinetics of P3 synthesis in protoplasts are consistent with P3 being an early viral protein. The implication is that cells infected with TMV or AlMV should be competent for virus movement at an early stage of their infection cycle. Indeed, not only is the transport of TMV to neighbouring cells an early viral function but it is impaired if the 30K is forced to be synthesized at the same time as the replicase and coat proteins (Lehto et al., 1990b).

That cell-to-cell movement is an early function in TMV and AlMV infections is consistent with immunogold electron microscopic observations. For both viruses, the movement protein was observed in areas of the tissue (tobacco leaf) which had just become infected and it was not observed in cells which had accumulated large amounts of virus. The 30K was found in plasmodesmata, as expected from its function, and P3 was seen in the middle lamella and occasionally in plasmodesmata. We do not know whether this slight difference is significant or not. Another difference between TMV- and AlMV-infected tissue is that, unlike TMV, AlMV seems to induce occasional tubular structures extending from plasmodesmata into the cytoplasm. We observed these tubules at the front of infection only. They had about the same diameter as that induced by cucumoviruses (e.g. tomato aspermy virus; Francki et al., 1985) but were much shorter and did not contain recognizable viral particles, although they were sometimes labelled with an anti-AlMV serum (Godefroy-Colburn et al., 1990).

Three more movement proteins have been detected in whole plants: those of a caulimovirus, CaMV, and of two comoviruses, CPMV and red clover mosaic virus (RCMV). Unlike AlMV and TMV, these viruses induce conspicuous modifications of the cell wall which persist throughout infection.

The movement proteins of CaMV (P1) and CPMV (48/58K protein) were detected in extracts of infected plants but their exact location was difficult to determine by subcellular fractionation. P1 was detected in several fractions (replication complex, inclusion bodies, cell walls) where it appeared to be differentially modified. A recent study (Maule et al., 1989) showed the amount of P1 to be maximum in young systemically-infected leaves which were just developing vein-clearing symptoms, and to decrease as the leaves were getting older and accumulated virus. This is quite consistent with the data on TMV and AlMV movement proteins, although the kinetics were much slower in the case of CaMV.

Immuno-electron microscopy confirmed the presence of P1 in cell walls, in the immediate vicinity of tubular structures crossing the wall between infected mesophyll cells and extending into the cytoplasm. By the same method, the 48/58K of CPMV was recently observed in bundle-sheath tissue, at the surface of virion-containing tubules which crossed the cell wall or were entirely included in the cytoplasm (Goldbach et al., 1990; van Lent et al., 1990). The movement protein was also seen in small electron-dense areas close to the end of the tubules. It is tempting to speculate that these areas are the site of tubule formation. Similar observations were made by Shanks et al. (1989) on RCMV-infected tissue. They detected the "43K" viral protein (equivalent to the 48K/58K of CPMV) in modified plasmodesmata and associated tubules. On the whole, electron microscopic observations of the movement protein seem more frequent (given a certain area of tissue section) for the caulimo and comoviruses than for either TMV or AlMV. This may indicate that P1 and the 48K/58K are more stable, or less transiently expressed, than the other two. If the movement proteins were indeed responsible for the ultrastructural modification of plasmodesmata, the absence or the transient nature of the modifications in TMV or AlMV infections would be explained by the rapid turnover of the movement protein in plasmodesmata.

## MECHANISM(S) OF VIRUS MOVEMENT

Many hypotheses have been made to explain how the movement proteins function and we are not going to discuss them all. Fortunately, several breakthrough reports were published in the last six months, which enable us to sort out some of them.

## Several Mechanisms of Virus Movement?

One of the characteristics of virus movement, which was among the arguments leading to the concept of a "viral transport function", is that a given virus can in some cases act as a helper for the deficient movement of an unrelated virus in a given host. For example, potato virus X (PVX) complements the deficient movement function of TMV mutant Ls1 in tomato. In the field, several cases of synergy between viruses may be explained by transport complementation. This tends to indicate that movement proteins are interchangeable between viruses and partly determine their host range (Atabekov and Dorokhov, 1984). This also implies that viruses from different taxonomic groups can be transported between cells by the same mechanism.

However, recent surveys of movement complementation (Malyshenko et al., 1989; Harrison et al. 1990; Atabekov and Taliansky, submitted) show that the story is not so simple. All the viruses are not equally able to complement the movement of other viruses. For instance, tobamoviruses and PVX can complement a number of other viruses but AlMV has only been reported to complement BMV in Vigna unguiculata (Malyshenko et al., 1989). This is particularly evident when one compares the ability of different sap-transmitted viruses to potentiate the movement of luteoviruses (normally phloem-limited) into non-phloem tissues (Barker, 1989; Harrison et al., 1990). Luteovirus movement is enhanced by tobamo, tobra, potex, poty and umbra (carrot mottle) viruses and is not enhanced by nepo, cucumo, parsnip yellow fleck or alfalfa mosaic viruses. Several viruses of the first group have been proven not to require their capsid protein for cell-to-cell spread and none has been reported to induce tubules containing virus particles. In contrast, several viruses of the second group induce such tubules and CPMV, a close cousin of nepoviruses, has been shown to require its coat protein to move between cells. The suggestion was made, therefore, that cell-to-cell movement could occur by at least two mechanisms: type A (Harrison et al., 1990), also called tobamovirus transport mechanism (Goldbach et al., 1990; van Lent et al., 1990), in which the entity being transported is the viral RNA (double-stranded RNA or RNA-protein complex) and type B, or comovirus transport mechanism, in which the virion itself moves through the cell wall via specialized structures. The tricorna and dianthoviruses have some of the characteristics of type B viruses (lack of complementation of luteoviruses and lack of free RNA mutants). However, with the exception of some cucumoviruses, they do not induce stable modifications of cell walls or plasmodesmata.

Type A mechanism is rather non-specific with respect to the virus being transported. Indeed, the movement proteins of type A viruses are able to complement the transport function of many unrelated (including type B) viruses (Malyshenko et al., 1989; Atabekov and Taliansky, submitted). Type B (or non-A) movement proteins on the contrary seem to be more specific, as would be expected if they recognized viral particles. Atabekov and Taliansky report few examples of complementation by viruses of this type: BMV complements TMV and PVX, CMV complements RCMV, arabis mosaic nepovirus (ArMV) complements TMV and RCMV, RCMV complements TMV and AlMV complements BMV at a low efficiency. The question then arises whether the coat protein of the helper plays a role in the complementation. We speculate that complementation by a coat-requiring virus should only be observed in two cases: (i) the two viruses are closely related and the helper movement protein is able to recognize the virions of the deficient virus; (ii) the coat protein of the helper is able to associate with the genome of the deficient virus into a complex (not necessarily a full capsid) that the movement protein can recognize. Indeed transcapsidation is well documented in a few cases and, in many more instances, the level of transcapsidation, although undetectable by physical methods, may be sufficient to promote the movement of a few molecules of the deficient genome.

## Functional Domains of the Movement Proteins: Some Speculations

The previous argument leads to the conclusion that movement proteins must recognize either the viral genome (type A) or a viral nucleoprotein containing the coat protein (type B). From subcellular localization data, we can also infer that they have a binding site for a plasmodesmatal component. Recognition of this receptor seems to require rather precise adjustment between the movement protein and the host component. This follows from two lines of evidence: (i) the Ls1 mutation, which affects the transport of TMV in tobacco and tomato, has no effect in *Amaranthus caudatus* (Mushegian *et al.*, 1989); (ii) in tomatoes carrying the Tm2 gene, resistance to TMV can be broken by mutations in specific regions of the 30K cistron (Meshi *et al.*, 1989).

How viral cell-to-cell movement functions exactly is a mystery, but common sense tells us that the transit of a particle through a tube having approximately the same diameter as the particle is not a simple diffusion process. This must require energy, i.e. ATP or GTP hydrolysis. By analogy with the movement of mRNPs through nuclear pores, Atabekov and Taliansky (submitted) propose that the movement protein might be a protein kinase (or might activate a protein kinase) regulating a cellular NTPase coupled to an energy-requiring process. It has indeed been observed that dibutyryl-cAMP partially restores the movement of Ls1 in tobacco leaves at non-permissive temperature (Mushegyan *et al.*, 1986). Moreover, the movement protein of CaMV has some homology to known kinases.

Thus we assume that movement proteins have at least two functional domains, binding respectively to a viral component and to a cellular receptor, and that they may also have a catalytic site.

The functional study of the TMV (type A) and AlMV (type B?) movement proteins has just started. Because these proteins are found at very low levels in infected plants, no attempt was made, to our knowledge, at purifying them from that source in large quantity. Instead, the movement protein genes were expressed in recombinant microorganisms. In addition to convenient purification techniques, recombinant DNA methods offer the possibility of site-directed mutagenesis. *In vitro* studies done with the mutant protein can then be compared with *in vivo* studies done with transgenic plants expressing the same protein.

## Functional Studies on the 30K of TMV

One of the most exciting breakthroughs in the field of virus movement was the report (Wolf *et al*., 1989) that plasmodesmata of transgenic plants expressing the 30K have a molecular exclusion limit of about 10000 Da, as compared with less than 1000 Da for normal plasmodesmata. These results prove that the 30K does interact with plasmodesmata in a functional way but the increase in porosity is still not sufficient to allow the transit of the infectious particles, assuming they were random-coiled viral RNA.

Another breakthrough report (Citovsky *et al*., 1990) may contribute to resolve this difficulty. It shows that the 30K (synthesized in recombinant *Escherichia coli* and purified) binds single-stranded nucleic acids, both DNA and RNA, cooperatively. The non-specificity of this binding explains quite well the results of the complementation studies. An interesting model was proposed by the authors whereby the transport form of TMV is a single-stranded RNA molecule coated with the 30K. Instead of assuming the rod-like shape of viral particles, this "transport RNP" could be very thin (the authors quote a diameter of less than 2 nm by analogy with the complexes of *Agrobacterium* VirE2 protein with ssDNA) and thus could pass through slightly enlarged plasmodesmata. Let us now push the argument further. If the RNA helix of TMV virions were completely extended, we calculate that its lengh would be several microns. This is one order of magnitude higher than the average thickness of mesophyll cell walls. It is likely that, *in vivo*, the viral RNA does not interact with free 30K but with a plasmodesma-associated protein. Thus the RNA molecules would be stretched only locally and would be pulled through the plasmodesma by a sort of scanning mechanism, reminiscent of mRNA translation by membrane-bound ribosomes.

## Mapping of the Functional Domains of the 30K

Sequence homology between the movement proteins of several tobamo and tobraviruses points to two evolutionary-conserved regions denoted I (aa 56 to 96 of the TMV 30K sequence) and II (aa 125 to 165) by Saito *et al*. (1988). We can now begin to assign functions to these regions.

By site-directed mutagenesis (Citovsky *et al*., 1990), the nucleic acid binding site of the 30K was mapped to a small domain (aa 65 to 86) which corresponds to the center of region I. If other type A movement proteins function like the 30K, they should have a similar site. Indeed, the N-terminal polypeptide (28K) of tobacco vein mottling potyvirus contains a domain with significant homology to region I (Hull, 1989).

In contrast to RNA binding, interaction of the 30K with its host receptor(s) seems to involve several regions of the molecule:

1. Meshi *et al*. (1989) have sequenced several mutants of TMV which overcome the Tm2 resistance gene and found mutations both in region II (aa 133) and in, or close to, the N-proximal portion of region I (aa 52 and 68).

2. The Ls1 and Ni2519 mutations of strain L, which confer temperature sensitivity to the movement function in some hosts, involve amino-acid changes at positions 154 and 144 respectively (region II).

3. In a recent study (A. Berna, R. N. Beachy and others, unpublished) the 30K gene (coding for 268 aa) was sequentially deleted from its 3' extremity and several constructions were selected, which coded for C-terminally deleted proteins with lengths of 258, 233, 213, 195 and 174 aa respectively. The constructions were used to transform tobacco and the resulting transgenic plants were studied from two points of view, ability to complement Ls1 and subcellular location of the deleted movement protein. It was found that the first three deletions affected neither the cell-wall location nor the function of the movement protein. The 195-aa protein, on the contrary, was found in the cytoplasm only and did not complement Ls1. The shortest protein seemed to be rapidly degraded because it could not be detected in plants.

These results indicate that the C-terminal portion of the 30K protein (aa 213 to 268) is not required for function, in tobacco at least. In contrast, the 195-212 portion includes some amino-acids which are necessary both for targeting the movement protein to its (plasmodesmatal?) receptor and for activity. Sequence comparison results (Saito *et al.*, 1988) support these findings. The C-terminal portion of the 30K is indeed the most variable part of the molecule but there is a stretch of 6 well-conserved amino-acids at positions 192-197, which belong to an acidic region denoted A (aa 182 to 200). These amino-acids seem important and may turn out as a real consensus. When examined by the standard methods of conformational analysis, region A appears to be an amphipathic alpha-helix, a good candidate for a hydrophobic binding site.

## Functional Studies on the AlMV Movement Protein

In comparison with the amount of data on the function of the 30K, information on the movement proteins of other viruses is rather sketchy. We have just started to apply the genetic engineering strategy explained above to P3 of AlMV and can only report preliminary results.

To investigate the function of P3 *in vitro*, we prepared the protein from recombinant *Saccharomyces cerevisiae*. The yeast was transformed with three different constructions, containing respectively the complete P3 cistron (coding for 300 aa) and two 5'-deleted cistrons (Godefroy-Colburn *et al.*, 1990). The ORFs of the deleted clones (designated 1 and 2) start respectively at the second and sixth AUG of the P3 ORF, hence code for proteins of 288 and 223 aa. The major proteins synthesized by yeast from the complete clone, from clone 1 and from clone 2 are called respectively P3-Y, P3-$Y_1$ and P3-$Y_2$. By Western blot, P3-Y was shown to have the same electrophoretic mobility as the movement protein made by AlMV-infected tobacco plants (Godefroy-Colburn *et al.*, 1990); P3-$Y_1$ and P3-$Y_2$ had the mobility expected from their theoretical molecular weight (Fig. 1A, lanes $Y_1$ and $Y_2$; compare with the AlMV-infected extract, $T_A$).

P3-Y and P3-$Y_1$ were purified to near-homogeneity as described (Godefroy-Colburn et al., 1990) and are now used for binding assays to the different viral components. By a method similar to that of Citovsky et al., the complete protein was found to bind viral RNA (F. Schoumacher, unpublished results). Further characterization of the binding is in progress.

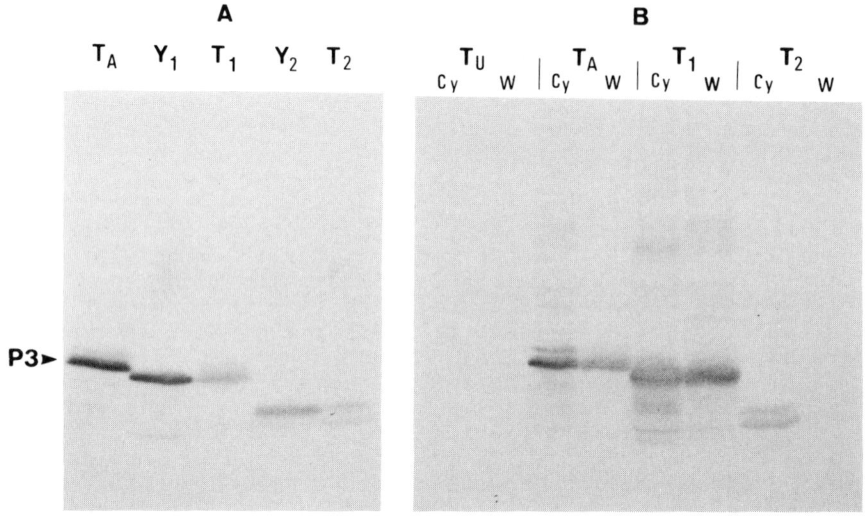


Figure 1. Expression of two N-terminally deleted mutants of P3 in yeast and in transgenic tobacco. (A) Western blot analysis of extracts from: AlMV-infected tobacco ($T_A$), yeast ($Y_1$ and $Y_2$) and transgenic tobacco ($T_1$ and $T_2$) expressing respectively the deleted genes 1 (coding for 288 aa) and 2 (223 aa). (B) Analysis of subcellular fractions (cy, cytoplasmic fraction; w, cell-wall fraction) from the same plants and from uninfected, untransformed tobacco ($T_U$); each lane of panel B contains an amount of material corresponding to 7 mg of leaf tissue. (Our unpublished results)

In parallel, we transformed tobacco (Nicotiana tabacum, cv. Xanthi and Xanthi nc) by Ti-agroinfection, with the complete P3 cistron and with each of the two deleted constructions, placed behind the 35S promoter. We have been unable so far to detect full-length P3 by Western blot in plants transformed with the complete cistron, although expression was seen at the callus stage (not shown). This may indicate that the presence of P3 interferes with regeneration and that, somehow, the plants eliminate the protein as they are regenerating. However, the transformations were quite successful with the two deleted clones. Fig. 1A shows Western blots of crude leaf extracts from plants $T_1$ and $T_2$ (cv. Xanthi nc transformed respectively with constructions 1 and 2) in comparison with the yeast extracts $Y_1$ and $Y_2$. Immunodetection was with an antiserum to purified P3-$Y_1$. It appears that $T_1$ and $Y_1$

synthesized the same protein. On the contrary, only a fraction of the recombinant protein made by $T_2$ comigrated with the yeast protein; most of it had a slightly higher mobility. This could have two explanations: either the protein was partially degraded, or it was initiated at a different AUG codon. Transformed plants derived from the other cultivar (Xanthi) made the same products as those derived from cv. Xanthi nc (not shown).

The localization of the deleted movement proteins was investigated by crude subcellular fractionation. Two fractions were made from the leaf extracts, containing respectively crude membranes and triton-washed cell walls. Material not sedimentable at 30000 g was discarded as it was found to contain insignificant amounts of the P3-related proteins. Fig. 1B shows a Western blot of the fractions made from plants $T_1$ and $T_2$, with fractions from uninfected ($T_U$) and AlMV-infected plants ($T_A$) as controls. It appears that a large proportion of the P3-related products fractionates with cell walls in $T_1$ plants, as they do in AlMV-infected plants. In $T_2$ plants on the contrary, the deleted proteins are not detectable in the cell-wall fraction. The absence of staining in lanes $T_U$ proves the specificity of the signal.

These data suggest that the first 12 N-terminal amino-acids of P3 are not required for targeting to the cell-wall (plasmodesmatal?) receptor. However we have not yet determined if the deleted protein is functional. On the other hand, removal of 77 amino-acids from the N-terminus abolishes cell-wall targeting and, presumably, biological activity. Part of the binding site of P3 to its receptor is therefore located between aa 13 and 77. More precise mapping of the site is needed before we can elaborate further on its possible structure. It is however intriguing that receptor binding should involve a C-proximal domain in the case of the 30K and an N-proximal domain in the case of P3. Either the proteins are not related at all, or the binding site includes C-proximal and N-proximal domains in both proteins. As we have seen, the second possibility is very likely in the case of the 30K. It would not be surprising to find the same thing for P3.

## CONCLUSION

From this rapid survey, it appears that our understanding of virus movement is progressing very rapidly.

1. The action of the TMV movement protein on plasmodesmata has been shown directly and is not a hypothesis anymore.

2. A useful distinction was made between type A (tobamovirus-like) viruses, which move from cell to cell as non-virion nucleoproteins and type B viruses (comovirus-like) which move as virion-like particles. Thanks to the work on TMV, we can start making educated guesses about the translocation of type A viruses through plasmodesmata but that of type B remains a mystery. Thus far, the properties of the AlMV movement protein have not proven fundamentally different from that of TMV, although AlMV has some of the characteristics of type B viruses. More experiments are needed to reveal the differences between AlMV and TMV movement mechanisms.

3. The functional domains of the TMV movement protein are slowly emerging and we hope that those of the AlMV protein will

follow. In addition to the sites involved in binding the viral RNA (or nucleoprotein) and the plasmodesmatal receptor(s), we suspect that the movement protein contains a catalytic site or an enzyme activator site. Which reaction this protein participates in is not known but we tend to think that it has to do with reorganizing the cytoskeleton, which is involved in cellular movements.

## ACKNOWLEDGEMENTS

We thank Drs. Atabekov and Taliansky for sending us their unpublished manuscript. Work carried out in Strasbourg was funded by the Centre National de la Recherche Scientifique. Some of the unpublished results were obtained during a two-year postdoctoral stay of A. Berna in the laboratory of Dr. Beachy, with a fellowship from the European Molecular Biology Organization and research funds, to Dr. Beachy, from the National Science Fundation and from Monsanto Co.

## REFERENCES

Atabekov, J. G., and Dorokhov, Y. L., 1984, Plant virus-specific transport function and resistance of plants to viruses, Adv. Virus Res., 29:313-364.
Barker, H., 1989, Specificity of the effect of sap-transmissible viruses in increasing the accumulation of luteoviruses in co-infected plants, Ann. Appl. Biol., 115:71-78.
Boulton, M. I., Steinkellner, H., Donson, J., Markham, P. G., King, D. I., and Davies, J. W., 1989, Mutational analysis of the virion-sense genes of maize streak virus, J. Gen. Virol., 70:2309-2323.
Briddon, R. W., Watts, J., Markham, P. G., and Stanley, J., 1989, The coat protein of beet curly top virus is essential for infectivity, Virology, 172:628-633.
Citovsky, V., Knorr, D., Schuster, G., and Zambryski, P., 1990, The P30 movement protein of tobacco mosaic virus is a single-strand nucleic acid binding protein, Cell, 60:637-647.
Dodds, J. A., and Hamilton, R. I., 1976, Structural interactions between viruses as a consequence of mixed infections, Adv. Virus Res., 20:33-86.
Francki, R. I. B., Milne, R. G., and Hatta, T., 1985, "Atlas of plant viruses," Vol. 1 and 2, CRC Press, Boca Raton.
Gibbs, A., 1976, Viruses and plasmodesmata, in: "Intercellular communications in plants: studies on plasmodesmata," B. E. S. Gunning and A. W. Robards, eds., p. 149-164, Springer-Verlag, Berlin, Heidelberg.
Godefroy-Colburn, T., Schoumacher, F., Erny, C., Berna, A., Moser, O., Gagey, M.-J., and Stussi-Garaud, C., 1990, The movement protein of some plant viruses, in: "Recognition and response in plant-virus interactions," R. S. S. Fraser, ed., NATO ASI Series H, Vol. 41, p. 207-231, Springer-Verlag, Berlin, Heidelberg.

Goldbach, R., Eggen, R., de Jager, C., van Kammen, A., van Lent, J., Rezelman, G., and Wellink, J., 1990, Genetic organization, evolution and expression of plant viral RNA genomes, in: "Recognition and response in plant-virus interactions," R.S.S. Fraser, ed., NATO ASI Series H, Vol. 41, p. 147-162, Springer-Verlag, Berlin, Heidelberg.

Harrison, B. D., Barker, H., and Derrick, P. M., 1990, Intercellular spread of potato leafroll luteovirus: effects of co-infection and plant resistance, in: "Recognition and response in plant-virus interactions," R. S. S. Fraser, ed., NATO ASI Series H, Vol. 41, p. 405-414, Springer-Verlag, Berlin, Heidelberg.

Hull, R., 1989, The movement of viruses in plants, Annu. Rev. Phytopathol., 27:213-240.

Lazarowitz, S. G., Pinder, A. J., Damsteegt, V. D., and Rogers, S. G., 1989, Maize streak virus genes essential for systemic spread and symptom development, EMBO J., 8:1023-1032.

Lehto, K., Bubrick, P., and Dawson, W. O., 1990a, Time course of TMV 30K accumulation in intact leaves, Virology, 174:290-293.

Lehto, K., Grantham, G. L., and Dawson, W. O., 1990b, Insertion of sequences containing the coat protein subgenomic promoter and leader in front of the tobacco mosaic virus 30K ORF delays its expression and causes defective cell-to-cell movement, Virology, 174:145-157.

Malyshenko, S. I., Kondakova, O. A., Taliansky, M. E., and Atabekov, J. G., 1989, Plant virus transport function: complementation by helper viruses is non-specific, J. Gen. Virol., 70:2751-2757.

Maule, A. J., Harker, C. L., and Wilson, T. M. A., 1989, The pattern of accumulation of cauliflower mosaic virus-specific products in infected turnips, Virology, 169:436-446.

Meshi, T., Motoyoshi, F., Maeda, T., Yoshiwoka, S., Watanabe, H., and Okada, Y., 1989, Mutations in the tobacco mosaic virus 30-kD protein gene overcome Tm-2 resistance in tomato, The Plant Cell, 1:515-522.

Mushegian, A. R., Malyshenko, S. I., Taliansky, M. E., and Atabekov, J. G., 1986, The role of cAMP in the transport of the virus genome in infected plants, Molekuliarnaia Biologia, 20:1371-1376, in russian.

Mushegian, A. R., Malyshenko, S. I., Taliansky, and Atabekov, J. G., 1989, Host-dependent suppression of temperature-sensitive mutations in tobacco mosaic virus transport gene, J. Gen. Virol., 70:3421-3426.

Saito, T., Imai, T., Meshi, T., and Okada, Y., 1988, Interviral homologies of the 30K proteins of tobamoviruses, Virology, 167:653-656.

Shanks, M., Tomenius, K., Clapham, D., Huskisson, N. S., Barker, P. J., Wilson, I. G., Maule, A. J., and Lomonossoff, G. P., 1989, Identification and subcellular localization of a putative transport protein from red clover mottle virus, Virology, 173:400-407.

Stratford, R., and Covey, S. N., 1989, Segregation of cauliflower mosaic virus symptom genetic determinants, Virology, 172:451-459.

Van Lent, J., Wellink, J., and Goldbach, R., 1990, Evidence for the involvement of the 58K and 48K proteins in the intercellular movement of cowpea mosaic virus, J. Gen. Virol., 71:219-223.

Wellink, J., and van Kammen, A., 1989, Cell-to-cell transport of cowpea mosaic virus requires both the 58/48K proteins and the capsid proteins, J. Gen. Virol., 70:2279-2286.
Wolf, S., Deom, C. M., Beachy, R. N., and Lucas, W. J., 1989, Movement protein of tobacco mosaic virus modifies plasmodesmatal size exclusion limit, Science, 246:377-379.
Zaitlin, M., and Hull, R., 1987, Plant virus-host interactions, Annu. Rev. Plant Physiol., 38:291-315.

# TOMATO SPOTTED WILT VIRUS: A BUNYAVIRUS INVADING THE PLANT KINGDOM

Richard Kormelink, Peter de Haan, Dick Peters and Rob Goldbach

Department of Virology, Agricultural University
Binnenhaven 11
6709 PD Wageningen, The Netherlands

## INTRODUCTION

Despite its economic importance tomato spotted wilt virus (TSWV) has gained only little popularity among plant molecular biologists and virologists so far. While the tomato spotted wilt disease has been described already in 1919 (Brittlebank) and shown to have viral etiology in 1930 (Samuel *et al.*, 1930) only during the past 10 years plant pathologists became aware of its worldwide spread and destructive properties. TSWV causes serious diseases in a great number of crops, including tomato, potato, pepper, peanut, lettuce and pea, and in many ornamental plant species, including chrysanthemum and impatiens. To date more than 400 plant species representing more than 50 botanical families, both mono- and dicots, are known to be susceptible to TSWV (Matthews, 1982; D. Peters, personal communication). Disease symptoms are often severe, frequently including necrosis of leaf and stem tissues. Hence, with its broad host range and virulence, TSWV ranks among the most agressive and pathogenic plant viruses currently known. The virus is exclusively transmitted by thrips species (*Thysanoptera*) (Sakimura, 1962). Vectors include *Thrips tabaci*, *Frankliniella occidentalis*, *F. fusca*, and *F. schultzei* (Cho *et al.*, 1989). Originally TSWV was a virus mainly restricted to (sub-) tropical climate zones but seems to have expanded its geographical range enormously in the past two decades. Nowadays the virus occurs commonly in the more temperate climate zones in both the New and Old World. In areas like Northern Europe and Canada TSWV has become primarily a greenhouse problem. Although little is known concerning the causal events that have led to its recent marked, geographical expansion a major reason for the current outbreak in Europe has been the introduction of *F. occidentalis* and the considerable resistance of this vector to chemical control.
Among plant viruses TSWV is unique in its particle morphology, genome structure and transmission. Hence, it has been plausible to classify this virus as the sole representative of a distinct plant virus group, the tomato spotted wilt virus group (Ie, 1970; Matthews, 1982). The virion of TSWV is a spherical enveloped particle, 80 to 110 nm in diameter, covered with surface projections that may consist of two different glycoproteins denoted G1 (molecular size of 78K) and G2 (58K), which are present in the lipid envelope (Fig. 1). The core consists of three ribonucleocapsid structures, each containing a single species of RNA, tightly complexed with nucleocapsid (N) proteins (molecular size 29K). The genomic RNAs are

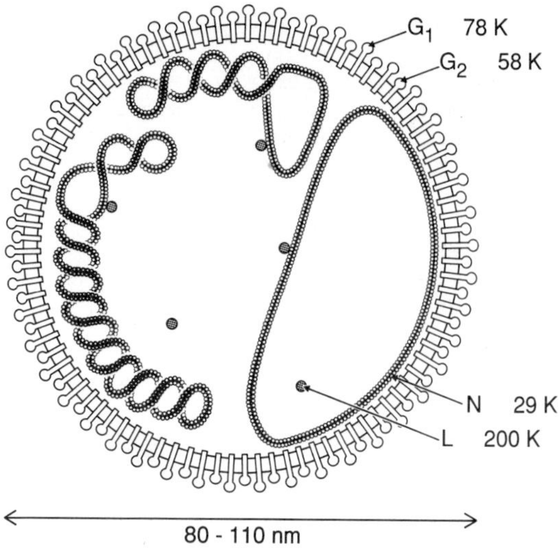


Fig. 1. Schematic representation of a TSWV particle. The three linear genomic RNA molecules are tightly associated with nucleocapsid (N) proteins and form circles, which may be coiled. Two species of glycoproteins, G1 and G2, are present in the lipid envelope. A large protein (L), present in minor amounts, may represent the viral replicase.

single-stranded and linear, but form pseudo-circles upon encapsidation (Figs. 1 and 2). The lengths of the genomic RNA species are 8000 nucleotides (L RNA), 5000 nucleotides (M RNA) and 3000 nucleotides (S RNA), respectively (Van den Hurk *et al.*, 1977; Verkleij and Peters, 1983; De Haan *et al.*, 1989a). Isolated nucleocapsids still retain their infectivity but purified RNAs are not infectious (Verkleij and Peters, 1983), suggesting that TSWV represents a negative strand virus. In that case a viral transcriptase activity should be present in virions, which

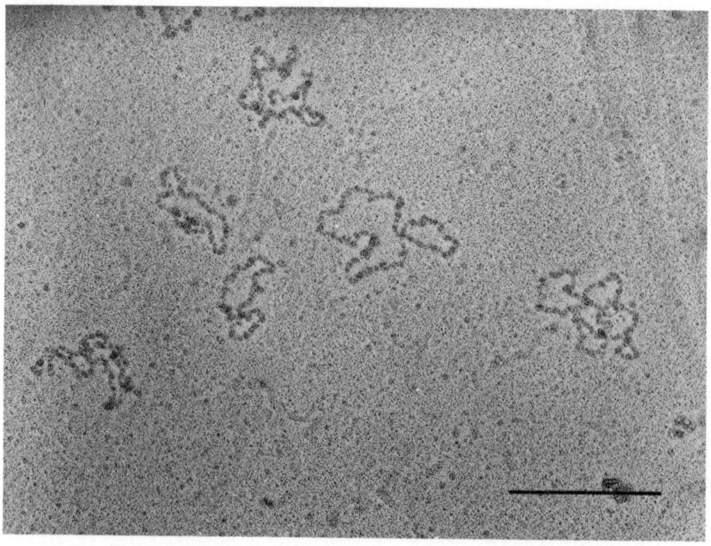

Fig. 2. Electron micrograph of isolated nucleocapsids of TSWV. The linear genomic RNAs in these nucleocapsids are folded into circles. The bar represents 300 nm.

could reside in the 200K protein (denoted L), found in minor amounts in TSWV particles (Fig. 1). Based on the morphological and structural properties it has been proposed (Milne and Francki, 1984; De Haan *et al.*, 1989a) that TSWV actually represents a member of the large family of arthropod-borne Bunyaviridae, being unique in its property to infect plants. To determine a possible evolutionary relationship to this family of exclusively vertebrates infecting viruses and to gain more insight in the molecular basis of its virulence to plants detailed information on the molecular properties, genomic expression and cytopathology of TSWV is required. Therefore we decided to clone and sequence the TSWV genome and to study its replication and translational expression in infected plant tissues.

## TSWV IS A NEGATIVE STRAND VIRUS

Using a series of clones, selected from pUC and λgt10 libraries a major part (6000 out of 8000 nucleotides) of the L RNA segment has been sequenced (Fig 3). From the sequence data the following conclusions can be drawn:

(a) There are no open reading frames (ORFs) present in the viral strand of L RNA, in stead the viral complementary (vc) strand contains an ORF which may be contiguous from position 34 to the UAA stop codon at a position 228 residues from the 3'terminus of the RNA.
(b) The termini are complementary over a region of approximately 60 residues. These termini can be folded into a stable panhandle structure with a free energy of $\Delta G$ = 235,4 kJ/mol. The formation of such panhandle may explain the pseudo-circular state of TSWV nucleocapsids (Fig. 2), as also found for all bunyaviruses analyzed thusfar.
(c) Only within the N-terminal regions the theoretical translation product of TSWV L RNA exhibits some sequence homology to the L protein of Bunyamwera virus (Fig. 4), the only bunyavirus from which the L RNA has been completely sequenced (Elliott, 1989). Two short sequence motifs (ARHDXFGXEL and NA/VTPDNY) may indicate a genetic inter-

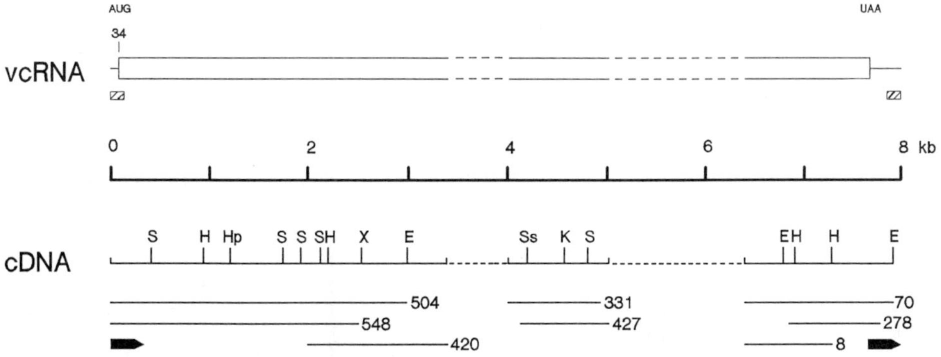


Fig. 3. Genetic organization of TSWV L RNA. Determination of a major part of its sequence, using the cDNA clones indicated, demonstrates that L RNA is a negative strand. The viral complementary (vc) strand contains open reading frames (indicated as open bars) for most of its sequence. It is anticipated that there is only one, long ORF in the vc strand. Symbols: ▨, complementary termini; ➡, primers used for primer extension sequencing; E, EcoRI; H, Hind III; Hp, HpaI; K, KpnI; S, SphI; Ss, SstI; X, XbaI.

relationship. At least the first motif is conserved between three bunyavirus members (Bunyamwera virus, La Crosse virus, Snowshoe Hare virus) and TSWV, for the second motif sequence information on La Crosse and Snowshoe Hare virus is lacking.

(d) If the ORFs in the sequenced regions of TSWV L RNA are parts of one, contiguous ORF, then this ORF would encode a protein with a size exceeding 250 kilodaltons. The L protein, with an apparent size of 200 kilodaltons and present in lower amounts in virions (Fig. 1) may therefore correspond to this ORF. Indeed, using riboprobes derived from L cDNA clones, no L-RNA-specific subgenomic mRNAs could be detected in infected *Nicotiana rustica* or tomato tissues. Hence, additonal, shorter ORFs seem unlikely to exist in L RNA.

```
TSWV      1 MNIQKIQKLIENGTTLLLSIEDCVGSNHDLALDLHKRNSDEIPEDVIINN 50

                     .         .         .         .         .
TSWV     51 NAKNYETMRELIVKITADGEGLNKGMATVDVKKLSEMVSLFEQKYLETEL 100
                            ::::.:. :  ::. :  ..:.. :....||
BUNYA     1 ................MEDQAYDQYLHRIQAARTATVAKDISADILE... 31

                     .         .         .         .         .
TSWV    101 ARHDIFG.ELISRHLRIKPKQRNEVEIEHALREYLDE..LNKKSCINKLS 147
            ||||.|| ||.           |.::||.  . .|||  |:  ..:| |
BUNYA    32 ARHDYFGRELC...........NSLGIEYKNNVLLDEIILDVVPGVNLL. 69

                     .         .         .         .         .
TSWV    148 DDEFERINKEYVATNATPDNYVIYKESKNSELCLIIYDWKISVDARTETK 197
                     :|  .|.|||||:       .:  |||.|:|:||:. ...
BUNYA    70 .........NYNIPNVTPDNYI......WDGHFLIILDYKVSVGNDSSEI 104

                     .         .         .         .         .
TSWV    198 TMEKYYKNIWKSFKGIKVNGKPFFYTFCLWGCASASINIYSMLPGEVNDS 247
            | .|| . |:. :.:: ::..  :.       |.: .  .|::.:|..:.
BUNYA   105 TYKKYTSLILPVMSELGIDTEIAIIR......ANPVTYQISIIGEEFKQR 148

                     .
TSWV    248 IR.......... 249
            :.
BUNYA   149 FPNIPIQLDFGR 160
```

Fig. 4. Alignment of the N-termini of L proteins from TSWV and Bunyamwera-virus, using the alignment program developed by the University of Wisconsin Genetics Computer Group (UWGCG).

The occurrence of an open reading frame exclusively in the antigenomic strand of L RNA and the presence of complementary termini provide the first direct proof that TSWV is a negative strand virus. Furthermore, the coding capacity of L RNA and the (weak) sequence homology of its theoretical translation product with the L protein of Bunyamwera virus, the prototype of the Bunyaviridae, give a first indication that TSWV is a genetic relative of the Bunyaviridae. This relationship is confirmed by the sequence information on the S RNA segment.

## TSWV S RNA HAS AN AMBISENSE CHARACTER

While the virion morphology and the sequence information on the L RNA already indicate that TSWV should be regarded as a member of the Bunyaviridae, this becomes even more clear when examining the genomic organization of the S RNA (Fig. 5).

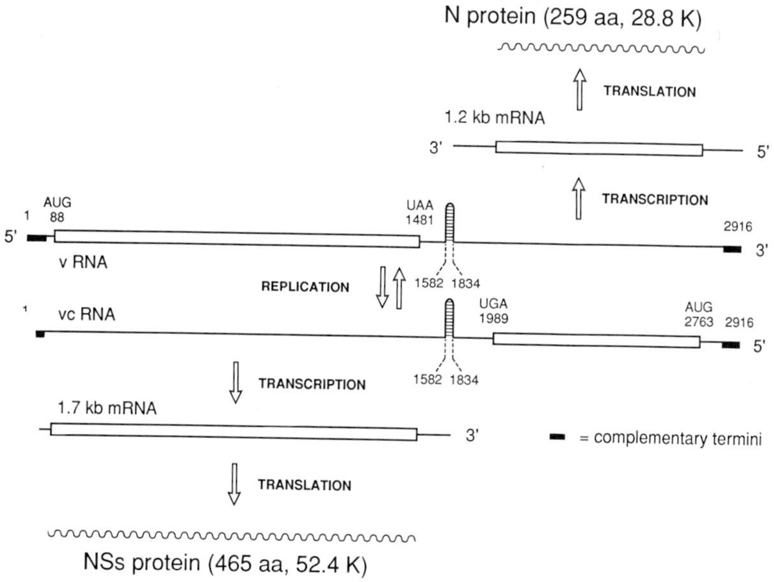


Fig. 5. Organization and expression of the S RNA of tomato spotted wilt virus. The RNA is 2916 nucleotides long and has an ambisense coding strategy. The ORF in the viral strand (vRNA) encodes a 52.4K non-structural protein (NSs), the ORF in the viral complementary strand (vcRNA) represents the nucleocapsid protein (N) gene. The ORFs are expressed from subgenomic mRNAs of 1.7 and 1.2 Kb respectively. Numbers refer to nucleotide positions, the black bars represent the complementary termini.

The complete sequence of S RNA is 2916 nucleotides long (De Haan *et al.*, 1990) and has some striking similarities in common with the S RNA species of uukuviruses and phleboviruses, two genera of the Bunyaviridae (Ihara *et al.*, 1984; Simons *et al.*, 1990):

(a) As found for L RNA the 3'- and 5'-terminal sequences of S RNA are complementary (over a stretch of 65 to 70 nucleotides) and can be folded into a stable panhandle structure ($\Delta G$ = -254.1 kJ/mol).

(b) The RNA has an ambisense coding strategy, containing one ORF in the viral strand, and a second ORF in the viral complementary (vc) strand (Fig. 5). This coding strategy is unique for plant viruses but is also found for the S RNA of phlebo- and bunyaviruses.

(c) As found for the ambisense S RNAs of phlebo- and uukuviruses, the ORF in the vc strand represents the nucleocapsid (N) gene. This was determined by *in vitro* transcription and translation experiments (De Haan *et al.*, 1990a).

(d) Again, similar to S RNA of phlebo- and uukuviruses, TSWV S RNA contains a long A-U rich hairpin in the intercistronic region (Fig. 5), which may act as a symmetric terminator during subgenomic mRNA synthesis.

## TSWV REPRESENTS A BUNYAVIRUS

Together with the morphological data the similarities in genomic structure and expression, not only for the S RNA but also for the L RNA, indicate that TSWV should be considered as a member of the arthropod-borne Bunyaviridae, which thusfar were known to infect vertebrates only (Matthews, 1982; Bishop, 1990; Elliott,1990). The ambisense gene organization in the S RNA of TSWV is a genome strategy unique for plant viruses, but is strikingly similar to the gene arrangement in the S RNA of phleboviruses and uukuviruses, two genera of the Bunyaviridae. When searching the EMBL protein and nucleotide sequence database no or hardly (see Fig. 4) any sequence homology could be detected between TSWV RNAs or proteins and bunyaviral RNA or protein sequences, which indicates that TSWV should be regarded as the sole member of a new genus within the Bunyaviridae, which could be named phytophlebovirus or maybe better, phythrivirus (siglum of phytos and thrips-transmitted).

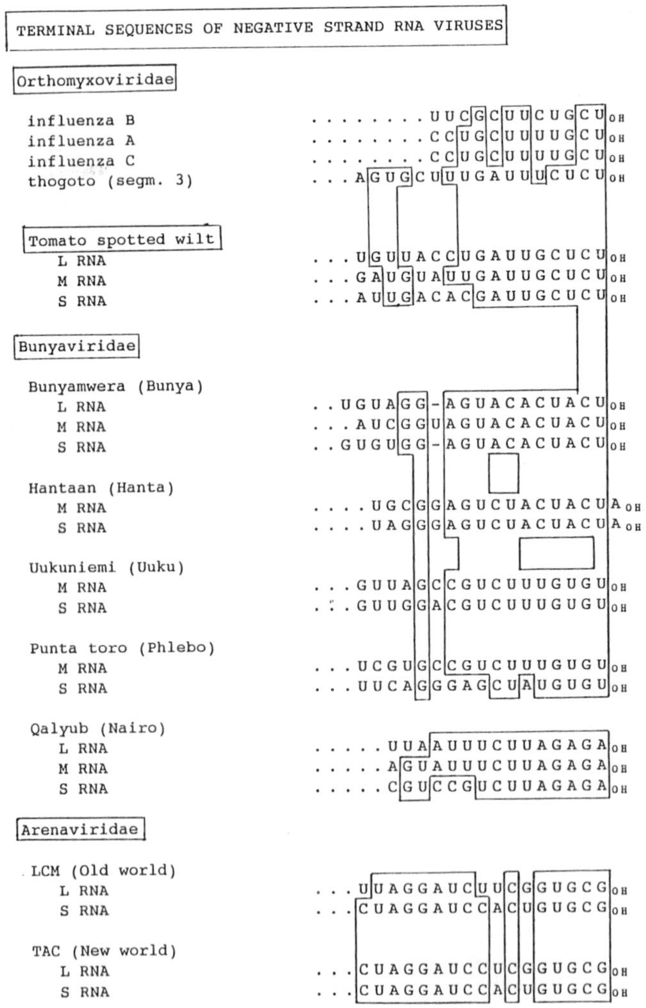


Fig. 6. Comparison of the 3' terminal sequences of the genomic RNA molecules of TSWV to those of members of the Bunyaviridae, Arenaviridae and Orthomyxoviridae. Identical sequences are boxed.

When comparing the ultimate terminal sequences of the RNAs of all negative strand viruses with segmented genomes the termini of TSWV RNAs appear to exhibit sequence homology only to those of thogoto virus, a tick-borne member of the Orthomyxoviridae (Staunton *et al.*, 1989), but not to e.g. bunyaviral RNA termini (Fig. 6). This latter finding is not in contradiction with the proposed taxonomic position of TSWV, since the different genera of the Bunyaviridae lack a consensus terminal sequence (Fig. 6). The significance of the sequence homology to an orthomyxovirus remains to be established.

## VIROLOGICAL AND EVOLUTIONARY IMPLICATIONS

Now, that it has been established that TSWV represents a member of the Bunyaviridae, a number of questions arise with respect to its potential to infect plants, instead of vertebrates, and with respect to the evolutionary pathway by which an insect-vertebrate virus has become an insect-plant virus.

The family Bunyaviridae contains more than 300 established members to date, while another 100 viruses are putative members (Bishop, 1990; Elliott, 1990). Hence, while the number of vertebrate-infecting bunyaviruses is quite high, TSWV appears to be the sole member of the Bunyaviridae capable to infect plants. Hence, it seems likely that TSWV has evolutionary departed from an animal bunyavirus, that became able to infect thrips species and, due to the very close ecological interactions between this insect group and (flowering) plants, subsequently adapted to plants instead of vertebrates. Also within another family of negative strand RNA viruses, that mainly infect animals, the Rhabdoviridae, a number of viruses are known to be adapted to plants (Baer *et al.*, 1990). While these rhabdoviruses infecting plants have, like their animal counterparts, rather restricted host ranges, TSWV seems to be capable to multiply in any plant species on which its thrips vector forages. Indeed, plant viruses differ from animal viruses in the fact that they don't require a cell receptor for successful infection. It is plausible that the envelope glycoproteins of bunyaviruses are involved in receptor recognition during infection of insects and vertebrates. Interestingly, when TSWV is maintained in plants only, and transmitted mechanically without the involvement of thrips, the virus appears to lose its envelope while the production of one or both glycoproteins is dramatically decreased or even ceased completely (Ie, 1982 ; Verkleij and Peters, 1983). Analysis of the genome of such defective isolates shows that the M RNA, i.e. the genome segment most probably encoding these glycoproteins, contains deletions (Verkleij and Peters, 1983). These observations suggest that the glycoprotein-containing envelope is not a prerequisite for multiplication in a host plant but is essential only during the insect stage of the TSWV multiplication cycle. In other words, receptor recognition is not required for successful TSWV infection of plants. In stead the feeding behaviour and host plant acceptance of its thrips vectors seem to be the major determinants of the (broad) host range of TSWV.

## ACKNOWLEDGMENTS

The authors wish to thank Annemieke van der Jagt for preparing the manuscript. This work was supported in part by Zaadunie, Enkhuizen, The Netherlands and by the Netherlands Foundation for Chemical Research (SON), with financial aid from the Netherlands Organization of Pure Research (NWO).

## REFERENCES


Baer, M.G., Bellini, W.J., and Fishbein D.B., 1990, Rhabdoviruses, in: "Virology", B.N. Fields, D.M. Knipe *et al.*, eds., second edition, Raven Press, Ltd., New York.

Bishop, D.H.L., 1990, Bunyaviridae and their replication, Part I: Bunyaviridae, in: "Virology", B.N.Fields, D.M. Knipe *et al.*, eds., Second edition, Raven Press, Ltd., New York.

Brittlebank, C.C., 1919, Tomato diseases, J. Agr. Victoria, 17:213.

Cho, J.J., Mau, R.F.L., German, T.L., Hartman, R.W. Yudin, L.S., Gonsalves, D., and Provvidenti, R., 1909, A multidisciplinary approach to management of tomato spotted wilt virus in Hawaii, Plant Disease, 73:375.

De Haan, P., Wagemakers, L., Goldbach, R., and Peters, D., 1989a, Tomato spotted wilt virus, a new member of the Bunyaviridae?, in: "Genetics and Pathogenicity of Negative Strand Viruses", D. Kolakofsky and B.W.J. Mahy, eds., Elsevier, Amsterdam.

De Haan, P., Wagemakers, L., Peters, D., and Goldbach, R., 1986, Molecular cloning and terminal sequence determination of the S and M RNAs of tomato spotted wilt virus, J. gen. Virol., 70:3469.

De Haan, P., Wagemakers, L., Peters, D., and Goldbach, R., 1990. The S RNA segment of tomato spotted wilt virus has an ambisense character, J. gen. Virol., 71:in press.

Elliott, R.M., 1989, Nucleotide sequence analysis of the large (L) genomic RNA segment of Bunyamwera virus, the prototype of the family Bunyaviridae, Virology, 173:426.

Elliott, R.M., 1990, Molecular biology of the Bunyaviridae, J. gen. Virol., 71:501.

Ie, T.S., 1970, Tomato spotted wilt virus, CMI/AAB Descriptions of Plant Viruses., 39.

Ie, T.S., 1982, A sap-transmissible, defective form of tomato spotted wilt virus, J. gen. Virol., 59:387.

Ihara, T., Akashi, K. and Bishop, D.H.L., 1984, Novel strategy (ambisense genomic RNA) revealed by sequence analysis of Punta Toro phlebovirus S RNA, Virology, 136:293.

Matthews, R.E.F., 1982, Classification and nomenclature of viruses, Intervirology, 17:1.

Sakimura, K., 1962, The present status of thrips-borne viruses, in: "Biological Transmission of Disease Agents", K. Maramorosch, ed., Academic Press, New York.

Samuel, G., Bald, J.G., and Pittman, H.A., 1930, Investigations on 'spotted wilt' of tomatoes, Counc. Sci. Ind. Res. Bull., 44

Simons, J.F., Hellman, U., and Petterson, R.F., 1990, Uukuniemi virus S RNA segment: Ambisense coding strategy, packaging of complementary strands into virions and homology to members of the genus phlebovirus. J. of Virol., 64:247.

Staunton, D., Nuttall, P.A., and Bishop, D.H.L., 1980, Sequence analysis of thogoto viral RNA segment 3: Evidence for a distant relationship between an arbovirus and members of the Orthomyxoviridae, J. gen. Virol., 70:2811.

Van den Hurk, J. Tas, P.W.L., and Peters, D., 1977, The ribonucleic acid of tomato spotted wilt virus, J. gen. Virol., 36:81.

Verkleij, F.N., and Peters, D., 1983, Characterization of a defective form of tomato spotted wilt virus, J. gen Virol., 64:677.

# ORIGIN AND EVOLUTION OF DEFECTIVE INTERFERING RNAS OF TOMATO BUSHY STUNT VIRUS

David A. Knorr and T. Jack Morris

Plant Pathology Department
University of California
Berkeley, CA 94720, USA

## INTRODUCTION

Viruses with defective genomes have been identified in association with virtually every major family of viruses and have been widely utilized as tools for investigating virus functions in animal cell culture systems (Perrault, 1981). It is generally thought that defective interfering viruses (DIs) arise through deletion, rearrangement, or recombination of a competent viral genome. DIs lack the ability for independent existence relying on their parental helper viruses to supply factors required for replication, maturation, and/or encapsidation (Huang and Baltimore, 1977). The interference attributed to DIs is thought to result from competition with the helper virus for factors required in *trans* for replication and/or encapsidation (Schlesinger, 1988). It is these features that have made DIs very useful tools for mapping viral signals required for replication and packaging (*e.g.* Levis, *et al.*, 1986). The DIs associated with animal viruses have usually been detected after serial passage of virus at high multiplicities of infection (m.o.i.) in cell cultures. It has also been noted that maintenance in continuous passage results in fluctuations of helper and DI which reflect the relative abilities of each to interfere or support the other (Huang, 1988). DIs have also been detected in natural virus infections and, although there is some debate, may function *in vivo* to modulate virus diseases and allow more persistent infections (Huang, 1988).

The majority of plant viruses have positive-sense, single-stranded RNA genomes. Additional RNA components capable of modulating symptoms have most commonly been associated with either satellite viruses or satellite RNAs. Satellite viruses differ from satellite RNAs by encoding their own capsid proteins, whereas satellite RNAs are encapsidated by their specific helper virus. Satellite RNAs are relatively common among plant viruses, having been found in association with at least 24 members in six virus groups (Francki, 1985). Unlike DIs, satellite RNAs in general share little sequence similarity with their specific helper virus. The one exception is the chimeric satellite RNA-C of turnip crinkle virus (Simon, 1988). Satellite viruses and satellite RNAs have been shown to both intensify as well as attenuate symptoms normally expressed by their specific virus helpers (Kaper, and Collmer, 1988, Simon, 1988).

In contrast to the many reports of satellites, few authentic DIs have been identified among plant viruses. These accounts include the presence of DI-like particles associated with the negative-stranded plant rhabdoviruses (Adam *et al.*, 1983; Ishmail and Milner, 1988) and the bunyavirus-like tomato spotted wilt virus (Verkleij and Peters, 1983). DI-like RNA components have also been reported for wound tumor

*Plant Molecular Biology* 2, Edited by R.G. Herrmann and
B. Larkins, Plenum Press, New York, 1991

virus, a plant reovirus (Nuss, 1988). Among the positive, single-stranded RNA plant viruses, authentic DIs have been identified for only two viruses: tomato bushy stunt virus (TBSV) (Hillman *et al.*, 1987; Morris *et al.*, 1989); and turnip crinkle virus (TCV)(Li *et al.*, 1989).

TBSV and TCV are members of two closely related groups of plant viruses, the tombusviruses and the carmoviruses respectively (see Martelli *et al.*, 1988 and Morris *et al.*, 1988 for reviews). Both viruses are good model systems for studying small RNA virus and DI RNA molecular biology because each has been well-defined biologically and detailed structural studies have been performed on both viruses. In addition, the genome of each virus has been cloned, entirely sequenced, and engineered to produce infectious RNA transcripts *in vitro* (Carrington *et al.*, 1989; Heaton, *et al.*, 1989; Hearne, *et al.*, 1990). The reader is referred to these papers for details on the genomic organizations of these two viruses. The features of the TBSV genome organization important in understanding the origins of associated DI RNAs are summarized in Figure 5.

In this paper we will focus on the derivation, molecular characterization, and evolution of TBSV-associated DIs. We will also discuss the possibilities for using plant virus DIs as a general method for controlling virus diseases.

## DISCOVERY OF DI RNAs IN TOMBUSVIRUSES

In studies with the cherry strain of TBSV, it was observed that plants inoculated with concentrated virus preparations tended to develop a less-severe, more persistent disease syndrome than plants inoculated with diluted virus. This interference was found to be associated with a novel small RNA species (400b) present in the virus isolate (Hillman, *et al.*, 1985). This observation was similar to the "autointerference" phenomenon in some animal virus systems attributed to the presence of DIs (Huang, 1988) and to the effects of some satellite RNAs associated with RNA plant viruses. Although a satellite RNA had been identified in association with another tombusvirus at that time (Gallitelli and Hull, 1985), it was not suspected to be involved in the TBSV-cherry system because northern hybridizations consistently indicated that the low molecular weight (LMW) RNA in our culture hybridized with the viral genome. Subsequent sequencing of clones of TBSV-associated small RNAs in conjunction with genomic sequencing clearly demonstrated that the interference-associated RNAs were collinear deletion mutants of the helper virus comprised of sequences derived from 5' proximal, internal, and 3' proximal regions of the genome (Hillman, *et al.*, 1987). Plant infection studies demonstrated that the DI RNA was responsible for the symptom attenuation as well as decreased viral replication and reduced virus accumulation in whole plants (Hillman, 1986). More recent studies using plant cell protoplasts have shown that TBSV DIs specifically interfere with the replication of helper virus as judged by reduced incorporation of $^{3}$H-uridine into viral and viral-specific, subgenomic RNA species (Jones, *et al.*, 1990). These features are consistent with the definition of defective interfering viruses and establish the first definitive proof for the existence of DIs in association with a plant virus.

## *DE NOVO* GENERATION OF TBSV DI RNAs

A previous report had suggested that the satellite RNA associated with CyRSV was generated upon serial passage of the virus in plants (Gallitelli and Hull, 1985). In our laboratory, however, LMW RNA species suggestive of DIs were readily detected after serial passage of several different tombusviruses including artichoke mottle crinkle virus, pelargonium leaf curl virus, petunia asteroid mosaic virus, and a field isolate of the BS3 strain of TBSV (Morris, unpublished). Curiosity about the possible *de novo* generation of these DIs prompted a more definitive set of passage experiments. Sixteen lines of a DI-free isolate of TBSV-cherry were generated by inoculating single local lesions from *Nicotiana glutinosa* onto each of 8 *N. clevelandii* and 8 *N. benthamiana* plants. Each of the independent lines of TBSV were passed at 7 day intervals, 6 lines for each host at high m.o.i. (undiluted sap) and 2 lines at low m.o.i. (buffer-diluted sap). This approach was similar to that routinely employed to generate DIs in animal cell lines. After each passage, infected plants were observed for alterations in symptom development, and inoculated leaves were analyzed for ssRNA, dsRNA, and virus production. An advantage in using an intact host rather than cell cultures was the ability to detect the presence of DIs by observing attenuation of disease symptoms. In the course of the passage experiment, symptom attenuation was evident in some of the high m.o.i. passed lines as early as passage 3 and in all such lines by passage 10 but not in any of the 4 lines passed at low m.o.i. The symptom attenuation was also correlated with LMW RNAs of approximately 400-600b present in viral RNA preparations analyzed by polyacrylamide gel electrophoresis (PAGE) of each of the independent virus lines passed at high m.o.i. and in none of lines passed at low m.o.i. (Figure 1). These results strongly support the contention that DIs arise spontaneously upon host passage because distinctively sized DI species appeared in each of the high m.o.i. lines at different stages of the passage experiment. More definitive proof for *de novo* generation of TBSV DI RNAs has since come from the identification of DI species in lines derived from clones of TBSV by inoculation of infectious transcripts of the viral genome (Knorr, in preparation.).

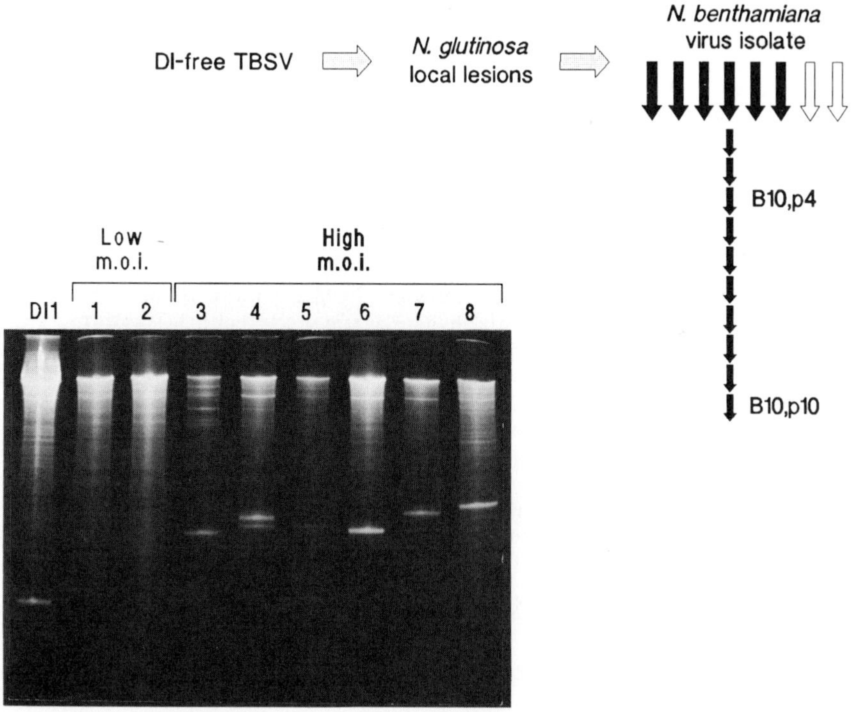


**Figure 1.** Strategy for *de novo* generation of TBSV DI RNAs. The diagram illustrates how independent TBSV isolates were obtained in *N. benthamiana*. Black vertical arrows indic ate high m.o.i. passage lines; white arrows indicate two lines passed at low m.o.i. Each independent line was passed ten times (indicated by the smaller black arrows for the B10 line). Results from the B10 passage are presented in Figure 4. The lower panel shows an ethidium bromide-stained gel of virion RNAs after the tenth passage from the TBSV isolates passed in *N. clevelandii.* DI1 is a 400b spontaneous DI originally described (Hillman, *et al.* (1987). Lanes 1 and 2 are from the low m.o.i. lines, 3-8 are from high m.o.i. lines.

## MOLECULAR CHARACTERIZATION OF A DI POPULATION

The TBSV DI RNAs generated in each of the high m.o.i. lines of the passage experiment were initially identified by northern analysis using a TBSV specific probe. Typical results for one of the lines generated by passage in *N. benthamiana* (designated DI B10) are presented in Figure 2. Consistent with the properties of the originally characterized TBSV DIs, the B10 DI species could be readily detected in the 2.5 M lithium chloride soluble fraction of infected tissue extracts as well as in the virion RNA preparations.

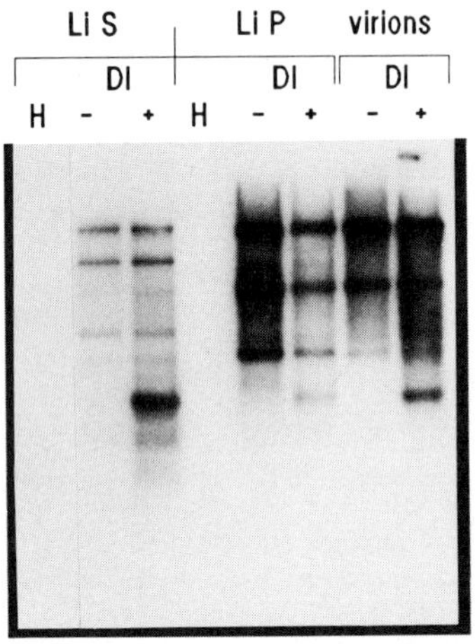


**Figure 2.** Northern analysis of DI RNAs. Total nucleic acids were extracted from healthy (H), TBSV (-), and TBSV-DI (+) infected plants, treated with 2.5 M LiCl and DNase, then separated into soluble (LiS) and pellet (LiP) fractions. The LiCl products and virions RNAs a TBSV isolate with (+) and without (-) DI RNAs were then separated by non-denaturing PAGE. The RNAs were transferred to a nylon membrane and probed with [$^{32}$P]-labelled TBSV-cDNA.

A series of cDNA clones of the B10 line then was produced in phagemid vectors by standard protocols using RNA templates purified from virions (Knorr, in prep). The complete nucleotide sequences for six apparently full-length cDNA clones of the B10 DI isolate were determined using the di-deoxynucleotide chain terminating method. The sequence analysis confirmed the DI nature of each of the clones which showed marked similarity. Each consisted of the same four regions of sequence derived entirely from the helper virus genome. The four regions of sequence maintained in each DI were: (I) the complete 168 nucleotide TBSV 5' leader sequence, including the initiator methionine codon; (II) a region of between 173 to 234 nucleotides (nt 1285 to 1481-1523) from the putative polymerase gene; (III) approximately 70 nt containing the 3' terminus of the viral P19 and P22 open reading frames (ORFs); and (IV) the extreme 3' terminal 130 nt of non-translated viral sequence (see the diagram in Fig. 5). Minor variations in each sequence consisting of single base changes and small deletions were detected. However, conservation of the deletion junctions were maintained between each region of retained sequence suggesting that this population evolved either from a common progenitor, or by some inflexible molecular mechanism. It is interesting that at each sequence junction there is an ambiguity of one to four nucleotides. At the junction between regions II and III, the sequence 'AGAA' could be derived from genomic sequences at either side of the deletion. The same final sequence motif (AGAA) was present between regions I and II, and between III and IV, in another DI population analyzed later. It is also worth mentioning here that the same general regions of conserved sequence were identified in the first DI to be sequenced (Hillman *et al.*, 1987) and in another of the *de novo* generated lines derived by passage in *N. clevelandii* that was also cloned and sequenced (see the C6-line identified in Fig. 5).

It is possible that DI RNAs translate proteins that are in some way involved in attenuating symptoms or virus replication. To examine this possibility, each sequence was analyzed for the presence of coding regions. The largest ORFs encoded by the B10 DIs are 31 and 34 amino acids respectively. They share no apparent similarity with virus, or other proteins, nor are they strictly conserved in each DI. Experiments with an additional set of DI clones indicates even more variation in these two ORFs. Also, previous attempts to translate DIs *in vitro* were negative (Brad Hillman, unpublished). Therefore, it seems unlikely that the mechanism by which DIs attenuate symptoms involves activity of a DI-encoded protein. It is possible, however, that sequestering of ribosomes by DI RNAs, or a similar activity, may affect disease outcome.

## INFECTIVITY OF CLONED TBSV DIs

Biological activity of DI sequences was assessed by making RNA transcripts *in vitro* from the cloned cDNAs and co-inoculating them with DI-free TBSV (Figure 3). In order to accomplish this, however, the inserts had to be modified. The 5' termini

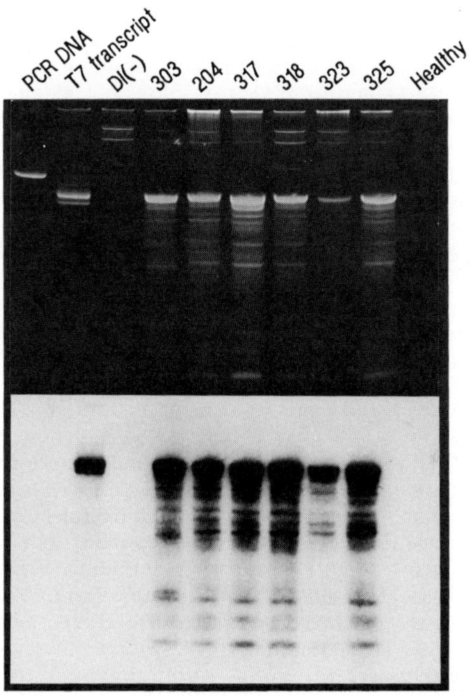


**Figure 3.** Infectivity of transcripts from TBSV DI B10 clones. PCR-amplified cDNAs were transcribed *in vitro* and inoculated onto *N. clevelandii* together with TBSV genomic RNA. Plants inoculated with DI transcripts survived, whereas those inoculated only with TBSV genomic RNA developed lethal systemic necrosis. Replication of the DIs was analyzed as in Figure 2. Upper panel shows an ethidium-stained gel, with the corresponding northern blot below. Numbered lanes refer to specific TBSV B10-DI clones. A typical PCR product and the T7 transcripts derived from it are indicated.

contained no promoter sequence and were also missing 2-4 nucleotides compared with the parental TBSV sequence. At the 3' end inserts lacked as few as none, but as many as 11 nucleotides. Polycytidine tails added prior to cloning also were an impediment. These problems were solved simultaneously by using the polymerase chain reaction (PCR) to correct and amplify cDNAs used for transcription. As outlined in Figure 4, the PCR strategy used one oligonucleotide primer containing the sequence for a bacteriophage T7 promoter fused to the 5' terminal 24 nt of TBSV and another primer specifying the compliment to the viral 3' terminal 28 nt. Conditions for PCR were identical to those used to modify and clone the termini of TBSV and have been described elsewhere (Hearne, *et al.*, 1990). Transcripts of B10 DI cDNAs were made from 1 µg of PCR-amplified product in 100 µl reactions essentially according to the method of Janda, *et al.* (1987), but omitting the GpppG 5' cap analogue. Transcripts replicated in plants co-inoculated with TBSV RNA or virions. DI RNAs accumulated to high titers in the presence of helper virus, but no accumulation of DI was observed when transcripts were inoculated without TBSV (Figure 3). No differences in symptom attenuation were observed in plants when transcripts from DI clones or the uncloned DI RNA was co-inoculated with helper virus.

## DI EVOLUTION

During the serial passage experiment it was noted previously that DIs appeared at different times in each of the individual high m.o.i. lines and that by passage 10 each had a distinctive size. We selected 2 lines (B10 and C6) in which the DIs appeared early to evaluate more thoroughly each step in the passage experiment. Interestingly, the predominant DI species present in each population after ten weeks of passage were noticeably smaller than that observed when the DI first appeared, suggesting evolution had occurred. Figure 4 (bottom panel) shows the DI RNA species present in the LiCl soluble fraction of RNA extracted from B10 infected plants at passages 3 (no DI), 4 and 10. The B10 DI first appeared during the $4^{th}$ passage in *N. benthamiana* and consisted of a largely heterogeneous group of RNA species from approximately 300-800 b, with a predominant band at ~780b. After 10 passes, there was less heterogeneity, and the predominant DI species was approximately 600b. Cloning and sequencing of the DI populations at each passage was accomplished directly from the LiCl fractions by initial cDNA synthesis followed by PCR amplification as outlined in Figure 4. Primers used for cloning were similar to those used for preparing transcription templates except that unique restriction sites were incorporated into each terminus to facilitate cloning. Clones were obtained from PCR amplification of cDNA from the equivalent of as little as 8 µg of infected tissue.

A summary of the results of the sequence analysis is presented in Figure 5 for both the original and evolved C6 and B10 DI populations. These results represent the consensus of 18 DI clones from 4 separate populations. It is notable that each comprised essentially the same regions of sequence with some minor variations. Only two clones (from C6, pass 3) were identical in sequence suggesting that they might have resulted from PCR amplification of the same template. Minor differences in each of the other sequences suggest each may have originated separately, rather than by duplication of an original cDNA template. A major difference between C6 and B10 was the location of the genomic deletion defining regions I and II. All of the C6 DIs had a precise deletion between nt 178 to 1285, whereas the comparable deletion for all B10 DIs was between nt 168 to 1285. The C6 population was also distinctive in that it contained a greater portion of the polymerase domain (region II) than B10. A comparison of all of the DIs sequenced shows that the 5'-proximal junction of region II is generated with precision while the 3'-proximal portion tolerates more extensive sequence heterogeneity. Deletions resulting in the creation of regions III and IV in the DI RNAs also appeared to occur with precision as the junctions in all but the least-evolved (B10, pass 4) and most evolved (DI1) sequences were highly conserved. In general, the sequence differences between pass 3 or 4 and pass 10 involve heterogeneity at the 3'-end of region II and the precise deletion defining the 5' terminus of region III. These observations strongly support the idea that the DIs continued to evolve by additional deletion events from previously existing DIs.

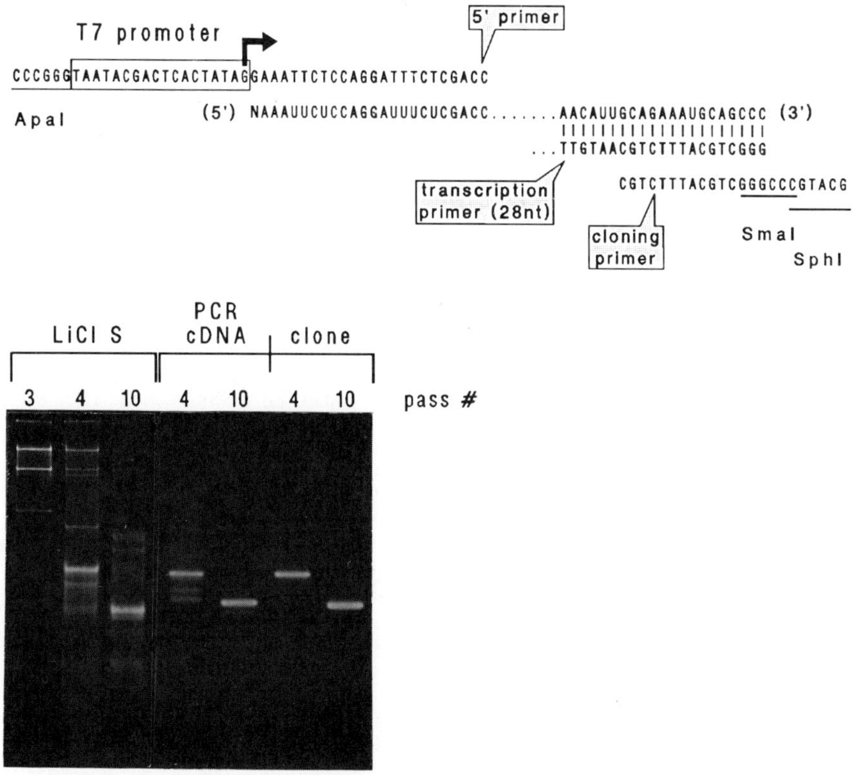


**Figure 4.** PCR strategy for cloning TBSV B10 DI RNAs. Upper panel shows sequences of primers used for cDNA synthesis and PCR amplification. The termini of TBSV template sequences are labelled (5' and 3'). Restriction sites, T7 promoter region, and transcription start site (solid arrow) are indicated. In the lower panel is an ethidium bromide-stained gel showing LiCl supernatant RNAs (LiS) isolated from passes 3, 4, and 10 (left 3 lanes). The center two lanes show PCR-amplified cDNAs resulting from passes 4 and 10. In the right two lanes are examples of inserts from clones generated from the amplified cDNAs. The identical strategy was employed to make clones from the TBSV C6 isolate after 3 and 10 passes.

## DISCUSSION

These experiments confirm the *de novo* generation of DIs during serial passage of TBSV. It appears that two relatively precise deletion events take place giving rise to DIs with a highly conserved sequence motif in which most of the sequence heterogeneity within a given population is confined to the 3'-end of region II. The slight variations in the sequences between and within independently-derived DI populations indicate they may arise from separate events. There are at least two possible explanations for the strong conservation of sequences observed in TBSV DIs. One is that the mechanism of recombination is rigidly controlled, resulting in the generation of only certain deletions. Another explanation is that random deletions may occur during replication, but by selection, only templates with optimum abilities for replication are maintained. In actuality, a combination of these two explanations may operate. For instance, in the coronavirus mouse hepatitis virus (MHV), it is thought that deletions leading to formation of DIs involve recombination at conserved elements of secondary structure (Makino, *et al.*, 1988). However, once generated, survival of the resulting recombinants depends upon maintaining the ability to be replicated. Although a similar mechanism may operate in creating TBSV DIs, searches for conserved elements of predicted secondary structure surrounding points of recombination resembling those of MHV, have so far been negative. Nevertheless, it seems clear that DIs contain elements required in *cis* for replication. That DIs maintain both viral termini, is not surprising since each is considered to be important in initiating complementary strand synthesis during replication. It is also not unprecedented for internal elements to be maintained. The intergenic region of brome mosaic virus RNA3 contains is important for efficient RNA replication (French and Ahlquist, 1987). In addition, it has been noted that animal virus DIs often contain internal regions of sequence (Makino, 1988; Schlesinger, 1988).

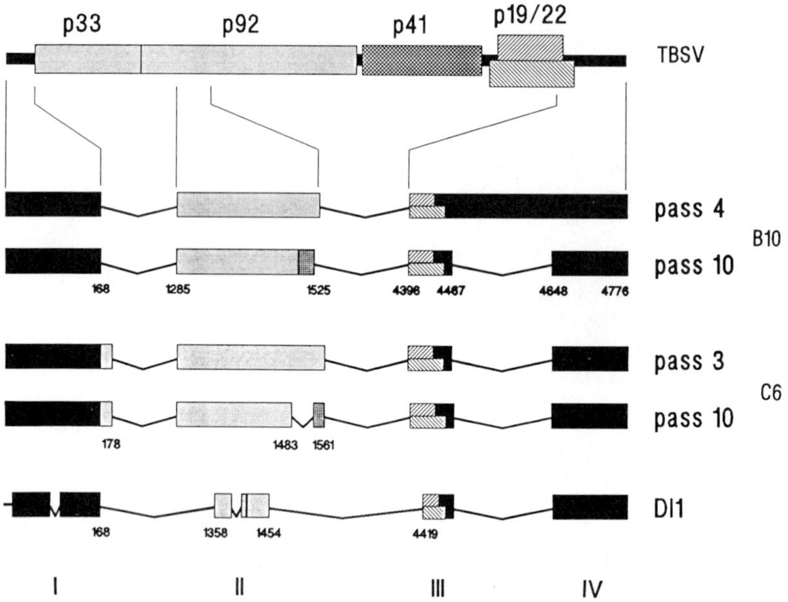


**Figure 5.** Evolution of *de novo* generated TBSV DI RNAs. A map of the TBSV genome is shown above with coding regions as shaded rectangles and the sizes of their protein products in kDa above; non-coding regions are black. Separate diagrams for the consensus sequence of each DI population in this study are given below with identical blocks of sequence in the same vertical position. Fine lines give the positions for each region of DI relative to the TBSV genome. Pass 3 and pass 4 represent sequences from initial populations of TBSV isolates C6 and B10 respectively, while pass 10 represents the passage at which the experiment was stopped. Precise definition of regions I-IV are given in the text. Numbers indicate consensus break points between different regions for some populations. A spontaneous DI characterized previously, (DI1) is provided for comparison.

The sequence motifs of TBSV DIs resemble in several ways those of the similarly-sized DIs recently characterized in association with TCV (Li *et al.*, 1989). In both viruses, DI populations have been generated *de novo* and have been shown to retain all of the 5' non-coding region and a precisely modified 3' non-coding region consisting of the terminal 130-150 nt fused to variable portions of the C-terminus of the 3'-most ORF. An interesting distinction between the two types is the significant modification of the 5'-terminal region in the evolved TCV DIs and the absence of a significant region of internal sequence from the viral polymerase gene. These comparisons provide useful clues about sequences important in replication. To emphasize this, we note that the TCV DI RNAs have recently been useful for defining possible encapsidation signals through the use of RNase protection assays to identify regions of coat protein affinity (Wei *et al.*, 1990).

Plant virus DIs may prove useful in developing novel strategies for controlling plant virus diseases. In California, significant economic losses occur from tomato decline, a disease in which TBSV infection appears to play a major role (J. Gerik, unpublished results). In an effort to control this disease we are currently engineering transgenic tomato plants to express DI RNAs. This strategy is plausible because preliminary trials have demonstrated that TBSV DIs can function in *trans* to attenuate virus symptoms in tomatoes. This approach is similar to other methods which have been advanced for controlling virus diseases by engineering expression of virus-derived sequences into plants such as coat protein-mediated protection (reviewed in Register, *et al.*, 1989), expression of satellite RNAs (reviewed in Baulcombe, *et al.*, 1989), and anti-sense RNA expression (Loesch-Fries, 1987; Hemenway, 1988). Potential advantages of using DIs are that expression of foreign proteins is not required and the levels of constitutive expression necessary to achieve protection are expected to be low. We expect that many more plant viruses will be shown to produce DIs. In the future, based on a more thorough understanding of the mechanisms by which DIs are generated and maintained, it may be possible to design DI-like molecules to control viruses that may not normally make DIs.

## REFERENCES


Adam, G., Gaedigk, K., and Mundry, K. 1983. Alterations of a plant Rhabdovirus during successive mechanical transfers. *Z. Pflanzenkr. Pflanzensch.* **90**:28-35.

Baulcombe, D., Devic, M., Jaegle, M., and Harrison, B. 1989. Control of viral infection in transgenic plants by expression of satellite RNA of cucumber mosaic virus. *In* B. Staskawicz, P. Ahlquist and O. Yoder (ed.), *Molecular Biology of Plant-Pathogen Interactions.* UCLA symposia on molecular and cellular biology, New series, Alan R. Liss., New York, N.Y. pg. 257-267.

Carrington, J.C., Heaton, L.A., Zuidema, D., Hillman, B.I., and Morris, T.J. 1989. The genome structure of turnip crinkle virus. *Virology* **170**: 219-226.

Francki, R.I.B. 1985. Plant virus satellites. *Ann. Rev. Microbiol.* **39**:151-174.

French, R., and Ahlquist, P. 1987. Intercistronic as well as terminal sequences are required for efficient amplification of brome mosaic virus RNA3. *J. Virol.* **61**:1457-1465.

Gallitelli, D. and Hull, R. 1985. Characterization of satellite RNAs associated with tomato bushy stunt virus and five other definitive Tombusviruses. *J. Gen. Virol.* **66**:1533-1543.

Hearne, P., Knorr, D., Hillman, B., and Morris, T. 1990. The complete genome structure and infectious RNA synthesized from clones of tomato bushy stunt virus. **Virology** (in press).

Heaton, L., Carrington, J. and Morris, T. 1989. Turnip crinkle virus infection from RNA synthesized in vitro. *Virology* **170**: 214-218.

Hemenway, C., Fang, R-X., Kaniewski, W., Chua, N-H., Tumer, N. 1988. Viral protection in transgenic plants expressing the cucumber mosaic virus coat protein or its antisense RNA. Bio/Technology **6**: 549.

Hillman, B., Schlegel, D.E., & Morris, T.J. 1985. Effects of low-molecular weight RNA and temperature on tomato bushy stunt virus symptom expression. *Phytopathology* **75**: 361-365.

Hillman, B. 1986. Genome organization, replication and defective RNAs of tomato bushy stunt virus, Ph.D. thesis, University of California, Berkeley.

Hillman, B., Carrington, J.C., and Morris, T.J. 1987. A Defective interfering RNA that contains a mosaic of a plant virus genome. *Cell* **51**: 427-433.

Huang, A. 1988. Modulation of viral disease processes by defective interfering particles. Vol. 3, p. 195-208, *In* E. Domingo, J.J. Holland and P. Ahlquist (ed.), *RNA Genetics*, CRC Inc., Boca Raton.

Huang, A., and Baltimore, D. 1977. Defective interfering animal viruses. *Compr. Virol.* **10**:73-116.

Ismail, I. and Milner, J. 1988. Isolation of defective interfering particles of sonchus yellow net virus from chronically infected plants. *J. Gen. Virol.* **69**:999-1006.

Janda, M., French, R., and Ahlquist, P. 1987. High efficiency T7 polymerase synthesis of infectious RNA from cloned brome mosaic virus cDNA and effects of 5' extensions on transcript infectivity. *Virology* **158**: 259-262.

Jones, R., Jackson, A.O. and Morris, T.J. 1990. Interaction of tomato bushy stunt virus and its defective interfering RNA in inoculated protoplasts. *Virology* (in press).

Kaper, J., and Collmer, C. Modulation of viral plant diseases by secondary RNA agents. Vol 3, pg. 171-194. *In* E. Domingo, J.J. Holland and P. Ahlquist (ed.), *RNA Genetics*, CRC Inc., Boca Raton.

Knorr, D., and Morris, T.J. Molecular characterization and infectious transcripts from clones of defective interfering RNAs of tomato bushy stunt virus. (in prep.).

Levis, R., Weiss, B., Tsiang, M., Huang, H., and Schlesinger, S. 1986. Deletion mapping of sindbis virus DI RNAs derived from cDNAs defines the sequences essential for replication and packaging. *Cell* **44**: 137-145.

Li, X. H., Heaton, L.A., Morris, T.J. and Simon, A.E. Defective interfering RNAs of turnip crinkle virus intensify viral symptoms. *Proc. Natl. Acad. Sci. USA* **86**:9173-9177.

Loesch-Fries, L., Halk, E., Merlo, D., Jarvis, N., Nelson, S., Krahn, K., and Burhop, L. 1987. Expression of alfalfa mosaic virus coat protein gene and antisense cDNA in transgenic tobacco tissue. *In* Arntzen, C., and Ryan, C. (eds.): *Molecular Strategies for Crop Protection*, Alan R, Liss, New York. p. 221

Makino, S., Sheih, C., Soe, L., Baker, S., and Lai,M. 1988. Primary structure and translation of a defective interfering RNA of murine coronavirus. *Virology* **166**: 550-560.

Martelli, G., Gallitelli, D., and Russo, M. 1988. Tombusviruses. *In* R. Konig, ed., *The Plant Viruses*, **vol.** 3: chapter 2. Polyhedral virions with monopartite RNA Genomes, Plenum, NY.

Morris, T., and Carrington, J. 1988. Carnation mottle virus and viruses with similar properties. p. 73-112, *In* R. Koenig (ed.), *The Plant Viruses*, Volume 3: Polyhedral Virions with Monopartite RNA Genomes, Plenum, NY.

Morris, T. and Hillman, B. 1989. Defective interfering RNAs of a plant virus, Volume 101, p.185-197, *In* B. Staskawicz, P. Ahlquist and O. Yoder (ed.), *Molecular Biology of Plant-Pathogen Interactions.* UCLA symposia on molecular and cellular biology, New series, Alan R. Liss., New York, N.Y.

Nuss, D. 1988. Deletion mutants of double stranded RNA genetic elements found in plants and fungi, Vol. 2, p. 188-210, *In* E. Domingo, J.J. Holland and P. Ahlquist (ed.), *RNA Genetics*, CRC Inc., Boca Raton.

Perrault, J. 1981. Origin and replication of defective interfering particles. *Curr. Top. Microbiol. Immunol.* **93**:151-207.

Register III, J., Powel, P., Nelson, R., and Beachy, R. 1989. Genetically engineered cross protection against TMV interferes with initial infection and long distance spread of the virus. *In* B. Staskawicz, P. Ahlquist and O. Yoder (ed.), *Molecular Biology of Plant-Pathogen Interactions.* UCLA symposia on molecular and cellular biology, New series, Alan R. Liss., New York, N.Y. pg.269-281.

Schlesinger, S. 1988. The generation and amplification of defective interfering RNAs, Vol. 2, p. 167-185, *In* E. Domingo, J.J. Holland and P. Ahlquist (ed.), *RNA Genetics*, CRC Inc., Boca Raton.

Simon, A. 1988. Satellite RNAs of plant viruses. *Plant Mol. Biol. Rep.* **6**:240-252.

Verkleij, F., and Peters, D. 1983. Characterization of a defective form of tomato spotted wilt virus. *J. Gen. Virol.* **64**: 677-686.

Wei, N., Heaton, L., Morris, T., and Harrison, S. 1990. Structure and assembly of turnip crinkle virus VI. Identification of coat protein binding sites on the RNA. *J. Mol. Biol.* **213**:(in press).

# SELF-CLEAVAGE ACTIVITIES FROM VIRAL SATELLITE RNAs

W.L. Gerlach and M.J. Young

CSIRO Division of Plant Industry
GPO Box 1600, Canberra, ACT 2601, Australia

## Satellite RNA to Ribozymes

We are studying aspects of the interaction of a satellite RNA with its plant host and helper virus. The satellite RNA of tobacco ringspot virus ('STobRV') is similar to a number of other low molecular weight satellite RNAs which have been described for a range of plant viruses (Francki, 1985). They are single stranded RNAs, often less than 400 bases, with no extensive sequence homology with either their plant host or supporting virus. They are dependent upon both the plant host and virus helper for their propagation and encapsidation during infection of the plant by the virus. Satellite RNAs can alter disease expression, with some exacerbating symptoms (eg. Cowpea CARNA5 satellite) while others (eg. STobRV) ameliorate disease.

Our interest in the STobRV molecule arose from its ability to protect plants from symptom development when infected with the helper virus. This is manifested when plants are infected with TobRV and satellite together, or when transgenic plants expressing satellite RNA are inoculated with virus alone (Schneider, 1971; Gerlach et al., 1986,1987). As part of this work we have begun to dissect the molecular properties of the satellite RNA. These properties include its ability to be recognised by the replication machinery in both plus (encapsidated) and minus strand forms, and to be encapsidated by the plant virus. Along with this, autolytic cleavage events which occur in multimeric replication intermediates have also been studied. The full nucleotide sequence of STobRV has been determined (Buzayan et al., 1986), making it ideal for molecular manipulations.

An unexpected outcome during mutagenesis experiments to investigate structure-function relationhips of the satellite RNA, was the ability to adapt the autolytic cleavage mechanism from the plus strand of the RNA to the design of new ribozyme RNAs. These have enzymatic ability to cleave new target RNAs according to the design rules shown in Figure 1 (Haseloff and Gerlach, 1988). There are three essential elements in the interaction of the substrate and ribozyme RNAs. First, the substrate can be any RNA sequence containing a GUC (or related, see below) trinucleotide sequence, termed the target site. Second, the ribozyme has a defined catalytic domain. Third, the ribozyme has sequences which base pair with the specific target sequence. Mutagenesis results have also shown the way to adapt the two component minus strand cleavage site (Haseloff and Gerlach, 1989; Feldstein et al., 1989) so that it can also cleave new target RNA sequences (Hampel et al.,1990).

*Plant Molecular Biology* 2, Edited by R.G. Herrmann and
B. Larkins, Plenum Press, New York, 1991

## DESIGN PARAMETERS FOR HAMMERHEAD RIBOZYMES

From our initial characterization of the design rules for new ribozymes based on the hammerhead cleavage structure present in the plus strand cleavage reaction of STobRV (see Symons et al. 1989 and Bruening 1990 for reviews) we have undertaken *in vitro* tests of various aspects of the design rules. These fall into three categories of modifications: the GUC target site, the catalytic domain and the extent of base pairing arms.

### Mutant Target Sites

A GUC trinucleotide cleavage site is chosen by most known naturally occurring hammerhead reactions. Through a systematic mutational analysis of a defined target sequence based on a GUC target in the chloramphenicol acetyl transferase (CAT) gene, and appropriate compensating changes in the ribozyme, we have shown that efficient cleavage is found with not only GUC but also GUA, GUU, UUC and CUC (Perriman *et al*, manuscript in preparation). Certain other target triplets can be cleaved, but at a slower rate, with efficiency at least an order of magnitude lower, while others have no activity. However, it is important to realise that there are at least five triplet target sequences which can be chosen for efficient cleavage in the design of these ribozyme RNAs. This has also been concluded by Koizumi et al. (1988) from a more limited series of mutational tests.

### Catalytic Domain

Although catalytic domains from known satellite RNas contain several conserved nucleotides, there are regions of variability (Symons, 1989; Bruening, 1990). For this reason we have tested the relative reaction rates of other catalytic domains in addition to the STobRV sequence. One of these involves a sequence derived from the satellite RNA of another plant virus, subterranean clover mosaic virus. Another involves the extension of the base paired stem of the STobRV catalytic domain (Figure 1) through direct repetition of the four base pairs to produce eight base pairs length. Both of these changes result in new catalytic domains with *in vitro* catalytic efficiencies comparable with the native STobRV sequence. An increased loop size on the stem-loop can also be tolerated. Other changes, such as alteration of the conserved GAAA boundary of the catalytic domain and a range of dideoxy substitutions (Perrault et al., 1990) result in inactivation of the ribozymes. They show that naturally occuring catalytic domain sequences from the autolytic cleavage regions of other plant satellite RNAs, or synthetic derivatives of these, are comparable to the STobRV domain and can therefore be used in the design of these ribozymes.

### Arm lengths

In simplest terms, the events that are required for successful reaction using these ribozymes are 1) a combination of the ribozyme with the target RNA sequence to form the reaction complex, 2) subsequent cleavage at the target site, and 3) disassociation to yield the ribozyme along with the cleaved reaction products. One of the important parameters in this sequence of events is the hybridization reaction which causes a ribozyme molecule to base pair with the target template through complementary base pairing. The rate of this base pairing is partly determined by the extent of complementary sequences between the ribozyme and target RNA sequences, but there is increasing evidence to suggest that secondary structure is important. Several studies have shown an increased reaction rate when ribozymes and substrates are denatured. Also, different reaction efficiencies of apparently similar ribozyme reactions are seen with with different substrate molecules (Fedor et al., 1989; Heus et al., 1989; also our unpublished results).

With complementary sequences involving four bases on each side of the ribozyme (ie. eight base pair total complementarity) there is insufficient energy to form the reaction complex through base pairing even at physiological temperatures and thus no reaction proceeds *in vitro*. With eight and twelve complementary bases either side of the catalytic domain there is sufficient base pairing for complex formation under our *in vitro* conditions (10 mM MgCl2, 10 mM Tris pH=7.5, 50oC or 37oC). With twelve base pairs the reaction proceeds much faster than with eight base pairs, ($T_{1/2}$ for 12 base pair complementarity = 8 minutes, while $T_{1/2}$ for 8 base pair complementarity = 70 minutes) presumably due to a more rapid rate of hybridization to form the complex. These ribozymes with eight and twelve base arms are capable of a true enzymatic reaction i.e. enzymatic turnover. We have extended this approach to produce 'catalytic antisense' ribozymes. One example is an 800 base ribozyme sequence targeted against the CAT (chloramphenicol acetyl transferase) mRNA, by embedding four catalytic domains within the antisense sequence . As expected, this ribozyme reacts rapidly to form active complex, but does not show enzymatic turnover in vitro. This most likely results from the stability of the reaction complexes which are unable to dissociate under standard reaction conditions due to the high degree of base pairing.

## PATHOGEN TARGETS - *IN VITRO*

One possible application of these ribozymes lies in their potential to specifically cleave, and thereby functionally inactivate, RNA *in vivo*. Amongst potential targets are mRNAs or genomic RNAs associated with viral and subviral pathogens. We have begun using two plant pathogen RNAs as model systems. These are the RNA genomes of tobacco mosaic virus (TMV) and citrus exocortis viroid (CEV). TMV was chosen because it is a well characterized plant virus with a linear single stranded RNA genome whose sequences have been determined for a number of isolates (Ohno et al., 1984). CEV was chosen as an example of a well characterized viroid for which, once again, the base sequences of different isolates are known (Visvader and Symons, 1985). Also, it has a high degree of internal base pairing and provides a test for the ability of ribozymes to cleave substrates with extensive secondary structure.

Catalytic antisense ribozymes were prepared against each of these pathogens. For TMV the extent of complementarity involved one kilobase of the RNA genome. Embedded within this antisense sequence were three catalytic domains to produce the ribozyme. For CEV, the complete 371 bases of the genome were targeted by the complementary sequence within which three catalytic domains were engineered. Both of these ribozymes were tested against their target RNAs *in vitro* and both were found to successfully cleave. Of particular note was the observation that the viroid RNA could be cleaved at the target sites by the ribozyme. This is relevant in view of the high degree of secondary structure of a viroid. It means that a ribozyme molecule can hybridize with a viroid, most likely initiated in short unpaired regions and then pair further through strand displacement.

However, the reaction rates for ribozyme mediated cleavage of these RNAs differed dramatically. The reaction against the viroid target was much slower compared with the reaction against the TMV virus RNA at all temperatures tested. We assume that this is due to the high secondary structure of the viroid molecule and its relative inaccessibility to the ribozyme. This was verified by an experiment in which the viroid and its ribozyme sequences were melted before the reaction commenced. This allowed a greater association of the ribozyme with the viroid RNA. It led to a thirty times increase in reaction rate, which we assume to be due to the greater accessibility of the target RNA. This points to the necessity for a consideration of secondary structure in target RNA sequences as a major limitation to catalytic efficiency of these ribozyme designs. Other studies have also shown that substrate and ribozyme structural effects can cause up to 70 times rate differences in ribozyme reactions (Fedor et al., 1989; Heus et al., 1989; also our unpublished results).

## PATHOGEN TARGETS - *IN VIVO*

Experiments have been conducted on transgenic tobaccos expressing TMV ribozymes. The gene construction which was used was straightforward, involving a CaMV 35S promotor driving the production of the ribozyme transcript and using the nopaline synthese 3' terminator sequence. Transgenic tobacco plants were produced and screened by northern analysis to determine those which were producing significant levels of the ribozyme transcript. The particular tobacco host chosen was a samsun NN genotype, which is a local lesion host for TMV. $F_2$ families of independent transformants for both the catalytic antisense ribozyme as well as the appropriate antisense and non-insert controls have been produced. Populations of plants for each of these $F_2$ families have been mechanically inoculated with TMV. In two separate experiments we have seen that the mean lesion numbers for different families varies, as would be expected for individual transformants expressing different amounts of ribozyme or control sequences. However, the mean lesion numbers for the ribozyme populations are generally lower than the lesion numbers for the antisense controls. Both types of gene constructions produced plants with reduced average lesion number compared with non-transgenic controls. We are pursuing this further by testing the effect on TMV infection in protoplasts from transgenic plants, and host plants which can be systemically infected with TMV while expressing ribozymes and appropriate control sequences.

We have begun production of transgenic tomatoes expressing the CEV ribozyme driven by the CaMV 35S promoter. Transformations have only recently been done, and tests will be done on the seed progeny of the regenerated transgenics.

## FUTURE DIRECTIONS

Our first constructions against plant pathogen targets and other genes have involved the use of relatively simple constructions. In these the ribozyme sequence is transcribed from a CaMV 35S promotor element. The next generation of constructions expressing ribozymes will have enhancements built into the gene casssettes. These will address (a) a balance between a rate of complex formation and dissociation, (b) modified expression cassettes capable of tissue specific, stable production of ribozymes, (c) a range of target gene sequences to assess the effects on different gene activities, and (d) an investigation of the relative cleavage ability compared with target sequence structure.

In considering the balance between rate of complex formation and rate of dissociation of the reaction we are examining ways of maximising the rate of formation through extended complementarity. Experiments to date suggest that complex formation rather than chemical cleavage is rate limiting. However, hybridization between substrate and ribozyme should not be so great that stable complexes are formed. It would be advantageous to provide for turnover (ie. dissociation of the complex) after cleavage has occurred. One way to achieve this may be through the interspersion of noncomplementary sequencese, so that extensive hybridization is maintained but there are not long uninterrupted tracts of complementarity.

Regarding expression cassettes, we are testing the production of ribozymes which could be stabilised within the cytoplasm, perhaps through their expression within the 3' noncoding region of a cytoplasmic mRNA gene. In a recent report this appears to increase ribozyme efficiency in animal cells (Cameron and Jennings, 1989). Other expression cassettes may involve the production of ribozymes from RNA polymerase III promoters by embedding them within a tRNA gene sequence (Cotten and Birnstiel, 1989), or the production of circular transcripts which may be more stable within the cell (Harland and Misher, 1988).

A range of target sequences are currently being tested within our laboratory and by others. Results from this work will provide insights into particular properties of target sequences which are important for the reactions. In particular, we need to know more about the accessibility of target sequence regions since, as mentioned above, it appears that the physical structure of either the target RNA or the ribozyme RNA may be a major factor in influencing the rate of reaction.

From the interest that has been shown in ribozyme reactions *in vivo* it appears that there will be a number of laboratories reporting the results of experiments in the near future. A recent report which is worthy of mention has been the observation of an effect on HIV infection by human cell cultures expressing a ribozyme from a transgene construction (Sarver et.al., 1990). Hela cells expressing a short ribozyme against the *gag* gene of the HIV virus were produced. When challenged with HIV virus it was found that the levels of p24 antigen and proviral DNA were significantly reduced. Although an antisense RNA control was not included in the experiments, it appears that this effect is due to the production of the ribozymes within the cells.

## CONCLUSIONS

Ribozyme systems are ubiquitious in nature. They have been found in animal, plant, fungal, protozoan, bacterial and viral systems. Several laboratories are successfully adapting their activities in new ways against new target sequences. We regard the results currently available as promising. There are a number of cases within both plant and animal systems where expression of ribozymes has been found to show an activity against gene or pathogen targets. This work provides the groundwork for second generation constructions for *in vivo* studies. Such constructs will involve refined ribozyme constructions and more sophisticated expression systems.

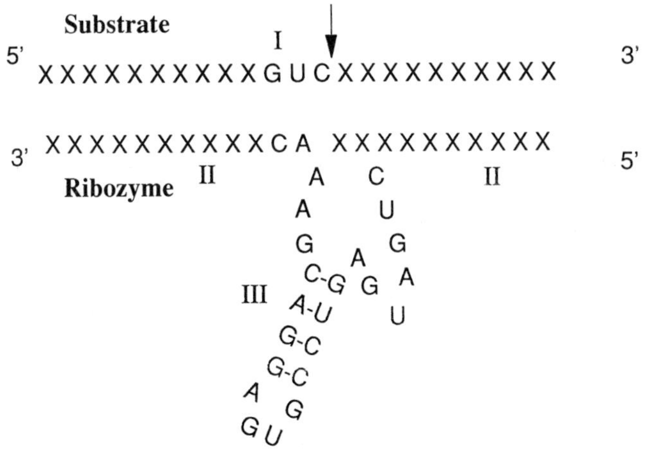


Figure 1. Ribozyme design according to Haseloff and Gerlach (1988). A substrate RNA is cleaved at a GUC site (arrowed) by the ribozyme. The ribozyme is composed of catalytic arms "II" which hybridize to the substrate, and a caalytic domain "III" which causes the cleavage of the substrate.

Acknowledgements: We would like to acknowledge that the interpretations described here come from the work of a number of our colleagues: Jim Haseloff, Rhonda Perriman, Lisa Kelly, Marianne Stapper, George Bruening, Lynda Graf and Christophe Robaglia. Rhonda Perriman made valuable comments on the manuscript.

## REFERENCES

Bruening, G. 1990. Compilation of self-cleaving sequences from plant virus satellite RNAs and other sources. Meth. Enzymol. 180: 546-558.

Buzayan, J.M. Gerlach, W.L., Bruening, G., Keese, P. and Gould, A.R. 1986. Nucleotide sequence of satellite tobacco ringspot virus RNA and its relationship to multimeric forms. Virology 151: 186-199.

Cameron, F.H. and Jennings P.A. 1989 Specific gene suppression by engineered ribozymes in monkey cells. Proc. Natl Acad. Sci. USA 86:9139-9143.

Cotten, M. and Birnstiel, M.L. 1989. Ribozyme mediated destruction of RNA in vivo. EMBO J. 8:3861-3866.

Fedor, M.J. and Uhlenbeck, O.C. 1990. Substrate sequence effects on "hammerhead" RNA catalytic efficiency. Proc. Natl Acad. Sci. USA 87:1668-1772.

Feldstein P.A., Buzayan, J.M. and Bruening, G. 1989. Two sequences participating in the autolytic processing of satellite tobacco ringspot virus complementary RNA. Gene 82:53-61.

Francki, R.I.B. 1985. Plant virus satellites. Ann. Rev. Microbiol. 39:151-174.

Gerlach, W.L., Buzayan, J.M. Schneider, I.R. and Bruening, G. 1986. Satellite tobacco ringspot virus RNA: biological activity of DNA clones and their *in vitro* transcripts. Virology 151: 172-185.

Gerlach, W.L. Llewellyn, D. and Haseloff, J. 1987. Construction of a plant disease resistance gene from the satellite RNA of tobacco ringspot virus. Nature 328:802-805.

Haseloff, J. and Gerlach, W.L. 1988. Simple RNA enzymes with new and highly specifc endoribonuclease activities. Nature 334:585-591.

Haseloff, J. and Gerlach, W.L. 1989. Sequences required for the self-catalysed cleavage of tobacco ringspot virus. Gene 82: 43-52.

Hampel, A., Tritz, R., Hicks, M. and Cruz, P. 1990. "Hairpin" catalytic RNA model: evidence for helices and sequence requirement for substrate RNA. Nucleic Acid Res. 18:299-304

Harland and Misher 1988 Development 102:837-852.

Heus, H.A., Uhlenbeck, O.C. and Pardi, A. 1990. Sequence-specific structural variations of hammerhead RNA enzymes.

Koizumi, M., Iwai, S., and Ohtsuka, E. 1988. Cleavage of specific sites of RNA by designed ribozymes. FEBS Lett. 239: 285-288.

Ohno, T., Aoyagi, M., Yamanashi, Y., Saito, H., Ikawa, s., Meshi, T. and Okada, Y. 1984. Nucleotide seqence of the tobacco mosaic virus (tomato strain) genome and comparison with the common strain genome. J. Biochem. 96:1915-1923.

Pereault, J-P., Wu, T., Cousineau, B., Ogilvie, K.K and Cedergren, R. 1990. Mixed deoxyribo- and ribo- oligonucleotides with catalytic activity. Nature 344:565-567.

Sarver, N., Cantin, E.M., Chang, P.S., Zaia, J.A., Ladne, P.A., Stephens, D.A. and ROSSI, J.J. 1990. Ribozymes as potential anti-HIV-1 therapeutic agents. Science 247:1222-1225.

Schneider, I.R. 1977. Defective plant viruses. pp201-219 in Beltsville Symposia in Agricultural research; Virology in Agriulture. Ed. J.A. Romberger. (Allenheld Osmun; Montclair, N.J.)

Symons, R.H. 1989. Self-cleavage of RNA in the replication of small pathogens of plant and animals. TIBS 14: 445-450.

Visvader, J.E. and Symons, R.H. 1985. Eleven new sequence variants of citrus exocortis viroid and the correlation of the sequence with pathogenicity. Nucleic Acids Res. 13:2907.

# VIROID STRUCTURES INVOLVED IN PROTEIN BINDING AND REPLICATION

Detlev Riesner, Jutta Harders, Rolf Hecker, Petra Klaff, Peter Loss, Noemi Lukacs, and Gerhard Steger

Institut für Physikalische Biologie, Heinrich-Heine-Universität Düsseldorf, Universitätsstr. 1, D-4000 Düsseldorf

## INTRODUCTION

Viroids are plant pathogens distinguished from viruses by the absence of a protein coat and by their small size. They are circular single-stranded RNA molecules consisting of a few hundred nucleotides, the smallest having about 240 and the largest about 600 nucleotides. Since viroids can have only a very limited coding capacity and as there is no experimental evidence for a viroid-coded translation product, one has to assume that viroid replication and pathogenesis depend completely on the enzyme systems of the host (reviews [1-5]). Thus, their genetic information is the RNA structure, the ability to undergo structural transitions and the capability to interact with host cell factors. Most results reported here were obtained from studies on the potato spindle tuber viroid (PSTVd)

The mechanism of viroid replication is known to a certain extent. The knowledge is summarized according to the literature in Fig. 1. The circular (by definition (+)strand) viroid is transcribed into an oligomeric (-)strand RNA. The (-)strand acts as template for synthesis of an oligomeric (+)strand RNA. Both transcription steps are catalyzed by an host enzyme, the DNA dependent RNA polymerase II [7,8]. This enzyme which normally transcribes mRNA from double-stranded DNA does accept viroid RNA as template. The (+)strand oligomeric RNA is cleaved enzymatically to unit length molecules which are then ligated to the mature viroid circles. Autocatalysis of cleavage and ligation was not found in viroids; a definite exception is, however, the avocado sunblotch viroid, which undergoes autocatalytic cleavage and ligation, and is also different in respect to other details of the replication cycle.

## STABLE AND METASTABLE STRUCTURES OF PSTVd

### Native Structure and Thermal Denaturation

Under physiological conditions viroids form a rod-like structure which may be described as an unbranched arrangement of short helices and small internal loops (cf. Fig. 2A). It was shown by a whole set of experiments that the viroid molecule is present in solution in the rod-like structure, i.e. it does not fold back and form a more globular structure. During thermal denaturation viroids undergo several structural transitions from the rod-like structure to the single-stranded circle without intramolecular base pairing (review [3,9]; cf. Fig. 2B). In a highly cooperative main transition

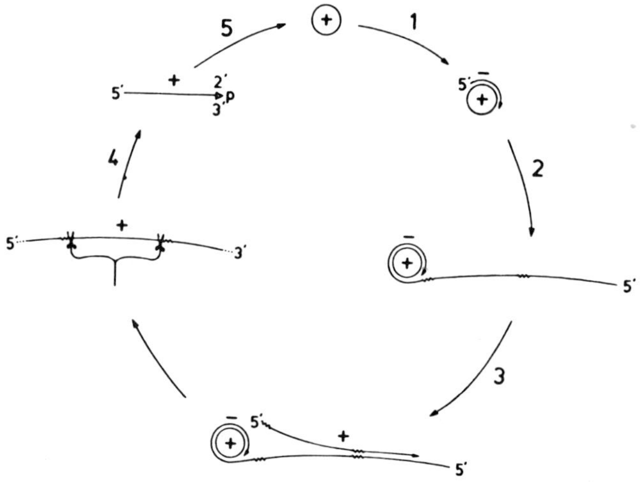


**Fig. 1.** Potato spindle tuber viroid (PSTVd) replication cycle. The infecting circular (+) molecule is transcribed into oligomeric (-)strands (1,2) followed by synthesis of (+)strand oligomers (3). The (+)strands are cleaved into monomeric units (4) and ligated to mature (+) molecules (5). The figure is reproduced from Branch and Robertson [6].

all base pairs of the rod-like structure are disrupted and particularly stable hairpins are newly formed. With other words, an extended structure switches over into a branched structure with a concommitant loss of base pairing. Due to thermodynamic competition the branched structure tends to destabilize the rod-like structure. At higher temperature the stable hairpins dissociate independently upon each other in the order of their individual thermal stabilities. Hairpin I is formed in a region of the molecule the sequence of which is highly homologous among all viroids (except avocado sunblotch viroid) and which is therefore called central conserved region (CCR). Hairpin II is present with slight sequence variations in all viroids of the PSTVd-class. Viroids of this group excert high sequence homology and do grow on tomato plants as host. Hairpin III was found only in PSTVd.

As outlined above, under physiological conditions the thermodynamically favoured secondary structure contains none of the stable hairpins. Thus, any functional role of these hairpins has to be attributed to metastable structures which contain the hairpins even under physiological conditions. One may consider two situations which favor the formation of stable hairpins. Firstly, the transition from the extended native structure to the branched, i.e. hairpin-containing structure may be facilitated by protein binding. At present, no experimental data are available supporting this possibility. Secondly, the stable hairpins may be formed during replication of the RNA - strand as an intermediate conformation before the synthesis of the whole strand is completed. The mechanism that a hairpin-containing structure is formed during synthesis and that it undergoes slow conformational transitions into more stable structures, could indeed be shown experimentally [10].

## Experimental Analysis of Stable and Metastable Structures

### Circular viroids

The structural transitions of both the circular PSTVd and of its oligomeric replication intermediates were studied with the recently developed method of temperature gradient gel electrophoresis (TGGE) [11,12]. This method

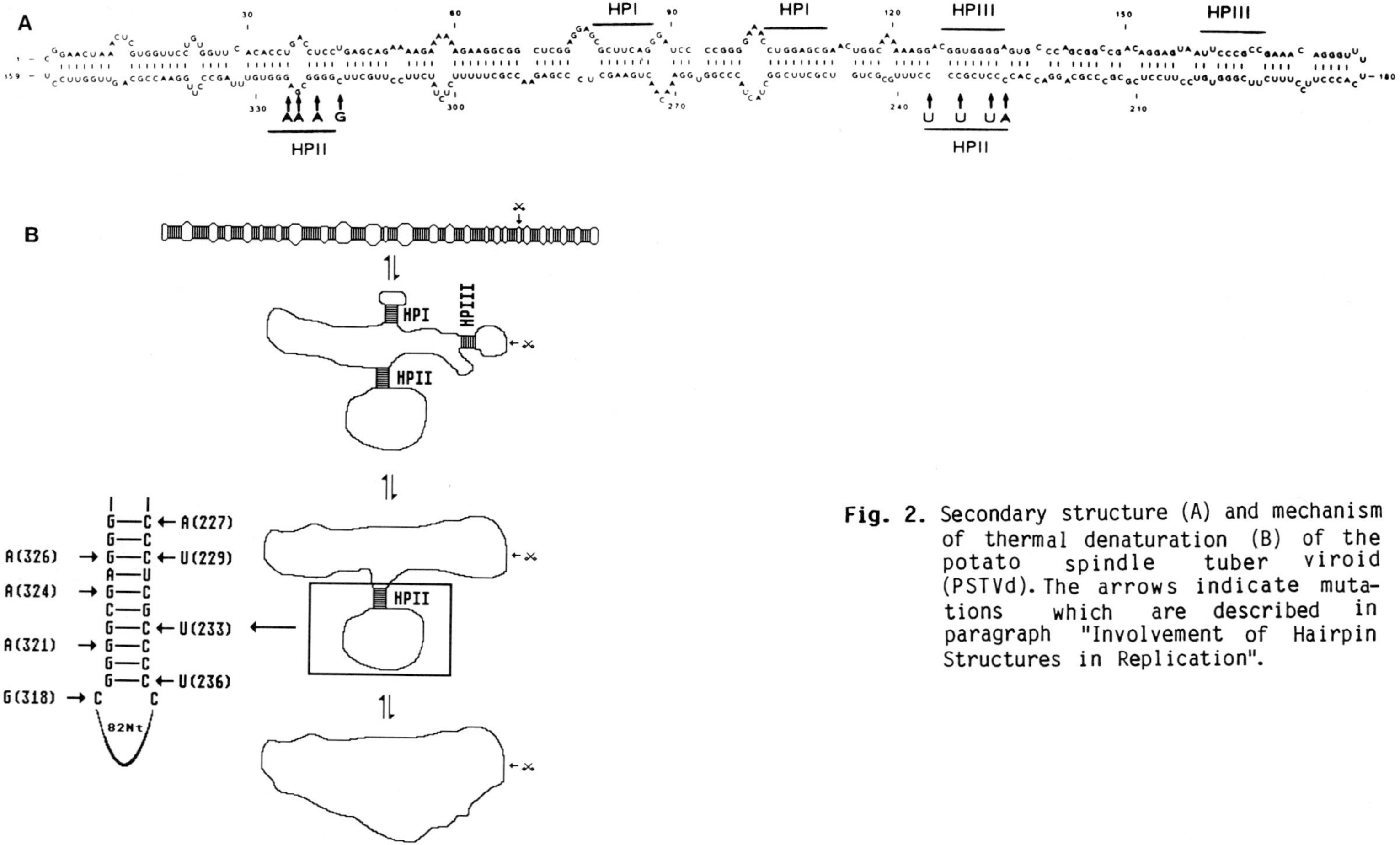


**Fig. 2.** Secondary structure (A) and mechanism of thermal denaturation (B) of the potato spindle tuber viroid (PSTVd). The arrows indicate mutations which are described in paragraph "Involvement of Hairpin Structures in Replication".

is explained most easily with the example of circular PSTVd (Fig. 3A). A linear temperature-gradient is established in a horizontal slab gel. The sample is applied in the broad sample slot which extends nearly over the whole width of the slab gel. The direction of the electrophoresis is from top to bottom, i.e. perpendicular to the temperature gradient. Thus, every individual molecule runs at constant temperature, those at the left side at low temperature, at the right side at high temperature, and in between at all intermediate temperatures. Since the instrument is presently industrially (DIAGEN-TGGE system, DIAGEN, Düsseldorf, FRG) manufactured, the method may be used as a routine procedure.

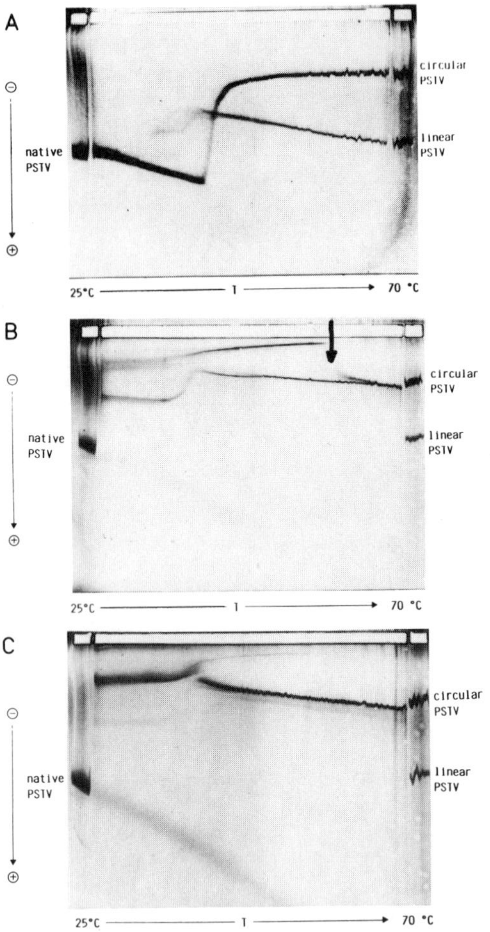


**Fig. 3.** Analysis of structural transisions of circular and linear PSTVd (A) and dimeric (+) strand RNA transcripts (B, C) by temperature-gradient gel electrophoresis (TGGE) in 5% polyacrylamide. Buffer condiions: 17.8 mM Tris, 17.8 mM boric acid, 0.4 mM EDTA. Direction of the electrophoresis is from top to bottom. The linear temperature-gradient is indicated. The gel was silver-stained. Marker slots (left and right side of the broad sample slot) contain natural circular PSTVd(C) and natural linear PSTVd(L). (Panel B) Transcript isolated from the transcription mixture, incubated in a high salt/urea buffer (1 mM Na-cacodylate, 500 mM NaCl, 4 M urea, 0.1 mM EDTA, pH 6.8) for 1 h at 53°C, cooled and analysed. (Panel C) Transcript analysed right after transcription without further treatment. Figure reproduced from [10].

As seen in Fig. 3A PSTVd migrates fastest in its native conformation and is drastically retarded within a narrow temperature range which corresponds to the main transition (cf. Fig. 2B). A further decrease of mobility is observed at higher temperatures. Some portion of the PSTVd-sample was in a linear form due to single nicks in the circle. Nicks at different sites lead to different stabilities of the native structure; correspondingly different transition curves are visible as faint bands all of which merge at high temperature into the curve of the completely denatured PSTVd-RNA. The advantages of TGGE over the conventional optical melting curves are obvious from Fig. 3A. A mixture of different molecules, e.g. circular and linear viroids, may be analysed in form of individual transition curves, whereas in optical melting curves only the superimposition of all curves is measurable (cf. [9]). The amount of material needed for one TGGE is as little as 100 ng or less if silver staining is used for detection. Radioactively labelled nucleic acids may be recorded as well, and if the detection is carried out by molecular hybridization with a specific probe, a single nucleic acid species may be analyzed even if it is present as a minor portion of a nucleic acid crude extract. The method of TGGE was originally developed for studies on viroids, but has been applied in the mean time for a large series of other problems of nucleic acids and proteins (cf. review[12]).

Linear oligomeric RNA transcripts

Structure formation during replication was simulated by synthesizing linear oligomeric RNA transcripts from PSTVd-cDNA clones with T7 polymerase [10]. An oligomeric RNA transcript is considered to represent a replication intermediate (cf. Fig. 1). Since the dimeric RNA (+)strand transcripts are the smallest model for oligomers, a series of studies was carried out on these molecules.

The secondary structures which dimeric molecules may adapt are depicted in Fig. 4.

Extended structure. An extended secondary structure without bifurcations (Fig. 4E) was calculated to be the most stable structure. It is described best as a doubling of the secondary structure of native circular viroids (cf. Fig. 4B).

Tri-helical structure. Because the region, which forms hairpin I in circular viroids, is present twice in the dimeric transcripts, both of them together may form a particularly stable double-helical region. This region shown in detail in Fig. 4A, contains three consecutive helices with alltogether 28 basepairs with a GC content of 71%. The central helix is formed only in dimeric transcripts because the sequence is located in the loop region of hairpin I. Once formed the other parts of the dimeric transcript will assume the structure as depicted in Fig. 4C. The structure is by 34 kJ/mol or 7% thermodynamically less favorable than the extended structure. Consequently the tri-helical structure is metastable.

Transient structures during synthesis (multi-hairpin-structures). During synthesis other than the extended or the tri-helical structure may be formed and may persist as metastable structures for some time. It was assumed that in a transient structure such as shown in Fig. 4D a hairpin is formed as soon as its sequence has been synthesized. In the particular model of Fig. 4D the well known stable hairpins of PSTVd and the left- and right-ended hairpins of the extended structure (designated by small arrows in their loops) are taken into account. Alternative models of a multi-hairpin structure could also be designed.

TGGE of dimeric transcripts. The analysis was carried out after a pretreatment of the sample in different solution conditions (for details see ref. [10]):

a) condition of thermal equilibrium between the extended structure (Fig. 4E) and its denatured conformation (as shown for PSTVd in Fig. 2B);
b) condition of synthesis, i.e. TGGE was carried out immediately after T7 transcription without further treatment.

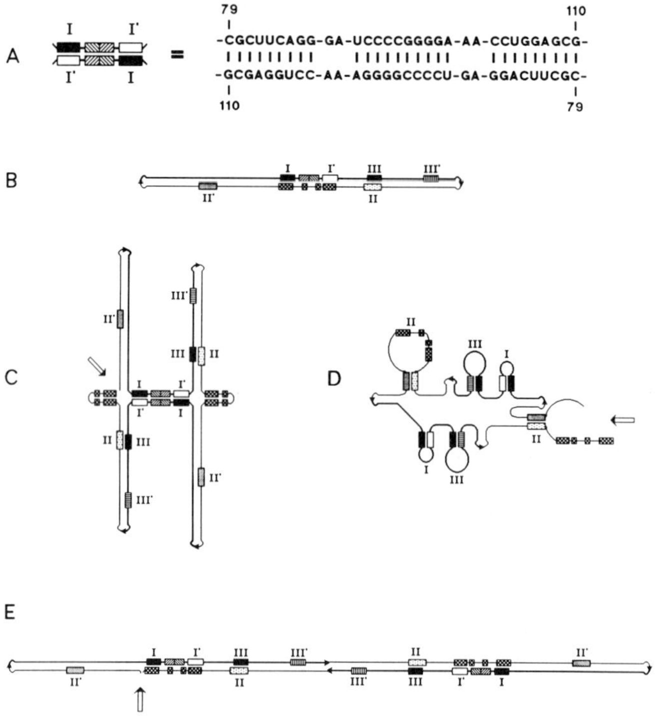


**Fig. 4.** Structural models of dimeric (+)strand transcripts. Complementary sequences are symbolized by boxes with corresponding graphic patterns. The 5' to 3' direction of the strand is indicated by arrows which are positioned in the hairpin loops of the left and the right end of the native structure. (Panel A) Schematic representation of the upper central conserved region of PSTVd (nt 79-110) and of the arrangement of two conserved regions in three successive helices. Segments I and I' form hairpin I during the thermal denaturation of the circular PSTVd (cf. Fig. 2). (Panel B) Secondary structure of mature circular viroid (see Fig. 2) showing the position of the segment from panel A and of the segment II and II' which form hairpin II, and III and III' which form hairpin III. The chequered boxes are complementary sequences (cf. Panel C) in the lower central conserved region.(Panel C) Structure of the dimeric transcript with the three successive helices, also called tri-helical structure. (Panel D) Transient or multi-hairpin structure with stable hairpins (I:nt 79-87/102-110; II:nt 227-236/319-328; III:nt 127-135/160-168; hairpin at th left end of the native structure: nt 3-28/334-357; hairpin at the right end of the native structure: nt 169-177/182-192). (Panel E) Extended structure similar to that of native circular viroid. The thick arrows with the open shafts in panel C-E represent the locations of the start and the stop of the linear transcript.

In Fig. 3B and C the results from TGGE after treating the dimers in solution conditions (a) and (b) are shown. Under equilibrium condition (Fig. 3B) two dominant bands and one faint band were visible, corresponding to three different structures. This is a graphical example for the potency of TGGE to analyse co-existing structures. The fastest migrating band was clearly identified as the extended structure because the transition was very similar to that of a circular PSTVd or a monomeric linear viroid (Fig. 3A). The band of slowest mobility has to represent the tri-helical structure because the three consecutive helices are the only structural elements which are stable enough to account for the transition at 58-60°C (designated in Fig. 3B by an arrow).

The faint band in Fig. 3B between the bands of the extended structure and the tri-helical structure was identical to that obtained under the conditions of synthesis (b) (Fig. 3C). Therefore, it has to be the transient structure (cf. Fig. 4D). One should note in Fig. 3C that to a minor extent also in the transcription mixture the extended structure and the tri-helical structure are present. Thus one has to conclude, that the transient structure is able to switch over into the other structures under conditions which are close to physiological.

In summary, the experiments and calculations have shown that viroids assume under physiological conditions a rod-like structure which switches over in a hairpin-containing structure at elevated temperature. The same hairpins were found, however, under physiological conditions, when viroid replication intermediates were analysed right after their synthesis.

## LOCATION OF VIROIDS IN NUCLEOLI

After describing the different structures of viroids we will concentrate on the relation between structure and protein interaction. Before describing molecular complexes we will summarize the cellular location of viroids.

The intracellular localization of viroids has been investigated by viroid-specific *in situ* hybridization and analysis by digital microscopy of the distribution of the fluorescent hybridization signal [13]. Isolated nuclei from green leaf tissue of tomato plants infected with potato spindle tuber viroid (PSTVd) were bound to microscope slides, fixed with formaldehyde and hybridized with biotinylated transcripts of cloned PSTVd cDNA. The bound probe was detected with lissamine-rhodamine conjugated streptavidin. Nucleoli were identified by immunofluorescence using the monoclonal antibody Bv96 [14] and a secondary FITC-conjugated antibody. In plants infected with either a lethal or an intermediate PSTVd strain, the highest intensity of fluorescence that arose from hybridization with the probe specific for the viroid (+)strand was found in the nucleoli, confirming results of previous fractionation studies [15]. A similar distribution was found for (-)strand replication intermediates of PSTVd using specific (+)strand transcripts as hybridization probes.

In order to determine if viroids are located at the surface or in the interior of the nucleoli, the distribution of the fluorescence hybridization signals was studied with a confocal laser scanning microscope (CLSM). In Fig. 5 the intensity distribution in an optical section with a focal depth of only 0.4 μ through an isolated nucleus is displayed as a three dimensional isometric projection, with pixel grey values represented by vertical displacement in form of a wire frame. The area in the crossection where the high viroid concentration was detected, is exactly the area of the nucleolus. It can be concluded that viroids are neither restricted to the surface of the nucleoli nor to a peripheral zone, but are instead homogeneously distributed throughout the nucleolus.

The particular subcellular localization of viroids clearly differentiates them from plant viruses. The latter generally accumulate and replicate in the cytoplasm and only rarely in the nucleus. No virus, viral RNA or any other form of a pathogen is known to be targeted to the nucleolus, PSTVd being the notable exception. However, since this is the only viroid species and infected tomato leaves the only tissue examined in this connection to

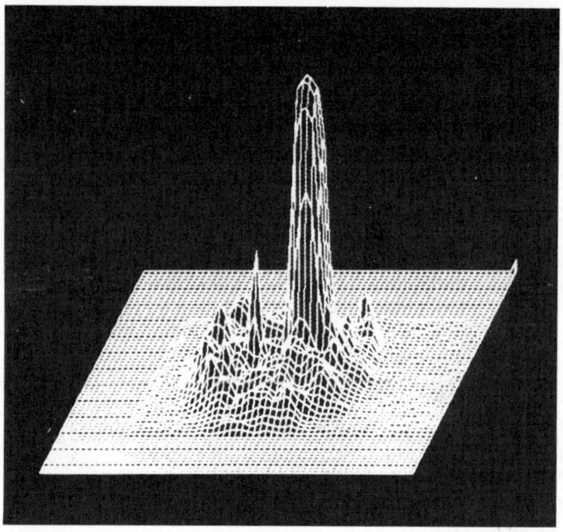

**Fig. 5.** Distribution of viroid concentration in a crossection through an isolated nucleus displayed as a three-dimensional isometric projection. Viroid concentration was measured as the fluorescence intensity after in situ hybridization with a fluorescence-labelled RNA probe; fluorescence intensity was scanned in an optical section of 0.4 μ focal depth in a confocal laser scanning microscope. Figure is reproduced from Harders et al. [13].

date, it remains to be established whether the involvement of the nucleolus in the mechanism of viroid infection is a general phenomenon.

## INTERACTIONS WITH "STRUCTURAL" PROTEINS

Viroid-host interactions were studied by two sets of experiments. Firstly, an attempt was made to bind isolated viroids to cellular proteins. Secondly, cellular viroid-protein complexes were analysed in situ by u.v. irradiation crosslinkig studies using isolated nuclei from infected plants.

### In vitro Reconstruction of Viroid-protein Complexes

Reconstitution of complexes with filter-bound proteins. Cellular proteins were separated by gel electrophoresis, blotted to nitrocellulose filters, renatured, and incubated with isolated viroids to form viroid-protein complexes. In these experiments the proteins to which viroids were bound, are well characterized with respect ot $M_r$ and/or charge. The conditions of the binding experiments, however, are far from being native. Those experiments were described in detail by Wolff et al. [16], and Klaff et al. [17]. A nuclear 43 KD protein was found to bind viroids even in the presence of a large excess of other RNA.

Reconstruction in solution. Reconstitution of viroid-protein complexes may also be carried out in solution before analysis by gel electrophoresis. Complex formation in solution occurs under native or near to native conditions, but analysis of the proteins involved will be more indirect.

Binding experiments in solution were performed using crude nuclear protein extracts from green leaf tissue (for details see [17]). Twenty μg of protein was incubated with 50 ng PSTVd, and the complexes were crosslinked by u.v. radiation. They were detected by molecular hybridization of the

PSTVd component after SDS-PAGE (7.5% polyacrylamide gels) and Western blotting onto nitrocellulose filters. If PSTVd was incubated with protein extracts without crosslinking by u.v. irradiation, no signals were detected indicating that the bands occurring in the crosslinked samples were due neither to the binding of the hybridization probe or free PSTVd nor to the transferred protein. Competition experiments showed that a 1000-fold excess of competitor RNA abolished viroid binding.

The position of the bands of viroid-protein complexes in SDS-PAGE was compared with marker proteins. In this way apparent $M_r$-values were determined: these were 72K, 56K, 47K, 42K, and 33K.

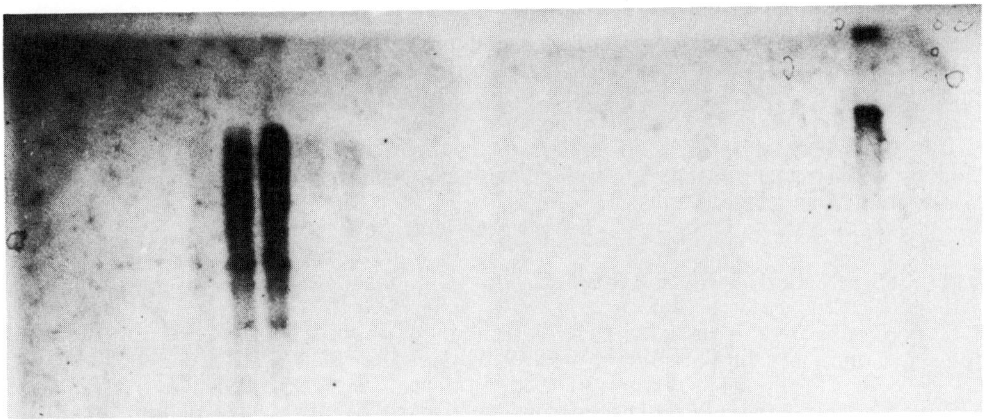


**Fig. 6.** Analysis of a sucrose gradient for cellular viroid-protein complexes after crosslinking by u.v. irradiation. The fractions from the gradient were analysed by 10% SDS-PAGE and tested for their viroid content by transfer to nitrocellulose and molecular hybridization using a PSTV-specific probe. Lanes 1 to 17 represent fractions 1 to 17; lane P, pellet. Fraction 1 refers to the top of the gradient, fraction 17 to the bottom. Figure is reproduced from [17].

## Crosslinking of Viroid-protein Complexes in Isolated Nuclei

Cellular viroid-protein complexes were identified in situ using the u.v. irradiation crosslinking method as reported by Klaff et al. [17]. Isolated nuclei (2 x $10^8$) from infected plants were irradiated with u.v. light to introduce covalent crosslinks between proteins and nucleic acids. Nucleoli were prepared from irradiated nuclei, digested with DNase I, sonicated according to the protocol to prepare 10S particles [16], and fractionated by a 5-30% (w/v) sucrose gradient. Fractions (2 ml) from a 25 ml gradient were collected and extracted with phenol. Covalent nucleic acid-protein complexes should be found in the phenolic phase, from which the proteins and nucleic acid-protein complexes were precipitated with ethanol, separated by SDS-PAGE, blotted to nitrocellulose filters and analysed for their content of viroids by hybridization. Binding of viroids to the filters was mediated solely by the protein crosslinking with the viroid. The distribution of viroid-protein complexes in the sucrose gradient fractions is shown in Fig. 6. Only in fractions 5 and 6 were hybridization signals clearly visible over the background. The sedimentation coefficients of these fractions are 7-8S and 10S, respectively. Thus, it has to be concluded that viroids are organized inside the cell in well defined complexes with proteins. Furthermore, it was seen from the SDS-PAGE of fractions 5 and 6 that after complete denaturation of the 10S particles with SDS, distinct bands of viroid-protein complexes were visible; these had the following apparent $M_r$-values as estimated form the marker proteins: 95K, 89K, 75K, 58K, 48K and 38K.

The apparant $M_r$-values of the complexes were correlated with those of the proteins bound to the viroid. This was achieved by combining a gel electrophoresis of the crosslinked nucleic acid-protein complexes with that of the protein alone after the nucleic acid moiety had been digested by nuclease treatment. From these experiments it could be concluded that the complex with an apparent $M_r$ of 75K contained a 56K protein, and the complex of $M_r$ of 58k contained a 43K protein. The 43K protein was identified as a viroid binding protein in the experiments mentioned above. The same protein could also be released from cellular viorid-protein complexes by fractionation with HPLC [17].

In summary, several lines of experiemental evidence showed that in nucleoli viroids are associated with proteins. Those cellular complexes showed several species which were characterized by their apparent $M_r$-values. Similar $M_r$-values were obtained if viorid-protein complexes had been reconstituted in solution. The reconstituted complexes with $M_r$ of 72K, 56K, 47K correspond within the limit of experimental error to cellular complexes with $M_r$ of 75K, 58K, and 48K. The cellular protein of 43K which could be identified as a major component of the complexes, will be purified and characterized in future experiments.

## COMPLEXES OF VIROIDS AND POLYMERASE

Considerable experimental evidence points to an involvement of the specific nuclear DNA-dependent RNA polymerase II in viroid replication [7,8,18]. Significantly, transcription studies with DNA-dependent RNA polymerase II, purified from wheat germ and tomato plants, have shown that the enzyme can synthesize full-length (-)viroid RNA copies *in vitro* [8].

The interaction of viroids and purified polymerase II from wheat germ has been studied quantitatively by *in vitro* transcription, determination of binding constants in the analytical ultracentrifuge and by electron microscopy of viroid-polymerase complexes [18]. Figure 7 shows electron micrographs of the complexes which can be observed between polymerase II and viroid (left panel), virusoid (middle panel), and a 187 bp DNA fragment (right panel). The latter two nucleic acids were used as controls for the specificity of binding under the given conditions. Virusoids are circular satellite RNAs which resemble viroids in size and structure but are not replicated by the enzyme. The 187 bp long fragment of *E. coli* DNA has the same size as PSTVd but no specific binding site for the eukaryotic RNA polymerase. The electron micrographs provided visual evidence for polymerase binding on all three nucleic acids; the binding occurred in most cases at the ends of the rod-like molecules. Whith the two-fold molar excess of enzyme used in the experiment, both 1:1 and 2:1 enzyme to nucleic acid stoichiometries were found.

From electron microscopic evidence only, it was not possible to distinguish between specific and nonspecific polymerase binding, so the binding constants of the polymerase-nucleic acid complexes were determined by quantitative evaluation of the sedimentation profiles measured in the analytical ultracentrifuge. The 1:1 polymerase-viroid complex showed a binding constant of $1.9 \times 10^7\ M^{-1}$ but the affinity of the 2:1 complex was markedly lower. This suggests that (a) the 2:1 polymerase-viorid complex is not physiologically significant and (b) the binding at one of the ends of the viroid RNA must be nonspecific. Since the virusoid and the 187 bp DNA fragment were bound to the polymerase with a distinctly lower affinity than viroid RNA it can be concluded that the polymerase binding to virusoid and 187 bp DNA seen on the micrographs must be nonspecific. Regarding the influence of viroid structure we have to conclude that circular viroids are present in their rod-like secondary structure when binding to the polymerase.

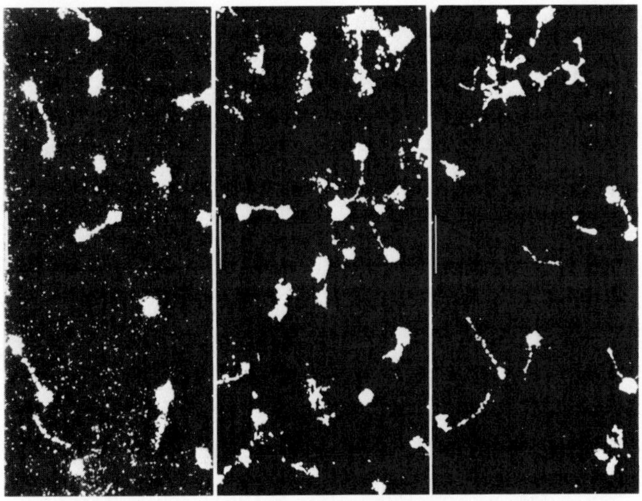

**Fig. 7.** Electron micrographs showing formation of complexes between DNA-dependent RNA polymerase II and PSTVd (left panel), vVTMoV (velvet tobacco mottle virusoid, middle panel), and 187 bp DNA fragment (right panel). Figure is reproduced from Goodman et al. [18].

## INVOLVEMENT OF HAIRPIN STRUCTURES IN REPLICATION

The functional relevance of hairpin I and hairpin II was studied by site-directed mutagenesis. The mutants were analysed in respect to their structure as well as to their functional changes. Preliminary data on hairpin I suggested that hairpin I is involved in processing the oligomeric (+)strand RNA to monomeric circles [19,20]. In this context the experiments on hairpin II will be described in some detail because more firm conclusions can be drawn at present. The details are described elsewhere [21,22].

### Experiments after Site-directed Mutagenesis in Hairpin II

Site-specific mutations were introduced into the regions of hairpin II. These muations are added to the presentation of the structure and structural transitions in Fig. 2. All mutations were single-site mutations. It is obvious that these mutations destabilize hairpin II, except the mutation at site 318 which stabilizes due to the formation of an additional base pair adjacent to the stem (cf. Fig. 2B). The sites of mutation were selected carefully in order to introduce as little perturbations as possible in the native structure (cf. Fig. 2A). Only 318 transforms a mismatch between two helical regions into a base pair.

Thermodynamic analysis. Using standard procedures the mutations were introduced in monomeric and dimeric cDNA clones, and from these templates monomeric and dimeric RNA transcripts were synthesized. Although the main interest was directed towards analysis of the biological effects of the mutations, the thermodynamics of the mutants was studied, too. Monomeric RNA transcripts were analysed by thermodynamic calculations and by experimental TGGE. Without describing the details it can be stated that the mechanism of the conformational transitions as depicted in Fig. 2 could be confirmed quantitatively: Destabilization of hairpin II by destroying a base pair led to a stabilization of the rod-like structure, i.e. the switch from the extended to the branched structure was shifted to higher temperature. Other details may be seen from the original literature [21].

Infectivity and genetic stability. All mutants were tested for their infectivity on tomato plants. For this purpose the plants were inocculated with the corresponding double-stranded DNA of dimeric length or with their linear RNA transcripts. Both types of inocculation gave identical results. To our big surprise, all mutants exept 321 were infectious. The sequences of all progeny-viroids were determined. Mutant 321 contained a second mutation due to a cloning artefact; it will therefcore not be discussed further. Only a portion of the other mutations were stable in the sequences of the progenies, the other portion was identified as revertants to the wild type sequence after the first passage through the host plant. In Fig. 8 the mutations in hairpin II are classified according to the results of sequencing as stable (S) or revertant (R). Reversion has to occur in the plant and cannot come from transcription errors during synthesis of the transcripts, because inocculation with cloned DNA or with RNA transcripts gave identical results.

## Hairpin II as a Regulatory Element for Replication

Indispensable core and variable periphery. The results allowed us to correlate the position of the mutation with the biological activity. Obviously, the stable mutations affect the structure in a way which is tolerated during replication. The reversions, however, have to be part of a structural element which is indispensable for replication. A high selection pressure in favour of the wild type sequence has to be assumed so that the revertants overgrow the mutants during the replication in the cell very fast.

Whereas a correlation of the genetic stability of the mutation with its influence on the native structure of the viroid could not be drawn, such a correlation with the position of the mutations in hairpin II is quite convincing. In Fig. 8 hairpin II is divided into a core and two peripheral segment. The mutations in the core are reverted to the wild type, the mutations in the peripheral segments remain stable. Thus, it may be generalized from the present data, that the structural integrity of the core is indispensable for replication whereas mismatches or wobble-pairs in the peripheral segments are tolerated.

Constant core and variable periphery in other viroids. As mentioned above hairpin II is present also in the other viroids of the PSTVd-class. PSTVd, TPMVd, TASVd, CEVd, CSVd, and CLVd belong to that class. The structure of hairpin II in the different viroids varies slightly as depicted in Fig. 9A. They may be aligned, however, so that the core is identical in all of them. This is demonstrated by the box in Fig. 9A. The only exception is an AU pair in CSVd compared to a GC pair in all other viroids. Thus, the core region of hairpin II which cannot be varied in PSTVd without losing the infectivity is also constant if different viroids are compared. The variable segments in hairpin II of PSTVd correspond to significant variations in hairpin II of other viroids. In summary, the variations in the structure of hairpin II, which were tolerated after site-directed mutagenesis of PSTVd are at very similar sites like the natural variations.

Homology with recognition sites of host genes. Structural elements similar to hairpin II also exist in the host. In fact, GC-rich segments are found in the 5' untranscribed regions of eukariotic genes. Unfortunately, those have been reported so far mostly for human [23], animal [24] and viral systems [25] whereas corresponding regions from plants have been analysed only in one case [26]. A striking similarity between hairpin II from PSTVd and a GC-rich segment from the 5'-upstream region of the superoxid-dismutase gene from man was found [23]. It is shown in Fig. 9B. Although depicted in Fig. 9B as a hairpin structure within one strand, the double-stranded GC-rich region may be formed in the host DNA also by the (+)- and the (-)strand. Thus, a hairpin structure in the viroid could correspond to a segment of the genomic double-stranded DNA of the host. Other examples of GC rich segments could be given with similar homology. Most of those GC-segments belong to so-called house-keeping genes and have been discussed to act as binding sites for transcription factors [24].

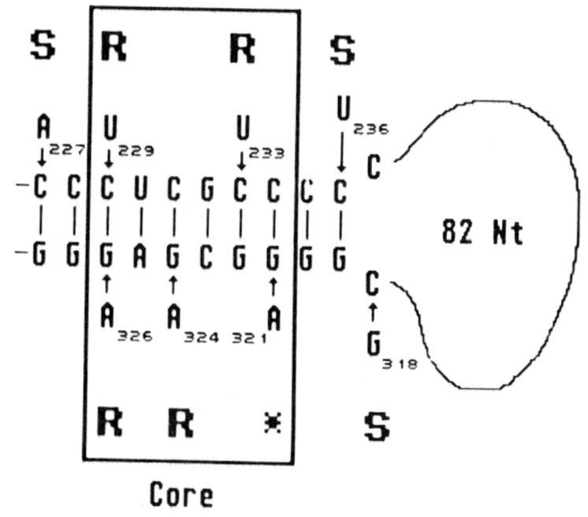


Fig. 8. Stable mutations (S) and reversions (R) in hairpin II of PSTVd. The core of the hairpin depicted in a box contains all reversions. For mutation 321 see text.

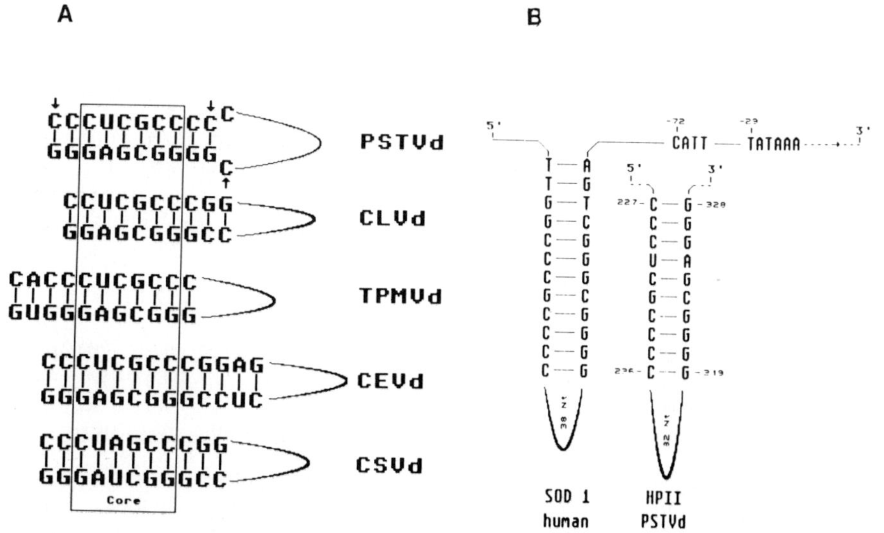


Fig. 9. Comparison of hairpin II in different viroids (A) and with a 5'-upstream GC-rich segment in the gene of superoxide-dismutase [23] (B).

Hairpin II in viroids and homologous DNA segments in host genes: Both form an A-double-helix. A RNA structural motif in viroids is compared with a DNA motif in host genes. Is this comparison legitimate? Recently several structures of double-stranded DNA oligomers with high GC-content were studied by X-ray analysis [27]. It was found that those fragments of DNA assume an A-form double-helix rather than the typical B-form. Although the A-form of DNA and the A-form of RNA are not identical, they are much more similar than A-form RNA and B-form DNA. If recognition of GC-rich segments would be interpreted merely on the basis of their A-form in an otherwise B-form DNA, one may speculate that viroids are taking their chance of utilizing a cellular recognition structure in which RNA and DNA assume nearly the same structure.

Hairpin II as a hypothetical binding site of transcription factors. The homology between a strucural element in viroids and in host genes suggests that both structures may act as a recognition structure for the same host protein. A priori recognition structures which are essential for transcription and/or replication are promotors, enhancers and binding sites for transcription factors. Because viroids are replicated by DNA-dependent RNA polymerase II (Pol II) one may consider that hairpin II acts as a promotor for this enzyme. On the other hand Pol II was shown by electron microscopy (cf. Fig. 7) to bind to the left- and right-end hairpins of the native rod-like structure. Thus, the electron micrographs of viroid-Pol II-complexes do not indicate an involvement of hairpin II in Pol II-binding. Inside the cell, however, the situation is more complex as presented in Fig. 7: Transcription factors bind to the template in addition to Pol II; not only the circular viroids in the most stable conformation but also the oligomeric viroid intermediates in metastable conformations may act as templates for Pol II. Taking into account that the host structural element which is homologous to hairpin II acts as binding site for transcription factors [24] it seems appropriate to assume that a not yet identified transcription factor of the host binds to hairpin II. Since in circular viroids the transcription factor could bind to hairpin II only after the native structure is denatured, it is more obvious to assume that the transcription factor binds to hairpin II in the (-)strand oligomeric viroid RNA. As shown above, this strand is synthesized in a multibranched structure and does not assume the rod-like structure during the replication cycle. Possibly, also the oligomeric (+)strand RNA may act as a template for further (-)strand RNA synthesis. Therefore, from the thermodynamic point of view the formation and thereby the functional relevance of hairpin II is more important in the oligomeric replication intermediates: These molecules are synthesized in a structure with hairpin II and require the recognition of hairpin II by a host factor for synthesis of their counter strands.

## REFERENCES


1. Diener, T.O. (1979) In: 'Viroid and Viroid Diseases', Wiley and Sons, N.Y.
2. Sänger, H.L. (1982 In: 'Encyclopedia of Plant Physiology', (Parthier, B. and Boulter, D., eds.), New Series, Vol 14 B, Springer-Verlag, Berlin/Heidelberg, N.Y., pp. 368-454.
3. Riesner, D. and Gross, H.J. (1983) "Viroids" Ann. Rev. Biochem. 54:531-564.
4. Diener, T.O. (ed.) (1987) "The Viroids", Plenum Publ. Corp.
5. Riesner, D. and Steger, G. In: 'Viroids and viroid-like RNAs' Landolt-Börnstein, New Series in Biophysics - nucleic acids -, Vol VII/1 (W. Saenger, ed.), Springer Verlag, Berlin, in press.
6. Branch, A.D. and Robertson, H.D. (1984) "A Replication cycle for viroids and other small infectious RNAs" Science 223:450-455.

7. Mühlbach, H.P. and Sänger, H.L. (1979) "Viroid replication is inhibited by $\alpha$- amanitin" Nature 278:185-187.
8. Rackwitz, H.R., Rohde, W., and Sänger, H.L. (1981) "DNA-dependent RNA polymerase II of plant-origin transcribes viroid RNA into full-length copies" Nature 291:297-301.
9. Riesner, D. (1987) "Structure formation" In: 'The Viroids' (T.O. Diener, ed.) Plenum, New York.
10. Hecker, R., Wang, Z., Steger, G., and Riesner, D. (1988) "Analysis of RNA structures by temperature-gradient gel electrophoresis: replication and processing" Gene 72:59-74.
11. Rosenbaum V. and Riesner, D. (1987) "Temperautre-gradient gel electrophoresis: thermodynamic analysis of nucleic acids and proteins in purified form and in cellular extracts" Biophys. Chem. 26:235-246.
12. Riesner, D., Steger, G., Zimmat, R., Owens, R.A., Wagenhöfer, M., Hillen, W., Vollbach, S., and Henco, K. (1989) "Temperature-gradient gel electrophoresis of nucleic acids: Analysis of conformational transitions, sequence variations, and protein-nucleic acid interactions" Electrophoresis 10:377-389.
13. Harders, J., Lukacs, N., Robert-Nicoud, M., Jovin, T.M., and Riesner, D. (1989) "Imaging of viroids in nuclei from tomato leaf tissue by in situ hybridization and confocal laser scanning microscopy" EMBO J. 8:3941-3949.
14. Frasch, M. (1985) "Charakterisierung chromatinassoziierter Kernproteine von Drosophila melanogaster mit Hilfe monoklonaler Antikörper" Thesis, Eberhard-Karls-Universität, Tübingen, F.R.G.
15. Schumacher, J., Sänger, H.L., and Riesner, D. (1983) "Subcellular localization of viroids in highly purified nuclei from tomato leaf tissue" EMBO J. 2:1549-1555.
16. Wolff, P., Gilz, R., Schumacher, J., and Riesner D. (1985) "Complexes of viroids with histones and other proteins" Nucleic Acids Res. 13:355-367.
17. Klaff, P., Gruner R., Hecker, R., Sättler, A., Theißen, G., and Riesner D. (1989) "Reconstituted and Cellular Viorid-Protein Complexes" J. Gen. Virol. 70:2257-2270.
18. Goodman, T.C., Nagel, L., Rappold, W., Klotz, G., and Riesner, D. (1984) "Viroid replication: Equilibrium association constants and comparative activity meassurements for the viroid-polymerase interaction" Nucleic Acids Res. 12:6231-6246.
19. Steger, G., Mörchen, M., Riesner, D., Tabler, M., Tsagris, M., and Sänger, H.L. (1989) "Secondary structure requirements for potato spindle tuber viroid processing" Cold Spring Harbor Mtg. on RNA Processing 1989, abstr. 236.
20. Sänger, H.L., Tsagris, M., Tabler, M., Schiebel, W., Anders, C., and Haas, B. (1989) "Processing of viroids in vitro and in vivo" Cold Spring Harbor Mtg. on RNA Processing 1989, abstr. 219.
21. Loss, P. (1990) "Untersuchungen zur Funktion der Haarnadelstruktur II des PSTVd mit Hilfe der in vitro-Mutagenese" Thesis, Heinrich-Heine-Universität Düsseldorf, F.R.G.
22. Loss, P., Schmitz, M., Steger, G., and Riesner, D., Manuscript in preparation.
23. Levanon, D., Lieman-Hurwitz, J., Dafni, N., Wigderson, M., Sherman, L., Bernstein, Y., Laver-Rudich, Z., Danciger, E., Stein, O., and Groner, Y. (1985) "Architecture and anatomy of the chromosomal locus in human chromosome 21 encoding the Cu/Zn superoxide dismutase" EMBO J. 4:77-84.
24. Gidoni, D., Dynan, W.S., and Tjian, R. (1984) "Multiple specific contacts between a mammalian transcription factor and its congnate promotors" Nature 312:409-413.
25. McKnight, S.L., Kingsbury, R.C., Spence, A., and Smith, M. (1984) "The distral transcription signals of the herpesvirus tk gene share a common hexanucleotide control sequence" Cell 37:253-262.
26. Matsuoka, M. and Minami, E. (1989) "Complete structure of the gene for phosphoenologyruvate carboxylase from maize" Eur. J. Biochem. 181:593-98

27. Heineman, K., Lauble, H., Frank, R., and Blöcker, H. (1987) "Crystal structure analysis of an A-DNA fragment of 1.8 Å resolution: d(GCCCGGGC)" Nucleic Acids Res. 15:9531-9550.

# ANALYSIS OF VIROID PATHOGENICITY BY GENOME MODIFICATION

Rosemarie W. Hammond and Robert A. Owens

Microbiology and Plant Pathology Laboratory
USDA-ARS, Beltsville, Maryland USA 20705

## INTRODUCTION

Viroids are the smallest autonomously replicating pathogenic agents yet described.[1] Viroids are low molecular weight RNAs (1.1-1.3 x $10^5$) that can be isolated from certain species of plants afflicted with specific diseases. Although viroids have been discovered because they cause symptoms in some hosts, they are often replicated in other hosts without causing obvious damage.

The single-stranded, circular RNA genomes of known viroids range in size from 246-375 nucleotides. Viroids differ from conventional viruses and satellite RNAs in their apparent lack of mRNA activity and the absence of a helper virus for replication and movement. They represent a novel class of subviral pathogens and, as the smallest known infectious agents, provide a minimal genetic and biochemical system for study of mechanisms controlling host-pathogen interactions and the control of gene expression in host cells. Two important aspects of viroid function - pathogenicity and host range - are amenable to study using the techniques of molecular biology.

Viroid pathogenicity is a complex process that is influenced by both the viroid and host genomes. As viroids lack an apparent mRNA activity, they may act as regulatory RNAs in the host cell and interact directly with host cellular constituents for the production of disease symptoms. At the biochemical level, little is known about the molecular interactions believed to be responsible for symptom expression. Comparison of field isolates which differ in severity of symptoms produced on tomato suggests that one particular region of the viroid molecule, the pathogenicity domain, may be important in the interaction of the molecule with host cellular components and influence symptom production.

The molecular mechanisms responsible for host range restriction are also obscure, but evidence accumulated from

Plant Molecular Biology 2, Edited by R.G. Herrmann and
B. Larkins, Plenum Press, New York, 1991

studies of virus/host interactions suggests that the transport function involved in systemic spread may control host range. We have begun to examine the the molecular signals which may be required for movement of viroid molecules from primarily infected cells.

The introduction of site-directed mutations into infectious cloned copies of viroid cDNAs and the construction of chimeric viroid molecules has allowed us to address these questions, among others. Mutational analysis has revealed the complex relationships between structure and function and has prompted us to examine the behaviour of selected mutants in systems other than the conventional systemic tomato bioassay.

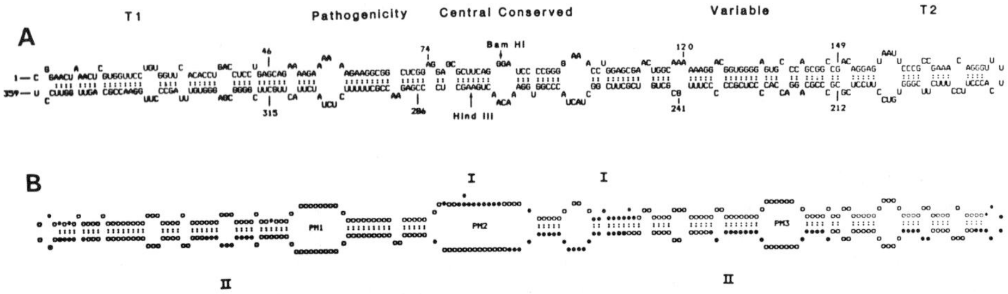


Figure 1. A. Nucleotide sequence and proposed secondary structure of the intermediate strain of potato spindle tuber viroid (PSTVd). The five structural domains proposed by Keese and Symons[6] are indicated above the native structure.
B. Premelting regions (PM) 1 - 3 and the stems of secondary hairpins I and II are indicated.

## MOLECULAR STRUCTURE AND BIOLOGICAL PROPERTIES OF VIROIDS

A secondary structure model for the first complete viroid sequence, the 359 nucleotide sequence of the type strain of potato spindle tuber viroid (PSTVd) was derived by Gross et al.[2] (Figure 1A). Unusual structural features of viroids include their single-stranded circularity, a highly base-paired secondary structure, and characteristic cooperative structural transitions during thermal denaturation.[3] Quantitative thermodynamic and kinetic studies of their denaturation indicate that viroids can assume alternative secondary structures that may possibly control various biological functions. Stable secondary hairpins involving the pairing of distant complementary bases form (eg. hairpins I and II, Figure 1A). Although the biological significance of these structures is unknown, mutational analysis of hairpin I[4] and hairpin II[5] suggest that they play a critical role in viroid replication.

When the stabilities of helices present in the native viroid molecule are calculated at increasing temperatures, three regions show lowest stability and have been designated as the premelting (PM) regions (1 - 3)[7] (see Figure 1B). As will be discussed later, correlations have been made between the decreased thermodynamic stability of premelting loop 1 and increasing virulence of PSTVd.[8]

Comparative analysis of viroid sequences has allowed Keese and Symons[6] to divide the native structure into five structurally distinguishable regions - the central conserved region (CCR), flanking pathogenicity (P) and variable (V) domains, and two terminal loops (T1 and T2)(illustrated in Figure 1A). Specific functions have been tentatively assigned to these regions, eg., the central conserved region has been proposed as the site for processing of (+) strand multimers in viroid replication[9] and comparative sequence analysis of different viroids and viroid strains strongly suggests that symptom expression is controlled by the pathogenicity domain. Structure/function correlations of this type provide a conceptual framework for site-directed mutagenesis studies of infectious cDNAs, and may lead to a better understanding of the requirements for biological functions, such as viroid replication and pathogenesis.

## ANALYSIS OF VIROID-HOST INTERACTIONS:PATHOGENICITY

Viroids differ from viral and viral satellite RNAs in several characteristics, most fundamentally in their apparent lack of an mRNA activity.[10] If they are indeed devoid of any mRNA activity, one would also speculate that there must be a direct interaction between the viroid RNA and host cellular constituents for the production of disease symptoms. Correlations between viroid pathogenicity and the ability of a certain region of the viroid molecule, the pathogenicity modulating region (Figure 1A), to interact with cellular constituents to produce disease symptoms have been made. In order to understand the subtleties of these interactions, we have employed the technique of site-directed mutagenesis of cloned, infectious copies of PSTVd cDNAs to introduce mutations into the pathogenicity modulating region of PSTVd.

### Site-directed and Random Mutagenesis in the Pathogenicity Modulating Region

Sequence comparisons between the intermediate (Type) and a severe (KF440) strain of PSTVd have shown that symptom expression is dramatically altered by a four-nucleotide difference with the P domain.[8] To test the plasticity of this region, we initially introduced one of these changes, a single G → C substitution at position 46 (Table 1), into cloned, infectious copies of PSTVd.[4] This mutation, which may disrupt base-pairing in the viroid native structure, resulted in a loss of cDNA infectivity. In a more recent series of studies, we have determined the effect of single and multiple nucleotide substitutions upon symptom expression in PSTVd. Successive site-directed mutations were introduced into the intermediate strain of PSTVd to create various combinations of the four nucleotide difference between the intermediate and severe strains (Table 1). The infectivity of each altered cDNA was determined and the progress of disease development was monitored.

Not all the mutant cDNAs produced were infectious when inoculated onto tomato. However, the severity of symptoms incited by two of the infectious variants (ie. nucleotide substitutions at residues 46 and 47 and at 46, 47, and 315) were milder than the intermediate strain parent. As expected, the newly created severe strain containing all four mutations incited severe symptoms on tomato. Determination of viroid titers in plants inoculated with the infectious cDNAs four weeks post-inoculation revealed no striking differences (data not shown).

The most stable secondary structures for this region were calculated using the algorithm of Zuker.[11] The predicted thermodynamic stabilities of the infectious variants are less than that of the intermediate strain, whereas those calculated for the lethal mutations are dramatically decreased (Table 1).

TABLE 1

Infectivity of PSTVd P domain cDNAs on tomato

| | Infectivity[a] | G kcal/mol)[b] |
|---|---|---|
| Site-directed mutations | | |
| wild-type | +,(int.) | 0 |
| $G_{46} \rightarrow C$ | - | +4.8 |
| $G_{46} \rightarrow C$, $C_{47} \rightarrow A$ | +,(mild) | +2.1 |
| $G_{46} \rightarrow C$, $C_{315} \rightarrow U$ | - | +4.6 |
| $G_{46} \rightarrow C$, $C_{315} \rightarrow U$, $U_{317} \rightarrow U$ | - | +6.5 |
| $G_{46} \rightarrow C$, $C_{47} \rightarrow A$, $C_{315} \rightarrow U$ | +,(mild-int.) | +2.1 |
| $G_{46} \rightarrow C$, $C_{47} \rightarrow A$, $C_{315} \rightarrow U$, $U_{317} \rightarrow C$ | +,(severe) | +2.3 |
| Random mutations | | |
| $A_{310} \rightarrow G$ | +,(mild) | -0.1 |
| $C_{311} \rightarrow U$ | +,(mild) | -1.1 |
| $C_{303} \rightarrow U$ | +,(int.) | -2.8 |

[a] Infectivity was monitored on 'Rutgers' tomato plants following inoculation with cDNA or RNA transcript inoculum. Symptoms observed after 3 to 4 weeks post-inoculation were mild, int.=intermediate or severe.
[b]The most stable thermodynamic structures were calculated using the algorithm of Zuker.[11] The free energy values are normalized to the intermediate strain of PSTV. A (+) value signifies a decrease in thermodynamic stability.

In a separate series of experiments, single and multiple mutations were introduced randomly throughout the entire sequence of PSTVd. Three of the resulting variants [ie. those containing C → U substitutions at positions 303 or 311 or an A → G substitution at position 310] produce mild to intermediate symptoms on tomato. Only one of the variant containing a C → U substitution at position 311 has a more stable P region than does the intermediate parent (Table 1). For all the mutations described above, sequence analysis of the progeny revealed no evidence of sequence reversion or instability.

The lowest free energy structures for the variants were calculated using the algorithm of Zuker;[11] a representative sample of these is shown in Figure 2. Minor structural

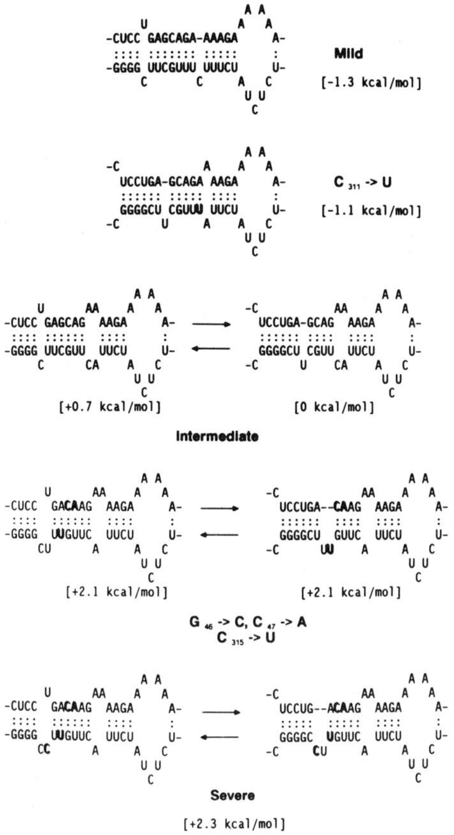


Figure 2. Sequence alterations within the pathogenicity modulating domain (nt. 39 - 60 and 300 - 322) of PSTVd. The lowest free energy structures were determined by the method of Zuker.[11] The structure of the mild strain was redrawn from Schnölzer et al..[8] Free energy values are expressed in kcal/mol and normalized to the value for the lowest free energy structure of the intermediate strain [arbitrarily expressed as 0 kcal/mol]. Where two structures were possible at the lowest free energy, both are shown.

rearrangements occur within the P domain. Although structural calculations suggest that only one of the variants shown has a more stable P domain ($C_{311} \rightarrow U$), all the variants, except $PSTVd_{severe}$ produce mild to intermediate symptoms on tomato.

The properties of the variants discussed above provide indirect support for the role of the P domain in pathogenicity. The inability to correlate the calculated thermodynamic stability of the region to symptom production, however, is in contrast to the observations made by Schnölzer et al..[8] Our results suggest that conformation rather than stability is important in symptom expression. In addition, the presence of as yet undiscovered structural interactions may play an important role in symptom development. To clarify the role of the P domain will require that we search for interactions of host proteins with this region and that we determine its conformation in vivo.

## ANALYSIS OF HOST RANGE AND MOVEMENT FUNCTIONS

The full expression of a plant virus or viroid infection in a susceptible host depends on the ability of the pathogen to spread to most parts of that host. The replicating viroid must be able to move within its plant host, both by cell-to-cell spread and by long distance movement. Viroids appear to move long distances with the flow of photosynthate,[12] but the mechanism of cell-to-cell spread is unknown.

Conventional bioassay techniques for cDNA infectivity are based upon the ability of viroid infections to rapidly become systemic. Bioassay systems where the Ti plasmid of Agrobacterium tumefaciens is used to mediate viroid infection, on the other hand, can provide large amounts of tissue in which viroid infection need not be systemic to be detectable. In its simplest form, Agrobacterium-mediated infection (agroinfection) protocols in which an infectious viroid cDNA is placed in the T-DNA of a virulent strain of Agrobacterium have been used to produce systemic infections in susceptible hosts.[13] In this case, the viroid first replicates in the gall tissue and then spreads systemically to the newly expanding leaves, presumably through the vascular tissue. If the inoculated plant is a non-host, systemic viroid replication does not occur.

Salazar et al.[14] and Yamaya et al.[15] used agroinfection to study the host range of PSTVd and hop stunt viroid (HSVd), respectively. Eight plant species believed to be resistant to mechanical inoculation with PSTVd were inoculated with virulent Agrobacterium tumefaciens strains containing infectious cDNA copies of PSTVd.[14] Small amounts of PSTVd-related RNAs were detected in the galls and roots (but not the leaves) of six of the species, suggesting that although PSTVd is passively translocated and stable in these species, RNA-directed replication does not occur. Because the previous reported resistance of Solanum acaule to PSTVd infection was broken by agroinfection, this resistance is actually resistance to mechanical inoculation rather than immunity to infection.

In a series of related experiments, Yamaya et al.[15] examined the sensitivity of tobacco to HSVd infection. The replication of HSVd was readily detected in plants agroinfected with HSVd, whereas mechanical inoculation resulted in barely detectable levels of replication. Obviously, care must be taken in defining the host range of viroids by mechnical inoculation.

We have used Agrobacterium-mediated inoculation of modified PSTVd cDNAs to examine the mechanisms responsible for restricting the cell-to-cell movement of viroids. These studies have also provided clues about determinants of host range. Some time ago, we reported that the introduction of multiple nucleotide substitutions in the left and right terminal loops of PSTVd abolished infectivity in a conventional tomato systemic bioassay,[4] but a conventional tomato bioassay may not be able to distinguish truly lethal mutations from those which severly restrict the viroids ability to replicate or move from cell to cell. To address these questions, monomeric copies of two mutant cDNAs were inserted behind the CaMV35S promoter and were introduced into the Ti plasmid of a virulent strain of Agrobacterium (Figure 3).

Wounds were made on the stems of 7-day old tomato seedlings, a solution containing bacteria was applied to the wound site, and viroid infection was monitored over a period of several weeks. Analysis of leaf, root, and gall tissues of tomato revealed systemic replication in plants inoculated with wild-type PSTVd constructions (pCGN160a and pCGN986:WT), whereas viroid-related RNAs were not detectable in any tissue inoculated with the pCGN986:P mutant. Mature circular viroid molecules as well as double-stranded replicative intermediates of the mutant in the right terminal loop (pCGN986:R) were confined to the gall and root (Table 2). Viroid related RNAs could not be detected in the foliage.

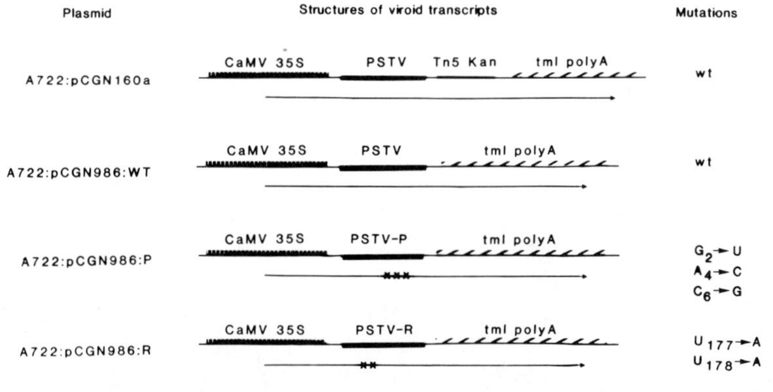


Figure 3. Agrobacterium Ti plasmid constructions containing PSTVd cDNAs. BamHI fragments (359 base-pairs) of wild-type (intermediate strain) PSTVd, PSTVd-P, or PSTVd-R were introduced into the BamHI site in the plasmid pCGN986 (Calgene, Inc.) to produce chimeric plasmids pCGN986:WT, pCGN986:P and pCGN986:R, respectively. pCGN160a has been described.[13] The inserts are in the PSTVd orientation (nt. 89 > 359-1 > 88). The resulting constructions were introduced into the pTiA6 plasmid of A. tumefaciens. The locations of mutations (x) and the transcripts predicted to occur in transformed cells are indicated.

Table 2

Distribution of PSTVd plus (+) strands in tissues collected from tomato plants inoculated with Agrobacterium tumefaciens carrying recombinant Ti plasmids

| Plasmid | A722 | pCGN160a | pCGN986:P | pCGN986:R | pCGN986:WT |
|---|---|---|---|---|---|
| Leaf | - | +[a] | - | - | ++ |
| Gall | - | + | - | + | ++ |
| Root | - | + | - | + | ++ |

[a] PSTV plus (+) strands were detected by nucleic acid hybridization. ++, wild-type signal; +, signal reduced 10 - 100-fold; -, no detectable signal. Samples were collected four weeks post-inoculation. A722 is a strain of Agrobacterium tumefaciens carrying the plasmid TiA6, and does not contain a viroid-related cDNA insert.

We and others routinely monitor the progress of viroid replication in infected hosts by measuring the increase in extractable viroid RNAs by dot blot hybridization. This method, however, cannot distinguish between a high rate of replication in comparatively few cells or a lower rate of replication in all cells of a particular tissue. Examination of the tissue collected from the lower portions of the stems (Figure 4) and roots (not shown) by in situ hybridization using PSTVd-specific RNA probes showed that the mutant (pCGN986:R) (+) strands are confined to the vascular cambium (Figure 4B), whereas wild-type PSTVd could be detected throughout the stem cross-section in all tissues (Figure 4A).

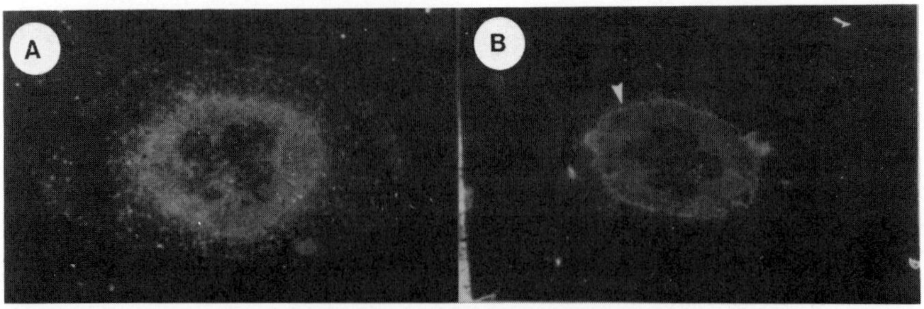


Figure 4. Dark-field micrographs of viroid-infected tomato stem sections collected four weeks post-inoculation and hybridized with $^{35}S$-labeled viroid-specific RNA probes. Tomato plants were inoculated with Agrobacterium tumefaciens constructions described in the text and illustrated in Figure 3. A. pCGN986:WT (wild-type) B. pCGN986:R (mutant). The arrow denotes the vascular cambium.

The pattern of viroid movement is postulated to be initial downward transport in the phloem followed by upward transport to the apical meristem via the xylem.[12] Therefore, this mutant appears to defective in cell-to-cell movement which may explain its confinement to the root system. Further study of this restriction in cell-to-cell movement in which the tomato host now behaves as a non-host will provide us with important clues about host range.

## CONCLUDING REMARKS

Aside from their obvious agricultural importance, viroids are of interest to biochemists and molecular biologists because of their unusual structural and biological properties. They are able to harness the host cells' machinery and induce dramatic alterations in plant development with what appears to be a minimal amount of sequence information. By introducing mutations into the viroid genome we have begun to address some of the questions raised about the ability of viroids to cause disease in certain hosts.

Contrary to initial indications, viroids appear able to undergo significant sequence alterations without losing their ability to replicate and move systemically in their host. Site-specific mutagenesis has already made a major contribution to our knowledge of structure/function relationships in viroid, primarily with respect to pathogenesis and cell-to-cell movement. This information can now be used to construct models of pathogenesis that contain testable predictions about these relationships.

## Acknowledgements

We thank T. O. Diener for continuing discussion of results and J. Hammond and K. Kamo for critical review of the manuscript. We thank S. M. Thompson and M. Hale for excellent technical assistance. Studies in our laboratories have been partially supported by Competitive Research Grants Program (Grants 85-CRCR-1738 and 88-37263-3990).

## REFERENCES

1. T. O. Diener, Potato spindle tuber "virus". IV. A replicating, low molecular weight RNA, Virology 45:411 (1971).
2. H. J. Gross, H. Domdey, C. Lossow, P. Jank, M. Raba, H. Alberty, and H. L. Sänger, Nucleotide sequence and secondary structure of potato spindle tuber viroid, Nature 273:203 (1978).
3. D. Riesner, K. Henco, U. Rokohl, G. Klotz, A. K. Kleinschmidt, H. Domdey, P. Jank, H. J. Gross, and H. L. Sänger, Structure and structure formation of viroids, J. Mol. Biol. 133:85 (1979).
4. R. W. Hammond and R. A. Owens, Mutational analysis of potato spindle tuber viroid reveals complex relationships between structure and infectivity, Proc. Natl. Acad. Aci. USA 84:3967 (1987).

5. P. Loss, M. Schmitz, G. Steger, and D. Riesner, Formation of a thermodynamically metastable structure containing hairpin II is critical for infectivity of potato spindle tuber viroid RNA, EMBO J. 10:719 (1991).
6. P. Keese and R. H. Symons, Domains in viroids: evidence of intermolecular RNA rearrangements and their contribution to viroid evolution, Proc. Natl. Acad. Sci. USA 82:4582 (1985).
7. G. Steger, H. Hofmann, J. Fortsch, H. J. Gross, J. W. Randles, H. L. Sänger, and D. Riesner, Conformational transitions in viroids and virusoids: comparison of results from energy minimization algorithm and from experimental data, J. Biomol. Struct. Dynam. 2:543 (1984).
8. M. Schnölzer, B. Haas, K. Ramm, H. Hofmann, and H.L. Sänger, Correlation between structure and pathogenicity of potato spindle tuber viroid (PSTV), EMBO J. 4:2181 (1985).
9. T. O. Diener, Viroid processing: a model involving the central conserved region and hairpin I, Proc. Natl. Acad. Sci. USA 83:58 (1986).
10. J. W. Davies, P. Kaesberg, and T. O. Diener, Potato spindle tuber viroid. XII. An investigation of viroid RNA as messenger for protein synthesis, Virology 61:281 (1974).
11. M. Zuker, On finding all suboptimal foldings of an RNA molecule, Science 244:48 (1989).
12. P. Palukaitis, Potato spindle tuber viroid: investigation of the long distance intra-plant transport route, Virology 158:239 (1987).
13. R. C. Gardner, K. Chonoles, and R. A. Owens, Potato spindle tuber viroid infections mediated by the Ti plasmid of Agrobacterium tumefaciens, Plant Mol. Biol. 6:221 (1986).
14. L. F. Salazar, R. W. Hammond, T. O. Diener, and R. A. Owens, Analysis of viroid replication following Agrobacterium-mediated inoculation of non-host species with potato spindle tuber viroid cDNA, J. Gen. Virol. 69:879 (1988).
15. J. Yamaya, M. Yoshioka, T. Sano, E. Shikata, and Y. Okada, Expression of hop stunt viroid from its cDNA in transgenic tobacco plants: identification of tobacco as a host plant, Molecular Plant Microbe Interactions 2:169 (1989).

# AN EXTRACELLULAR OLIGOSACCHARIDE SYMBIOTIC SIGNAL PRODUCED BY *RHIZOBIUM MELILOTI*

**Philippe Roche, Patrice Lerouge and Jean-Claude Promé**

Centre de Recherche de Biochimie et de Génétique Cellulaires, CNRS-UPS, 118 route de Narbonne, 31062 Toulouse Cedex, France

**David G. Barker, Catherine Faucher, Fabienne Maillet, Georges Truchet and Jean Dénarié**

Laboratoire de Biologie Moléculaire des Relations Plantes-Micro-organismes, CNRS-INRA, BP27, 31326 Castanet-Tolosan Cedex, France

## INTRODUCTION

The infection of leguminous plants by *Rhizobium* and the subsequent co-differentiation of both organisms leads ultimately to the formation of the unique nitrogen-fixing plant organ known as the root nodule. Within the nodule the *Rhizobium* are furnished with photosynthate-derived energy and, in exchange, the endosymbiotic bacterium reduces atmospheric nitrogen to a form which can be assimilated by the host plant. This remarkable symbiosis has been the subject of extensive research for many years, but it is only recently that it has become possible to study, at the molecular level, the role of diffusible factors in plant-bacterial signalling and recognition, with the exciting prospect that such molecules might also be responsible for triggering plant morphogenesis.

In the case of temperate legumes, bacterial infection proceeds via the root hair, and involves curling of the hair tip, followed by the formation of an infection thread within the root hair. While these early stages of infection are proceeding, nodule meristematic activity is induced in the inner cortical cell layers of the root. By means of the infection thread the *Rhizobium* penetrate into the root cortex, invade the newly-divided cells and subsequently differentiate into the nitrogen-fixing bacteroid form within the central tissue of the nodule. The use of various types of *Rhizobium* mutants has revealed that the micro-symbiont is capable of eliciting nodule organogenesis at a distance (Truchet *et al.*, 1980; Finan *et al.*, 1985). Recent studies have even demonstrated that, for certain alfalfa clones, nodulation can occur in the total absence of *Rhizobium* (Truchet *et al.*, 1989). This suggests that the plant possesses the entire genetic programme for nodule organogenesis and that, under normal circumstances, the *Rhizobium* provides only the trigger to switch on this programme.

Fast growing *Rhizobia* such as *R. meliloti* and *R. leguminosarum* have a narrow

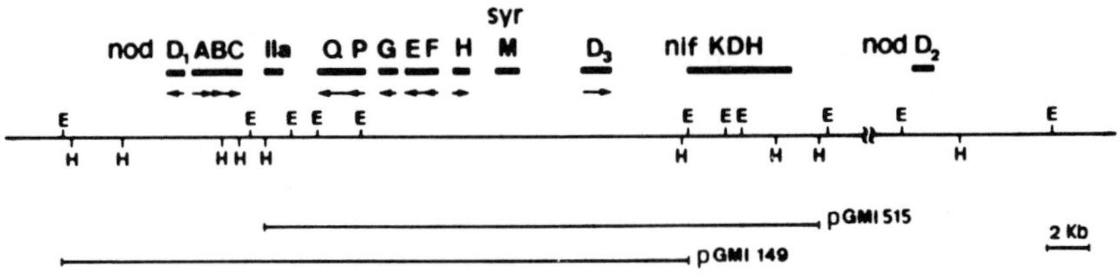


Figure 1. Genetic and physical map of the nodulation region of *R. meliloti* 2011. The horizontal line represents the restriction map (E, *Eco*R1; H, *Hin*dIII). The plasmids pGMI149 and pGMI515 are shown below the map, and the arrows indicate the direction of transcription of *nod* genes (Long, 1989).

host range, with the former able to nodulate *Medicago*, *Melilotus* and *Trigonella* species and the latter *Pisum* and *Vicia*. Figure 1 shows the genes of *R. meliloti* which are necessary for nodulation (*nod* genes) and which can be classified as either "common" (i.e. genes such as *nodABC* which can complement mutations in analagous genes of another *Rhizobium*), "regulator" (i.e. *nodD*) or "host-specific" (such as *nodH* and *nodQ* in *R. meliloti*). For example, a mutation in the *R. meliloti nodH* or *nodQ* genes causes a shift in host-range from *Medicago* to *Vicia* (Faucher *et al.*, 1988; Cervantes *et al.*, 1989).

It is now clear that part of the early interaction between the plant host and *Rhizobium* involves the transcriptional activation of bacterial nodulation genes by the combined action of specific flavonoid compounds present in plant exudates and the regulator *nodD* gene products (Long, 1989). However it is also clear that this very early stage of the interaction cannot account for the specificity of infection and nodulation (Faucher *et al.*, 1989). Recently, it has been shown that when *R. meliloti nod* genes are induced in batch culture by the flavone luteolin, soluble factors can be recovered in the supernatant which are able to provoke plant reactions such as root hair deformations (Faucher *et al.*, 1988; Banfalvi and Kondorosi, 1989; Faucher *et al.*, 1989). Most significantly, the specificity of this response mirrors the specificity of infection and nodulation by the *Rhizobium* itself.

In this article we would like to present the approach that we have undertaken in order to identify *Rhizobium* extracellular signals which elicit a host-specific response in legume root hairs. One of these factors has been purified and its structure determined, and in the light of this new information we will discuss the possible role of such molecules in bacterial-plant recognition and nodule organogenesis.

## RESULTS

### Bioassays for *Rhizobium* Extracellular Signals

An important advance in our understanding of how *Rhizobium nod* genes operate came with the discovery that sterile supernatants of *R. leguminosarum* cultures are able

Table 1. Correlation between the specificity of symbiotic behaviour and the specificity of extracellular factors

| | Alfalfa | | | Vetch | | |
|---|---|---|---|---|---|---|
| | Hac | Nod | Had | Hac | Nod | Had/Tsr |
| *R. meliloti* | + | + | + | - | - | - |
| *Rm. nodA*⁻ | - | - | - | - | - | - |
| *Rm. nodC*⁻ | - | - | - | - | - | - |
| *Rm. nodH*⁻ | - | - | - | + | + | + |
| *Rm. nodQ*⁻ | + | + | + | + | ND | + |
| *R. leguminosarum* | - | - | - | + | + | + |
| *Rl.* (pGMI515) | + | + | + | - | - | - |
| *Rl.* (pGMI515/*nodH*⁻) | - | - | - | + | + | + |
| *Rl.* (pGMI515/*nodQ*⁻) | - | - | - | + | + | + |

Effects on the specificity of both the symbiotic response and extracellular symbiotic signal production were examined in parallel for a variety of *R. meliloti nod* gene mutations, and also after the introduction of plasmids carrying the host specificity genes of *R. meliloti* into *R. leguminosarum* (see text for details). Symbiotic behaviour was assessed by both hair curling (Hac) and nodulation (Nod) tests, and extracellular signal activity by either hair deformation (Had) or thick short root (Tsr) bioassays (from Faucher *et al.*, 1988; Faucher *et al.*, 1989). ND = not determined.

to elicit a thick and short root (Tsr) reaction on seedlings of the *R. leguminosarum* host common vetch (*Vicia sativa* sp. nigra) (Van Brussel *et al.*, 1986; Zaat *et al.*, 1987). The authors showed that the *nodD* and *nodABC* genes are required for the Tsr response, and furthermore, that the flavonoid *nod* gene inducer naringenin must also be present in the bacterial growth medium.

Unfortunately, the Tsr reaction could not be detected on any of a range of *R. meliloti* hosts tested (Faucher *et al.*, 1988), but it was observed that treating plants with the sterile supernatants of flavonoid-activated cultures can induce a generalised deformation of root hairs (Had). This bioassay can be used not only for common vetch (Zaat *et al.*, 1987), but also for alfalfa (Faucher *et al.*, 1988) and white clover (Bhuvaneswari and Solheim, 1985). Furthermore, it is more rapid than the Tsr assay and significantly more sensitive (Faucher *et al.*, 1989). As with Tsr, the Had reaction is dependent on functional *nodDABC* genes. The role of host-specific genes in the production of hair deformation factors will now be discussed.

### *R. meliloti nodH* and *nodQ* Genes Determine the Host-specific Modification of Extracellular Signals

The observation that supernatants of *R. meliloti* were able to induce root hair deformations on alfalfa but not on common vetch, and *vice versa* for the supernatants of *R. leguminosarum* cultures (Faucher *et al.*, 1988) demonstrated clearly that there is a correlation between host specificity and extracellular factor activity. To investigate the possible role of host specificity genes of *R. meliloti* in the production of this symbiotic signal, we examined the Had activities of various *R. meliloti* mutants on both homologous and heterologous hosts. As shown in Table 1 mutations in *nodH* lead to a shift in signal activity (Had$^-$ on alfalfa (HadA$^-$) and Had$^+$ on vetch (HadV$^+$)), and mutations in *nodQ* lead to an extended activity (HadA$^+$ HadV$^+$). These modifications precisely mirror the changes in host specific nodulation of the corresponding mutated strains (Faucher *et al.*, 1988; Cervantes *et al.*, 1989). It should be pointed out that in all the above experiments the same flavonoid inducer, luteolin, was used to induce *nod* activity.

The introduction of the *R. meliloti* host range genes into *R. leguminosarum* (plasmid pGMI515 containing *nodPQ*, *nodFEG*, *nodH*, *nodD*3 and *syrM*, Fig. 1) results in the production of an extracellular factor that is now able to deform alfalfa hairs, and at the same time has decreased Tsr activity on vetch (Faucher *et al.*, 1989). Most significantly, mutations in either *nodH* or *nodQ* restores the original *R. leguminosarum* phenotype (Had$^-$ on alfalfa and Tsr$^+$ on vetch, see Table 1). This clearly shows that the introduction of *R. meliloti nodH* and *nodQ* into *R. leguminosarum* results in a modification of the specificity of the extracellular Had factors. The additional observation that a sterile filtrate of a *R. meliloti* strain carrying only *nodD*1 and the common *nodABC* genes is able to deform root hairs on vetch, but not on alfalfa (Faucher *et al.*, 1988) has led us to propose the model shown in Fig. 2. Both *R. leguminosarum* and *R. meliloti* common *nod* genes lead to the synthesis of a factor which is Had$^+$ on vetch, but Had$^-$ on alfalfa. The *nodH* and *nodQ* genes of *R. meliloti*

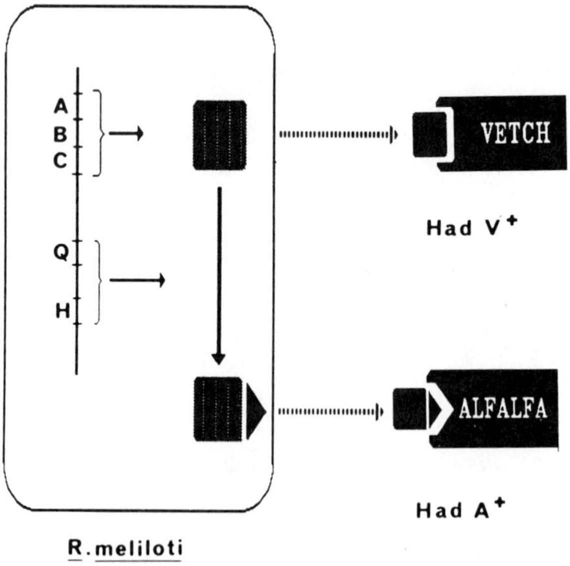


Figure 2. A model for the role of the *R. meliloti* common (*nodABC*) and host specific (*nodH* and *nodQ*) *nod* genes in the production of symbiotic extracellular signals which are able to deform root hairs on either alfalfa (HadA$^+$) or vetch (HadV$^+$). See text for explanation.

then convert this common factor to an alfalfa specific signal, which is no longer recognised by vetch root hairs. The reason why the *nodQ* gene is apparently required for the production of the alfalfa signal in *R. leguminosarum*, but not in *R. meliloti* itself could be due to the presence of more than one *nodQ* gene in *R. meliloti* (Schwedock and Long, 1989), and that this second gene has weak activity.

**Purification and Structure of a *R. meliloti* Extracellular Symbiotic Signal**

Preliminary analysis of the hair deformation factors present in filtrates of luteolin-induced *R. meliloti* cultures had shown that most of the activity could be extracted in a butanol-soluble fraction. Unfortunately, the concentration of the active molecule(s) proved to be insufficient for further structural analysis. In order to amplify factor production by *R. meliloti*, we introduced into the strain Rm 2011 the plasmid pGMI149 (Fig.1), carrying all the common and host specific *nod* genes, as well as the three regulatory genes *nodD*1, *nodD*3 and *syrM*. The presence of this plasmid led to an increase of more than a hundred-fold in the Had activity on alfalfa. As a result of this amplification it was now possible to observe two far-UV (220 nm) absorbing peaks when the butanol soluble fraction was further fractionated by HPLC on a $C_{18}$ reverse-phase column (Lerouge *et al.*, 1990). As had been previously observed for both nodulation and Had signal production (Faucher *et al.*, 1988), Tn5 insertions into either the *nodA* or *nodC* genes resulted in the simultaneous disappearance of these two UV-absorbing peaks.

Subsequent large-scale culture supernatant processing was carried out with the exopolysaccharide-deficient *R. meliloti* strain EJ355 (containing plasmid pGMI149) by virtue of its non-mucoid characteristics (Finan *et al.*, 1985). The signals produced by Rm 2011 (pGMI149) and Rm EJ355 (pGMI149) are identical as assessed by both chemical criteria (HPLC, NMR and mass spectrometry) and the biological Had assays. Filtrates were extracted and extracellular factors purified by a combination of reverse-phase $C_{18}$ HPLC, gel permeation on a Sephadex LH20 column and by ion-exchange chromatography on a DEAE column (Lerouge *et al.*, 1990). Ten litres of a luteolin-induced *R. meliloti* culture yielded around 4 mg of purified factors, which showed Had activity on alfalfa (at concentrations between $10^{-8}$ and $10^{-11}$ M), but not on common vetch.

Mass spectrometry, chemical modification and NMR spectroscopy were used to establish the structure of the factors as N-acyl tri N-acetyl ß-1,4 D-glucosamine tetrasaccharides, bearing a sulphate group on C-6 of the reducing sugar. The aliphatic chain carried by the non-reducing terminal sugar residue is a 2,9-hexadecadienoic N-acyl group (Lerouge *et al.*, 1990 and Fig 3). The two factors correspond to $\alpha$ and ß anomers of the same molecule at the C-1 position of the reducing end sugar, and we propose that this molecule be called NodRm-1. The choice of this nomenclature is based rather on the *nod* genes responsible for its production rather than the effects (such as hair deformation) that this factor can elicit on the plant host (see below). The fact that we have always observed a strict correlation both qualitatively and quantitatively between this molecule and the specific Had activity (Had$^+$ on alfalfa and Had$^-$ on vetch) is strong evidence that NodRm-1 is indeed the plant specific symbiotic signal previously characterised in the hair deformation assays.

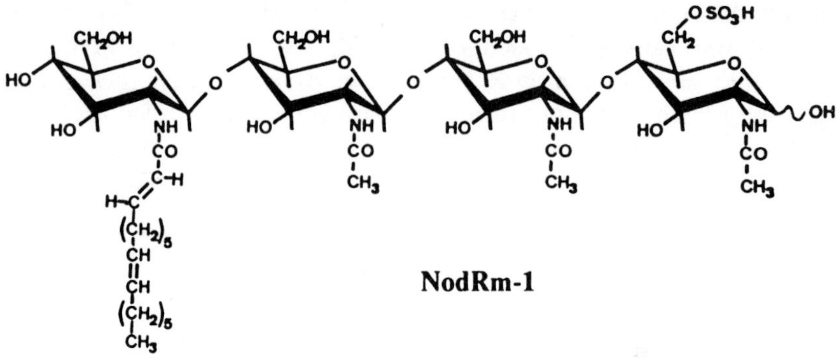


Figure 3. Structure of the sulphated and acylated glucosamine oligosaccharide symbiotic signal (NodRm-1) which has been purified from supernatants of flavonoid-induced cultures of *R. meliloti*. This structure has been established from a combination of NMR spectroscopic, mass spectrometric, $^{35}$S-labelling and methylation analyses (Lerouge *et al*, 1990).

## DISCUSSION AND PERSPECTIVES

Current research into the earliest stages of the interaction between *Rhizobium* and its legume host is now giving us a clearer view of how these two organisms exchange information at the molecular level even before they come into contact with each other. Firstly, plant signals in the form of flavonoids, and present in the root exudates, activate bacterial nodulation genes in conjunction with an appropriate regulator *nodD* gene. Efficient *nod* gene activation does require a correct matching between the *nodD* gene product and the flavonoid content of the plant exudate (Long, 1989).

The second, and highly specific stage of the bacteria-plant interaction occurs with the production of extracellular signals by a combination of common and host specific *Rhizobium nod* gene activities. Using the hair deformation assay as a means of identifying and purifying these signals we have always observed a strict correlation between the specificity of the bioassay and the symbiotic behaviour of the intact *Rhizobium meliloti.* The structure of the signal molecule for *R. meliloti* (NodRm-1) has been determined to be a sulphated lipo-oligosaccharide (Fig 3), which is active in the root hair deformation assay at extremely low concentrations (down to $10^{-11}$ M). It is worth noting that most phytohormones are only active in the $10^{-7}$ M range when added exogenously (Zeroni and Hall, 1980), and only certain fungal oligosaccharide eliciters which trigger the plant defense response are known to be active in the nanomolar range (Darvill and Albersheim, 1984).

These results immediately pose a certain number of questions concerning symbiotic signalling between *Rhizobium* and its legume host which we are now in a position to investigate. (i) What is the modification in the structure of NodRm-1 which is caused by *nodH* or *NodQ* mutations, and which causes a shift or extension of host range? (ii) How does this modified signal compare in structure with the wild-type signal produced by *R. leguminosarum*, both of which are able to deform root hairs of common vetch? (iii) Does the production of this extracellular molecule involve *de novo* synthesis from N-acetyl D-glucosamine or rather the degradation and/or modification of pre-existing cell wall macromolecules such as peptidoglycan? (iv) Does NodRm-1 also have the potential to induce mitotic activity in cortical root cells of alfalfa? If so, it will be important to establish to what extent NodRm-1 resembles the *R. meliloti* factors identified by Schmidt *et al.*, (1988), and which stimulate mitosis in plant protoplasts. The structure of these molecules are likely to be different since the production of these latter factors does not require either *nodC* or the host specificity genes. (v) Finally, could this molecule be responsible for the triggering of nodule organogenesis? An elegant series of experiments carried out with combinations of nodulation and infection mutants of *R. meliloti* have recently demonstrated that a diffusible bacterial signal can trigger nodulation even when the bacteria and the plant are separated by a filter (Kapp *et al.*, 1990). Furthermore, only mutations in the *nodABC* and *nodH* genes of *R. meliloti* give rise to a Nod$^-$ phenotype on alfalfa.

The identification and purification of a bacteria-derived molecule which can initiate hair branching on legume hosts in a specific manner is also extremely important for future research on the plant host itself. It is intriguing that the NodRm-1 molecule has structural features which are typical of lectin ligands such as hexosamine oligosaccharides (Lis and Sharon, 1986). The proposition that plant lectins play a role in the specificity of the plant-*Rhizobium* interaction has long been the subject of debate (for review see Dazzo and Gardiol, 1984). Further direct evidence in favour of such an idea has come from recent experiments in which the introduction of a pea lectin gene into white clover led to successful infection and nodulation by a pea-specific *Rhizobium* (Diaz *et al.*, 1989). The same authors (Kijne *et al.*, 1989) go on to suggest that a plant root lectin located in the cytoplasmic membrane of susceptible root hairs might be part of a Nod signal receptor. With purified signal molecules such as NodRm-1 now available it should be possible to examine the role of lectin or other receptors present at the root tip. Radioactive labelling of NodRm-1 can easily be carried out via the sulphate moiety, and this should also facilitate the subsequent purification of such receptors.

Finally, Scheres *et al.*, (1990) have reported that supernatants of flavonoid-induced *R. leguminosarum* cultures are able to elicit expression of an infection-specific plant gene ENOD12 in pea root hairs. The transcription of this gene is normally detected in root hairs only 24 h after infection with *Rhizobium* (Scheres *et al.*, 1990) and the fact that its expression is triggered by hair deformation factors suggests that it is one of the earliest plant genes to be activated in the interaction with *Rhizobium*. It will be very interesting to examine if this gene responds to Had factors in a host specific manner, and to use reporter gene/transgenic plant technology to monitor the expression of ENOD12 after factor induction. By this means we should gain access to the signal transduction pathway within the root hair and perhaps also to the pathway which leads ultimately to the triggering of nodule organogenesis.

Thus new approaches towards studying the molecular basis of specific plant-microbe interactions are opening up and the availability of molecules such as NodRm-1 will surely prove invaluable in our effort to understand some of the mechanisms by which signalling can take place between a prokaryotic and a eukaryotic organism.

## REFERENCES


Banfalvi, Z. and Kondorosi, A., 1989, Production of root hair deformation factors by *Rhizobium meliloti* nodulation genes in *Escherichia coli*: *HsnD* (*NodH*) is involved in the plant host-specific modification of the *NodABC* factor, *Plant Mol. Biol.*, 13:1.

Bhuvaneswari, T.V. and Solheim, B., 1985, Root hair deformation in the white clover/*Rhizobium trifolii* symbiosis, *Physiol. Plant.*, 63:25.

Cervantes, E., Sharma, S.B., Maillet, F., Vasse, J., Truchet, G. and Rosenberg, C., 1989, The *Rhizobium meliloti* host range *nodQ* gene encodes a protein which shares homology with translation elongation and initiation factors, *Molec. Microbiol.*, 3:745.

Darvill, A.G. and Albersheim, P., 1984, Phytoalexins and their elicitors - a defense against microbial infection in plants, *Ann. Rev. Plant. Physiol.*, 35:234.

Dazzo, F.B. and Gardiol, A.E., 1984, Host specificity in *Rhizobium*-legume interactions, *in* Genes involved in microbe-plant interactions, Verma, D.P.S. and Hohn, T.H. eds., Springer-Verlag, Wien/New York.

Debellé, F., Rosenberg, C., Vasse, J., Maillet, F., Martinez, E., Dénarié, J. and Truchet, G., 1986, Assignment of symbiotic developmental phenotypes to common and specific nodulation (*nod*) genetic loci of *Rhizobium meliloti*, *J. Bacteriol.*, 168:1075.

Diaz, C.L., Melchers, L.S., Hooykaas, P.J.J., Lugtenberg, B.J.J. and Kijne, J.W., 1989, Root lectin as a determinant of host-plant specificity in the *Rhizobium*-legume symbiosis, *Nature*, 338:579.

Faucher, C., Maillet, F., Vasse, J., Rosenberg, C., van Brussel, A.A.N., Truchet, G. and Dénarié, J., 1988, *Rhizobium meliloti* host-range *nodH* gene determines production of an alfalfa-specific extracellular signal, *J. Bacteriol.*, 170:5489.

Faucher, C., Camus, S., Dénarié, J. and Truchet, G., 1989, The introduction of the *Rhizobium meliloti* host-range *nodH* and *nodQ* genes into *R. leguminosarum* results in changes in the production of an alfalfa-specific extracellular signal, *Mol. Plant-Microbe Interact.*, 2:291.

Finan, T.M., Hirsch, A.M., Leigh, J.A., Johansen, E., Kuldau, G.A., Deegan, S., Walker, G.C. and Signer, E.R., 1985, Symbiotic mutants of *Rhizobium meliloti* that uncouple plant from bacterial differentiation, *Cell*, 40:869.

Kapp, D., Niehaus, K., Quandt, J., Müller, P. and Pühler, A., 1990, Cooperative action of *Rhizobium meliloti* nodulation and infection mutants during the process of forming mixed infected alfalfa nodules, *The Plant Cell*, 2:139.

Kijne, J.W., Diaz, C.L. and Lugtenberg, B.J.J., 1989, Role of lectin in the pea-*Rhizobium* symbiosis, *in*: Signal molecules in Plant and Plant-Microbe Interactions, B.J.J. Lugtenberg, ed., Springer-Verlag, Berlin/Heidelberg.

Lerouge, P., Roche, P., Faucher, C., Maillet, F., Truchet, G., Promé, J.C. and Dénarié, J., 1990, Symbiotic host-specificity of *Rhizobium meliloti* is determined by a sulphated and acylated glucosamine oligosaccaride signal, *Nature*, 344:781.

Lis, H. and Sharon, N., 1986, Lectins as molecules and as tools, *Ann. Rev. Biochem.* 55:35.

Long, S.R., 1989, *Rhizobium*-legume nodulation: Life together in the underground, *Cell*, 56:203.

Scheres, B., Van De Wiel, C., Zalensky, A., Horvath, B., Spaink, H., Van Eck, H., Zwartkruis, F. Wolters, A-M. Gloudemans, T. Van Kammen, A. and Bisseling, T., 1990, The ENOD12 gene product is involved in the infection process during the pea-*Rhizobium* interaction, *Cell*, 60:281.

Schmidt, J., Wingender, R., John, M., Wieneke, U. and Schell, J., 1988, *Rhizobium meliloti nodA* and *nodB* genes are involved in generating compounds that stimulate mitosis of plant cells, *Proc. Nat. Acad. Sci. USA*, 85:8578.

Schwedock, J. and Long, S.R., 1989, Nucleotide sequence and protein products of two new nodulation genes of *Rhizobium meliloti*, *nodP* and *nodQ*, *Molec. Plant-Microbe Interact.*, 2:181.

Truchet, G., Michel, M., Dénarié, J., 1980, Sequential analysis of the organogenesis of lucerne (*Medicago sativa*) root nodules using symbiotically-defective mutants of *Rhizobium meliloti*, *Differentiation*, 16:163.

Truchet, G., Barker, D.G., Camut, S., de Billy, F., Vasse, J. and Huguet, T., 1989, Alfalfa nodulation in the absence of *Rhizobium*, *Mol. Gen. Genet.*, 219:65.

Van Brussel, A.A.N., Zaat, S.A.J., Canter Cremers, A.C.J., Wijfellman, C.A., Pees, E., Tak, T. and Lugtenberg, B.J.J., 1987, Role of plant exudate and Sym plasmid-localised nodulation genes in the synthesis by *Rhizobium leguminosarum* of Tsr factor, which causes thick and short roots on common vetch *J. Bacteriol.*, 165:517.

Zaat, S.A.J., Van Brussel, A.A.N., Tak, T., Pees, E.E. and Lugtenberg, B.J.J., 1987, Flavonoids induce *Rhizobium leguminosarum* to produce *nodABC*-gene-related factors that cause thick, short roots and root hair responses on common vetch, *J. Bacteriol.*, 169:3388.

Zeroni, M. and Hall, M.A., 1980, Molecular effects of hormone treatment on tissue, *in* Hormonal regulation of development 1. Molecular aspects, MacMillan, J., ed., Springer-Verlag, Berlin/Heidelberg.

# TISSUE-SPECIFIC EXPRESSION OF EARLY NODULIN GENES

Clemens van de Wiel and Ton Bisseling

Dept. of Molecular Biology, Agricultural University
Dreijenlaan 3, 6703 HA Wageningen
The Netherlands

## INTRODUCTION

Root nodule development in legumes has been extensively described from a morphological/anatomical point of view (Dart, 1977; Newcomb, 1981; Bergersen, 1982). Members of the bacterial genera *Rhizobium* and *Bradyrhizobium* induce root cortical cells to resume meristematic activity at specific sites. The meristematic activity leads to the establishment of a nodule primordium, to which the bacteria are carried from the primary infection site, a root hair, by way of an infection thread deposited by the plant. Subsequently, the primordium develops into a consistently organized root nodule. The mature nodule consists of a central tissue, containing infected and uninfected cells, surrounded by a peripheral tissue. The latter tissue is supplemented with the vascular traces that connect the nodule with the central cylinder of the subtending root. Thus, the root is provided with a new organ that is specialized for supporting nitrogen fixation and exhibits unique anatomical features. These features contrast with the situation in other nitrogen-fixing root-based associations, such as the cyanobacterial ones with cycads (Pate et al., 1988) and the actinorrhizal ones with members of various angiospermous families (Becking, 1977), where the nodules can be regarded as modified lateral roots.

In the last few years, a set of nodulin cDNA clones has been isolated representing genes that are expressed before the onset of nitrogen fixation: the early nodulin genes. These clones and their characteristics are listed in Table 1. Recently, the location of the expression of the different early nodulin genes has been determined with the aid of *in situ* hybridization in pea (*Pisum sativum*) and soybean (*Glycine max*), in order to gain more insight into their roles in nodule development. The expression of each of the early nodulin genes turned out to be restricted to a specific developmental stage and appeared closely linked with the formation of specific nodule tissues (Van de Wiel et al., 1990; Scheres et al., 1990; Scheres et al., in press). In this paper we shall

*Plant Molecular Biology*, Edited by R.G. Herrmann and
B. Larkins, Plenum Press, New York, 1991

examine to which extent the localization data of early nodulin gene products can contribute to the characterization of nodule tissues and the distinction of developmental stages during nodule formation. In addition, the possible functions of the different early nodulins will be discussed in view of the spatial and temporal distribution in nodule development. For that purpose, the expression of the early nodulin genes will be discussed in sections dealing with the nodule tissues in which they are expressed specifically, i.e. the meristem, the central, the peripheral and the vascular tissue.

Table 1 Cloned early nodulin sequences

| cDNA clone | site of expression | characteristics | reference |
|---|---|---|---|
| PsENOD2 | nodule parenchyma ("inner cortex") | HRGP-like (cell wall protein?) | Van de Wiel et al., 1990 |
| PsENOD3 | early symbiotic zone | cluster of 4 cysteines (metal-binding protein?) | Scheres et al., in press |
| PsENOD5 | root cortex cells with IT, invasion zone, early symbiotic zone | rich in Pro, Ala, Ser, Gly (arabinogalactan protein?) | Scheres et al., in press |
| PsENOD12 | root cortex cells with IT, nodule primordium, invasion zone | HRGP-like (cell wall protein?) | Scheres et al., 1990 Scheres et al., in press |
| PsENOD14 | early symbiotic zone | homologous to PsENOD3 | Scheres et al., in press |
| GmENOD2 | nodule parenchyma ("inner cortex") | homologous to PsENOD2 | Franssen et al., 1987 Van de Wiel et al., 1990 |
| GmENOD40 | pericycle around nodule vascular bundle | no resemblance to other known proteins | Franssen et al., in prep. |

PsENOD clones were isolated from a pea (*Pisum sativum*) nodule cDNA library, GmENOD clones from a soybean (*Glycine max*) nodule cDNA library. Abbreviations: IT, infection thread tip; HRGP, hydroxiproline-rich glycoprotein.

## THE NODULE MERISTEM

The pattern of meristematic activity during root nodule development forms the basis for the main division of legume nodules into determinate and indeterminate growth types (for review, Sprent, 1980). In determinate nodules, of which the soybean nodule is an example, meristematic activity gradually diminishes during development and finally all cells differentiate into mature nodular tissue. On the other hand, in the indeterminate type, of which the pea nodule is an example, meristematic activity persists at the distal end of the nodule.

In the indeterminate type of nodule, two successive stages of meristematic activity can be distinguished: the primary meristem that leads to the formation of a nodule primordium, and then the persistent meristem that establishes itself in a subsequent stage at the distal end of the primordium, enabling the nodule to sustain indeterminate growth. The meristematic cells in these two distinct stages characteristically differ in the expression of the PsENOD12 genes (Scheres et al., 1990). These genes are expressed in the meristematic cells of the first stage, in which large, highly vacuolated cells of the inner root cortex resume meristematic activity and several cell divisions occur within the framework of the original cortex cells. Hence, increasingly smaller cells arise and in the course of this process the originally large central vacuole becomes more and more fragmented, while the cells become richer in cytoplasm. Soon, the primary meristematic cells in the centre of the primordium become infected by bacteria, carried there by an infection thread. These cells subsequently differentiate into the central tissue. At the same time the lateral and proximal cells of the primordium differentiate into the peripheral tissues. A group of cells at the distal end of the primordium continue to generate small cytoplasmic-rich cells that retain division activity. These then form the persistent distal meristem of the future nodule. At this second stage of meristematic activity, the expression of the PsENOD12 genes becomes restricted to the invasion zone, i.e. the zone immediately proximal to the distal meristem where infection threads penetrate cells produced by the meristem.

When the nodule primordial cells are induced to express their PsENOD12 genes, the infection thread, containing the bacteria, is still in the outer cortical cell layers. Hence, the nodule primordial cells must be amenable to react to a rhizobial signal carried over several cell layers. In the invasion zone, all the cells contain PsENOD12 transcripts, whereas only a part of the cells is penetrated by a branch of the infection thread (Scheres et al., 1990; Scheres et al., in press). So, most likely also in the nodule apex, a rhizobial signal is operating that is effective outside the cells actually containing bacteria and that should therefore be able to reach the distal meristem as well. Nevertheless, the PsENOD12 genes are not expressed in the distal meristem, whereas the level of PsENOD12 transcripts is already maximal in the invasion zone cell layer directly adjacent to it, indicating that the distal meristem is not susceptible to such a signal. Taken together, these observations strongly suggest that the distal meristem is not an unchanged remnant of the primary nodule meristem that partially differentiated into other nodular tissue. At the molecular level, there must be a significant difference between the cells of the nodule distal meristem and the primary meristematic cells of the nodule primordium.

Besides in the already mentioned sites, namely the nodule primordial cells and the nodule invasion zone, PsENOD12 transcripts are also detected in the root cortical cells through which the infection thread grows on its way to the nodule primordium. Interestingly, these root cortical cells and the primordial cells start the redirection of their developmental pathways in a comparable manner. In both cell types cytoplasmic strands are intersecting the large central vacuole of the original root cortical cells. In the inner cortical cells these cytoplasmic strands serve their usual role in positioning the nucleus in the centre of the cell prior to cell division; in the outer root cortical cells, however, these strands attain a new function, namely providing a sort of bridge for guiding the infection thread through the central vacuole of the cell. In this connection, Bakhuizen et al. (1988) suggested that the root outer cortical cells through which the infection thread will migrate start a cell division cycle in which they become arrested.

What these data mean in terms of the function of the PsENOD12 nodulin is not immediately apparent. Its amino acid sequence, as deduced from the cDNA sequence, indicates that it might be a cell wall protein (cf. Table 1). This would be consistent with a role in nodule meristematic activity as well as in infection thread growth, both being processes in which new cell walls are produced and with which PsENOD12 gene expression appears to be associated, as can be deduced from the localization data discussed above. In addition, the outer cortical cells through which the infection thread makes its way to the nodule primordium, deposit an additional wall layer.

Further studies have shown that PsENOD12 transcripts can also be detected, albeit in smaller amounts than in the root nodule, in other parts of the plant, namely in the flower and in older parts of the stem. Strictly speaking, PsENOD12 can therefore no longer be regarded as a true nodulin. In the stem, PsENOD12 mRNA is only found in the cells of the inner cortex immediately adjacent to the central ring of vascular bundles (Scheres et al., 1990). These parenchymatic cells have relatively large central vacuoles and show no signs of incipient meristematic activity, such as cytoplasmic strands through their vacuoles. It would be worthwhile to check whether these cells deposit an additional cell wall layer and if so, to compare the structure and the composition of this layer with those of the wall layers deposited in nodule development. If no additional clues to the function of PsENOD12 have come from the studies of its presence in other parts of the plant than the nodule, it still can be concluded that PsENOD12 must have a role in normal plant developmental processes and that most probably it also proved useful for specific functions in nodule development in the course of the evolution of the symbiosis.

The function of PsENOD12 in the nodule appears most likely to be related to the infection process. Even in the nodule, the relationship of PsENOD12 gene expression with meristematic activity is not consistent, as is evident from its absence in the distal meristem. Also, PsENOD12 transcripts are not detectable in root and shoot meristems. On the other hand, the sites in nodule development where PsENOD12 gene expression is found, sooner or later all become involved in the infection process. Yet the function of PsENOD12 will not necessarily be in formation of the infection thread itself, since not all cells of the nodule primordium expressing the PsENOD12 genes will eventually be penetrated by one. We would rather hypothesize a more indirect function for ENOD12 than in infection thread growth itself, i.e. preparing the walls of cells that *might* receive an infection thread, to enable these cells to sustain a controled infection process. In the root outer cortical cells this preparation involves depositing an additional cell wall layer, in the root inner cortical cells it is executed concomitantly with the formation of new cells and in the cells of the nodule invasion zone it is attained only after the production of new cells by the mitotic activity of the nodule distal meristem.

## THE CENTRAL TISSUE

The central tissue containing the infected cell type with the nitrogen-fixing bacteroids is the most specific tissue of the nodule. For that reason, it is hardly surprising that the majority of nodulin genes, studied in this respect, were shown to be expressed here during the development of this tissue (for review, Nap and Bisseling, 1990). Several of the nodulin genes indeed encode proteins that are integrated into the peribacteroid membrane, which surrounds the bacteroids in the infected cells. Besides the infected cells, the central tissue contains uninfected, highly vacuolated parenchymatous cells that also appear to have a nodule-specific character. In soybean, this is shown by the exclusive presence of the nodulin Ngm-35 in peroxisomes of the uninfected cells (Bergmann et al., 1983). The Ngm-35 nodulin is a nodule-specific uricase involved in the assimilation pathway of the bacterial-derived ammonia. In pea, the specialized character of the uninfected cells of the central tissue is illustrated by the exclusive presence of the Nps-40' early nodulin (Van de Wiel et al., in prep.).

As indicated above, the nodule central tissue develops after infection of the primordium by the bacteria. The sequential steps in this developmental process can simultaneously be observed in indeterminate pea nodules, since the distal meristem continuously provides the nodule with new cells, which subsequently differentiate into the various nodular cell types. Thus, Newcomb (1981) distinguished the following developmental stages in the central tissue, going from the distal meristem towards the base of the nodule: the invasion zone where infection threads penetrate the cells deposited by the meristem and subsequently release their bacteria; the early symbiotic zone where the bacteria proliferate and develop into pleiomorphic shapes; the late symbiotic zone where the infected cells are completely filled with bacteroids that, for some time, continue enlargement concomitantly with the host cells themselves; the senescent zone where bacteroids and host cells degenerate.

These different developmental stages have been merely defined by morphological criteria. With the availability of clones representing both bacterial and plant genes involved in the nodulation process, it becomes possible to define developmental stages in a more precise way, by using molecular markers in combination with morphological criteria. A relevant example with regard to the nitrogen fixation process comes from our recent studies on the location of the transcripts of the bacterial *nif*H gene, which encodes the nitrogenase component II. We found that in the first two cell layers of the late symbiotic zone *nif*H transcripts are not yet detectable, while in the subsequent cell layer *nif*H transcripts are already present at a maximal level (Yang et al., in prep.). So, in the first few cell layers of the late symbiotic zone, as defined by the morphological criteria mentioned above, nitrogen fixation will not be possible.

Most of the pea early nodulin cDNA clones characterized so far, represent genes the expression of which marks specific developmental zones of the central tissue (Scheres et al., in press). Some of the early nodulin transcripts are specifically present in zones coinciding with the ones defined morphologically above, while others are not. Transcripts of the PsENOD12 gene, for example (see the discussion in the previous section), are only detected in the invasion zone, while transcripts of the PsENOD5 gene are found in infected cells of the invasion as well as in the early symbiotic zone. Similarly, the PsENOD3 and PsENOD14 genes are expressed in the infected cells of the late part of the early symbiotic zone and the early part of the late symbiotic zone. Therefore, the localization of these early nodulin mRNAs enables to refine the distinction of developmental zones in the indeterminate root nodule. This, in turn, will facilitate the unravelling of the successive steps in nodule development and the identification of the signals, both on the plant and the bacterial side, involved in effecting these steps.

The possible function of the PsENOD12 protein has already been discussed in the section on the nodule meristem. The PsENOD5 gene is expressed in all cells with growing infection thread tips and, after the release of the bacteria from the infection threads, the amount of transcript increases concomitantly with the proliferation of bacteroids in the infected cells. The amino acid composition of PsENOD5 is reminiscent of that of arabinogalactan proteins. Hence, PsENOD5 might be an extracellular factor occurring in the infection thread matrix as well as in the peribacteroid space. On the other hand, there are some hydrophobic domains in the amino acid sequence of PsENOD5 indicating that the PsENOD5 protein might be inserted in a membrane. If the latter is the case, PsENOD5 will most probably be part of both the infection thread membrane and the peribacteroid membrane. Both PsENOD3 and PsENOD14 contain four cysteines with a spatial distribution indicating that they might be involved in binding of a metal ion. Transcripts for these proteins are present in cells with developing bacteroids. Since the bacteroids need relatively large amounts of iron and molybdenum for the synthesis of active nitrogenase, it is conceivable that PsENOD3 and PsENOD14 have a function in metal ion transport.

## THE PERIPHERAL TISSUE

The peripheral tissue surrounds the central tissue and is separated from the latter by a boundary layer consisting of uninfected cells. The peripheral tissue is divided into an outer and an inner part by an endodermal layer, the so-called nodule or common endodermis, which links up with the endodermis of the subtending root. Nodulin genes characteristic for the outer part of the peripheral tissue have not been described up till now, but one early nodulin gene, ENOD2, is exclusively expressed in the inner part of the peripheral tissue in both the determinate nodules of soybean (Van de Wiel et al., 1990) and the indeterminate nodules of pea (Van de Wiel et al., 1990) and alfalfa (*Medicago sativa*) (Van de Wiel et al., in prep.).

The peripheral tissue has usually been called the nodule cortex (cf. Newcomb, 1981; Bergersen, 1982). The arrangement of tissues in this "cortex", however, is different from that of the cortex of the subtending root. The nodule endodermis forms the inner boundary of a tissue (the "outer cortex") that morphologically resembles the root cortex. Both these tissues are next to each other at the base of the nodule, separated from the rest of the nodule and the root central cylinder by a continuous endodermal layer. The other part of the peripheral tissue, located inside the endodermal layer (the "inner cortex"), consists of highly vacuolated, parenchymatous cells that are smaller and more densely packed, i.e. with fewer and smaller intercellular spaces, than those of the peripheral tissue outside the endodermal layer (Witty et al., 1986; Van de Wiel et al., 1990). Moreover, it is traversed by vascular traces bounded by a second endodermal layer, named bundle endodermis, and in which no ENOD2 transcript was detected. Thus, the organization of tissues inside of the nodule endodermis is unique for nodules, if compared to other plant organs. The unique character of the inner part of the peripheral tissue, both positionally and morphologically, is emphasized by the specific presence of the early nodulin ENOD2 transcript in this tissue in soybean, pea and alfalfa nodules.

These observations taken together prompted us to reconsider the terminology of the peripheral tissue. Indeed, one might loosely distinguish a cortex and a medulla (the central tissue) in nodules. However, according to strict appliance of plant-anatomical terminology, the term "inner cortex" is highly misleading. By definition, the root endodermis delimits the root cortex from the central cylinder with the vascular tissue (cf. Esau, 1977). The nodule endodermis, on the contrary, delimits only the nodule "outer cortex" from the rest of the nodule. Thus, only this "outer cortex" has a position comparable to that of the root cortex and might be called a cortex proper, in agreement with its morphology that is similar to that of the root cortex. Contrary to the situation in the root, however, no epidermal layer is discernable around this nodule cortex. On the other hand, the "inner cortex" in the nodule has a position that in the root is occupied by the pericycle of the central cylinder. However, the "inner cortex" is traversed by vascular traces that are bounded by a bundle endodermis and that have a sheath of one to several cell layers, depending on the species (see the next section), continuous with the root pericycle. Hence, one must conclude that it is a tissue for which no counterpart can be found in the root. For that reason, we have proposed to rename this tissue to avoid the potentially misleading term "inner cortex", which might be taken to imply a relationship with the root cortex that actually does not exist (Van de Wiel et al., 1990). For reasons of clarity, the proposed term, nodule parenchyma, could be extended with the term "inner peripheral" to define more precisely where the tissue is located in the nodule.

The nodule parenchyma ("inner cortex") has been shown to function as a barrier towards the penetration of free oxygen into the central tissue (Witty et al., 1986). The flow of free oxygen into the central tissue needs to be regulated to protect the extremely oxygen-sensitive nitrogen-fixing enzyme nitrogenase. The scarcity of intercellular air-filled spaces, through which oxygen can diffuse most easily, might enable the nodule parenchyma to perform a function as oxygen barrier. From the cDNA nucleotide sequence it has been deduced that ENOD2 might be a cell wall protein (Table 1) and as such it appears conceivable that ENOD2 contributes to a specific cell wall conformation with limited intercellular space.

## THE VASCULAR TISSUE

Contrary to the situation in the root, the nodule vascular tissue is organized into discrete bundles. The outer part of these bundles is formed by a parenchymatous sheath that is continuous with the root pericycle. In soybean, the early nodulin gene GmENOD40 is expressed specifically in this sheath (Franssen et al., in prep.), which consists of three layers of cells with dense cytoplasm and little vacuolation (Walsh et al., 1989). There is some variation in the organization of this sheath among different species of legume. In pea, the sheath consists of one layer of transfer cells (Pate, 1976), a cell type with an elaborate network of wall ingrowths that is associated with intensive transport processes (Pate and Gunning, 1972).

The sheath is designated as pericycle because of its position between the vascular tissue proper and the bundle endodermis. The sheath is morphologically clearly different from the root pericycle. This difference might be related to a specialized function of the sheath in the loading of the nodule vascular bundles, against a concentration gradient, with high levels of nitrogenous compounds (Walsh et al., 1989). In the case of soybean, this is highlighted by the specific presence of the GmENOD40 gene product. We have not been able to establish whether GmENOD40 might have a role in such a loading function of the sheath, since the amino acid sequence of GmENOD40, as derived from the nucleotide sequence of its cDNA did not offer any clue to a possible function, neither was there any homology identified with a known protein.

## CONCLUDING REMARKS

During root nodule development, a range of tissues differentiates that can be distinguished by a combination of morphological and positional criteria. The localization of several early nodulin transcripts, of which cDNA clones have become available, turns out to be helpful in addition to the existing morphological/positional criteria to distinguish more precisely specific nodular tissues. Moreover, in pea nodules, the localization data of several such nodulin and rhizobial transcripts have proven to be useful in refining the characterization of successive developmental stages in the formation of the central tissue. Future research should be directed to the localization and further characterization of the corresponding proteins, in an attempt to understand their specific functions. This should be of help in the further understanding of the specific functions of the various nodule tissues in the symbiosis.

## REFERENCES


Bakhuizen, R., Van Spronsen, P.C., Díaz, C.L., and Kijne, J.W., 1988, The plant cytoskeleton in the *Rhizobium*-legume symbiosis, Ph.D. thesis, Leiden University, The Netherlands.

Becking, J.H., 1977, Dinitrogen-fixing associations in higher plants other than legumes, *in*: A treatise on dinitrogen fixation, vol. III, R.W.F. Hardy and W.S. Silver, eds., Wiley, New York.

Bergersen, F.J., 1982, Root nodules of legumes: structure and functions, Research studies Press Wiley, Chichester.

Bergmann, H., Preddie, E., and Verma, D.P.S., 1983, Nodulin-35: a subunit of specific uricase (uricase II) induced and localized in the uninfected cells of soybean nodules, *EMBO J.*, 2: 2333.

Dart, P., 1977, Infection and development of leguminous nodules, *in*: A treatise on dinitrogen fixation, vol. III, R.W.F. Hardy and W.S. Silver, eds., Wiley, New York.

Esau, K., 1977, Anatomy of seed plants, Wiley, New York.

Franssen, H.J., Nap, J.-P., Gloudemans, T., Stiekema, W., Van Dam, H., Govers, F., Louwerse, J., Van Kammen, A., and Bisseling, T., 1987, Characterization of cDNA for nodulin-75 of soybean: a gene product involved in early stages of root nodule development, *Proc. Natl. Acad. Sci. USA*, 84: 4495.

Nap, J.-P., and Bisseling, T., 1990, Nodulin function and nodulin gene regulation in root nodule development, *in*: Molecular biology of symbiotic nitrogen fixation, P.M. Gresshoff, ed., CRC Press, Boca Raton.

Newcomb, W., 1981, Nodule morphogenesis and differentiation, *Int. Rev. Cytol. Suppl.*, 13: 247.

Pate, J.S., 1976, Transport in symbiotic systems fixing nitrogen, *in*: Transport in plants, Encyclopedia of plant physiology, N.S., vol. 2, pt. B, Tissues and organs, U. Lüttge and M.G. Pitman, eds., Springer, Heidelberg.

Pate, J.S., and Gunning, B.E.S., 1972, Transfer cells, *Ann. Rev. Plant Physiol.*, 23: 173.

Pate, J.S., Lindblad, P., and Atkins, C.A., 1988, Pathways of assimilation and transfer of fixed nitrogen in coralloid roots of cycad-*Nostoc* symbioses, *Planta*, 176: 461.

Scheres, B., Van de Wiel, C., Zalensky, A., Horvath, B., Spaink, H., Van Eck, H., Zwartkruis, F., Wolters, A.-M., Gloudemans, T., Van Kammen, A., and Bisseling, T., 1990, The ENOD12 gene product is involved in the infection process during the pea-*Rhizobium* interaction, *Cell*, 60, 281.

Scheres, B., Van Engelen, F., Van der Knaap, E., Van de Wiel, C., Van Kammen, A., and Bisseling, T., 1990, Sequential induction of nodulin gene expression in the developing pea nodule, *Plant Cell*, in press.

Sprent, J.I., 1980, Root nodule anatomy, type of export product and evolutionary origin in some *Leguminosae*, *Plant Cell Env.*, 3: 35.

Van de Wiel, C., Scheres, B., Franssen, H., Van Lierop, M.-J., Van Lammeren, A., Van Kammen, A., and Bisseling, T., 1990, The early nodulin transcript ENOD2 is located in the nodule parenchyma (inner cortex) of pea and soybean root nodules, *EMBO J.*, 9: 1.

Walsh, K.B., McCully, M.E., and Canny, M.J., 1989, Vascular transport and soybean nodule function: nodule xylem is a blind alley, not a throughway, *Plant Cell Env.*, 12: 395.

Witty, J.F., Minchin, F.R., Skøt, L., and Sheehy, J.E., 1986, Nitrogen fixation and oxygen in legume root nodules, *Oxford Surv. Plant Mol. Cell Biol.*, 3: 275.

# INTERNALIZATION OF RHIZOBIUM BY PLANT CELLS: TARGETING AND ROLE OF PERIBACTEROID MEMBRANE NODULINS

Desh Pal S. Verma, Guo-Hua Miao, Chandrashekhar P. Joshi, Choong-Ill Cheon and Ashton Delauney

Department of Molecular Genetics and Biotechnology Center
The Ohio State University
Columbus, OH 43210 USA

## INTRODUCTION

*Rhizobium* enters the root cells of legume plants through a process resembling endocytosis. To reach appropriate host cells and to avoid any pathogenic reaction, both organisms have evolved an elaborate mechanism involving many signals (Lerouge *et al.*, 1990) that induce specific genes leading to the formation of the infection thread and eventual release of the bacteria inside the plant cell (see for reviews, Verma and Nadler, 1984; Verma and Fortin,1989; Long,1989). During this state bacteria remain enclosed in a membrane envelope (peribacteroid membrane, PBM) which is constituted by the components of the plasma membrane of the host cell (Verma *et al.*, 1978) and forms a novel "extracellular compartment". Physiological "internalization" of this compartment is a key step in acceptance of the foreign organism inside the plant cell. Success of this process allows bacteria to reduce dinitrogen for the benefit of the host plant and establish an endosymbiosis. *Rhizobium* enclosed in PBM behaves as an "organelle" having many properties in common with mitochondria. The PBM, while keeping bacteria "outside" the host cell cytoplasm and avoiding any pathogenic interaction, allows development of a close contact between the plant cell and bacteria for rapid and efficient exchange of metabolites.

In order to perform this specialized role, many specific host gene products (nodulins) are integrated into the PBM. Since formation of this membrane compartment is vital for symbiosis, a large amount of PBM is produced during the early phase of nodule development before the commencement of nitrogen fixation, and consequently many nodulins are induced at this stage (Verma and Delauney, 1988). We have identified several PBM nodulins from soybean (Fortin *et al.*, 1985) and have characterized nodulin-24 and nodulin-26 in detail (Katinakis and Verma, 1985 and Fortin *et al.*, 1987) including determination of the function of nodulin-26 (G.-H. Miao and D. P. S. Verma, unpublished data). PBM nodulins have also been identified from Pea nodules (Katinakis *et al.*, 1988). Our data suggests that nodulins may form novel channel(s) in this membrane for rapid equilibration of the peribacteroid compartment. This sheds new light on the manner in which endosymbiotic bacteria comes in close contact with their hosts.

## PERIBACTEROID MEMBRANE NODULINS IN SOYBEAN

### Targeting of Nodulin to the Peribacteroid Membrane

Release of bacteria from the infection thread requires proliferation of PBM during early stages of the infection process and redirecting some of the plasma membrane proteins and all of the PBM nodulins to this *de novo* synthesized subcellular compartment. In nodules formed by a *Bradyrhizobium japonicum* strain defective at the stage of endocytosis, a large number of membrane vesicles are found to fuse at the tip of the infection thread (Morrison and Verma, 1987). The PBM nodulins are found to be induced at this stage of development prior to the formation of the PBM compartment. Apparently, transduction of all the signals necessary for development and differentiation of nodules preceed the release of bacteria from the infection thread (Verma *et al.*, 1988). Soybean PBM contain at least six PBM nodulins. We studied two PBM nodulins of soybean in detail in order to determine any common features that direct these proteins to this newly formed subcellular compartment. The mechanism of targeting of these nodulins to PBM differ for each peptide. Nodulin-24 was found to have an amino terminus signal sequence which is cleaved co-translationally (Katinakis and Verma, 1985) and the cleaved peptide is further processed post-translationally, which increases its molecular weight by several kilo Daltons (Fortin et al., 1985). The nature of this modification, however, is not known.

Nodulin-24 is a highly hydrophobic peptide, with three repeated domains which form an $\alpha$–helix which may be embeded in the lipid bilayer. The last one-third of the protein which is also hydrophobic appear to contain a membrane spanning region (Fortin et al., 1987). *In vitro* co-translational processing and membrane protection experiments using trypsin have suggested that no part of this protein is exposed outside the vesicle (C.-I. Cheon and D. P. S. Verma, unpublished data). The amphipathic $\alpha$-helix of this protein thus may face the bacteroid side of the PBM. It has recently been shown that amphipathic molecules may act as antibody paratope (antigenic recognition site, Kauvar *et al.*, 1990.). It is possible that such recognition may allow the attachment of bacteroid to the PBM surface. An attachment is apparently necessary to coordinate the specific amount of PBM synthesis with bacterial cell division. A physical attachment of bacteroid with PBM has often been observed (see, Brewin et al., 1985).

A sequence encoding soybean nodulin-26 was isolated by immuno-precipitation of polysomes using PBM-specific antibodies followed by screening of a nodule cDNA library (Fortin et al., 1987). Structure analysis of this sequence suggested it to be a trans-membrane protein with potentially six membrane-spanning regions. This nodulin was found to have homology with an intrinsic protein of the bovine eye lens membrane, MIP-26 (Gorin et al., 1984). The latter is known to form a gap junction type channel (Gooden et al., 1985). A longer cDNA clone was subsequently isolated and cloned downstream of a T7 RNA polymerase promoter. Transcripts made *in vitro* from this construct were translated in a rabbit reticulocyte lysate both in presence and absence of dog pancreatic membranes, followed by determination of sensitivity of the nodulin-26 polypeptide to trypsin. While the entire peptide is inserted in to microsomes, a truncated sequence containing amino terminal 91 amino acids failed to be inserted indicating that this protein contains non-cleavable internal signal sequence down stream of amino acid 91. Thus nodulin-26 differs in its mode of membrane targeting in the PBM from nodulin-26. The membrane trans-located nodulin-26 is glycosylated co-translationally as evidenced by the Con-A binding.

Based on these results and in comparison with MIP-26, nodulin-26 can be folded as shown in Figure 1. The orientation of the carboxy and amino termini was confirmed by chemical cleavage mapping at cystein residues of trypsinized and untreated *in vitro* synthesized peptide. A single glycosylation site exists in this protein and this seems to face the bacteroid surface of the PBM. The carboxy end contains several phosphorylation sites which may interact with the acidic lipids present in this membrane.

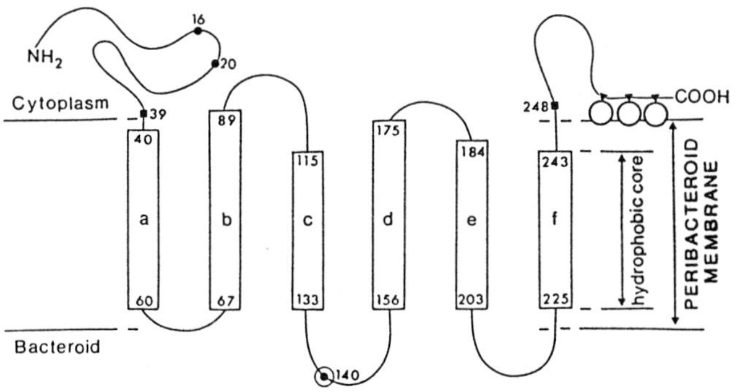


Figure 1. Diagrammatic representation of nodulin-26 structure in the PBM. The membrane spanning amino acids were deduced from the computer analysis while orientation of the protein was determined experimentally (G.-H. Miao and D. P. S. Verma, unpublished data).

Nodulin-26 also shows sequence similarity with a Pea protein induced as a result of change in turgor pressure (Guerrero et al., 1990) and the glycerol facilitator protein (glpF) of *Escherichia coli* (Baker and Saier, 1990). The latter was suggested to be a common precursor of MIP and nodulin-26. Figure 2 shows the extent of direct sequence homology between these four proteins. The hydropathy profile of these proteins clearly show six membrane spanning regions, despite the extensive sequence divergence. All four proteins appear to lack signal sequence at the amino terminal end. Secondary structure analysis of these proteins (Figure 3) showed extensive similarity. The nodulin-26 and glpF have a similar isoelectric point (PI 6.9) while MIP-26 and Pea turgor regulated protein show similarity(PI 9.9 and 9.6, respectively). All of these proteins form novel ion channels which are different from other known channels (Maelicke, 1988) These are involved in translocation of a group of compounds, some of which may be common to each channel. MIP and glycerol facilitator are known to form channels (Zampighi et al., 1985, in the membrane; however, the nature of compounds translocating through the MIP channel is not known.

Function of Nodulin-26 in PBM

In order to decipher the function of this nodulin in PBM, we attempted to integrate this protein into *E. coli* membrane and compare the properties of the channel formed by this protein with that formed by the glycerol facilitator of *E. coli* A *glp*F$^-$ strain was used for these studies (the *glp*F gene was kindly provided by G. Sweet). The results suggest that nodulin-26 does not transport glycerol and is unable to complement *glp*F mutation in *E. coli*. However, we discovered that it does form a channel analogous to glpF that facilitates the translocation of succinate, an important dicarboxylic known to be present in nodule cytoplasm during symbiotic nitrogen fixation (Streeter et al., 1987). Nodulin-26 has no homology with the dctA or dctB genes (Ronson et al., 1984). This channel is also able to translocate glycine, but not proline, and thus seems to be very specific. Up to 10mM of glycine has little effect on the uptake of succinate, suggesting the non-competitive nature of this channel. Interestingly, *glp*F also allows uptake of succinate, albeit at a slower rate. The uptake of succinate is passive and differs from active transport in bacteroids. Thus, recent results obtained on the uptake of succinate using peribacteroid units (Udvardi et al., 1988) must be reassessed in light of this data.

Succinate is essential for symbiosis since dct$^-$ *Rhizobium* often make ineffective nodules (Finan et al., 1983). This compound also seems to be preferentially utilized by *Rhizobium* (Humbeck et al., 1985) and after sucrose it occurs in highest concentration in nodules (Streeter, 1987). comparison was made by using the University of Wisconsin Genetic Computer Group programs. The boxed areas represent a minimum of three homologies in aligned sequences including replacement substitutions.

```
          1                                                    50
Nod26.   .......... .......... .MADYSAGTE SQEVVVNVTK NTSETIQRSD
 Jm7a.   MEAKEQDVSL GANKFPERQP LGIAAQSQDE PKDYQEPPPA PLFEPS....
Mip26.   .......... .......... .......... ........MW ELRSAS....
 GlpF.   .......... .......... .......... .......... .MSQTS....

          51                                                  100
Nod26.   .SLVSVPFLQ. ....KLVAEA VGTYFLIFAG CASLVVNENY YNMITFPGI.
 Jm7a.   .....ELTSW SFYRAGIAEF IATFLFLYIT VLTVMGVVRE SSKCKTVGIQ
Mip26.   .....FWRA. ....IC.AEF FASLFYVFFG LG..ASLRWA PGPLHVLQV.
 GlpF.   .....TLKG. ....QCIAEF LGTGLLIFFG VGCVAALKVA GASFGQWEI.

          101                                                 150
Nod26.   AIVWGLVLTV ..LVYTVGHI SGGHFNPAVT IAFASTRRFP LIQVPAYVVA
 Jm7a.   GIAWAFGGMI FALVYCTAGI SGGHINPAVT FGLFLARKXS LTRAIFYMVM
Mip26.   ALAFGLALAT ..LVQAVGHI SGAHVNPAVT FAFLVGSQMS LLRAICYMVA
 GlpF.   SVIWGLGVAM ..AIYLTAGV SGAHLNPAVT IALWLFACFD KRKVIPFIVS

          151                                                 200
Nod26.   QLLGSILASG .......TLR LLFMGNHDQF SGTV...... ........PN
 Jm7a.   QVLGAICGAG VVKGF..... .......... EGKQRFGDLN GG.ANFVAPG
Mip26.   QLLGAVAGAA VLYSV..... .....TPPAV RGNLALNTL. .......HPG
 GlpF.   QVAGAFCAAA LVYGLYYNLF FDFEQTHHIV RGSVESVDLA GTFSTYPNPH

          201                                                 250
Nod26.   GTNLQAFVFE FIMIFFLMFV ICGVATDNRA V..GE....L AGIAIGSTLL
 Jm7a.   YTKGDGLGAE IMGTFILVYT VFSAT.DAKR SARDSHVPIL APLPIG..FA
Mip26.   VSVGQATIVE IFLT...LQF VLCIFATYDE RRNGR....L GSVALA..VG
 GlpF.   INFVQAFAVE MVITAILMGL ILALTVDGNG VPRGP....L APLLIG..IL

          251                                                 300
Nod26.   LNVIIG...G PV.TGASMNP ARSLGP.... ....AFVHGE YEG......I
 Jm7a.   VFLVHLAT.I PI.TGTGINP ARSLGAAIVF NKKIGWND.. .........H
Mip26.   FSLTLGHLFG MYYTGAGMNP ARSFAPA.IL .......... ...TRNFTNH
 GlpF.   IAVIGASM.G PL.TGFAMNP ARDFGPK.VF AWLAGWGNVA FTGGRDIPYF

          301                                                 350
Nod26.   WIYLLAPVVG AIAGAWVYN. .IVRYTDKPL SEI.TKSASF LKGRAASK..
 Jm7a.   WIFWVGPFIG AALAAL.YHQ VVIRAIPFKS K......... ..........
Mip26.   WVYWVGPVIG AGLGSLLYDF LLFPRLKSVS ERLSILKGSR PSESNGQPEV
 GlpF.   LVPLFGPIVG AIVGAFAYRK LIGRHLPCDI CVVEEKETTT PSEQKASL..

          351       362
Nod26.    ......... ..
 Jm7a.   .......... ..
Mip26.   TGEPVELKTQ AL
 GlpF.   .......... ..
```

Figure 2. DNA sequence comparison of soybean nodulin-26 (nod-26) which contains 66 extra amino acids on the 5' end (as compared to previously published sequence [Fortin et al., 1987]); JM7a, a turgor-regulated gene from Pea (Guerrero et al., 1990); Mip26, bovine eye lens intrinsic membrane protein (Gorin et al., 1984); *glp*F, *E. coli* glycerol facilitator protein (Muramatsu and Mizuno, 1989). The sequence

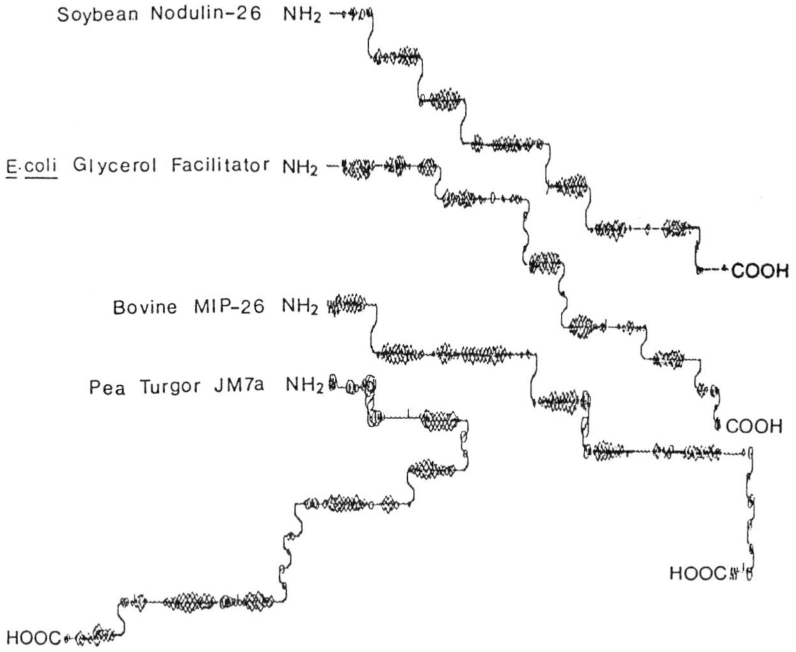


Figure 3. Secondary structure comparison of nodulin-26, *E. coli* glycerol facilitator, MIP-26 and a turgor-regulated sequence from Pea (see Figure 2). Note the general similarity in these four proteins.

## Ionic equilibration of the "Extracellular" Compartment Enclosing the Bacteroid with the Host Cell Cytoplasm

In light of the homology of of nodulin-26 with MIP-26 and glycerol facilitator, it suggests that nodulin-26 forms a gap junction-type pore in the PBM for equilibrating the concentration of metabolites such as succinate in the peribacteroid fluid from where it can be taken up actively by the bacteroid. The presence of this type of channel in the PBM sheds new light on symbiosis where the "extracellular compartment" enclosing the microsymbiont must be internalized with respect to its ionic equilibrium with the host cell cytoplasm and thus bring nitrogen-fixing bacteria closer to the cellular metabolism of the host. Since the peribacteroid space occupies as much or more space as the total cytoplasm in an infected cell such an equilibration becomes even more essential . It is also possible that this channel or other similar channels in PBM translocate $NH_4$ from the bacteroid, since an increase in the concentration of this compound in the peribacteroid fluid would directly inhibit nitrogen fixation and must be rapidly diffused to the host cell, where it is converted to glutamine by the host glutamine synthetase. An ammonia-induced glutamine synthetase is present in soybean root nodules (Hirel et al., 1987; G-H. Miao et al., manuscript in preparation).

Regulation of Turgor and Osmoticum in Root Nodules

The similarity of nodulin-26 with a turgor-regulated protein in Pea (Guerrero et al., 1990) is very interesting in light of the fact that osmoticum in infected cells of nodules is higher than uninfected cells. This is evident from the observation that there is a high level of proline in nodules (Kohl et al., 1988) and high concentrations of sucrose is translocated to this tissue as photosynthate. Direct measurement of osmoticum in nodules (Verma et al., 1978; D. Layzal, personal communication) give a value of 2-4 times higher than uninfected root tissue. We have isolated a soybean gene encoding $\Delta^1$-pyrroline-5-carboxylate reductase (P5CR) and found its expression to be higher in nodule than in root (Delauney and Verma, 1990). A high osmoticum may be necessary to protect bacteria since they alter their membrane structure during endosymbiosis. Furthermore, in order to prevent buildup of carbon and nitrogen in the infected cells, the turgor pressure needs to be controlled. The nodulin-26 channel may be regulated by these processes to provide proper flux of metabolites.

## CONCLUSIONS

Discovery of pore-type channels such as the one formed by nodulin-26 translocating specific metabolites across the peribacteroid membrane in root nodules suggest that equilibration of the peribacteroid fluid (extracellular), is essential for bringing bacteria in close proximity of the host cellular metabolism. Passive transport of succinate would make this metabolite available to the bacteroid for active uptake avoiding the need for coupling the plant and bacterial membranes for active transport. The peribacteroid compartment must be tightly regulated by the appropriate osmoticum which could control the flow of specific metabolites. The presence of such channels in other symbioses between eucaryotic and procaryotic cells thus seems essential and sheds a new light on these associations.

## ACKNOWLEDGEMENTS

This work was supported by research grants from NSF (#DCB 8819399 and DCB 8904101) and the Ohio State University. Our sincere thanks are to Dr G. Sweet who made the *E. coli* glpF$^-$ mutant and the glycerol facilitator gene availble to us. We would like to thank John Mullett for making sequence data of turgor regulated sequence available prior to its publication and Ruth-Hsu for computer analysis.

## References


Baker, M.E., and M.H. Saier, 1990, A common ancestor for bovine lens fiber major intrinsic protein, soybean nodulin-26 protein, and *E. coli* glycerol facilitator, Cell 60:185-186 (1990).

Bok, D., J. Dockstader, and J. Horwitz, Immunocytochemical localization of the lens main intrinsic polypeptide (MIP26) in communicating junctions, J. Cell. Biol. 92: 213-220 (1982).

Brewin, N.J., J.G. Robertson, E.A. Wood, B. Wells, A.P. Larkins, G. Galfre, and G.W. Butcher, Monoclonal antibodies to antigens in the peribacteroid membrane from *Rhizobium*-induced root nodules of pea cross-react with plasma membranes and Golgi bodies, EMBO J. 4:605-611 (1985).

Delauney, A.J. and D.P.S. Verma, A soybean gene encoding $\Delta^1$-pyrroline-5-carboxylate reductase was isolated by functional complementation in *Escherichia coli* and is found to be osmoregulated, Mol. Gen. Genet. in press (1990).

Finan, T.M., J.M. Wood, and D.C. Jordan, Symbiotic properties of C4-dicarboxylic acid transport mutants of *Rhizobium leguminosarum*, J. Bact. 154: 1403-1413 (1983).

Fortin, M.G., M. Zelechowska, and D.P.S. Verma, Specific targetting of the membrane nodulins to the bacteroid enclosing compartment in soybean nodules, EMBO J. 4:3041-3046 (1985).

Fortin, M.G., N.A. Morrison, and D.P.S. Verma, Nodulin-26, a peribacteroid membrane nodulin is expressed independently of the development of the peribacteroid compartment, Nucl. Acids Res. 15: 813-824 (1987).

Gooden, M., D. Rintoul, M. Takehana, and L. Takemoto, Major intrinsic polypeptide (MIP26K) from lens membrane: Reconstitution into vesicles and inhibition of channel forming activity by peptide antiserum, Biochem. Biophys. Res. Comm. 128: 993-999 (1985).

Gorin, M.B., S.B. Yancey, J. Cline, J.-P. Revel, and J. Horwitz, The major intrinsic protein (MIP) of the bovine lens fiber membrane: Characterization and structure based on cDNA cloning, Cell 39: 49-59 (1984).

Guerrero, F.D., J.T. Jones, and J.E. Mullet, Turgor responsive gene transcription and RNA levels increase rapidly when pea shoots are wilted. Sequence and expression of three inducible genes, Plant Mol. Biol. (in press).

Humbeck, C., H. Thierfelder, P.M. Gresshoff, and D. Werner, Competitive growth of slow growing *Rhizobium japonicum* against fast growing *Enterobacter* and *Pseudomonas* species at low concentrations of succinate and other substrates in dialysis culture, Arch Microbiol. 191: 1-6 (1985).

Katinakis, P. and D.P.S. Verma, Nodulin-24 gene of soybean codes for a peptide of the peribacteroid membrane and was generated by tandem duplication of an insertion element, Proc. Natl. Acad. Sci. USA 82:4157-4161 (1985).

Katinakis, P., R.M.K. Lankhorst, J. Louwerse, A. van Kammen, and R.C. van den Bos, Bacteroid-encoded proteins are secreted into the peribacteroid space by *Rhizobium leguminosarum*, Plant Mol. Biol. 11: 183-190 (1988).

Kauvar, L. M., P. V. K. Cheung, R. H. Gomer and A. A. Fleischer Paralog chromatography. Biochromatography 5: 22-26 (1990).

Kohl DH, Schubert KR, Carter MB, Hagedorn CH, Shearer G: Proline metabolism in N2-fixing root nodules: Energy transfer and regulation of purine synthesis. Proc. Natl. Acad. Sci. USA 85: 2036-2040 (1988).

Lerouge, P., P. Roche, C. Faucher, F. Maillet, G. Truchet, J.C. Promé, and J. Dénarié, Symbiotic host-specificity of *Rhizobium meliloti* is determined by a sulphated and acylated glucosamine oligosaccharide signal, Nature 344: 781-784 (1990).

Long, S.R., *Rhizobium*-legume nodulation: Life together in the underground, Cell 56: 203-214 (1989).

Maelicke A. Structural similarities between ion channel proteins. TIBS 13: 199-202.(1988).

Morrison, N. and D.P.S. Verma, A block in the endocytotic release of *Rhizobium* allows cellular differentiation in nodule but affects the expression of some peribacteroid membrane nodulins, Plant Mol. Biol. 7:51-61 (1987).

Muramatsu, S. and T. Mizuno, Nucleotide sequence of the region encompassing the *glpKF* operon and its upstream region containing a bent DNA sequence of *Escherichia coli*, Nucl. Acids Res. 17;4378 (1989).

Ronson, C.W., P.M. Astwood, and J.A. Downie, Molecular cloning and genetic organization of C4-dicarboxylate transport genes from *Rhizobium leguminosarum*, J. Bact. 160: 903-909 (1984).

Streeter, J. G. Carbohydrate, organic acids, and amino acid compsition of bacteroid and cytosol from soybean nodules. Plant Phys. 85: 768-773 (1987).

Udvardi, M.K., G.D. Price, P.M. Gresshoff, and D.A. Day, A dicarboxylate transporter on the peribacteroid membrane of soybean nodules, FEB Letters 231: 36-40 (1988).

Verma D. P. S. and A. Delauney (1988) Root nodule symbiosis: nodulins and nodulin genes. In Plant Gene Research Vol. V: Temporal and Spatial Regulation of Plant Genes, eds. D. P. S. Verma and R. Goldberg, Springer-Verlag pp. 169-199.

Verma, D. P. S., A. J. Delauney, M. Guida, B. Hirel, R. Schafer and S. Koh (1988) Control of expression of nodulin genes. In Molecular Genetics of Plant-Microbe Interactions 1988, eds. R. Palacios and D. P. S. Verma, APS Press, St Paul, Minnesota.

Verma, D.P.S. and M. Fortin, 1989, Nodule development and formation of the endosymbiotic compartment, in: "Cell Culture and Somatic Cell Genetics of Plants, Vol. 6," I.K. Vasil, ed., Academic Press, New York.

Verma, D.P.S. and K. Nadler, 1984, Legume-*Rhizobium* symbiosis: Host's point of view, in: "Genes Involved in Plant-Microbe Interactions", D.P.S. Verma and T. Hohn, eds., Springer-Verlag, Wien/New York.

Verma, D.P.S., V. Kazazian, V. Zogbi, and A.K. Bal, Isolation and characterization of the membrane envelope enclosing the bacteroids in soybean root nodules, J. Cell Biol. 78: 919-936 (1978).

Zampighi, G.A., J.E. Hall, and M. Kreman, Purified lens junctional protein forms channels in planar lipid films, Proc. Natl. Acad. Sci. USA 82: 8468-8472 (1985).

# REGULATION OF NODULE SPECIFIC GENES

Peter Lauridsen, Niels Sandal, Astrid Kühle, Kjeld Marcker and Jens Stougaard

Department of Molecular Biology and Plant Physiology, Laboratory of Gene Expression University of Aarhus, DK-8000 Denmark

## INTRODUCTION

One of the unique developmental processes studied in plants is the induction and formation of nitrogen fixing root nodules on leguminous plants. Several plant genes encoding "nodulins" specifically involved in different stages of nodulation has been characterised (Listed in Delauney and Verma 1988). Early nodulin genes induced shortly after infection with the rhizobial microsymbiont respond to signal molecules excreted by the bacteria, while later nodulin genes like leghemoglobin responsible for oxygen transport in the mature nodule seem to require additional factors for induction (Govers et al. 1986; Scheres et al. 1990). In an approach to study the regulatory circuits directing expression of nodulin genes we have used a *Lotus corniculatus* transformation regeneration system (Petit et al. 1987; Stougaard et al. 1987; Hansen et al. 1989) to define *cis*-acting regulatory DNA elements on both early and late nodulin promoters. In this communication results from analysis of the soybean leghemoglobin *lbc*$_3$ gene promoter will be presented, together with attempts to characterise *trans*-acting protein factors linking the developmental status of the cell to regulatory events at the promoter.

## IDENTIFICATION OF REGULATORY DNA ELEMENTS

Organ-specific and correct developmental expression of a chimeric soybean *lbc*$_3$ 5´3´-CAT gene in *Lotus corniculatus* and *Trifolium repens* nodules formed after infection with their respective microsymbionts, *Rhizobium loti* and *R. trifolii* demonstrated extensive conservation of regulatory mechanisms for nodulin genes. Nodulins from different symbiotic systems can therefore be analysed in a heterologous legume host (Stougaard et al. 1986;Jørgensen et al.1988). The soybean leghemoglobin *lbc*$_3$ promoter was consequently analyzed in *Lotus corniculatus*.

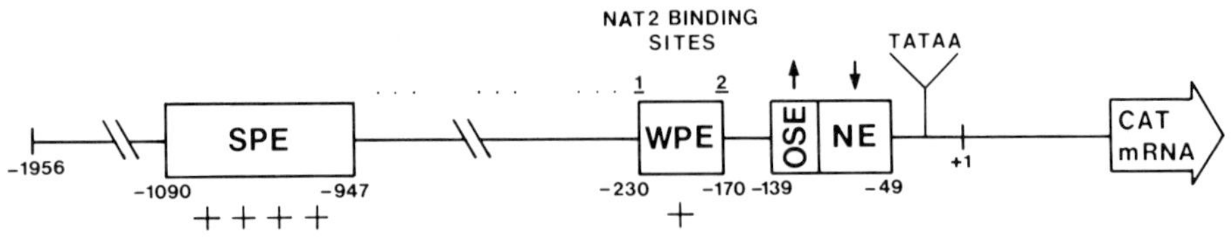


Figure 1. Organisation of regulatory DNA elements in the 2 kb 5´region of the soybean leghemoglobin *lbc3* gene. The mapped binding sites for the NAT2 *trans*-acting factor are indicated by 1 and 2, further upstream sites by dots. SPE; strong positive element. WPE; weak positive element. OSE; organ-specific element NE; negative element.

A 2 kb 5´region of *lbc3* 5´3´-CAT was initially analysed in a classical deletion analysis. Two upstream quantitative promoter components were defined by this approach. A strong positive element (SPE) located between positions -1090, -947 and a weak positive element (WPE) located between -230,-170. Two qualitative elements were characterised after reactivation of silent deletions using the "constitutive" CaMV 35S enhancer. An organ-specific element (OSE) containing putative nodulin consensus sequences (AAAGAT; CTCTT) , and a negative element (NE) were found overlapping the basic promoter. See figure 1. The qualitative and quantitative contributions of these elements were also analysed using internal deletions removing one or more of the defined regulatory regions (Stougaard et al. 1990). Low levels of expression resulting from removal of the OSE element indicated that this element also acts as a quantitative element possibly interacting with the upstream SPE. The WPE element previously found responsible for 2 % of full level expression could on the other hand be removed without effecting organ specificity and with only minor effect on the expression level. Nodule specific expression of deletions with only the TATA box region and strong positive element present indicates the presence of specific control sequences on the putative enhancer (SPE) itself. Full level expression seems to depend on the strong positive element, the organ specific element as well as basic promoter elements.

The regulatory capacity of the region carrying the OSE and NE elements was also investigated in hybrid promoter studies using the " neutral constitutively" expressed CaMV 35S promoter. To facilitate the promoter constructs a *Hpa*I linker was inserted in the *Eco*RV site at -9o between the enhancer and basic promoter of the 35S promoter. Ligation of oligonucleotides corresponding to the -139,-35 OSE,NE region of the *lbc3* promoter formed the hybrids. Figure 2 shows the constructs. Insertion of the complete -139, -35 region carrying the organ specific and the negative element of the *lbc3* gene promoter change the expression of the 35S promoter towards a higher level in nodules. This effect was even more pronounced when a dimer of the -139,-35 region was used. Low level expression was only detected in roots from a few plants and no activity was detected in leaves. Figure 2.

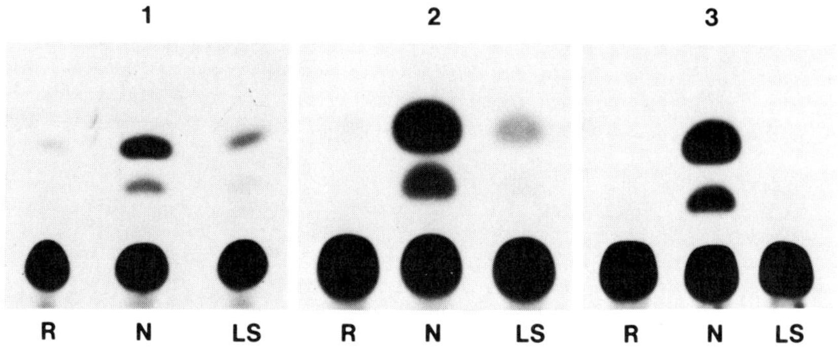


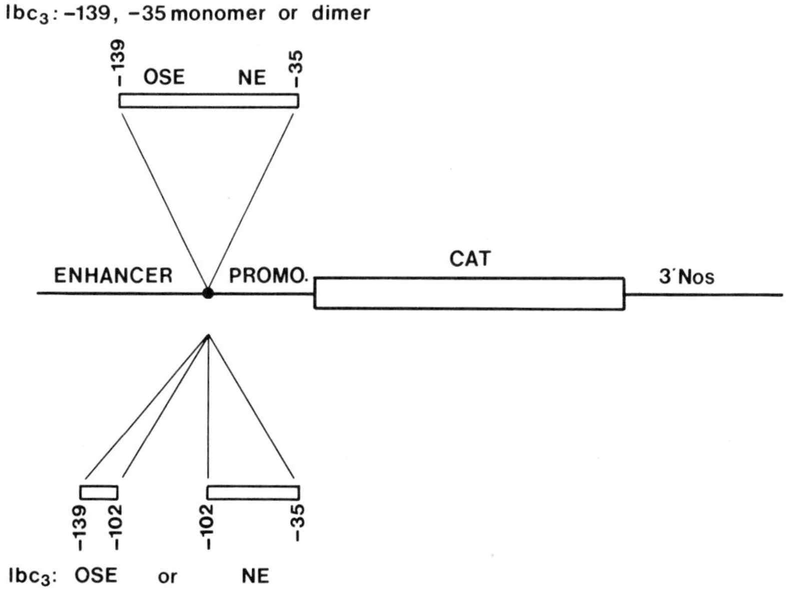


Figure 2. (A) Hybrid promoter constructs of the CaMV 35S promoter. Oligonucleotides representing the -139,-35 region of soybean leghemoglobin *lbc3* promoter region were inserted at -90 between the enhancer and basic promoter of the 35S promoter. (B) Chloramphenicol acetyl transferase activity expressed from 35S hybrid promoter constructs. 1; 35S itself. 2; Activity from hybrid promoter with *lbc3* OSE;NE region (-139,-35) inserted at -90. 3; As 2 with -139,-35 dimer.

Strong preference for nodule expression was therefore conferred on the 35S promoter by the OSE and NE elements acting together. See fig. 2B. Insertion of the NE element alone (-102,-35) seems to lower expression from the 35S in roots, leaves and nodules. The OSE alone does not confer any nodule specificity on the 35S promoter, but might elevate expression.(Data not shown). Presence of both OSE and NE seems therefore required to give preferential expression of the 35S promoter in root nodules. The possibility that individual elements cannot interact with 35S promoter elements due to for example position effects can however not be excluded. Interdependence between various elements of the *lbc3* promoter seems nevertheless to be an important feature of the regulation leading to high level organ specific expression (Stougaard et al. 1990).

In an attempt to speed up the analysis of *cis*-acting elements and to avoid the labour intensive plant transformation regeneration procedures, a protoplast system was tested for transient expression. Purified DNA (plasmids) was $Ca^{++}$/PEG precipitated onto protoplasts isolated from soybean root nodules or from a soybean callus culture. Activity from the CAT reporter gene was determined after three days of culture. Results from a set of parallel experiments using promoter constructs outlined in figure 3A are shown in figure 3B. The results demonstrate that the CAT activity obtained was dependent on addition of DNA and the level depends on the amounts of DNA (Lane 1,2 and 8). Expression of the *lbc*$_3$ promoter in protoplasts from nodules and lack of expression in protoplasts from callus further indicates a tissue dependent expression of genes in the protoplasts system (Lane 3). The effect of removing the 35S enhancer from the 35S promoter is shown in lane 2 and 4, while lane 6 and 7 show the elevated level of CAT activity from the -90 35S minimal promoter with the enhancer replaced by the -978,-919 or -1049,-730 regions from the SPE of the *lbc*$_3$ promoter.

Based on these experiments transient expression systems seem to be capable of faithful expression of various types and strengths of promoters and to reproduce organ specific expression patterns. An interesting difference between protoplast expression and plant expression was however observed with the -1049,-730 and -978,-919 reactivations of the -90 35S promoter. (Lanes 6 and 7) Reactivation was observed in protoplasts but not in plants. One explanation of this could be a different conformation of plasmid DNA compared to chromosomal integrated DNA. Weaker elements might therefore reactivate in protoplasts and not in plants. Another interesting speculation could be interaction between elements in *trans* mimicking the interaction of a complete promoter.

Although very interesting observations were obtained with transient expression in protoplasts the system proved difficult to work with. Some undefined property of protoplast preparation determines whether expression is possible or not. Preparations of protoplasts would either reproduce previous results or express no activity at all from any plasmids. As this problem was not overcome easily transient expression analysis was no faster than analysis in transgenic plants and the system was not developed further.

## IDENTIFICATION AND CHARACTERIZATION OF NODULE SPECIFIC REGULATORY PROTEINS INVOLVED IN GENE INDUCTION

Three different organ specific nuclear proteins recognizing AT rich DNA sequences were identified in soybean by gel retardation assays. One protein (NAT2) is present in mature nodules, another protein (NAT1) is detected in roots and nodules, and a third protein (LAT1) is only observed in leaves. Footprinting, deletion and point mutation analysis reveal different binding properties for all three proteins and further show that even single base pair substitutions have dramatic effects on binding affinity. The NAT1 and LAT1 proteins are released from chromatin by low salt extraction and are soluble in 2% TCA, indicating a relationship to high mobility group proteins (HMG), (Jensen et al. 1988; Jacobsen et al. 1990).

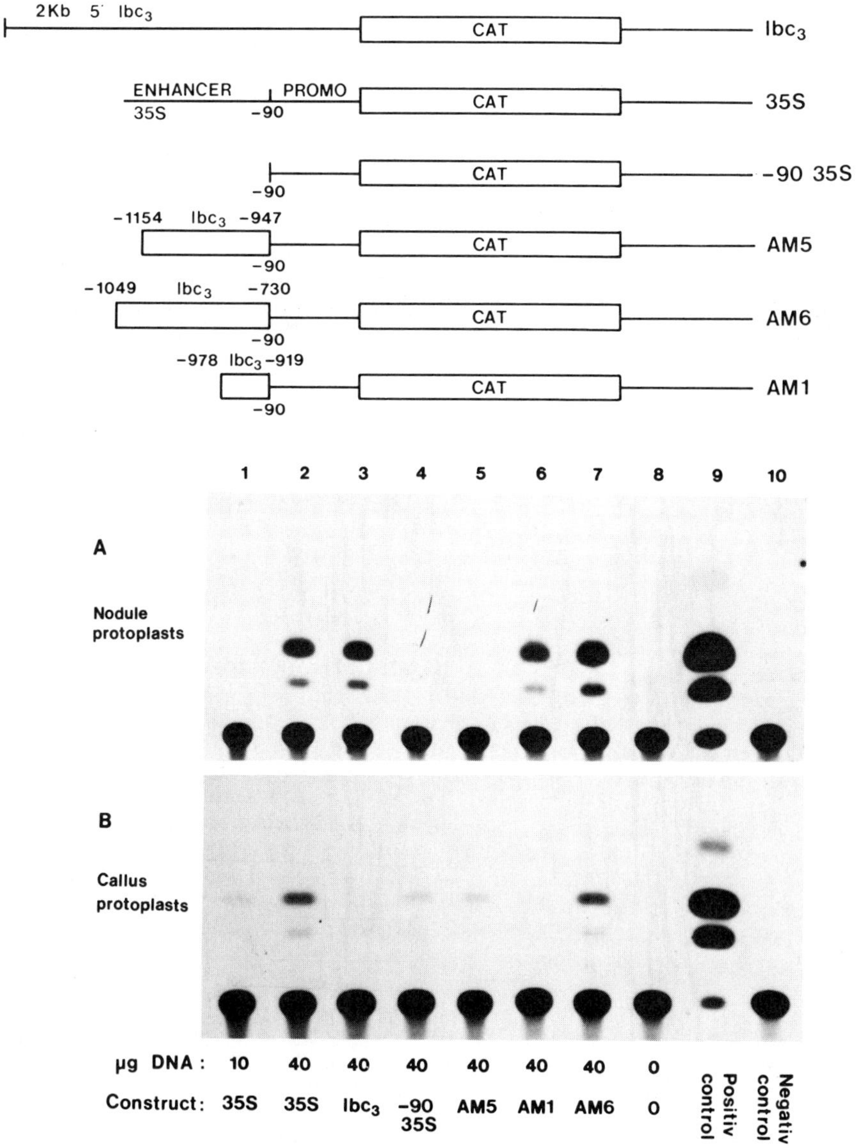


Figure 3. (A) Chimeric gene constructs used for transient expression in protoplasts from soybean nodules or soybean callus. All constructs use CAT as reporter gene. (B) Transient expression of CAT activity in protoplasts from soybean nodules or callus. Plasmid DNA of constructs was calcium precipitated onto protoplasts and CAT measured 3 days later.

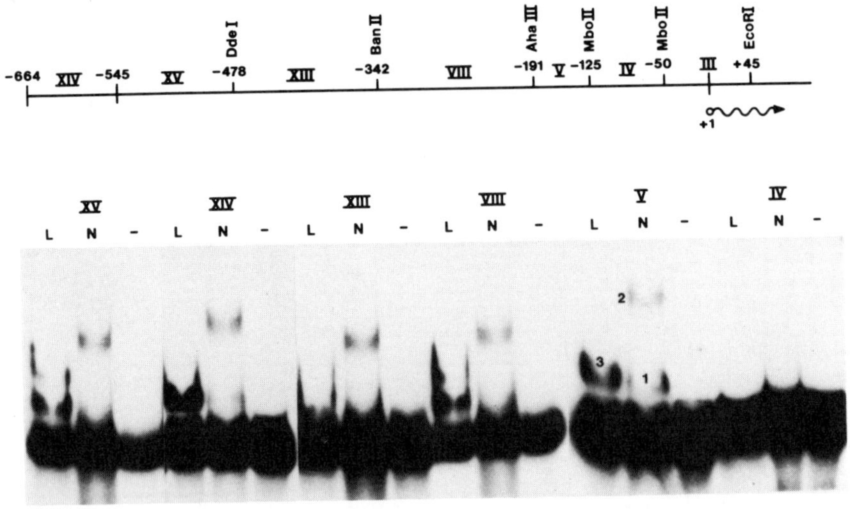


Figure 4. Gelretardation assay of soybean nodule (N) and leaf (L) nuclear extracts using leghemoglobin $lbc_3$ promoter fragments. Restriction map of the $lbc_3$ 5′region with numbered fragments is shown above the retardation assays of these fragments. Protein-DNA complexes corresponding to the NAT2, NAT1 and LAT1 binding factors are indicated as (1) NAT1, (2) NAT2, (3) LAT1.

The *in vivo* role of NAT2 binding was investigated using the $lbc_3$ promoter. Two distinct bindings sites overlap the weak positive element of this promoter, see fig 1. Internal deletions removing one or both of these binding sites from an otherwise intact (2kb) promoter region linked to a CAT reporter sequence were tested in plants. Removal of the weak positive element and the two NAT2 binding sites resulted in only minor reduction of the nodule specific expression (Stougaard et al. 1990). One explanation is that NAT2 is a chromatin associated protein with a more structural role. Another possibility is that NAT2 binding at the WPE sites can be compensated by binding at alternative binding sites. Figure 4 demonstrate the presence of other NAT2 binding sites in the $lbc_3$ promoter region between the SPE and WPE elements. These sites might bind NAT2 with lower affinity since they were not discovered in the original characterization of NAT2 sites, (Jensen et al. 1988) but they may nevertheless compensate the loss of sites at the WPE. Binding sites for the LAT1 protein are located at the -947 border of the strong positive element.See fig.4 The functional role of these sites and their relation to the SPE has not yet been investigated in any detail. Binding of NAT1 occurs on at least two fragments of the $lbc_3$ promoter, but the precise position of binding sites has not been determined.

## CONCLUSION

Several *cis*-elements and *trans*-acting factors have been identified in the soybean leghemoglobin $lbc_3$ promoter region. Strong correlation between *in vivo* effects of DNA regulatory elements and factor binding can however not be established on present knowledge. Improved methods of extracting and purifying nuclear proteins is needed for such correlative studies. Progress in procedures has already allowed identification of more binding sites for known factors, such as NAT2 and has recently made it possible to detect interaction on the organ specific OSE element. Previously no interactions was found in this part of the promoter (Jensen et al. 1988). Refined studies of factor binding at defined sites of a complete promoter taking account of possible protein to protein interactions will furthermore depend on improved footprinting techniques. Binding of factors at mutated binding sites compared to *in plant* effects of these sites can however be performed and results of such studies will be presented.

REFERENCES

Govers, F., Moerman, M., Downie, J.A., Hooykaas, P., Franssen, F.J., Louwerse, J., Van Kammen, A., and Bisseling, T., 1986, Rhizobium *nod* genes are involved in inducing an early nodulin gene. Nature, 323: 564-566.

Hansen, J., Jørgensen J-E., Stougaard, J., and Marcker, K.A., 1989, Hairy roots - a short cut to transgenic root nodules. Plant Cell Rep., 8: 12-15.

Jacobsen, K., Laursen N.B., Jensen E.Ø., Marcker, A., Poulsen, C., and Marcker, K.A.,1990, HMG I like proteins from leaf and nodule nuclei interact with different AT motifs in soybean nodulin promoters. The Plant Cell, 2: 85-94.

Jensen, E.Ø., Marcker, K.A., Schell, J., and de Bruijn, J., 1988, Interaction of a nodule specific, trans-acting factor with distinct DNA elements in the soybean leghemoglobin *lbc*$_3$ 5´upstream region. EMBO J., 7: 1265-1271.

Jørgensen, J-E., Stougaard J., Marcker, A., and Marcker, K.A., 1988, Root nodule specific gene regulation: Analysis of the soybean nodulin N23 gene promoter in heterologous symbiotic systems. Nucleic Acids Res. 16: 39-50.

Petit, A., Stougaard J., Kühle, A., Marcker, K.A., and Tempe´, J., 1987, Transformation and regeneration of the legume *Lotus corniculatus*: A system for molecular studies of symbiotic nitrogen fixation. Mol. Gen. Genet. 207: 245-250.

Scheres, B., Van De Wiel, C., Zalanski, A., Horvath, B., Spaink, H., Van Eck, H., Zwartkruis, F., Wolters, A-M., Gloudemans, T., Van Kammen, A., and Bisseling, T., 1990, The ENOD12 product is involved in the infection process during the pea-*Rhizobium* interaction. Cell 60::: 281-294.

Stougaard, J., Marcker, K.A., Otten, L., and Schell, J., 1986, Nodule-specific expression of a chimeric soybean leghemoglobin gene in transgenic *Lotus corniculatus*. Nature 321: 669-674.

Stougaard, J., Abildsten, D., and Marcker, K.A., 1987, The *Agrobacterium rhizogenes* pRi TL-segment as a gene vector system for transformation of plants. Mol. Gen. Genet. 207: 251-255.

Stougaard, J., Sandal, N.N., Grøn, A., Kühle, A., and Marcker, K.A., 1987, 5´analysis of the soybean leghemoglobin *lbc*$_3$ gene: Regulatory elements required for promoter activity and organ specificity. EMBO J., 6: 3565-3569.

Stougaard, J., Jørgensen, J-E., Christensen, T., Kühle, A., and Marcker, K.A., 1990, Interdependence and nodule specificity of *cis*-acting regulatory elements in the soybean leghemoglobin *lbc*$_3$ and N23 promoters. Molec. Gen. Genet. 220: 353-360.

Delauney, A.J., and Verma D.P.S., 1988, Cloned nodulin genes for symbiotic nitrogen fixation.Pl. Mol.Biol. Rep. 6:279-285.

# REGULATION OF GENES FOR ENZYMES ALONG A COMMON NITROGEN METABOLIC PATHWAY[1]

Gloria M. Coruzzi, Janice W. Edwards, Elsbeth L. Walker, Fong-Ying Tsai, and Timothy Brears

Laboratory of Plant Molecular Biology, The Rockefeller University, New York, N.Y. 10021-6399, USA

## ABSTRACT

We have characterized the multigene families encoding glutamine synthetase (GS) and asparagine synthetase (AS) in *Pisum sativum*. The isolated GS and AS genes have been used as probes to study the expression of individual members of these gene families during various aspects of plant development. These studies have shown that chloroplast GS2 and cytosolic GS are encoded by homologous nuclear genes which are differentially expressed *in vivo* (18,19). The nuclear gene for chloroplast GS2 is regulated by light, phytochrome, and photorespiration (4). In contrast, two nearly identical genes for cytosolic GS (GS3A and GS3B) are expressed at highest levels in developmental contexts where large amounts of nitrogen are mobilized in plants (22). Analysis of transgenic tobacco plants containing the pea GS promoters fused to a GUS reporter gene has shown that the genes for chloroplast GS2 and cytosolic GS3A are expressed in distinct cell types (5). These transgenic experiments have demonstrated that the chloroplast GS2 and cytosolic GS3A isoforms serve distinct, non-overlapping roles in plant nitrogen metabolism. Parallel studies on the gene family for plant AS have shown that peas contain two AS genes (AS1 and AS2) (20). The AS1 gene shows a dramatic dark-induced expression, which reflects the role of asparagine as the preferred nitrogen transport compound in dark-grown plants (20). Both AS1 and AS2 are expressed coordinately with genes for cytosolic GS during germination and nitrogen-fixation (20). Our combined studies on the gene families for GS and AS should uncover the molecular basis for the coordinate regulation of genes for enzymes along a common nitrogen metabolic pathway in plants.

## INTRODUCTION

In plants, amino acids serve not only as building blocks of proteins but also as nitrogen transport compounds. Two primary nitrogen transport amino acids in most higher plants are glutamine and asparagine. Ammonia

[1] This work was supported by NIH grant GM32877 and DOE grant DEFG0289ER14034. J.W.E is the recipient of a National Science Foundation Fellowship in Plant Biology. T.B. is the recipient of an EMBO long-term fellowship.

*Plant Molecular Biology 2*, Edited by R.G. Herrmann and B. Larkins, Plenum Press, New York, 1991

assimilated into organic form by glutamine synthetase (GS) may be transported directly, or may be converted into asparagine by asparagine synthetase (AS). Molecular studies on the gene families for these two enzymes involved in generating the nitrogen transport amino acids glutamine and asparagine has shown that the expression of individual GS and AS genes are dramatically regulated both by internal factors such as cell-type or developmental stage, and by external factors such as light or dark growth conditions. In addition to providing insight into how genes for enzymes along a common nitrogen metabolic pathway are regulated, these studies have also provided insight into the function of multiple isoenzymes in plants.

For glutamine synthetase (GS), previous biochemical studies demonstrated the existence of multiple isoforms which were present in the chloroplast or cytosol (10). However, since GS functions in a number of contexts to assimilate ammonia, the relative role of each isoform in each process was difficult to assess biochemically due to inadequate cell fractionation, overlapping activity profiles, and immunological cross reactivity (11,12). The ability to monitor the expression of individual genes encoding each GS isoform has provided insight into the role of each encoded isoenzyme in plant nitrogen metabolism. Furthermore, transgenic plant technology and "reporter" gene systems have enabled the localization of the expression of an individual GS gene to a particular cell type. These cell-specific expression studies have indicated that chloroplast and cytosolic GS perform non-redundant functions in plants (5), as outlined below.

For asparagine synthetase (AS), previous biochemical studies have been hampered by problems with *in vitro* enzyme assays which include; instability of AS *in vitro* (7,14), presence of contaminating asparaginase activity (16), and the presence of non-protein inhibitors of AS in partially purified preparations (8). Biochemical questions remain as to the substrate specificity of plant AS (e.g. for glutamine or ammonia), and also as to its subcellar localization (1). We have isolated and characterized AS cDNA clones from plants in order to answer questions about the nature of the AS enzyme, and to examine how asparagine biosynthesis is regulated at the level of gene expression.

Since GS and AS genes encode enzymes along a common nitrogen metabolic pathway, we hope to gain insight not only into the function of individual genes, but should also uncover the molecular mechanisms which coordinate the expression of GS and AS genes during plant development.

## Glutamine synthetase is encoded by a diverse multigene family

We initially isolated GS cDNAs from plants using a DNA probe encoding animal GS (17). Full length cDNA clones corresponding to the entire GS gene family of *Pisum sativum* were shown to encode either chloroplast GS2 or cytosolic GS by nucleotide sequence analysis, characterization of *in vitro* translation products, and *in vitro* chloroplast uptake experiments (18,19). One cDNA clone GS185 corresponds to the single nuclear gene for chloroplast GS2 (GS2ct) (18).

Cytosolic GS is encoded by several different cDNAs which correspond to homologous but distinct genes. One cDNA for cytosolic GS called GS299, corresponds to a single nuclear gene (GS1) (18). Two other cDNAs for cytosolic GS called GS341 and GS132, correspond to two nearly identical or "twin" GS genes (GS3A and GS3B, respectively) (18,22). Gene-specific experiments described below were used to correlate the

Table 1. Genes for GS and AS in *Pisum sativum*

| | cDNA | genomic | reference |
|---|---|---|---|
| chloroplast GS2 | GS185 | GS2 | (5,18) |
| cytosolic GS1 | GS299 | GS1 | (18) |
| cytosolic GSn | GS341 | GS3A | (5,19) |
| | GS132 | GS3B | (22) |
| cytosolic AS1 | cAS1 | gAS1 | (20) |
| cytosolic AS2 | cAS801<br>cAS201 | gAS2 | (20) |

expression of individual members of the GS gene family (see Table 1) with the distinct roles that GS serves in primary assimilation, photorespiration, nitrogen-fixation, and during germination.

<u>The nuclear gene for chloroplast glutamine synthetase is regulated by light, phytochrome, and photorespiration</u>

Expression studies have shown that the mRNA for chloroplast GS2 accumulates preferentially in leaves, in a light-dependent fashion (18). The light-induced expression of chloroplast GS2 is shown to be due in part to the chromophore phytochrome, and in part to light-induced changes in chloroplast metabolism (4,18). In particular, it has been shown that GS2 mRNA accumulates preferentially in pea plants grown under photorespiratory conditions, while mRNA for cytosolic GS remains unchanged (4).

These molecular studies in pea complement genetics studies in barley which demonstrated that photorespiratory mutants lacked chloroplast GS2, but contained normal levels of cytosolic GS (23). The patterns of GS2 gene regulation observed in pea, underscore the role of chloroplast GS2 in primary ammonia assimilation in plastids and in the reassimilation of photorespiratory ammonia.

<u>"Twin" cytosolic GS genes are induced in developmental contexts of increased nitrogen transport</u>

In pea, the three cDNAs for cytosolic GS (GS299, GS341, GS132) encode highly homologous polypeptides (86-99%) (18,19,22). Gene-specific probes from the 3' non-coding regions of these cDNAs were used to monitor the expression of individual genes for cytosolic GS in two contexts where large amounts of nitrogen are mobilized, in nitrogen-fixing root nodules and in cotyledons of germinating seedlings (22). A 3' S1 nuclease technique was used to discriminate between the mRNAs of the "twin" genes for cytosolic GS3A and GS3B, which correspond to cDNAs GS341 and GS132, respectively.

These studies showed that the GS3A gene is expressed at highest levels in all contexts examined, and that mRNAs for both GS3A and GS3B are coordinately induced to accumulate in nitrogen-fixing root nodules and in cotyledons of germinating seedlings (22). In contrast, the mRNA for cytosolic GS1 (cDNA GS299) accumulates in nodules, but not in cotyledons of germinating seedlings (22). These differences in expression patterns amongst the genes for cytosolic GS suggest that the GS3A gene product serves a major role in generating glutamine for intercellular nitrogen transport (22).

## Transgenic analysis demonstrates that chloroplast and cytosolic GS are expressed in distinct cell types and serve non-overlapping functions

To determine the cell-specific expression patterns for chloroplast and cytosolic GS, the promoters for chloroplast GS2 or cytosolic GS3A were fused to the beta-glucuronidase (GUS) reporter gene and introduced into *Nicotiana tabacum* using an *Agrobacterium* based Ti transformation system (5). GUS activity was monitored *in situ* in tissue sections of the transgenic tobacco plants. This analysis demonstrated that the promoter for chloroplast GS2 directs high level GUS expression predominantly in photosynthetic cells (e.g. palisade parenchymal cells of the leaf blade, chlorenchymal cells of the midrib and stem, and in photosynthetic cells of tobacco cotyledons) (5). The promoter for chloroplast GS2 is also able to confer light-regulated expression to the GUS reporter gene in transgenic tobacco (5). These expression patterns reflect the physiological role of chloroplast GS2 in photosynthetic cells, namely in the assimilation of ammonia in plastids from reduced nitrate and from photorespiration.

In contrast, the promoter for cytosolic GS3A directs expression of GUS specifically within the phloem cells of all organs of mature plants (5). This phloem-specific expression pattern of the gene for cytosolic GS3A, suggests that this cytosolic GS isoenzyme functions to generate glutamine for intercellular nitrogen transport. The nitrogen "transport" role for cytosolic GS3A is particularly evident in cotyledons of germinating seedlings where the intense expression of the GS3Acy-GUS transgene in the vasculature of cotyledons correlates with the mobilization of nitrogen from the seed storage reserves (5).

The distinct cell-specific expression patterns conferred by the promoters for chloroplast GS2 and cytosolic GS3A indicate that the corresponding GS isoforms perform separate non-overlapping metabolic functions. In light of the fact that chloroplast GS2 and cytosolic GS3A are expressed in distinct cell types, it is now clear why cytosolic GS is unable to compensate for the loss of chloroplast GS2 in photorespiratory mutants of barley (23).

## Peas contain two homologous but distinct genes for asparagine synthetase

Previous biochemical studies on AS were hampered by problems of enzyme instability in partially purified preparations (7,14). To circumvent biochemical problems in the purification of plant AS, we have directly isolated AS cDNAs from higher plants using an animal AS DNA probe (20). A characterization of the AS cDNAs clones from pea has begun to answer some questions about the encoded AS enzymes, their substrate specificities and their subcellular localization.

Molecular studies on AS have revealed that there are at least two genes for AS in *P. sativum* (AS1 and AS2) which are 86% homologous to each other at the level of encoded amino acids (20). Full length cDNAs for

Table 2. Organ-specific and regulated expression of GS and AS genes in *Pisum sativum*.

| Gene | Leaves Light | Leaves Dark | Roots | Nodules | Cotyledons | Ref. |
|---|---|---|---|---|---|---|
| chloroplast GS2 | +++++ | + | + | ++ | - | (5,18) |
| cytosolic GS1 | + | + | ++ | +++ | - | (18,19,22) |
| cytosolic GS3A | + | + | ++ | +++++ | +++++ | (19,22) |
| cytosolic GS3B | + | + | ++ | ++++ | ++++ | (22) |
| cytosolic AS1 | + | +++++ | + | +++++ | +++++ | (20) |
| cytosolic AS2 | + | ++ | ++ | +++++ | +++++ | (20) |

AS1 and AS2 each encode cytosolic enzymes whose four amino terminal amino acid residues correspond to the glutamine binding domain of AS in animals (6). The non-homologous 3′ non-coding regions of cAS1 and cAS2 were used as DNA probes to monitor the expression of each AS gene in various developmental contexts, as outlined below.

Dark-induced expression of AS1

While asparagine and glutamine both serve as nitrogen transport amino acids in plants, the ratio of asparagine to glutamine transported varies according to growth conditions. When plants are grown in the dark, asparagine is the preferred, more economical nitrogen transport amino acid since it contains a higher ration of nitrogen to carbon than glutamine. Early biochemical studies showed that AS enzyme activity increases with dark treatment (8), as do the levels of asparagine in the phloem (21).

Using AS cDNAs as molecular probes, we have shown that AS1 mRNA accumulates to high levels specifically in response to dark-treatment, both in etiolated plants as well as in dark-adapted mature green plants (20). In contrast, AS2 mRNA accumulates only to low levels in dark-treated plants. The fold-induction of AS1 mRNA in leaves of dark- verses light-grown plants is as high as 30-fold (20). Further studies have shown that the repression of AS1 gene expression in the light is due to a phytochrome-mediated response (20).

Induced expression of AS1 and AS2 in developmental contexts of increased nitrogen transport

Early biochemical studies showed that AS activity was induced in nitrogen-fixing root nodules (13,15), and in cotyledons of germinating seedlings (2,3,9). Northern blot analysis of AS mRNAs revealed that both AS1 and AS2 mRNAs accumulate to high levels in both nodules and cotyledons of germinating seeds (20). The accumulation of AS mRNAs in these contexts where plants need to transport large amounts of nitrogen is coordinate with the induction of cytosolic GS mRNA in these contexts, albeit with slightly different kinetics.

CONCLUSIONS

Our molecular studies on the GS and AS gene families in pea have shown a dramatic correlation between the expression of individual genes with developmental contexts involving glutamine or asparagine biosynthesis. In order to define the cis-acting DNA elements which are responsible for the differential regulation of individual GS and AS genes, we have begun to dissect the promoter elements of the GS and AS genes in transgenic plants. Cis-acting sequences important for gene regulation *in vivo* are currently being used to define the transcription factors which interact with these elements, using *in vitro* approaches. Such studies should uncover the molecular mechanisms which regulate the differential transcription of chloroplast GS2 and AS1 by light, or for the coordinate regulation of genes for cytosolic GS and AS in cotyledons and nodules.

REFERENCES


1. Boland, M.J., Hanks, J.F., Reynolds, P.H.S., Blevins, D.G., Tolbert, N.E., Schubert, K.R. 1982. Subcellular organization of ureide biogenesis from glycolytic intermediates in nitrogen-fixing soybean nodules. *Planta* 155:45-51

2. Capdevila, A.M., Dure, L., III 1977. Developmental biochemistry of cottonseed embryogenesis and germination. *Plant Physiol.* 59:268-273

3. Dilworth, M.F., Dure, L. 1978. Developmental biochemistry of cotton seed embryogenesis and germination. X. Nitrogen flow from arginine to asparagine in germination. *Plant Physiol.* 61:698-702

4. Edwards, J.W., Coruzzi, G.M. 1989. Photorespiration and light act in concert to regulate the expression of the nuclear gene for chloroplast glutamine synthetase. *Plant Cell* 1:241-248

5. Edwards, J.W., Walker, E.L., Coruzzi, G.M. 1990. Cell-specific expression in transgenic plants reveals nonoverlapping roles for chloroplast and cytosolic glutamine synthetase. *Proc. Natl. Acad. Sci. USA.* 87:3459-3463

6. Heeke, G.V. Schuster, M. 1989. Expression of human asparagine synthetase in *Escherichia coli. J. Biol. Chem.* 264:5503-5509

7. Huber, T.A., Streeter, J.G. 1985. Purification and properties of asparagine synthetase from soybean root nodules. *Plant Science* 42:9-17

8. Joy, K.W., Ireland, R.J., Lea, P.J. 1983. Asparagine synthesis in pea leaves, and the occurrence of an asparagine synthetase inhibitor. *Plant Physiol.* 73:165-168

9. Kern, R., Chrispeels, M.J. 1978. Influence of the axis in the enzymes of protein and amide metabolism in the cotyledons of mung bean seedlings. *Plant Physiol.* 62:815-819

10. McNally, S.F., Hirel, B., Gadal, P., Mann, F., Stewart, G.R. 1983. Glutamine synthetases of higher plants. *Plant Physiol.* 72:22-25

11. Miflin, B.J., Lea, P.J. 1980. ammonia assimilation. In *The Biochemistry of Plants, Vol. 5*, 169-202, New York: Academic Press

12. Miflin, B.J., Lea, P.J. 1982. Biosynthesis and metabolism of protein amino acids and proteins. In *Nucleic acid and proteins in plants I: Structure, biochemistry and physiology of proteins*, 5-64, eds. D. Boulter, B. Parthier. Berlin Heidelberg New York: Springer-Verlag

13. Reynolds, P.H.S., Blevins, D.G., Boland, M.J., Schubert, K.R., Randal, D.D. 1982. Enzymes of ammonia assimilation in legume nodules: A comparison between ureide- and amide-transporting plants. *Physiol. Plant* 55:255-260

14. Rognes, S.E. 1975. Glutamine-dependent asparagine synthetase from *Lupinus luteus Phytochemistry* 14:1975-1982

15. Scott, D.B., Farnden, K.J.F., Robertson, J.G. 1976. Ammonia assimilation in lupin nodules. *Nature* 263:703-705

16. Streeter, J.M. 1977. Asparaginase and asparagine transaminase in soybean leaves and root nodules. *Plant Physiol.* 60:235-239

17. Tingey, S.V , Coruzzi, G.M. 1987. Glutamine synthetase of *Nicotiana plumbaginifolia:* Cloning and *in vivo* expression. *Plant Physiol.* 84:366-373

18. Tingey, S.V., Tsai, F.-Y., Edwards, J.W., Walker, E.L., Coruzzi, G.M. 1988. Chloroplast and cytosolic glutamine synthetase are encoded by homologous nuclear genes which are differentially expressed *in vivo. J. Biol. Chem.* 263:9651-9657

19. Tingey, S.V , Walker, E.L., Coruzzi, G.M. 1987. Glutamine synthetase genes of pea encode distinct polypeptides which are differentially expressed in leaves, roots and nodules. *EMBO J.* 6:1-9

20. Tsai, F.-Y., Coruzzi, G.M. 1990. Dark-induced and organ-specific expression of two asparagine synthetase genes in Pisum sativum. *EMBO J.* 9:323-332

21. Urquhart, A.A., Joy, K.W. 1981. Use of phloem exudate technique in the study of amino acid transport in pea plants. *Plant Physiol.* 68:750-754

22. Walker, E.L., Coruzzi, G.M. 1989. Developmentally regulated expression of the gene family for cytosolic glutamine synthetase in *Pisum sativum.* *Plant Physiol.* 91:702-708

23. Wallsgrove, R.M., Turner, J.C., Hall, N.P., Kendally, A.C., Bright, S.W.J. 1987. Barley mutants lacking chloroplast glutamine synthetase: Biochemical and genetic analysis. *Plant Physiol.* 83:155-158

# MOLECULAR BASIS OF PLANT DEFENSE RESPONSES TO FUNGAL INFECTIONS

Klaus Hahlbrock, Petra Groß, Christiane Colling and Dierk Scheel

Max-Planck-Institut für Züchtungsforschung
Abteilung Biochemie
D - 5000 Köln 30
F.R.G.

## INTRODUCTION

Plant diseases caused by fungal infections are the results of the relatively rare cases of compatible plant-fungus interactions, where a susceptible plant is invaded by a virulent fungus. Much more frequent are incompatible interactions between the two types of organism. In these cases, the plant is resistant and the fungus avirulent. If all known genotypes of the interacting organisms are incompatible with each other, the plant is by definition nonhost resistant; if one or more genotype combinations of the two species lead to a compatible interaction, the plant is a host which can be susceptible or resistant. The various possible types of interaction are depicted in a general scheme in Figure 1. It should be noted that in most plant-fungus combinations, no interaction takes place at all.

The interest of a growing number of scientists in molecular plant pathology is to a large extent caused by the increasing problems arising from the spread of plant diseases, that is, of compatible plant-pathogen interactions, in agriculture. In several cases, breeders have been successful with the introduction of resistance genes directed against particular genotypes of pathogens. However, pathogens mutate rapidly and form new virulent genotypes. Whether the result of deliberate breeding efforts or of natural inheritance, host resistance is a constantly endangered trait. If a particular plant interacts at all with a potential pathogen, the more desirable trait is nonhost resistance which cannot be overcome by any genotype of the pathogen.

With the aim of understanding the mechanism of nonhost resistance in plants, we are investigating the interaction of parsley (*Petroselinum crispum*) with the soybean-pathogenic fungus *Phytophthora megasperma* f.sp. *glycinea* (*Pmg*). Under standardized conditions, *Pmg* grows efficiently on a parsley leaf surface and occasionally enters the plant tissue, preferentially through stomata (Hahlbrock et al 1985). The attempted penetration is counteracted by rapid, highly localized plant cell death, the "hypersensitive response" of the cells that come in direct contact with the fungus. Immediately thereafter, the area surrounding these cells undergoes a drastic change in metabolic activity, including the activation of numerous genes. By *in situ* RNA hybridization, we have shown that several of the genes tested were transiently activated by 4 h after inoculation of parsley leaves with *Pmg* zoospores (Schmelzer et al 1989).

*Plant Molecular Biology* 2, Edited by R.G. Herrmann and
B. Larkins, Plenum Press, New York, 1991

FUNGUS- OR ELICITOR-INDUCED GENE ACTIVATION

Among the genes that are activated locally around fungal infection sites are those encoding the enzymes of general phenylpropanoid metabolism. This pathway converts a product of primary metabolism, phenylalanine, to the common substrate of various phenylpropanoid branch pathways, 4-coumaroyl-CoA and several related CoA esters of cinnamate derivatives. Two enzymes catalyzing key steps in this conversion are phenylalanine ammonia-lyase (PAL) and 4-coumarate:CoA ligase (4CL) (Hahlbrock and Scheel 1989). They are encoded in parsley by small families of four and two genes, respectively (PAL: Lois et al 1989, Schulz et al 1989; 4CL: Douglas et al 1987).

The triggering of gene activation in *Pmg*-infected parsley leaves can be mimicked by treatment of suspension-cultured cells with *Pmg*-derived elicitor (Somssich et al 1989). From a collection of cDNA probes generated for the transcripts of elicitor-responsive genes in parsley cells, several were tested by *in situ* RNA hybridization and all demonstrated the same kind of local, rapid and transient gene activation in infected leaf tissue. Among the activated genes were those encoding PAL and 4CL, suggesting a

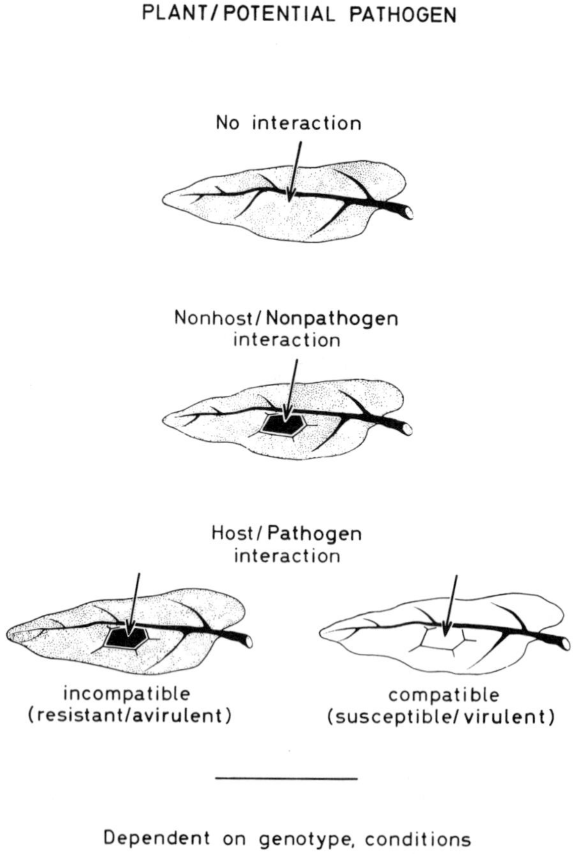


**Figure 1.** Schematic representation of plant-microbe interactions: No detectable interaction (majority of cases), nonhost-nonpathogen interaction (frequent), host-pathogen interaction (rare).

causal connection with the accumulation of various types of soluble and wall-bound phenolics in both elicitor-stimulated cells and leaf areas around infection sites (Scheel et al 1986, Jahnen and Hahlbrock 1988, Schmelzer et al 1989).

Transcriptional activation in cultured cells involved at least three of the four PAL genes (Lois et al 1989) and both 4CL genes (Douglas et al 1987). High transcription rates were reached within about 2 h of the elicitor treatment and remained at this level for at least another 8 h (Lois et al 1989). The response was essentially the same whether elicitor was applied to cultured cells or to protoplasts derived from them (Dangl et al 1987), and whether a crude elicitor preparation from *Pmg* mycelial cell walls or its purified, most active component, a 42 kDa glycoprotein, was used for stimulation (Parker et al 1988; J.E. Parker and D. Scheel, unpublished results). This and a few other proteins from *Pmg*, which are potent phytoalexin elicitors in parsley, are inactive in soybean, the host plant for *Pmg* (Parker et al 1988). The recognition mechanism for the glycoprotein elicitor and the signal-transduction chain in parsley cells are presently being investigated. One aspect is the involvement of ion fluxes.

## INVOLVEMENT OF ION FLUXES

When the transcriptional activities of various elicitor-responsive as well as control genes were measured in the presence or absence of $Ca^{2+}$ in the protoplast culture medium, we observed strong differential effects. All tested elicitor-responsive genes were activated at greatly reduced rates in the absence of appreciable concentrations of $Ca^{2+}$. Control genes that were constitutively transcribed at high rates in the presence of $Ca^{2+}$ were equally active in $Ca^{2+}$-depleted medium, and another group of control genes that were activated by UV light responded strongly to the signal regardless of whether $Ca^{2+}$ was present or absent. Direct measurement of $Ca^{2+}$ in elicitor-treated parsley cells revealed a rapid decrease of $Ca^{2+}$ in the culture medium, an equally rapid appearance of $^{45}Ca^{2+}$ in the cells, both within a few minutes, and a simultaneous efflux of $K^+$ and increase in the extra-cellular pH (Scheel et al 1989; C. Colling and D. Scheel, unpublished results).

With similar timing, several large and rapid changes in the protein phosphorylation patterns occurred in elicitor-treated parsley cells (Dietrich et al 1990). These changes appeared to be signal-specific. They were dependent on the continued presence of elicitor, as demonstrated by a washing experiment, and were triggered by the same proteinaceous elicitor as mentioned above. The effect was again greatly reduced when $Ca^{2+}$ was omitted from the medium.

Taken together, these results suggest that ion fluxes play a major role in the elicitation process. We are presently testing the hypothesis that ion channels are affected by binding of the elicitor at the membrane. The resulting influx of $Ca^{2+}$ may then have the observed effects on protein phosphorylation, which in turn may be directly or indirectly related to gene activity. To see whether all these phenomena are causally connected, we have also begun to analyze the *cis*-acting elements and *trans*-acting factors involved in the activation of elicitor-responsive genes, including PAL and 4CL (Lois et al 1989).

## LOCALIZED STRUCTURAL CHANGES IN INFECTED CELLS

As a first step towards elucidating the sequence of various molecular events at the more complex level of true plant-pathogen interactions, we have started looking at some microscopically visible effects of the attempted fungal penetration into cultured parsley cells. Fungal cysts were fixed to the gelatin-coated surface of a microscope slide and were allowed to germinate for 1 h. Cultured parsley cells were then added, those not interacting with the fungus were washed off, and those remaining attached to fungal hyphae were examined under the microscope. Two striking observations are shown in Figure 2 and Figure 3 (P. Groβ, unpublished results).

Figure 2 shows that the nucleus of an attached parsley cell moved towards the site of attempted penetration and reached this point within approximately 4 h postinoculation. Visualization of the cytoskeleton of another challenged cell, by indirect FITC-labelling of a tubulin-specific antibody, demonstrated a highly localized disintegration of microtubules precisely around the site of interaction of the two organisms. One example is shown in Figure 3. It will be interesting to see whether the observed structural changes in the cell are accompanied by local increases in the intracellular $Ca^{2+}$ concentrations, and whether they also occur in infected whole-plant tissue.

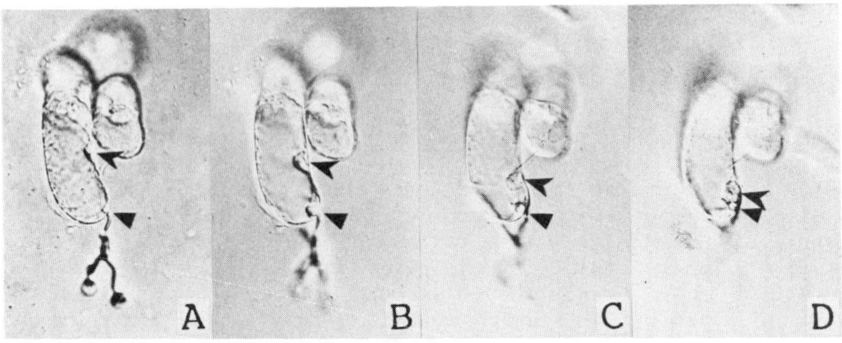


**Figure 2.** Migration of parsley cell nucleus (arrow head) towards infection site (triangular arrow). A) ca. 2 h after incubation of germinating fungus with cultured cells; B-D) 22,94 and 136 min later.

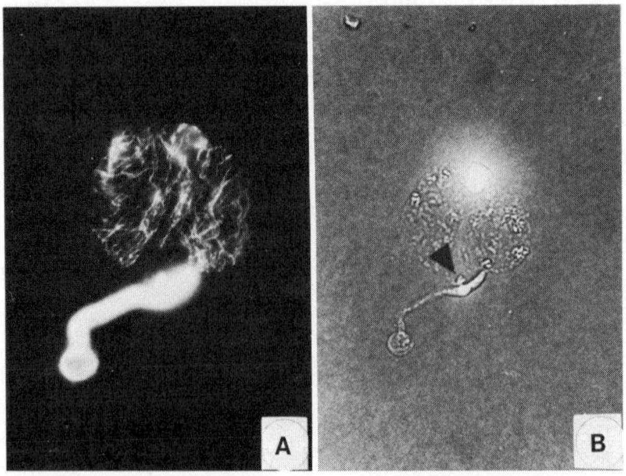


**Figure 3.** Visualization of microtubule arrangement in infected, cultured parsley cell. A) Staining with $\alpha$-tubulin-specific antibodies (upper part) and with antibodies specific for fungal surface material (lower part). B) Same specimen stained with Hoechst dye to localize plant (upper part) and fungal (lower part) nuclei. Note small fungal penetration peg (arrow) and absence of microtubular structures in its vicinity.

We hope that a better understanding of such early events in nonhost-pathogen interactions may help in identifying new target sites for improved crop protection as well as genes responsible for non-self recognition in plants.

## REFERENCES


Dangl JL, Hauffe KD, Lipphardt S, Hahlbrock K, Scheel D (1987) Parsley protoplasts retain differential responsiveness to UV light and fungal elicitor. EMBO J **6**:2551-2556

Dietrich A, Mayer JE, Hahlbrock K (1990) Fungal elicitor triggers rapid, transient, and specific protein phosphorylation in parsley cell suspension cultures. J Biol Chem **265**:6360-6368

Douglas C, Hoffmann H, Schulz W., Hahlbrock K (1987) Structure and elicitor or u.v.-light stimulated expression of two 4-coumarate:CoA ligase genes in parsley. EMBO J **6**:1189-1195

Hahlbrock K, Chappell J, Jahnen W, Walter M (1985) Early defense reactions of plants to pathogens, in: "Molecular form and function of the plant genome," L. van Vloten-Doting, G.S.P. Groot, T.C. Hall, eds., Plenum Publishing Corporation, pp. 129-140

Hahlbrock K, Scheel D (1989) Physiology and molecular biology of phenylpropanoid metabolism. Annu Rev Plant Physiol Plant Mol Biol **40**:347-369

Jahnen W, Hahlbrock K (1988) Cellular localization of nonhost resistance reactions of parsley (*Petroselinum crispum*) to fungal infection. Planta **173**:197-204

Lois R, Dietrich A, Hahlbrock K, Schulz W (1989) A phenylalanine ammonia-lyase gene from parsley: structure, regulation and identification of elicitor and light responsive cis-acting elements. EMBO J **8**:1641-1648

Parker JE, Hahlbrock K, Scheel D (1988) Different cell wall components from *Phytophthora megasperma* f.sp. *glycinea* elicit phytoalexin production in soybean and parsley. Planta **176**:75-82

Scheel D, Hauffe KD, Jahnen W, Hahlbrock K (1986) Stimulation of phytoalexin formation in fungus-infected plants and elicitor-treated cell cultures of parsley, in: "Recognition in Microbe-Plant Symbiotic and Pathogenic Interactions," B Lugtenberg, ed., NATO ASI Series H 4, pp. 325-331

Scheel D, Colling C, Keller H, Parker JE, Schulte W, Hahlbrock K (1989) Studies on elicitor recognition and signal transduction in host and non-host plant/fungus interactions, in: "Signal molecules in plants and plant-microbe interactions," B.J.J. Lugtenberg, ed., NATO ASI Series H 36, pp. 212-218

Schmelzer E, Krüger-Lebus S, Hahlbrock K (1989) Temporal and spacial patterns of gene expression around sites of attempted fungal infection in parsley leaves. The Plant Cell **1**:993-1001

Schulz W, Eiben HG, Hahlbrock K (1989) Expression in *Escherichia coli* of catalytically active phenylalanine ammonia-lyase from parsley. FEBS Letters **258**:335-338

Somssich IE, Bollmann J, Hahlbrock K, Kombrink E, Schulz W (1989) Differential early activation of defense-related genes in elicitor-treated parsley cells. Plant Mol Biol **12**: 227-234

# PLANT GENES INVOLVED IN RESISTANCE TO VIRUSES

Bernard Dumas, Estelle Jaeck, Annick Stintzi, Jacques Rouster, Serge Kauffmann, Pierrette Geoffroy, Marguerite Kopp, Michel Legrand, and Bernard Fritig

Institut de Biologie Moléculaire des Plantes, CNRS
12 Rue du Général Zimmer 67000 Strasbourg, France

## INTRODUCTION

The model system that will be considered here is the hypersensitive response (HR) since it is one of the most efficient natural mechanisms of defence of plants against pathogens. This is also true in the case of virus infections. In hypersensitivity to viruses the same characteristics are found as in incompatible interactions between plants and fungi or bacteria: i) necrosis at and around each point at which the leaf tissue was infected ; ii) localization of the pathogen to the region of each initiated infection; iii) induction of marked metabolic changes in the cells surrounding the necrotic area, these changes being believed to cause, or at least to contribute to, the resistance observed. Furthermore, the hypersensitive response to viruses involves a cascade of events and signals (Fritig *et al.*, 1987) similar to that proposed for active defence against fungi and bacteria (Lamb *et al.*, 1989): the response would be initiated by a specific gene-for-gene recognition between the plant and the virus that would lead to cell damage (death of a number of cells) and the release of intermediary signals which in turn would enter a reception-transduction pathway leading to major changes in gene expression (usually gene activation) responsible for the resistance observed.

The present paper will deal mainly with the downstream part of this cascade, i.e. the changes in host gene expression and their possible contribution to the defence response. Major emphasis will be put on a group of strongly induced proteins with typical properties, the PR proteins ("pathogenesis-related") also called "stress proteins". They were initially discovered (Van Loon and Van Kammen, 1970 ; Gianinazzi *et al.*, 1970) in tobacco reacting hypersensitively to tobacco mosaic virus (TMV) and, therefore, thought to be anti-viral. It will be shown that most of the yet characterized major PR proteins induced by virus infection are apparently not directed specifically against viruses but rather against fungi, bacteria and perhaps insects

*Plant Molecular Biology* 2, Edited by R.G. Herrmann and B. Larkins, Plenum Press, New York, 1991

and thus are part of the defence mechanims which would be useful against secondary invaders or predators.

## PLANT GENE(S) INVOLVED IN THE EARLY RECOGNITION TRIGGERING RESISTANCE TO VIRUS INFECTION

In the HR, resistance results from an active defence mechanism that is induced by the invading pathogen itself. We have demonstrated such an induction of resistance against virus infection by using two almost isogenic lines of *Nicotiana tabacum*, one sensitive and the other containing the N gene for resistance to TMV. The leaves of the two cultivars were micro-inoculated with the U1 strain of TMV. The micro-inoculation procedure yielded infection sites at predetermined locations on the leaves of the two host cultivars (Konate and Fritig, 1984). At various intervals after the micro-inoculations the individual infection sites were assayed for virus content by ELISA or for incorporation of $^{3}$H-uridine into the viral RNA after pulse-labelling by injection around the micro-inoculated areas at the end of the infection period (Konate and Fritig, 1983). Analysis of more than 3000 individual infection sites showed that, up to the time (33-36 hours after inoculation) visible necrosis appeared, the rates of virus multiplication did not differ significantly between the two hosts. These results resemble those mentioned by Fraser (1988) for the accumulation of viral RNA in the case of leaves of White Burley tobacco inoculated either by a necrotic (avirulent) virus strain of TMV forming local lesions, or by a systemically-spreading strain (virulent). These examples show that shortly after virus inoculation, virus multiplication is not impaired despite the presence of a gene for resistance. Hypersensitive resistance to virus infection does not preexist but is induced later during the infection. In our model system, the tobacco Samsun NN-TMV interaction, hypersensitive resistance appears at about the same time as the necrotic symptoms and its efficiency increases with time after infection (Konate, 1984). Around the necrotic lesions there is a ring of cells about 1.0-1.5 mm in width containing detectable virus but in which virus multiplication is inhibited or from which it cannot spread. Thus, it appears that in Samsun NN a period of 30-36 hours of interaction between plant components and viral components is required to produce the necrotic stress accompanied by the metabolic alterations responsible for resistance.

Gene-for-gene complementarity is a widely occurring phenomenon in which single dominant alleles in plants and their pathogens determine whether the plant will be resistant or susceptible to attack (Flor, 1942; Ellingboe, 1982). If the dominant, matching alleles of a pathogen avirulence gene and a plant disease resistance gene are present, recognition occurs and the plant responds to infection by invoking the HR. However, if dominant alleles are not present in both partners, recognition does not occur and the pathogen develops extensively, causing disease. While some aspects concerning expression of the HR are understood (Lamb *et al*., 1989), little is known about the molecular functions of avirulence genes and resistance genes, particularly in how they lead to specific recognition of incompatible races or strains. One of the major hypothesis that is currently tested is the elicitor-receptor model, which holds that pathogen biotypes carrying a certain avirulence gene produce an elicitor substance which is specifically recognized by a receptor present in plants carrying the matching resistance gene (Keen, 1985). A prediction from this model is that elicitor-receptor binding initiates the subsequent biochemical events associated with the expression of the HR. While disease resistance genes have not been isolated from higher plants, several avirulence genes have been cloned and characterized from bacterial pathogens (for a review, see Keen and Staskawicz, 1988). Very recently these authors showed that bacteria expressing avirulence gene avr D from Pseudomonas syringae pv tomato produce a specific elicitor of the soybean HR (Keen et al., 1990).Even though specific HR phytoalexin elicitors from several pathogens have been previously reported (for a review, see Lamb et al., 1989), this is the first case where the production of a cultivar-specific elicitor of the HR-phytoalexin mechanism has been causally linked to the action of a defined microbial avirulence gene. The structure of the elicitor (a rather low molecular weight component) is currently being investigated.

Analysis of the genetics of plant-virus interactions is less advanced. Clearly, in plant-virus interactions, the pathogen half of the interaction is not a cell, and might not be expected to be capable of the same degree of complexity of signals as those emanating from a bacterium or fungus. But recognition and signalling events between plants and viruses do nevertheless involve the universal principle of interactions between host- and pathogen-specified molecules, just as happens with microbial pathogens and plants. However, for a long time it has been difficult to do genetical analysis with viral genomes : recombination analysis has been impossible, and most plant viruses have RNA genomes which are inherently unstable. The concepts of dominance and recessiveness, which have proved so useful in analysis of resistance and virulence in plant-microbial interactions, are also difficult to apply in the viral context.

But recently, the use of recombinant DNA techniques has allowed fine genetic analysis of the genomes of RNA viruses, and determinants which effect interactions with host resistance genes can now be located with great precision (Saito et al., 1987; Meshi et al., 1988). For instance, avirulence in TMV matching the hypersensitivity N' gene of Nicotiana sylvestris has been located on the coat protein gene of the virus (Culver and Dawson, 1990 ; Mundry et al., 1990). But removal of the entire TMV coat protein gene does not effect the induction of HR in NN genotype plants. Thus, a viral component resulting from a gene other than the coat protein gene is the elicitor of HR in NN genotype plants (Culver and Dawson, 1990) that are used in our model system.

As with microbial pathogens, the plant component of the early recognition events between plant and virus is poorly understood. A survey of the genetics of resistance to a large number of viruses in cultivated crops has been reported by Fraser (1986, 1988, 1990). Most resistance to viruses in crops is monogenic. Dominant alleles are associated with virus-localization mechanisms that accompany the HR. It has been observed that many resistance genes have been overcome by virulent isolates of viruses;  however, a significant number of resistance genes have proved exceptionally durable. So it can be expected that in the near future, efforts will be devoted to the isolation of such genes with the aim to understand the recognition event but also to use them in practical crop protection. It is likely that precise and detailed information on how matching virulence and avirulence is determined in the virus will help in devising strategies for the isolation of host resistance genes.

## OVERALL METABOLIC CHANGES DURING THE HYPERSENSITIVE REACTION TO VIRUSES

Even though the triggering of the hypersensitivity reaction results from a very specific gene-for-gene recognition between plant and pathogen, what the plant does thereafter to defend itself appears to be specific only to the host. Most of the metabolic changes that have been reported as associated to active defence in incompatible plant-fungi and plant-bacteria interactions (for reviews, see : Bailey and Deverall, 1983; Darvill and Albersheim, 1984; Goodman et al., 1986; Collinge and Slusarenko, 1987; Lamb et al., 1989) have also been found during the hypersensitive reaction of plants to viruses (Ponz and Bruening, 1986; Fritig et al., 1987, 1990). In summary, the alterations include cell wall thickening resulting from production and deposition of various macromolecules, and the production of defence enzymes and proteins. Among the cell wall macromolecules are lignins and other phenolic polymers, polysaccharides such as callose, and proteins such as the hydroxyproline rich glycoproteins (HRGP). Defence enzymes fall into two classes : enzymes that catalyze the production of various metabolites participating in resistance (ethylene, phytoalexins, aromatic compounds, oxidized metabolites) ; and direct defence enzymes (hydrolases such as chitinases and glucanases). The defence proteins include inhibitors of microbial proteases or of polygalacturonases and "pathogenesis-related" (PR) proteins, the latter being rather abundant proteins. In a previous review, we (Fritig et al., 1987) indicated that if most of these changes had been reported in the case of HR to viruses, a few of these changes were only suspected and not unequivocally proven : for instance the production of HRGP, of microbial proteinase inhibitors and of inhibitors of polygalacturonases. If the latter are still not well documented, even in infections by microbial pathogens (with the exception of Cervone et al., 1989), HRGP (Benhamou et al., 1990) and inhibitors of microbial proteases (see below) have now been demonstrated in HR of tobacco to TMV.

## CHANGES IN PLANT GENE EXPRESSION ASSOCIATED TO THE HYPERSENSITIVE RESPONSE TO VIRUSES

In this paragraph we shall discuss the major changes in gene expression that accompany the HR to viruses. One strategy, used by the group of J. Bol, consisted in following induction of host mRNAs in hypersensitively reacting plants by a differential screening of cDNA libraries with probes made to poly(A) RNA from either infected or healthy leaves. This approach led to groups of clones corresponding to seven TMV-induced host mRNAs of Samsun NN and were called "clusters A to G" (Hooft van Huijsduijnen et al., 1986; Cornelissen et al., 1987). Several of these clusters were found to correspond to already known major PR proteins. Later these cDNA clones were used as probes to isolate the corresponding genes and in turn to study regulatory sequences involved in the induction of the genes during the HR (Bol et al., 1989 ; Linthorst et al., 1990).

In our group, we used another strategy which turned out to be complementary to that of the group of J. Bol. It consisted in isolating and characterizing the major proteins that had been detected in the HR of tobacco Samsun NN to TMV (Van Loon, 1982). Enzymatic activities were then demonstrated for some of these proteins. The antibodies raised against them helped to find other proteins of the PR-type and, again, for some of them, a biological activity. During the fractionation used to purify the latter, novel minor induced proteins were detected and were also isolated. Thus, this strategy using biochemical approaches led to the characterization of a higher number of HR-induced proteins than the differential hybridization, and to the identification of their biological activity (including comparison of specific activities).

## The Major Defence-Related Proteins Induced during the Hypersensitive Response

The PR-proteins represent the major changes in protein production during the hypersensitive response to viruses as well as to other pathogens (Van Loon, 1985; Fritig et al., 1987; Lamb et al., 1989). These PR-proteins have characteristic properties: they are selectively extractable at low pH, highly resistant to exogenous and endogenous proteases, easily resolved by electrophoresis on polyacrylamide gels under native conditions.

The occurrence of these proteins in many plant species has been reviewed (Van Loon, 1985; Carr and Klessig, 1989). Tobacco is indeed a reference system for which the highest number of PR-proteins has been described after isolation from TMV-infected leaves with a very performant purification procedure (Fritig et al., 1989). The identification of some of the major tobacco PR-proteins as glycanhydrolases has been described previously (Legrand et al., 1987; Kauffmann et al., 1987; Fritig et al., 1990). Here we will merely summarize the present stage of knowledge concerning the isolated proteins.

We have isolated up to now 25 defence-related proteins from tobacco reacting hypersensitively to TMV (see Table 1). These include the ten acidic PR-proteins already detected by L.C. Van Loon (1982) and named PR-1a, 1b, 1c, 2, N, O, P, Q, R and S in order of decreasing mobility on native basic gels. These 10 latter proteins belong to 4 distinct serological groups: the PR-1 group with PR-1a, 1b, 1c whose biological function is not known; PR-2, N and O are in fact 1,3-ß-glucanases (Kauffmann et al., 1987) that are serologically related to another acidic glucanase (Q') and to a basic glucanase (gluc.b.), these latter glucanases exhibiting typical behaviours of PR-proteins. Another 1,3-ß-glucanase has been isolated (O') that has not all of the typical properties of PR-proteins and is not serologically related to the 5 other glucanases. Furthermore the 6 glucanases exhibit very different specific enzyme activities towards laminarin as the substrate. PR-P and Q are chitinases serologically related to two other chitinases (chi 32 and chi 34) with again a behaviour typical of PR-proteins even though they exhibit

Table 1. Properties of tobacco PR-proteins induced by infection with TMV

| Proteins | Elution pH (chromato-focusing) | Molecular weight[a,b] ($x10^{-3}$) | Serological group[c] | Localization[d] | Biological activity |
|---|---|---|---|---|---|
| 1a | ND | 15.5[a] | 1 | ext | ? |
| 1b | 4.5 | 15.5[a] | 1 | ext | ? |
| 1c | 4.7 | 15[a] | 1 | ext | ? |
| 2 | 4.4 | 35[b] | 2 | ext | glucanase[f] |
| N | 4.7 | 36[b] | 2 | ext | glucanase[f] |
| O | 4.8 | 37[b] | 2 | ext | glucanase[f] |
| O'[e] | 4.9 | 25[b] | | int | |
| P | 5.3 | 25[a] | 3 | ext | chitinase[g] |
| Q' | 5.3 | 35[b] | 2 | ext | glucanase[f] |
| Q | 5.8 | 25[a] | 3 | ext | chitinase[g] |
| R | 6.9 | 24[b] | 4 | ext | ? |
| r1 | 7.0 | 14.5[b] | - | ext | ? |
| r2 | 7.0 | 13[b] | - | ext | ? |
| s1 | 7.4 | 14.5[b] | - | ext | ? |
| s2 | 7.4 | 13[b] | - | ext | ? |
| S | 7.5 | 22[a] | 4 | ext | ? |
| gluc.b | > 7 | 33[a] | 2 | int | glucanase[f] |
| ch32 | > 7 | 32[a] | 3 | int | chitinase[g] + lysozyme |
| ch34 | > 7 | 34[a] | 3 | int | chitinase[g] + lysozyme |
| Lys1 | > 7 | 28[b] | - | ND | lysozyme |
| Lys2 | > 7 | 29[b] | - | ND | lysozyme |
| Rb | > 7 | 22[a] | 4 | int | ? |
| 16 K | > 7 | 16[b] | ND | ND | ? |
| 42 K | > 7 | 42[b] | | ND | |
| InhI[e] | > 7 | 6[b] | | ND | microbial proteinase inhibitor |

a Molecular weight calculated from sequence data.
b Molecular weight determined on SDS-polyacrylamide gels.
c The 10 tobacco PR-proteins initially detected by Van Loon (1982) are members of 4 distinct serological groups (with high cross-reactivity between the members within a given group).
d Localization refers to extracellular (ext) or intracellular (int).
e Proteins O' and InhI do not have all of the typical properties of PR-proteins.
f endo-1,3-ß-glucanase.
g endo-chitinase.
ND Not determined.

basic isoelectric points (Legrand et al., 1987). Some chitinases are known to have both lysozyme and chitinase activities (Métraux et al., 1989). We found that P and Q had only chitinase activity, chi 32 and chi 34 had both chitinase and lysozyme activities and we found 2 proteins (lys 1 and lys 2) with only high lysozyme activity and with a behaviour of PR proteins (Stintzi et al., unpublished). Proteins R and S are serologically related to each other (Kauffmann et al., 1990) and to a basic protein of 22 K (Rb) which has recently been identified as the salt stress-induced protein osmotin (Kauffmann et al., unpublished). These 3 proteins are of unknown function, even though they show sequence homology to the sweet-tasting protein thaumatin (Cornelissen et al., 1986 ; Pierpoint et al., 1987) and to a bifunctional enzyme (amylase and proteinase) inhibitor from maize (Richardson et al., 1987). We found no significant α-amylase or proteinase inhibitor activity in isolated proteins R, S, and Rb. Two couples of acidic proteins, though migrating in native gels like R (r1, r2) and S (s1, s2) are apparently not related to R and S (Kauffmann et al., 1990), and are also of unknown function. There is also no known function for two other proteins with basic isoelectric points (42 K and 16 K), the 16 K is perhaps the basic counterpart of acidic proteins of the PR-1 group (Bol et al., 1989).

We examined other potential activities for these proteins, namely proteinase and proteinase inhibitor activities. In tobacco, none of the known PR-proteins (Van Loon, 1985) or of the newly identified PR-proteins (Table 1) has proteolytic activity. This contrasts with tomato where one of the PR-proteins induced by infection with citrus exocortis viroid, P-69, was shown to be an alkaline endoproteinase (Vera and Conejero, 1988), whose role in pathogenesis and defence is still unclear.

Polypeptide inhibitors of proteases are widely distributed in all plant tissues and are thought to have a role in defence against herbivores since they tend to be active against animal and not endogenous proteases (Ryan, 1981). Microorganisms produce serine proteinases resembling animal proteinases and, therefore, proteinase inhibitors produced by plants are thought to be involved in defence against attacking pests. Indeed, the fact that inhibitors from plants are able to inhibit proteinases from bacteria and fungi is well documented. In melon plants infected by *Colletotrichum lagenarium* an increase of activity inhibiting the protease produced by the fungus has been demonstrated (Roby et al., 1987). It is apparently one of the rare reports on increase of proteinase inhibitory activity in plants infected by pathogens. Concerning virus infections, there is a report on induction of proteinase inhibitor activity in tobacco reacting hypersensitively to TMV (Pierpoint et al., 1981), but this activity does not correspond to PR-proteins 1a, 1b and 1c (Pierpoint et al., 1981) or to any of the other well known PR-proteins. From this plant material we have recently isolated an inhibitor (Inh I of Table 1) that is highly active against 4 different serine endoproteinases of fungal and bacterial origin but that inhibits poorly 2 serine endoproteinases of animal origin, trypsin and chymotrypsin (Geoffroy et al., 1990). The inhibitor behaves upon electrophoresis under denaturing conditions as a small polypeptide of about 6 Kd. From its amino acid composition and N-terminal amino acid sequence analysis it appears that the inhibitor belongs to the potato inhibitor I family. It is highly induced in inoculated leaves and weakly in uninfected upper leaves.

Most of the proteins with acidic isoelectric points were found to be extracellular (Van Loon, 1985) while most basic proteins were found to be intracellular (Grosset et al., 1990), some being localized in the vacuolar compartment (Van der Bulcke et al., 1989). cDNAs encoding a number of these proteins have been cloned (for reviews, see Bol et al., 1989; Linthorst et al., 1990). Recently cDNAs encoding chitinases P and Q (Payne et al., 1990) and a gene encoding a basic chitinase (Shinshi et al., 1990) have been cloned. The sequence data indicate in all cases reported the presence of a N-terminal signal peptide.

For two of the serological groups of proteins with unknown function(s), transgenic plants have been obtained expressing constitutively PR-1a or PR-S. This constitutive expression had no effect on virus infection (Linthorst et al., 1989), suggesting that these proteins are not directly anti-viral. As indicated in Table 1, among the 25 defence-proteins are 10 glycanhydrolases (4 chitinases and 6 1,3-ß-glucanases) with an endo-catalytic activity. Thus, they may play a central role in the release of oligosaccharides from the walls of some plant pathogens or of the plant cells themselves, some of these oligosaccharides being known as elicitors of defence reactions (Darvill and Albersheim, 1984; Ryan, 1988). Thus, these glycanhydrolases are probably direct anti-bacterial and anti-fungal enzymes but not direct anti-viral enzymes. They might be anti-viral indirectly by releasing polysaccharidic or oligosaccharidic fractions, some of which are anti-viral (Kopp et al., 1989).

Stimulation of Aromatic Metabolism and Resistance to Virus Infection

Since the major induced proteins appear not involved directly in anti-viral resistance, we looked at other changes in gene expression. We have described previously a striking increase in ethanol-soluble aromatic metabolites and in lignin (Massala et al., 1987) which correlated well with a sharp increase in the activities of phenylalanine ammonia-lyase (PAL), cinnamic acid hydroxylase, cinnamoyl CoA ligases, ortho-diphenol-O-methyl-transferases and peroxidases (Legrand et al., 1976). Calculations revealed high intensities of enzymic stimulation per cell (more than hundredfold for all the enzymes studied) which correlated well with the efficiencies of localizing different TMV strains (Legrand et al., 1976).

We also studied the comparative effects on hypersensitive resistance of two competitive inhibitors of PAL, the first enzyme on the pathway. The inhibitors used were α-aminooxyacetic acid (AOA) and a structural analogue of phenylalanine, α-aminooxy-ß-phenylpropionic acid (AOPP). When supplied to tobacco leaves they increased the size of the lesions for several tobacco-TMV combinations examined (Massala et al., 1980, 1987), AOA being the more effective compound in weakening hypersensitive resistance. We concluded from these comparative studies that insoluble polymers such as lignin are likely to be the phenylpropanoid derivatives that participate in the mechanism of hypersensitive resistance to virus infection.

Density labelling studies have shown an increase in *de novo* synthesis of some of these enzymes, for instance phenylalanine ammonia-lyase and 3 isoforms of o-diphenol-O-methyltransferases during the HR. An approach using the tools of molecular biology is under way to establish whether there is a transcriptional or translational control of the increased enzyme synthesis. Within the frame of an ECC ECLAIR programme, we are trying to modulate (increase by gene overexpression, decrease by anti-sense RNA expression) lignification, the 3 target enzymes being cinnamyl alcohol dehydrogenase, cinnamoyl-CoA reductase and our system of o-diphenol-O-methyltransferases. This approach should shed some light on the role in virus resistance of increased deposition of aromatic polymers in the cell walls.

## OTHER INHIBITORS OF VIRUS REPLICATION ASSOCIATED WITH THE RESISTANCE RESPONSES

It is likely that the above mentioned cell wall thickening (deposition of aromatic polymers and other macromolecules) is acting in virus resistance by inhibiting cell-to-cell movement rather than by inhibiting intracellular virus replication. But some inhibitors of intracellular virus replication have been detected and found to be proteinaceous in nature (for a review, see Loebenstein and Gera, 1990). It was found that protoplasts from tobacco, carrying the NN gene responsible for resistance (local lesions) to TMV, release after their inoculation with TMV a compound that inhibits several plant viruses, the effect being dose responsive (Loebenstein and Gera, 1981). This substance termed inhibitor of virus replication (IVR) inhibited TMV in protoplasts, TMV, potato virus X (PVX), potato virus Y (PVY), and cucumber mosaic virus (CMV) in leaf disks from different hosts, indicating that IVR is neither host nor virus-specific (Gera and Loebenstein, 1983). IVR also inhibited TMV replication in intact leaves when applied by spraying to tobacco and tomato plants and CMV in cucumber. Subsequently, IVR was also obtained from the intercellular spaces of TMV-infected tobacco NN plants and from induced-resistant tissue (Spiegel *et al*., 1989). IVR has a proteinaceous nature, and its induction is sensitive to actinomycin D. Antibodies were prepared (Gera and Loebenstein, 1989) which neutralized IVR's anti-viral activity and enabled its detection in immunoblots. It will be interesting to know more about the biochemistry of this type of protein(s), its mechanism of action on virus replication, its occurrence in other plant-virus interactions or even in other plant-pathogen interactions.

## CONCLUSION

The hypersensitive reaction is triggered by a very specific (gene-for-gene) recognition between a pathogen and the host plant. It is likely that the modern tools of molecular biology, that have already lead to the identification of avirulence genes in pathogens, will also lead in the near future to the isolation of major resistance genes, including resistance genes against viruses. Even if the practical use of these genes in biotechnology for crop protection is questionable (unstable resistance as discussed by Fraser, 1988), they probably will help in understanding the triggering of the cascade of biochemical events responsible for active defence.

As discussed above, active defence of plants is associated with metabolic alterations that are apparently only host specific and confer a universal type of resistance. It is noteworthy that some of the tobacco PR-proteins, which were first detected in viral infections and were thought to be related to anti-viral defence, are chitinases, lysozymes and glucanases and are therefore defence enzymes apparently directed against insects, fungi and bacteria. Perhaps, it is not one only change but the coordinated superposition of many changes which is efficient against viruses, or specific anti-viral metabolites or proteins have still to be identified. The major induced proteins being glycanhydrolases could also act indirectly by releasing polysaccharidic elicitors, some of which have been shown to be anti-viral (Kopp et al., 1989). It follows that controlling the production of regulatory molecules such as elicitors could perhaps confer resistance to various pathogens. Hypersensitivity to viruses might be a simplified model in which to identify elicitors of host origin. These elicitors could then be used to induce metabolic alterations, and the elicitors that proved most effective in inducing resistance against challenge infection by various pathogens could be selected. The gene(s) of the corresponding hydrolase(s) releasing the most efficient elicitor(s) would be useful for plant genetic engineering. If adequate regulatory sequences were used, the expression of these genes could be targeted to the sites of preferential attack by pathogens.

## REFERENCES


Bailey, J. A., Deverall, B. J., 1983, "The dynamics of Host Defence", Academic Press, Sydney New York London.

Benhamou, N., Mazau, D., Esquerré-Tugayé, M. T., and Asselin, A., 1990, Immunogold localization of the hydroxyproline-rich glycoproteins in necrotic tissue of Nicotiana tabacum L. cv. Xanthi-nc infected by tobacco mosaic virus, Physiol. Mol. Plant Pathol., 36:129.

Bol, J. F., van de Rhee, M. D., Kan, J. A. L., Gonzales Jaen, M. T., and Linthorst, M. J. M., 1989, Characterization of two virus-inducible plant promoters, in: "Signal Molecules in Plants and Plant-Microbe Interactions," B. J. J. Lugtenberg, ed., NATO ASI Series, Cell Biology, Vol. 36, Springer-Verlag, Berlin, 169.

Carr, J. P., and Klessig, D. F., 1989, The pathogenesis-related proteins of plants, in: "Genetic Engineering ; Principles and Methods," J. K. Setlow, ed., vol. 11, Plenum Press, New York and London, 65.

Cervone, F., Hahn, M. G., De Lorenzo, G., Darvill, A., and Albersheim, P., 1989, Host-pathogen interactions XXXIII. A plant protein converts a fungal pathogenesis factor into an elicitor of plant defense responses, Plant Physiol., 90:542.

Collinge, D. B., Slusarenko, A. J., 1987, Plant gene expression in response to pathogens, Plant. Mol. Biol., 9:389.

Cornelissen, B. J. C., Hooft van Huijsduijnen, R. A. M., and Bol, J. F., 1986, A tobacco mosaic virus-induced tobacco protein is homologous to the sweet-tasting protein thaumatin, Nature, 321:531.

Cornelissen, B. J. C., Horowitz, J., Van Kal, J. A. L., Goldberg, R. B., and Bol, J. F., 1987, Structure of tobacco genes encoding pathogenesis-related proteins from the PR-1 group, Nucleic Acids Res., 15:6799.
Culver, J. N., and Dawson, W. O., 1990, Modifications of the coat protein gene of tobacco mosaic virus resulting in the induction of necrosis, in: "Recognition and Response in Plant-Virus Interactions," R. S. S. Fraser, ed., NATO ASI Series, Cell Biology, Vol. 41, Springer-Verlag, Berlin, 337.
Darvill, A. G., Albersheim, P., 1984, Phytoalexins and their elicitors- a defense against microbial infection in plants, Annu. Rev. Plant Physiol., 35:243.
Ellingboe, A. H., 1982, Genetical aspects of active defence, in: "Active Defence Mechanisms in Plants," R. K. S. Wood, ed., Plenum Press, New York, 179.
Flor, H. H., 1942, Inheritance of pathogenicity in Melampsora lini, Phytopathology, 32:653.
Fraser, R. S. S., 1986, Genes for resistance to plant viruses, CRC Crit. Rev. Plant Sci., 3:257.
Fraser, R. S. S., 1988, Virus recognition and pathogenicity : implications for resistance mechanisms and breeding, Pestic. Sci., 23:267.
Fraser, R. S. S., 1990, The genetics of plant-virus interactions: mechanisms controlling host range, resistance and virulence, in: "Recognition and Response in Plant-Virus Interactions," R. S. S. Fraser, ed., NATO ASI Series, Cell Biology, Vol. 41, Springer-Verlag, Berlin, 71.
Fritig, B., Kauffmann, S., Dumas, B., Geoffroy, P., Kopp, M., Legrand, M., 1987, Mechanism of the hypersensitivity reaction of plants, in: "Ciba Foundation Symposium 133 : Plant Resistance to Viruses," D. Evered and S. Harnett, eds., John Wiley & Sons, Chichester, 92.
Fritig, B., Rouster, J., Kauffmann, S., Stintzi, A., Geoffroy, P., Kopp, M., and Legrand, M., 1989, Virus-induced glycanhydrolases and effects of oligosaccharide signals on plant-virus interactions, in: "Signal Molecules in Plants and Plant-Microbe Interactions," B. J. J. Lugtenberg, ed., NATO ASI Series, Cell Biology, Vol. 36, Springer-Verlag, Berlin, 161.
Fritig, B., Kauffmann, S., Rouster, J., Dumas, B., Geoffroy, P., Kopp, M., and Legrand, M., 1990, Defence proteins, glycanhydrolases and oligosaccharide signals in plant-virus interactions, in: "Recognition and Response in Plant-Virus Interactions," R. S. S. Fraser, ed., NATO ASI Series, Cell Biology, Vol. 41, Springer-Verlag, Berlin, 375.
Geoffroy, P., Legrand, M., and Fritig, B., 1990, Isolation and characterization of a proteinaceous inhibitor of microbial proteinases induced during the hypersensitive reaction of tobacco to tobacco mosaic virus, Mol. Plant-Microbe Interac., 3: (in the press).
Gera, A., and Loebenstein, G., 1983, Further studies of an inhibitor of virus replication from tobacco mosaic virus-infected protoplasts of a local lesion-responding tobacco cultivar, Phytopathology, 73:111.
Gera, A., and Loebenstein, G., 1989, Evaluation of antisera to an inhibitor of virus replication, J. Phytopathol., 124:366.
Gianinazzi, S., Martin, C., and Vallée, J. C., 1970, Hypersensibilité aux virus, température et protéines solubles chez le Nicotiana Xanthi nc. Apparition de nouvelles macromolécules lors de la répression de la synthèse virale,

C. R. Acad. Sci. (Paris), D270:2383.
Goodman, R. N., Kiraly, Z., and Wood, K. R. (eds), 1986, The Biochemistry and Physiology of Plant Disease, University of Missouri Press, Columbia, Missouri.
Grosset, J., Meyer, Y., Chartier, Y., Kauffmann, S., Legrand, M., and Fritig, B., 1990, Tobacco mesophyll protoplasts synthesize 1,3-ß-glucanase, chitinases and "osmotins" during in vitro culture, Plant Physiol., 92:520.
Hooft van Huijsduijnen, R. A. M., Van Loon, L. C., and Bol, J. F., 1986, cDNA cloning of six mRNAs induced by TMV infection of tobacco and a characterization of their translation products, EMBO J., 5:2057.
Kauffmann, S., Legrand, M., Geoffroy, P., and Fritig, B., 1987, Biological function of 'pathogenesis-related' proteins : four PR proteins of tobacco have 1,3-ß-glucanase activity, EMBO J. 6:3209.
Kauffmann, S., Legrand, M., and Fritig, B., 1990, Isolation and characterization of six pathogenesis-related (PR) proteins of Samsun NN tobacco, Plant Mol. Biol., 14:381.
Keen, N. T., 1985, Progress in understanding the biochemistry of race-specific interactions, in: "Genetic Basis of Biochemical Mechanisms of Plant Disease," J. V. Groth and W. R. Bushnell, eds., APS Press, Saint Paul, MN, 85.
Keen, N. T., and Staskawicz, B., 1988, Host range determinants in plant pathogens and symbiots, Annu. Rev. Microbiol., 42:421.
Keen, N. T., Tamaki, S., Kobayashi, D., Gerhold, D., Stayton, M., Shen, H., Gold, S., Lorang, J., Thordal-Christensen, H., Dahlbeck, D., and Staskawicz, B., 1990, Bacteria expressing avirulence gene D produce a specific elicitor of the soybean hypersensitive response, Mol. Plant-Microbe Interac., 3:112.
Konate, G., and Fritig, B., 1983, Extension of the ELISA method to the measurement of the specific radioactivity of viruses in crude cellular extracts, J. Virol. Meth., 6:347.
Konate, G., and Fritig, B., 1984, An efficient microinoculation procedure to study plant virus multiplication at predetermined individual infection sites on the leaves, Phytopathol. Z., 109:131.
Kopp, M., Rouster, J., Fritig, B., Darvill, A., and Albersheim, P., 1989, Host-pathogen interactions XXXII. A fungal glucan preparation protects Nicotianae against infection by viruses, Plant Physiol., 90:208.
Lamb, C. J., Lawton, M. A., Dron, M., and Dixon, R. A., 1989, Signals and transduction mechanisms for activation of plant defenses against microbial attack, Cell, 56:215.
Legrand, M., Fritig, B., and Hirth, L., 1976, Enzymes of the phenylpropanoid pathway and the necrotic reaction of hypersensitive tobacco to tobacco mosaic virus, Phytochemistry, 15:1353.
Legrand, M., Kauffmann, S., Geoffroy, P., and Fritig, B., 1987, Biological function of "pathogenesis-related" proteins : four tobacco PR-proteins are chitinases, Proc. Natl. Acad. Sci. USA, 84:6750.
Linthorst, H. J. M., Meuwissen, R. L. J., Kauffmann, S., and Bol, J., 1989, Constitutive expression of pathogenesis-related proteins PR-1, GRP, and PR-S in tobacco has no effect on virus infection, The Plant Cell, 1:285.
Linthorst, H. J. M., Cornelissen, B. J. C., van Kan, J. A. L., van de Rhee, M., Meuwissen, R. L. J., Gonzalez Jaen, M. T., and Bol, J. F., 1990, Induction of plant genes by compatible

and incompatible virus-plant interactions, in: "Recognition and Response in Plant-Virus Interactions," R. S. S. Fraser, ed., NATO ASI Series, Cell Biology, Vol. 41, Springer-Verlag, Berlin, 361.

Loebenstein, G., and Gera, A., 1981, Inhibitor of virus replication released from tobacco mosaic virus-infected protoplasts of a local lesion-responding tobacco cultivar, Virology, 114:132.

Loebenstein, G., and Gera, A., 1990, Inhibitor of virus replication associated with resistance responses, in: "Recognition and Response in Plant-Virus Interactions," R. S. S. Fraser, ed., NATO ASI Series, Cell Biology, Vol. 41, Springer-Verlag, Berlin, 395.

Massala, R., Legrand, M., and Fritig, B., 1980, Effect of α-aminooxyacetate, a competitive inhibitor of phenylalanine ammonia-lyase, on the hypersensitive resistance of tobacco to tobacco mosaic virus, Physiol. Plant Pathol., 16:213.

Massala, R., Legrand, M., and Fritig, B., 1987, Comparative effects of two competitive inhibitors of phenylalanine ammonia-lyase on the hypersensitive resistance of tobacco to tobacco mosaic virus, Plant Physiol. Biochem., 25:217.

Meshi, T., Motoyoshi, F., Adachi, A., Watanabe, Y., Takamatsu, N., and Okada, Y., 1988, Two concomitant base substitutions in the putative replicase genes of tobacco mosaic virus confer the ability to overcome the effects of a tomato resistance gene, Tm-1, EMBO J., 7:1575.

Métraux, J. P., Burkhart, W., Moyer, M., Dincher, S., Middlesteadt, W., Williams, S., Payne, G., Carnes, M., and Ryals, J., 1989, Isolation of a complementary DNA encoding a chitinase with structural homology to a bifunctional lysozyme/chitinase, Proc. Natl. Acad. Sci. USA, 86:896.

Mundry, K. W., Schaible, W., Ellwart-Tschürtz, M., Nitschko, H., and Hapke, C., 1990, Hypersensitivity to tobacco mosaic virus in N'-gene hosts : which viral genes are involved ? in: "Recognition and Response in Plant-Virus Interactions," R. S. S. Fraser, ed., NATO ASI Series, Cell Biology, Vol. 41, Springer-Verlag, Berlin, 345.

Payne, G., Ahl, P., Moyer, M., Harper, A., Beck, J., Meins, F., and Ryals, J., 1990, Isolation of complementary DNA clones encoding pathogenesis-related proteins P and Q, two acidic chitinases from tobacco, Proc. Natl. Acad. Sci. USA, 87:98.

Pierpoint, W. S., Robinson, N. P., Leason, M. B., 1981, The pathogenesis-related proteins of tobacco : their induction by viruses in intact plants and their induction by chemicals in detached leaves, Physiol. Plant Pathol., 19:85.

Pierpoint, W. S, Tatham, A. S., and Pappin, D. J. C., 1987, Identification of the virus-induced protein of tobacco leaves that resembles the sweet-protein thaumatin, Physiol Mol. Plant Pathol., 31:291.

Ponz, F., and Bruening, G., 1986, Mechanisms of resistance to plant viruses, Annu. Rev. Phytopathol., 24:355.

Richardson, M., Valdes-Rodriguez, S., and Blanco-Labra, A., 1987, A possible function for thaumatin and a TMV-induced protein suggested by homology to a maize inhibitor, Nature, 327:432.

Roby, D., Toppan, A., and Esquerré-Tugayé, M. T., 1987, Cell surfaces in plant-microorganism interactions. VIII Increased proteinase inhibitor activity in melon plants in response to infection by Colletotrichum lagenarium or to treatment with an elicitor fraction from this fungus, Physiol. Mol. Plant Pathol., 30:453.

Ryan, C. A., 1981, Proteinase inhibitors, in: "Biochemistry of Plants," P. K. Stumpf and E. E. Conn, eds., vol. 6, Academic Press, London, 351.

Ryan, C. A., 1988, Oligosaccharides as recognition signals for the expression of defensive genes in plants, Biochemistry, 27:8879.

Saito, T., Meshi, T., Takamatsu, N., and Okada, Y., 1987, Coat protein gene sequence of tobacco mosaic virus encodes a host response determinant, Proc. Natl. Acad. Sci. USA, 84:6074.

Shinshi, H., Neuhaus, J. M., Ryals, J., and Meins, F., 1990, Structure of a tobacco endochitinase gene : evidence that different chitinase genes can arise by transposition of sequences encoding a cysteine-rich domain, Plant Mol Biol., 14:357.

Spiegel, S., Gera, A., Salomon, R., Ahl, P., Harlap, S., and Loebenstein, G., 1989, Recovery of an inhibitor of virus replication from the intercellular fluid of hypersensitive tobacco infected with tobacco mosaic virus and from uninfected induced-resistant tissue, Phytopathology, 79:258.

Van den Bulcke, M., Bauw, G., Castresana, C., Van Montagu, M., and Vandekerckove, J., 1989, Characterization of vacuolar and extracellular ß(1,3)-glucanases of tobacco : Evidence for a strictly compartmentalized plant defense system, Proc. Natl. Acad. Sci. USA, 86:2673.

Van Loon, L. C., 1982, Regulation of changes in proteins and enzymes associated with the active defence against virus infection, in: "Active Defence Mechanisms in Plants," R. K. S. Wood, ed., Plenum, New York, 247.

Van Loon, L. C., 1985, Pathogenesis-related proteins, Plant Mol. Biol., 4:111.

Van Loon, L. C., and van Kammen, A., 1970, Polyacrylamide disc electropheresis of the soluble leaf proteins from Nicotiana tabacum var. "Samsun" and "Samsun NN". Changes in protein constitution after infection with TMV, Virology, 40:199.

Vera, P., and Conerejo, V., 1988, Pathogenesis-related proteins of tomato. P-69 as an alkaline endoproteinase, Plant Physiol., 87:58.

# REPEATED DNA SEQUENCES AND THE ANALYSIS OF HOST SPECIFICITY IN THE RICE BLAST FUNGUS

Forrest G. Chumley, Barbara Valent, Marc J. Orbach, James A. Sweigard, Leonard Farrall and Anne Walter

E. I. Du Pont de Nemours and Co., Inc.
Central Research and Development Department
Experimental Station, P. O. Box 80402
Wilmington, Delaware 19880-0402 U.S.A.

## Introduction

Field isolates of the heterothallic Ascomycete, *Magnaporthe grisea* Barr (anamorph, *Pyricularia oryzae* Cav. or *P. grisea*), include pathogens of many grasses, with rice (*Oryza sativa*) being the most important example from an economic perspective. Individual field isolates have a limited host range, parasitizing one or a few grass species (MACKILL and BONMAN 1986). We have recently shown that rice pathogens from around the world contain a family of repeated DNA sequences, MGR sequences, that appear to be absent from or present in low copy number in field isolates that infect grasses other than rice (HAMER *et al.* 1989a). We have hypothesized that the correlation between MGR sequence conservation and rice pathogenicity is due to genetic isolation and independent evolution of rice pathogens descended from a small ancestral population. As we will discuss, MGR sequences have become useful as tools for strain identification and as genetic markers for cloning host specificity genes of interest.

Many genes in the pathogen must act together to execute the metabolic and developmental pathways that comprise the disease cycle, the ordered series of events leading from spore germination through penetration, lesion development, and conidiation. Genes that control the development and function of infection

*Plant Molecular Biology* 2, Edited by R.G. Herrmann and B. Larkins, Plenum Press, New York, 1991

structures are examples of general pathogenicity determinants, those required for the pathogen to infect any host. Genes that determine host specificity, however, appear to provide important signals for a complex recognition system in the host that can trigger an effective disease resistance response (FLOR 1971; DAY 1974; CRUTE 1986).

Strains of the fungus that parasitize rice are subdivided into races, depending on the rice cultivars they can successfully infect. The rice blast fungus shows a high degree of variability in the field; new races frequently appear with the ability to attack previously resistant rice cultivars (OU 1985). Cultivars of rice that differ from one another by the presence or absence of dominant resistance genes have been developed (YAMADA *et al.* 1976). The resistance of any one of these cultivars is effective only against certain races of the pathogen. The genetic basis for differences in cultivar specificity between races of the rice blast pathogen has been previously unknown due to the infertility of field isolates of *M. grisea* that infect rice. Fertile *M. grisea* rice pathogens have now been developed. We have identified pathogen genes that control rice cultivar specificity, as well as genes that control pathogenicity toward a second host, weeping lovegrass (*Eragrostis curvula*). In this analysis, a gene was termed an "avirulence gene" if alleles determine virulence or avirulence in an all-or-nothing, cultivar specific fashion. Genes with no cultivar specific effects were termed "pathogenicity genes." Because obstacles to genetic analysis of rice pathogens have been overcome, the *M. grisea* system has properties that recommend it for undertaking a detailed analysis of the molecular basis for host specificity (VALENT 1990).

**MGR Sequences May Be Derived From a Retrotransposon**

MGR sequences, repeated DNA sequences identified in the genome of *M. grisea* (HAMER *et al.* 1989a) show conservation of sequence homology according to host species specificity. MGR sequences are distributed among all six *M. grisea* chromosomes. Genomic segments with MGR sequences showed significant polymorphism in the arrangement of restriction sites and blocks of DNA. However, Southern hybridization analysis indicated that a 2.0 kb *Bam*H I fragment is highly conserved among MGR sequences in almost all *M. grisea* isolates. MGR sequences hybridized to polyA$^+$ RNAs ranging up to 7.6 kb in length.

Although MGR sequences show significant polymorphism, we now report that rice pathogens contain some highly conserved MGR sequences, detected as *Hind* III fragments of 4.0 and 5.6 kb in Southern hybridization with the conserved 2.0 kb *Bam*H I fragment as probe (M.J. Orbach and L. Farrall, unpublished). We have isolated Lambda clones that contain each of the conserved *Hind* III fragments. Subclones that overlap the 2.0 kb *Bam*H I fragment were isolated from each Lambda clone. The DNA sequences of these subclones and of the 2.0 kb *Bam*H I fragment have been determined. The sequenced region contains a large open reading frame ("ORF886") that could encode a protein of more than 886 amino acids. The deduced amino acid sequence of ORF886 shares extensive homology with the "ORF2" reverse transcriptase (RT) domains of several polyA-type retrotransposons (FINNEGAN 1989). Thus, MGR sequences may be fungal polyA-type retrotransposons, a hypothesis that seems to explain some features of the biology of MGR sequences. We are now attempting to isolate 3'-terminal MGR clones to determine if they contain sequences typical of polyA-type retrotransposons. We are also checking for the presence of RT in *M. grisea* cell-free extracts, using immunohybridization and enzyme assays.

## Avirulence Genes Identified in a Backcrossing Scheme

The interfertility of *M. grisea* strains with different host species specificities has permitted genetic analysis of those differences. A backcrossing regime was initiated by crossing the weeping lovegrass (*Eragrostis curvula*) pathogen, 4091-5-8, a highly fertile hermaphrodite, and the rice pathogen, O-135, a female sterile field isolate that also infects weeping lovegrass (VALENT, FARRALL and CHUMLEY 1990). All progeny from this cross infected weeping lovegrass, as expected, because both parents are weeping lovegrass pathogens. However, only 6 of 59 progeny were pathogens of rice, indicating that the two parental strains differ by several genes that control the ability to infect rice.

In the backcrossing regime, the rice pathogen was the recurrent parent. Female parents for the backcross generations were progeny strains chosen for having fertility comparable to that of 4091-5-8. The ratios of pathogenic to nonpathogenic (and virulent to avirulent) progeny through the backcross generations suggested that O-135 and 4091-5-8 differ in two types of genes that control the ability to infect rice. First, polygenically-inherited factors determine the extent of lesion development on rice. Second, single avirulence genes govern, in an all-or-nothing fashion, virulence toward specific cultivars of rice. Several crosses confirmed the segregation of three unlinked genes, *Avr1-CO39*, *Avr1-M201* and *Avr1-YAMO* that determine avirulence on rice cultivars CO39, M201, and Yashiro-mochi, respectively. *Avr1-M201* may be linked to the mating type locus, *Mat1*.

The avirulence alleles of the cultivar specificity genes identified in these crosses appear to have been inherited from the parent that is a nonpathogen of rice, because the rice pathogen parent is virulent on all three cultivars. We identified avirulence genes for each cultivar used in these experiments, suggesting a high potential for identification of other avirulence genes derived from the nonpathogen of rice. YAEGASHI and ASAGA (1981) reported similar results suggesting that a finger millet pathogen, WGG-FA40, carries an avirulence gene corresponding to the rice blast resistance gene, *Pi-a*.

**Avirulence Genes Identified in Other Crosses**

A separate series of crosses also identified single genes that control cultivar specificity (B. Valent, L. Farrall and F. Chumley, unpublished). Both parents in Cross 4360 (Table 1) were virulent on rice cultivars CO39, M201 and Sariceltik, and all progeny were virulent on these three cultivars. The parents differed in specificity toward rice cultivars Yashiro-mochi and Maratelli. Data from five complete tetrads (a representative tetrad is shown in Table 1) and from 30 random ascospore cultures suggested that two unlinked avirulence genes segregated among the progeny of this cross: *Avr2-YAMO* (alleles determine virulence or avirulence on cultivar Yashiro-mochi) and *Avr1-MARA* (alleles determine virulence or avirulence on cultivar Maratelli). The segregation of both genes was confirmed in at least two subsequent crosses. These genes were not linked to the mating type locus. The parents of Cross 4360 differ in specificity toward five additional cultivars; further analysis of the progeny will probably reveal the segregation of additional avirulence genes.

Genetic crosses were performed to determine if *Avr2-YAMO* is allelic to *Avr1-YAMO*, described in the previous section. Data

TABLE 1

Lesion Types Produced by Parents of Cross 4360 and Progeny from One Tetrad

| STRAIN | MATING TYPE | LESION TYPE ON RICE CULTIVAR | | | LESION TYPE ON WEEPING LOVEGRASS |
|---|---|---|---|---|---|
| | | SARICELTIK | YASHIRO-MOCHI | MARATELLI | |
| PARENTS: | | | | | |
| 4224-7-8 | 1 | 5[a] | 0[a] | 0 | 5 |
| 6043 | 2 | 5 | 5 | 5 | 0 |
| PROGENY FROM TETRAD 1: | | | | | |
| 4360-1-1 | 2 | 5 | 0 | 5 | 0 |
| 4360-1-4 | 2 | 5 | 0 | 5 | 0 |
| 4360-1-2 | 2 | 5 | 5 | 0 | 5 |
| 4360-1-8 | 2 | 5 | 5 | 0 | 5 |
| 4360-1-3 | 1 | 5 | 5 | 5 | 0 |
| 4360-1-7 | 1 | 5 | 5 | 5 | 0 |
| 4360-1-5 | 1 | 5 | 0 | 0 | 5 |
| 4360-1-6 | 1 | 5 | 0 | 0 | 5 |

[a]A scale of lesion types has been defined according to the size of lesions formed, ranging from Type 0, no visible symptoms, to Type 5, the largest spreading lesions characteristic of the host variety (VALENT et al. 1986; VALENT et al. 1990).

from three different crosses indicated that the two genes are not linked to one another.

### Genes for Pathogenicity To Weeping Lovegrass

The parents of Cross 4360 differed in ability to infect weeping lovegrass (see Table 1). Single gene segregation of pathogenicity toward weeping lovegrass was observed among the random ascospore progeny and in the tetrads (Table 1). Subsequent crosses confirmed simple segregation of the gene now named *Pwl2*. Preliminary data suggest that *Pwl2* is not linked to the weeping lovegrass pathogenicity determinant, *Pwl1*, previously identified in a cross between a weeping lovegrass pathogen and a goosegrass pathogen (VALENT *et al.* 1986; VALENT and CHUMLEY 1987).

The genes *Pwl1* and *Pwl2* must be considered "pathogenicity genes," with no cultivar specific effects, because cultivars of weeping lovegrass are not available. However, it remains an intriguing possibility that these genes may function in a manner analogous to avirulence genes.

### Unstable Genes in *M. grisea*

Host specificity genes we have identified are described in Table 2. Two of these genes, *Pwl2* and *Avr2-YAMO*, appear to be unstable. That is, spontaneous pathogenic or virulent mutants derived from strains carrying the nonpathogenic or avirulent alleles of these genes appear frequently in standard assays (B. Valent, L. Farrall and F. Chumley, unpublished). Two other *M. grisea* genes appear to be unstable in several strains: $SMO^+$, a gene that controls spore morphology (HAMER *et al.* 1989b), and $BUF^+$, a gene that encodes an enzyme involved in biosynthesis of the gray pigment, melanin (CHUMLEY and VALENT 1990). However, many other genes appear to mutate at normal frequencies, including $ALB^+$ and $RSY^+$, which encode melanin biosynthetic enzymes. Cloning stable and unstable genes will yield clues to the molecular basis for genetic instability in *M. grisea*. The mechanism by which new races of a pathogen arise is an intriguing problem. Although the extent of instability in race determinants has been debated (LATTERELL 1975; OU 1985), it is clear that new races rapidly appear in the field when new blast resistant rice cultivars are introduced.

TABLE 2

Genes For Host Specificity Identified in *M. grisea*

| Gene | Host | Source[a] | Stability[b] |
|---|---|---|---|
| *Avr1-CO39* | CO39 | Weeping Lovegrass Pathogen | Stable |
| *Avr1-M201* | M201 | Weeping Lovegrass Pathogen | Stable |
| *Avr1-YAMO* | Yashiro-mochi | Weeping Lovegrass Pathogen | Stable |
| *Avr2-YAMO* | Yashiro-mochi | Rice Pathogen | Unstable |
| *Avr1-MARA* | Maratelli | Unknown | Unknown |
| *Pwl1* | weeping lovegrass | Goosegrass Pathogen | Stable |
| *Pwl2* | Weeping lovegrass | Unknown | Unstable |

[a]The "source" listed for each gene is based on the infectivity of the parents in the cross in which the gene was identified. The source indicated is the parent strain that carried the allele for nonpathogenicity or avirulence.

[b]A host specificity determinant is listed as "unstable" if virulent or pathogenic mutants appear at a high frequency, as described in the text.

## Cloning Genes for Host Specificity

We are pursuing three independent approaches for cloning *M. grisea* genes that govern cultivar or host species specificity. The first effort is to clone avirulence genes by complementation of function. A gene library has been prepared in the cosmid vector pMOcosX (M.J. Orbach, unpublished) using DNA from strain 4392-1-6, which carries *Avr1-CO39, Avr1-M201, Avr1-YAMO, Avr2-YAMO, and Avr2-MARA*. The library has been stored as approximately 4400 individual clones. We are screening individual cosmid clones for the presence of avirulence genes by introducing them one-by-one into recipient strain CP983 (which is virulent on all 4 cultivars) and screening transformants for loss of virulence on the 4 rice varieties. For the purpose of these experiments, we have assumed dominance of the alleles for avirulence (see Discussion below); we expect to isolate one of the avirulence genes per 600 to 1000 clones screened.

The second and third approaches rely on cloning avirulence and pathogenicity genes by chromosome walking from physical genetic markers linked to the gene of interest. The second approach utilizes classical RFLP mapping (MICHELMORE and HULBERT 1987); the third approach utilizes the mapping of MGR sequence polymorphisms (HAMER *et al.* 1989a). The parents for the RFLP mapping population (74 random progeny) are strains 4224-7-8 [*Avr2-YAMO, Avr1-MINE* (cultivar Minehikari), *Avr1-TSUY* (cultivar Tsuyuake), *Avr1-MARA*] and 6043 (*Pwl2;* LEUNG et al. 1988). These are also the parents for the cross shown in Table 1. Cosmid clones are being used as probes for detecting RFLPs; about 30% of the cosmids yield useful markers with at least one of the 8 restriction enzymes tested. With about 71 RFLP markers mapped, linkage has been detected between *Avr2-YAMO* and RFLPs that lie 5.8 cM from the gene (J. Sweigard, A. Walter, unpublished). MGR sequence polymorphisms have served as reliable genetic markers in the backcrossing scheme described above (VALENT, FARRALL and CHUMLEY 1990). One such marker maps approximately 8.5 cM from *Avr1-CO39;* single copy DNA adjacent to the MGR sequence has been cloned to serve as the starting point for a chromosome walk to clone the Avr gene (M.J. Orbach, unpublished).

## Wheat Blast

Over the past 5 years, Brazilian wheat farmers have experienced an increasingly severe outbreak of blast disease on wheat caused by the fungus *M. grisea* (*Pyricularia* spp.) (IGARASHI *et al.* 1986). The disease has occurred in Paraná, a state where rice is cultivated in rotation with wheat. We have explored the possibility that wheat blast pathogens appeared as variants in the population of rice blast pathogens (F. Chumley, B. Valent and L. Farrall, unpublished). We characterized a number of wheat blast pathogens with regard to fertility, pathogenicity and MGR sequence "fingerprints." The MGR fingerprints have been generated by Southern hybridization using MGR sequences as probes. The results of these studies clearly indicate that the Brazilian wheat pathogens are not related to rice pathogens from the same area. Thus, wheat blast and rice blast represent diseases of independent origin, and efforts to control rice blast can be expected to have little effect on the occurrence of wheat blast disease. The wheat pathogens appear to be closely related to pathogens of feral triticale collected in Parana.

**Discussion**

Characterization of conserved components of MGR sequences has shown these elements may encode the structure of a protein with reverse transcriptase activity. Amino acid sequence homologies suggest that MGR sequences may be derived from a retrotransposon with similarities to "polyA" (or "non-LTR") eukaryotic transposable elements. Further characterization of the termini of MGR sequences will be needed before such a conclusion can be made with certainty. MGR sequence polymorphisms can be used to generate distinctive DNA fingerprints for *M. grisea* isolates from nature. MGR fingerprints have been useful for epidemiological studies, demonstrating, for example, that wheat blast pathogens in Brazil did not arise from the local population of the rice blast fungus. MGR sequence polymorphisms segregate as Mendelian determinants in genetic crosses of *M. grisea*, and linkage has been detected between one particular MGR sequence and a gene that governs compatibility with the rice cultivar CO39.

Host specificity genes of *M. grisea* that have been confirmed by segregation in three or more crosses are listed in Table 2. LEUNG *et al.* (1989) report evidence for additional genes. Nonpathogens of rice contain genes that function as cultivar specific avirulence genes when they are introduced into strains that infect rice. Similar observations have been published concerning bacterial plant pathogens (WHALEN, STALL and STASKAWICZ 1988; KOBAYASHI, TAMAKI and KEEN 1989). The ease with which single genes for cultivar specificity have been identified in the crosses reported here suggests that these genes are numerous.

Whether or not the cultivar specificity genes we have defined may be classical avirulence genes remains to be determined. We can provide no information on dominance relationships between alleles of these genes because *M. grisea* does not form stable vegetative diploids or heterokaryotic conidia (CRAWFORD *et al.* 1986). Further genetic analysis of both the host and the pathogen will be required to define any functional correspondence that may exist between a particular pathogen gene and a particular resistance gene in rice. The rice cultivar Yashiro-mochi, one of the ten Japanese differential cultivars (YAMADA *et al.* 1976), is reported to have the blast resistance gene, *Pi-ta*. Genetic crosses of Yashiro-mochi will determine if the pathogen genes *Avr1-YAMO* and *Avr2-YAMO* interact with the blast resistance gene, *Pi-ta*, with previously unidentified resistance genes, or with any identifiable gene for resistance. The cultivars CO39, M201 and Maratelli have not previously been reported to carry blast resistance genes. Genetic crosses involving these cultivars will determine if they contain unidentified blast resistance genes.

Our efforts now are focused on cloning avirulence genes as a first step toward understanding the molecular mechanisms that determine cultivar specificity. With cloned avirulence genes in hand, we plan to address the following questions. What gene products are encoded by avirulence genes? How might these gene products interact with the products of host resistance genes? How is the expression of avirulence genes controlled? What molecular genetic events accompany the appearance of new races in the field? What is the difference between stable and unstable avirulence genes? How do rice pathogen avirulence genes differ from those present in nonpathogens of rice? The goal of understanding the molecular basis for race-cultivar specificity remains a major challenge in the study of host-pathogen interactions.

**Literature Cited**

Chumley FG, and Valent B (1990) Genetic analysis of melanin deficient nonpathogenic mutants of *Magnaporthe grisea*. Molecular Plant-Microbe Interactions 3:135-143

Crawford MS, Chumley FG, Weaver CG, Valent B (1986) Characterization of the heterokaryotic and vegetative diploid phases of *Magnaporthe grisea*. Genetics 114:1111-1129

Crute IR (1986) The genetic basis of relationships between microbial parasites and their hosts. In: Fraser RSS (ed) *Mechanisms of Resistance to Plant Diseases*. Martinus Nijhoff and W. Junk, Dordrecht pp 80-142

Day PR (1974) *Genetics of Host-Parasite Interaction*. WH Freeman and Co., San Francisco

Finnegan DJ (1989) Eukaryotic transposable elements and genome evolution. Trends in Genetics 5:103-107

Flor HH (1971) Current status of the gene-for-gene concept. Annu Rev Phytopathol 9:275-296

Hamer JE, Farrall L, Orbach MJ, Valent B, and Chumley FG (1989a) Host-species specific conservation of a family of repeated DNA sequences in the genome of a fungal plant pathogen. Proc. Nat. Acad. Sci. U.S.A. 86:9981-9985

Hamer JE, Valent B, Chumley FG (1989b) Mutations at the *SMO* genetic locus affect the shape of diverse cell types in the rice blast fungus. Genetics 122:351-361

Igarashi S, Utiamada CM, Igarashi LC, Kazuma AH, Lopes RD (1986) Pyricularia sp. em trigo. I. Ocorrencia de Pyricularia sp. no estado do Paraná. In: resumo do 19° Congreso Brasileiro de Fitopatologica, DF, 11(2):351-352

Kobayashi DY, Tamaki SJ, Keen NT (1989) Cloned avirulence genes from the tomato pathogen *Pseudomonas syringae* pv. *tomato* confer cultivar specificity on soybean. Proc Natl Acad Sci USA 86:157-161

Latterell FM (1975) Phenotypic stability of pathogenic races of Pyricularia oryzae, and its implication for breeding blast resistant rice varieties. In *Proceedings of the Seminar on Horizontal Resistance to Blast Disease of Rice*. Colombia Series CE-No. 9 Centro Internacional de Agricultura Tropical, Cali pp 199-234

Leung H, Borromeo ES, Bernardo MA, Notteghem JL (1988) Genetic analysis of virulence in the rice blast fungus *Magnaporthe grisea*. Phytopathology 78:1227-1233

Mackill AO, Bonman JM (1986) New hosts of *Pyricularia oryzae*. Plant Disease 70:125-127

Michelmore RW and Hulbert SH (1987) Molecular markers for genetic analysis of phytopathogenic fungi. Annual Review of Phytopathology 25:383-404

Ou SH (1985) In *Rice Diseases*, Commonwealth Agricultural Bureaux, Slough, UK.pp. 109-201

Valent B (1990) Rice blast as a model system for plant pathology. Phytopathology 80:33-36

Valent B, Chumley FG (1987) Genetic analysis of host species specificity in *Magnaporthe grisea*. UCLA Symp Mol Cell Biol (New Series) 48:83-93

Valent B, Crawford MS, Weaver CG, Chumley FG (1986) Genetic studies of pathogenicity and fertility of *Magnaporthe grisea*. Iowa State J Res 60:569-594

Valent B, Farrall L, Chumley FG (1990) *Magnaporthe grisea* genes for pathogenicity and virulence identified through a series of backcrosses. Genetics (in press)

Whalen MC, Stall RE, Staskawicz B (1988) Characterization of a gene from a tomato pathogen determining hypersensitive resistance in non-host species and genetic analysis of this resistance in bean. Proc Nat Acad Sci USA 85:6743-6747

Yaegashi H, Asaga K (1981) Further studies on the inheritance of pathogenicity in crosses of *Pyricularia oryzae* with *Pyricularia* sp. from finger millet. Ann Phytopathol Soc Jpn 47:677-679

Yamada M, Kiyosawa S, Yamaguchi T, Hirano T, Kobayashi T, Kushibuchi K, Watanabe S (1976) Proposal of a new method for differentiating races of *Pyricularia oryzae* Cavara in Japan. Ann Phytopathol Soc Jpn 42:216-219

# MOLECULAR GENETICS OF THE TOMATO PATHOGEN *CLADOSPORIUM FULVUM*

Richard P. Oliver, Nick J. Talbot, Mark T. McHale and Alan Coddington

Norwich Molecular Plant Pathology Group
University of East Anglia
School of Biological Sciences

Cladosporium fulvum Cke (syn. Fulvia fulva) is a fungal pathogen of the tomato, Lycopersicon esculentum, causing the disease leaf mould. A variety of natural features of the interaction together with techniques and knowledge built up over past few years have established these organisms as a near-ideal systems for the study of pathogenicity, host specificity and race cultivar specificity. In this review I will emphasise two aspects of the genetics of the pathogen which have received attention in the recent past. The first is genetic analysis by parasexual analysis and protoplast fusion coupled with pulsed field gel separation of chromosomes. The second is the discovery and analysis of long-terminal repeat retrotransposons.

## GENETIC AND GENOMIC ANALYSIS

C. fulvum is an imperfect fungus with no known sexual stage. This makes genetic analysis by conventional means impossible. To overcome this barrier our approach has been to utilise protoplast fusion of mutant strains to encourage chromosome mixing followed by analysis of haploid progeny (Talbot et al 1988a). The first step was the isolation of a range of mutants. All our results to date suggest that wild-type strains of C. fulvum are haploid. Thus the isolation of at least some classes of mutants was straightforward. Mutants resistant to chlorate were isolated after UV mutagenesis or spontaneously and assigned to the phenotype classes nia, nir or cnx by growth tests. C. fulvum is efficient at scavenging substrates in solid media so these growth test were best performed in liquid culture and assayed by measuring dry weights. This scavenging efficiency made the isolation of auxotrophic mutants by selection on minimal media inefficient although we were able to isolate a methionine and a biotin requiring mutant. A number of mutants resistant to various fungicides and heavy metals were also isolated but not utilised further. The other main source of marked strains was to use transformation. Two chimeric antibiotic resistance genes pAN7-1 and pAN8-1 giving resistance to hygromycin and phleomycin were used (Punt et al 1987, Oliver et al 1987, Mattern et al 1988). Transformation in filamentous fungi is via chromosomal integration. As these vectors have no C. fulvum sequences the sites of integration are of necessity non-homologous. Transformed strains show a high degree of phenotypic

*Plant Molecular Biology* 2, Edited by R.G. Herrmann and
B. Larkins, Plenum Press, New York, 1991

stability through mitosis and during sporulation. They are therefore suitable parent strains for use in protoplast fusion experiments.

The marked strains are protoplasted and fused using polyethylene glycol solutions (Talbot et al 1988a, Anné & Peberdy). The fusion products were then plated on media selective for the markers in both parental strains. In the case of the chlorate resistant mutants the selection comprises using nitrate as sole nitrogen source. Colonies are recovered from the selection plates at frequencies $10^{-6}$ and $10^{-4}$ with respect to the imput number of protoplast. This compares to a regeneration frequency of ca. 0.4 and $10^{-2}$ before and after PEG treatment respectively on non-selective plates.

The analysis of these fusion products was dependent on the development of a rapid method of determining ploidy (Talbot et al 1988b). DAPI- stained slides of mycelium and spores of fusion-products and parental stains were viewed in a fluorescence microscope. The video image of individual nuclei was analysed by computer to assess DNA content. The system was established using the well characterised yeast Schizosaccharomyces pombe. Haploid and diploid strains were shown to be easily distinguishable. This system allows large numbers of nuclei to be examined giving statistical certainty to the assignment of ploidy despite variation in condensation state. Using this system we have demonstrated that primary fusion products contain nuclei of diploid and higher ploidy. These nuclei are unstable and spontaneously breakdown to give various haploid products. The frequent presence of selectable markers from both parental strains in haploid products in evidence for chromosome reassortment. It supports our contention that true diploids are formed rather than heterokaryons.

Analysis of the segregation of markers in the progeny strain allows the construction of linkage maps. In addition to the markers used in the fusion process, the segregation of other phenotypic markers and RFLPs can be followed. C. fulvum exists in a number of races distinguishable by their ability to infect certain cultivars of the host tomato carrying different resistance genes. A particular aim of this project was to investigate the dominance relationships of this cultivar specificity and to detect linked RFLP markers. The parental races in a trial fusion were derived from Race 4 (able to infect tomato with resistance genes (Cf0) and tomato with the Cf4 resistance gene) and Race 5 (able to infect Cf0 and Cf5 lines). The fusion products are unstable diploids making their analysis by pathogenicity testing difficult. Nevertheless, none of the fusion products were able to infect both Cf4 and Cf5 lines. Rather they were unable to infect these lines whilst retaining full pathogenicity of CfO lines. This result indicates that Race 4 and Race 5 carry dominant genes for avirulence on Cf5 and Cf4 respectively. The paradoxical possession of avirulence genes confirms the biochemical isolation of peptides inducing hypersensitive necrosis in resistant tomato cultivar (de Wit & Spikman, 1982).

After one or a few generations the unstable diploids resolved into haploids. Pathogenicity testing of haploid progeny identified products carrying no avirulence genes (Race 4,5) either Avr4 or Avr5 (Race 5 and Race 4) and both Avr4 & Avr5 (Race 0). The segregation pf cnx, nia, nir, 2 RFLPs revealed by cloned cosmids (4318 and 3470), a variation in the non-transcribed spacer of the rDNA repeat and the presence of the pAN7-1 DNA was followed. Despite the low number of progeny and of markers a number of conclusions are possible. Each of the progeny are different indicating that chromosome reassortment and possibly even non-sister chromatid exchange are frequent. There is evidence of linkage in a number of cases: between nir and nia, genes which are clustered in many filamentous fungi; between p3470 and the rDNA; and between the Avr5 gene and the nitrate/nitrite reductase gene cluster. However the most important conclusion is that genetic analysis is not only possible but efficient and informative in an imperfect fungus: the lack of sexual cycle should not hinder the analysis of a fungal pathogen of interest.

An alternative approach to the genetic analysis is to utilise pulsed field gel electrophoresis to physically separate the chromosomes. The chromosomes of filamentous fungi are generally in the range 0.5 - 8 Mbp, a size which is within the range resolved by current pulsed field gel techniques. The analysis of chromosomes by these techniques allied to Southern blotting and hybridisation techniques complements the fusion strategies and has provided further evidence for the preliminary linkage data mentioned above.

Protoplasts of C. fulvum races were embedded in agarose blocks and lysed. The electrophoresis was carried out in a Pharmacia Pulsaphor apparatus using the hexagonal electrode. Using a switching time of 4500s and a voltage of 45v resolution of ten bands is achieved in 14 days. The largest two bands (bands 1 & 2) are probably doublets. Comparison with Aspergillus nidulans chromosomes suggests that the bands vary in size from 1.8 to 5.4 Mbp giving a total genome size of 44 Mbp. In addition there is evidence for a minichromosome of about 500 kbp. Three RFLP probe hybridise to the doublet band two. This supports the linkage of 3470 and the rDNA. 4318 may either be distant or on the other chromosome represented in this band.

A particularly interesting finding was the detection of chromosome-length polymorphism between races of C. fulvum. In two races one of the chromosomes comprising the band 2 doublet has deleted about 500kbp. This may be a translocation as the band 6 chromosome may be larger by a similar figure in these races. The possibility that this putative translocation is related to the loss of Avr9 in these strains is under investigation. We anticipate that the combination of pulsed-field gels and protoplast fusion with RFLP technology will enable us to create rapidly an RFLP map aiding as in the cloning of interesting genes.

RETROTRANSPOSONS IN THE C. FULVUM GENOME

C. fulvum is an intercellular pathogen of the leaves of tomato. The success or failure of the interaction, dependent on the interaction of avirulence and resistance genes, is decided by molecular interactions in the apoplastic fluid. A strategy designed to isolate genes for fungal protein exported into the apoplast was devised. Proteins isolated from the apoplast of infected leaves were used to raise antisera in rabbits (by P de Wit). Genomic DNA of C. fulvum was cloned into λgtll after cleavage with EcoR1 under star conditions. The resulting library was screened with the antiserum and one positive clone, P5, has proved of considerable interest. The sequence of the 225bp insert was determined and compared to DNA sequences in the databases. The amino acid sequences, inferred from the expressed open-reading frame was also used to interrogate the databases. Both analyses revealed a strking similarity to reverse transcriptase genes of LTR retrotransposons of the Gypsy class and, to a lesser extent, retroviruses (Doolittle et al 1989).

Not only was the homology and similarity highly significant (45% and 80% respectively) but also conserved amino acids were also present (McHale et al 1989). This suggested that a LTR retrotransposon might be present in the genome of a filamentous fungus. Hybridisation analysis revealed that the sequences were present in multiple copies (50-100) in the genome. Several full-length copies have been isolated in cosmid and analysed in detail. The element named Cft1-712 contains identical long-terminal repeats of 425bp. The putative target site duplication is TATAG and the terminal inverted duplication is 5'TGTTA..ACAAT5'. Putative first and second strand primer binding sites were identified. Reverse transcriptase enzyme activity was detected. Sucrose gradient fractionation of fungal homogenates copurified the RT activity, RNA homologous the retroelements and a virus-like particles similar to Ty particles. These results suggest that the transposable element is actively expressed into proteins which are packaged.

The findings have manifest significance. The presence of retrotransposon may explain the notorious variability of plant pathogens. The transposon may either directly inactivate genes or it may by providing dispersed sites of homology around the genome promoting translocations and deletions. Secondly this or similar transposons may provide genetic tools helping to overcome the difficulties of working with genetically uncharacterised organisms.

## ACKNOWLEDGEMENTS

The work was supported by the AFRC, SERC and the Gatsby Charitable Foundation.

## REFERENCES

Anné, J. and Peberdy, J.F., 1976, Induced fusion of fungal protoplasts following treatment with polyethylene glycol, J. Gen. Microbiol., 92: 413-417.

De Wit, P.J.G.M. and Spikman, G., 1982, Evidence for the occurence of race and cultivar-specific elicitors of necrosis in intercellular fluids of compatible interactions of Cladosporium fulvum and tomato, Physiological Plant Pathology, 21: 1-11.

Doolittle, R.F., Feng, D-F., Johnson, M.S. and McClure, M.A., 1989, Origins and evolutionary relationships of retroviruses, Quart. Rev. Biology, 64: 1-30.

Mattern, I.E., Punt, P.J. and van den Hondel, C.A.M.J.J., 1988, A vector of Aspergillus transformation conferring phleomycin resistance, Fungal Genetics Newsletter, 35: 25.

McHale, M.T., Roberts, I.N., Talbot, N.J. and Oliver, R.P., 1989, Expression of Reverse Transcriptase genes in Fulvia fulva, Mol. Plant Microbe Int., 2: 165-168.

Oliver, R.P., Roberts, I.N., Harling, R., Kenyon, L., Punt, P.J., Dingemanse, M.A. and van den Hondel, C.A.M.J.J., 1987, Transformation of Fulvia fulva, a fungal pathogen of tomato, to hygromycin B resistance, Curr. Genet., 12: 231-233.

Punt, P.J., Oliver, R.P., Dingemanse, M.A., Pouwels, P.H. and van den Hondel, C.A.M.J.J., 1987, Transformation of Aspergillus based on the hygromycin B resistance marker from Escherichia coli, Gene, 56: 117-124.

Talbot, N.J., Coddington, A., Roberts, I.N. and Oliver, R.P., 1988a, Diploid construction by protoplast fusion in Fulvia fulva (syn. Cladosporium fulvum): Genetic analysis of an imperfect plant pathogen, Current Genetics, 14: 567-572.

Talbot, N.J., Rawlins, D. and Coddington, A., 1988b, A rapid method for ploidy determination in fungal cells, Curr. Genet., 14: 51-52.

# TRANSGENIC POTATO CULTIVARS RESISTANT TO POTATO VIRUS X

André Hoekema, Marianne J. Huisman, Dinie Posthumus-Lutke Willink, Erik Jongedijk, Peter van den Elzen and Ben J.C. Cornelissen

MOGEN International nv
Einsteinweg 97
2333 CB Leiden The Netherlands

## INTRODUCTION

Extensive potato breeding programs over the years have yielded improved varieties with a whole set of proven valuable traits. The process of potato breeding, however, is laborious and time-consuming due to the tetraploid character of the potato genome. Techniques for the genetic engineering of plants present a new tool to improve potato via the introduction of traits presently missing in the existing cultivars. An essential element in the successful application of these new techniques is the preservation of the existing, desired traits of the cultivars. So far, no systematic studies on the actual performance of engineered plants have been performed. In this paper we report the engineering of resistance to potato virus X in the two commercial potato cultivars Bintje and Escort. Also, the results of two years of field trials testing the performance of these transgenic lines are discussed.

## DESIRABLE TRAITS IN POTATO CULTIVARS

Many of the major potato cultivars lack several important characteristics. Examples of desirable traits are resistance to several

Plant Molecular Biology 2, Edited by R.G. Herrmann and
B. Larkins, Plenum Press, New York, 1991

potato viruses, to fungal pathogens like *Phytophthora infestans* and *Alternaria solani*, to *Pseudomonas* and *Erwinia* species, and to the cyst nematodes *Globodera rostochiensis* and *G. pallida*. The high incidence of diseases in potato cultivars (as compared to their wild relatives) may be due to the selection for yield, processing quality etc. rather than for disease resistance in the breeding programs. Also the use of monocultures may select for strains of pathogens that can overcome resistances. As a consequence, the farmer relies heavily on chemical protectants for the control of insects, fungi and nematodes.

Elimination of viruses has hardly been possible sofar. Several factors have contributed to the control of viral diseases. Traditional breeding for resistance, and the agricultural practice of spraying the foliage to prevent aphids from infecting the potato crop, have helped to diminish damage. Furthermore, certification schemes have been developed for potato in order supply virus-free seed potatoes. The affection of potato yields by the occurrence of viruses, however, still is a recognized problem.

### Virus resistance

Recent developments in molecular biology have stimulated research into the development of genetically engineered virus resistance. The strategies reported sofar to obtain protection against viral infections, are based on the interference with virus propagation, either via the expression of viral sequences or sequences derived from viral satellites in transgenic plants (for recent reviews on this subject see 1,2). Other more general mechanisms for virus resistance based on the use of plant genes are being pursued as well (1-4).

The first report of engineered virus resistance is from 1986, concerning protection of tobacco against tobacco mosaic virus (TMV), mediated by the expression of the TMV coat protein gene (5). Since then, resistance has been obtained against alfalfa mosaic (6-8), tobacco streak (9), tobacco rattle (10), and cucumber mosaic virus (11) as well as against potato virus X (12-14) and the potyviruses potato virus Y (14,15) and tobacco etch virus (15).

## ENGINEERED VIRUS RESISTANCE IN POTATO

The majority of the studies on coat protein mediated virus resistance concern tobacco as a model plant system. We have chosen to focus on the introduction of PVX-resistance in commercial potato cultivars. Seed potatoes of two PVX sensitive cultivars, Escort and Bintje, were used in this work. Bintje is the most widely grown cultivar in Europe. A cDNA fragment corresponding to the PVX coat protein cistron was placed under the control of a cauliflower mosaic virus (CaMV) 35S promoter. A transformation system for potato, based on regeneration from tuber discs was used (13).

Fifty transgenic Escort lines and 80 Bintje lines were obtained, and tested for the expression level of the coat protein. The expression levels in different lines ranged from undetectable to 0.3 % of total soluble leaf protein (13).

To assess the level of protection against infection with PVX, we challenged two transgenic Escort lines, three transgenic Bintje lines as well as Bintje and Escort control plants with buffer (mock) or with PVX (1μg/ml), and performed ELISA at various days post infection to monitor the amounts of PVX. The transgenic plants showed up to a 100-fold reduction in the accumulation of PVX coat protein. CP-levels reflect the amounts of infectious virus in the plants. This was confirmed by inoculation of a local lesion host for PVX (*Chenopodium quinoa*) with extracts of the inoculated potato plants.

These experiments show that the introduced coat protein gene mediates a high level of protection against infection of potato plants with PVX.

**PVX resistance in the next generation**

We were interested to test for the presence of the PVX resistance trait in potato plants grown from transgenic tubers. To this aim tubers were harvested from a control regenerant Bintje line (Br), three transgenic Bintje lines (B13, B33 and B66), an Escort control regenerant line (Er), and two transgenic Escort lines (E13 and E32). Of each line eight tubers were planted. At four weeks after potting six of the plants were inoculated with PVX (1 μg/ml) and two with buffer. At weekly intervals the PVX coat protein accumulation was analysed by ELISA. At 21 days post inoculation the PVX accumulation in the tuber grown transgenic Bintje plants was about 20-30-fold lower than that in the regenerant ones. For Escort the reduction in accumulation was upto a 100-fold as compared to the regenerant Escort plants (Fig. 1).

**The role of coat protein in protection**

The constitutive expression of the gene encoding the PVX coat protein has resulted in protection against infection with PVX. In the transgenic plant the coat protein may interfere with the multiplication cycle of the virus. Such interference is also believed to underlie the phenomenon of cross-protection, the protection of a plant against a virulent strain of a virus by prior inoculation with a mild strain of the same virus.
Coat protein-mediated protection may exemplify one aspect of the cross-protection

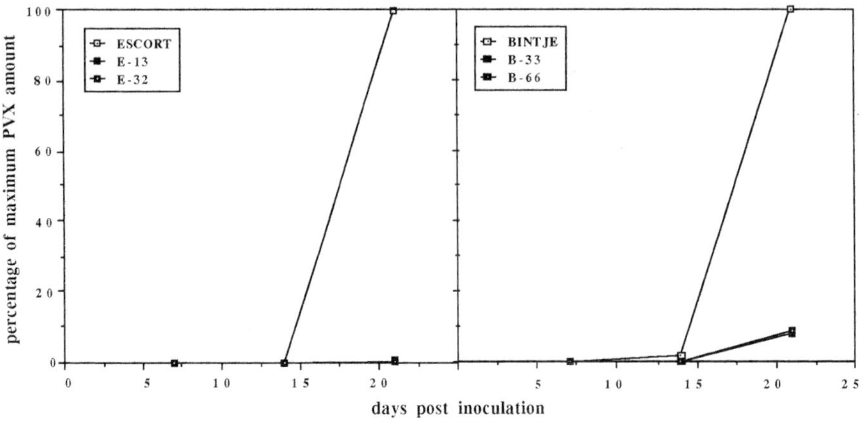

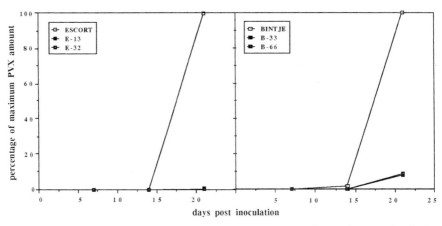

Fig.1 Accumulation of PVX in tuber-grown transgenic and control plants

phenomenon. The coat protein itself is believed to play an important role in both cross protection (16,17) and engineered CP-mediated expression: a transgenic plant expressing a CP-gene with a frameshift mutation produces normal amounts of RNA but does not confer protection (8), showing that protection involves the coat protein itself and not its mRNA. This was confirmed recently for protection against tobacco mosaic virus (18). Furthermore, both types of protection can be overcome by high concentrations of the inoculum and both are less effective against inoculations with RNA, with maybe one exception (12). However, the coat protein model cannot explain the cross-protection observed with viroids, CP-less mutants (19), or with a tobacco rattle virus RNA 1 infection (20). This data suggest that the coat protein plays a role in protection, but that it is not the only factor.

## EVALUATION OF CHARACTERISTICS OF THE POTATO CULTIVARS

A crucial prerequisite for the successful application of genetic modification, is that the intrinsic properties of the cultivars are preserved. In one report in literature the potential reduction in yields of two tomato lines transgenic for the TMV coat protein gene was investigated. Yields from one line were equal to that of control plants, yields of the other were depressed (21). Potato in known to be very susceptible to somaclonal variation, especially after regeneration from callus or protoplasts (22,23). This can often be correlated with variations in the number of chromosomes, and in structural chromosome aberrations. Although we avoided an intervening callus phase in the regeneration process of potato plants, we analyzed to what extent the identity of the original cultivars had been preserved in the transgenic lines.

**Testing of growthroom material**

In total 62 independently transformed lines (39 Bintje lines and 23 Escort lines) were analyzed. To assess possible karyotype changes in the transgenic lines chromosome numbers were determined in root tip cells from tissue culture grown plantlets. Sixty plants showed the normal tetraploid number of chromosomes (2n=4x=48). Phenotypically all these plants were normal under growthroom conditions. One Bintje line with an apparently normal phenotype contained the aneuploid number of chromosomes (2n=47), one other Bintje line was definitely abnormal and contained approximately 96 chromosomes (4n~96).
No major chromosome aberrations were found. However, this does not imply that no cultivar characteristics were changed.

**Performance of potato cultivars under field conditions**

In a more detailed study, tests were carried out according to the criteria used in the official potato variety registration procedure. In these official tests new cultivars are judged for 50 morphological characteristics as described in the UPOV guidelines to test for stability, homogeneity and distinctiveness (24). In our analysis the 50 official morphological characteristics were scored, among which were 21 plant, 7 tuber, 10 flower and 12 light-sprout characteristics. We also tested yield, and grading of the tubers. Field tests were carried out in Dronten, The Netherlands, by permission under the Dutch Nuisance Act.

In the summer of 1988 axenically grown plantlets were planted in the field and compared to control Bintje and Escort plants which were propagated axenically as well. From these preliminary studies we concluded that approximately 60% of the Bintje transgenic lines and 87% of the Escort lines were true to type. Plant lines that were found deviant generally differed in at least five characteristics. Some of the tubers from the field were sprouted under low light conditions and were judged for all light-sprout characteristics. This revealed that the sprout characteristics were indicative of the plant phenotype observed.

In the 1989 field trial, tubers from plants with a low, medium, or high expression level for the PVX coat protein were planted. This time a more complete observation and scoring of the lines was carried out. In order to allow statistical tests for the occurrence of differences in varietal characteristics, a so-called randomized complete block experiment was carried out. In this experiment seed tubers of a uniform size were planted in 2 replicates, using 5 plants per plot. Statistical data were obtained on most plant characteristics, plant vigour, tuber yield and grading. Flower characteristics were scored in the growthroom.

During this trial leaf samples were collected throughout the season to follow the expression patterns of the PVX coat protein. This revealed the stability of CP-expression during the growing season. Figure 2 shows a summary of the results obtained in these field trials. This illustrates the main conclusion of our work, namely that in the majority of the transgenic lines, none of the original cultivar characteristics had been affected. The results of the statistical analysis of all combined data will be published elsewhere (Jongedijk *et al.*, in preparation).

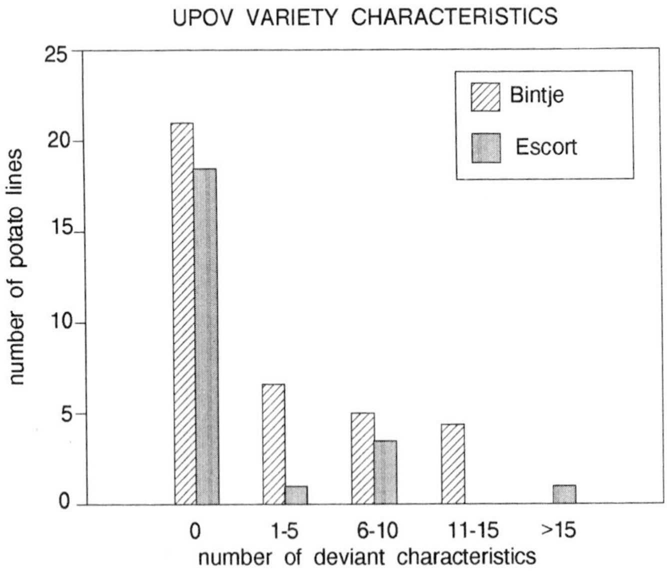


Fig.2 Summary of morphological performance data of transgenic Escort and Bintje lines

**DISCUSSION**

**PVX-resistant potato cultivars**

In this paper we have described the recent progress towards the development of potato cultivars carrying a virus resistance trait. We have shown that the new trait, resistance against PVX, is expressed both in tissue-culture grown plants, and in the next generation of tuber-grown plants. We have demonstrated that commercial cultivars of potato can be engineered for a new trait with the apparent preservation of all cultivar characteristics. The transgenic cultivars are now in the process of cultivar registration. Furthermore, tests will be carried out in 1990/1991 to establish the PVX resistance level in the field.

It is interesting to note that registration on the cultivar list requires a morphologically distinct phenotype of the new cultivar. Evidently, with the availability of techniques to improve existing cultivars without affecting varietal characteristics, there is a need for new guidelines. Also, new tests to show the identity of newly developed cultivars will be needed.

### Other virus resistance strategies

At present, the coat protein-mediated protection strategy seems to be the most promising one. Other strategies involving expression of viral genes have been tested in several plant species. Constitutive expression in tobacco of the AlMV RNA 1 and 2 encoded genes or the so-called movement protein gene of the same virus or TMV, did not lead to resistance (25-27). The expression of the 13K, 16K and 29K genes of tobacco rattle virus gave no detectable protection against TRV (28). Also, the expression of an antisense AlMV coat protein gene does not result in protection (25). Plants transgenic for an antisense cucumber mosaic virus and PVX coat protein gene show a low level of resistance to cucumber mosaic virus and PVX (11,12).

Some strains of certain plant RNA viruses have satellites, small RNA molecules which are replicated by the helper virus polymerase and encapsidated in helper virus CP. The production of satellite RNA in transgenic plants results in a reduction of helper virus particles and symptoms as compared to non-transformed plants (29,30).

Finally, one can envisage virus resistance strategies based on the phenomenon of induced resistance. With the increase of our knowledge of PR proteins the chances to come to a broad virus resistance via the expression of PR proteins may grow as well. It has been speculated that PR proteins may be responsible for induced resistance against viruses (31).

### REFERENCES


1. Baulcombe D: Strategies for virus resistance in plants. Trends in Genetics 5: 56-60 (1989).
2. Van den Elzen PJM, Huisman MJ, Posthumus-Lutke Willink D, Jongedijk E, Hoekema A and Cornelissen BJC: Engineering virus resistance in agricultural crops. Plant Mol. Biol. 13: 337-346 (1989).
3. Linthorst HJM, Meuwissen RLJ, Kauffman S, Bol JF: Constitutive expression of pathogenesis-related protein PR-1, GRP, and PR-S in tobacco has no effect on virus infection. The Plant Cell 1: 285-291 (1989).

4. Spiegel S, Gera A, Salomon R, Ahl P, Harlap S, Loebenstein G: Recovery of an inhibitor of virus replication from the intercellular fluid of hypersensitive tobacco infected with tobacco mosaic virus and from uninfected induced-resistant tissue: Phytopathology 79:258-259 (1989).
5. Powell Abel P, Nelson RS, De B, Hoffman N, Rogers SG, Fraley RT, Beachy RN: Delay of disease development in transgenic plants that express the tobacco mosaic virus coat protein gene. Science 232: 738-743 (1986).
6. Tumer NE, O'Connell KMO, Nelson RS, Sanders PR, Beachy RN, Fraley RT, Shah DM: Expression of alfalfa mosaic virus coat protein gene confers cross-protection in transgenic tobacco and tomato plants. EMBO J 6: 1181-1188 (1987).
7. Loesch-Fries LS, Merlo D, Zinnen T, Burhop L, Hill K, Krahn K, Jarvis N, Nelson S, Halk E: Expression of alfalfa mosaic virus RNA 4 in transgenic plants confers virus resistance. EMBO J 6: 1845-1987 (1987).
8. Van Dun CMP, Bol JF, Van Vloten-Doting L: Expression of alfalfa mosaic virus and tobacco rattle virus coat protein genes in transgenic tobacco plants. Virology 159: 299-305 (1987).
9. Van Dun CMP, Overduin B, Van Vloten-Doting L, Bol JF: Transgenic tobacco expressing tobacco streak virus or mutated alfalfa mosaic virus coat protein does not cross-protect against alfalfa mosaic virus infection. Virology 164: 383-389 (1988).
10. Van Dun CMP, Bol JF: Transgenic tobacco plants accumulating tobacco rattle virus coat protein resist infection with tobacco rattle virus and pea early browning virus. Virology 167: 649-652 (1988).
11. Cuozzo M, O'Connell KM, Kaniewski W, Fang R-X, Chua N-H, Tumer, NE: Viral protection in transgenic tobacco plants expressing the cucumber mosaic virus coat protein or its antisense RNA. Bio/Technology 6: 549-557 (1988).
12. Hemenway C, Fan R-X, Kaniewski WK, Chua N-H, Tumer NE: Analysis of the mechanism of protection in transgenic plants expressing the potato virus X coat protein or its antisense RNA. EMBO J 7: 1273-1280 (1988).
13. Hoekema A, Huisman MJ, Molendijk L, Van den Elzen PJM, Cornelissen BJC: The genetic engineering of two commercial potato cultivars for resistance to potato virus X. Bio/Technology 7: 273-278 (1989).
14. Lawson C, Kaniewski W, Haley L, Rozman R, Newell C, Sanders P, Tumer NE: Engineering resistance tomixed virus infection in a commercial potato cultivar: resistance to potato virus X and potato virus Y in transgenic Russet Burbank. Bio/Technology 8: 127-134 (1990).
15. Stark DM, Beachy RN: Protection against potyvirus infection in transgenic plants: evidence for broad spectrum resistance. Bio/Technology 7: 1257-1262 (1989).

16. DeZoeten GA, Fulton RW: Understanding generates possibilities. Phytopathology 65: 221-222 (1975).
17. Sherwood JL, Fulton RW: The specific involvement of coat protein in tobacco mosaic virus cross-protection. Virology 119: 150-158 (1982).
18. Powell PA, Sanders PR, Tumer N, Fraley RT, Beachy RN: Protection against tobacco mosaic virus infection in transgenic plants requires accumulation of coat protein rather than coat protein RNA sequences. Virology 175: 124-130 (1990).
19. Zaitlin M: Viral cross-protection; more understanding is needed. Phytopathology 66: 382-383 (1976).
20. Cadman CH, Harrison BD: Studies on the properties of soil-borne viruses of the tobacco rattle type occurring in Scotland. Ann Appl Bil 47: 542-556 (1959).
21. Nelson RS, McCormick SM, Delannay X, Dubé P, Layton J, Anderson EJ, Kaniewska M, Proksch RK, Horsch RB, Rogers SG, Fraley RT, Beachy RN: Virus tolerance, plant growth, and field performance of transgenic tomato plants expressing coat protein from tobacco mosaic virus. Bio/Technology 6: 403-409 (1988).
22. Karp A, Nelson RS, Bright SWJ: Chromosome variation in protoplast derived potato plants. Theor Appl Genet 63: 265-272 (1982).
23. Gill BS, Kam-Morgan LMW, Shepard JF: Cytogenetic and phenotypic variation in mesophyll cell derived tetraploid potatoes. J Heredity 78: 15-20 (1987).
24. UPOV, 1986. International union for the protection of new varieties of plants. Guidelines for the conduct of tests for distinctness, homogeneity and stability of potato (*Solanum tuberosum*).
25. Van Dun CMP: Expression of viral cDNA in transgenic tobacco. Thesis, University of Leiden (1988).
26. Van Dun CMP, Van Vloten-Doting L, Bol JF: Expression of alfalfa mosaic virus cDNA1 and 2 in transgenic tobacco plants. Virology 163: 572-578 (1988).
27. Deom CM, Oliver MJ, Beachy RN: The 30-kilodalton gene product of tobacco mosaic virus potentiates virus movement. Science 237: 389-394 (1987).
28. Angenent GC, Van den Ouweland JMW, Bol JF: Susceptibility to virus infection of transgenic tobacco plants expressing structural and nonstructural genes of tobacco rattle virus. Virology 175, 191-198 (1990).
29. Harrison BD, Mayo MA, Baulcombe DC: Virus resistance in transgenic plants that express cucumber mosaic virus satellite RNA. Nature 328: 799-802 (1987).

30. Gerlach WL, Liewellyn D, Haseloff J: Construction of a plant disease resistance gene from the satellite RNA of tobacco ringspot virus. Nature 328: 802-805 (1987).

31. Kassanis B, Gianninazzi S, White RF: A possible explanation of the resistance of virus-infected tobacco plants to second infection. J Gen Virol 23: 11-16 (1974).

# THE AGROBACTERIUM VIRULENCE SYSTEM

Paul J.J. Hooykaas, Leo S. Melchers,
Kees W. Rodenburg and Stefan C.H. Turk

Dept. Plant Molecular Biology,
Clusius Laboratory, Leiden University
Wassenaarseweg 64
2333 AL Leiden, The Netherlands

## INTRODUCTION

The soil bacterium Agrobacterium tumefaciens causes the plant tumor crown gall at wound sites on dicotyledonous plants (Fig.1). During crown gall induction, agrobacteria transfer a segment of their tumor-inducing (Ti)-plasmid to plant cells at the wound sites. This transferred (T)-DNA, which becomes integrated in one or more of the chromosomes of the host plant cells either in one copy or in a limited number of copies, is responsible for the tumorous character of crown gall cells. This is due to the overproduction of an auxin (IAA) and a cytokinin (isopentenyl-AMP) in these cells via enzymes that are encoded by genes in the T-DNA. Besides these well-characterized onc-genes the T-DNA contains some other onc-genes (e.g. gene 6b) that are important for tumorigenicity of particalar plant species (Hooykaas et al., 1988). A characteristic feature of crown gall cells is that they produce compounds called opines which are mostly built from an amino acid and a keto acid. The synthases catalyzing the formation of these opines are encoded by genes located in the T-DNA. Agrobacterium can be classified on the basis of the opines that are produced in the tumor they induce. Thus octopine, nopaline, succinamopine and leucinopine strains have been described. A number of recent reviews describe several aspects of the Agrobacterium-plant interaction in detail (Hooykaas and Schilperoort, 1984; Nester et al., 1984; Koukolikova-Nicola et al., 1987; Melchers and Hooykaas, 1987; Binns and Thomashow, 1988).

### Virulence genes and T-DNA processing

It is not known how the T-DNA is transferred from agrobacteria to plant cells, but numerous genes have been identified in the bacterium which are involved in this process. Some of these are located in the chromosome (chromosomal virulence (chv)-genes), while other are present on the Ti plasmid (vir-genes). Together these genes determine the transfer apparatus, which specifically mediates the transfer of only the T-region from the bacterium to plant cells. When present in the Ti plasmid the T-region is surrounded by a 24bp imperfect direct repeat, which is called the border repeat (Yadav et al., 1982). The deletion of the right border sequence but not the left border sequence leads to avirulence of the host bacterium. Virulence can be restored by the re-introduction of either a left border sequence (Van Haaren et al., 1987[a]), or right border sequence (Wang et al., 1984) into the Ti plasmid at the site from which the right border sequence had originally been deleted. This work also showed that the border repeat is active only in one particular orientation (Wang et al., 1984; Peralta and Ream, 1985; Van Haaren et al., 1987a). Border repeats from heterologous Ti plasmids

Plant Molecular Biology 2, Edited by R.G. Herrmann and
B. Larkins, Plenum Press, New York, 1991

Fig.1 Tumors induced by Agrobacterium tumefaciens strain 15955

could functionally replace the right border sequence of the octopine Ti plasmid, showing that the transfer apparatus determined by this latter plasmid can cope with these heterologous border repeats (Van Haaren et al., 1988). When a 24bp border repeat is used to replace a right border region, virulence is only partially restored. This is due to the fact that in the wild-type situation right border repeats are accompanied by a sequence called overdrive which is essential for full virulence (Peralta et al., 1986; Van Haaren et al., 1987[a]). The overdrive sequence is active in both orientations, and at a much larger distance than in the wild-type situation versus the right border repeat and acts as a T-DNA transfer enhancer (Van Haaren et al., 1987b).

After induction of the Agrobacterium virulence system (see below) T-DNA processing can be observed. Initially excised circular T-molecules were detected (Koukolikova-Nicola et al., 1985; Machida et al., 1986). Later on single stranded molecules (T-strands) were found, which represented the bottom strand of the T-region (Stachel et al., 1986). The presence of the overdrive sequence next to the right border repeat led to a stimulation of the production of T-strands in the bacterium (Van Haaren et al., 1987b). The formation of T-strands is probably preceded by the nicking of both the left and right border repeats in the bottom strand (Albright et al., 1987; Wang et al., 1987). These nicks might act as the start sites for DNA-synthesis leading to T-strand production. Whether single stranded DNA molecules as such are transferred to plant cells via Agrobacterium is not known, but recent experiments have shown that single stranded DNA molecules that are introduced into plant cells via a method of direct DNA transfer can transform plant cells efficiently (Rodenburg et al., 1989; Furner et al., 1989). Probably, the single stranded molecules are converted into double stranded forms shortly after uptake.

An unexpected finding by Buchanan-Wollaston et al (1987) was that certain plasmids can be transferred to plant cells via the Agrobacterium virulence system even if they lack a 24 bp border repeat. The explanation for this phenomenon was that in such plasmids the border repeat is functionally replaced by the origin of conjugative transfer (oriT) with the accompanying mob-genes which code for proteins that can nick at this sequence. This system is used by bacterial plasmids for the synthesis of single stranded copies of itself that are subsequently transferred via the conjugative transfer system to recipient bacteria. It is thus possible that Agrobacterium uses a conjugation-like system to transfer DNA into recipient plant cells.

## The chromosomal virulence genes

The Agrobacterium T-DNA transfer apparatus is determined by a combination of chromosomal (chv)- and Ti plasmid (vir)-genes. Mutants with defects in either the chvA or chvB gene do not produce extracellular ß1,2-glucan and show a limited ability to attach to plant cells (Douglas et al.,1985). The chvB gene codes for a large innermembrane protein that is probably directly involved in the synthesis of ß1,2-glucan (Zorreguieta et al., 1988), whereas the chvA gene determines an innermembrane protein with homology to a number of export proteins such as HlyB and Mdr (Cangelosi et al.,1989). In chvA mutants ß1,2-glucan is formed, but this is not exported, but remains in the cytoplasm (Cangelosi et al., 1989; Inon de Ianino and Ugalde, 1989). This strongly suggests that the chvA encoded protein is essential for transport of ß1,2-glucan to the periplasm. It was supposed that chvA and chvB mutants were avirulent on all plant species (Douglas et al., 1985); however, on potato (Hooykaas et al., 1987) and Solanum hougasii (Hawes and Pueppke, 1989) these mutants were later found to be virulent, probably because these chv genes are not necessary for attachment to cells of these plant species (Hawes and Pueppke, 1989). A number of (exo) loci have been identified that are involved in the production of extracellular polysaccharides in Agrobacterium (Thomashow et al., 1987). Mutations in one of these loci (exoC or pscA) leads to avirulence, possible because ß1,2-glucan is not produced bij these mutants (Thomashow et al., 1987).Besides these well-known loci a few other chromosomal mutations have been described that affect virulence or host range for tumor induction, i.e. chvC (Hooykaas et al., 1984), chvD (Winans et al.,1988) and chvE (Huang et al., 1990). The proteins encoded by the latter two loci are necessary for an optimal expression of the Ti plasmid vir-genes (see below). Agrobacterium produces cellulose fibrils during the initial phases of tumor induction, which leads to the formation of large aggregates (Matthysse, 1983). The cel genes involved are located in the chromosome and have been mapped in a region of the chromosome which also embraces the chvA and chvB genes (Robertson et al., 1988) on the chromosomal map that was constructed for strain C58 (Hooykaas et al., 1982). Mutants that are defective in cellulose production are normally virulent in standard laboratory assays, but are much less effective than the wild-type when the inoculation sites are washed with water (Matthysse, 1983). This suggests that in nature the ability to form cellulose fibrils is an asset for Agrobacterium useful to avoid removal by heavy rainfall for instance. It is important to realize that chromosomal (background) genes may affect the efficiency or host range of the T-DNA delivery from Agrobacterium to plant cells. This is probably the reason that the socalled supervirulent phenotype mediated by the pTi542 plasmid is expressed in the C58 cured background, but less well in the natural 542 chromosomal background (Hood et al., 1987). Heterologous bacteria such as Rhizobium trifolii (Hooykaas et al., 1977) and Phyllobacterium myrsinacearum (Van Veen et al., 1988) are converted into tumorigenic bacteria after receipt of the Ti plasmid, but not related bacteria such as R.meliloti or P.rubiacearum (Melchers and Hooykaas, 1987; Van Veen et al., 1989). This is probably due to the absence of certain chromosomal genes affecting virulence in these bacteria, since vir-expression and T-strand formation occur normally in them (Van Veen et al., 1989).

## The Ti plasmid virulence genes

The vir-genes are located in one particular segment of the Ti plasmid, the Vir(ulence) region, which is about 40 kbp in size (Fig.2). For the octopine type Ti plasmid 24 vir-genes have been identified in seven operons, which together belong to one regulon (see below). The operons may contain from one up to 11 genes, and have been given the names virA-virH. The Vir-regions of other types of Ti plasmids may be slightly different in their gene composition (see below).

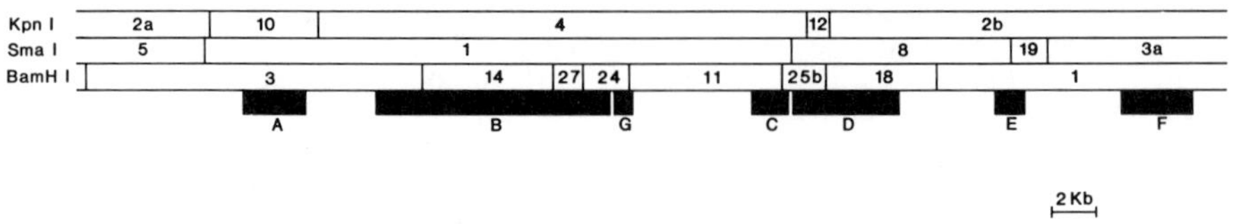


Fig.2 Map of the Vir-region of the Ti plasmid

Four of the vir-operons are essential for virulence, i.e. virA, virB, virD and virG. Mutants in any of these loci lead to avirulence on all plant species. However, mutations in any of the other loci (virC, virE, virF, virH,) lead to an attenuated tumorigenicity or a restricted host range for tumor induction. The functions determined by some of the vir-genes have been established. The genes virA and virG encode a socalled two component regulatory system that controls the expression of the whole vir-regulon (see next paragraph). Proteins encoded by the first two genes virD1 and virD2 of the virD operon are the only vir-genes that are involved in nicking of the border repeats anex in the production of T-strands after induction of the virulence system (Stachel et al., 1987). The VirD1 protein has a helix-turn-helix consensus sequence and because of that may have DNA-binding capacity. This protein has topoisomerase-activity and can relax covalently closed circular DNA molecules (Ghai and Das, 1989). During nicking the VirD2 protein becomes covalently attached to the 5'end of the nick sites, and T-strands still have this protein covalently attached at their 5'end (Ward and Barnes, 1988; Young and Nester, 1988; Dürrenberger et al., 1989). This is the reason that protease treatment is an essential step in the procedure for isolation of T-strands. The protein encoded by the virD4, but not the virD3 gene, is essential for virulence, but its function is unknown. The essential virB operon encodes a large number of proteins, many of which are hydrophobic or have a potential signal peptide (Thompson et al., 1988; Ward et al., 1988). These are therefore likely to be membrane proteins or secreted proteins. For some of these a location in the bacterial membrane has indeed been demonstrated (Engström et al., 1987). It can therefore be speculated that the virB-operon determines proteins that together form a structure (pilus, pore?) that is essential for transfer of the T-DNA from the bacterium to the plant cell. Interestingly, recently a protein involved in the uptake of single stranded DNA by competent Bacillus subtilis cells (the ComG protein) turned out to have significant homology with one of the virB encoded proteins, viz VirB11 (Albano et al., 1989). This VirB11 protein has consensus ATP-binding domains (Thompson et al., 1988) and has indeed ATPase activity (Christie et al., 1989). It may help to provide the energy that is necessary for the transport of T-strands. Both the non-essential virC and virE operons code for two proteins. For VirCI it was shown that it binds specifically to the overdrive sequence and promotes T-strand production (Toro et al., 1989), and it can therefore be speculated that after binding to overdrive it may promote T-strand formation by "opening" the right border repeat for the VirD proteins or by attracting VirD proteins to this area. The phenotype of virC mutants as to tumorigenicity is similar to that of overdrive lacking mutants. The VirE2 protein binds efficiently and cooperatively to single stranded DNA and may protect T-strands against nucleases (Gietl et al., 1987; Citovsky et al., 1989). If cooperative binding would also occur in vivo this would convert the T-strand into a virus-like particle (Citovsky et al., 1989). The host range gene virF (Hooykaas et al., 1984) is absent from certain other types of Ti plasmids such as the nopaline Ti plasmids (Otten et al., 1985). This gene, which encodes a 22 Kd hydrophilic protein is important for tumor induction on certain plant species (e.g. Nicotiana glauca). Its absence from the nopaline Ti plasmid explains why nopaline strains are almost avirulent on N.glauca (Melchers et al., 1990). The recently described virH operon encodes two proteins that have homology to cytochrome P450 enzymes (Kanemoto et al., 1989). When diluted inoculants are used virH mutants are less virulent than the wild-type, possibly because the proteins encoded by the vir operon play a role in the detoxification of toxic compounds that are exuded by the wounded and infected plants. While octopine strains specifically have the virF gene, nopaline strains have a gene called tzs. This latter gene forms part of the vir-regulon, and encodes a protein which is similar to that encoded by the ipt- or cyt-gene in the T-region, and this determines production and secretion of trans-zeatin (a cytokinin) by the bacterium (Akiyoshi et al., 1985; John and Amasino, 1988). The presence of this gene may make nopaline strains more effective on certain plant species than octopine strains. When planning to use Agrobacterium for the genetic engineering of a new plant species, it is therefore advisable to consider the use of a set of different helper strains in order to find out which is the optimal strain for the transformation of that particular plant species.

## Regulation of the vir-genes

It was shown that plant exudates contain compounds that can induce the vir-genes. In wounded tobacco the responsible vir-inducers were identified as acetosyringone and α-hydroxyacetosyringone (Stachel et al., 1985). However, besides these compounds a variety of other phenolic compounds can be used for the induction of vir-genes, the most simple inducer being guiacol, a derivative of phenol with a methoxyl-group in the ortho-position (Spencer and Towers, 1988; Melchers et al., 1989[b]). Substitutions in the para-position are tolerated to some extent, and can lead to compounds with either enhanced or reduced inducing activity. The absence of sufficient amounts of inducer molecules in the exudates of certain plant species can be the major reason that these are not effiiciently transformed by Agrobacterium (Schäfer et al., 1987).

Even in the presence of an inducer particular conditions have to be met in order to obtain an optimal induction of the vir-genes: a) the pH of the medium must be adjusted to about 5.3, (b) the temperature must be below 30°C, (c) a high sugar (preferably sucrose) concentration must be present in the medium. Although the vir-genes of different types of Ti plasmids can be induced in similar acetosyringone containing media, there are small differences in their requirement for optimal induction. For instance, while octopine and leucinopine Ti plasmids require pH5.3 for maximal vir-gene induction, for the vir-genes of nopaline Ti plasmids the preferred medium is at pH5.8. This probably reflects small differences in the VirA regulatory proteins determined by these plasmids (see below).

The virA and virG gene products together are necessary for the induction of the other vir-operons (Stachel and Zambryski, 1986). Sequence analysis in our laboratory (Melchers et al., 1986; 1987) as well as by others showed that these proteins have homology with proteins comprising other two-factor regulatory systems such as cheA/cheY, envZ / ompR, ntrB / ntrC, phoR / phoB (Ronson et al., 1987).

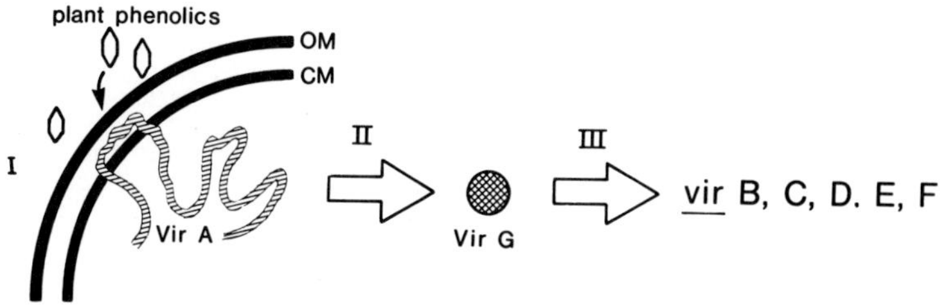


Fig.3 Regulation of the virulence genes

The model (Fig.3) is that the virA-like component of the system is a sensor (for phenolic compounds in the case of the VirA protein), and the virG-like component a DNA-binding activator protein capable of specifically activating a particular set of genes (the other vir-operons in the case of the VirG protein).
For the ntrB/ntrC (Keener and Kustu, 1988; Weiss and Magasanik, 1988), cheA/cheY (Hess et al., 1988) and envZ/ompR (Igo et al., 1989) systems in enterobacteria evidence has been obtained that signal transmission from sensor to activator occurs via phosphorylation. Recently it was shown that the VirA protein like other sensor proteins has autophosphorylating activity (Jin et al., 1990[a], Huang et al., 1990), while the VirG protein has the capacity to bind to specific sequences in the vir-promoters (Jin et al., 1990[b]; Pazour and Das, 1990). As described above some chromosomal genes are involved in the operation of the VirA-VirG signalling system, which adds to the complexity of this system (Winans et al., 1988; Huang et al., 1990).

## Structure of the VirA protein

For the VirA protein the sequence data show that it has two long stretches of hydrophobic amino acids capable of forming membrane-crossing α-helices like most sensor proteins. Using antibodies against the C-terminal part of the VirA protein we could show (after membrane separation) that the VirA protein is present in the inner membrane of *Agrobacterium* indeed (Melchers *et al.*, 1987). Further work in which we used *in vitro* constructed fusions between different N-terminal parts of VirA and PhoA devoid of its signal peptide revealed that the topology of the VirA protein is as predicted from the DNA-sequence, *i.e.* having in the N-terminus a periplasmic domain followed by a membrane spanning domain (TM2) and a cytoplasmic C-terminal domain (Melchers *et al.*, 1989[a]). On the basis of this it was postulated that - like with other types of receptor proteins - the periplasmic (outside) domain would form the sensor (signal receiver) domain and the cytoplasmic (inside) domain the signal transmitter domain. In order to find out whether this was indeed the case, we made hybrids between VirA and the *E.coli* chemoreceptor Tar, which activates the chemotaxis apparatus after sensing the presence of aspartate or loaded maltose binding protein. The Tar-protein has a basic structure similar to that of the VirA protein, and it has been shown that the periplasmic domain of Tar is the sensor domain, while the cytoplasmic domain contains the signal transmitter region. When replacing the periplasmic part of VirA by that of Tar, we obtained sensor proteins that were active as vir-regulators. Unexpectedly, however, they still responded to phenolic compounds such as acetosyringone and not to aspartate or maltose (Melchers *et al.*, 1989[a]). Certain of these hybrid proteins were even more active in mediating the activation of vir-genes than the wild-type VirA protein, and rendered the system less sensitive to the negative influence of a relatively high pH or temperature. Significant induction took place even at pH6.8 and at temperatures as high as 33°C and 37°C. Thus our data indicate that 1) the sensor domain is not located in the periplasmic region of VirA 2) the amino acid composition of the periplasmic domain is one element conferring pH-and temperature-sensitivity on the system. In line with this we found that the deletion of the periplasmic domain from the VirA protein did not render the protein inactive. In fact such deleted VirA proteins behaved like the wild-type VirA protein in acetosyringone-dependent signal transduction. Further work showed that the deletion of the second transmembrane domain (TM2) from VirA or the substitution of this domain by the corresponding domain of Tar led to proteins that no longer responded to plant phenolics, but rather mediated a low, signal-independent level of vir-gene induction. We conclude that the second transmembrane domain of VirA is an essential element of VirA, which possibly contains the sensor (signal receiver) function of this protein. That the location of the sensor domain in VirA is different from that in the more classically studied chemoreceptors does not have to be too surprising in light of the fact that the signals in the case of VirA are formed by lipophilic phenolic compounds. The only other example, where the signal receiver function is located in a membrane domain of a chemoreceptor protein is given by the ß-adrenergic receptors of eukaryots (Strader et al., 1987). The ligands in this case are formed by catecholamines, which can interact with particular serine, cysteine and residues in the membrane spanning domains of the receptor (Fraser, 1989; Strader et al., 1989). When the TM2 areas of different VirA proteins are depicted in a helical wheel configuration, a conserved pocket is seen which contains besides strictly hydrophobic amino acids a cysteine and a serine residue, which might play an important role in the binding of the phenolic inducers. Deletion analysis of the cytoplasmic C-terminal domain of VirA showed that small deletions already render the protein inactive as a signal transducer, which is in line with a possible function of this part of the protein in signal transmission to the VirG protein. Future research will hopefully shed more light on the exact molecular details of T-DNA transfer between Agrobacterium and plant cells and the signal communication between plant cells and agrobacteria that forms part of this process.

## REFERENCES

Akiyoshi, D. E., Regier, D. A., and Gordon, M. P. (1989) Nucleotide sequence of the tzs gene from Pseudomonas solanacearum strain K60. Nucleic Acids Res. 17, 8886

Albano, M., Breitling, R., and Dubnau, D. A. (1989) Nucleotide sequence and genetic organization of the Bacillus subtilis comG operon. J.Bacteriol. 171, 5386-5404

Albinger, C., and Beiderbeck, R. (1977) Ubertragung der Fahigkeit zur Wurzelinduktion von Agrobacterium rhizogenes auf A.tumefaciens. Phytopath.Z. 90, 306-310

Albright, L. M., Yanofsky, M. F., Leroux, B., Ma, D., and Nester, E. W. (1987) Processing of the T-DNA of Agrobacterium tumefaciens generates border nicks and linear, single stranded T-DNA. J.Bacteriol. 169, 1046-1055

Binns, A. N., and Thomashow, M. F. (1988) Cell biology of Agrobacterium infection and transformation of plants. Ann. Rev.Microbiol. 42, 575-606

Buchanan-Wollaston, V., Passiatore, J. E., and Cannon, F. (1987) The mob and oriT mobilization function of a bacterial plasmid promote its transfer to plants. Nature 328, 172-175

Cangelosi, G. A., Martinetti, G., Leigh, J. A., Lee, C. C., Theines, C., and Nester, E. W. (1989) Role of Agrobacterium tumefaciens ChvA protein in export of ß-1,2 glucan. J. Bacteriol. 171, 1609-1615

Citovsky, V., Wong, M. L., and Zambryski, P. (1989) Cooperative interaction of Agrobacterium VirE2 protein with single-stranded DNA: implications for the T-DNA transfer process. Proc.Natl.Acad.Sci.USA 86, 1193-1197

Douglas, C. J., Staneloni, R. J., Rubin, R. A., and Nester, E. W. (1985) Identification and genetic analysis of an Agrobacterium tumefaciens chromosomal virulence region. J. Bacteriol. 161, 850-860

Dürrenberger, F., Crameri, A., Hohn, B., and Koukolikova-Nicola, Z. (1989) Covalently bound VirD2 protein of Agrobacterium tumefaciens protects the T-DNA from exonucleolytic degradation. Proc.Natl.Acad.Sci.USA 86, 9154-9158

Engström, P., Zambryski, P., Van Montagu, M., and Stachel, S. (1987) Characterization of Agrobacterium tumefaciens virulence proteins induced by the plant factor acetosyringone. J.Mol.Biol. 197, 635-645

Fraser, C. M. (1989) Site-directed mutagenesis of ß-adrenergic receptors. Identification of conserved cysteine residues that independently affect ligand binding and receptor activation. J.Biol.Chem. 264, 9266-9270

Furner, I. J., Higgins, E. S., and Berrington, A. W. (1989) Single-stranded DNA transforms plant protoplasts. Mol.Gen. Genet. 220, 65-68

Ghai, J., and Das, A. (1989) The virD operon of Agrobacterium tumefaciens Ti plasmid encodes a DNA-relaxing enzyme. Proc. Natl.Acad.Sci.USA 86, 3109-3113

Gietl, C., Koukolikova-Nicola, Z., and Hohn, B. (1987) Mobilization of T-DNA from Agrobacterium to plant cells involves a protein that binds single-stranded DNA. Proc.Natl.Acad.Sci.USA 84, 9006-9010

Hawes, M. C., and Pueppke, S. G. (1989) Variation in binding and virulence of Agrobacterium tumefaciens chromosomal virulence (chv) mutant bacteria on different plant species. Plant Physiol. 91, 113-118

Hess, F. J., Oosawa, K., Kaplan, N., and Simon, M. I. (1988) Phosphorylation of three proteins in the signaling pathway of bacterial chemotaxis. Cell 53, 79-87

Hood, E. E., Fraley, R. T., and Chilton, M. - D. (1987) Virulence of Agrobacterium tumefaciens strain A281 on legumes. Plant Physiol. 83, 529-534

Hooykaas, P. J. J., Klapwijk, P. M., Nuti, M. P., Schilperoort, R. A., and Rörsch, A. (1977) Transfer of the Agrobacterium tumefaciens Ti plasmid to avirulent agrobacteria and to Rhizobium ex planta. J.Gen.Microbiol. 98, 477-484

Hooykaas, P. J. J., Peerbolte, R., Regensburg-Tuïnk, A. J. G., De Vries, P., and Schilperoort, R. A. (1982) A chromosomal linkage map of Agrobacterium tumefaciens and a comparison with the maps of Rhizobium spp. Mol.Gen.Genet 188, 12-17

Hooykaas, P. J. J., and Schilperoort, R. A. (1984) The molecular genetics of crown gall tumorigenesis. In: "Molecular Genetics of Plants". Advances in Genetics 22, 209-283, Academic Press, Orlando, USA (J.G.Scandalios, Ed.)

Hooykaas, P. J. J., Hofker, M., Den Dulk-Ras, H., and Schilperoort, R. A. (1984) A comparison of virulence determinants in an octopine Ti plasmid, a nopaline Ti plasmid, and an Ri plasmid by complementation analysis of Agrobacterium tumefaciens mutants. Plasmid 11, 195-205

Hooykaas, P. J. J., and Schilperoort, R. A. (1986) The molecular basis of the Agrobacterium plant interaction. Characteristics of Agrobacterium virulence genes and their possible occurrence in other plant-associated baccteria. In "Recognition in Microbe-Plant Symbiotic and Pathogenic Interactions" ( B. Lugtenberg,Ed), pp 189-202. Springer Verlag, Berlin.

Hooykaas, P. J. J., Den Dulk-Ras, H., and Schilperoort, R. A. (1988) The Agrobacterium tumefaciens T-DNA gene 6b is an onc gene. Plant Mol.Biol. 11, 791-794

Huang, M. - L. W., Cangelosi, G. A., Halperin, W., and Nester, E. W. (1990) A chromosomal Agrobacterium tumefaciens gene required for effective plant signal transduction. J.Bacteriol. 172, 1814-1822

Huang, Y., Morel, P., Powell, B., and Kado, C. L. (1990) VirA, a coregulator of Ti-specified virulence genes, is phosphorylated in vitro. J.Bacteriol. 172, 1142-1144

Igo, M. M., Ninfa, A. J., and Silhavy, T. J. (1989) A bacterial environmental sensor that functions as a protein kinase and stimulates transcriptional activation. Genes and Development 3, 598-605

Inon de Iannino, N., and Ugalde, R. A. (1989) Biochemical characterization of avirulent Agrobacterium tumefaciens chvA mutants : synthesis and excretion of ß-(1,2) glucan.J. Bacteriol. 171, 2842-2849

Jin, S., Roitsch, T., Ankenbauer, R. G., Gordon, M. P., and Nester, E. W. (1990) The VirA protein of Agrobacterium tumefaciens is autophosphorylated and is essential for vir gene regulation. J.Bacteriol. 172, 525-530

Jin, S., Roitsch, T., Ankenbauer, R. G., Gordon, M. P., and Nester, E. W. (1990) The regulatory VirG protein specifically binds to a cis acting regulatory sequence involved in transcriptional activation of Agrobacterium tumefaciens virulence genes. J.Bacteriol. 172, 531-537

John, M. C., and Amasino, R. M. (1988) Expression of an Agrobacterium Ti plasmid gene involved in cytokinin biosynthesis is regulated by virulence loci and induced by plant phenolic compounds. J.Bacteriol. 170, 790-795

Kanemoto, R. H., Powell, A. T., Akiyoshi, D. E., Regier, D. A., Kerstetter, R. A., Nester, E. W., Hawes, M. C., and Gordon, M. P. (1989) Nucleotide sequence and analysis of the plant-inducible locus pinF from Agrobacterium tumefaciens. J. Bacteriol. 171, 2506-2512

Keener, J., and Kustu, S. (1988) Protein kinase and phosphoprotein phosphatase activities of nitrogen regulatory proteins NTRB and NTRC of enteric bacteria: roles of the conserved amino-terminal domain of NTRC. Proc.Natl.Acad.Sci. USA 85, 4976-4980

Koukolikova-Nicola, Z., Shillito, R. D., Hohn, B., Wang, K., Van Montagu, M., and Zambryski, P. (1985) Involvement of circular intermediates in the transfer of T-DNA from Agrobacterium tumefaciens to plant cells. Nature 313, 191-196

Koukolikova-Nicola, Z., Albright, L., and Hohn, B. (1987) The mechanism of T-DNA transfer from Agrobacterium tumefaciens to the plant cell. In: "Plant-DNA Infectious Agents" ( T. Hohn,and J. Schell,Eds), pp 109-148. Springer Verlag, Vienna.

Machida, Y., Usami, S., Yamamoto, A., Niwa, Y., and Takebe, I. (1986) Plant-inducible recombination between the 25 bp border sequences of T-DNA in Agrobacterium tumefaciens. Mol.Gen. Genet. 204, 374-382

Matthysse, A. G. (1983) Role of bacterial cellulose fibrils in Agrobacterium tumefaciens infection. J.Bacteriol. 154, 906-915

Melchers, L. S., Thompson, D. V., Idler, K. B., Schilperoort, R. A., and Hooykaas, P. J. J. (1986) Nucleotide sequence of the virulence gene virG of the Agrobacterium tumefaciens octopine Ti plasmid: significant homology betweeen virG and the regulatory genes ompR, phoB and dye of E.coli. Nucleic Acids Res. 14, 9933-9942

Melchers, L. S., Thompson, D. V., Idler, K. B., Neuteboom, S. T. C., De Maagd, R. A., Schilperoort, R. A., and Hooykaaas, P. J. J. (1987) Molecular characterization of the virulence gene virA of the Agrobacterium tumefaciens octopine Ti plasmid. Plant Mol.Biol. 9, 635-645

Melchers, L. S., Regensburg-Tuïnk, A. J. G., Schilperoort, R. A., and Hooykaas, P. J. J. (1989) Specificity of signal molecules in the activation of Agrobacterium virulence gene expression. Mol.Microbiol. 3, 969-977

Melchers, L. S., Regensburg-Tuïnk, T. J. G., Bourret, R. B., Sedee, N. J. A., Schilperoort, R. A., and Hooykaas, P. J. J. (1989) Membrane topology and functional analysis of the sensory protein VirA of Agrobacterium tumefaciens. EMBO J. 8, 1919-1925

Melchers, L. S., Maroney, M. J., Den Dulk-Ras, A., Thompson, D. V., Van Vuuren, H. A. J., Schilperoort, R. A., and Hooykaas, P. J. J. (1990) Octopine and nopaline strains of Agrobacterium tumefaciens differ in virulence; molecular characterization of the virF-locus. Plant Mol. Biol. 14, 249-259

Nester, E. W., Gordon, M. P., Amasino, R. M., and Yanofsky, M. F. (1984) Crown gall: a molecular and physiological analysis. Ann.Rev.Plant Physiol. 35, 387-413

Otten, L. A. B. M., Piotrowiak, G., Hooykaas, P. J. J., Dubois, M., Szegedi, E., and Schell, J. (1985) Identification of an Agrobacterium tumefaciens pTiB6S3 vir region fragment that enhances the virulence of pTiC58. Mol.Gen.Genet. 199, 189- 193

Pazour, G. J., and Das, A. (1990) VirG, an Agrobacterium tumefaciens transcriptional activator, initiated translation at a UUG codon and is a sequence-specific DNA-binding protein. J.Bacteriol. 172, 1241-1249

Peralta, E. G., and Ream, L. W. (1985) T-DNA border sequences required for crown gall tumorigenesis. Proc.Natl.Acad.Sci.USA 82, 5112-5116

Peralta, E. G., Hellmiss, R., and Ream, W. (1986) Overdrive, a T-DNA transmission enhancer on the A.tumefaciens tumour- inducing plasmid. EMBO J. 5, 1137-1142

Robertson, J. L., Holliday, T., and Matthysse, A. G. (1988) Mapping of Agrobacterium tumefaciens chromosomal genes affecting cellulose synthesis and bacterial attachment to host cells. J.Bacteriol. 170, 1408-1411

Rodenburg, C. W., De Groot, M. J. A., Schilperoort, R. A., and Hooykaas, P. J. J. (1989) Single stranded DNA used as an efficient new vehicle for plant protoplast transformation. Plant Mol.Biol. 13, 711-720

Ronson, C. W., Nixon, B. T., and Ausubel, F. M. (1987) Conserved domains in bacterial regulatory proteins that respond to environmental stimuli. Cell 49, 579-581

Schäfer, W., Görz, A., and Kahl, G. (1987) T-DNA integration and expression in a monocot crop plant after induction of Agrobacterium. Nature 327, 529-532

Spencer, P. A., and Towers, G. H. N. (1988) Specificity of signal compounds detected by Agrobacterium tumefaciens. Phytochemistry 27, 2781-2785

Stachel, S. E., Timmerman, B., and Zambryski, P. (1986) Generation of single-stranded T-strand molecules during the initial stages of T-DNA transfer from Agrobacterium tumefaciens to plant cells. Nature 322, 706-712

Stachel, S. E., and Zambryski, P. C. (1986) virA and virG control the plant-induced activation of the T-DNA transfer process of Agrobacterium tumefaciens. Cell 46, 325-333

Stachel, S. E., Timmerman, B., and Zambryski, P. (1987) Activation of Agrobacterium tumefaciens vir gene expression generates multiple single-stranded T-strand molecules from the pTiA6 T-region: requirement for 5' virD gene products. EMBO J. 6, 857-863

Strader, C. D., Sigal, I. S., Register, R. B., Candelore, M. R., Rands, E., and Dixon, R. A. F. (1987) Identification of residues required for ligand binding to the ß-adrenergic receptor. Proc.Natl.Acad.Sci. USA 84, 4384-4388

Strader, C. D., Candelore, M. R., Hill, W. S., Sigal, I. S., and Dixon, R. A. F. (1989) Identification of two serine residues involved in agonist activation of the ß-adrenergic receptor. J.Biol.Chem. 264, 13572-13578

Thomashow, M. F., Karlinsey, J. E., Marks,J.R., and Hurlbert, R.E. (1987) Identification of a new virulence locus in Agrobacterium tumefaciens that affects polysaccharide composition and plant cell attachment. J.Bacteriol. 169, 3209-3216

Thompson, D. V., Melchers, L. S., Idler, K. B., Schilperoort, R. A., and Hooykaas, P. J. J. (1988) Analysis of the complete nucleotide sequence of the Agrobacterium tumefaciens virB operon. Nucleic Acids Res. 16, 4621-4636

Toro, N., Datta, A., Carmi, O. A., Young, C., Prusti, R. K., and Nester, E. W. (1989) The Agrobacterium tumefaciens virC1 gene product binds to overdrive, a T-DNA transfer enhancer. J.Bacteriol. 171, 6845-6849

Van Haaren, M. J. J., Pronk, J. T., Schilperoort, R. A., and Hooykaas, P. J. J. (1987) Functional analysis of Agrobacterium tumefaciens octopine Ti-plasmid left and right T-region border fragments. Plant Mol.Biol. 8, 95-104

Van Haaren, M. J. J., Sedee, N. J. A., Schilperoort, R. A., and Hooykaas, P. J. J. (1987) Overdrive is a T-region transfer enhancer which stimulates T-strand production in Agrobacterium tumefaciens. Nucleic Acids Res. 15, 8983- 8997

Van Haaren, M. J. J., Sedee, N. J. A., Krul, M., Schilperoort, R. A., and Hooykaas, P. J. J. (1988) Function of heterologous and pseude border repeats in T region transfer via the octopine virulence system of Agrobacterium tumefaciens. Plant Mol.Biol. 11, 773-781

Van Veen, R. J. M., Dulk-Ras, H., Bisseling, T., Schilperoort, R. A., and Hooykaas, P. J. J. (1988) Grown gall tumor and root nodule formation by the bacterium Phyllobacterium myrsinacearum after the introduction of an Agrobacterium Ti plasmid or a Rhizobium Sym plasmid. Mol.Plant-Microbe Interactions 1, 321-234

Van Veen, R. J. M., Den Dulk-Ras, H., Schilperoort, R. A., and Hooykaas, P. J. J. (1989) Ti plasmid containing Rhizobium meliloti are non-tumorigenic on plants, despite proper virulence gene induction and T-strand formation. Arch. Microbiol. 153, 85-89

Wang, K., Herrera-Estrella, L., Van Montagu, M., and Zambryski, F. (1984) Right 25 bp terminus sequence of the nopaline T-DNA is essential for and determines direction of DNA transfer from Agrobacterium to the plant genome. Cell 38, 455-462

Wang, K., Stachel, S. E., Timmerman, B., Van Montagu, M., and Zambryski, P. C. (1987) Site-specific nick in the T-DNA border sequence as a result of Agrobacterium vir gene expression. Science 235, 587-591

Ward, E. R., and Barnes, W. M. (1988) VirD2 protein of Agrobacterium tumefaciens very tightly linked to the 5'end of T-strand DNA. Science 242, 927-930

Ward, J. E., Akiyoshi, D. E., Regier, D., Datta, A., Gordon, M. P., and Nester, E. W. (1988) Characterization of the virB operon from an Agrobacterium tumefaciens Ti plasmid. J. Biol.Chem. 263, 5804-5814

Weiss, V., and Magasanik, B. (1988) Phosphorylation of nitrogen regulator I (NRI) of Escherichia coli. Proc.Natl.Acad.Sci, USA 85, 8919-8923

Winans, S. C., Kerstetter, R. A., and Nester, E. W. (1988) Transcriptional regulation of the virA and virG genes of Agrobacterium tumefaciens. J.Bacteriol. 170, 4047-4054

Yadav, N. S., Vanderleyden, J., Bennet, D. R., Barnes, W. M., and Chilton, M. - D. (1982) Short direct repeats flank the T- DNA on a nopaline Ti plasmid. Proc.Natl.Acad.Sci.USA 79, 6322-6326

Young, C., and Nester, E. W. (1988) Association of the VirD2 protein with the 5'end of T strands in Agrobacterium tumefaciens. J.Bacteriol. 170, 3367-3374

Zorreguieta, A., Geremia, R. A., Cavaignac, S., Cangelosi, G. A., Nester, E. W., and Ugalde, R. A. (1988) Identification of the product of an Agrobacterium tumefaciens chromosomal virulence gene. Mol. Plant-Microbe Interactions 1, 121-127

# T-DNA GENE-FUNCTIONS

Csaba Koncz[1,2], Thomas Schmülling[1], Angelo Spena[1] and Jeff Schell[1]

1- Max-Planck-Institut für Züchtungsforschung, Abt. Genetische Grundlagen der Pflanzenzüchtung 5000 Köln 30, FRG

2- Institute of Genetics, Biological Research Center Hungarian Academy of Sciences, 6701 Szeged, Hungary

## INTRODUCTION

The products of genes located on T-DNA, or Transferable DNA segments, of Ti and Ri plasmids from plant-pathogenic *Agrobacteria* are the immediate cause of the abnormal growth known as "Crown-gall's" and "Hairy-roots". (For recent review see Weising et al., 1988.) "T-DNA" segments from Ti and Ri plasmids are transferred from the bacterial pathogen to the nucleus of plant cells by a mechanism which is remarkably analogous to a bacterial conjugation system (see Zambryski, 1989). T-DNA's carry a number of genes that are transcribed and translated in plant cells. Genetic studies have demonstrated that the products of these T-DNA genes are responsible for the abnormal growth patterns of plant cells in "Crown-galls" (Ti-T-DNA) and "Hairy roots" (Ri-T-DNA). Two general categories of T-DNA linked genes have thus far been characterized:

a) oncogenes: the products of these genes are directly involved in causing the abnormal growth pattern and

b) opine synthase genes that code for enzymes involved in the synthesis, by transformed plant cells, of a number of organic compounds (opines) that can serve as C and N sources for the growth of free-living (or symbiotic?) *Agrobacteria* that are genetically endowed with the capacity to specifically catabolize defined opines.

In this short review we plan to discuss the function of some T-DNA linked genes involved in plant growth control that had not been previously understood.

## $T_L$-DNA-GENES CARRIED BY Ti PLASMIDS

It has been documented that the major mechanism responsible for the proliferation of largely undifferentiated cells in Crown-gall tumors is the production by the transformed cells of abnormal levels of two of the major plant growth hormones: auxins and cytokinins. (For recent review see Zambryski et al., 1989). The abnormal production of auxins was

*Plant Molecular Biology* 2, Edited by R.G. Herrmann and B. Larkins, Plenum Press, New York, 1991

shown to be the consequence of the activity of the T-DNA genes 1 (iaaM) and 2 (iaaH) (Inze et al., 1984; Schröder et al., 1984; Thomashow et al., 1984) which code respectively for a tryptophan 2-monooxygenase catalizing the formation of indol-3-acetamide from tryptophan and an indol-3-acetamide hydrolase, catalizing the conversion of indol-3-acetamide into indol-3-acetic acid. Similarly the abnormal production of cytokinin was shown to be controlled by the product of gene 4 (iptZ), which is an isopentenyl transferase which converts 5'AMP and isopentenylpyrophosphate into the active cytokinin isopentenyladenosine-5-monophosphate (Akiyoshi et al., 1984; Barry et al., 1984; Buchmann et al., 1985). One can readily demonstrate that the introduction in plant cells of a subsegment of T-DNA only carrying the genes iaaM, iaaH and ipt results in the formation of Crown-gall-like proliferations. Wild-type T-DNA's from many A. tumefaciens strains however carry some other genes in addition to these three essential oncogenes, such as gene 5 which is located to the left of genes iaaM and iaaH on the $T_L$-DNA genetic map and gene 6 which is located to the right of the iptZ gene. The elucidation of the function of gene 5 was not straightforward because mutant T-DNA's from which the function of gene 5 was eliminated by deletion or transposon insertion, are still capable of producing Crown-gall tumors essentially undistinguishable from W.T. Crown galls (Garfinkel et al., 1981; Leemans et al., 1982; Joos et al., 1983) and tobacco plants transgenic for gene 5 or for a chimeric gene, with the strong CaMV35S promoter driving its expression, exhibit a normal growth habit (i.e. no abnormal phenotype was observed that correlated with gene 5 expression). Two observations led to the elucidation of the function of this gene. 1) Koncz and Schell (1986) found that the natural promoter of gene 5 was activated by the presence of auxins and 2) transgenic tobacco plantlets that contain and express gene 5 are resistant to toxic levels of different exogeneously supplied auxins such as IAA, NAA and 2,4-D (our unpublished data). This phenotype was to some extent explained when it was found that gene 5 codes for an enzyme that catalizes the synthesis of indol-lactate (an auxin analogue). Apparently the synthesis of indol-lactate can counterbalance the toxic effect of exogeneously supplied auxins. In wild-type Crown-galls gene 5, induced by the presence of auxins synthesized by the products of the iaaM and iaaH genes, might protect the transformed cells from toxic effects resulting from the accumulation of auxins in Crown-gall tissue. In transgenic plants carrying a Gus gene driven by the promoter of gene 5, Gus (ß-glucuronidase) activity was observed primarily in phloem cells. The auxin induced function of the gene 5 promoter was shown to be mediated by a cis-regulatory element that was previously identified in auxin activated soybean and Arabidopsis genes. (5' CXAXCATCACAXXTXTGTCGGCXXC 3'). One would expect that an enzyme involved in the synthesis of an auxin analogue that can somehow block auxin-activity (in this case auxin-toxicity) would be detrimental to normal plant growth. However transgenic tobacco plants, overexpressing the gene 5 product and in which indol-lactate was detected, exhibit a normal growth phenotype except in the presence of toxic levels of auxins. This observation could be explained if one assumes that indol-lactate blocks an auxin-carrier or an auxin-receptor specifically involved in the movement or signal-transduction of exogeneously supplied auxins. This hypothesis is presently tested experimentally. The function of gene 6b appears to be similar and possibly reciprocal to that of gene 5. It was indeed observed (Spanier et al., 1989) that the activity of gene 6b reduces a concentration dependent activity of cytokinins. The molecular mechanism responsible for this effect has however not yet been elucidated.

## *rol* GENES CARRIED BY $T_L$-DNA of Ri PLASMIDS

*A. rhizogenes* is the causative agent of the hairy root disease, consisting of adventitious roots growing at the site of bacterial infection. While the underlying mechanisms to transform plant cells are apparently identical for *A. tumefaciens* and *A. rhizogenes*, the latter one harbors genes on its T-DNA that use a different mechanism to alter the developmental faith of transformed cells. At least three genes (*rol*A, B and C) of the $T_L$-DNA are individually capable of stimulating root formation in competent plant tissues. The detailed action of these genes is unknown, but they probably affect the auxin sensitivity of transformed cells.

The idea that the *rol*B gene can enhance the sensitivity to auxins of *rol*B transformed cells stems from the observation that some plant tissues (e.g. kalanchoe leaves, carrot disks) that do not react to exogeneously added auxin, react by root formation if they are transformed by an active *rol*B gene, which by itself does not induce a morphogenetic reaction (White et al., 1985; Estramareix et al., 1986; Spena et al., 1987; Capone et al., 1989). To further investigate the mode of action of this gene it was cloned in a binary plant vector cassette and delivered via *A. tumefaciens* to plant cells. Its coding region was brought under the transcriptional control of the strong 35S RNA promoter of the cauliflower mosaic virus ($P_{35S}$-*rol*B) to study the consequences of overexpression of the *rol*B gene. Tobacco calli transgenic for the chimeric $P_{35S}$-*rol*B gene display an increased sensitivity towards auxins in tissue culture. These calli are necrotic when grown on MS medium containing 0.2 mg/l kinetin and 0.6 mg/l NAA, a medium usually required to support normal callus growth. Lowering the NAA concentration to 0.1 mg/l led to the disappearance of necrosis in $P_{35S}$-*rol*-B transgenic calli. Comparison of growth of $P_{35S}$-*rol*B calli to normal tobacco calli on media with varying concentrations of several substances with auxin activity (NAA, IAA, IBA, 2,4-D) revealed a 5-10 fold difference in the sensitivity towards the various auxins tested. Also *rol*B transgenic arabidopsis calli display a higher sensitivity towards auxin in tissue culture. Tobacco plants regenerated from $P_{35S}$-*rol*B transgenic calli display a number of phenotypic traits that are reminiscent of auxin-mediated effects (e.g. leaf necrosis). Interestingly Northern blot analysis has shown that the *rol*B gene, even when its transcription is controlled by the 35S promoter, is expressed at a rather low level in transgenic plants. Overexpression of this gene is probably lethal for plant cells, one might therefore select for transformants with a low level of *rol*B expression due to position effects. Further support for the viewpoint that the *rol*B gene product somehow controls auxin sensitivity comes from electrophysiological studies on the effect of NAA on the transmembrane electrical potential (Em) difference of tobacco mesophyll protoplasts. The Em variations have been measured as a function of auxin concentration. The NAA concentration inducing the maximum hyperpolarization gives an estimate of the sensitivity to auxin. These investigations revealed that protoplasts from *rol*B transgenic plants are 1000-fold more sensitive than normal protoplasts (Maurel et al., 1989). However, the increased sensitivity of *rol*B transgenic tobacco tissue towards auxin was not observed when the sensitivity of *rol*B transgenic seeds towards auxin was compared to that of seeds from normal tobacco plants. In either case inhibition of germination and growth was obtained at similar auxin concentrations in the medium. These results indicate that the putative auxin signal transduction chain that is somehow activated by the *rol*B gene product, is not active or present at all developmental stages or in all tissues or cells.

In contrast tobacco calli containing and expressing the rolC gene under the control of the 35S promoter ($P_{35S}$-rolC) were less sensitive to auxin when compared to normal tobacco calli. Necrosis, as an indication of toxic auxin effects, was observed in these calli at auxin concentrations two to four fold higher than in control calli. Also plants regenerated from these calli display phenotypical traits that are indicative of a reduced auxin activity (e.g. reduced apical dominance). Seeds of $P_{35S}$-rolC transgenic plants were able to germinate and grow on media containing auxin concentrations toxic for control seeds. For example $P_{35S}$-rolC transgenic seedlings grow well on MS medium containing 5 μM NAA, an auxin concentration inhibiting growth of control seedlings. The opposite reactions of rolC transformed plant tissues towards auxin might explain why the combined expression of these genes in transgenic plants can in part reverse the altered growth characteristics caused by the expression of either these genes separately (Schmülling et al., 1988) and why they work synergistically to induce root formation (Spena et al., 1987).

## CONCLUSIONS

Ti and Ri plasmids of *Agrobacteria* have acquired, in their T-DNA's, a remarkable set of genes with which they modify plant cell growth and differentiation. In addition to genes involved in growth hormone synthesis, they also carry various genes involved in the modulation of hormone activity. It will be most fascinating to find out whether similar functions, involved in the modulation of plant growth hormone activity, are operative in non-transformed plant tissues.

## REFERENCES


Akiyoshi, D. E., Regier, D. A., Jen, G., and Gordon, M. P., 1985, Cloning and nucleotide sequence of the tzs gene from *Agrobacterium* strain T37, *Nucl. Acids Res.*, 13:2773-2788.

Barry, G. F., Rogers, D. A., Fraley, R. T., and Brand, L., 1984, Identification of a cloned cytokinin biosynthetic gene, *Proc. Natl. Acad. Sci. USA.*, 81:4776-4780.

Buchmann, I., Marner, F. J., Schröder, G., Waffenschmidt, S., and Schröder, J., 1985, Tumor genes in plants: T-DNA encoded cytokinin biosynthesis, *EMBO J.*, 4:853-859.

Capone, I., Cardarelli, M., Trovato, M., and Costantino, P., 1989, Upstream non-coding region which confers polar expression to Ri plasmid root inducing gene rolB, *Mol. Gen. Genet.*, 216:239-244.

Estramareix, C., Ratet, P., Boulanger, F., and Richaud, F., 1986, Multiple mutations in the transferred regions of the *Agrobacterium rhizogenes* root-inducing plasmids, *Plasmid*, 15:245-247.

Garfinkel, D. J., Simpson, R. B., Ream, L. W., White, F. F., Gordon, M. P., and Nester, E. W., 1981, Genetic analysis of Crown gall: fine structure map of the T-DNA by site-directed mutagenesis, *Cell*, 27:143-153.

Inze, D., Follin, A., Van Lijsebettens, M., Simoens, C., Genetello, C., Van Montagu, M., and Schell, J., 1984, Genetic analysis of the individual T-DNA genes of *Agrobacterium tumefaciens*; further

evidence that two genes are involved in indole-3-acetic acid synthesis, Mol. Gen. Genet., 194:265-274.
Joos, H., Inze, D., Caplan, A., Sormann, M., Van Montagu, M., and Schell, J., 1983, Genetic analysis of T-DNA transcripts in nopaline Crown galls, Cell, 32:1057-1067.
Koncz, C., and Schell, J., 1986, The promoter of $T_L$-DNA gene 5 controls the tissue-specific expression of chimeric genes carried by a novel type of Agrobacterium binary vector, Mol. Gen. Genet., 204:383-396.
Leemans, J., Deblaere, R., Willmitzer, L., De Greve, H., Hernalsteens, J.-P., Van Montagu, M., and Schell, J., 1982, Genetic identification of functions of TL-DNA transcripts in octopine Crown galls, EMBO J., 1:147-152.
Maurel, C., Spena, A., Barbier-Brygoo, H., Tempe, J., and Guern, J., 1989, Single genes from Agrobacterium rhizogenes increase the sensitivity of tobacco protoplasts to auxin, J. Cell. Biochem., Suppl. 13, CD, p 321.
Schmülling, T., Schell, J., and Spena, A., 1988, Single genes from Agrobacterium rhizogenes influence plant development, EMBO J., 7:2621-2629.
Schröder, G., Waffenschmidt, S., Weiler, E. W., and Schröder, J., 1984, The T-region of Ti plasmids codes for an enzyme synthesizing indole-3-acetic acid, Eur. J. Biochem., 138:387-391.
Spanier, K., Schell, J., and Schreier, P. H., 1989, A functional analysis of T-DNA gene 6b: the fine tuning of cytokinin effects on shoot development, Mol. Gen. Genet., 219:2o9-216.
Spena, A., Schmülling, T., and Schell, J. S., 1987, Independent and synergistic activity of rol A, B and C loci in stimulating abnormal growth in plants, EMBO J., 6:3891-3899.
Thomashow, L. S., Reeves, S., and Thomashow, W. F., 1984, Crown gall oncogenesis: evidence that a T-DNA gene from the Agrobacterium Ti plasmid pTiA6 encodes an enzyme that catalyzes synthesis of indoleacetic acid, Proc. Natl. Acad. Sci. USA, 81:5071-5075.
Weising, K., Schell, J., and G. Kahl, 1988, Foreign genes in plants: transfer, structure, expression, and applications, Annu. Rev. Genet., 22:421-77.
White, F. F., Taylor, B. H., Huffmann, G. A., Gordon, M. P., and Nester, E. W., 1985, Molecular and genetic analysis of the transferred DNA regions of the root-inducing plasmid of Agrobacterium rhizogenes, J. Bacteriol., 164:33-44.
Zambryski, P., 1989, Agrobacterium-plant cell DNA transfer, in: "Mobile DNA", D. E. Berg, and M. M. Howe, eds., American Society for Microbiology, Washington D. C., pp. 3o9-333.
Zambryski, P., Tempe, J., and Schell, J., 1989, Transfer and function of T-DNA genes from Agrobacterium Ti and Ri plasmids in plants, Cell 56:193-201.

# MORPHOGENETIC GENES IN THE T-DNA OF Ri PLASMIDS

P. Costantino, M. Cardarelli, I. Capone,
A. De Paolis, P. Filetici, M. Pomponi and
M. Trovato

Dip. Genetica e Biologia Molecolare, Università
"La Sapienza" and Centro Acidi Nucleici, CNR
P.le A. Moro 5, 00185 Roma, Italy

## INTRODUCTION

*Agrobacterium rhizogenes* is responsible for the hairy root syndrome of dicotyledonous plants which consists in an abundant proliferation of roots at the wounded site of bacterial infection (Elliot, 1951). Hairy roots contain a portion (T-DNA) of a large bacterial plasmid (Ri plasmid), which directs growth and differentiation of the transformed plant cells (Chilton et al., 1982; Spanò et al., 1982b; White et al., 1982; Willmitzer et al.,1982). Hairy roots grow very actively *in vitro* in the absence of hormones with a characteristic highly branched and plagiotropic pattern (David et al., 1984); they synthesize specific opines of which three major types have been so far identified, agropine, mannopine and cucumopine (Petit and Tempè., 1985), corresponding to different families of Ri plasmids (Costantino et al., 1981). Whole, fertile plants of different species have been regenerated from hairy root tissues in several laboratories (Chilton et al., 1982; Spanò et al., 1982a; Tepfer, 1984; Spanò et al., 1987).

## THE T-DNA OF Ri PLASMIDS

As mentioned above, three families of Ri plasmids, with distinctive restriction endonuclease maps (Costantino et al., 1981; Pomponi et al., 1983; Koplow et al., 1984), have been so far identified. As far as their T-DNA they show differences in that, while mannopine and cucumopine type Ri plasmids possess only one transferred region (Koplow et al., 1984; Combard et al., 1987), the agropine type harbours two indipendent T-DNAs, denominated TL and TR (De Paolis et al., 1985). Southern blot hybridizations (Filetici et al., 1987) as well as electron microscope heteroduplex analysis (Brevet and Tempè, 1988) indicates that two highly conserved regions exist in mannopine and cucumopine T-DNAs which are also present in the TL-DNA of the agropine type. Fig.1 represents the DNA region of this latter plasmid (pRi1855) encompassing both the TR and TL-DNA, with the

*Plant Molecular Biology 2*, Edited by R.G. Herrmann and
B. Larkins, Plenum Press, New York, 1991

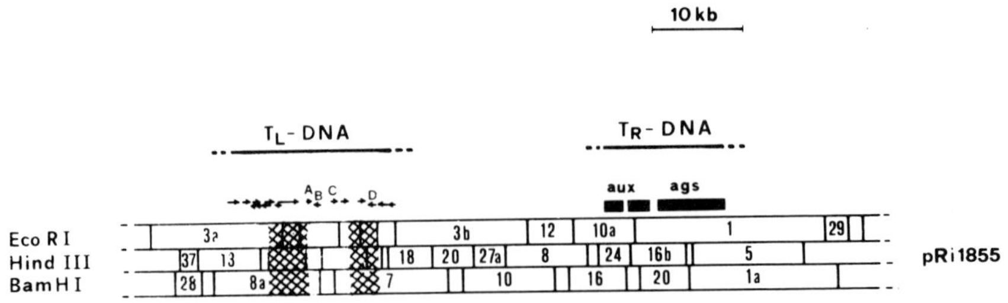


Fig.1. Restriction endonuclease map of the T-regions of pRi1855. Arrows below the TL-DNA represent open-reading frames. A, B, C, D: *rol* genes. *aux*, *ags*: auxin and agropine biosynthetic genes, respectively.

regions homologous to the other Ri plasmids' T-DNAs indicated by shaded areas; also shown are the genetic loci localized in the TR-region consisting in the biosynthetic genes for agropine (*ags*) and for the plant hormone auxin (*aux*). Opine biosynthesis functions have been localized in mannopine and cucumopine T-regions (Lahners et al., 1984 ; Brevet et al., 1988), while *aux* genes are not present on these plasmids (Cardarelli et al., 1985). Also shown in Fig.1 (arrows) are the genes identified on the TL-DNA on the basis of the complete nucleotide sequence (Slightom et al., 1986); four of these genes correspond to the four genetic loci (*rolA,B* and *C*) previously shown by transposon mutagenesis to affect hairy root growth and development (White et al., 1985).

## THE ROLE OF TL AND TR-DNA AND OF AUXIN

Since auxin is the plant hormone involved in root formation, it may seem at first sight obvious the presence of *aux* genes on the root inducing T-DNA of *A.rhizogenes*. However, mannopine and cucumopine T-DNAs lack auxin biosynthetic genes but induce hairy root symptoms in most hosts indistinguishable from those due to the agropine strains which harbour *aux* genes on their TR-DNA.
On carrot discs, however, a major difference in the virulence properties between $aux^+$ and $aux^-$ strains can be observed, as the former are capable to induce abundant rooting on both the apical (auxin rich) and the basal (auxin depleted) surface of the cut discs while the latter strains do not elicit any symptoms on the basal surface (Ryder et al., 1985; Cardarelli et al., 1985). The polar infectivity of mannopine and cucumopine strains could be ascribed to the lack of *aux* genes (Cardarelli et al., 1985) and by means of basal infections with recombinant *Agrobacterium* strains carrying different configurations of TR and TL-DNA and addition of exogenous auxin we could demonstrate the relative role of the TL and TR-DNA and of auxin in hairy root induction (Cardarelli et al., 1987a). Thus, the TR-DNA *aux* genes merely provide auxin, whenever endogenous auxin is lacking, which triggers root differentiation in plant cells whose response to the hormone is amplified by the presence of TL-DNA genes (Cardarelli et al., 1987a). Hairy root induction is therefore not the result of a gross auxin unbalance

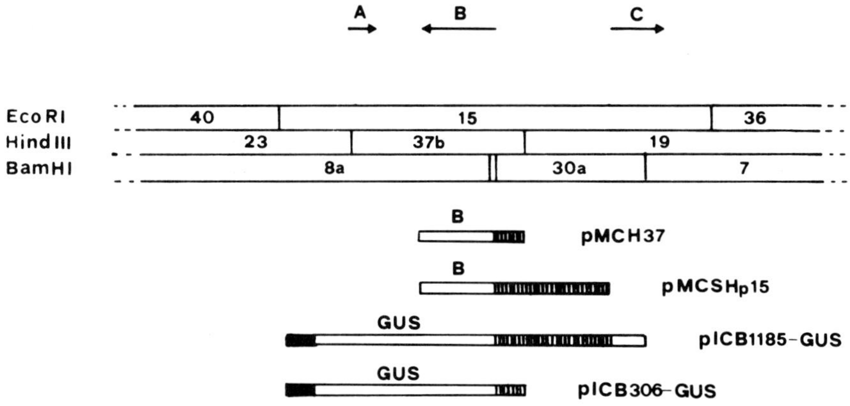


Fig.2. The segment of the TL-DNA of pRi1855 encompassing *rol* genes *A*, *B* and *C*. Bars below the map represent constructions of the *rolB* promoter cloned in Bin19 upstream the coding sequence of either *rolB* or the GUS reporter gene.

in the transformed cells but rather of the alteration, due to genes on the TL-DNA, of the capability of the cells to respond to the hormone. The much increased sensitivity to auxin conferred by Ri TL-DNA was subsequently demonstrated by comparing the levels of auxin needed to differentiate roots from leaf explants of normal vs. hairy root tobacco regenerants (Spanò et al., 1988) and, more quantitatively, by determining the dose/response curves for auxin on the elongation of apical root segments and on protoplast proton secretion and induction of transmembrane potential (Shen et al., 1988). Hairy root cells were shown to be two to three orders of magnitude more responsive to auxin than normal cells (Shen et al., 1988).
The TR-DNA plays thus a rather trivial and ancillary role in hairy root induction while gene(s) relevant to morphogenesis and hormonal response are present on the TL-DNA.

## TL-DNA GENES NECESSARY FOR (HAIRY) ROOT INDUCTION

The availability of the nucleotide sequence of the whole TL-DNA (Slightom et al., 1986) allowed cloning in plant vectors of different DNA segments containing single or groups of genes to be tested for their capability of directing root differentiation on various plant hosts. In our laboratory all of the TL-DNA genes normally found in hairy root tissues (Cardarelli et al., 1987b, Capone et al.,1989a) were cloned in the binary vector Bin19 (Bevan, 1984). The crucial region of the TL-DNA resulted to be the segment encompassing the *rol* genes *A*, *B* and *C*, reported in Fig.2. Abundant rooting was in fact elicited on all tested hosts by recombinant *Agrobacterium* strains harbouring fragment EcoRI 15 (Cardarelli et al., 1987; Capone et al., 1989a). An extensive analysis carried out by transforming carrot disc cells with each TL-DNA gene individually and in combination singled out *rolB* as the only gene capable alone to trig-

ger rooting, provided that an adequate auxin supply is available and that a long (1185 bp ) 5' upstream region is included in the *rolB* clone (Capone et al., 1989a) (Fig.2, construction pMCSHp15). Interestingly, the region corresponding to *rolA, B* and *C* is the less conserved in the Ri T-DNAs (see above); the conseved regions to the left (ORF 8) and to the right (ORF 13 and 14) of the *rol* genes enhance the effect of *rolB* and relieve the auxin requirement but are not sufficient *per se* for root induction. No diffusible compound is produced by genes in these regions since even when coinoculated from a different strain, these genes are normally found in the same roots as *rolB* (Capone et al., 1989a).
More recently, the DNA segments responsible for root induction have been identified and cloned from the T-DNAs of mannopine (pRi8196, M.D. Chilton, personal communication) and cucumopine (pRi2659, Failla et al., submitted) Ri plasmids. In both cases they fall within the central non conserved region of the T-DNAs. Curiously, among Ri plasmids the "core" rooting functions seem to have more widely diverged, at least at the nucleotide sequence level, than the apparently more accessory functions encoded by the strongly conserved flanking regions.

## GROWTH PROPERTIES OF ROOTS CONTAINING *rolB*

Roots derived from carrot disc infections with various TL-DNA constructions and from coinoculations with multiple strains were cultured and the growth properties described (Capone et al., 1989a). Compared to untransformed carrot roots which grow rather slowly and with a limited branching of lateral roots in the absence of plant hormones, roots containing *rolB* grow more rapidly and branched, in a fashion typical of hairy roots. The hairy root nature of *rolB* roots is further confirmed by their pronounced lack of geotropic response. Since induction of secondary meristems and suppression of geotropism are among the most noticeable effects of addition of auxin to untransformed carrot root cultures (Capone et al., 1989a), the growth pattern of *rolB* roots can be regarded as a manifestation of the abnormal auxin sensitivity of these cells. Recently, the much higher responsiveness to auxin of *rolB* vs. untransformed protoplasts has been convincingly demonstrated (Barbier-Brygoo et al., 1990).

## REGULATED EXPRESSION OF *rolB*

The presence of auxin and of a long non-coding region at the 5' of *rolB* (construction pMCSHp15 in Fig.2) are necessary for this gene to induce root differentiation: *rolB* constructions encompassing 306 bp of its 5', including the TATA and CCAAT boxes (construction pMCH37 in Fig.2) are ineffective in root induction on carrot discs (Capone et al., 1989a). The question was addressed of the possible presence of an upstream regulatory sequence and of the role of auxin in its activation. Initially, constructions with the GUS reporter gene placed under the control of the short (306 bp) and long (1185 bp) version of the *rolB* promoter (Fig.2, constructions pICB306-GUS and, respectively, pICB1185-GUS) were utilized in carrot disc infections. It was shown that the 1185 bp construction is capable to drive high GUS expression only on the auxin-rich (apical) side of the discs, while the 306 bp deletion only allows a

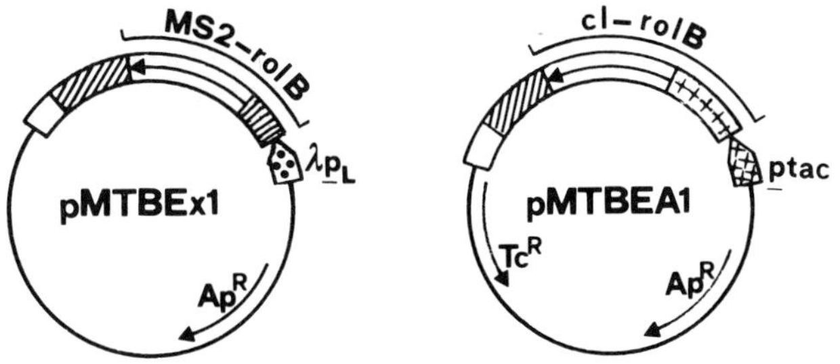


Fig.3. Plasmids utilized for expression in *E.coli* of *rolB* gene fusions. pMTBEx1: *rolB*-MS2 fusion; pMTBEA1: *rolB*-cI fusion.

low level of gene expression, undetectable by the histochemical staining analysis utilized in that work (Capone et al., 1989b). It was thus suggested that auxin controls the level of *rolB* expression via upstream activating sequence(s) localized between position -306 and -1185 (Capone et al., 1989b). Subsequently, the GUS gene was placed under the control of several deleted fragments of the *rolB* promoter in order to localize the regulatory regions and their responsiveness to auxin. Carrot disc infections and the fluorimetric and histochemical analysis of transgenic tobacco plants allowed the identification of different regulatory domains responsible for the level and tissue specificity of expression of *rolB*. A CCAAT box-proximal upstream DNA segment is responsible for *rolB* expression in the vascular tissue (phloem) of the aerial organs of the plant while a more distal element controls expression in the (root) meristems (Capone et al., submitted). Auxin regulation of the *rolB* promoter was also demonstrated by culturing protolplasts from transgenic plants under various hormonal conditions (ibidem). Binding to these DNA regions of distict protein regulatory factors was also demonstrated (Filetici et al., in preparation).

## ANTIBODIES AGAINST RolB

So far nothing is known concerning the biochemical role of the *rolB* gene product in determining root differentiation. At this stage, given the profound effect of the still elusive RolB protein on the responsiveness of transformed cells to auxin, it is tempting to suggest a role for it in auxin signal reception/transduction/amplification mechanisms.
Antibodies specific against the *rolB* gene product represent a major tool in the clarification of its function.
Recently anti-RolB antibodies were raised by means of expression of gene fusions in *E.coli* (Trovato et al., 1990). About 90% of the *rolB* coding sequence from the C-terminus was fused in frame with either the first 297 bp of the gene encoding the DNA polymerase of phage MS2, under the control of phage lambda *p*L promoter (plasmid pMTBEx1 in Fig.3) or to the first 480 bp of the gene encoding the lambda repressor, under the control of

the *p*tac promoter (plasmid pMTBEA1). Antiserum specific against the fusion proteins were raised by injecting rabbits with these latter. The antisera recognise specifically the unfused, complete RolB protein obtained by *in vitro* translation of a transcript derived by *in vitro* transcription of *rolB* (Trovato et al., 1990).

## CONCLUSIONS

Achieving knowledge and control of growth, development and morphogenesis of higher plants is among the most formidable challenges of modern biology and major goals for the agriculture of the future. The *rol* genes of Ri plasmids affect deeply growth and morphogenetic potential of plant cells: they represent thus privileged "windows" through which a unique possibility is offered to investigate on these domains of plant biology and to elaborate powerful tools to manipulate on them.

As described in the previous paragraphs, expression of *rolB*, which is dependent upon the presence of auxin, determines in plant cells a state of competence to differentiate roots, manifested by an enormous increase of responsiveness to auxin. Clarifying the biochemical function of the RolB protein in altering the response to this hormone and the interaction of *rolB* with endogenous plant genes will be of paramount relevance in shedding light on the molecular mechanisms of hormonal response in higher plants.

## REFERENCES


Barbier-Brygoo H, Maurel C, Shen WH, Ephritikhine G, Delbarre A, Guern J, 1990, Use of mutants and transformed plants to study the action of auxins, in: "Hormone perception and signal transduction in animals and plants", JA Roberts, C Kirk, M Venis, eds, Company of Biologists, Cambridge

Bevan M, 1984, Binary *Agrobacterium* vectors for plant transformation, Nucleic Acids Res, 12:8711-8721

Brevet J, Tempè J, 1988, Homology mapping of T-DNA regions of three *Agrobacterium rhizogenes* Ri plasmids by electron microscope eteroduplex studies, Plasmid, 19:75-83

Brevet J, Borowski D, Tempè J, 1988, Identification of the region encoding opine synthesis and of a region involved in hairy root induction on the T-DNA of cucumber-type Ri plasmid, Mol Plant-Micr Interact, 1:75-79

Capone I, Spanò L, Cardarelli M, Bellincampi D, Petit A, Costantino P, 1989a, Induction and growth properties of carrot roots with different complements of *Agrobacterium rhizogenes* T-DNA, Plant Mol Biol, 13:43-52

Capone I, Cardarelli M, Trovato M, Costantino P, 1989b, Upstream non-coding region which confers polar expression to Ri plasmid root inducing gene *rolB*, Mol Gen Genet, 216:239-244

Cardarelli M, Spanò L, De Paolis A, Mauro ML, Vitali G, Costantino P, Identification of the genetic locus responsible for non-polar root induction by *Agrobacterium rhizogenes* 1855, Plant Mol Biol, 5:385-391

Cardarelli M, Spanò L, Mariotti D, Mauro ML, Costantino P,

1987a, The role of auxin in hairy root induction, Mol Gen Genet, 208:457-463
Cardarelli M, Mariotti D, Pomponi M, Spanò L, Capone I, Costantino P,1987b, *Agrobacterium rhizogenes* T-DNA genes capable of inducing hairy root phenotype, Mol Gen Genet, 209.475-480
Chilton MD, Tepfer DA, Petit A, Casse-Delbart F, Tempé J, 1982, *Agrobacterium rhizogenes* inserts T-DNA into the genome of host plant root cells, Nature, 295.432-434
Combard A, Brevet J, Borowski D, Cam K, Tempé J, Physical map of the T-DNA region of *Agrobacterium rhizogenes* strain NCPPB 2659, 1987, Plasmid 18:70-75
Costantino P, Mauro ML, Micheli G, Risuleo G, Hooykaas PJJ, Schilperoort RA, 1981, Fingerprinting and sequence homology of plasmids from different virulent strains of *Agrobacterium rhizogenes*, Plasmid, 5:170-182
David C, Chilton MD, Tempé J, 1984, Conservation of T-DNA in plants regenerated from hairy root cultures, Biotechnology, 2:73-76
De Paolis A, Mauro ML, Pomponi M, Cardarelli M, Spanò L, Costantino P, 1985, Localization of agropine synthesizing functions in the TR-region of the root inducing plasmid of *Agrobacterium rhizogenes* 1855, Plasmid, 13:1-7
Filetici P, Spanò L, Costantino P, 1987, Conserved regions in the T-DNA of different *Agrobacterium rhizogenes* root-inducing plasmids, Plant Mol Biol, 9:19-26
Koplow J, Byrne MC, Jen G, Tempè J, Cilton MD, 1984, Physical map of the *Agrobacterium rhizogenes* strain 8196 virulence plasmid, Plasmid, 11:17-27
Lahners K, Byrne MC, Chilton MD, 1984, T-DNA fragments of hairy root plasmid pRi8196 are distantly related to octopine and nopaline Ti plasmid T-DNA, Plasmid, 11:130-140
Petit A, Tempé J, The function of T-DNA in nature, 1985, in: "Molecular form and function of the plant genome", L Van Vloten-Doting, G Groot, T Hall, eds, Plenum Press, New York
Pomponi M, Spanò L, Sabbadini MG, Costantino P, 1983, Restriction endonuclease mapping of the root-inducing plasmid of *Agrobacterium rhizogenes* 1855, Plasmid, 10:119-129
Shen WH, Petit A, Guern J, Tempé J, 1988, Hairy roots are more sensitive to auxin than normal roots, Proc Natl Acad Sci USA, 85:3417-3421
Slightom JL, Durand-Tardif M, Jouanin L, Tepfer D, 1986, Nucleotide sequence analysis of *Agrobacterium rhizogenes* agropine type plasmid: identification of open-reading frames, J Biol Chem, 261:108-121
Spanò L, Costantino P, 1982a, Regeneration of plants from callus cultures of roots induced by *Agrobacterium rhizogenes* on tobacco, Z Pflanzenphysiol, 106.87-92
Spanò L, Pomponi M, Costantino P, Van Slogteren GMS, Tempè J, 1982b, dentification of T-DNA in the root inducing plasmid of the agropine-type *Agrobacterium rhizogenes* 1855, Plant Mol Biol, 1:291-300
Spanò L, Mariotti D, Pezzotti M, Damiani F, Arcioni S, 1987, hairy root transformation in alfalfa (*Medicago sativa L.*), Theor Appl Genet, 73:523-530
Spanò L, Mariotti D, Cardarelli M, Branca C, Costantino P, 1988, Morphogenesis and auxin sensitivity of transgenic tobacco with different complements of Ri T-DNA, Plant Physiol, 87:479-483

Tepfer D, 1984, Transformation of several species of higher plants by *Agrobacterium rhizogenes*: sexual transmission of the transformed genotype and phenotype, Cell, 37:959-967

Trovato M, Cianfriglia M, Filetici P, Mauro ML, Costantino P, 1990, Expression of *Agrobacterium rhizogenes rolB* gene fusions in *Escherichia coli*: Production of antibodies against the RolB protein, Gene, 87:139-143

White FF, Ghidossi G, Gordon MP, Nester EW, 1982, Tumor induction by *Agrobacterium rhizogenes* involves the trnsfer of plasmid DNA to the plant genome, Proc Natl Acad Sci USA, 79:3193-3197

White FF, Taylor BH, Huffmann GA, Gordon MP, Nester EW, 1985, Molecular and genetic analysis of the transferred DNA regions of the root-inducing plasmid of *Agrobacterium rhizogenes*, J Bacteriol, 164:33-44

Willmitzer L, Sanchez-Serrano J, Buschfeld E, Schell J, 1982, DNA from *Agrobacterium rhizogenes* is transferred to and expressed in axenic hairy root plant tissue, Mol Gen Genet, 186:16-22

# TRANSIENT EXPRESSION AND STABLE TRANSFORMATION OF MAIZE USING MICROPROJECTILES

Michael Fromm, Theodore M. Klein[1], Stephen A. Goff, Brad Roth, Fionnuala Morrish, and Charles Armstrong[2]

USDA/U.C. Berkeley, Plant Gene Expression Center, Albany, California 94710; [1]E.I. Du Pont De Nemours & Co, Medical Products Dept., Glasgow Site, Newark, Delaware, 19714; [2]Monsanto Co., 700 Chesterfield Village Pkwy, St. Louis, Missouri, 63198 USA

## Introduction

High-velocity microprojectiles provide a direct means of introducing DNA into plant cells (Klein *et al.*, 1987, 1988a, 1988b; Wang *et al.*, 1988). In principle, this method could be used to introduce DNA into any accessible plant tissue, although the majority of DNA transfer seems to occur in the exposed cells on the surface of the tissue. This article describes progress in transient gene expression in intact plant tissues and in obtaining stably transformed maize plants.

Although there are a number of gene transfer techniques available (reviewed in Davey *et al.*, 1989; Potrykus, 1989), high-velocity microprojectiles seem uniquely suited for performing transient gene expression studies in intact plant tissues. The ideal transient assay technique would efficiently deliver DNA into a large number of the cells in a tissue. Additionally, the transferred gene should show the same physiological, tissue-specific, and genetic regulation of expression as that of the equivalent endogenous gene in the cells' chromosomes.

The two most common DNA delivery techniques, *Agrobacterium*-mediated gene transfer and protoplast gene transfer techniques, fail to meet the ideal transient assay conditions. *Agrobacterium*-mediated gene transfer (Hooykaas, 1989) is host-specific, of unknown efficiency for whole tissues, and the analysis of transient gene expression is complicated by expression of the gene in *Agrobacterium* itself, which persists in the tissue. (The use of a reporter gene with introns in the reporter gene coding region should solve the latter problem). Direct gene transfer techniques that require the use of protoplasts (Fromm *et al.*, 1985, 1986; Potrykus, 1989) suffer from the fact that the isolation of the protoplasts from the plant tissue is a dramatic change in the cells' physiology and thus is often inappropriate for the experimental objective. Therefore, microprojectiles are unique in their ability to deliver free DNA to a large number of cells in an intact plant tissue.

There have been a number of experiments performed to ascertain how accurately the transient expression of the introduced gene reflects that of the endogenous genes of the plant tissue. The most detailed studies have used the known tissue-specificity and different regulatory genotypes of the maize anthocyanin biosynthetic pathway to evaluate this. One of the maize anthocyanin structural genes that has been cloned and sequenced is the *Bz1* gene, encoding the enzyme UDP glucose:flavonol 3-O-glucosyl transferase (Furtek *et al.*, 1988). In the aleurone of the maize seed, the expression of anthocyanins is known to be regulated by two genes, *C1* and *R*. Additionally, the appearance of the **Bz1** enzyme and mRNA has been shown to be dependent on the presence of functional *C1* and *R* genes (Dooner and Nelson 1979; Cone *et al.*, 1986). *C1* (Cone *et al.*, 1986; Paz-Ares *et al.*, 1987) and *R* (Ludwig *et al.*, 1989) also have been cloned and sequenced, and encode proteins with homology to the *myb* and *myc* families of transcription factors.

A chimeric construct pBzL, which uses the *Bz1* promoter to express the firefly luciferase coding region, was introduced into maize aleurones using microprojectiles. The *Bz1* promoter was only expressed in *C1, R* aleurones: there was 100 to 200 fold less expression in aleurones mutant for either *C1* or *R* (Klein *et al.*, 1989b). Additionally, there was little expression in *C1, R* embryos (Klein *et al.*, 1989b), as expected from genetic studies. The high level of expression in the aleurones and low levels of expression in the embryos of the same genotype demonstrates tissue-specific expression of the introduced *Bz1* gene. Similar studies using microprojectiles have demonstrated red/far red light control of a phytochrome gene (Bruce *et al.*, 1989), and pollen-specific transient expression of pollen specific genes of tomato (Twell *et al.*, 1989). These studies indicate that genes introduced into the cells of intact plant tissues by microprojectiles are properly regulated. Thus, many studies of promoter structure and function can be rapidly performed using transient assays. We have performed a mutation analysis of the *Bz1* promoter in maize aleurones and identified a 38 bp region that responds to *C1* and *R* control (Brad Roth and Michael Fromm, unpublished data). Transient gene expression studies can also be used to test the function of regulatory genes. We have shown that an *r* mutation in maize aleurones can be complemented by transient expression of an introduced *B* regulatory gene (Goff *et al.*, 1990). The *B* gene is essentially a duplicate gene of *R* (Chandler *et al.*, 1989). Another group has shown that an *R* gene can complement *r* tissues (Ludwig *et al.*, 1990). In a separate report we showed that a *C1* cDNA could be expressed and complement *c1* aleurones. Further, we found that this transient assay system could be used to dissect the functional domains of the **C1** protein (Goff *et al.*, manuscript in preparation).

In addition to being useful for transient gene expression assays, microprojectiles are also useful for stably transforming plant cells. Microprojectile gene transfer has been used to obtain stably transformed maize cells (Klein *et al.*, 1989a), transgenic tobacco plants (Klein *et al.*, 1988c; Tomes *et al.*, 1990) and transgenic soybean plants (McCabe *et al.*, 1989). However, unlike the case for transient assays, microprojectiles do not necessarily have clear advantages over other gene transfer techniques for stable transformation. Indeed, given the low frequency of stable transformation obtained with microprojectiles, both *Agrobacterium*-mediated transformation and protoplast related direct gene transfer techniques are often more efficient (Potrykus, 1989; Davey *et al.*, 1989). The need for microprojectile gene transfer arises in the cases where the plant species is not susceptible to *Agrobacterium*-mediated transformation and is not easily regenerated from protoplasts. This is the case for maize, where despite indications that *Agrobacterium* can infect maize (Grimsley *et al.*, 1987, 1989) and other cereals (Marks *et al.*, 1989), it has not been possible to obtain stably transformed plants with this technique. Likewise, although maize protoplasts have been regenerated into fertile plants recently (Rhodes, 1988; Shillito *et al.*, 1989; Prioli *et al.*, 1989), fertile transgenic maize plants have not been reported. Therefore, for maize, microprojectiles are very useful since they allow genes to be transferred into cells that are easily regenerated, as described below.

## Stable Transformation of Embryogenic Maize Cells

### Structure of the Marker Plasmid pALSLUC

The selectable marker plasmid, pALSLUC (Figure 2), uses the CaMV 35S promoter to express a chlorsulfuron-resistant form of the maize acetolactate synthase (ALS) cDNA (M. Fromm, in preparation). Chlorsulfuron (CS) inhibits ALS thus preventing the synthesis of the branched chain amino acids leucine, isoleucine and valine (Chaleff, 1984, 1987).

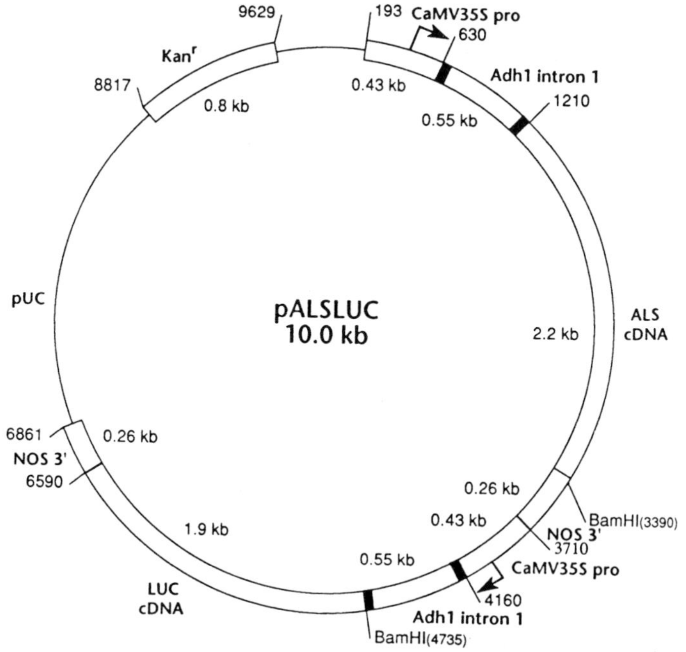


Fig. 1. Structure of pALSLUC. Symbols:CaMV 35S pro, Cauliflower mosaic virus 35S promoter; Adh1 intron 1, maize alcohol dehydrogenase 1 intron 1; ALS cDNA, chlorsulfuron-resistant form of the acetolactate synthase cDNA; NOS 3', polyadenylation region of the nopaline synthase gene; LUC cDNA, cDNA of the firefly luciferase coding region; $Kan^r$, kanamycin resistance marker for selection in *E. coli*.

pALSLUC also contains a chimeric gene that uses the CaMV 35S promoter to express the firefly luciferase coding region (De Wet *et al.*, 1987). Luciferase expression provides a convenient assay (De Wet *et al.*, 1987; Callis *et al.*, 1987) for determining if cells are transformed.

## Gene Transfer Using Microprojectiles

The target cells for the microprojectile bombardment were maize suspension culture cells (B73 X A188 genotype; Hodges, 1986) that are capable of regenerating into fertile maize plants. Maize embryogenic suspension cultures were bombarded with tungsten microprojectiles coated (Klein *et al.*, 1989a) with pALSLUC DNA. After 12 days in non-selective liquid media, the bombarded suspension culture cells were plated on agarose solidified N6 media (Chu *et al.*, 1975) containing 50 nM chlorsulfuron. 6 to 8 weeks after bombardment, the growth of chlorsulfuron-resistant calli could be distinguished from the limited growth of the non-transformed cells.

## Luciferase Expression in the Transgenic Callus and Plants

A total of 8 bombardments in two different experiments produced three independently derived transformed calli that expressed luciferase. A portion of each callus was subjected to standard regeneration protocols (Kamo *et al.*, 1985; Armstrong and Green, 1985; Hodges *et al.*, 1986). The first of these calli could not be regenerated into plants. However, the other two transformed calli contained somatic embryos that could be regenerated into plantlets that developed into mature plants in the greenhouse. These plants were tested for the presence of luciferase. Additionally, DNA was isolated from many of the plants for PCR

**Table 1**

*Luciferase Expression in the R1 Progeny of R0 Transgenic Maize Plants*

| Parent | Luciferase Expression[a] | Total |
|---|---|---|
| pALSLUC-3-CC | 3+; 9- | 25 % |
| pALSLUC-3-CE | 7+; 2- | 78 % |
| pALSLUC-3-CG | 11+; 18- | 38 % |
| pALSLUC-3-CH | 9+; 5- | 64 % |
| TOTAL | 30+; 34- | 47 % |

[a]Luciferase expression was in light units (l.u.) as measured using a luminometer. Positive values ranged from 5,000 to 760,000 l.u., while negative values were between 100 and 200 l.u.

analysis (Lassner *et al.*, 1989). The two PCR primers were specific for the CaMV 35S promoter and Adh1 intron portion of the expression cassettes. The second transformed callus produced plants that did not contain luciferase enzyme activities (data not shown). This callus evidently consisted of chimeras of transformed and untransformed cells. Apparently, there was preferential regeneration of the non-transformed cells in the chimeric embryogenic tissue.

The third callus produced 15 luciferase positive plants after regeneration (data not shown). Southern blots of BamHI digested DNA from these plants showed a constant ratio of the bands from the endogenous ALS gene to those from the introduced ALS gene from pALSLUC (data not shown), indicating they consisted of entirely transformed cells. All of these plants produced ears and partially feminized tassels. The tassels contained mature anthers but did not shed noticeable pollen. However, the ovules in the tassel and ear produced seeds when fertilized with pollen from a non-transformed plant.

Sixteen day old embryos from immature seeds were dissected out of the kernels, germinated *in vitro*, and the plantlets transferred back to the greenhouse. These seedling progeny were examined for luciferase expression (Table 1). Assays of the leafs showed luciferase expression in 47 % of the progeny as would be expected from seeds fertilized with pollen from a non-transformed plant. Luciferase positive plants contained pALSLUC DNA as determined by PCR analysis, while luciferase negative plants did not contain pALSLUC DNA (data not shown).

## Discussion

We have found that high-velocity microprojectiles have reproducibly transformed embryogenic maize cells. The chlorsulfuron marker gene provides a convenient selection for recovering the stable transformation events. The selected cells retained their embryogenic qualities as all three transformed calli still formed globular embryos, and two of the three calli could be regenerated to plants. However, one of these gave rise to non-transformed plants, indicating the original callus consisted of transformed and non-transformed cells. Only one of three transformed calli gave rise to multiple fertile plants, that when fertilized with non-transgenic pollen, showed about 1:1 segregation of luciferase expression and the transforming DNA in its progeny. The transferred genes were stably maintained and expressed in the progeny as has been found in other plant transformation systems (Christou,

1989; Shimamoto, 1989; Hooykaas, 1989). The availability of a reproducible, general system for producing transgenic cereal plants will greatly facilitate the analysis and manipulation of gene expression in cereals.


## Acknowledgments

We thank Rosalind Williams and Andrew Watson for technical assistance in this work. A Biolistics (Du Pont, DE) high-velocity microprojectile gun was used in this research. The use of this product does not constitute an endorsement or advertisement by the USDA. This work was supported by the US Department of Agriculture-Agricultural Research Service.


## References


Armstrong, C. L. and Green, C. E., 1985, Establishment and maintenance of friable, embryogenic maize callus and the involvement of L-proline. *Planta*, **164**:207-214.

Bruce, W. B., Christensen, A. H., Klein, T., Fromm, M., and Quail, P., 1989, Photoregulation of a phytochrome gene promoter from oat transferred into rice by particle bombardment, *Proc. Natl. Acad. Sci. USA.*, **86**:9692-9696.

Callis, J., Fromm, M., and Walbot, V., 1987, Introns increase gene expression in cultured maize cells, *Genes and Develop.*, **1**:1183-1200.

Chaleff, R. S., and Mauvais, C. J., 1984, Acetolactate synthase is the site of action of two sulfonylurea herbicides in higher plants, *Science*, **224**:1443-1445

Chaleff, R. S., and Bascomb, N. F., 1987, Genetic and biochemical evidence for multiple forms of acetolactate synthase in *Nicotiana tabacum, Molec. Gen. Genet.*, **210**:33-38.

Chandler, V. L., Radicella, J. P., Robbins, T. P., Chen, J., and Turks, D., 1989, Two regulatory genes of the maize anthocyanin pathway are homologous: Isolation of *B* utilizing *R* genomic sequences, *Plant Cell*, **1**:1175-1183.

Christou, P., Swain, W. F., Yang, N-S., McCabe, D. E., 1989, Inheritance and expression of foreign genes in transgenic soybean plants, *Proc. Natl. Acad. Sci. USA*, **86**:7500-7504.

Chu, C. C., Wang, C. C., Sun, C. S., Hsu, C., Yin, K. C., Chu, C. Y., and Bi, F. Y., 1975, Establishment of an efficient medium for anther culture of rice through comparative experiments on the nitrogen sources, *Sci. Sin. Peking*, **18**:659-688.

Cone, K. C., Burr, F. A., and Burr, B., 1986, Molecular analysis of the maize anthocyanin regulatory locus *C1, Proc. Natl. Acad. Sci. USA*, **83**:9631-9635.

Davey, M. R., Rech, E. L., and Mulligan, B. J., 1989, Direct DNA transfer to plant cells, *Plant Mol. Biol*, **13**:273-285.

De Wet, J. R., Wood, K. V., Deluca, M., and Helinski, D. R., 1987, Firefly luciferase gene: structure and expression in mammalian cells, *Molec. Cell. Biol*, **7**:725-737.

Dooner, H. K., and Nelson, O. E., 1979, Interaction among *C, R* and *Vp* in the control of the *Bz* glucosyltransferase during endosperm development in maize, *Genetics*, **91**:309-315.

Fromm, M., Taylor, L. P., and Walbot, V., 1985, Expression of genes transferred into monocot and dicot plant cells by electroporation, *Proc. Natl. Acad. Sci. USA*, 82:5824-5828.

Fromm, M. E., Taylor, L. P., and Walbot, V., 1986, Stable transformation of maize after gene transfer by electroporation, *Nature*, **319**:791-793.

Furtek, D., Schiefelbein, J. W., Johnston, F., and Nelson, O. E., 1988, Sequence comparisons of three wild-type *Bronze-1* alleles from *Zea mays, Plant Mol. Biol*, **11**:473-481.

Goff, S. A., Klein, T. M., Roth, B. A., Fromm, M. E., Cone, K. C., Radicella, J. P., and Chandler, V. L., 1990, Transactivation of anthocyanin biosynthetic genes following transfer of *B* regulatory genes into maize tissues, *EMBO J.*, in press

Grimsley, N., Hohn, T., Davies, J. W., and Hohn, B., 1987, *Agrobacterium*-mediated delivery of infectious maize streak virus into maize plants, *Nature*, **325**:177-179.

Grimsley, N., Hohn, B., Ramos, C., Kado, C., and Rogowsky, P., 1989, DNA transfer from *Agrobacterium* to *Zea mays* or *Brassica* by agroinfection is dependent on bacterial virulence functions, *Mol. Gen. Genet.*, **217**:309-316.

Hodges, T. K., Kamo, K. K., Imbrie, C. W., and Becwar, M. R., 1986, Genotype specificity of somatic embryogenesis and regeneration in maize, *Biotechnology*, **4**:219-223.
Hooykaas, P. J. J., 1989, Transformation of plant cells via *Agrobacterium, Plant Mol. Biol.*, **13**:327-336.
Kamo, K. K., Becwar, M. R., and Hodges, T. K, 1985, Regeneration of Zea Mays L. from embryogenic callus, *Bot. Gaz.*, **146**:327-334.
Klein, T. M., Wolf, E. D., Wu, R., and Sanford, J. C., 1987, High-velocity microprojectiles for delivering nucleic acids into living cells, *Nature*, **327**:70-73.
Klein, T. M., Fromm, M. E., Weissinger, A., Tomes, D., Schaaf, S., Sletten, M., and Sanford, J. C., 1988a, Transfer of foreign genes into intact maize cells with high-velocity microprojectiles, *Proc. Natl Acad. Sci. USA*, **85**: 4305-4309.
Klein, T. M., Gradziel, T., Fromm, M. E., and Sanford, J. C., 1988b, Factors influencing gene delivery into *Zea mays* cells by high-velocity microprojectiles, *Bio/Technology*, **6**:559-563.
Klein, T. M., Harper, E. C., Svab, Z., Sanford, J. C., Fromm, M. E., and Maliga, P., 1988c, Stable genetic transformation of intact *Nicotiana* cells by the particle bombardment process, *Proc. Natl. Acad. Sci. USA*, **85**: 8502-8505.
Klein, T. M., Kornstein, L., Sanford, J. C., and Fromm, M. E., 1989a, Genetic transformation of maize cells by particle bombardment, *Plant Physiol*, **91**:440-444.
Klein, T. M., Roth, B. A., and Fromm, M. E., 1989b, Regulation of anthocyanin biosynthetic genes introduced into intact maize tissues by microprojectiles, *Proc. Natl. Acad. Sci. USA*, **86**:6681-6685.
Lassner, M. W., Peterson, P. and Yoder, J. I., 1989, Simultaneous amplification of multiple DNA fragments by polymerase chain reaction in the analysis of transgenic plants and their progeny, *Plant Mol. Biol. Rep.*, **7**:116-128.
Ludwig, S. R., Habera, L. F., Dellaporta, S. L., and Wessler, S. R., 1989, Lc, a member of the maize R gene family responsible for tissue-specific anthocyanin production, encodes a protein similar to transcription activators and contains the myc-homology region, *Proc. Natl. Acad. Sci. USA*, **86**:7092-7096
Ludwig, S. R., Bowen, B., Beach, L., and Wessler, S., 1990, A regulatory gene as a novel visible marker for maize transformation, *Science*, **247**:449-450
Marks, M. S., Kemp, J. M., Woolston, C. J., and Dale, P. J., 1989, Agroinfection of wheat: A comparison of Agrobacterium strains, *Plant Sci.*, **63**:247-256.
McCabe, D. E., Swain, W. F., Marinell, B. J., and Christou, P., 1988, Stable transformation of soybean (Glycine Max) by particle acceleration, *Bio/Technology*, **6**:923-926.
Paz-Ares, J., Ghosal, D., Wienand, U., Peterson, P. A., and Saedler, H., 1987, The regulatory c1 locus of Zea mays encodes a protein with homology to *myb* proto-oncogene products and with structural similarities to transcriptional activators, *Embo J.*, **6**:3553-3558.
Potrykus, I., 1989, Gene transfer to cereals: An Assessment, *Trends Biotechnol.*, **7**:269-272.
Prioli, L. M., and Sondahl, M. R., 1989, Plant regeneration and recovery of fertile plants from protoplasts of maize Zea mays L., *Bio/technology*, **7**:589-594.
Rhodes, C. A., Lowe, K. S., and Ruby, K. L., 1988, Plant regeneration from protoplasts isolated from embryogenic maize cell cultures, *Bio/Technology*, **6**:56-60.
Shillito, R. D., Carswell, G. K., Johnson, C. M., DiMaio, J. J., and Harms, C. T., 1989, Regeneration of fertile plants from protoplasts of elite inbred maize, *Bio/technology*, **7**:581-587.
Shimamoto, K., Terada, R., Izawa, T., and Fujimoto, H, 1989, Fertile transgenic rice plants regenerated from transformed protoplasts, *Nature*, **338**:274-276.
Tomes, D., Weissinger, A. K., Ross, M., Higgins, R., Drimmond, B. J., Schaaf, S., Malone-Schoneberg, J., Staebell, M., Flynn, P., Anderson, J., and Howard, J., 1990, *Plant Molec. Biol.*, **14**:261-268.
Twell, D., Klein, T. M., Fromm, M. E., and McCormick, S., 1989, Transient expression of chimeric genes delivered into pollen by microprojectile bombardment, *Plant Physiol.*, **91**:1270-1274.
Wang, Y. C., Klein, T. M., Fromm, M., Cao, J., Sanford, J. C., and Wu, R., 1988, Transformation of rice, wheat and soybean by the particle bombardment method, *Plant Mol. Biol.*, **11**:433-439.

# AGROINFECTION AS A TOOL FOR THE INVESTIGATION OF PLANT-PATHOGEN INTERACTIONS

Nigel Grimsley*, Elke Jarchow, Juerg Oetiker** Michael Schlaeppi and Barbara Hohn

Friedrich Miescher-Institut, PO Box 2543, CH-4002 Basel, Switzerland. *Laboratoire de Biologie Moleculaire, Auzeville BP 27, 31326 Castanet-Tolosan, France. **NIAR, Department of Molecular Biology, 2-1-2 Kunnondai, Tsukuba Ibaraki 305, Japan

## INTRODUCTION

The fascinating way by which *Agrobacterium* invades a host plant has been the subject of intensive investigations by various laboratories, and has opened new avenues of research (reviewed by other authors in this volume, and in references quoted in their articles). Most of these avenues of research utilize the uncanny ability of this bacterium to effect an inter-kingdom type of exchange of genetic information, namely to transfer sequences of nucleic acids from bacterium to plant. This genetic exchange begins with the detection of substances released by wounded plant cells, to which the bacterium reacts by chemotaxis and induction of the bacterial virulence genes necessary for DNA transfer, and ends with the integration and expression of bacterial genes in the plant, that direct production of a plant tumour.

We will focus this presentation on the use of the bacterium as a route for the introduction of viral genetic information to plants, a process referred to as "agroinfection" (Grimsley et al., 1986a), or "agroinoculation" (Elmer et al., 1988). The production of viral symptoms in a recipient plant can be used as a sensitive marker for T-DNA transfer, and, since it is independent of T-DNA integration and tumour formation, adds another dimension to research on the *Agrobacterium*-plant interaction. The addition of a third biological entity, the virus or viroid, to interactions that normally involve two of these components (either plant-bacterium or plant-virus), provides the opportunity to study phenomena that lie outside of the scope of those that can be studied in the two-component systems. Since general reviews about agroinfection have appeared recently (Grimsley and Bisaro, 1987; Grimsley, 1990), we prefer to present some recent results obtained in our laboratory, preceded by a short introduction concerning the systems that we have been using.

*Plant Molecular Biology* 2, Edited by R.G. Herrmann and B. Larkins, Plenum Press, New York, 1991

Cauliflower mosaic virus (CaMV), the subject of many molecular investigations (see Gronenborn, 1987 for a review) is a double-stranded DNA virus that is transmitted between crucifer host plants by aphids. *Agrobacterium*, however, can be used as an alternative vector for viral transmission, by the introduction of tandemly repeated copies of the virus into the T-DNA. Inoculation of *Brassica campestris* var. *rapa*, a host plant for both bacterium and virus, with such bacterial strains leads to systemic viral infection (Grimsley et al., 1986a)

Surprisingly, maize streak virus (MSV), a monopartite single-stranded DNA geminivirus (reviewed by Davies et al., 1987) that is transmitted in nature by leafhoppers, could also be introduced to maize (*Zea mays*) plants using agroinfection (Grimsley et al., 1987). Isolated DNA of MSV or cloned MSV DNA were never previously shown to be infectious. Our experiment, and others, using agroinfection, (Lazarowitz, 1988; Donson et al., 1988; Woolston et al, 1988), also demonstrated the ability of *Agrobacterium* to transfer DNA to cereal plants, a remarkable result considering that *Agrobacterium* could not incite tumours in these cases.

We describe below the use of agroinfection to study the nature of the T-DNA intermediate and the specificity of the plant-bacterium interaction, and include some preliminary information about virulence gene inducing activity found in extracts of maize seedlings.

## RESULTS AND DISCUSSION

### The nature of the T-DNA intermediate

During the process of infection, following the attachment of bacteria to a plant cell, a piece of the Ti plasmid DNA, the T-DNA, is transferred from bacteria to plants. The nature of this intermediate has been the subject of many investigations, but it is not known whether a single or a double strand of DNA enters the plant cell. Although the balance of evidence would favour the former hypothesis, both single and double-stranded T-DNA molecules have been found within *Agrobacterium* cells that have been grown under appropriate inducing conditions. Biochemical analyses of the kinds of intermediate that are present have been conducted by various laboratories, but these experiments are hampered by the low concentrations of putative intermediates and the inability to distinguish true intermediates from possible side-products. We have attempted to use agroinfection to resolve this question. Geminiviruses, such as maize streak virus, are single-stranded DNA viruses, and the sense (+) strand is packaged in the virion. During the course of infection in nature (+) strands are transferred from plant to plant by the leafhopper vector, and might therefore be more infectious than (-) strands. This feature of the natural system tempted us to use it for analysis of T-DNA transfer, since the intermediate may contain all of its genetic information encoded on a single strand of DNA.

In a series of plasmid constructions using MSV genomes, we produced binary vectors containing the viral sequences in different orientations with respect to the T-DNA borders. If a single strand of T-DNA would be transferred to the plant, either a (+) or a (-) strand of virus would be transferred, depending on the orientation of the MSV sequences (Figure 1). Although we have previously observed (Grimsley et al., 1987) that both orientations of MSV cloned in *Agrobacterium* binary vectors are agroinfectious, we have now extended these experiments to a much more careful comparison, using a serial dilution technique that has previously been shown to be effective for comparison of

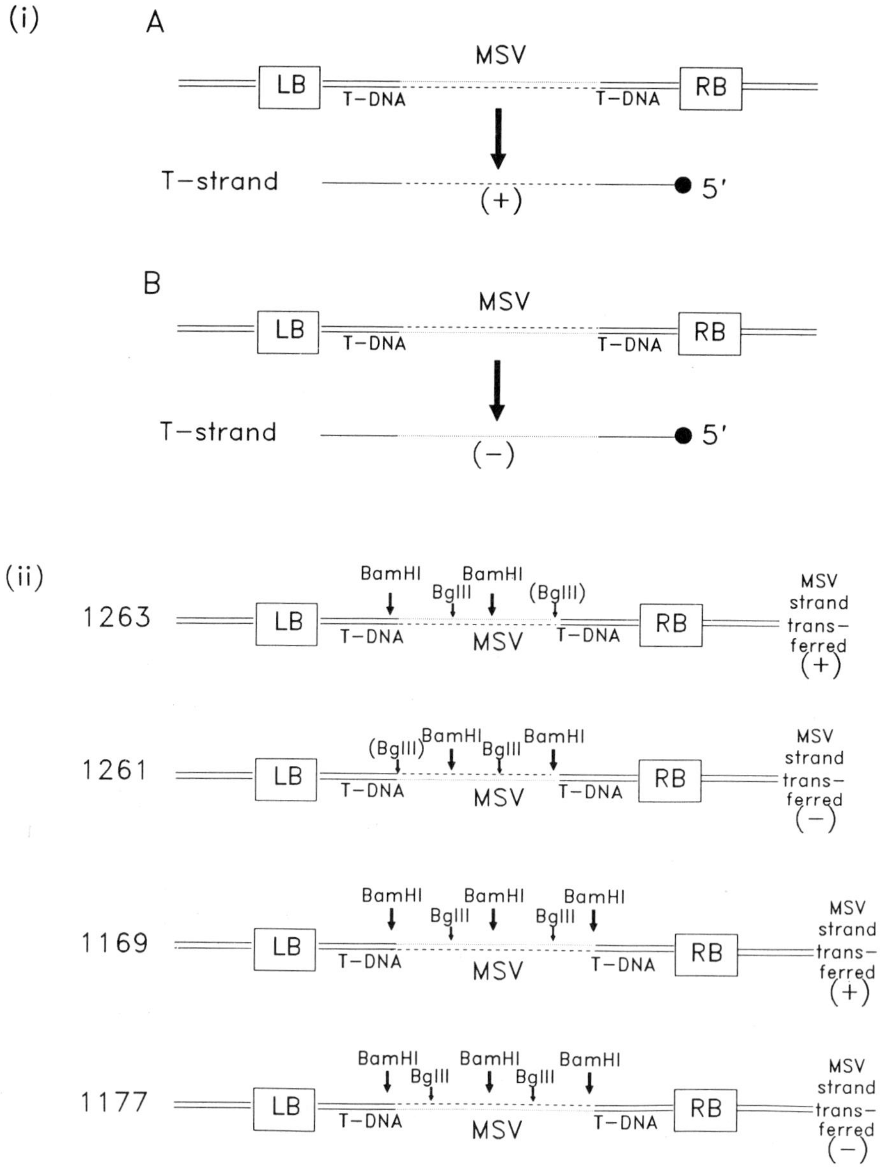


**Figure 1.** (i). The orientation of viral sequences within the T-DNA may affect the nature of the molecule that is transferred to the plant. If a single strand is transferred, then either a sense (+) (case A, dashed line) or an anti-sense (-) strand (case B, dotted line) of viral DNA is transferred.
(ii). Four constructs in *Agrobacterium* were used to test DNA transfer by agroinfection; (1) strain NGB1169 [C58(pEAP37), Grimsley et al., 1987], (2) NGB1177, a similar strain, but carrying MSV sequences in the opposite orientation with respect to the binary plasmid border sequences, (3) strain NGB1263, carrying an MSV 1.6mer (unpublished data) and (4) NGB1261, MSV sequences in the opposite orientation to 1261. Assuming that a single strand of DNA is transferred to the plant, the polarity of the strand that would be transferred is indicated on the right. Inoculation of plants was carried out as previously described (Grimsley et al., 1988), and the appearance of viral symptoms was checked regularly (Table 1)

**Table 1.** The time of appearance of viral symptoms following agroinfection of maize with different orientations of MSV.

| Inoculum | Lg no. bacteria | No. of symptomatic plants per 2-day interval | | | | | | | | | Infected /total no. plants (%) | |
|---|---|---|---|---|---|---|---|---|---|---|---|---|
| | | 4 | 6 | 8 | 10 | 12 | 14 | 16 | 18 | 20 | | |
| 1263 | 7 | 0 | 0 | 7 | 0 | 0 | 1 | 0 | 2 | 0 | 10/10 | 100 |
| | 6 | 0 | 3 | 1 | 4 | 0 | 0 | 0 | 0 | 0 | 8/9 | 89 |
| | 5 | 0 | 1 | 8 | 5 | 6 | 1 | 0 | 0 | 2 | 23/26 | 88 |
| | 4 | 0 | 0 | 6 | 3 | 0 | 1 | 0 | 0 | 0 | 10/23 | 43 |
| | 3 | 0 | 0 | 0 | 2 | 0 | 0 | 0 | 1 | 0 | 3/27 | 11 |
| 1261 | 7 | 0 | 3 | 4 | 0 | 1 | 0 | 0 | 0 | 0 | 8/8 | 100 |
| | 6 | 0 | 4 | 4 | 1 | 0 | 1 | 0 | 0 | 0 | 10/10 | 100 |
| | 5 | 2 | 1 | 14 | 5 | 0 | 0 | 0 | 0 | 0 | 22/26 | 85 |
| | 4 | 0 | 3 | 1 | 3 | 0 | 0 | 0 | 6 | 0 | 13/26 | 50 |
| | 3 | 0 | 0 | 2 | 2 | 1 | 0 | 0 | 0 | 0 | 5/27 | 19 |
| 1169 | 7 | 0 | 5 | 7 | 2 | 0 | 0 | 0 | 0 | 0 | 19/19 | 100 |
| | 6 | 0 | 0 | 2 | 6 | 5 | 0 | 1 | 0 | 0 | 14/16 | 88 |
| | 5 | 0 | 0 | 4 | 2 | 5 | 0 | 4 | 0 | 0 | 15/19 | 79 |
| | 4 | 0 | 0 | 0 | 1 | 0 | 0 | 1 | 1 | 0 | 3/18 | 17 |
| | 3 | 0 | 0 | 0 | 0 | 1 | 0 | 0 | 0 | 1 | 2/20 | 10 |
| 1177 | 7 | 0 | 4 | 8 | 4 | 2 | 0 | 0 | 0 | 0 | 18/18 | 100 |
| | 6 | 0 | 0 | 4 | 12 | 4 | 0 | 0 | 0 | 0 | 20/20 | 100 |
| | 5 | 0 | 2 | 6 | 2 | 4 | 0 | 0 | 2 | 1 | 17/19 | 89 |
| | 4 | 0 | 0 | 4 | 1 | 1 | 0 | 0 | 0 | 0 | 6/19 | 31 |
| | 3 | 0 | 0 | 0 | 2 | 1 | 0 | 0 | 1 | 0 | 4/20 | 20 |
| TOTALS | | 2 | 26 | 82 | 60 | 33 | 4 | 6 | 13 | 4 | 230/380 | |

Please refer to the legend of Figure 1 for a description of the bacterial strains used in the inoculum. The logarithm of the number of bacteria injected by a standard method (Grimsley et al., 1989) is given in column 2.

*Agrobacterium* virulence mutations (Grimsley et al., 1989). However, this careful comparison of the efficiencies of agroinfection with the different virus orientations, using dilution series' of inocula and monitoring symptom appearance on a daily basis, failed to reveal any large differences in the efficiencies of agroinfection (Table 1).

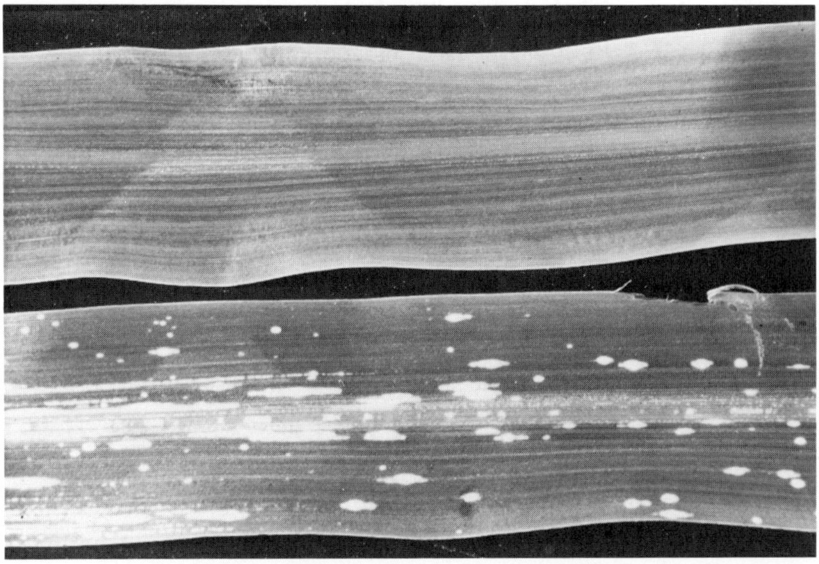

**Figure 2.** The appearance of maize streak virus symptoms. Above: healthy leaf. Below: infected leaf; the distal tip of the leaf is off to the right. As the leaf grows from the base (to the left), the spots usually appear first, these move out towards the tip as the leaf grows, and then the chlorotic streaks develop.

The appearance of MSV symptoms on maize seedlings is a clear indicator for DNA transfer, the transition between non-symptomatic and symptomatic plants occurring within a day. Symptoms first appear as yellow spots at the base of a new leaf, and are easily distinguished from the surrounding green tissue (Figure 2). Exceptionally symptoms appear only 4 days after inoculation, but usually after about 9 days, on the second or the third true leaf. Clearly, there is no difference in efficiency between orientations (NGB1169 vs NGB1177 and NGB1261 vs NGB1263), and consequently we cannot draw any conclusion about the single or double stranded nature of the T-DNA intermediate from this experiment. In case a single strand is transferred, synthesis of a complementary strand might proceed using a bacterial gene product (for example the attached *vir*D2 protein [Herrera-Estrella et al., 1988, Ward et al., 1988, Dürrenberger et al., 1989]), or by host-encoded functions, since it is known that single-stranded DNA can be used to transform isolated protoplasts (Rodenburg et al., 1989); alternatively, double stranded DNA might be transferred.

Since the time of symptom appearance is independent of, or only very weakly dependent on, inoculum size, we can sum all of the data (Table 1). A variety of factors that may influence the plant-bacterium interaction are already known to exist (Figure 3), and it was therefore not obvious *a priori* that the timing of symptom formation would be independent of inoculum size. The dilution series' show that a minimum number of bacteria are required for successful agroinfection (inoculation of $10^4$ bacteria leads to infection in about 50% of cases). We thus think it highly unlikely that the majority of the bacteria can grow in the wound, since smaller (subthreshold) inocula do not usually give rise to symptoms at a later time. Maize is known to produce a compound, DIMBOA (2,4-dihydroxy-7-methoxy-2H-1,4-benzoxazin-3(4H)-one) that hinders bacterial growth (Corcuera et al.,1978), including growth of *Agrobacterium* (S. Sahi et al., 1990 and our own unpublished results), and also strongly inhibits the transformation process itself (Sahi et al., 1990). Extracellular complementation experiments (Otten et al., 1984, 1985; Christie et al., 1988) show that the products of some virulence genes are released from the bacteria, and increased inoculum levels may thus lead to increased levels of these substances. However, our results suggest that the plants' effective defense response against the bacterium does not differ with differing inoculum densities, indicating that any factors that may be produced by the bacteria do not aid or abet the DNA transfer in this system.

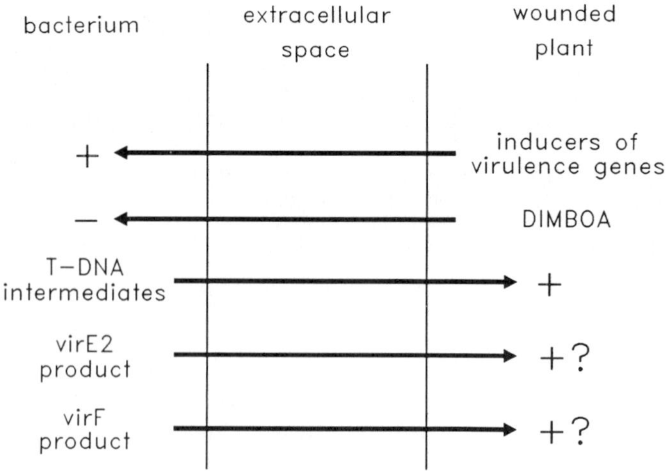


**Figure 3.** Positive (+) or negative (-) influences on the compatibility of the interaction between maize and *Agrobacterium.* Maize does not form tumours after inoculation with *Agrobacterium*, but many steps of the interaction nevertheless occur.

The summed data show a surprising second small peak in the time of symptom formation at about 18 days, that we are currently unable to explain. The overall mean time of symptom appearance is 9.92 days, and 9.04 days if the second peak is excluded by limiting the observation time to 14 days (this excludes only 10% of observations).

**The specificity of the plant-bacterial interaction**

Prior to our studies using agroinfection with CaMV to test DNA transfer in a series of experiments (Grimsley et al., 1989), we tested a variety of crucifers, normally hosts for both bacterium and virus. Locally available varieties of the host species radish (*Raphanus sativus*) were included in these tests. Surprisingly, strong differences in the appearance of CaMV symptoms were observed between different varieties of radish; thus, under our experimental conditions, "Zuericher" showed no chlorotic lesions, "Fruehwunder" developed only faint symptoms, while "Eiszapfen" and "Muenchener Bierrettich" both showed clear symptoms (see also Figure 4).

**Figure 4**. Cauliflower mosaic virus (CaMV) symptoms on radish (*Raphanus sativus*) four weeks after agroinfection. Left: variety "Carnita," a healthy and an infected leaf. Right: "Münchener Bierrettich," infected leaf. In the former case, mostly isolated chlorotic lesions developed, while in the latter case, mosaic vein-clearing developed. Plants were germinated and grown in soil for two weeks prior to inoculation at the crown with a toothpick that had been dipped in the pellet of bacteria obtained from centrifugation of a saturated (2-day old) suspension, and incubated in the P3 laboratory under continuous light for a further 4 weeks (about 10000 lux) at about 25$^{o}$C before scoring viral symptoms.

Using "Muenchener Bierrettich" as an indicator plant, various strains of *Agrobacterium* were tested for their ability to cause tumours and viral symptoms (Table 2). Control inoculations gave high levels of both tumour formation (86%) and symptom formation (91%). Using crown inoculations, production of viral symptoms did not seem to be a much more sensitive indicator for T-DNA transfer than tumour formation, since in 6% of cases plants developed tumours but no symptoms, and *vice versa* in 11% of cases. Similar results were obtained using a miniature variety of *Brassica campestris* (Grimsley et al., 1989). The simplest interpretation of the similarity of the efficiencies of tumour formation and viral symptom formation is that once a T-DNA molecule has entered the nucleus, it has equal probabilities for integration and expression, and/or production of viral replicative intermediates. Intuitively, one might expect that the production of a viral replicative intermediate by transcription/reverse transcription would proceed more efficiently than integration, since the same piece of DNA might be used repeatedly as a template for replication (Grimsley et al., 1986b), whereas there may be only one opportunity for the T-DNA molecule to integrate. Our data therefore suggest that the integration step might be very efficient. However, there could be other more complex reasons for the parallel between these efficiencies of infection; perhaps, for example, a minimum number of viral replicative intermediates are required for the onset of a viral infection, or perhaps the T-DNA molecule itself can replicate prior to integration. Investigations of more virus-plant combinations by agroinfection might help to resolve this question.

**Table 2.** Agroinfection of *Raphanus sativus* "Muenchner Bierrettich" with cauliflower mosaic virus using different strains of *Agrobacterium*.

| Plasmid | Number of plants | | | |
|---|---|---|---|---|
| | Symptoms + tumours | Symptoms only | Tumours only | Healthy |
| pEAP42 | 0 (0%) | 0 (0%) | 0 (0%) | 39 (100%) |
| pGV3850::Ca305 | 0 (0%) | 27 (79%) | 0 (0%) | 7 (21%) |
| pTiC58,pEAP42 | 28 (80%) | 4 (11%) | 2 (6%) | 1 (3%) |
| pTiBo542,pEAP42 | 0 (0%) | 32 (94%) | 0 (0%) | 2 (6%) |

Plants were agroinfected with *Agrobacterium* strain C58C1 carrying the plasmids shown and scored as described in the legend to Figure 3.

The strain C58(pEAP42) produced both symptoms and tumours, but C58C1(pTiBo542), a wide host-range (WHR) supervirulent strain (Hood et al., 1984; Komari et al., 1986), did not produce tumours on radish, even though it rapidly incited tumour formation on tabacco (latter data not shown). However, T-DNA transfer to this strain occurs, since CaMV symptoms appeared! Evidently, the block in response of radish to C58C1(TiBo542) occurs at a stage after DNA transfer, but is probably not due to a hypersensitive type of response, since if this were the case systemic spread of the virus from the point of inoculation would also be arrested. We must conclude either that T-DNA integration does not occur, or that some other late stage of the interaction, such as expression of the T-DNA genes or some aspect of the plants' response, is altered.

Although we have no evidence for integration of the T-DNA in the host genome in this case, it seems unlikely that there is a block in the integration process itself, because this bacterial strain can integrate its T-DNA into the genome of many other host plants, and because other WHR strains of *Agrobacterium* form tumours with radish. In addition, the enzymes that are involved are so conserved that a single species of *Agrobacterium* can attack host plants from widely differing genera. Similarly, sequences governing the constitutive expression of plant genes are conserved; indeed, there is strong selection on the pathogen to maintain promoters that enable gene expression in a wide range of host species.

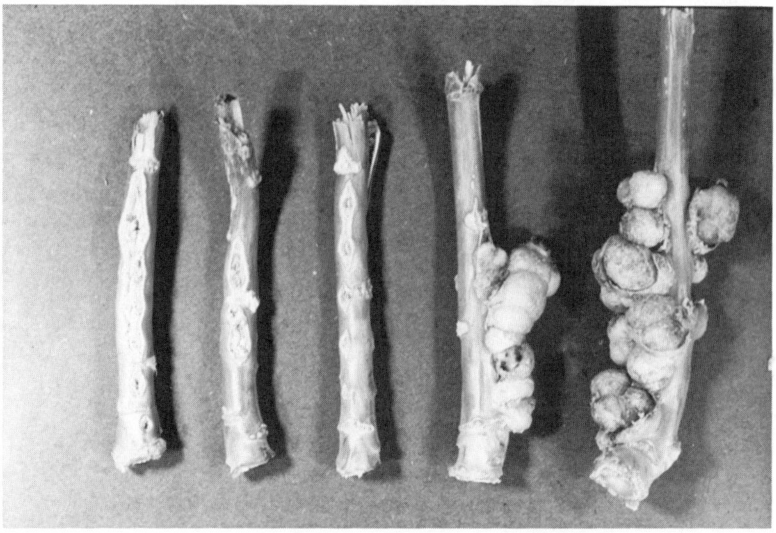

**Figure 5.** Response of rape (*Brassica napus*) to inoculation with different strains of *Agrobacterium*. The two stems on the right were inoculated with C58, the two on the left with C58C1(pTiBo542). A mock-inoculated stem is shown in the centre.

The strain C58C1(pTiBo542) also shows a reduced ability to form tumours on rape (*Brassica napus*) plants (Figure 5, and Charest et al., 1989). There is thus a group of related plants which do not form tumours in response to this WHR strain that might be related to differences between the T-DNAs of this plasmid and other WHR strains (Hood et al.,1984, Komari et al.,1986). It would be very interesting to check integration of the T-DNA of C58C1(TiBo542) in these plants by another independent route, such as the selection of kanamycin-resistant cell lines in culture, or perhaps examining the T-DNAs found in callus of galls after mixed inoculations with C58 and C58C1(pTiBo542).

Bacterial strain-dependent differences also exist in the interaction of maize, a graminaceous plant, with *Agrobacterium* (the data for various plant species are summarized in Table 3). In general, agroinfection of maize with MSV using nopaline-type strains of *Agrobacterium* produces viral symptoms, indicating DNA transfer, whereas octopine-type strains do not transfer their DNA in this way. We are currently investigating the reasons for this difference. The reason for the reduced ability of C58C1(pTiBo542) to agroinfect maize also remains mysterious, especially since this strain is supervirulent on many dicotyledonous plants. We have no idea whether it might be due to some similarity with the octopine type plasmids, or, in contrast to the cases of crucifers discussed above, be due to a hypersensitive response of maize to this particular strain. The plasmid pTiBo542 is known to direct production of elevated levels of certain virulence gene products (Jim et al., 1987, Engstroem et al., 1987). We attempted to test the effect of increasing the level of virG protein in agroinfection by introducing a plasmid carrying *vir*G into a strain capable of agroinfecting maize (Grimsley et al., 1989), but this neither augmented nor diminished the efficiency of agroinfection. This may indicate that the overproduction of the virG protein alone, at least in conjunction with pTiC58, is not eliciting a hypersensitive response in maize. Surprisingly, in rice, another graminaceous plant, gene transfer worked better with strains carrying pTiBo542, again highlighting the complex nature of the *Agrobacterium*-plant interaction (Raineri et al., 1990).

**Table 3.** T-DNA transfer to various plants, measured by tumour formation or agroinfection.

| Plant | Transformation marker | Type of bacterial strain: octopine | nopaline | super-virulent |
|---|---|---|---|---|
| *Nicotiana tabacum* | agroinfection | NT | NT | NT |
| | tumour | + | + | ++ |
| *Brassica napus* | agroinfection | NT | + | NT |
| | tumour | (−) | + | (−) |
| *Raphanus sativus* | agroinfection | + | + | + |
| | tumour | + | + | − |
| *Zea mays* | agroinfection | − | + | (−) |
| | tumour | − | − | − |

NT: not tested. ++: supervirulent. +: positive. (-) weakly negative. -: negative. Comparison of agroinfection with tumour formation using different species of plants and different strains of bacteria reveals bacterial strain specificities, and shows that DNA transfer can occur in some cases where tumour formation does not. (Data summarised from our unpublished observations and Grimsley et al., 1987, Charest et al., 1989; Boulton et al., 1989)

## Bacterial inducers are produced by plants that do not form tumours

Although a range of chemical substances have now been found that are able, to varying extents, to induce the bacterial virulence genes (Stachel et al., 1985; Bolton et al., 1986; Spencer and Towers, 1988), relatively few investigations have actually led to characterisation of the substances responsible for induction *in vivo* (Stachel et al., 1985,

Zerback et al., 1989). We decided to attempt to characterise the inducer(s) produced by maize tissues, after finding that DNA transfer from *Agrobacterium* to maize could proceed in the absence of exogenously supplied inducers (Grimsley et al.,1989). Our experimental protocol is outlined in Figure 6.

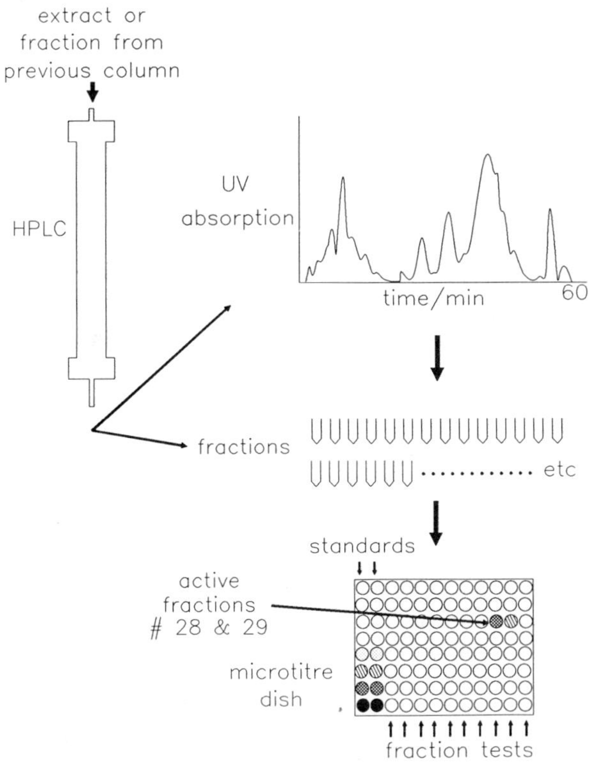


**Figure 6.** Experimental design for isolation of inducers from maize. Organic extracts are passed down a series of different reverse phase HPLC columns, each set of (usually 80) fractions being assayed using a ß-galactosidase test in microtitre dish wells.

Until now, we have found several different activities, but we have not yet characterised these compounds chemically. Since the most important inducers found in dicotyledonous plants may be intermediates of lignin biosynthesis (Stachel, 1985), we might expect the compounds from maize to be different, since there are some differences in the pathways of lignin biosynthesis between monocotyledonous and dicotyledonous plants (Dehne and Kreysig, 1982). However, the molecules most likely resemble those from dicotyledonous plants in some way, since the *vir*A gene product is the receptor molecule, and mutations in *vir*A also block DNA transfer to maize (Grimsley et al., 1989).

## SUMMARY AND PERSPECTIVES

We have used agroinfection as a springboard to begin investigation of several different aspects of the interaction between *Agrobacterium* and plants. In biological systems, we must look for the simplest hypotheses to account for experimental observations. In the case of the *Agrobacterium*-plant interaction, we know that a piece of DNA is transferred from bacterium to plant, an event directed by a set of bacterial genes, and that once integrated in the plant cell nulear DNA, the T-DNA can direct synthesis of plant hormones. Our findings, however, lead to questions that suggest that there may be some quite fundamental variations of this system that still remain a mystery.

Are small hydrophobic molecules, such as acetosyringone (AS)(Stachel et al., 1985), the only kind of molecule involved in the induction of the virulence genes? There is one report of a hydrophilic inducer (Usami et al., 1988). In maize, it is surprising that DNA transfer occurs at all because of the high concentrations of substances inhibitory to bacterial growth (Sahi et al., 1990), particularly since molecules such as AS seem to induce the bacterial virulence genes only after several hours of interaction with the bacteria (Stachel et al., 1985). However, we should add that the inducing activities that we have found from maize all partition to the organic phase in our extractions.

Although circumstancial evidence favours the hypothesis that a single strand of DNA is transferred from bacterium to plant, we still lack direct proof that this is the case. Since the concentration of this intermediate in a mixture of plant cells and bacteria is most likely low relative to the amount of unprocessed Ti plasmid, approaches based on biochemical separations are likely to be difficult; perhaps *in vivo* approches, such as the one we have outlined, or *in vitro* reconstitutions of the system, should yield more information.

Which components are responsible for the specificity of the host-pathogen interaction? In other systems, gene for gene interactions (see Keen and Staskowicz, 1988, for a review) have been identified, but so far not in *Agrobacterium*, although intriguing specificities do exist. "Narrow host range" strains of *Agrobacterium tumefaciens* show differences in the kinds of oncogenes that they carry, even deletions in some cases, (Paulus et al., 1989, Huss et al 1990), but they nevertheless incite tumours on certain species; conversely, in the case discussed in this paper, C58C1(pTiBo542) transfers a full set of plant hormone genes to radish, but a tumour does not develop. We still have a great deal to learn about these interactions.

**Acknowledgements**: We are grateful for technical assistance provided by Andrea Crameri and Cynthia Ramos.

## REFERENCES

Bolton G W, Nester E W, Gordon M P 1986 Plant phenolic compounds induce expression of the *Agrobacterium tumefaciens* loci needed for virulence. Science 232:983-985

Boulton M I, Buchholz W G, Marks M S, Markham P G, Davies J W 1989 Specificity of *Agrobacterium*-mediated delivery of maize streak virus DNA to members of the gramineae. Plant Molec. Biol. 12:31-40

Charest P J, Iyer V N, Miki B L 1989 Virulence of *Agrobacterium tumefaciens* strains with *Brassica napus* and *Brassica juncea*. Plant Cell Reports 8:303-306

Christie P J, Ward J E, Winans S C, Nester E W 1988 The *Agrobacterium tumefaciens vir*E2 gene product is a single-stranded-DNA-binding protein that associates with T-DNA. J. Bacteriol. 170:2659-2667

Corcuera L J, Woodward M D, Helgeson J P, Kelman A, Upper C D 1978 2,4-Dihydroxy-7-methoxy-2H-1,4-benzoxazin-3(4H)-one, an Inhibitor from *Zea mays* with differential Activity against soft rotting *Erwinia* species. Plant Physiol. 61:791-795

Davies J W, Townsend R, Stanley J 1987 The structure, expression, functions and possible exploitation of geminivirus genomes. In: (eds) Hohn, T, Schell, J. Plant Gene Research: Plant Plant DNA Infectious Agents. Springer, New York and Vienna. pp.31-52

Dehne H, Kreysig D 1982 Natürliche organische Makromolüle. In: (eds) Kempter G, Kasper F, Kresig D, Uhlemann E, Welsch F. Chemie für Lehrer Volume 15. VEB Deutscher Verlag der Wissenschaften, Berlin

Donson J, Gunn H V, Woolston C J, Pinner M S, Boulton M I, Mullineaux P M, Davies J W 1988. *Agrobacterium*-mediated infectivity of cloned digitaria streak virus DNA. Virology 162:248-250

Dürrenberger F, Crameri A, Hohn B, Koukl olikova-Nicola Z 1989 Covalently bound VirD2 protein of *Agrobacterium tumefaciens* protects the T-DNA from exonucleolytic degredation. Proc. Natl. Acad. Sci. USA. 86:9154-9158

Elmer J S, Sunter G, Gardiner W E, Brand L, Browning C K, Bisaro D M, Rogers S G 1988 *Agrobacterium*-mediated inoculation of plants with tomato golden mosaic virus DNAs. Plant Mol. Biol. 10:225-234

Engstroem P, Zambryski P, Van Montagu M, Stachel S 1987 Characterisation of *Agrobacterium tumefaciens* virulence proteins induced by the plant factor acetosyringone. J. Mol. Biol. 197:635-645

Grimsley N H 1990 Agroinfection. Physiol. Plant. 79:147-153

Grimsley N H, Hohn B, Hohn T, Walden R 1986a Agroinfection, an alternative route for viral infection of plants by using the Ti plasmid. Proc.Natl.Acad.Sci.USA. 83:3282-3286

Grimsley N H, Hohn T, Hohn B 1986b Recombination in a plant virus: template-switching in cauliflower mosaic virus. EMBO J. 5:641-646

Grimsley N H, Bisaro D 1987. Agroinfection. In: (eds) Hohn, T, Schell, J. Plant Gene Research: Plant Plant DNA Infectious Agents. Springer, New York and Vienna. pp.87-107

Grimsley N H, Hohn T, Davies J W, Hohn B 1987 *Agrobacterium*-mediated delivery of infectious maize streak virus into maize plants. Nature 325:177-179

Grimsley N H, Ramos C, Hein T, Hohn B, 1988 Meristematic tissues of maize plants are most susceptible to agroinfection with maize streak virus. Biotechnology 6:185-189

Grimsley N H, Hohn B, Ramos C, Kado C, Rogowsky P 1989 DNA transfer from *Agrobacterium* to *Zea mays* or *Brassica* by agroinfection is dependent on bacterial virulence functions. Mol. Gen. Genet., 217:309-316

Gronenborn B 1987 The molecular biology of cauliflower mosaic virus and its application as a plant gene vector. In: (eds) Hohn, T, Schell, J. Plant Gene Research: Plant DNA Infectious Agents. Springer, New York and Vienna. pp.1-29

Herrera-Estrella A, Chen Z, Van Montagu M, Wang K 1988 VirD proteins of *Agrobacterium tumefaciens* are required for the formation of a covalent DNA-protein complex at the 5' terminus of T-strand molecules. EMBO J. 7:4055-4062

Hood E, Jen G, Kayes L, Kramer J, Fraley R T, Chilton M-D, 1984 Restriction endonuclease map of pTi Bo542, a potential Ti plasmid vector for genetic engeneering of plants. Biotechnology 2:702-708

Huss B, Bonnard G, Otten L 1989 Isolation and functional analysis of a set of auxin genes with low root-inducing activity from an *Agrobacterium tumefaciens* biotypeIII strain. Plant Molecular Biology 12:271-283

Jin S, Komari T, Gordon M P, Nester E W 1987 Genes responsible for the supervirulence phenotype of *Agrobacterium tumefaciens* A281 J. Bacteriol. 169:4417-4425

Keen N T, Staskawicz B 1988 Host range determinants in plant pathogens and symbionts. Ann. Rev. Microbiol. 42:421-440

Komari T, Halperin W, Nester E W 1986 Physical and functional map of supervirulent *Agrobacterium tumefaciens* Tumor-inducing plasmid pTiBo542. J. Bacteriol. 166:88-94

Lazarowitz S G 1988 Infectivity and complete nucleotide sequence of the genome of a South African isolate of maize streak virus. Nucl. Acids Res. 16:229-249

Otten L, De Greve H, Leemans J, Hain R, Hooykaas P, Schell J 1984 Restoration of virulence of Vir region mutants of *Agrobacterium tumefaciens* strain B6S3 by coinfection with normal and mutant *Agrobacterium* strains. Mol.Gen.Genet. 195:159-163

Otten L, Piotrowiak G, Hooykaas P, Dubois M, Szegedi E, Schell J 1985 Identification of an *Agrobacterium tumefaciens* pTiB6S3 vir region fragment that enhances the virulence of pTiC58. Mol Gen Genet 199:189-193

Paulus F, Huss B, Bonnard G, Ride M, Szegedi E, Tempe J, Petit A, Otten L 1989 Molecular systematics of biotype III Ti plasmids of Agrobacterium tumefaciens. Molec. Plant-Microbe Interact. 2:64-74

Raineri D M, Bottino P, Gordon M P, Nester E W 1990 *Agrobacterium*-mediated transformation of rice (*Oryza sativa* L.) Biotechnology 8:33-38

Rodenburg C W, De Groot MJA, Schilperoort R A, Hooykaas P Y J 1989 Single-stranded DNA used as an efficient new vehicle for transformation of plant protoplasts. Plant Mol. Biol. 13:711-719

Sahi S V, Chilton M D, Chilton W S 1990 Corn metabolites affect growth and virulence of *Agrobacterium tumefaciens*. Proc. Natl. Acad. Sci. USA 87:3879-3883

Spencer P A, Towers G H N 1988 Specificity of signal compounds detected by *Agrobacterium tumefaciens*. Phytochemistry 27:2781-2785

Stachel S E, Messens E, Van Montagu M, Zambryski P 1985 Identification of the signal molecules produced by wounded plant cells that activate T-DNA transfer in *Agrobacterium tumefaciens.* Nature 318:624-629

Usami S, Okamoto S, Takebe I, Machida Y 1988 Factor inducing *Agrobacterium tumefaciens* vir gene expression is present in monocotyledonous plants. Proc. Natl. Acad. Sci. USA 85:3748-3752

Ward E R, Barnes W M 1988 VirD2 protein of *Agrobacterium tumefaciens* very tightly linked to the 5' end of T-strand DNA. Science 242:927-930

Woolston C J, Barker R, Gunn H, Boulton M I, Mullineaux P M 1988 Agroinfection and nucleotide sequence of cloned wheat dwarf virus DNA. Plant Mol.Biol. 11:35-43

Young C, Nester E W 1988 Association of the VirD2 Protein with the 5' End of T Strands in *Agrobacterium tumefaciens.* J of Bacteriol. 170:3367-3374

Zerback R, Dressler K, Hess D 1989 Flavonoid compounds from pollen and stigma of *Petunia hybrida*: inducers of the *vir* region of the *Agrobacterium tumefaciens* Ti plasmid. Plant Sci. 62:83-91

# PHYSICAL MAPPING OF THE *ARABIDOPSIS* GENOME AND ITS APPLICATIONS

Brian M. Hauge[1], Jerome Giraudat[1,2], Susan Hanley[1], Inhwan Hwang[1], Takayuki Kohchi[1], and Howard M. Goodman[1]

1-Department of Genetics, Harvard Medical School and the Department of Molecular Biology, Massachusetts General Hospital, Boston, MA, 02114, USA.
2-Institut Des Sciences Vegetales, C.N.R.S., 91198 Gif Sur Yvette,Cedex, France

## INTRODUCTION

We are engaged in a project to construct a complete physical map of the *Arabidopsis thaliana* genome which will ultimately consist of a fully overlapping collection of cloned DNAs encompassing the five *Arabidopsis* linkage groups. There are several reasons for using *Arabidopsis* as a model system for the study of plant biology. Its short life cycle, small size and large seed output make it well suited for classical genetic analysis (reviewed in Meyerowitz, 1987). Mutations have been described affecting a wide range of fundamental developmental and metabolic processes (reviewed in Estelle and Somerville, 1986). A genetic linkage map consisting of some 100 loci has been assembled (Koornneef, 1987) and an increasing number of cloned genes and restriction fragment length polymorphisms (RFLPs) are available for correlation of the genetic map with the physical map (Chang et. al., 1988; Nam et. al., 1989). For molecular biological studies, *Arabidopsis* offers the additional advantages of having a very small genome (70,000 kb) and a remarkably low content of interspersed repetitive DNA (Pruitt and Meyerowitz, 1986). Both of these features are highly unusual among higher plants and suggest that the *Arabidopsis* genome is ideally suited for a physical mapping project.

The benefits of having a physical map are two-fold. First, the map provides access to any region of the genome which can be genetically identified. In other words, the physical map serves as a cloning tool by facilitating the movement from the genetic locus to the cloned gene. Given a mutation of known genetic map location, the physical map can be used to easily isolate a collection of overlapping clones encompassing the locus of interest. By eliminating the need for labor intensive steps such as chromosome walking, researchers are free to focus their efforts on the isolation and characterization of the gene of interest. Conversely, the physical map allows for the movement from the cloned gene to the genetic locus. Cloned genes can be easily integrated into the physical map and consequently aligned with the genetic linkage map. In some cases it may be possible to assign the clone to a defined genetic locus, thereby establishing its biological function. Clearly, in most cases assignment of clones to known loci will not be possible. Nonetheless, as an increasing number of genes are cloned and molecular biological information is accumulated, the map can be used to investigate the physical linkage of cloned genes, study the organization and distribution of repetitive elements and address questions such as how physical distance and genetic distance are correlated. In this context, the map provides the framework for cataloging and integrating molecular biological information, thereby serving as a starting point to investigate large scale genomic organization. Ultimately, genome organization will be investigated at the nucleotide level. Clearly, physical maps are the logical substrates for genome sequencing projects.

The first stage of the mapping project involves the characterization of random clones using the "fingerprinting" strategy of Coulson, Sulston and co-workers (1986). At present more than 20,000 random cosmid clones have been "fingerprinted" (Hauge et. al., 1988). Our working data set consists of 17,000 clones representing an 8-10 fold sampling redundancy. Using computer matching programs, these clones have been aligned into overlapping groups, referred to as contigs. The 17,000 clones fall into some 750 contigs representing approximately 90-95% of the *Arabidopsis* genome. Ideally, as the random clone analysis proceeds and the sample size grows, the contigs will be expanded and ultimately merge to produce a map of the five *Arabidopsis* linkage groups. However, having fingerprinted 8-10 genomic equivalents we have reached the practical limit of the procedure. In other words we are at the stage of the project where the rate of finding new joins is unacceptably low due to the scarcity of the remaining clones. Success in completing the map will depend on the ability to close the gaps. Several approaches are being used to complete the physical map; software development to automate assembly of the random clones; multi-enzyme fingerprinting of clones residing at the end of the contigs and the unattached clones to enhance the statistical detection of overlaps; selecting linking clones by hybridization using end-probes from clones residing at the ends of contigs and using YACs as probes to bridge the gaps.

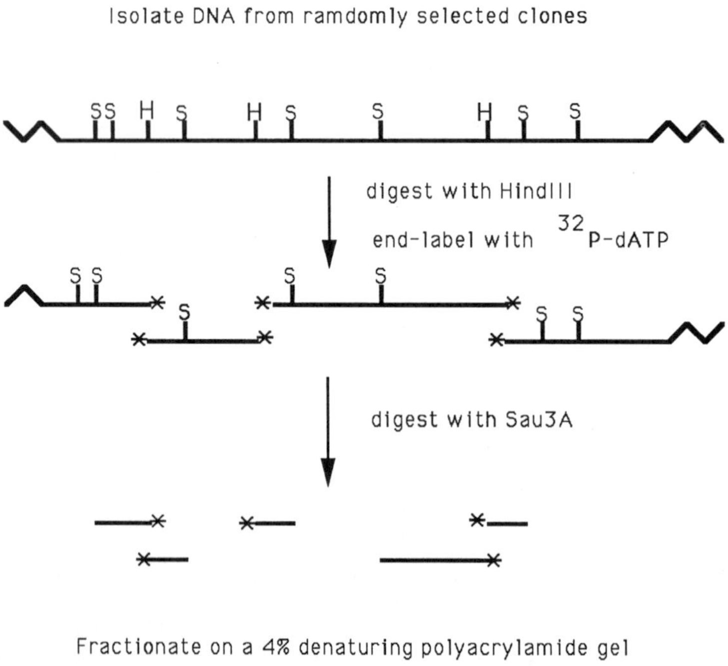


FIGURE 1. Fingerprinting procedure. The solid lines depict the cloned insert while the vector is represented by the zig-zag line. H HindIII recognition sequence: S Sau3A recognition sequence. The stars represent the end-labeled digestion products.

If the physical map is to be of any utility as a cloning tool, it is necessary to align the physical map with the genetic map. Toward this goal, we have assembled a restriction fragment length polymorphism (RFLP) map of the *Arabidopsis* genome (Nam et. al., 1989). The RFLP map, which currently consists of 158 markers, provides the contact points for alignment of the genetic and physical maps.

One of our major objectives for engaging in the mapping project is to simplify the cloning of genes when only the locus and not the product of the gene is known. In particular, we are using the maps to clone genes involved in plant hormone metabolism and their corresponding signal transduction pathways. We are focusing on two of the hormones, abscisic acid (ABA)(Giraudat et. al., 1989) and the gibberellins (GA)(Hauge et. al., 1989). Both hormones have long been implicated as important regulators affecting cell growth and development, seed germination, stem elongation and flower development; however, their precise role in regulating these processes is presently unclear. In all likelihood, the magnitude of pleiotrophic effects reflect the complex interplay between the various plant hormones. Consequently, it is difficult to establish the primary mode of action. As an approach to investigate the mechanism of hormone action, we are using the maps to clone genes involved in ABA and GA biosynthesis and their signal transduction pathways. When the molecular probes have been isolated, the pattern of expression of these genes will be examined during plant development and in response to alterations in hormone levels achieved either by the use of mutants or exogenous application. These experiments will be a starting point to investigate the role of hormones in specific developmental processes.

TOWARD COMPLETION OF THE PHYSICAL MAP

The strategy that we are using to construct the map is the random clone approach developed by Coulson, Sulston and co-workers (1986; Figure 1). Briefly, DNA is isolated from randomly selected clones. We are working with random shear cosmid libraries having a mean insert size of 40 kb. The DNAs are digested with a restriction enzyme having a 6 bp specificity which leaves staggerd ends. The ends are simultaneously labeled with $^{32}P$ and the appropriate nucleoside triphoshates. The reactions are then terminated and the fragments are subject to a second round of cleavage with an enzyme having a 4 bp specificity. The resultant fragments are of suitable size for fractionation on a 4% denaturing polyacrylamide gel. Figure 2 shows an example of a typical gel containing the reaction products from 48 clones. The band which is common to all of the clones corresponds to the vector band. The intensity of the vector band reflects the viability of the clone. The main point which is illustrated in figure 2, is that individual clones give a distinct banding pattern or "fingerprint" which serves as a unique signature for a given clone.

For subsequent analysis, the banding patterns are inputted into the computer using a scanning densitometer and an image-processing package (Sulston et. al., 1989; B.M. Hauge, W.D.B. Loos and H.M. Goodman, unpublished). Once the banding patterns are entered into the computer, they are compared in a pairwise fashion against the entire data set. Using clone matching programs the regions of probable overlap are determined and the clones are assembled into contigs. Figure 3 shows an example of a contig. It should be noted that the computer does not actually assemble the contigs. Assembly is performed interactively with a computer program (Coulson et. al., 1986). Before the clones are joined, the reliability of the match is assessed by visually aligning the films and the overlap must be logically consistent.

There are two limitations to this strategy; the combination of enzymes HindIII and Sau3A only permit the detection of overlaps in the range of 35-50%. So in general, smaller overlaps will go undetected and clones containing few or no HindIII sites will be unattached since a minimal number of bands is required for the statistical detection of overlapping regions. To circumvent these problems, the clones residing at the end of contigs and the unattached clones will be fingerprinted using several different combinations of enzymes. In doing so, we will be able to establish joins which were previously undetected due to the statistical limitations since the match probability is now the product of the individual probabilities. In addition, by using several

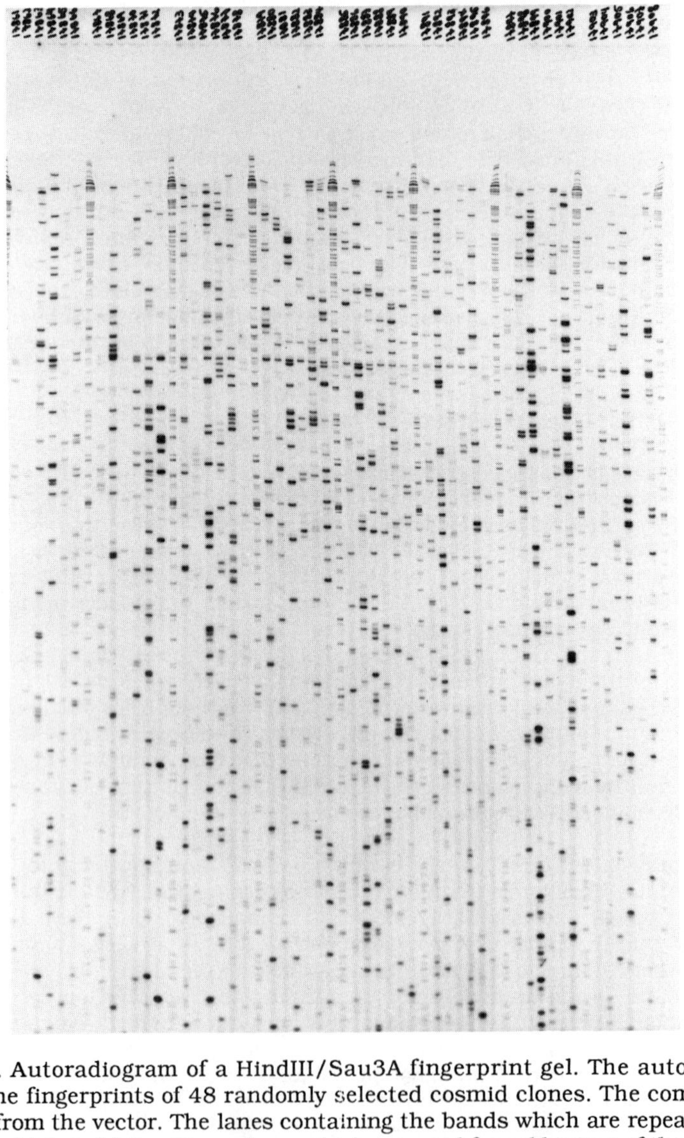

FIGURE 2. Autoradiogram of a HindIII/Sau3A fingerprint gel. The autoradiogram contains the fingerprints of 48 randomly selected cosmid clones. The common band originates from the vector. The lanes containing the bands which are repeated every 7 lanes are end-labeled λ Sau3A markers, which are used for calibration of the films.

combinations of enzymes, it is unlikely that a given clone will have a non-random distribution of restriction sites for all of the enzymes. Therefore, we can establish joins with the clones which were unattached as a result of having a minimal number of HindIII sites.

Once the practical limit of random clone mapping is reached, success in completing the map depends largely on the ability to bridge the remaining gaps. The most viable option is to select the missing clones by hybridization. One approach is to make end-probes from unattached clones and clones residing at the ends of the contigs, to select the linking clones. For this approach to be practical it is important that the end-probes can be made with minimal effort. Therefore, our cosmid libraries were constructed in vectors containing convergent bacteriophage *T7* and *Sp6* promoters flanking the inserts. The end-clones and unattached clones are plated out as ordered grids using a standard 96-well microtiter dish configuration. Probes are then made from pools of clones corresponding to rows and columns of the microtiter dish (Figure 4; Evans and Lewis, 1989). The mixed-probes are then hybridized to ordered arrays of *Arabidopsis* YAC clones. This procedure is repeated with probes derived from each of the rows and columns. To determine which of the cosmids hybridizes to a given YAC, the common signals are identified. The hybridizing clone is identified by the intersection of a row and a column. For example, in Figure 4, the cosmid clone at position 1A hybridizes to the YAC indicated by the arrows in panels A and B. By using the multiplex strategy outlined in Figure 4, the number of hybridizations can be reduced by 4.8 fold. Clearly, by using larger pools of probes the number of hybridizations can be further reduced. We have established the conditions for hybridization with mixed-probes and the feasibility of using this approach has been assessed by using characterized contigs.

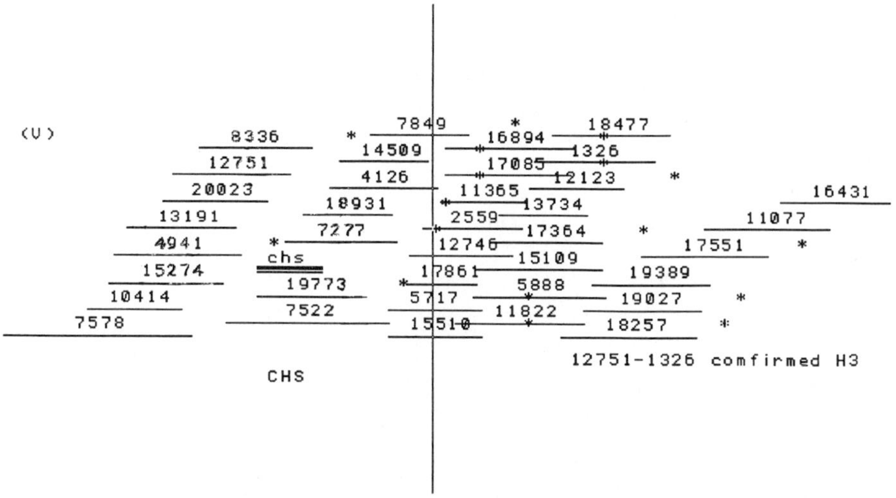


FIGURE 3. Computer display of a contig. Clones are depicted by the solid lines, the length of which is proportional to the number of bands in the fingerprint. The asterisk denotes buried clones which are not displayed. Contig assembly is essentially as described by Sulston et. al., (1988).

There are two reasons why we are using YAC libraries to select the missing clones. The obvious advantage of the yeast cloning system is that the large size of YAC clones means that fewer clones need be examined. Secondly, YACs offer the potential to give a random, or at least a different representation of clones than are obtained using bacterial host/vector systems. This is important since is likely that many of the gaps in the contig map can be attributed to cloning bias, due to the inability to clone certain sequences in cosmids or non-uniform growth of individual clones leading to sampling bias.

Another approach which we are using to bridge the gaps is to use YACs as hybridization probes. The strategy is to prepare two sets of ordered grids; one of 1536 random YAC clones, representing about 3 genomic equivalents, and one of cosmids, which is representative of our contigs and the unattached clones. YACs are then isolated and used as hybridization probes. The YACs may be picked at random, or alternatively selected from the YAC grid by probing with cosmids as described above. The hybridization pattern of the cosmid grid is then used to establish linkage, as well as its position with respect to the cosmids. This approach has been successfully used to reduce the number of contigs for the *C. elegans* map from 700 to 170 (Coulson et. al., 1988; J. Sulston and A. Coulson, personal communication). It is significant, that Coulson and co-workers (1988) find that sequences which are poorly represented in the cosmid libraries are represented in the YAC clones.

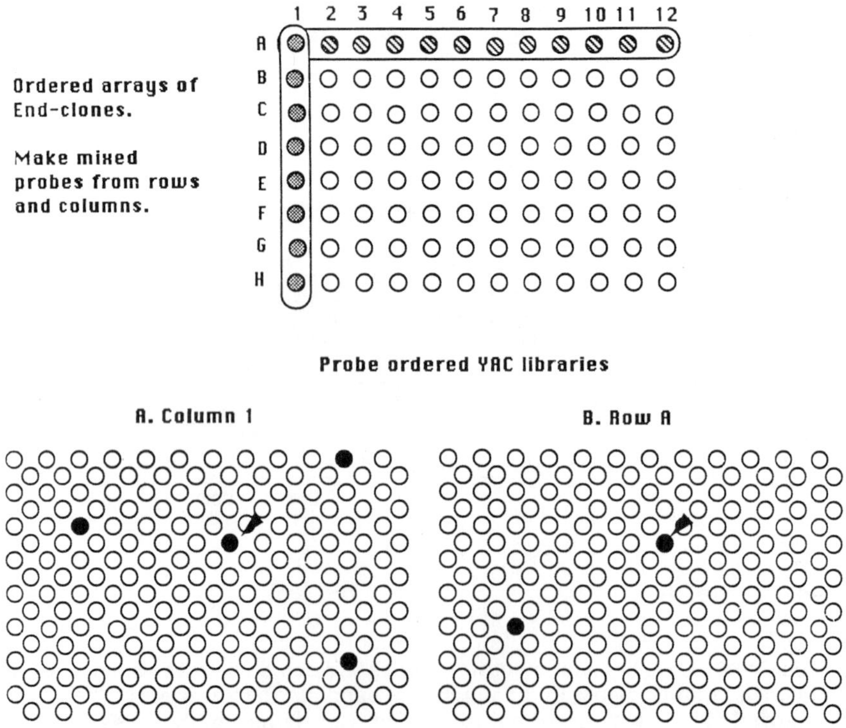


FIGURE 4. Strategy for linking contigs with multiplex hybridization. Cosmid clones from the ends of contigs are archived in microtiter dishes and held as glycerol stocks at -80$^{o}$ C. Individual clones are assigned a unique position. Probes are prepared from pools of 8 and 12 clones, corresponding to columns and rows respectively. The mixed probes are hybridized to ordered arrays of YAC clones and the procedure is repeated for each row and column. The clone at the intersection (1A) hybridizes to the YAC indicated by the arrow (Panels A and B).

## RFLP MAPPING

An important component of the mapping project is the alignment of the physical map with the established genetic map. We have previously published an RFLP map consisting of 94 randomly distributed molecular markers (Nam et. al., 1989). In addition, we have incorporated 17 markers from an independently constructed *Arabidopsis* RFLP map (Chang et. al., 1988). The combination of the two maps provides good coverage of the genome; however, several large gaps still exist. In an effort to close the gaps, we are adding additional markers to the map. Currently, our RFLP map contains 158 markers, several of the new markers map to previously unmarked regions of the genome. This analysis will be continued until coverage of the genome is complete. RFLPs are being identified using both random clones as well as cloned genes.

## CLONING OF HORMONE BIOSYNTHETIC AND RESPONSE LOCI

One of the main goals of this project is to utilize the maps to clone genes involved in the synthesis and response to plant hormones. We are focusing our efforts on two of the hormones, ABA and GA. Numerous mutants which are defective in the synthesis of GA have been characterized in *Arabidopsis* and genetically mapped to five loci (*ga-1, ga-2, ga-3, ga-4, ga-5*). These mutants are all phenotypic dwarfs with reduced apical dominance and for many alleles GA is required for germination ( Koornneff et. al., 1982; Koornneff and van der Veen, 1989). Normal growth of these mutants is restored by exogenous application of GA. In addition, a sixth locus (*Gai*) which is GA-insensitive has been identified. By genetic criteria, the phenotypes are consistent with the possibility that *Gai* encodes a GA receptor or a second messenger.

Several "ABA-insensitive" mutants (loci *abi-1, abi-2, abi-3*) have been identified which produce normal levels of endogenous ABA but display reduced sensitivity to ABA. Mutations at all three loci display reduced dormancy and ABA-inhibition of germination; however, with respect to the other responses, the effects of abi-1 and abi-2 are confined to vegetative growth, while abi-3 primarily effects seed development (Koornneef et. al., 1984; Koornneef et. al., 1989). In contrast to ABA-deficient mutants (*aba* locus), the wildtype phenotype cannot be restored by addition of exogenous ABA. Like the GA-insensitive mutant (*Gai*), the decreased sensitivity to ABA probably results from an alteration in the number or properties of the receptor(s) or other elements of the ABA signal transduction pathway.

Using the combined RFLP and physical maps, experiments are in progress to clone several of the loci described above. Clearly, the RFLP map provides the starting points for cloning. However, it is difficult to assign with confidence a precise position on the genetic map to an RFLP probe since only a few morphological markers per chromosome were used to align the genetic map with the RFLP map. In addition, due to the necessarily limited number of F2 progeny used for the segregation analysis, we estimate that the resolution of the RFLP map is on the order of 2cM (Nam et. al., 1989). So for each gene that we are interested in cloning, it is first necessary to map the locus with respect to the RFLP map. This is achieved using a "fine structure mapping" approach (Figure 5; Giraudat et. al., 1989; Hauge et. al., 1989). To enhance the resolution, we have constructed lines containing the mutation of interest flanked by linked markers. For the RFLP mapping, we examine the segregation of markers only among the F2 progeny which have undergone a recombination event within a small genetically defined interval (i.e., recombinant between the flanking markers and the gene of interest). Using this approach, we enhance the resolution of the RFLP map in the vicinity of the locus while minimizing the number of progeny to analyze. Once a linked RFLP has been identified, the physical map can be used to localize the gene to one or to several clones. The gene can then be identified by complementation or alternatively, physical means.

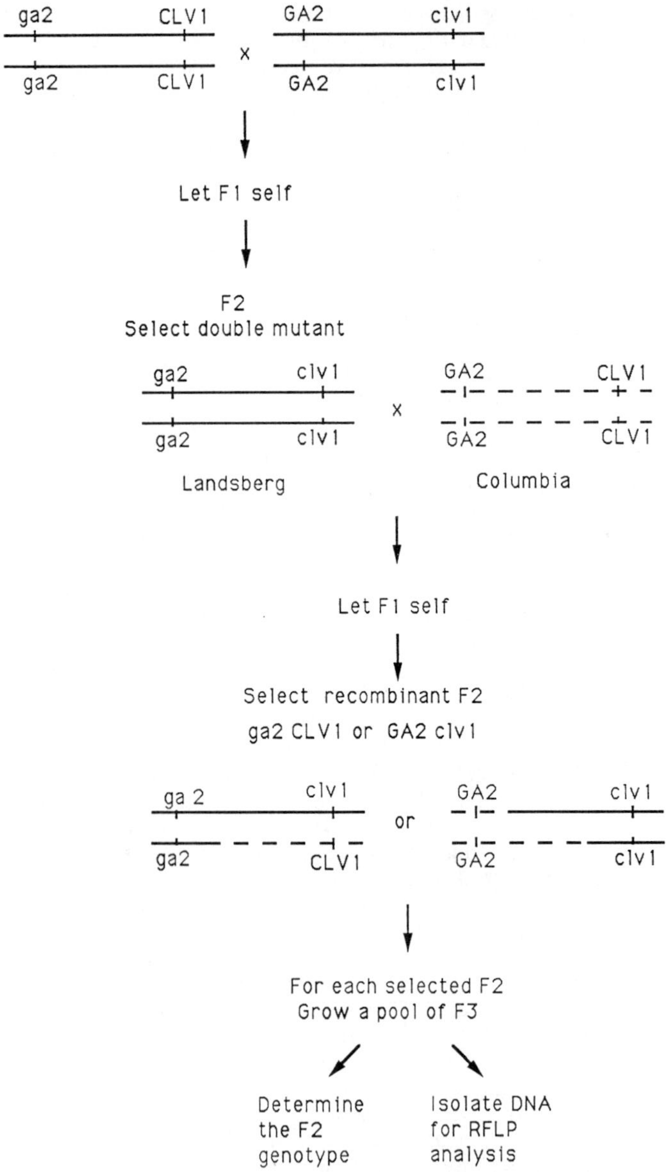


FIGURE 5. Strategy for "fine structure" mapping. For each locus that we are interested in cloning, lines are constructed containing the mutation of interest flanked by a linked marker. For example, we construct a *ga-2* and *clv-1* double mutant in a *Landsberg erecta* background. The double mutant is crossed to a wild type *Columbia* line which shows a high level of polymorphism with *Landsberg* (Nam et. al., 1989). The F1 progeny are allowed to self. In the F2 we select recombinants between *ga-2* and *clv-1*. The F2s are selfed and the resultant F3s are scored to uncover the F2 heterozygotes and pooled to make DNA for RFLP analysis.

The fine structure mapping is essentially complete for two loci, *abi-3* (Giraudat et. al., 1989) and *ga-2* (Hauge et. al., 1989) In both cases, using the combined RFLP/physical map we were able to localize the genes to a single cosmid. Experiments to further localize the respective genes within the cosmids are in progress. We are also attempting to clone several of the other *ga* loci and the *Gai* locus. These experiments are at various stages of progress. The cloning of the *aba*, *abi-1* and *abi-2* loci is being done in collaboration with J. Giraudat (CNRS, Gif sur Yvette, France). Once the genes have been isolated they will be characterized using the various genetic and molecular biological tools which are available. Of particular interest, the genes provide the starting point to elucidate the molecular mechanisms of plant hormone action.

DNA sequence analysis of the cosmid clones harboring abi-3 and ga-2 is currently underway. In addition to the obvious goal which is to determine the nucleotide sequence of the respective genes, we are using the clones to investigate strategies for large scale sequencing projects. We are pursuing several approaches; direct sequencing of cosmids in solution; sequencing of cosmids which have been immobilized onto solid supports; shotgun sequencing; and multiplex sequencing.

REFERENCES


Chang,C., Bowman,J.L., DeJohn,A.W., Lander,E.S. and Meyerowitz,E.M. (1988). Restriction fragment length polymorphism linkage map for *Arabidopsis thaliana*. Proc. Natl. Acad. Sci. USA, **85**, 6856-6860.

Coulson,A., Sulston,J., Brenner,S. and Karn,J. (1986). Toward a physical map of the nematode *Caenorhabditis elegans*. Proc. Natl. Acad. Sci. USA, **83**,7821-7825.

Coulson,A., Waterston,R., Kiff,J., Sulston,J. and Kohara,Y. (1988). Genome linking with yeast artificial chromosomes. Nature, **335**, 184-186.

Estelle, M.A. and Somerville, C.R. (1986). The mutants of *Arabidopsis*. Trends Genet., **2**, 89-93.

Evans,G.A. and Lewis,K.A. (1989). Physical mapping of complex genomes by multiplex analysis. Proc. Natl. Acad. Sci. USA, **86**, 5030-5034.

Giraudat,J., Hauge,B.M, and Goodman,H.M. (1989). Progress toward the cloning of the abi3 locus of *Arabidopsis*. Abstract; The Genetics and Molecular Biology of Arabidopsis. Bloomington, Indiana.

Hauge,B.M., Giraudat,J. Nam,H.-G. and Goodman,H.M. (1989) Progress toward a physical map of the *Arabidopsis thaliana* Genome. in Development and Application of Molecular Markers to Problems in Plant Genetics (ed. Helentjaris,T. and Burr,B.) Cold Spring Harbor Laboratory Press, Cold Spring Harbor, New York. 149-152.

Hauge,B.M., Gallant,P., Malolepszy,J. and Goodman,H.M. (1989). Toward the cloning of the *ga-2* locus from *Arabidopsis*. Abstract; The Genetics and Molecular Biology of Arabidopsis. Bloomington, Indiana.

Koornneef,M., Hanhard,C.J., Hilhorst,H.W.M. and Karssen,C.M. (1989). In vivo inhibition of seed development and reserve protein accumulation in recombinants of abscisic acid biosynthesis and responsiveness mutants in *Arabidopsis thaliana*. Plant Physiol. **90**, 463-469.

Koornneef,M. (1987). Linkage map of *Arabidopsis thaliana*(2n=10). In O'Brien,S.J. (ed.), Genetic Maps. Cold Spring Harbor Laboratory Press, Cold Spring Harbor, NY, pp. 742-745.

Koornneef,M., Elgersma,A., Hanhart,E.P., van Loenen-Martinet,C., van Rijn,L. and Zeevaart,J.A.D. (1985). A gibberellin insensitive mutant of *Arabidopsis thaliana*. Physiol. Plant., **65**, 33-39.

Koornneef,M., Reuling,G. and Karssen,C.M. (1984). The isolation and characterization of abscisic acid-insensitive mutants of *Arabidopsis thaliana*. Physiol. Plant., **61**, 377-383.

Koornneef,M and van der Veen,J.H. (1980). Induction and analysis of gibberellin sensitive mutants in *Arabidopsis thaliana* (L.) Heynh. Theor. Appl. Genet., **58**, 257-263.

Leutwiler,L.S., Hough-Evans,B.R. and Meyerowitz,E.M. (1984). The DNA of *Arabidopsis thaliana*. Mol.Gen.Genet., **194**, 15-23.

Meyerowitz,E.M. (1987). *Arabidopsis thaliana*. Ann. Rev. Genet., **21**, 93-111.

Nam,H-G., Giraudat,J., den Boer,B., Moonan,F., Loos,W.D.B., Hauge,B.M. and Goodman,H.M. (1989). Restriction fragment length polymorphism linkage map of *Arabidopsis thaliana*. Plant Cell, **1**, 699-705.

Pruitt,R.E. and Meyerowitz,E.M. (1986). Characterization of the genome of *Arabidopsis thaliana*. J.Mol.Biol., **187**, 169-183.

Sulston,J., Mallett,F., Durbin,R. and Horsnell,T. (1989). Image analysis of restriction enzyme fingerprint autoradiograms. Cabios, **5**, 101-106.

APPLICATION OF RESTRICTION FRAGMENT LENGTH POLYMORPHISM TO MAIZE BREEDING*

Michael G. Murray, Yan San Chyi, Jane H. Cramer, Sandra DeMars, Jane Kirschman, Yu Ma, Jan Pitas, Jeanne Romero-Severson, Jennifer Shoemaker, David P. West, and Dave Zaitlin

Agrigenetics Company, Madison WI 53716, USA

## INTRODUCTION

Restriction fragment length polymorphism (RFLP) is attracting considerable attention in the plant breeding community. RFLP-based genetic maps either have been or are being developed for a wide variety of crops (e.g. Apuya *et al.*, 1988; Bernatzky and Tanksley, 1985; Bonierbale *et al.*, 1988; Burr *et al.*,1988; Gebhart *et al.*, 1988; Havey *et al.*, 1989; Helentjaris, 1987; Landry *et al.*, 1987; McCouch *et al.*, 1988; Murray *et al.*, 1988a; Nam *et al.*, 1989) and the technology is already being applied to real-world breeding problems. The development of new cultivars depends on genetic recombination followed by selection. However, historically breeders have lacked an efficient system for monitoring recombination. This deficiency coupled with the apparent genetic complexity of the most economically valuable traits, has necessitated the use of large populations and extensive field evaluation. In conjunction with reliable field evaluation techniques, a well-saturated RFLP map now makes it possible to tag chromosomal segments carrying loci that contribute to traits which are difficult and/or expensive to score in the field. It then becomes possible to trace these segments using molecular techniques in lieu of field evaluation and to accurately assess the genetic composition of progenies. RFLP, thus, makes the old idea of marker-based breeding truly practical (see Tanksley *et al.*, 1989). We discuss here several examples of how RFLP technology can be applied to basic and applied problems in maize breeding and consider some future prospects for the technology.

## THE AGRIGENETICS MAIZE RFLP LINKAGE MAP

The construction of Agrigenetics' Maize RFLP map (Figure 1) has been described in detail previously (Murray *et al.*,1988a). The map is derived from linkage data observed in an A619Ht X Mangelsdorf's tester F2 population consisting of 87 individuals. Genetic linkage was detected using the method of Mather (1938) and recombination frequencies were calculated by the method of maximum likelihood (Allard, 1956). Recombinant DNA libraries representing different classes of sequences were screened for useful single copy clones. These included two cDNA libraries prepared from etiolated coleoptile total poly(A) and seedling root polysomal poly(A) RNA and two libraries made from 1-2 kBp size-selected *Pst*1 or *Xho*1 genomic DNA fragments. The genomic DNA libraries exploited the observation of Burr *et al.* (1988) that under-methylated sequences appear to be enriched for single copy sequences in maize. Although clones marking a single locus were preferred, the map does include markers that hybridize to more than one locus. We have also used a large number of markers from other researchers in the construction of our map (see Figure 1 legend).

*In memory of Dr. Craig Cowley, a friend, colleague, and an excellent corn breeder.

*Plant Molecular Biology* 2, Edited by R.G. Herrmann and
B. Larkins, Plenum Press, New York, 1991 

## AGRIGENETICS MAIZE LINKAGE MAP

May, 1990

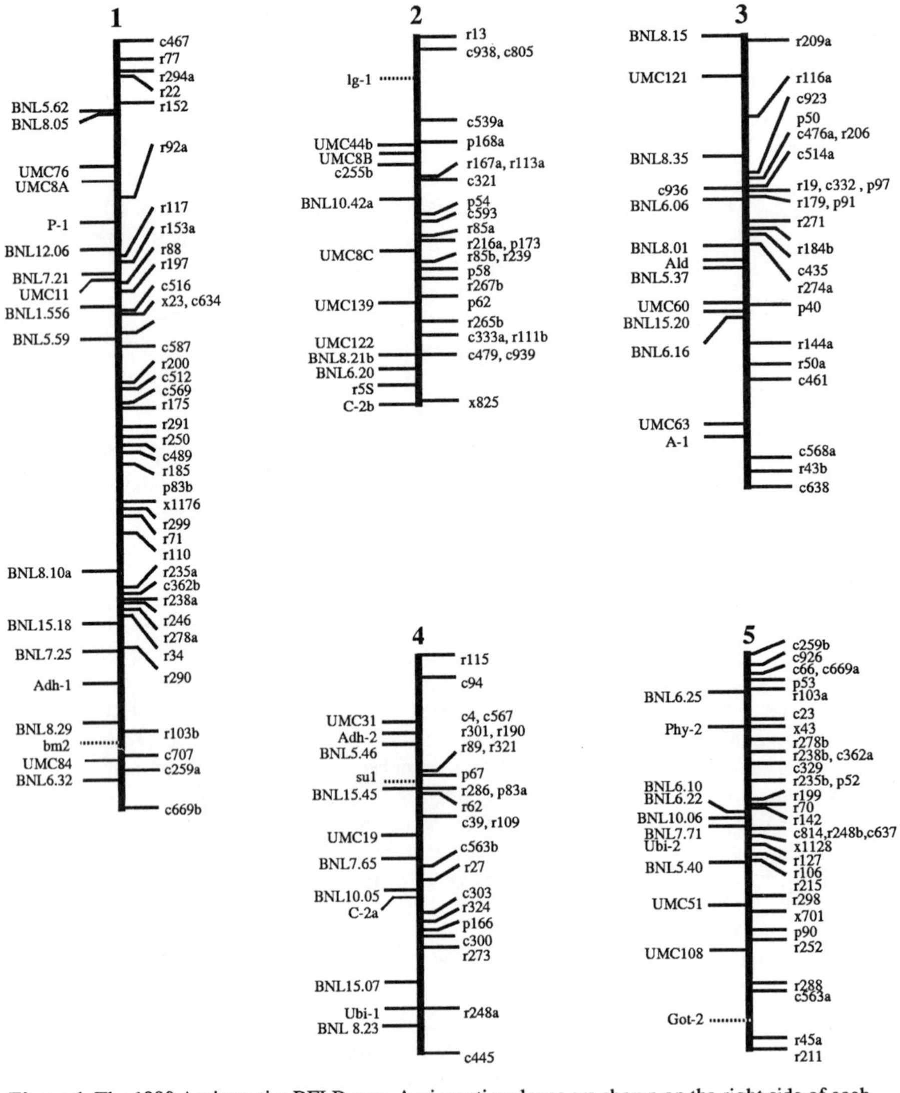


**Figure 1.** The 1990 Agrigenetics RFLP map. Agrigenetics clones are shown on the right side of each chromosome. Established genetic markers and clones obtained from other researchers are shown on the left.Codes are: BNL=Brookhaven National Laboratory; UMC=University of Missouri-Columbia; c=coleoptile cDNA clones; r=root cDNA clones; p=PstI genomic clones; x=XhoI genomic clones; a or b (suffix)=duplicated sequences recognized by a single clone. Genes of known identity obtained from other researchers are: A-1, S. Schwarz-Sommer; Adh-1, Adh-2 and Ald, E. Dennis; bz-1, O.E. Nelson; C-2, V. Weinand; P-1, T. Peterson; r5S, E. Zimmer; sh-1 and Sus, L.C. Hannah; Phy-2, Ubi-1 and Ubi-2, A. Christensen and P. Quail; wx-1, S. Wessler; 15b zein, B. Larkins; and 27g zein, A. Esen.

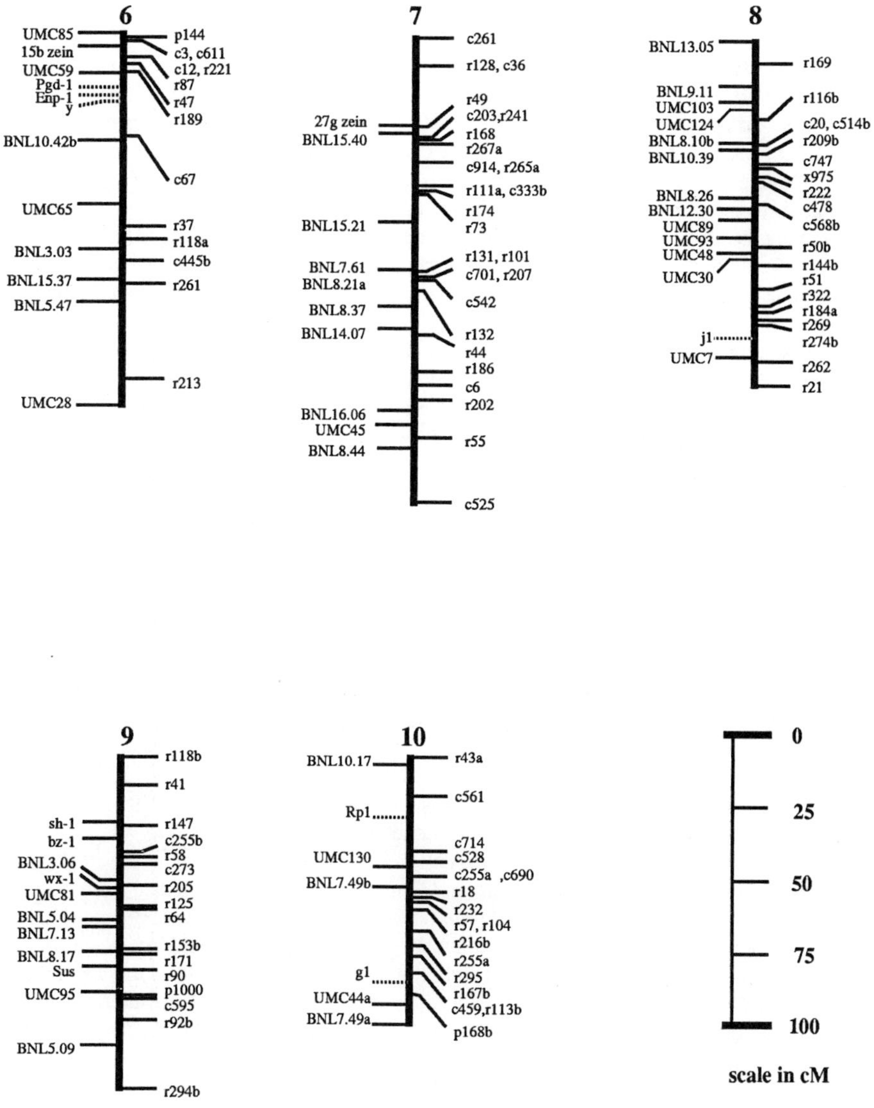
6
UMC85
15b zein
UMC59
Pgd-1
Enp-1
y
BNL10.42b
UMC65
BNL3.03
BNL15.37
BNL5.47
UMC28
p144
c3, c611
c12, r221
r87
r47
r189
c67
r37
r118a
c445b
r261
r213
7
27g zein
BNL15.40
BNL15.21
BNL7.61
BNL8.21a
BNL8.37
BNL14.07
BNL16.06
UMC45
BNL8.44
c261
r128, c36
r49
c203,r241
r168
r267a
c914, r265a
r111a, c333b
r174
r73
r131, r101
c701, r207
c542
r132
r44
r186
c6
r202
r55
c525
8
BNL13.05
BNL9.11
UMC103
UMC124
BNL8.10b
BNL10.39
BNL8.26
BNL12.30
UMC89
UMC93
UMC48
UMC30
j1
UMC7
r169
r116b
c20, c514b
r209b
c747
x975
r222
c478
c568b
r50b
r144b
r51
r322
r184a
r269
r274b
r262
r21
9
sh-1
bz-1
BNL3.06
wx-1
UMC81
BNL5.04
BNL7.13
BNL8.17
Sus
UMC95
BNL5.09
r118b
r41
r147
c255b
r58
c273
r205
r125
r64
r153b
r171
r90
p1000
c595
r92b
r294b
10
BNL10.17
Rp1
UMC130
BNL7.49b
g1
UMC44a
BNL7.49a
r43a
c561
c714
c528
c255a ,c690
r18
r232
r57, r104
r216b
r255a
r295
r167b
c459,r113b
p168b
0
25
50
75
100
scale in cM

The current map consists of 343 independent chromosomal loci designated by 299 probes. Of the 299 probes, 201 are proprietary, 55 are from Brookhaven National Laboratory (Burr *et al.*, 1988), 29 are from the University of Missouri (see Coe *et al.*, 1988) and 15 are cloned genes obtained from a number of researchers. The loci cover 1750 cM with an average spacing of 6.5 cM. The linear order has been confirmed in a subsequent F2 population of 103 individuals and a recombinant inbred population consisting of 200 individuals.

## HISTORICAL ANALYSIS OF CURRENT GERMPLASM

With RFLP, it will ultimately be possible to trace the ancestry of specific chromosomal segments from maize progenitors through the domestication of modern commercial lines and varieties. Helentjaris (1988) has demonstrated that RFLP analysis of mummified corn tissue (800 to 1200 AD) can be used to evaluate breeding practices among the Anasazi culture of the American Southwest, and one may expect increasing use of this technology in archaeology and systematics. RFLP markers are useful for evaluating relative genetic distances among modern maize inbreds as well as revealing the broad-scale selections that have resulted from intensive plant breeding (Murray *et al.*, 1988b). Combined with known pedigree records, RFLP technology will allow precise evaluation of the contributions of the individual progenitors to commercially significant germplasm.

Reid Yellow Dent is an example of a maize cultivar that was grown throughout much of the U.S. corn belt in the latter half of the last century, and it still features in many modern commercial hybrids through the inbreds that came out of the Iowa Stiff Stalk Synthetic [(BSSS); Baker (1984)]. During the late 19$^{th}$ and early 20$^{th}$ centuries, Reid Yellow Dent was the source of a number of selected varieties or subpopulations such as Wilson Farm Reid (Purdue 11) and Funk's Yellow Dent, which in turn served as sources of germplasm in the development of many of the earliest corn inbreds (see Crabb, 1947). A stand-out among these early inbreds was Wf9, a 1$^{st}$ cycle selection from Wilson Farm Reid. Wf9 featured prominently in many public double-cross hybrids into the 1950's, and even in some of the single crosses of the 1960's (Baker, 1984). A large number of maize inbreds are descended from Wf9 (see Figure 2), although none were ever as completely embraced by the hybrid corn industry as was Wf9 itself.

In the process of developing a comprehensive data base for major maize inbreds, we observed that the RFLP marker UMC6 failed to hybridize with DNA from the lines Wf9, A554, A654, and Pa91 (Zaitlin, manuscript in preparation). The most reasonable explanation is that the corresponding chromosomal region has been deleted in these lines. Because pedigree records (Stringfellow, Jenkins, personal communications) revealed that A554, A654 and Pa91 share Wf9 as a common ancestor, the analysis was extended to a large number of lines related to Wf9 (Figure 2). Of the inbreds examined, the UMC6 deletion was detected in 8 of 11 linear descendants of Wf9, 3 of 5 that came directly out of Funk's Yellow Dent (and Funk 176A) and 4 of 5 selected from a double-cross hybrid, Funk's G10. The deletion is also found in Pa91 and 1 of 3 of its descendents (NC254), as well as in 2 of 4 inbreds derived from a complex Wf9-based synthetic population. A survey of another 75 maize inbreds, representing at least eight major genetic groups, reveals that this DNA sequence is deleted solely in material that traces its ancestry to Reid Yellow Dent, and not in inbred lines derived from other open-pollinated midwestern dent varieties such as Midland Yellow Dent, Wooster Clarage, Lancaster Surecrop, Yellow Surecropper or Jarvis Prolific.

Additional analysis of 51 related S3 families descended directly from Reid Yellow Dent has shown that the UMC6 deletion is the most prevalent allele of 8 detected at this locus. Since Reid Yellow Dent itself arose by the combination of two diverse types of maize (dent and flint), it is relevant here to consider the possible source of this deletion. Northern Flint varieties in general have short generation times. One Northern Flint, Gaspé Flint, flowers in 35 days after emergence. Isozyme analysis suggests extreme genetic distance from dent corn (Doebley *et al.*, 1986). It is intriguing to speculate that the UMC6 deletion arose in this distinct type of maize as a result of the overall reduction in genome size that accompanied the adaptation of Northern Flints to the environment of northeastern North America. The deletion has also been detected in the Canadian inbred CQ205, a selection out of (Gaspé Flint x A495)A495. It will, therefore, be interesting to determine whether this deletion exists in Gaspé Flint, which has the smallest genome of the Northern Flint corns examined (Rayburn *et al.*, 1985).

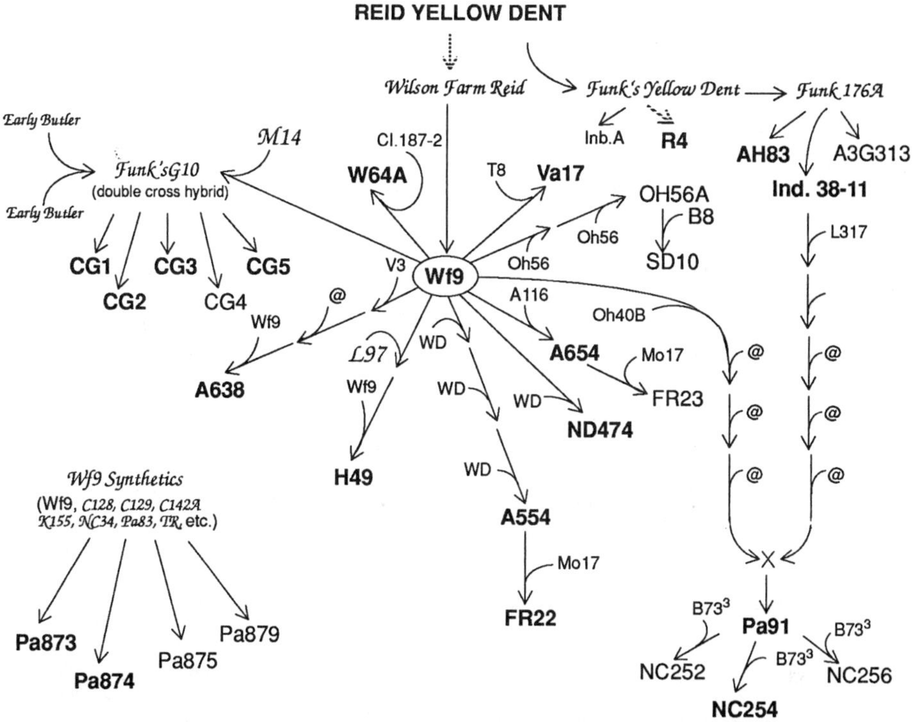


Figure 2. Documented pedigrees of Wf9-derived and related inbred lines of maize. Fully established relationships are indicated by solid lines. The arrows emanating from Wf9 and other parental inbreds represent a single generation of hybridization. It is inferred that as many as 6 generations of selfing follow the final hybridization event prior to the release of named inbred lines. Dashed arrows indicate some uncertainty in the events that led to the development of a maize inbred or variety. Reid Yellow Dent, Funk's Yellow Dent, Wilson Farm Reid and Funk 176A are open-pollinated Corn Belt Dent cultivars. The lines known to carry the chromosomal deletion at UMC6 are shown in boldface type, and the inbreds and varieties that have not been analyzed with UMC6 are italicized. Generations in which progenitors were selfed during the development of an inbred are indicated by an "@". The pedigree information contained in this diagram was obtained from the 16th edition of The Seedsman's Handbook (Mike Brayton Seeds, Inc., Ames, Iowa), published historical sources and communications from academic and industrial corn breeders.

While the existence of this genomic deletion is of some interest by itself, the phenomenon is not unknown in maize (Turner *et al.*, 1988). We have observed that several other probes occasionally reveal null alleles in maize, and RFLP-detected deletions have also been found in other crop species (e.g. rice, McCouch *et al.*, 1988 and tomato, Young *et al.*, 1988). However, UMC6 is the only probe to detect a locus that is entirely absent in maize inbreds of known lineage. What is particularly intriguing about the UMC6 deletion is its high level of genetic persistence. In one case (A554), it was retained through two additional generations of backcrossing with the inbred WD. We speculate that the deleted allele must have conveyed some advantage to have been independently selected over many years in separate maize breeding programs. This theory is supported by the fact that there is no consistent pattern of inheritance for the other markers on 2S which flank UMC6 (data not shown).

## USE OF RFLP IN BACKROSSING

Backcross conversion is a routine plant breeding procedure used to move individual traits into established inbreds. Theoretically, six backcrosses to the recurrent parent followed by selfing should produce a "near isogenic line" containing about 99.2% of the recurrent genome. The fidelity of recurrent genome recovery in backcross breeding programs has customarily been judged by plant appearance, performance, and combining ability patterns. In reality, breeders often find the performance of the recurrent parent to be difficult if not impossible to recover. Brinkman and Frey (1977) suggested that chromosomal segments "dragged" into the genotype of the recurrent parent may undermine the recovery of the line's potential, but breeders have had no means of monitoring the actual genomic composition. Accurate estimation of the true extent of "linkage drag" is now quite practical with RFLP (e.g. Young and Tanksley,1989).

Extensive analysis of backcrossed-derived maize lines indicates that more than just the region surrounding the selected locus can be dragged into the progeny (Ma *et al.*, 1990; manuscript in preparation). Figure 3 shows an RFLP analysis of the inbred A632Ht, the presumptive isogenic conversion of the line A632 to race 1 Northern Corn Leaf Blight resistance [*Helminthosporium turcicum* Pass. race 1; currently *Excerohilum turcicum* (Pass.) Leonard + Suggs]. The original conversion involved the use of GE440 as the trait donor and 6 cycles of backcrossing (the $BC_6$ generation) (A. L. Hooker, personal communication). DNA samples of A632Ht and the recurrent parent (A632) were digested separately with *Dra*I, *Eco*RI, and *Eco*RV prior to RFLP analysis with 156 different RFLP markers. Because DNA of the original trait donor was not available at the time, we have defined only marker loci derived from the recurrent A632 parent. Loci were considered identical if no polymorphism was detected with any of the three restriction endonucleases. When polymorphisms were revealed, they were detected by all three enzymes in 98% of the comparisons made. Consequently, we think it unlikely that analysis of DNA digested with additional restriction enzymes would have provided more information. RFLP-based genome mosaic patterns or "riflotypes" showing the recurrent and non-recurrent segments were constructed by arbitrarily placing boundaries midway between the relevant markers.

The *Ht1* locus has been mapped to chromosome 2L (Hooker, 1963; Coe *et al.*, 1988) and, accordingly, the isogenic conversion A632Ht1 shows a large (approximately 50 cM) non-recurrent chromosomal segment in the expected region. However, additional non-recurrent segments are apparent on chromosomes 4L, 8L, and 9L. Nonrecurrent segments account for a total of about 11% of the genome and thus recurrent parent recovery is significantly lower than the 99% recovery predicted after 6 cycles of backcrossing. It is unlikely that retention of these unselected non-recurrent segments may have resulted merely from chance. They may result from inadvertent or deliberate selection applied by the breeder, or could reflect events which can plague breeding and defy theory explanation. Regardless, in this particular isogenic conversion, A632Ht never did recover the performance of A632 (numerous personal communications). Prior to the onset of RFLP technology, breeders could only speculate as to why they had difficulty recovering the performance of the recurrent parent. With the type of information shown in Figure 3, it is now straightforward to directly address the effect of each of the non-recurrent segments.

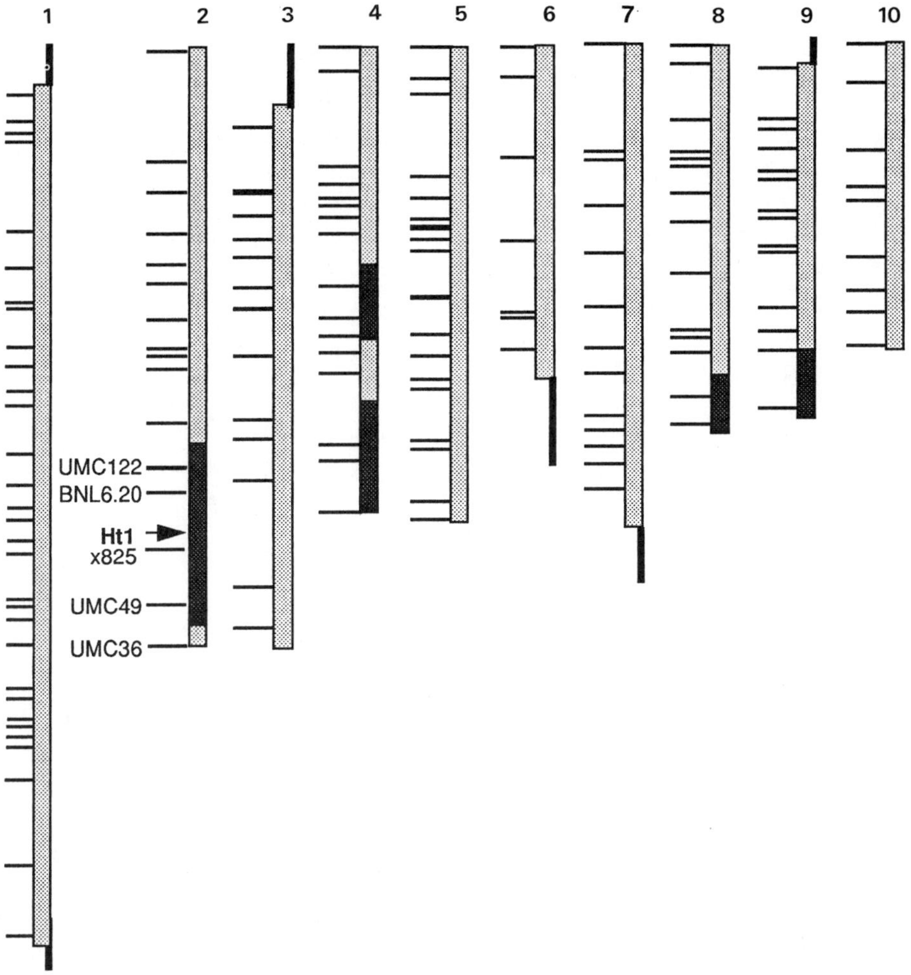


Figure 3. Riflotype of the A632Ht. Horizontal lines indicate the positions of DNA markers used to compare A632Ht with the recurrent A632 parent. Segments which match the recurrent parent are shown in grey and those presumed to be derived from the Ht1 donor (see text) are shown in black. The thin lines represent terminal chromosomal segments not assayed by RFLP analysis.

The significance of this new capability is demonstrated in Figure 4. The histogram shows the distribution of genetic contribution from both parents in an F2 population. As predicted the mean is centered about 50% of the A genome. This particular population was part of an experiment to map the *rhm* locus relative to a DNA marker and subsequently move it from line B (the donor) to A (the recipient). The dark blocks in the bar graph represent individuals which were resistant to Southern Corn Leaf Blight and are thus presumed homozygous for the *rhm* gene. This gene confers resistance to *Bipolaris maydis* (Nisik) Shoemaker and has long been thought to reside on chromosome 6S (Coe *et al.*, 1988). Riflotypes for two such resistant individuals of greatly differing overall parental constitution are also shown. It is obvious that individual 19 is a much better choice than individual 90 for further backcrossing if the objective is to recover the A genome. Thus, the ability to examine the chromosomal mosaic of various progeny allows a more efficient and precise conversion.

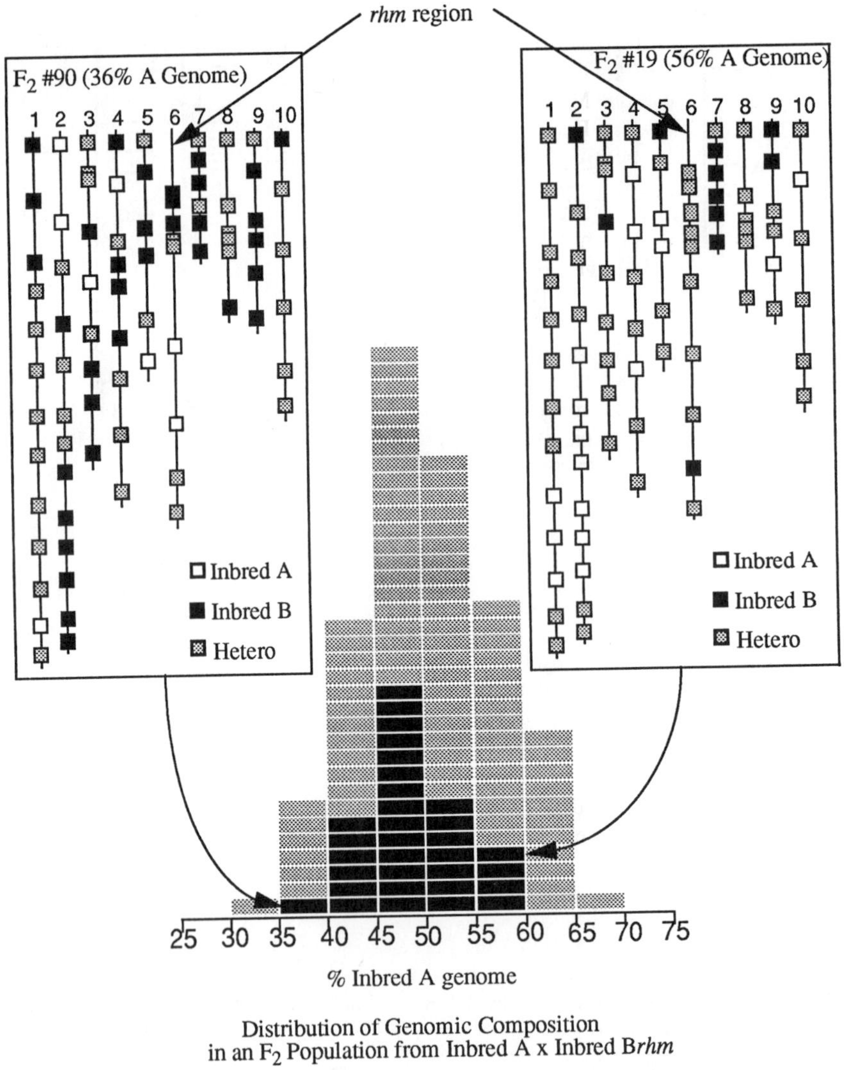


Figure 4. Use of RFLP to identify the most appropriate offspring for further breeding in a complex population. 120 $F_2$ individuals were riflotyped with 83 well distributed probes and the distribution of genetic composition plotted. Each block represents a single individual, with black blocks representing those plants resistant to *B. maydis* upon field inoculation. Riflotypes of two selected individuals are shown.

## PRACTICAL MARKER-FACILITATED BREEDING

One of the greatest potentials of RFLP technology is the understanding and manipulation of polygenic traits which have been difficult to manage in traditional breeding programs. One must first map the genes underlying these traits and then develop efficient strategies to exploit this information in a breeding program.

Resistance to Maize Dwarf Mosaic Virus (MDMV) is an example of one such trait. Breeders have historically had some difficulty incorporating effective MDMV resistance into elite lines, particularly in sweet corn. Existing literature suggests that anywhere between 1 and 5 genes are involved in resistance (reviewed in McMullen & Louie, 1989). We developed an F2 population consisting of 103 individuals using B68 as the resistance donor and B73 as the sensitive line. F2 plants were riflotyped and the $F_3$ progeny from each were screened for MDMV resistance using standard field practices. Analysis of these results identified 7 independent chromosome segments which together accounted for more than 95% of the phenotypic variance for resistance. Some of the segments were substantially more important than others. The accuracy of this identification has been subsequently proven in several different genetic backgrounds.

The identification of RFLP markers linked to loci governing MDMV resistance permits a rapid breeding program which integrates a number of technologies to convert elite maize lines to MDMV resistance while retaining their agronomic performance (Figure 5). The basic scheme involves alternating cycles of backcrossing and self-pollination combined with strong phenotypic (i.e. MDMV resistance) and genotypic (i.e. RFLP marker) selection. Field disease inoculation of the first backcross ($BC_1$) population was used to select for the maximal level of adult resistance and thereby reduce the population by an order of magnitude. It then became quite practical to riflotype this resistant subset using 90 well-spaced RFLP markers to find the most useful genotype in terms of both elite genome recovery and persistence of resistance regions.

A $Bc_1S_1$ population was generated by selfing the most ideal $BC_1$. A second cycle of disease screening (greenhouse seedling) and riflotyping was applied to determine fixation of resistance loci and the extent of recurrent genome recovery. However, on this cycle, it was only necessary to assay those markers which were heterozygous in the selected $BC_1$ parent.The best $BC_1S_1$ was backcrossed again and the disease screen and RFLP analysis again applied to the $BC_2$ progeny. The resulting best $BC_2$ individual was self-pollinated for seed increase and backcrossed a third time. $BC_2S_1$ seed was used to prepare hybrid seed for an early test of the selected line's yield potential in hybrid combination (i.e. combining ability) during the summer of 1990. RFLP analysis showed that selected individual had recovered 87% of the elite recurrent genome while maintaining a high level of MDMV resistance.

In addition to the gains achieved by RFLP-mediated genotypic selection, we were able to save one month per generation by direct germination of 14 day-old immature embryos thus eliminating lag time due to endosperm development, dry-down and germination. This integrated approach allowed the first yield trial one year from the screening of the first backcross. In contrast, a conventional breeding program would normally take from about 3 years to reach the same point. In a practical sense, the recovery of the elite parent's combining ability while maintaining practical levels of resistance will be the most crucial tests for this technology. Without knowledge of the genotypic constitution of the select individual, such early testing would have been futile.

## USE OF RFLP IN COMMERCIALIZING TRANSGENIC PLANTS

RFLP-facilitated breeding will be an essential component in commercializing transgenic plants. From a practical perspective, transgenic plants are merely trait donors. The potential value of a trait cannot be realized until it is incorporated into competitive germplasm. The present technology in maize does not permit routine transformation of elite commercial inbreds. Regenerable lines such as Black Mexican Sweet and A188 may have laboratory utility but few redeeming virtues in the field. RFLP analysis has identified several chromosomal regions of A188 origin that were associated with enhanced regenerability when transferred into elite germplasm (Armstrong *et al.*, in preparation). Thus, it may be possible to develop more elite regenerable lines through selective breeding.

Until such time as every current elite line can be transformed and regenerated at will, substantial breeding will be needed to commercialize transgenic traits. In theory, the transfer into an elite line seems to be a simple case of backcross breeding. The foreign gene is easy to detect so

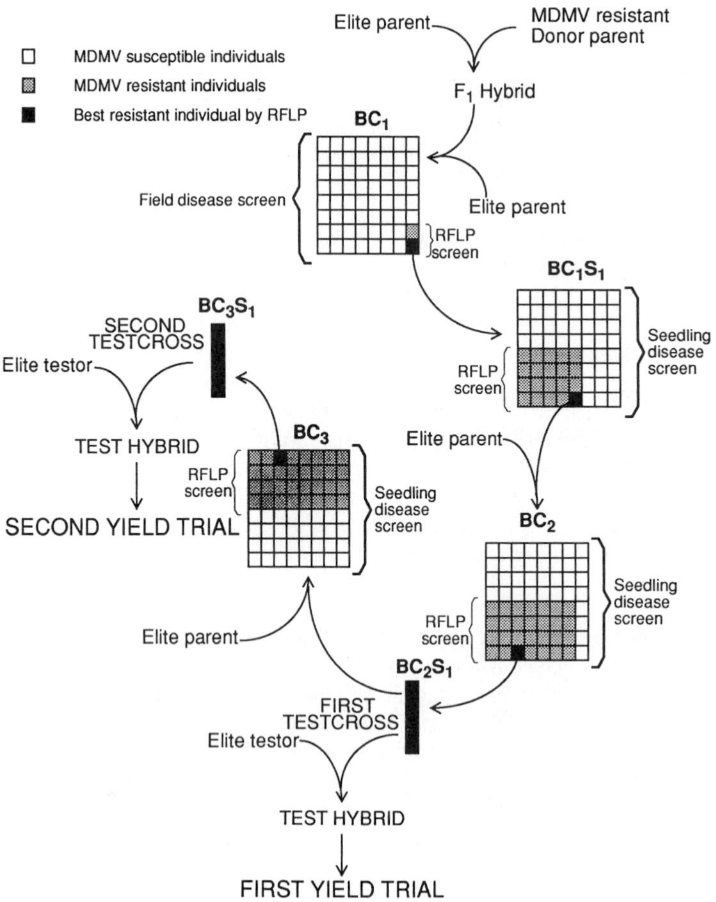


Figure 5. RFLP-facilitated breeding scheme. Grids represent populations resulting from the indicated backcrossing and self pollination steps. Shaded areas denote individuals which were resistant to MDMV and black boxes represent those resistant individuals with the most auspicious riflotypes.

numbers of plants at each generation can be kept quite low. The time needed is simply a function of how many generations can be cycled per year. Given the right maturity and good greenhouse and field capabilities, 3 generations per year are possible with maize (see Figure 5). It is clear that RFLP-based backcrossing can make this process much more efficient. However, more serious complications arise if the gene is inserted into an unfavorable donor segment. In such a case, substantial extra effort will be required to break deleterious linkages. Thus, prudence dictates that we should know the map position of the introduced gene before embarking on the effort. In reality, one may need to map a large number of independent transgenic plants to choose the most effective donor. RFLP-assisted breeding will be essential to bring the product to market as fast as possible.

## FUTURE CONSIDERATIONS

In less than ten years, the developing science of RFLP technology has expanded from an interesting theory (Botstein *et al.*, 1980) to an industrial practice in several crops. RFLP technology is powerful but costly when compared to traditional plant breeding methods. Improvements in probe detection systems, lab automation, data acquisition, and computer software will be essential before RFLP can have its full impact in routine breeding. Fortunately the medical diagnostics industry is spawning tremendous innovation (e.g., Polymerase Chain Reactions) some of which will have direct application to the plant breeding issues.

## ACKNOWLEDGEMENT

We wish to acknowledge Jack Gardiner, Dave Hoisington, and Ben and Frances Burr for the UMC and BNL maize markers and Prof. Dale Steffensen for Reid Yellow Dent seed samples. We thank Lynette Bausch, Diana Mefford and Mary Hoenecke for the preparation of the final manuscript.

## REFERENCES

Allard, R. W., 1956, Formulas and tables to facilitate the calculation of recombination values in heredity, Hilgardia 24:235.

Apuya, N. R., Frazier, B. L., Keim, P., Roth, E. J., and Lark, K. G., 1988, Restriction fragment length polymorphisms as genetic markers in soybean, *Glycine max* (L.) Merril, Theor. Appl. Genet. 75:889.

Baker, R., 1984, Some of the open-pollinated varieties that contributed most to modern hybrid corn, in: "Proc. 20th Annual Corn Breeders School", Champaign, Ill. pp. 1-19.

Bernatzky, R. and Tanksley, S. D., 1985, Toward a saturated linkage map in tomato based on isozymes and random cDNA sequences, Genetics 112:887.

Bonierbale, M.W., Plaisted, R.W., and Tanksley, S. D., 1988, RFLP maps based on a common set of clones reveal modes of chromosomal evolution in potato and tomato, Genetics 120:1095.

Botstein, D., R. L. White, M. Skolnick, and R.W. Davis, 1980, Construction of a genetic linkage map in man using restriction fragment length polymorphism. Am. J. Human Genet. 32:314-331.

Brinkman, M. A. and Frey, K. J., 1977, Yield-component analysis of oat isolines that produce different grain yields, Crop. Sci. 17:165.

Burr, B., Burr, F.A., Thompson, K. H., Albertson, M. C., and Stuber, C.W., 1988, Gene mapping with recombinant inbreds in maize, Genetics 118:519.

Coe, E. H., Neuffer, M.G., and Hoisington, D.A., 1988, The Genetics of Corn, in: "Corn and Corn Improvement", Third Edition, American Society of Agronomy, Sprague, G. F. and Dudley, J. W., eds., Madison, WI, pp. 81-258.

Crabb, A. Richard, 1947, "The Hybrid Corn Makers: Prophets of Plenty." Rutgers University Press, New Brunswick, NJ, pp. 331.

Doebley, J. F., Goodman, M. M., and Stuber, C.W., 1986, Exceptional genetic divergence of Northern Flint Corn, Amer. J. Bot. 73:64.

Gebhardt, C., Ritter, E., Debener, T., Schachtschabel, U., Walkemeir, B., Uhrig, H., and Salamini, F., 1989, RFLP analysis and linkage mapping in *Solanum tuberosum*, Theor. Appl. Genet. 78:65.

Havey, M. J. and Muehlbauer, F. J., 1989, Linkages between restriction fragment length, isozyme, and morphological markers in lentil, Theor. Appl. Genet. 77:395.

Helentjaris, T., 1987., A genetic linkage map for maize based on RFLPs, Trends in Genetics, 3:217.

Helentjaris, T., 1988., Does RFLP analysis of Ancient Anasazi samples suggest that they utilized hybrid maize, Maize Genetics Cooperation Newsletter, 62:104.

Hooker, A. L., 1963, Monogenic resistance to *Zea mays* L. *Helminthosporium turcicum*, Crop. Sci. 3:381.

Landry, B. S., Kesseli, R.V., Farrara, B., and Michelmore, R.W., 1987, A genetic map of lettuce (*Latuca sativa* L.) with restriction fragment length polymorphism, isozyme, disease resistance, and morphological markers, Genetics 116:331.

Mather, K., 1938, The measurement of linkage in heredity, Metheun & Co. Ltd., London.

McCouch, S. R., Kochert, G., Yu, Z. H., Wang, Z.Y., Khush, G. S., Coffman, W. R., and Tanksley, S. D., 1988, Molecular mapping of rice chromosomes, Theor. Appl. Genet. 76:815.

McMullen, M. D. and Louie, R., 1989, The linkage of molecular markers to a gene controlling the symptom response in maize to maize dwarf mosaic virus, Molec. Plant-microbe Interact. 6:309.

Murray, M., Cramer, J., Ma, Y., West, D., Romero-Severson, J., Pitas, J., DeMars, S., Vilbrandt, L., Kirschman, J., McLeester, B., Schilz, J., and Lotzer, J., 1988a, Agrigenetics maize RFLP linkage map, Maize Genetics Cooperation Newsletter 62:89.

Murray, M. G. , Ma, Y., Romero-Severson, J., West, D. P., and Cramer, J. H., 1988b, What are restriction fragment length polymorphisms and how can breeders use them. Proceedings of the 43rd Annual Corn and Sorghum Industry Research Conference 43:72.

Nam, H.-G., J. Giraudat, B. den Boer, F. Moonan, W.D.B. Loos, B. M. Hauge, and H. M. Goodman, 1989, Restriction fragment length polymorphism linkage map of *Arabidopsis thaliana*.The Plant Cell 1:699-705.

Rayburn, A. L., Price, H. J., Smith, J. D., and Gold, J. R., 1985, C-band heterochromatin and DNA content in *Zea mays*, Amer. J. Bot. 72:1610.

Tanksley, S. D., Young, N. D., Patterson, A. H., and Bonierbale, M.W., 1989, RFLP mapping in plant breeding: New tools for an old science, Bio/Technology 7:257.

Turner,V., Wright, S., Suzuki, J., and Helentjaris, T., 1988, Deletions of Loci detected by RFLP, Maize Genetics Cooperation Newsletter 62:103.

Young, N. D., Zamir, D., Ganal, M.W., and Tanksley, S. D., 1988, Use of isogenic lines and simultaneous probing to identify DNA markers tightly linked to the Tm-2a gene in tomato, Genetics 120:579.

Young, N. D. and Tanksley, S. D., 1989, RFLP analysis of the size of chromosomal segments retained around the Tm-2 locus of tomato during backcross breeding, Theor. Appl. Genet. 77:353.

# SOYBEAN GENOME ANALYSIS: DNA POLYMORPHISMS ARE IDENTIFIED BY OLIGONUCLEOTIDE PRIMERS OF ARBITRARY SEQUENCE

Scott V. Tingey, J. Antoni Rafalski, John G.K. Williams
and Scott Sebastian

E.I. du Pont de Nemours and Company
P.O. Box 80402
Wilmington, DE 19880-0402

## Abstract

We have recently constructed a complete linkage map of *Glycine max* using restriction fragment length polymorphism technology. The map contains more than 550 markers which describe 23 linkage groups and 2700 cM of DNA. Given the level of polymorphism between soybean cultivars, for a single genetic map to have general utility in a breeding program, the map must contain over 1300 markers. To reach this goal, we have developed new technology for genetic mapping that promises to automate genotype determination. The new assay is called a RAPD assay after **R**andom **A**mplified **P**olymorphic **D**NA. It is based on the observation that single oligonucleotide primers of arbitrary sequence can be used to amplify discrete loci in a complex genome.

## Soybean Genome Analysis

Soybean, one of the major U.S. and world crops, has received relatively little attention from the field of genetics. This is because of the fairly narrow genetic base of modern cultivated soybean. The classical genetic map of soybean contains only 50 loci. These loci define 17 linkage groups, many comprised of only two markers (Palmer and Kiang, 1990). Researchers were forced to rely on a large number simple two or three point crosses, leaving great doubt as to the relative position of any two genetic markers mapped in separate crosses.

To define the soybean genome in more detail we have created a complete genetic linkage map of soybean, within a single cross, using restriction fragment length polymorphism (RFLP) technology. This map, to be described elsewhere (Rafalski and Tingey, 1990) was constructed by screening over three thousand low copy number soybean genomic DNA clones to identify polymorphisms between Bonus, a cultivar of Glycine max and PI 81762 an accession of Glycine soja. Approximately 550 polymorphic probes have been mapped in a population of 68 F2 individuals and their F2-derived F3 families. The total length of the soybean genome as measured by recombination is 2700

*Plant Molecular Biology* 2, Edited by R.G. Herrmann and
B. Larkins, Plenum Press, New York, 1991

cM. The map contains 23 linkage groups and an average distance between markers of 7 cM. This results in a map resolution that will yield 95% confidence in finding a RFLP marker within 7.2 cM of any locus in the soybean genome.

For any single genetic map to have utility in a plant breeding program it must contain a sufficient number markers to be able to distinguish between any two genomes of interest, at a minimum map resolution of 20 cM. The total number of markers required will be a function of the degree of polymorphism within the gene pool, and to some extent will be effected by the organization (ploidy level) of the genome. In 1983, Delannay et al. completed a survey of the pedigrees of 158 USA and Canadian cultivars of soybean. This study indicated that only 10 introductions contributed, collectively, more than 80% of the northern gene pool, while only 7 contributed the same amount to the southern gene pool. This would suggest that the current gene pool available to soybean breeders is in fact quite narrow. We have recently completed a RFLP survey of these same introductions in part to confirm these suggestions, and to determine the degree of polymorphism that exists in the breeding gene pool. This also allowed us to estimate the total number of RFLP markers that would be necessary to effectively describe the soybean breeding gene pool. The survey was accomplished by assaying genotype with over 200 RFLP markers representing an average map resolution in soybean of 20 cM. Genotype was determined for each of the 12 introductions implicated above (ancestral lines), 13 elite cultivars representing a conservative selection of the current breeding gene pool, and 8 principle introductions (PIs) that have contributed to several gene introgression programs (see Table 1) and represent a somewhat random sample of the G. max gene pool. Comparisons were made between these three populations to determine the extent of polymorphism, and the degree genetic identity among alleles from each population.

The results of this survey suggest that cultivated soybean indeed does have a fairly narrow genetic base. Greater than 46% of all loci tested were found to be monomorphic across all populations. Of the remaining loci, nearly 25% had a frequency of the predominant allele exceeding 90% in each population. Of the RFLP markers that had detected polymorphism between Bonus and PI81762 (the parents used to create the soybean genetic map), only 64% detected polymorphism within the elite gene pool. Most markers detected only two alleles at each locus, and the average frequency of the predominant allele was 0.24. The probability of drawing two individuals from the elite gene pool, each containing a different allele is $2pq = 0.36$. This means that on average, only 23% of the markers previously mapped between Bonus and PI81762 will be infor-

Table 1. Soybean Germplasm Survey.

| Elite | Ancestral | Principle Introduction |
|---|---|---|
| Bonus | A.K. Harrow | A5 |
| Essex | CNS | A6 |
| Williams 82 | Manchu | N85-2176 |
| A1937 | Mandarin | Pando |
| A3127 | Mukden | Peking |
| A3205 | PI 240664 | PI 84946-2 |
| A3307 | PI 54610 | PI 90763 |
| A3966 | Richland | PI 437654 |
| A4271 | Roanoke | |
| A4595 | S-100 | |
| A4906 | Strain 171 | |
| A4997 | Tokyo | |
| A5474 | | |

mative for any two elite inbred soybean lines. It was our experience that approximately 300 random markers must be mapped in the soybean genome before the average map resolution reaches 20 cM. If the goal is to have a standard genetic map that will provide an average map resolution of 20 cM between markers, for any two inbreds, then the total number of mapped markers must exceed 1300.

Between each population allele identity is high; Nei's similarity index (IN, Nei, 1965) shows that 93% and 96% of the alleles can be considered identical between elite and PI or elite and ancestral populations, respectively, and that they are present at similar frequencies in each population. This means that polymorphic markers detected in one population of soybean can reliably assay another.

In order to increase the utility of a standard soybean map (ie. to define a 20 cM map for any two elite cultivars) approximately another 750 markers must be mapped in the soybean genome. To accomplish this task, and to increase the chance that molecular markers will be useful in soybean breeding programs, we have begun to discover and implement technology that will automate the process of determining genotype.

**Technology Development**

For the past ten years researchers have enjoyed the power and utility of RFLP analysis. These techniques have led to the development of complete genetic maps in several diverse organisms, many of which did not exist before the advent of this technology. The procedure that allows one to observe a RFLP, Southern blot analysis, suffers in that it does not lend itself to automation. High levels of sample processing are limited and the techniques become repetitive and mundane. Many new assays have appeared in the past few years employing the polymerase chain reaction (PCR). The advantage of these techniques lie in the fact that they do lend themselves to automation. Unfortunately, the major cost of automating a PCR based assay is the requirement of target sequence information. For example, in order to determine the genotype of a single soybean genome using a PCR based assay, you would have to determine the target sequence of 100 to 200 loci, each of which must describe an assayable polymorphism. In any genome, especially those that have a relatively low level of sequence diversity, this task becomes prohibitive.

We have recently described a new assay (Williams *et al.*, 1990), distinct from the PCR, that can assay discrete loci without the overhead of sequence determination. The concept of the assay, described in Figure 1, depends on the observation that a single oligonucleotide primer of random sequence will initiate the reproducible amplification of discrete loci (2-7) in a complex genome. This new assay, which we have named the RAPD assay after **R**andom **A**mplified **P**olymorphic **D**NA, provides a means for automating the tedious process of genotype determination. Figure 2 shows the detection of polymorphism between Bonus and PI81762 with several different primers of arbitrary sequence. When RAPD markers are identified and scored in a segregating population they can be placed in a genome with high confidence (see Figure 3 and 4). This technology will make a new level of genome analysis available to the molecular geneticist. The determination of genotype can essentially be automated. Genetic maps of high density can be created for many different organisms using the same set of random primers. The sequence and molecular weight information that uniquely define a RAPD marker can be maintained electronically doing away with the storage and maintenance of large clone banks, while providing a rapid means of information dissemination.

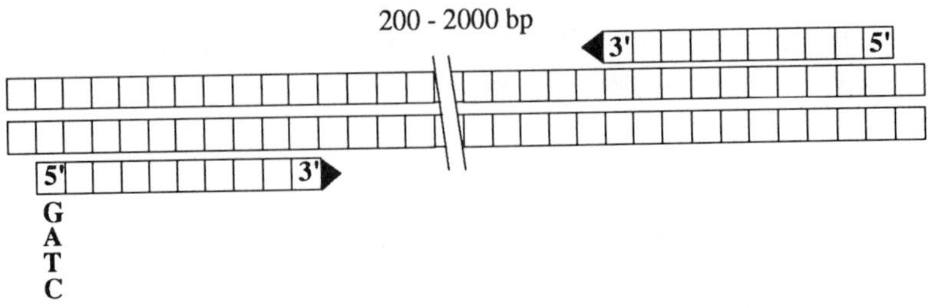


<u>**Figure 1.**</u>

A single primer of arbitrary nucleotide sequence is used to detect polymorphism in a complex genome. The chance that a single primer of arbitrary sequence will produce an effective amplification is the product of the probability of finding the 10 base sequence justaposed in the genome (1/4E20), the size of the genome (3x10E9 bases), and the number of bases that could separate the primers (~1800). The result is that in soybean one would expect approximately 5 amplification products from a single oligonucleotide primer 10 bases long.

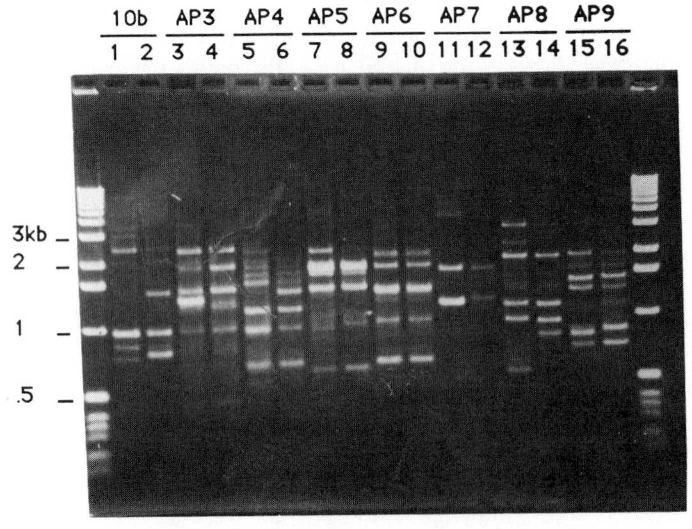


<u>**Figure 2.**</u>

Genomic DNA isolated from G. max (odd numbered lanes) or from G. soja (even numbered lanes) was amplified and compared to discover RAPD polymorphisms. Amplification reactions were in a volume of 25 ul containing 10 mM Tris-HCl pH 8.3, 50 mM KCl, 1.9 mM MgCl2, 0.001% gelatin, 0.1 mM each of dATP, dCTP, dGTP, and TTP (Pharmacia), 0.2 uM primer, 15 ng of genomic DNA, and 1 unit of Taq DNA polymerase (Perkin Elmer-Cetus). Amplification was performed in a Perkin Elmer-Cetus DNA Thermal Cycler programmed for 35 cycles of 1 min. at 94C, 1 min. at 35C, 2 min. at 72C. The products of amplification were separated by electrophoresis in a 1.4% agarose gel and subsequently stained with ethidium bromide and photographed. The sequence of the primers are as follows: 10b, GTGATGAACG; AP3, TCGTAGCCAA; AP4, TCACGATGCA; AP5, CATCGTAGTC; AP6, GCAAGTAGCT; AP7, CTGATACGGA; AP8, TGGTCACTGA; AP9, ACGGTACACT.

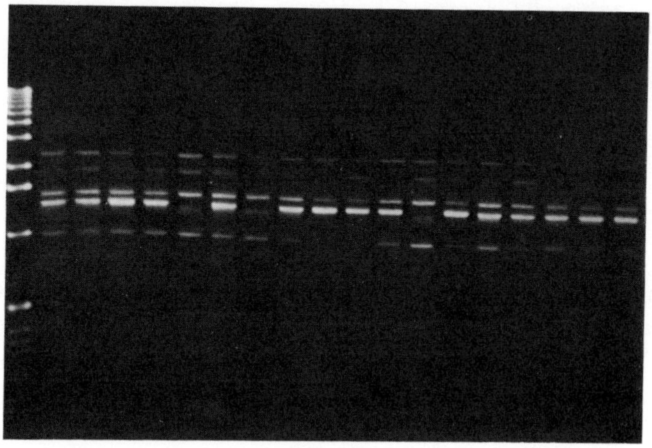

**Figure 3.**

Primer AP3 was used to detect a 1.5 Kbp polymorphism in G. max (see Figure 2). This polymorphism was detected in the amplification products of 68 F2 individuals segregating in a cross between Bonus and PI81762 (see text). Eighteen of these individuals are shown above.

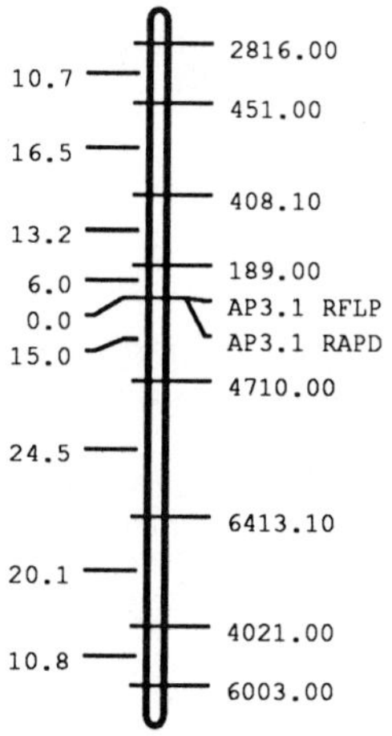


**Figure 4.**

The segregation data obtained in Figure 3 was used to map the polymorphism detected by primer AP3 (AP3-RAPD) using the MapMaker program. Primer AP3-RAPD mapped to linkage group 14 with a LOD score in support of linkage of 10.1. To confirm this map position, the genomic DNA amplified by primer AP3 was used to detect a RFLP by classical Southern blot hybridization. The AP3-RFLP marker was shown to co-segregated with the AP3 RAPD polymorphism. The figure shows the distance in centimorgans between markers on the left, and the marker names on the right.

## References

Delannay, X., Rodgers, D.M., and Palmer, R.G., 1983, Relative Genetic Contributions Amoung Ancestral Lines to North American Soybean Cultivars, Crop Science 23:944-949.

Nei, M., 1965, Genetic Distance Between Populations. Am. Nat. 106:283- 292.

Palmer, R.G., and Kaing, Y.T., 1990, Linkage Map of Soybean (Glycine max L. Merr.), in: "Genetic Maps," S. J. O'Brien, ed., Cold Spring Harbor Laboratory Press, Cold Spring Harbor, NY.

Rafalski, J.A., and Tingey, S.V., 1990, A Complete Genetic Map of the Soybean Genome, maunscript in preparation.

Williams, J.G.K., Kubelik, A.R., Livak, K., Rafalski, J.A., Tingey, S.V., 1990, DNA Polymorphisms Amplified by Arbitrary Primers Are Useful as Genetic Markers, manuscript submitted.

# GENOME ANALYSIS IN BRASSICA USING RFLPs

Thomas C. Osborn[1], Keming M. Song[1], Wayne C. Kennard[1], Mary K. Slocum[2], Scott Figdore[2], Jon Suzuki[2] and Paul H. Williams[3]

[1]Department of Agronomy, University of Wisconsin, Madison, WI 53706, USA; [2]NPI, 417 Wakara Way, Salt Lake City, UT 84108, USA; [3]Department of Plant Pathology, University of Wisconsin, Madison, WI 53706, USA

## INTRODUCTION

The analysis of plant genomes has been advanced greatly by using genetic information derived from restriction fragment length polymorphism (RFLP) of nuclear DNAs. RFLPs can be detected by digesting plant DNAs with restriction endonucleases and hybridizing Southern blots of the DNAs with labeled clones of low copy nuclear sequences. Homologous restriction fragments which vary in length can be distinguished as different molecular weight bands, and in segregating progenies, these RFLPs behave as simple Mendelian markers. RFLPs have been used to construct detailed linkage maps in several plant species, providing information on the organization and relationships of plant genomes. They also have been used to locate and study the effects of genes controlling traits of interest. They are particularly useful for analyzing quantitative traits which are under complex genetic control and influenced by the environment. These and other applications of RFLPs have great potential for assisting plant breeders in the improvement of crop species (Beckmann and Soller, 1986; Tanksley et al., 1989).

The genus *Brassica* represents an interesting and useful group of plants for genome analysis using RFLPs. Six cultivated species of *Brassica* are grown worldwide for vegetables, oil and fodder. Three of the species, *B. rapa* (syn. *campestris*) (AA genome, n = 10), *B. nigra* (BB, n = 8) and *B. oleracea* (CC, n = 9) are diploids, and these have hybridized naturally to form three amphidiploid species, *B. juncea* (AABB, n =18), *B. napus* (AACC, n = 19) and *B. carinata* (BBCC, n = 17). *Brassica* species, particularly *B. oleracea* and *B. rapa*, include accessions with very divergent morphotypes, such as cabbage, broccoli, kales, cauliflower, Chinese cabbage, Pak choi, and oil seed types. The genetic control of this morphological variation is complex, involving many genes that influence different components of the morphotypes.

Although cytogenetic studies have provided information on the chromosome relationships of *Brassica* species (Prakash and Hinata, 1980), there is very little information on genome organization and relationships

Plant Molecular Biology 2, Edited by R.G. Herrmann and B. Larkins, Plenum Press, New York, 1991 from studies with genetic markers. Also, little is known about the genetic control of morphological variation in these species. We have used RFLP markers to analyze genomes of two species, B. oleracea and B. rapa. We have constructed detailed linkage maps of RFLP loci and these markers were used to dissect the genetic control of complex morphological traits. The organization of RFLP and trait loci within and between genomes has provided substantial information on the evolution of these species.

## BRASSICA RFLP LINKAGE MAPS

### Brassica oleracea map

A detailed linkage map of B. oleracea was constructed based on segregation of 258 RFLP loci in a broccoli x cabbage $F_2$ population. The details of this map have been published (Slocum et al., 1990), and only a summary of its features will be presented here. The mapped loci were identified by probing Southern blots of 96 $F_2$ individuals with 197 random genomic DNA clones. The DNA clones were selected from a PstI library for those that hybridized to low copy number, polymorphic sequences in the B. oleracea parents (Figdore et al., 1988). The 258 loci mapped into nine linkage groups, corresponding to the haploid chromosome number of this species, and covered a total of 820 recombination units. Linkage groups have been designated 1C to 9C, and they ranged in size from 38 to 136 recombination units (Table 1).

Table 1 Characteristics of RFLP linkage maps of B. oleracea and B. rapa.

| Linkage map characteristic | Species: B. oleracea | Species: B. rapa |
|---|---|---|
| No. of loci mapped | 258 | 273 |
| No. of linkage groups | 9 | 10 |
| Map distance[a] | | |
| Total genome | 820 | 1455 |
| Range for linkage groups | 38(9C)-136(1C) | 85(10A)-209(1A) |
| RFLP probes | | |
| Total no. used for mapping | 197 | 184 |
| No. detecting single polymorphic loci | 145 | 116 |
| No. detecting multiple polymorphic loci | 52 | 68 |
| Conserved duplicated regions[b] | 1C-3C (5) | 1A-3C (4) |
| | 2C-2C (4) | 1A-8A (7) |
| | 3C-5C (4) | 3A-8A (10) |
| | 3C-8C (6) | 5A-6A (4) |
| | | 5A-10A (5) |

[a]Map distances are in units of percent recombination and linkage group designations are in parentheses.
[b]Pairs of linkage groups having at least four duplicated loci in common with the same linkage arrangements are shown. The numbers of duplicated loci in these conserved regions are in parentheses.

A majority (92.5%) of the DNA clones used as probes hybridized to more than two restriction fragments in the $F_2$ population. For many of these probes only one segregating RFLP locus was observed; however, 52 probes (26% of the total) hybridized to multiple segregating loci. Many of these mapped duplicated loci were distributed as clusters between pairs of linkage groups and the relative linkage arrangement of the loci between and within linkage groups was conserved (Table 1).

These results agree with previous cytological evidence for the presence of duplicated regions and for the evolution of B. oleracea from a lower chromosome number progenitor (Prakash and Hirata, 1980). However, based on our results, duplicated regions do not appear to span entire pairs of linkage groups, and thus, if B. oleracea has evolved by duplication of whole chromosomes, there has been considerable subsequent chromosome rearrangement. Evidence for genome rearrangements between the parents used to construct the mapping population also was obtained.

Brassica rapa map

A linkage map of B. rapa was constructed based on segregation of 273 RFLP loci in a $F_2$ population of Chinese cabbage x Spring broccoli. This map was constructed in a manner similar to that of the B. oleracea map using 184 PstI genomic DNA clones as probes on Southern blots containing DNAs from 96 $F_2$ individuals. The 273 loci mapped into 10 linkage groups, corresponding to the haploid chromosome number of this species. The total linkage distance of this map, 1455 recombinational units, was substantially larger than that of the B. oleracea map. Linkage groups have been designated 1A to 10A and they ranged in size from 85 to 209 recombination units (Table 1).

As in B oleracea, most of the DNA probes hybridized to more than two restriction fragments in the $F_2$ population, and a large proportion of the probes (37%) hybridized to multiple segregating loci. Conservation in the linkage arrangement of duplicated loci between pairs of linkage groups was even more evident in this map as compared to the B. oleracea map (Table 1). Five pairs of linkage groups had four or more duplicated loci in the same linear order, and one pair shared 10 duplicated loci that spanned large regions of the linkage groups. The conserved duplicated regions involved only portions of linkage groups and there was evidence of rearrangements, such as insertions or deletions, between these regions. The level of sequence duplication in B. rapa and B. oleracea detected in these studies is much higher than the levels detected by RFLP mapping in other species (Bernatzky and Tanksley, 1986; Bonierbale et al., 1988; Helentjaris et al., 1988; Landry et al., 1987; McCouch et al., 1988) and probably reflects a unique feature of evolution in Brassica compared to other species examined.

Comparison of B. oleracea and B. rapa maps

A common set of 165 DNA clones was used to generate the B. oleracea and B. rapa linkage maps. This has allowed us to compare linkage arrangement of loci in both populations. More than 100 RFLP loci mapped to regions including 3 or more loci with the same map order in the two species. These conserved regions spanned more than two thirds of the map distance of each genome. Between the two species, there were three pairs of linkage groups which had a single large tract of loci in common. Although loci within these conserved regions had the same linear order, the relative map distances for some adjacent loci differed between the two genomes. The remaining linkage groups included large tracts of common loci; however, each of these linkage groups included conserved regions that were present on two or three linkage groups in the other species. These observations suggest that evolutionary divergence of these species has included several chromosome translocation events.

The level of conservation in linkage order of RFLP loci between species has been assayed for tomato and pepper (Tanksley et al., 1988) and for tomato and potato (Bonierbale et al., 1988). Tomato and pepper had little conservation in linkage arrangements, and although tomato and potato are completely cross-incompatible, they had highly conserved linkage arrangements. The degree of conserved locus order observed for B. oleracea and B. rapa appears to be intermediate to that of the previous two comparisons. Based on taxonomic studies using RFLP data (Song et al., 1988 and 1990), these two Brassica species are closely related, and when crossed, they form viable but sterile amphihaploid progeny. Cytological analysis of meiosis in B. oleracea x B. rapa amphihaploids has revealed numerous multivalents which were inferred to include allosyndetic (between species) pairs of chromosomes (Attia and Robbelen, 1986). Our results are consistent with these observations and they suggest a high degree of homeology between chromosomes of the two species. However, the rearrangements between genomes that we have observed would preclude normal meiotic segregation of chromosomes in species hybrids resulting in sterility.

## ANALYSIS OF GENES CONTROLLING MORPHOLOGICAL VARIATION

### Brassica oleracea

The two parents used to develop the $F_2$ population for RFLP mapping represent extremes of morphotypes within B. oleracea. The cabbage parent 'Wisconsin Golden Acre' is a biennial, requires low temperatures for flower induction (vernalization), forms short internodes and has round sessile leaves that clasp the plant apex. The broccoli parent 'Packman' is an annual, does not require vernalization to flower, forms long internodes and has oblong, petiolate leaves that turn away from the apex of the plants. Single plants of these cultivars were crossed and a single $F_1$ plant was self-pollinated to create an $F_2$ population. The $F_1$ plant was phenotypically intermediate to the parents with a late annual flowering habit, and the $F_2$ population segregated widely for plant morphotype. Ninety-two $F_2$ plants which were used for RFLP mapping were measured for 37 morphological traits, and 10 additional traits were derived from these measurements. The traits included various flowering, leaf and stem characteristics.

The locations and effects of genes controlling these traits were determined by analyzing the variation in trait expression associated with genotypes at marker loci. Genotypic data for 72 RFLP loci scattered throughout the genome were used for this analysis. Significant associations between marker genotypes and trait expression were interpreted as indicating linkage between the marker and trait loci. This analysis allowed us to determine the minimum number of trait loci, the location of trait loci with respect to marker loci, and the magnitudes of gene effects ($R^2$ values) on trait expression. For each trait, several genomic regions were found to be associated with trait expression; however, one or a few regions usually explained much of the phenotypic variation.

The most interesting group of traits analyzed were those related to flowering. A summary of the genetic information on these traits derived from marker analysis is shown in Table 2. Most of the $F_2$ plants initiated flowering in less than 300 days without vernalization; however 10 plants did not flower during this time. These plants were placed at $4^{o}C$ for six weeks and subsequently flowered. RFLP loci in four genomic regions were significantly associated with this vernalization requirement. The main trait loci effects were found on linkage group 5, and markers linked to these loci were not associated with other flowering traits. Thus, the primary gene(s) involved in flowering response to cold temperatures appear to be located on linkage group 5. As expected, the cabbage parent contributed alleles associated with the requirement for vernalization. Trait loci for days to flower and for days to bud were localized to several common regions on linkage groups 1, 2, 3, and 7. The major gene effects for both traits were located on linkage group 7 and cabbage contributed alleles that increased days to bud or flower. An allele(s) from cabbage that decreased days to flower was detected for a minor gene(s) on linkage group 1, demonstrating the power of marker analyses to uncover hidden gene effects. A delay between bud formation and flowering is a characteristic of broccoli morphotypes, and the two regions associated with this trait had alleles from broccoli that increased the days from bud formation to flowering. Many of the same genomic regions were associated with height to first flower and days to flower, and similar allele effects were observed in these regions, suggesting that a delay in flowering also resulted in an increased height of plants at flowering time.

Table 2. The number, linkage group position, and magnitude of gene effects ($R^2$ values) for trait loci related to flowering in B. oleracea.

| Flowering Trait | Minimum no. of trait loci[a] | Linkage groups[a] | Range of $R^2$ values[b] |
|---|---|---|---|
| Requirement for vernalization[c] | 4 | 3, 5*, 7 | 0.09 - 0.20 |
| Days to bud[d] | 5 | 1, 2, 3, 7*, 9 | 0.10 - 0.39 |
| Days to flower[d] | 6 | 1, 2, 3, 7* | 0.10 - 0.43 |
| Days from bud to flower[e] | 2 | 3*, 8 | 0.10 - 0.14 |
| Height to first flower | 8 | 1, 2, 3, 5, 7*, 8 | 0.09 - 0.18 |

[a]The minimum number of trait loci and their linkage group positions were determined by counting clusters of unlinked marker loci that were significantly ($p < 0.05$) associated with the trait. Linkage groups with marker loci having the largest effects ($R^2$ values) are marked with *.
[b]The minimum and maximum $R^2$ values for clusters of unlinked marker loci that were significantly associated with traits are shown.
[c]Plants were scored as flowering or not flowering during a 299 day period without vernalization.
[d]The number of days from planting to bud formation or first open flower was measured for plants that did not require vernalization.
[e]Determined by subtracting days to bud from days to flowers.

Since many RFLP loci are duplicated in the B. oleracea genome, trait loci also may be duplicated and this could account for some multiple locus effects we observed. To investigate this possibility, we analyzed all pairs of mapped duplicated RFLP loci for significant associations with each trait. Associations of a trait with pairs of duplicated loci did not occur more frequently than expected by chance for all pairs of loci. However, in conserved duplicated regions, some pairs of duplicated loci were significantly associated with days to flower, stem length and leaf dimensions. These associations may reflect trait locus duplication in these portions of the genome. Our ability to detect potential duplication of morphological trait loci was dependent on our marker coverage of [duplicated regions in] the genome, the way in which we measured traits, the effects of the environment on trait expression, and the parents we used to develop the test population. Other experiments utilizing better marker coverage of duplicated regions and/or variation in the other conditions listed above would provide additional information on possible duplication of trait loci in the B. oleracea genome.

Brassica rapa

The B. rapa mapping population was developed using parents having analogous morphotypes to those used for the B. oleracea mapping population. The Chinese cabbage parent 'Michihili' is a heading type of B. rapa which will flower when vernalized for a short period. This vernalization requirement is weak, and plants will sometimes flower without a prolonged cold treatment. 'Michihili' has short internodes, entire leaf margins and sessile leaves which form cylindrical heads. Spring broccoli is a rapid flowering annual that forms small whorls of buds without any vernalization treatment. It has long internodes, lyrate leaf margins and petiolate leaves.

The B. rapa population was developed and analyzed in a manner similar to that of the B. oleracea population. Many of the same morphological traits were measured in this population and they were analyzed by the same methods using genotypic data from 79 to 132 RFLP loci distributed throughout the B. rapa genome.

A summary of genetic information on flowering traits derived from marker analysis of this population is presented in Table 3. All individuals in this $F_2$ population flowered within 190 days without vernalization treatment. However, the $F_2$ plants were kept outdoors into the fall season, and plants with weak vernalization requirements may have received enough cold treatment to flower. Trait loci for days to bud and for days to flower were localized to several common regions on linkage groups 3, 4, 6, and 8. Genes with major effects on flowering time, located on linkage groups 6 and 8, had alleles contributed from Michihili which delayed the time to bud and flower formation. However, genes with the largest effects on days to flower, located on linkage group 3, had alleles for delayed flowering from Spring broccoli. These alleles for rapid flowering from Michihili normally may be masked by alleles for vernalization requirement which were not detected in this study. Trait loci for days from bud to flower and height to first flower also were located in same position as other flowering loci on linkage group 3, and Spring broccoli contributed alleles associated with increased values of these measurements.

Data from the B. rapa population also were analyzed for potential duplication of trait loci. Although more extensive regions of conserved

Table 3. The number, linkage group position, and magnitude of gene effects ($R^2$ values) for trait loci related to flowering in B. rapa.[a]

| Flowering Trait | Minimum no. of trait loci | Linkage groups | Range of $R^2$ values |
|---|---|---|---|
| Days to bud | 7 | 3, 4, 6*, 7, 8, 10 | 0.08 - 0.19 |
| Days to flower | 5 | 3*, 4, 6, 8, 9 | 0.07 - 0.21 |
| Days from bud to flower | 4 | 1*, 3, 5, 8 | 0.08 - 0.11 |
| Height to first flower | 5 | 3, 4, 5, 8* | 0.08 - 0.13 |

[a]See Table 2 for explanation of table headings and flowering traits.

duplicated loci were detected in B. rapa compared to B. oleracea, there was little evidence for duplicated trait loci in these regions. The same considerations discussed for this type of analysis in B. oleracea also apply to B. rapa, and further studies are needed to determine the extent of potential trait locus duplication in B. rapa.

Comparative mapping of trait loci in B. oleracea and B. rapa

The extensive conservation in linkage arrangement of RFLP loci observed between B. oleracea and B. rapa raises an interesting question regarding the extent of conservation in trait loci between the two species. In each species, we used parents with analogous morphotypes for population development and we measured many of the same morphological traits. We compared the two populations for conservation in the map position of loci governing 14 traits which were measured identically in each study. Trait effects associated with marker loci from conserved regions of both species were compared to identify genes for morphological variation that are potentially conserved in the two species.

We did not find extensive conservation in the map position of genes controlling morphological variation in the two populations we analyzed. However, there were a few examples where conserved morphotype genes appear to exist. Genes controlling petiole and lamina dimensions, flowering characteristics and stem length were associated with common marker loci in conserved regions of the two genomes.

Although B. oleracea and B. rapa are closely related, they diverged as separate species before the various morphotypes were selected. Thus, different genes controlling morphological variation could have been selected in each species to derive analogous morphotypes. Our results from different accessions representing specific morphotypes suggest that many of the genes controlling morphological variation have been selected as alleles at different loci in the two species. However, gene effects at some common loci appear to have been maintained during the selection of morphotypes within these species. Additional studies with different morphotypes are needed to further determine the extent of conservation in genes controlling morphological variation in B. oleracea and B. rapa.

REFERENCES


Attia, T., and Robbelen, G., 1986, Cytogenetic relationship within cultivated Brassica analyzed in amphidiploids from the three diploid ancestors, Can. J. Genet. Cytol., 28:323.

Beckmann, J.S. and Soller, M., 1986, Restriction fragment length polymorphism in plant genetic improvement, in: "Oxford Survey of Plant Molecular Biology Vol. 3," B.J. Mifflin, ed., Oxford Press, Oxford.

Bernatzky, R., and Tanksley, S. D., 1986, Toward a saturated linkage map in tomato based on isozymes and random cDNA sequences, Genetics, 112:887.

Bonierbale, M. W., Plaisted, R. L., and Tanksley, S. D., 1988, RFLP maps based on a common set of clones reveal modes of chromosomal evolution in potato and tomato, Genetics, 120:1095.

Figdore, S. S., Kennard, W. C., Song, K. M., Slocum, M. K., and Osborn, T. C., 1988, Assessment of the degree of restriction fragment length polymorphism in Brassica, Theor. Appl. Genet., 75:833.

Helentjaris, T., Weber, D., and Wright, S., 1988, Identification of the genomic locations of duplicated sequences in maize by analysis of restriction fragment length polymorphisms, Genetics, 118:353.

Landry, B. S., Kesseli, R. V., Farrara, B., and Michelmore, R. W., 1987, A genetic map of lettuce (Lactuca sativa L.) with restriction fragment length polymorphism, isozyme, disease resistance and morphological markers, Genetics, 116:331.

McCouch, S. R., Kocher, G., Yu, Z. G., Wang, Z. Y., Khush, G. S., Coffman, W. R., and Tanksley, S. D., 1988, Molecular mapping of rice chromosomes, Theor. Appl. Genet., 76:815.

Prakash, S., and Hinata, K., 1980, Taxonomy, cytogenetics and origin of crop Brassica, a review, Opera Bot., 55:1.

Slocum, M. K., Figdore, S. S., Kennard, W. C., Suzuki, J. Y., and Osborn, T. C., 1990, Linkage arrangement of restriction fragment length polymorphism loci in Brassica oleracea. Theor. Appl. Genet., in press.

Song, K. M., Osborn, T. C., and Williams, P. H., 1988, Brassica taxonomy based on nuclear restriction fragment length polymorphisms (RFLPs).2. Preliminary analysis of subspecies within B. rapa (syn. campestris) and B. oleracea, Theor. Appl. Genet., 76:593.

Song, K. M., Osborn, T. C., and Williams, P. H., 1990, Brassica taxonomy based on unclear RFLPs. 3. Genome relationships in Brassica and related genera and the origin of B. oleracea and B. rapa (syn. campestris), Theor. Appl. Genet., in press.

Tanksley, S. D., Young, N. D., Paterson, A. H., and Bonierbale, M. W., 1989, RFLP mapping in plant breeding: New tools for an old science, Bio/Technology, 7:257.

# PHYSICAL MAPPING OF DNA SEQUENCES ON PLANT CHROMOSOMES BY LIGHT MICROSCOPY AND HIGH RESOLUTION SCANNING ELECTRON MICROSCOPY

Harald Lehfer, Gerhard Wanner and Reinhold G. Herrmann

Botanisches Institut der Ludwig-Maximilians-Universität München
Menzinger Str. 67, D-8000 München, Federal Republic of Germany

## INTRODUCTION

The development of the field emission source enables a sensational resolution of 0.5-2 nm in field emission scanning electron microscopy (FESEM). In combination with advances in procedures for metal coating, scanning electron microscopy has reached a new dimension of ultrastructural research in biology (Tanaka, 1990; Müller and Hermann, 1990), including investigations on plant chromosomes (Wanner et al., 1990; 1991). The current excitement concerning FESEM is not only based on the expectation that it may contribute to uncover fundamental chromosomal detail but also that it may provide "the missing link", i.e., to fill the gap between the DNA thread and the three-dimensional structure of the chromosome.

Primarly due to technical difficulties, chromosomes of plants have not yet been investigated at high resolution. We have recently described a preparation technique for high resolution scanning electron microscopy of plant chromosomes (Wanner et al., 1990; 1991), which was optimized for the use of routine standard squash preparations of mitotic as well as meiotic chromosomes in the size range from 5-50 $\mu$m from root tips and pollen mother cells of various cereals and *Lilium*. This technique permits, after light microscopic observation and documentation, to investigate the *same* specimen with a 100-fold higher resolution using a field emission scanning electron microscope. Furthermore, tilting of the specimen can provide a three-dimensional insight into both chromosomal structure and the arrangement of chromosomes within the nuclear matrix (Fig. 1). With this technique it has been possible to document - at high resolution - structures such as centromeric regions, spindle apparatus, spindle fibre attachment, nucleolus organizing regions, or stages of condensation and decondensation (Figs. 2 and 3). The smallest unit of DNA package that can be resolved in both meiotic and mitotic chromosomes appears to be the "beads on a string"-like structure of nucleosomes (15 nm fibre), followed by chromatin condensation into solenoids and supersolenoids. However, only parallel arrays of strands are typically observed during further condensation instead of helical and superhelical arrangements (Wanner et al., 1990; 1991).

The mapping of genes or of anonymous DNA fragments (RFLPs) on chromosomes represents doubtlessly one of the future developments of the outlined technique. A crucial prerequisite towards this aim rests in the availability of procedures for locating DNA sequences on chromosomes by *in situ* hybridization with satisfactory preservation of structural detail, specifically at high resolution. While repetitive sequences can be quite readily detected *in situ* at the light microscopic level

*Plant Molecular Biology 2*, Edited by R.G. Herrmann and
B. Larkins, Plenum Press, New York, 1991

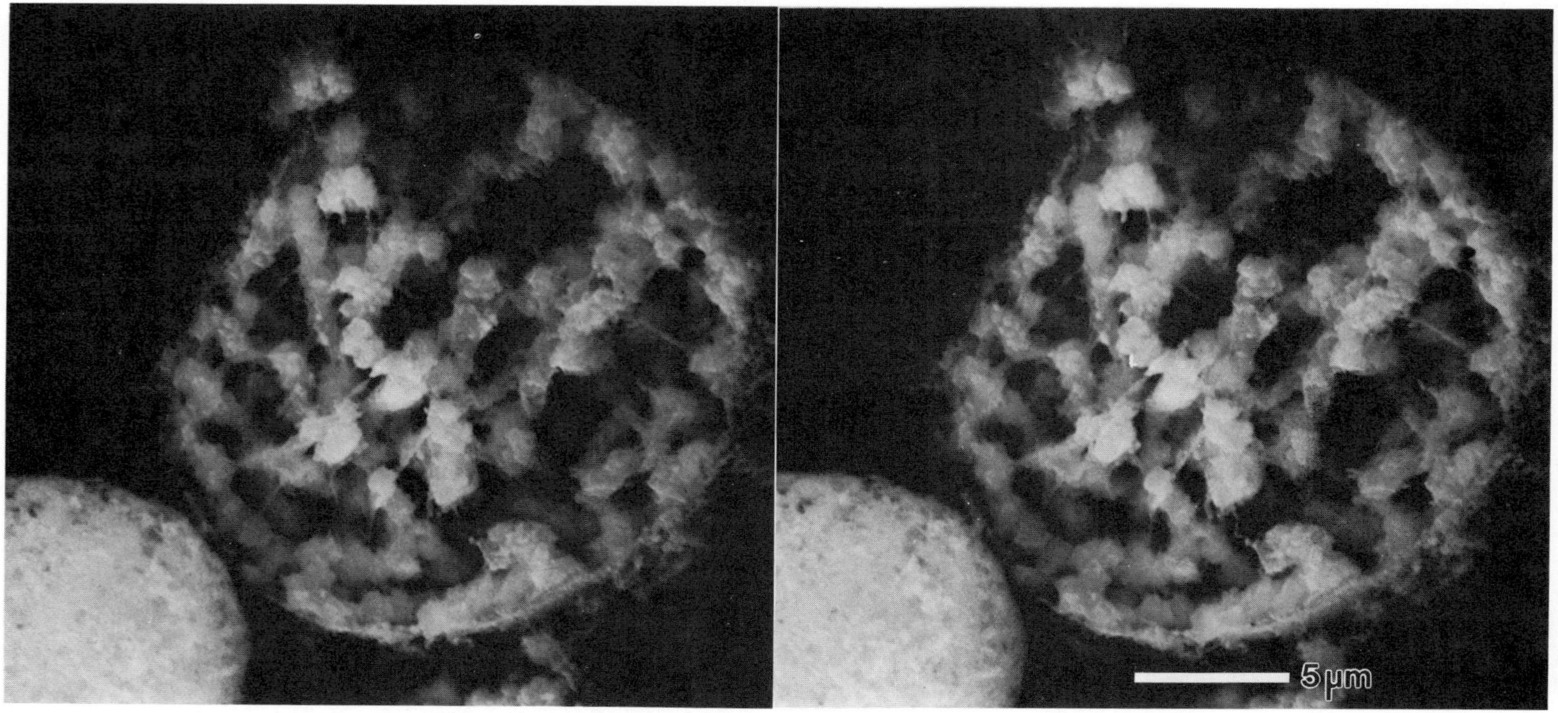


**Fig. 1.** Scanning electron micrograph of a fractured meiotic (microspore) nucleus of *Lilium henryi*, fixed with glutaraldehyde prior to EtOH/acetic acid-fixation. The stereo pair can be viewed by stereo viewers and gives an impressing insight into the three-dimensional arrangement of prophase chromosomes within the nuclear matrix. Lower left: part of an intact interphase nucleus.

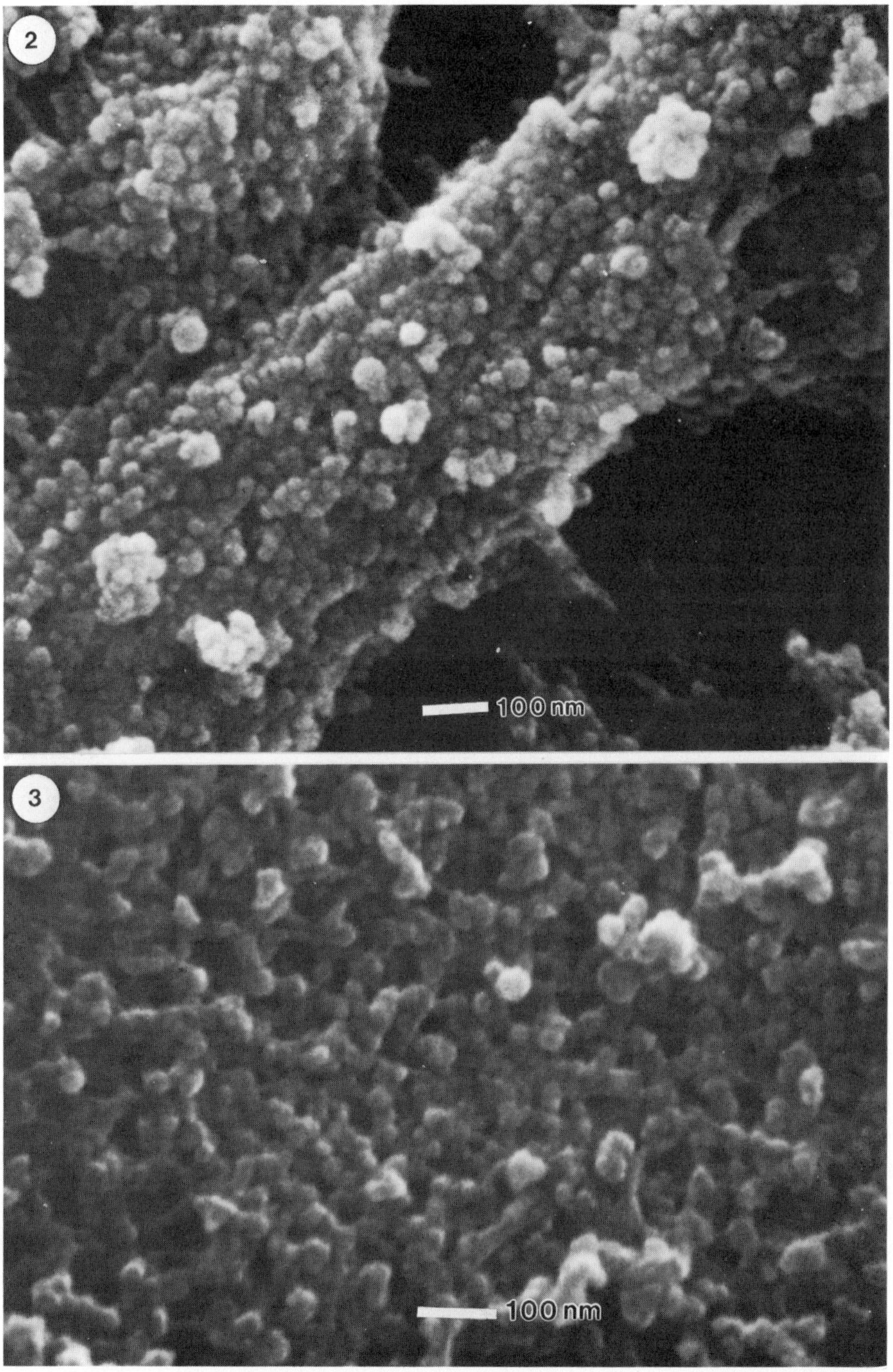


**Figs. 2 and 3.** High resolution scanning electron micrographs of cereal chromosomes prepared by the squash technique and impregnated by osmiumtetroxide/thiocarbohydrazide (Wanner et al., 1991). The resolution of FESEM allows investigation of chromosome structure down to the level of solenoid (10-15 nm) and supersolenoid (100-300 nm) organization. Fig. 2: Detail of a rye chromosome in meiotic prophase. Fig. 3: Detail of a mitotic metaphase chromosome surface of barley.

(Hutchinson et al., 1981; Rayburn and Gill, 1985; Rayburn and Gill, 1987; Lapitan et al., 1989), only few attempts were successful in the demonstration of low- and single-copy sequences using $^3$H-labeled and biotinylated probes (Ambros et al., 1986; Mouras et al., 1987; Huang et al.; 1988, Simpson et al., 1988). In animal chromosomes, biotinylated probes allowed the detection of either enzymatically amplified signals (Langer-Safer et al., 1982; Manuelidis et al., 1982) or, preferably, fluorochrome fluorescence with its advantage of higher resolution of the hybridizing locus (Langer-Safer et al., 1982; Pinkel et al., 1986; Nederlof et al., 1990). This approach also included the use of different haptens for DNA-labeling with the view to discriminate, by such fluorescence techniques, several probes on the same specimen (Nederlof et al., 1989; 1990). However, the application of hapten-labeled DNA and fluorescence detection is limited by the size of the probes. Short fragments (< 5 kb) require several amplification steps, which frequently cause a substantial increase of background signals and prevent clear results (cf. Lawrence et al., 1988). To bypass this obstacle, either low-copy sequences or cosmid-inserts with sizes up to 30 kb are generally used as probe-DNA, especially if several probes need to be discriminated on the same specimen (Landegent et al., 1987; Lichter et al., 1990; Nederlof et al. 1989; Nederlof et al., 1990). Additional technical difficulties will certainly be encountered with the adaption from light microscopic scale to ultrastructural level. To our knowledge, no report has yet appeared locating DNA fragments with FESEM at high resolution.

The localization of specific sequences to plant chromosomes has proven to be more difficult than to animal chromosomes. Major drawbacks reside in the presence of the rigid cell wall, which adversely affects the accessibility of the probe to the chromosome, and the generally higher background emerging after signal amplification. Only one of the outlined novel procedures has yet been employed successfully in plant cytogenetics, namely for the detection of (repetitive) rDNA-sequences on the chromosomes of *Crepis capillaris* (Maluszynska and Schweizer, 1989).

We have begun to explore the potential of various *in situ* hybridization and probe-labelling techniques on barley chromosomes both at the light microscopic and electron microscopic level of detection. In initial experiments, we have used three kinds of probes, the highly repetitive barley sequence plApa1 (insert length 570 bp), two probes (clones pBSC4: 1.2 kb, pBSC21: 3.5 kb) encoding different segments of one of the members of the clustered, small B1 hordein gene family, and a 1.8 kb single-copy DNA (MWG7H507), derived from barley chromosome 7. We have employed biotin-labeled probes detected via FITC fluorescence, or digoxigenin-labeled probes visualized by TRITC-labeled antibodies for light microscopic examinations.

**Fig. 4.** Fluorescence detection of a biotinylated repetitive DNA-probe (plApa1) on barley metaphase chromosomes. Fig. 4a shows the propidiumiodide counterstain of a barley metaphase. Fig. 4b demonstrates the widespread distribution of the probe (yellow-green) over barley chromosomes (red) after *in situ* hybridization using the avidin-FITC/anti-avidin technique. Fig. 4c illustrates the results obtained by the same probe but using a different hapten (digoxigenin) and detection procedure. Chromosome counterstain in this case was done with DAPI (blue). The signals appear in red (TRITC-labeled antibodies). The arrows mark chromosomal satellites.

**Fig. 5 a, b.** Localization of the B-hordein locus on barley metaphase chromosomes with biotinylated probes and avidin-FITC/anti-avidin-technique. The signals can be seen on both chromatides at the distal end of the short arm of chromosome 5 (arrows).

**Fig. 6.** Visualization of a 1.8 kb single-copy probe (MWG7H507) on barley interphase nuclei. Either single (Fig. 6 a), double (Fig. 6 c), or four signals (Fig. 6 b) can be seen, which might reflect G1- (6a, c) and G2-phase (b) nuclei, respectively.

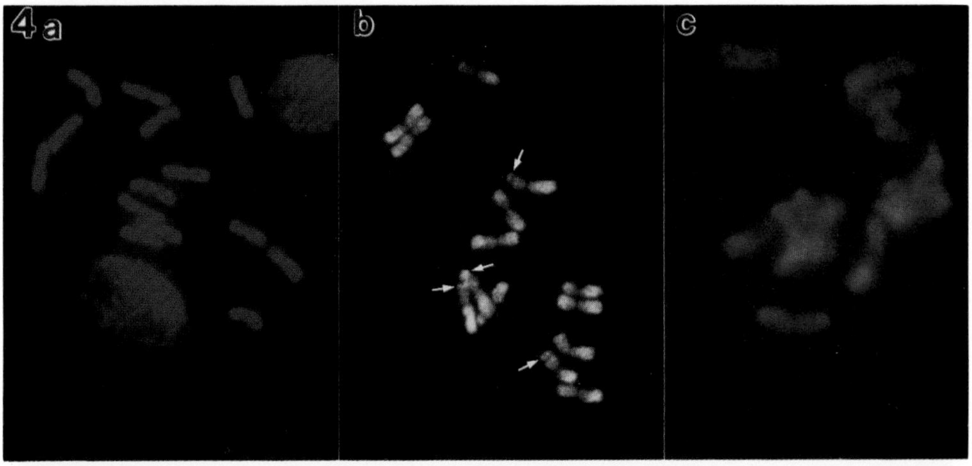
4a
b
c

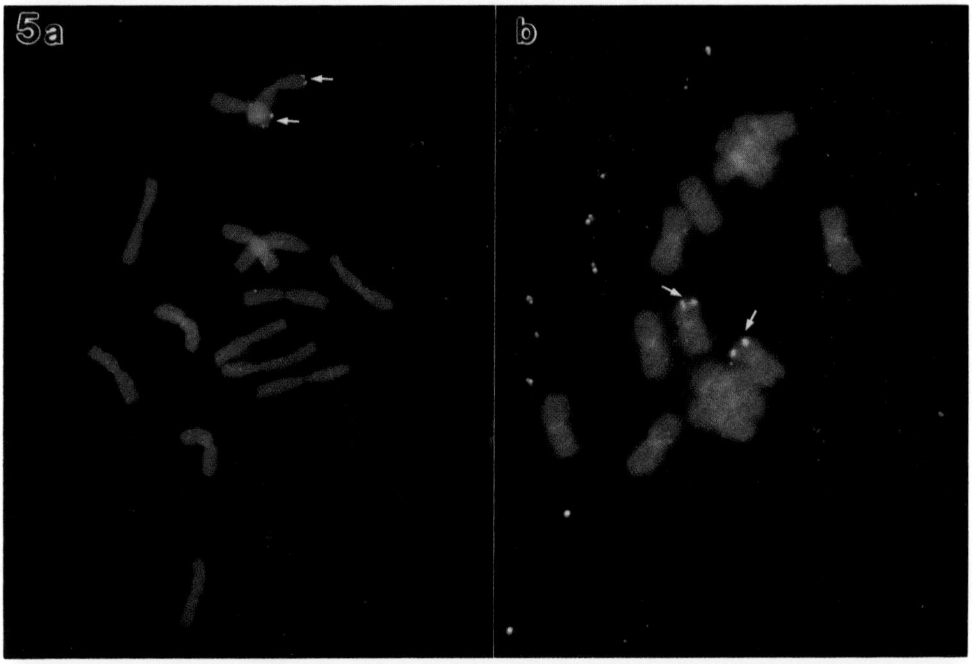
5a
b

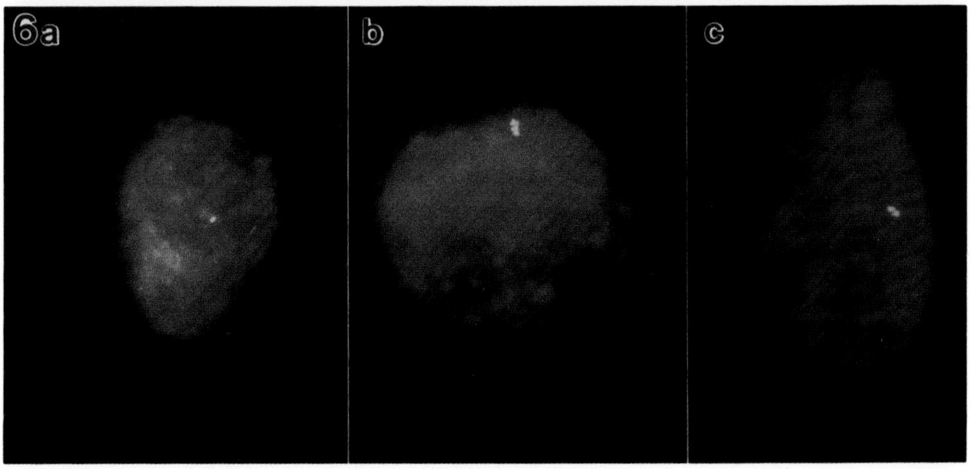
6a
b
c

Streptavidin-gold complexes (diameter 15 nm) along with biotin-labeled probes, in turn, were used for high-resolution scanning electron microscopy.

Metaphase chromosomes of *Hordeum vulgare* cv. Igri of appropiate quality were prepared with a modified version of the Linde-Laursen (1975) procedure. Prior to *in situ* hybridization, all slides were incubated with DNase-free RNase A (Boehringer Mannheim) in 2 x SSC, briefly treated with Proteinase K (Merck, Darmstadt) at pH 7.5 and 37°C in the presence of 2 mM $CaCl_2$, and washed. The specimen were then fixed with acid-free formaldehyde, and dehydrated in an ethanol-series for hybridization. The hybridization buffer contained 50% - 60% deionized formamide, 2 x SSC, 15-40 ng of biotinylated DNA or DNA labeled with digoxigenin, 1 $\mu$g sonicated herring sperm DNA, 1 $\mu$g of *E. coli* tRNA (Boehringer Mannheim), and, in the case of pBSC4-, pBSC21- or MWG7H507-DNA, 10% dextrane sulfate (Sigma Chemie, Deisenhofen). The outlined procedure will be detailed elsewhere.

Fig. 4b illustrates the hybridization of the repetitive probe plApa1 to a metaphase plate of barley, counterstained with the red fluorescent dye propidiumiodide (Fig. 4a). Signal detection was performed with avidin-FITC, biotinylated anti-avidin, and an additional layer of avidin-FITC. This probe stains all chromosomes, with slightly brighter signals in the distal regions, indicating that the sequence element is spread over both arms of all chromosomes. The chromosomal satellites of the chromosomes 6 and 7 are stained as well but all centromeres and the secondary constrictions of the satellite chromosomes do not or only weakly display FITC signals. Fig. 4c shows *in situ* hybridization with the same probe using the digoxigenin/anti-digoxin detection system. The result is comparable to that presented in Fig. 4b though with somewhat lower resolution which is predominantly caused by the relatively low affinity of the commercially available primary antibody against digoxigenin, a cross-reacting antibody against digoxin, and the lower efficiency of the amplifying system.

Fig. 5 presents barley metaphases hybridized with the two different avidin/FITC labelled B-hordein sequences. Signals are reproducibly found at the distal ends of the short arms of two homologous chromosomes. The twin signal on two chromosomes demonstrates hybridization to both chromatides. Determination of arm ratios of the chromosomes suggests that the signals are located on the short arm of chromosome 5 ($L/S = 0.69 \pm 0.065$) as expected. The twin signal at these sites is relatively bright and approximately 30% of ca. 100 well-spread, complete metaphases displayed such signals on both chromatides or on all four chromatides of the chromosomes. The position of the B-hordein locus at the more distal end of the short arm of chromosome 5 is not consistent with a recent report which, in *Hordeum vulgare* cv. Betzes, places that locus on the short arm of chromosome 5, at a distance of 46% from the centromere (Clark et al., 1989).

Fig. 6 illustrates fluorescence signals obtained by the 1.8 kb single-copy probe MWG7H507 in interphase nuclei of barley. The probe fluorescence in Fig. 6a appears to consist of two closely positioned signals, as is more evident in Fig. 6c at higher resolution. Fig. 6b, in turn, displays four resolvable signals, again in remarkable vicinity. This might imply that the loci on both chromosome homologues are in direct neighbourhood during interphase in G1 (Fig. 6a and c) and in G2 (Fig. 6b).

The example of Figs. 7 and 8 serves to demonstrate the efficiency of using high resolution SEM to locate sequences on mitotic barley chromosome structures. The secondary image in (a) indicates a satisfactory structural preservation of the chromosomes inspite of the denaturation and renaturation of the specimen during *in situ* hybridization. When examining the imaging of backscattered electrons (b), the colloidal gold particles of the pBSC21 probe can be detected as bright spots on the chromosome-surface. It is worth noting that this deposit is specific, that the arrangment of the probes on the chromatin surface seems to form helical-like structures, and that a distinction between hybridized and non-hybridzed signals is generally possible (not shown).

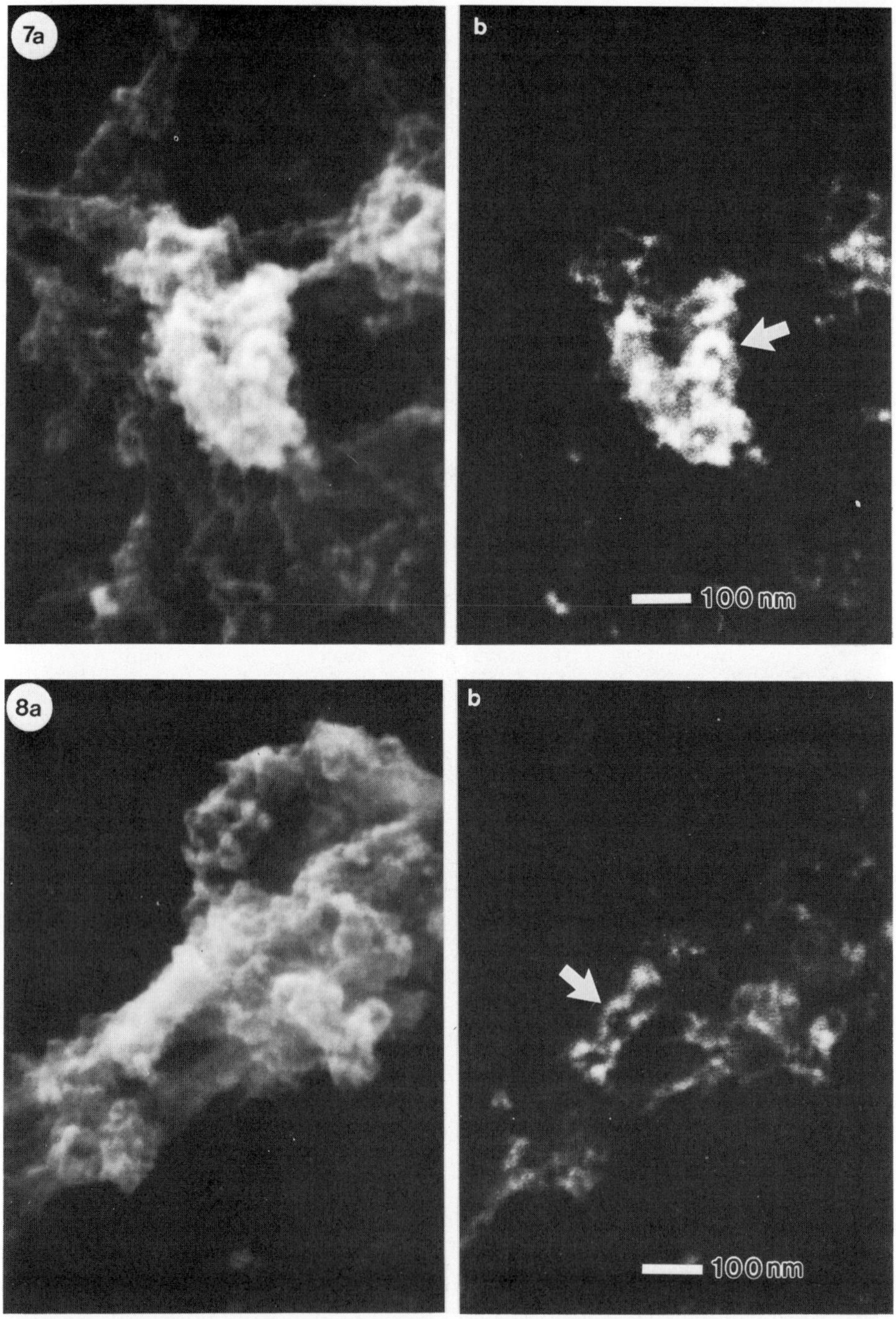


**Figs. 7 and 8.** Localization of the pBSC21-probe, visualized by a high-resolution scanning electron microscope, using the biotin/streptavidin-gold (15 nm) technique on mitotic chromosomes (prometaphase) of barley. The secondary electron images (a) indicate the sites of hybridization by higher brightness (enhancement of secondary-electron-number by gold-label). Backscatter-scanning-electron-images (b) prove bright spots to be gold. The hybridized probes show a remarkable "helical" arrangement of the gold-label (arrows).

## CONCLUSION

Collectively, the results demonstrate that it is technically feasible to map relatively small single-copy DNA sequences on plant chromosomes. The limit of light microscopic resolution of DNA probes resides probably in the order of 200 -300 bp. Using a novel "insert amplification/sandwich" technique for signal detection, we have successfully localized single-copy DNA fragments of 200 bp (to be published elsewhere), and even smaller fragments may be detected at the ultrastructural level. Work currently in progress aims to exploit this approach further by using probes to each of the barley chromosomes and double or triple labelling techniques with a view to construct topographical gene (RFLP) maps and to correlate them with genetic and physical maps (e.g., Hauge et al., this Volume). We expect that FESEM opens avenues not only for studies of chromosome structure or of chromosome arrangements within nuclei, but also to aspects of gene location and expression patterns *in situ*, or to chromosome evolution using strategies of chromosomal *in situ* suppression (CISS) hybridization and "chromosome painting" (Lichter et al., 1988; Wienberg et al., 1990).


## ACKNOWLEDGEMENTS

We are grateful to Dr. Andreas Graner (Grünbach) for the barley clones. This work was supported by grant nr. 0318990 of the Bundesministerium für Forschung und Technologie (BMFT).


## REFERENCES


Ambros PF, Matzke MA, Matzke AJM (1986) Detection of a 17 kb unique sequence (T-DNA) in plant chromosomes by in situ hybridization. Chromosoma 94: 11-18

Clark M, Karp A, Archer S (1989) Physical mapping of the B-hordein loci on barley chromosome 5 by *in situ* hybridization. Genome 32: 925-929

Hauge BM, Giraudat J, Hanley S, Hwang I, Kohchi T, Goodman HM (1991) Physical mapping of the *Arabidopsis* genome and its applications (this Volume)

Huang P-L, Hahlbrock K, Somssich I (1988) Detection of a single-copy gene on plant chromosomes by in situ hybridization. Mol Gen Genet 211: 143-147

Hutchinson J, Lonsdale DM (1982) The chromosomal distribution of cloned highly repetitive sequences from hexaploid wheat. Heredity 48: 377-381

Landegent JE, Jansen in de Wal N, Dirks RW, Baas F, van der Ploeg M (1987) Use of whole cosmid cloned genomic sequences for chromosomal localization by non-radioactive in situ hybridization. Hum Genet 77: 366-370

Langer-Safer P, Levine M, Ward DC (1982) Immunological method for mapping genes on *Drosophila* polytene chromosomes. Proc Natl Acad Sci USA 79: 4381-4385

Lapitan NVL, Ganal MW, Tanksley SD (1989) Somatic chromosome karyotype of tomato based on in situ hybridization of the TGRI satellite repeat. Genome 32: 992-998

Lawrence JB, Villnave CA, Singer RH (1988) Sensitive, high-resolution chromatin and chromosome mapping in situ: presence and orientation of two closely integrated copies of EBV in a lymphoma line. Cell 52: 51-61

Lichter P, Cremer T, Borden J, Manuelidis L, Ward DC (1988) Delineation of individual human chromosomes in metaphase and interphase cells by in situ suppression hybridization using recombinant DNA libraries. Hum Genet 80: 224-234

Lichter P, Chang Tang CJ, Call K, Hermanson G, Evans GA, Housman D, Ward DC (1990) High resolution mapping of human chromosome 11 by *in situ* hybridization with cosmid clones. Science 247: 64-69

Linde-Laursen IB (1975) Giemsa C-banding of the chromosomes of 'Emir' barley. Hereditas 81: 285-289

Maluszynska J, Schweizer D (1989) Ribosomal RNA genes in B chromosomes of *Crepis capillaris* detected by non-radioactive *in situ* hybridization. Heredity 62: 59-65

Manuelidis L, Langer-Safer P, Ward DC (1982) High resolution mapping of satellite DNA using biotin-labeled DNA probes. J Cell Biol 95: 619-625

Mouras A, Saul MW, Essad S, Potrykus I (1987) Localization by in situ hybridization of a low copy chimaeric resistance gene introduced into plants by direct gene transfer. Mol Gen Genet 207: 204-209

Müller M and Hermann R (1990) Towards high resolution SEM of biological objects. Hitachi Instrument News 19: 50-57

Nederlof PM, Robinson D, Abuknesha R, Wiegant J, Hopman AHN, Tanke HJ, Raap AK (1989) Three-color fluorescence in situ hybridization for the simultaneous detection of multiple nucleic acid sequences. Cytometry 11: 20-27

Nederlof PM, van der Flier S, Raap AK, Tanke HJ, Ploem JS, van der Ploeg M (1990) Multiple fluorescence in situ hybridization. Cytometry 11: 126-131

Pinkel D, Straume T, Gray JW (1986) Cytogenetic analysis using quantitative, high-sensitivity fluorescence hybridization. Proc Natl Acad Sci USA 83: 2934-2938

Rayburn AL, Gill BS (1985) Use of biotin-labeled probes to map scientific DNA sequences on wheat chromosomes. J Hered 76: 78-81

Rayburn AL, Gill BS (1987) Molecular analysis of the B-genome of the Triticae. Theor Appl Genet 73: 385-388

Simpson PR, Newman MA, Davies DR (1988) Detection of legumin gene sequences in pea by in situ hybridization. Chromosoma 96: 454-458

Tanaka K (1990) High resolution scanning electron microscopy of the cell. Hitachi Instrument News 19: 21-31

Wanner G, Formanek H und Herrmann RG (1990) Ultrastructure of plant chromosomes by high-resolution scanning electron microscopy. Plant Mol Biol Rep 8: 224 - 236

Wanner G, Formanek H, Martin R, Herrmann RG (1991) High-resolution scanning electron microscopy of plant chromosomes. Chromosoma 100:103-109

Wienberg J, Jauch A, Stanyon R, Cremer T (1990) Molecular cytotaxonomy of primates by chromosomal *in situ* suppression hybridization. Genomics 8: 347-350

# STRUCTURE AND FUNCTION OF THE MAIZE TRANSPOSABLE ELEMENT *ACTIVATOR* (*AC*)

Reinhard Kunze, George Coupland[1], Heidi Fußwinkel, Siegfried Feldmar, Ulrike Courage, Sylvia Schein, Heinz-Albert Becker, Shivani Chatterjee, Min-gang Li[2], and Peter Starlinger

Institut für Genetik, Universität zu Köln, Weyertal 121, 5000 Köln 41, FRG

[1] Institute of Plant Science Research, Maris Lane, Trumpington, Cambridge CB2 2JB, United Kingdom

[2] Department of Genetics and Cell Biology, University of Minnesota, 1445 Gortner Avenue, St. Paul MN 55108-1095, USA

## INTRODUCTION

The discovery of transposable elements in maize by Barbara McClintock was greatly facilitated by the thorough genetic characterization of this species since the early years of this century. In particular, the identification of several colour genes was helpful to detect unstable mutants. Meanwhile, at least ten independent mobile element systems are known in corn (Petersen 1986). However, only four of these families are molecularly characterized in more detail (recent reviews: Döring and Starlinger, 1986; Fedoroff, 1989).

*Activator* (*Ac*) is the autonomous element in the *Ac/Ds* transposable element family. It was the first transposable element that was detected by McClintock (1947), and it is now, beside the *En-Spm* maize transposable element, the best understood plant transposon. Genetic data indicate that it can transpose in nonreplicative manner and that transposition can be coupled to the replication process (Greenblatt and Brink 1963). A functional *Ac* can mobilize itself as well as *Ds* elements, which are stable insertions in the absence of *Ac*.

Like many other transposable elements, *Ac* and *Ds* have terminal inverted repeats (11 bp) and create target site duplications (8 bp) upon insertion. The *Ac* elements that have been isolated from several genes and were characterized molecularly turned out to be structurally identical (Fedoroff et al., 1984; Pohlman et al., 1984; Müller-Neumann et al., 1984; Lechelt et al., 1989). Unlike *Ac*, *Ds* elements generally seem to be present in the maize genome in many copies. Estimates vary between ~30 and several hundred copies.

*Ac* elements can loose their transpositional activity in different ways. Internal deletions can irreversibly mutate an *Ac* into a *Ds* element. Deletion of sequences required *in cis* for excision can generate an immobilized element that is still capable of transactivation of *Ds* elements. *Ac* can also be inactivated in reversible way: it can switch transiently into an inactive state during which it can neither transpose itself, nor transactivate a *Ds* element, but still can be mobilized if an additional active *Ac* is introduced.

## Structure and Transcription of *Ac*

Extensive transcription analyses were performed with various *Ac*-containing and *Ac*-free maize lines, and transgenic tobacco plants, respectively. In addition, cDNA clones were isolated and sequenced. The results of these experiments led to the conclusion that *Ac*-activity is correlated with the synthesis of a 3.5 kb mRNA (Kunze et al., 1987). This transcript spans most of the 4565 bp long *Ac* element, leaving only about 300 bp at the 5′-end and 264 bp at the 3′-end of *Ac* untranscribed. Four introns with a combined length of 654 bp are spliced from the primary transcript. The 3.5 kb mRNA has some unusual features:

Transcription does not start at a single position, but at more than ten positions that are located between 290 bp and 380 bp inside of *Ac*. The lack of a TATA box upstream of the transcription initiation sites might explain this phenomenon, as similar observations were made with the SV40 early promoter (Benoist and Chambon, 1981), a H2A histone gene (Grosschedl and Birnstiel, 1980) and some mammalian housekeeping genes.

The *Ac*-mRNA starts with a ~650 nucleotides long leader sequence, which is devoid of AUG start codons. The leader sequence contains several direct repeats that are between 13 bp and 20 bp long, and the G+C content in its first ~400 nucleotides is considerably higher than in the following open reading frame. It is not yet known if the leader sequence has any function on the DNA or on the RNA level. We know, however, that it is dispensable for the function of the transposable element: Deletions of the leader sequence down to about 70 bp upstream of the methionine codon do not cause a decreased transposition frequency in transgenic tobacco plants (Coupland et al., 1988).

Figure 1 summarizes the most important structural features of *Ac*.

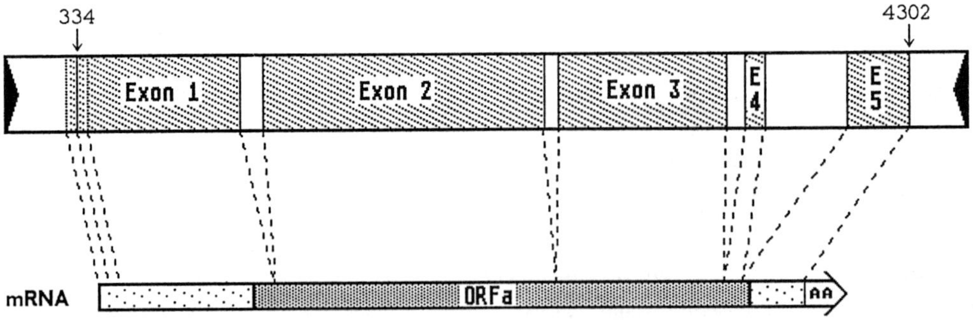


**Figure 1.** Schematic representation of *Ac* and its transcript. *Ac* is 4565 bp long. The filled triangles at its ends symbolize the 11 bp inverted repeats. The two positions marked by arrows indicate the major transcription start site and the poly(A)-addition site, respectively. The *Ac* mRNA is ~3.5 kb long.

## *Ac*-Encoded Protein in Maize

The first AUG codon on the *Ac* mRNA opens an 807 amino acids long open reading frame (ORFa). To search in maize for the protein encoded by ORFa, the putative transposase of *Ac*, polyclonal antisera were raised against different parts of this reading frame (Figure 2). Since transposition of *Ac* can be frequently observed in developing endosperm, and we expected that a transposase has to be a DNA-binding protein, nuclear protein extracts were prepared from developing endosperm tissue of *Ac*-carrying and *Ac*-free maize plants, respectively. All five anti-ORFa-antisera detect a protein with an apparent molecular weigth of 112 kD exclusively in *Ac*-carrying material (Fußwinkel et al., 1988; H. Fußwinkel et al., in preparation). This protein is probably the full-length ORFa protein.

Preliminary results indicate that, depending on the composition of the extraction buffer, a second, smaller protein co-segregates with the *Ac* element in the plants. We do not yet know if this protein has any function in the transposition process or for its regulation.

The first AUG codon of the *Ac* protein is not located within the sequence contexts frequently observed in plants or eukaryotes in general (Kozak, 1986; Lütcke et al., 1987), which raised the question to which extent this methionine codon can serve as a translational start point. *In vitro* translation experiments demonstrated that, despite the unusual sequence context, the first AUG codon is preferentially used as the initiation site (Kunze et al., 1987).

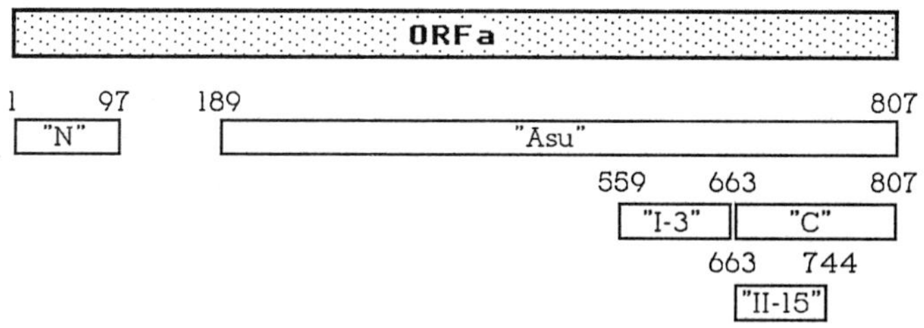


**Figure 2.** Polyclonal antisera were raised against different segments of the *Ac* protein. "N": antiserum recognizing amino acids 1-97; "Asu": antiserum recognizing amino acids 189-807; "I-3": antiserum recognizing amino acids 559-663; "C": antiserum recognizing amino acids 663-807; "II-15": antiserum recognizing amino acids 663-744.

## The migration anomaly of the *Ac* protein

The large *Ac* protein that is detected by Western blotting in maize migrates in SDS-containing polyacrylamide gels only slightly faster than the 116 kD ß-galactosidase; its electrophoretic mobility indicates a molecular weigth of ~112 kD. However, the calculated molecular weight of the 807 amino acids long ORFa protein is only 92 kD. This unexpected migration in denaturing gels could have several reasons:
1) The *Ac* protein could be modified, i.e. glycosylated and/or phosphorylated.
2) An unusual secondary structure could account for the migration anomaly.

→1) Glycosylation is unlikely to be the reason for the slow migration, because the ORFa-protein overexpressed in E. coli migrates only slightly faster (~110 kD) than the *Ac* ORFa-protein from maize.

We have no direct evidence if the *Ac* ORFa-protein in maize is phosphorylated, but it was shown that the ORFa-protein synthesized in insect cells is phosphorylated (Hauser et al., 1988). Although we would consider it unlikely that phosphorylation could account for the observed degree of migration anomaly, the nuclear protein extract from maize was incubated with alkaline phosphatase prior to electrophoresis. The dephosphorylated *Ac* protein migrates only slightly faster (at about 110 kD) than the untreated protein, at approximately the same position as the ORFa-protein synthesized in E. coli.

The cDNA-fragment that was taken as a template for the *in vitro* translation experiment encodes the first 230 amino acids of ORFa. The resulting polypeptide was calculated to have a molecular weight of ~25 kD. Instead, it migrates at approximately 39 kD, although the *in vitro* translation extract is probably incapable of authentic protein modification.

Taken together, these data demonstrate that the abnormally slow migration of the *Ac* protein is not (only) caused by modification.

→2) To answer the question if the migration anomally can be ascribed to a certain segment of the *Ac* protein, N-terminal deletions of ORFa were expressed in E. coli and the apparent molecular weights of the polypeptides were determined. The results are summarized in table 1 below. The *in vitro* translated protein, which is a C-terminally deleted *Ac* protein, was also included in this table.

**Table 1.** Plasmid pcAcN was *in vitro* translated, pRK20 expresses unfused *Ac* ORFa-protein, all other plasmids express unfused, N-terminally deleted ORFa-protein. The apparent molecular weights were determined by SDS-PAGE.

| Plasmid | amino acids | molecular weight [kD] calculated | apparent |
|---|---|---|---|
| pcAcN | 1 - 230 | 25.4 | 39 |
| pRK20 | 1 - 807 | 92.0 | 112 |
| pET13-PvuII | 76 - 807 | 84.2 | 100 |
| pRK22 | 189 - 807 | 71.5 | 72 |
| pRK24 | 465 - 807 | 39.6 | 40 |
| pRK25 | 663 - 807 | 16.0 | 16 |

The protein derivatives containing amino acids 77 - 188 migrate too slow, whereas the deletion derivatives starting at or behind amino acid 188 have the expected electrophoretic mobility. These result are in agreement with the assumption that a secondary structure between amino acids 77 and 188 is responsible for the migration anomally.

A conspicuous amino acid sequence is located between positions 108 and 133 of the *Ac* protein: the dipeptides Pro-Gln and Pro-Glu are eight and three times repeated, respectively, and in total six glutamic acid residues are located in this region.

Due to the high proline-content, this segment of the protein could be rather unstructured. Eventually, in combination with the high local glutamic acid content this causes the unusually slow migration in SDS-PAGE. Such a phenomenon was reported for some other proteins having either a high proline content (Franssen et al., 1987; Freytag et al., 1979), or a high overall or local glutamic acid content, respectively (Benedum et al., 1986; Kleinschmidt et al., 1986; Dingwall et al., 1987; Ollo and Maniatis, 1987; Benson and Pirotta, 1987).

The DNA-Binding Properties of *Ac*-Encoded Protein

The 112 kD *Ac* protein putatively is the *Ac* transposase and as such should be a DNA-binding protein. To test this hypothesis, the *Ac* cDNA was inserted in a baculovirus expression vector. Insect cells that were infected with the recombinant baculovirus expressed the *Ac* protein and accumulated it in their nuclei (Hauser et al., 1988). Nuclear protein extracts from cells infected with the recombinant baculovirus were incubated with various *Ac* DNA fragments and synthetic oligomers, respectively, and formation of complexes was assayed on gels (Kunze and Starlinger, 1989). Table 2 summarizes the results of the gel mobility shift assays:

**Table 2.** *Ac* fragments tested in gel mobility shift assays. The pairs of numbers denote the *Ac* nucleotides present in the fragment. + weak binding; ++ moderate binding; +++ strong binding; - no binding; ND bot determined.

| *Ac* fragment | Binding to *Ac* protein |
|---|---|
| 1- 75 | + |
| 76- 181 | +++ |
| 182- 251 | ++ |
| 252- 305 | - |
| 306- 588 | - |
| 589- 738 | + |
| 739- 961 | - |
| 962-1051 | - |
| 1053-1320 | - |
| 1321-1785 | - |
| 1786-3390 | ND |
| 3391-3556 | - |
| 3557-3844 | - |
| 3845-4194 | - |
| 4195-4419 | + |
| 4420-4565 | +++ |

The *Ac* protein binds specificially to subterminal sequences of the *Ac* DNA. It does not bind to DNA fragments containing the isolated inverted repeats, regardless if one to six copies of the inverted repeats are present in direct and/or inverted orientation. Apparently, not a single site in either terminus of *Ac* is recognized, but the protein can bind at several positions along certain fragments.

This interpretation was corroborated by a Bal 31 dele[illegible]periment (Østergaard Jensen et al., 1988): Both the 5′ terminal and the 3′ terminal fragment of *Ac* were labelled at either end, respectively, and deleted with Bal 31 to varying extents from the other end. By gel mobility shift assays, subsequent elution of the complexes and analysis on sequencing gels, the minimal length that is required for binding of the *Ac* protein was determined. This technique allowed the definition of major *Ac* protein binding segments in either end of the *Ac* element (Kunze et al., 1989) (Figure 3).

As all fragments that are bound by the *Ac* protein, and in particular the major *Ac* protein binding segments, contain several copies of the sequence motif WWWCGG (or CGGWWW), which is frequently AAACGG (or CGGAAA) in the fragments strongly bound by the *Ac* protein, synthetic AAACGG-hexamers were ligated, cloned into pUC19, and used as targets in mobility shift assays. The *Ac* protein binds specificially to the AAACGG-motif. The binding affinity increases strongly with the copy number of these hexamer motifs: A fragment containing three hexamers is only very weakly bound, whereas fragments containing eight or more copies are good targets. The central C-residue of the AAACGG-motif is absolutely required for the binding reaction: a mutation into an AAAGGG-motif abolishes binding of the *Ac* protein completely.

The recognition of subterminal sequences (and not the inverted repeats) by the transposase protein has also been described for the *Drosophila* transposable P element (Kaufman et al., 1989), whereas data for other eukaryotic transposases are not yet available. The tnpA-protein, an *En/Spm*-encoded protein that possibly participates in the regulation of transposition of the *En/Spm* transposable element from maize, binds to subterminal repeated motifs of *En/Spm* (Gierl et al., 1988)

### Sequences Required *in cis* for Excision of *Ac*

To answer the question how the *in vitro* DNA binding studies correlate with the *in vivo* sequence requirements for excision, a *Ds* excision assay was designed (Baker et al., 1987). Protoplasts from a tobacco line containing a stably integrated active *Ac* element were transformed with constructs carrying *Ac*-deletion derivatives in the leader sequence in front of the NPT II gene. Only after excision of the *Ac*-derivative the NPT II gene can be expressed and give rise to the development of kanamycin-resistent calli (Coupland et al., 1988).

The results of these experiments are summarized in Figure 4. Not surprisingly, the inverted repeats of the transposable element are absolutely required for excision: A deletion of the four outermost bases at the 3′ end of *Ac* generates a stable insertion, but does not prevent the expression of transposase activity from the element (Hehl and Baker, 1989). However, the inverted repeats are not sufficient for excision: deletion of the subterminal nucleotides 44-92 and 75-181, respectively, at the 5′ end of *Ac* abolishes transposability completely. In contrast, the deletion of *Ac* internal sequences between *Ac* positions 246 and 3381 does not inhibit excision (Coupland et al., 1988).

To achieve a higher resolution in the determination of the *Ac* sequences required *in cis* for excision, at both *Ac* ends internal deletions were created, that leave decreasing portions of the *Ac* termini intact (Coupland et al., 1989) (Figure5). At the 5′ end of *Ac* the excision frequency drops approximately 10fold when the deletions leave less than 238 bp of the terminus intact. Such deletions extend into the DNA region that is bound by the *Ac* protein. When the deletions extend into the major *Ac* protein binding segment as delineated by the Bal 31 deletion analysis, that is, leaving less than 132 bp of the terminus intact, the excision frequency drops to zero. Equivalent effects are observed at the 3′ end of *Ac*: deletions reaching into the major *Ac* protein binding segment abolish transposability completely.

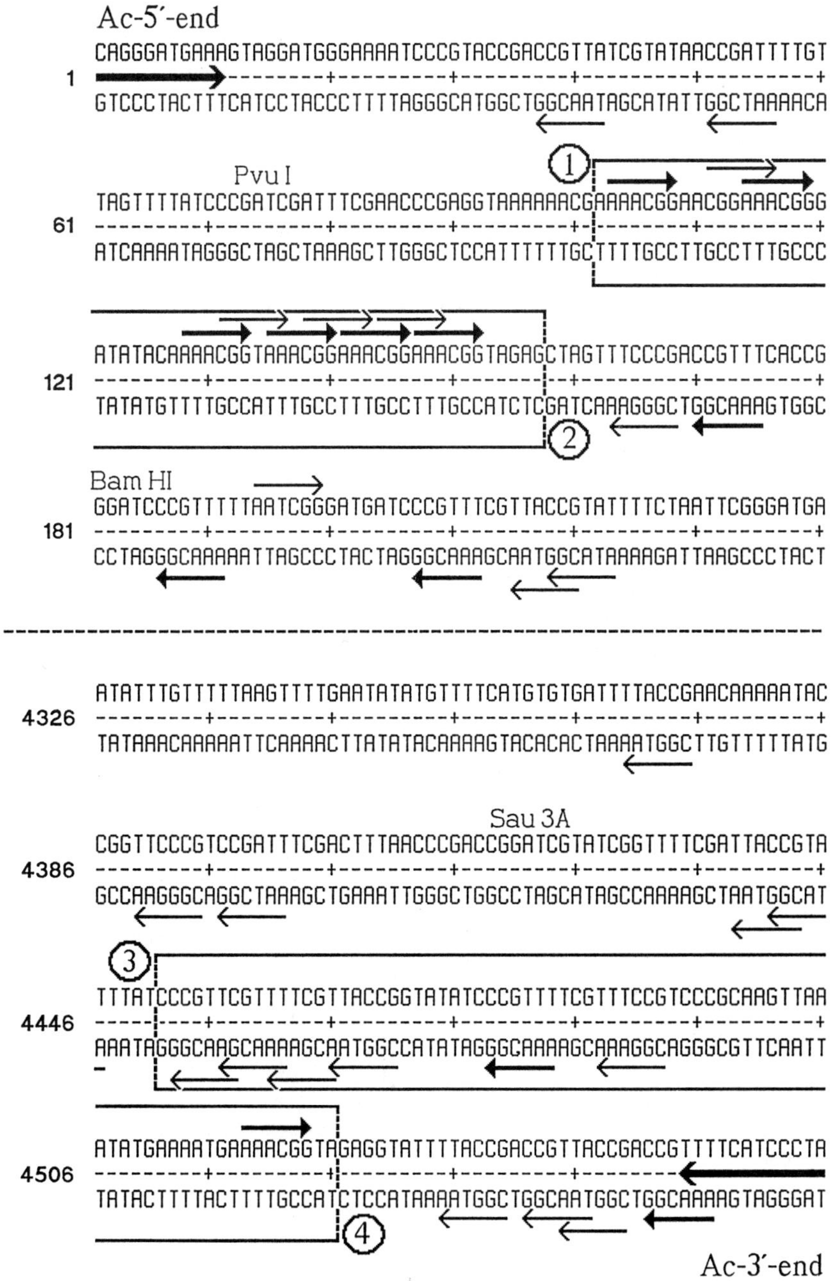


**Figure 3.** *Ac* protein binding sites in the termini of *Ac*. The upper block shows the 240 5′-terminal bases of *Ac*; the lower block shows the 240 3′-terminal bases of *Ac*. The 11 bp inverted repeats are symbolized by the thick arrows between the DNA strands. Arrows with filled heads indicate AAACGG motifs; arrows with open heads indicate hexamer sequences containing the central CG residues and deviating in no more than one nucleotide from AAACGG or its permutation CGGAAA. The boxed sequence blocks contain the major *Ac* protein binding segments as delineated by Bal 31 deletion analysis.

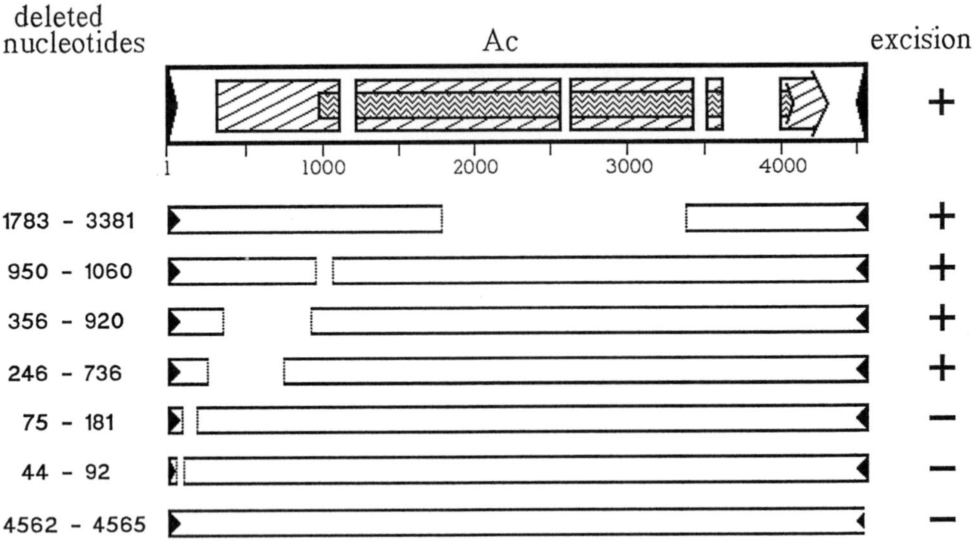


**Figure 4.** Determination of sequences required *in cis* for excision of *Ac*. Within the *Ac* element the localization of its mRNA and ORFa are shown. To the left from the scheme of the derivatives the deleted nucleotides are indicated. + : the derivative can excise if transposase is provided *in trans*; −: the derivative is stable.

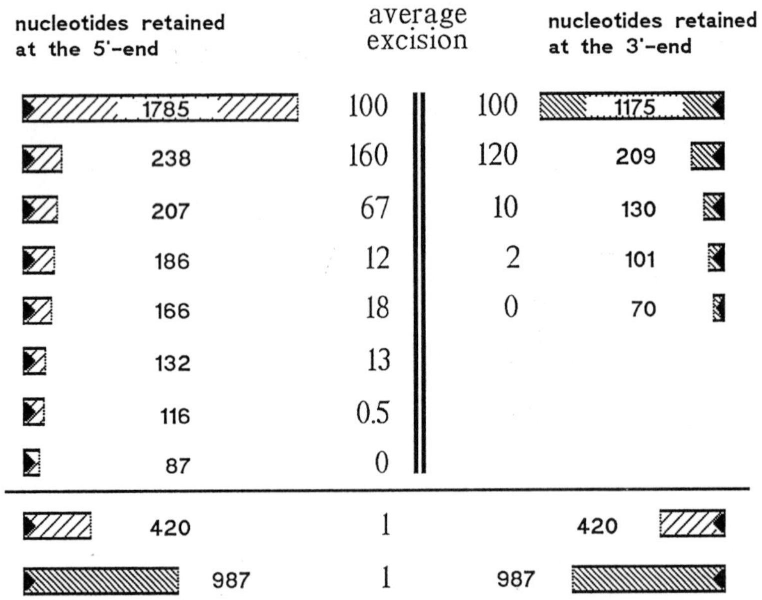


**Figure 5.** Subterminal *Ac* sequences are required for excision. The left half of the figure shows the 5′ endpoints of internal *Ac* deletions and the corresponding average excision frequency of the derivative. The right half of the figure shows the equivalent data for internal *Ac* deletions at the 3′ end of the element. Below the divider line the structures and results for two constructs are shown that contain two *Ac* 5′ ends (420 bp) and two *Ac* 3′ ends (987 bp), respectively.

A possible interpretation of the results of the DNA binding-studies and the deletion analyses is that several *Ac* protein molecules have to bind to subterminal motifs in both ends of *Ac* to achieve the normal excision rate. Deletion of a small number of the binding motifs is tolerable, but leads to a sharp decrease of the excision frequency.

But not only the length of the remaining termini determines the transposability: elements having either two 5′ ends or two 3′ ends of *Ac* do not excise (Figure 5). We do not yet know if the perfect, several hundred nucleotides long terminal inverted repeats with their potential to form very stable stem and loop-structures are inhibitory for the excision reaction. An alternative interpretation could be that accessory host factors are involved in the transposition process, that distinguish between the *Ac* ends.

Binding of Host Factors to *Ac*

The 11 bp terminal inverted repeats of *Ac* are the sites at which the excision reaction has to take place. These sites have to be recognized by some factor, probably by a protein. Since the *Ac* protein does not (at least not *in vitro*) bind to the inverted repeats, we began to search for cellular maize factor(s) with this property. Again, gel mobility shift assays with various DNA fragments of *Ac* were performed. The labelled DNA fragments this time were incubated with maize nuclear protein extracts. Intense complexes with slightly different mobilities are formed between fragments from the 3′ end of *Ac* and a 5′ subterminal fragment containing the *Ac* positions 75-181 (H.-A. Becker, unpublished). Preliminary results indicate that the DNA fragments from the two ends can cross-compete for complex formation. Surprisingly, however, no (strong) binding to the inverted repeats of *Ac* was detected. By DNase I footprinting experiments at least two protected sites in each terminus of *Ac* were detected. These sites are partly overlapping with the major *Ac* protein binding segments. The isolated AAACGG-hexamer motifs, however, are not complexed by a maize host factor.

Apparently, one or more proteins, including the putative *Ac* transposase, can bind to several positions in the ends of *Ac*. The functions and interactions of these proteins are not yet understood. If the host factor(s) binding to *Ac* play a crucial role for the transposition process, their functions must be strongly conserved among plants, because in all seven plant species that were transformed with *Ac* the element can transpose. The participation of host proteins during excision/transposition of a transposable element is well documented for some prokaryotic elements (Craigie et al., 1985; Phadnis and Berg, 1987; Yin and Reznikoff, 1987; Wiater and Grindley, 1988), and is assumed for the *Drosophila* P element. Here, the binding of a host protein to the inverted repeats was found (Rio and Rubin, 1988).

We have initiated studies that are intended to find out if these host factors are important during the excision event.

Effects of Methylation

Occasionally, the *Ac* element can switch from the normal, active state into a transiently inactivated state (McClintock, 1964; 1965). During that inactive state the *Ac* DNA is methylated at particular restriction sites (Schwartz and Dennis, 1986; Chomet et al., 1987), it is not transcribed (Kunze et al., 1988), and it behaves like a *Ds* element.

The analysis of the methylation status is practically limited by the fact that only a small proportion of the potential methylation sites (CpG and CpNpG) is located in restriction sites of methylation-sensitive endonucleases. Figure 6 shows the distribution of CpG dinucleotides in the *Ac* element. It is interesting that the ~240 bp at either end of *Ac* contain a high number of CpG, but nearly no GpC. Many of these CpG dinucleotides are located in the AAACGG-hexanucleotide motifs, but these are not accessible by restriction analysis. Between *Ac* positions 240 and ~1000 the G+C-content is still very high, but here no bias in the distribution of CpG and GpC is observed (Kunze et al., 1988). Most of the Hpa II restriction sites that are methylated during the inactive state of *Ac* (Schwartz and Dennis, 1986) are located in the untranslated leader sequence.

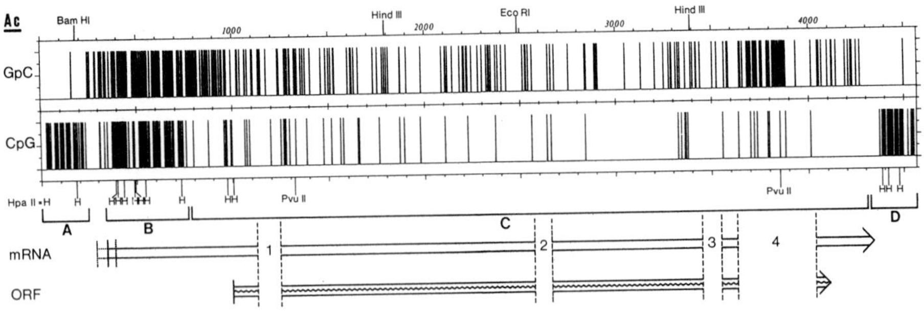


**Figure 6.** Distribution of CpG and GpC dinucleotides along the *Ac* element. CpG and GpC: each vertical line indicates the occurrence of the dinucleotide in the *Ac* sequence. The methylation-sensitive Hpa II and Pvu II restriction sites are marked below. The relative localization of the *Ac* mRNA and ORFa are shown at the bottom. A, B, C, and D are sequence segments with distinct CpG/GpC-distribution.

The analysis of the fate of an *Ac* element that was inserted in the *P* locus of maize after transposition (Greenblatt and Brink, 1962; 1963) lead to the hypothesis that transposition of *Ac* can be correlated with the replication process. Such a phenomenon was described in great detail for IS10, and it was demonstrated that the methylation state of the DNA is responsible for it (Roberts et al., 1985). We therefore tested if the affinity of the *Ac* protein to the AAACGG-hexamers is dependent on their C-methylation state (Kunze and Starlinger, 1989; R. Kunze, unpublished results). The results are summarized in Table 3.

We conclude from these experiments that methylation of the two C-residues in the lower strand of these motifs enhances the binding of the *Ac* protein to them. In contrast, methylation of the C-residue in the upper strand strongly decreases complex formation, and methylation of the cytosines in the lower strand can only weakly, if at all, compensate for this effect.

A hemimethylated state of the *Ac* DNA would appear for a certain time immediately after replication if prior to methylation the DNA was fully methylated. However, since the majority of the hexamer motifs in the 5′ end have the opposite orientation to those in the 3′ end, after replication in one sister chromatid a situation would result which favors *Ac* protein binding at the 5′ end, but inhibits *Ac* protein binding at the 3′ end, whereas the opposite would be true for the other sister chromatid.

We believe it is likely that methylation of the *Ac* protein recognition sites is involved in the coupling of transposition to the replication process. To confirm this assumption *in vivo* data about the methylation state of the *Ac* DNA are required. Since the AAACGG-motifs are not accessible by restriction analysis, we have initiated genomic sequencing experiments to obtain such data.

We hypothesize that the methylation-correlated inactivation of *Ac* is a different effect, perhaps similar to the "classical" transcription inhibition by promoter methylation.

**Table 3.** DNA fragments containing 18 copies of the hexamers in tandem were terminally labelled and tested as targets in gel mobility shift assays. The relative intensities of the complexes were estimated. $^{m}$C : 5-methylcytosine.

| target DNA | relative intensity of complex |
|---|---|
| AAACGG<br>TTTGCC | + |
| AAAC G G<br>TTTG$^{m}$C$^{m}$C | +++ |
| AAA$^{m}$CGG<br>TTT GCC | – |
| AAA$^{m}$C G G<br>TTT G$^{m}$C$^{m}$C | (+) |
| AAAC GG<br>TTTG$^{m}$CC | ++ |
| AAACG G<br>TTTGC$^{m}$C | ++ |
| AAA$^{m}$C GG<br>TTT G$^{m}$CC | – |
| AAA$^{m}$CG G<br>TTT GC$^{m}$C | – |

## Analysis of *Ac*-Protein Functions

A mutational analysis of the ORFa-encoded *Ac* protein was initiated in transgenic tobacco plants (Li, 1989; Li and Starlinger, in preparation). In a series of experiments it was tested wether the N-terminal amino acids of the *Ac* protein in front of the (Pro-Gln/Glu)$_{11}$-segment are required for transposition. A construct with a deletion of the 101 N-terminal amino acids is still able to excise itself as well as an unlinked *Ds* element. Removing 236 amino acids from the N-terminus, however, abolishes the activity of the derivative. The N-terminus of the *Ac* protein in front of the (Pro-Gln/Glu)$_{11}$-segment obviously is dispensable for the excision/transposition reaction. We do not yet know what function this part of the protein has. A possibility would be a regulatory function. Since the regulation of transposition is different in tobacco than in maize (Jones et al., 1989), however, the assay system might be inadequate for the detection of such a function.

To complement the *in vivo* analysis of the *Ac* protein mutants with biochemical studies we began to establish an expression system for the *Ac* protein and derivatives in E. coli. The ATG codon of ORFa was mutated into a Nco I site, and at several other sites close to the N-terminus Nco I sites were inserted in frame into ORFa. Subsequently, ORFa and N-terminal deletion derivatives of ORFa were inserted as Nco I-Bam HI fragments into the expression vector pET-8c which has a Nco I site behind the Shine-Dalgarno sequence (W. Studier, personal communication). This system allows the expression of unfused foreign proteins.

The *Ac* protein accumulates as insoluble inclusion bodies in the bacteria and has no DNA binding activity. The inclusion bodies can be solubilized in guanidiniumhydrochloride and renatured by dilution (Jaenicke and Rudolph, 1989). After this treatment, the *Ac* protein synthesized in E. coli binds with the same specificity to DNA as the protein derived from insect cells (S. Feldmar, unpublished).

Preliminary results were obtained from experiments with the complete *Ac* protein, a N-terminal deletion of 75 amino acids, and a N-terminal deletion of 188 amino acids: The 75 amino acid deletion derivative binds as well to the ends of *Ac* as the complete *Ac* protein. The 188 amino acid deletion derivative, however, has lost the DNA binding property. These results are in agreement with the *in vivo* studies which demonstrate that the deletion of 101 N-terminal amino acids does not prevent transposition activity. At present, experiments with a higher resolution are under way, that will allow the localization of the DNA binding domain(s) and, if existing, the oligomerization domain(s) of the *Ac* protein.

## Acknowledgements

This work was supported by the Deutsche Forschungsgemeinschaft through SFB 274.

## References

Baker, B., Coupland, G., Fedoroff, N., Starlinger, P., and Schell, J., 1987, Phenotypic assay for excision of the maize controlling element *Ac* in tobacco, *EMBO J.*, **6**:1547.

Benedum, U.M., Baeuerle, P.A., Konecki, D.S., Frank, R., Powell, J., Mallet, J., and Huttner, W.B., 1986, The primary structure of bovine chromogranin A: a representative of a class of acidic secretory proteins common to a variety of peptidergic cells, *EMBO J.*, **5**:1495.

Benoist, C., and Chambon, P., 1981, In vivo sequence requirements of the SV40 early promoter region, *Nature*, **290**:304.

Benson, M., and Pirotta, V., 1987, The product of the *Drosophila zeste* gene binds to specific DNA sequences in *white* and *Ubx*, *EMBO J.*, **6**:1387.

Chomet, P.S., Wessler, S., and Dellaporta, S.L., 1987, Inactivation of the maize transposable element *Activator* (*Ac*) is associated with DNA modification, *EMBO J.*, **6**:295.

Coupland, G., Baker, B., Schell, J., and Starlinger, P., 1988, Characterization of the maize transposable element *Ac* by internal deletions, *EMBO J.* **7**:3653.

Coupland, G., Plum, C., Chatterjee, S., Post, A., and Starlinger, P., 1989, Sequences near the termini are required for transposition of the maize transposon *Ac* in transgenic tobacco plants, *Proc. Natl. Acad. Sci. USA*, **86**:9385.

Craigie, R., Arndt-Jovin, D.J., Mizuuchi, K., 1985, A defined system for the DNA strand-transfer reaction at the initiation of bacteriophage Mu transposition: protein and DNA substrate requirements, *Proc. Natl. Acad. Sci. USA*, **82**:7570.

Dingwall, C., Dilworth, S.M., Black, S.J., Kearsey, S.E., Cox, L.S., and Laskey, R. A., 1987, Nucleoplasmin cDNA sequence reveals polyglutamic acid tracts and a cluster of sequences homologous to putative nuclear localization signals, *EMBO J.*, **6**:69.

Döring, H.-P., and Starlinger, P., 1986, Molecular genetics of transposable elements in plants, *Ann. Rev. Genet.*, **20**:175.

Fedoroff, N., 1989, Maize transposable elements, in: Mobile DNA, M.M. Howe and D.E. Berg, eds., American Society for Microbiology, Washington, D.C., pp. 377-411.

Fedoroff, N., Furtek, D., and Nelson Jr., O., 1984, Cloning of the *Bronze* locus in maize by a simple and generalizable procedure using the transposable controlling element *Ac*, *Proc. Natl. Acad. Sci. USA*, **81**:3825.

Franssen, H.J., Nap, J.-P., Gloudemans, T., Stiekema, W., van Dam, H., Govers, F., Louwerse, J., van Kammen, A., and Bisseling, T., 1987, Characterization of cDNA for nodulin-75 of soybean: A gene product involved in early stages of root nodule development, *Proc. Natl. Acad. Sci. USA*, **84**:4495.

Freytag, J.W., Noelken, M.E., Hudson, B.G., 1979, Physical properties of collagen-sodium dodecyl sulfate complexes, *Biochemistry*, **18**:4761.

Fußwinkel, H., Müller-Neumann, M., Both, C., Doerfler, W., and Starlinger, P., 1988, Studies on the *Ac* protein, *Maize Genet. Coop. Newslett.*, **62**:47.

Gierl, A., Lütticke, S., and Saedler, H., 1988, *TnpA* product encoded by the transposable element En-1 of *Zea mays* is a DNA binding protein, *EMBO J.*, **7**:4045.

Greenblatt, I.M., and Brink, R.A., 1962, Twin mutations in medium variegated pericarp maize, *Genetics*, **47**:489.

Greenblatt, I.M., and Brink, R.A., 1963, Transposition of Modulator in maize into divided and undivided chromosome segments, *Nature*, **197**:412.

Grosschedl, R., and Birnstiel, M.L., 1980, Identification of regulatory sequences in the prelude sequences in an H2A histone gene by the study of specific deletion mutants in vivo, *Proc. Natl. Acad. Sci. USA*, **77**:1432.

Hauser, C., Fußwinkel, H., Li, J., Oellig, C., Kunze, R., Müller-Neumann, M., Heinlein, M., Starlinger, P., and Doerfler, W., 1988, Overproduction of the protein encoded by the maize transposable element *Ac* in insect cells by a baculovirus vector, *Mol. Gen. Genet.*, **214**:373.

Hehl, R., and Baker, B., 1989, Induced transposition of *Ds* by a stable *Ac* in crosses of transgenic tobacco plants, *Mol. Gen. Genet.*, **217**:53.

Jaenicke, R., and Rudolph, R., 1989, in: Protein structure - a practical approach, T.E. Creighton, ed., IRL Press, Oxford, pp 191-223.

Jones, J.D.G., Carland, F., Maliga, P., and Dooner, H., 1989, Visual detection of transposition of the maize element *Activator* (*Ac*) in tobacco seedlings, *Science*, **244**:204.

Kaufman, P., Doll, R., and Rio, D., 1989, Drosophila P element transposase recognizes internal P element DNA sequences, *Cell*, **59**:359.

Kleinschmidt, J.A., Dingwall, C., Maier, G., and Franke, W.W., 1986, Molecular characterization of a karyophilic, histone-binding protein: cDNA cloning, amino acid sequence and expression of nuclear protein N1/N2 of *Xenopus laevis*, *EMBO J.*, **5**:3547.

Kozak, M., 1986, Point mutations define a sequence flanking the AUG initiator codon that modulates translation by eukaryotic ribosomes, *Cell*, **44**:283.

Kunze, R., Stochaj, U., Laufs, J., and Starlinger, P., 1987, Transcription of transposable element *Activator* (*Ac*) of *Zea mays* L., *EMBO J.*, **6**:1555.

Kunze, R., Starlinger, P., and Schwartz, D., 1988, DNA methylation of the maize transposable element *Ac* interferes with its transcription, *Mol. Gen. Genet.*, **214**:325.

Kunze, R., and Starlinger, P., 1989, The putative transposase of transposable element *Ac* from *Zea mays* L. interacts with subterminal sequences of *Ac*, *EMBO J.*, **8**:3177.

Lechelt, C., Peterson, T., Laird, A., Chen, J., Dellaporta, S.L., Dennis, E., Peacock, W.J., and Starlinger, P., 1989, Isolation and molecular analysis of the maize *P* locus, *Mol. Gen. Genet.*, **219**:225.

Li, M., 1989, Funktionelle Ausprägung von mutierten Derivaten des transponierbaren Elementes *Ac* aus Mais, PhD thesis, Köln.

Lütcke, H.A., Chow, K.C., Mickel, F.S., Moss, K.A., Kern, H.F., and Scheele, G.A., 1987, Selection of AUG initiation codons differs in plants and animals, *EMBO J.*, **6**:43.

McClintock, B., 1947, Cytogenetic studies of maize and neurospora, *Carnegie Inst. Washington Yearbk.*, **46**:146

McClintock, B., 1964, Aspects of gene regulation in maize, *Carnegie Inst. Washington Yearbk.*, **63**:592.

McClintock, B., 1965, Components of action of the regulators Spm and Ac, *Carnegie Inst. Washington Yearbk.*, **64**:527.

Müller-Neumann, M., Yoder, J.I., and Starlinger, P., 1984, The DNA sequence of the transposable element *Ac* of *Zea mays* L., *Mol. Gen. Genet.*, **198**:19.

Ollo, R., and Maniatis, T., 1987, *Drosophila* Krüppel gene product produced in a baculovirus expression system is a nuclear phosphoprotein that binds to DNA, *Proc. Natl. Acad. Sci. USA*, **84**:5700.

Peterson, P.A., 1986, Mobile elements in maize: A force in evolutionary and plant breeding processes, in: Genetics, Development, and Evolution, J.P. Gustafson, G. L. Stebbins, and F.J. Ayala, eds., Plenum Publishing Corporation, pp 47-78.

Phadnis, S.H., and Berg, D.E., 1987, Identification of base pairs in the outside end of insertion sequence IS*50* that are needed for IS*50* and Tn*5* transposition, *Proc. Natl. Acad. Sci. USA*, **84**:9118.

Pohlman, R.F., Fedoroff, N.V., and Messing, J., 1984, The nucleotide sequence of the maize controlling element Activator, *Cell*, **37**:635.

Rio, D.C., and Rubin, G.M., 1988, Identification and purification of a *Drosophila* protein that binds to the terminal 31-base-pair inverted repeats of the *P* transposable element, *Proc. Natl. Acad. Sci. USA*, **85**:8929.

Roberts, D., Hoopes, B.C., McClure, W.R., and Kleckner, N., 1985, IS10 transposition is regulated by DNA adenine methylation, *Cell*, **43**:117.

Schwartz, D., and Dennis, E., 1986, Transposase activity of the *Ac* controlling element in maize is regulated by its degree of methylation, *Mol. Gen. Genet.*, **205**:476.

Wiater, L.A., and Grindley, N.D.F., 1988, Gammadelta transposase and integration host factor bind cooperatively at both ends of gammadelta, *EMBO J.*, **7**:1907.

Yin, J.C.P., and Reznikoff, W.S., 1987, dnaA, an essential host gene, and Tn5 transposition, *J. Bacteriol.*, **169**:3714.

# EXPRESSION AND REGULATION OF THE MAIZE *Spm* TRANSPOSABLE ELEMENT

N.V. Fedoroff[1], P. Masson[1,3], and J.A. Banks[2]

[1]Department of Embryology, Carnegie Institution of Washington, 115 West University Parkway, Baltimore, Maryland, 21210
[2]Biology Department, McGill University, 1205 Docteur Penfield Avenue, Montreal, Canada H3A 1B1
[3]Faculty of Agronomy and Station for Plant Breeding, B-5800 Gembloux, Belgium

## INTRODUCTION

There is both genetic and molecular evidence that the maize *Suppressor-mutator (Spm)* element encodes gene products necessary for transposition and autoregulation (reviewed in Fedoroff, 1983, 1989, Gierl *et al.*, 1989). Identification of the element-encoded gene products that participate in transposition and regulation has been hindered by the pleiotropy of spontaneous mutations in the element and a poor understanding of how the element is regulated. Here we review recent progress in our laboratory in the identification and characterization of element-encoded transcripts, as well as in understanding both positive and negative regulation of *Spm* activity.

## THE *Spm* ELEMENT CODES FOR TRANSPOSITION AND REGULATORY FUNCTIONS

Evidence that the *Spm* element encodes a gene product necessary for transposition is provided by the observation that internally deleted *dSpm* elements transpose only in the presence of a full-length *Spm* element. The *Spm* and the virtually identical *Enhancer* (*En*) elements are automutagenic, giving rise to intraelement deletions at a relatively high frequency (Fedoroff, 1989a; Gierl *et al.*, 1989). Mutant *Spm* elements of two types have been identified. These are weak *Spm* (*Spm-w*) elements, which transpose and trans-activate transposition at a reduced frequency relative to a standard *Spm* (*Spm-s*), and transposition-defective *Spm* (*dSpm*) elements, which do not transpose autonomously. Transposition-defective modifying *Spm* and *En* elements that either increase or decrease transposition frequency have also been identified (reviewed in Fedoroff, 1989a; Gierl *et al.*, 1989). The ability of *dSpm* elements to move in the presence of an *Spm-s* or *Spm-w* implies that the *Spm* element encodes one or more gene products required for transposition.

*Plant Molecular Biology 2*, Edited by R.G. Herrmann and
B. Larkins, Plenum Press, New York, 1991

The existence of a positive autoregulatory gene product is implicit in the interactions between active and inactive *Spm* elements, as well as those between an *Spm* element and the *a-m2* alleles of the *a* locus involved in anthocyanin biosynthesis (McClintock, 1958, 1962; Masson *et al.*, 1987). McClintock noted that transposable elements can be inactivated by a genetic mechanism that is heritable, but reversible (reviewed in Fedoroff, 1983 and 1989a). She also reported that an inactive element can be transiently reactivated by an active element, implying the existence of a *trans*-acting element-encoded positive regulatory function (McClintock, 1958). Our recent genetic studies on the epigenetic mechanism that affects *Spm* activity, summarized below, have provided evidence that the same mechanism can either stably inactivate an element or impose one of a variety of heritable developmental programs of expression on the element (Fedoroff and Banks, 1988; Fedoroff, 1989). The results of molecular studies on epimutant and programmed elements have revealed a close correlation between the C-methylation pattern of the element's 5' end and its genetic properties (Banks *et al.*, 1988; Banks and Fedoroff, 1989). They have also provided evidence that element-encoded regulatory gene products promote both the transient and heritable reactivation of an inactive *Spm* element.

## GENE PRODUCTS ESSENTIAL FOR *Spm* ACTIVITY ARE ENCODED BY ALTERNATIVELY SPLICED TRANSCRIPTS

Until recently, only one element-encoded transcript had been characterized (Pereira *et al.*, 1986). This transcript, designated *tnpA* (Figure 1), consists of 11 exons and its transcription unit extends almost the full length of the 8.3-kb element. Of the *tnpA* transcription unit's 10 introns, 9 are less than 0.2 kb in length. Its first intron is 4.2 kb long and contains 2 adjacent long open reading frames (ORF1 and 2). In order to determine whether ORFs 1 and 2 encode gene products required for transposition, frameshift mutations were introduced into the intron 1 ORFs, as well as into intron 4. The *in vitro* mutagenized elements were tested in transgenic tobacco for their ability to promote transposition of a *dSpm* element from the leader sequence of a bacterial ß-glucuronidase (GUS) gene expressed from a cauliflower mosaic virus (CaMV) 35S promoter (Masson and Fedoroff, 1989, Masson *et al.*, 1989). It was found that mutations in either ORF1 or ORF2 of the *tnpA* transcript's first intron disrupted the element's ability to *trans*-activate excision of the *dSpm* element, while a comparable mutation in its fourth intron did not (Figure 2). This observation suggested that the *tnpA* intron 1 encodes gene products that participate in transposition.

Analysis of element-homologous transcripts present in *Spm*-containing maize and tobacco cells revealed the presence of two size classes of large transcripts (5-6 kb) with homology to both the *tnpA* transcript and *tnpA* intron 1 sequences (Masson *et al.*, 1989). The structure of the large transcripts was investigated using the polymerase chain reaction (PCR, Saiki *et al.*, 1985). Three different alternatively spliced large transcripts were identified (Masson *et al.*, 1989). The new transcripts have been designated *tnpB*, *tnpC*, and *tnpD* (Figure 3). All of the newly identified transcripts contain all of the *tnpA* transcript's 11 exons and an intron 1 exon comprising most of ORF1. In *tnpB*, a splice donor site near the end of ORF1 is spliced to the acceptor site at the beginning of the *tnpA* transcript's second exon, giving rise to a transcript that encodes a 171-kd ORF1-ORFA fusion polypeptide (Figure 4). *TnpC* and *tnpD* contain two and one additional exons, respectively, comprising ORF2 sequences (Figure 3). *TnpD* contains all of ORF2, while an additional small intron is removed from the ORF2 sequence in *tnpC*. In both *tnpC* and *tnpD*, the same ORF1 splice

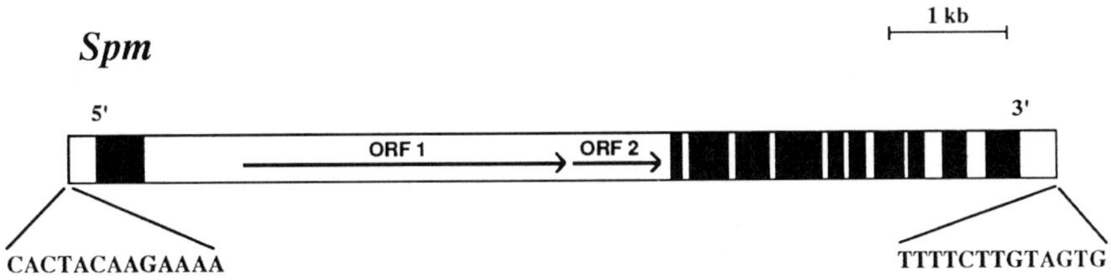


Fig. 1. The structure of the standard *Spm* element. The standard *Spm* element is 8.3 kb in length and is bounded by the 13-bp inverted terminal repetitions shown below the diagram. The exons of the element's most abundant transcript, designated *tnpA*, are represented by the filled boxes and the direction of transcription is left to right. The *tnpA* transcript's first intron contains two almost contiguous long open reading frames, designated ORFs 1 and 2 and represented by the arrows within the intron.

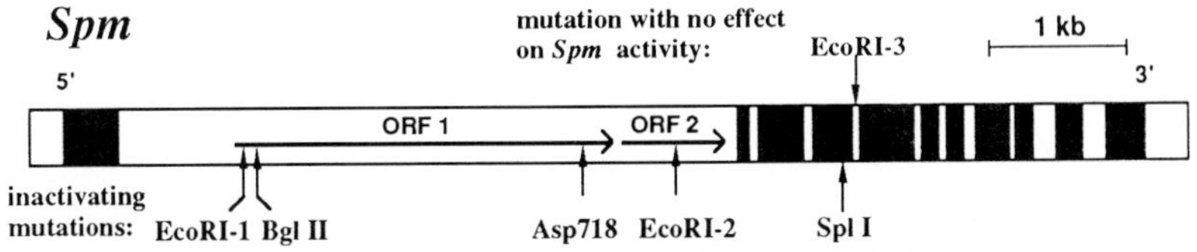


Fig. 2. The effect of frameshift mutations on the activity of the *Spm* element in transgenic tobacco. The *Spm* element is represented as in Figure 1. Vertical arrows indicate the sites at which frameshift mutations were introduced *in vitro* by filling in restriction endonuclease cleavage sites (Masson *et al.*, 1989; Masson and Fedoroff, unpublished). The biological activity of the mutated *Spm* elements was tested as described by Masson and Fedoroff (1989). Mutations which reduce or abolish its activity are shown below the diagram; the single mutation tested which had no effect on the element's activity is shown above the diagram.

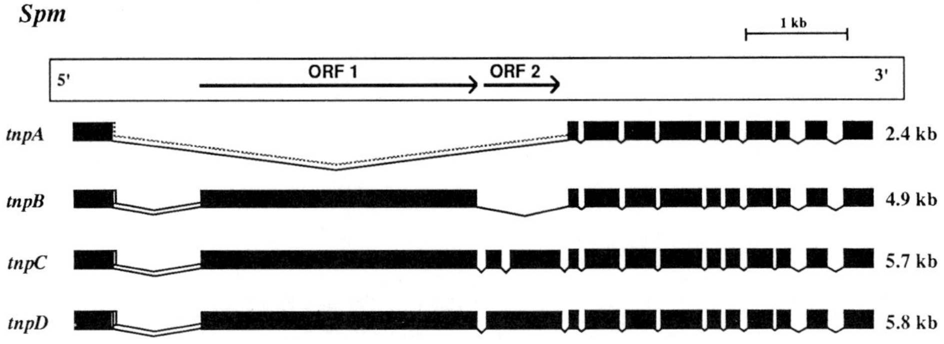


Fig. 3. The structure of the several large alternatively spliced *Spm* transcripts. The *Spm* element is represented by the box as in previous figures. The exons of the previously identified *tnpA* transcript are shown immediately below the element. The exon structure of the newly identified *tnpB*, *tnpC*, and *tnpD* transcripts is shown on succeeding lines. As discussed in the text, two different first exons have been identified and the two lines connecting the 1st and 2nd exons of each transcript indicate that both 1st exons may be present in the population of each transcript. The sizes of the transcripts containing the shorter and more abundant 1st exon and exclusive of the polyA tract are shown at the right of the diagram.

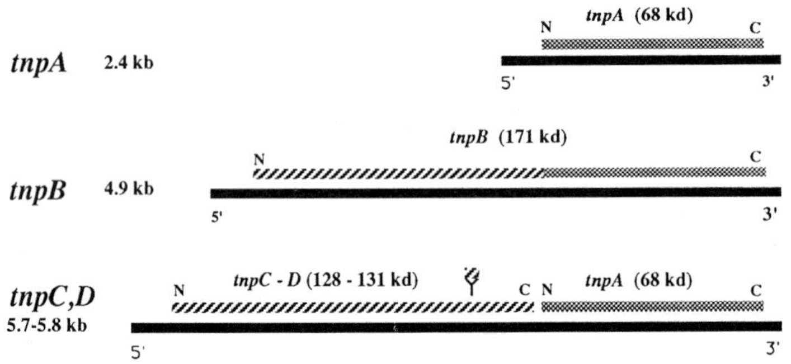


Fig. 4. A diagrammatic representation of the polypeptides encoded by the large ORFs of the *Spm* transcripts. The solid bars represent the transcripts described in figure 3. The lengths of the transcripts are given at the left. The hatched and stippled bars show the lengths of the polypeptides encoded by the largest ORFs of each transcript.

donor site used in *tnpB* is spliced to an acceptor site at the beginning of ORF2 and a splice donor site 19 bp downstream from the end of ORF2 is spliced to the acceptor site at the beginning of the *tnpA*'s second exon. There are translational stop codons in all three reading frames just downstream from the end of ORF2. Both *tnpC* and *tnpD* therefore encode two different polypeptides, a large one of 128 kd (*tnpC*) or 131 kd (*tnpD*) and the ORFA polypeptide encoded by the *tnpA* transcript (Figure 4).

In addition, two alternative exon 1 splice donor sites were identified in this study (Masson *et al.*, 1989). In both maize and tobacco, *Spm* mRNAs can have one of two 1st exon sequences which end either at a donor site 578 or 704 bp from the element's 5' end. Almost all *tnpA* transcripts commence with the shorter first exon in maize and the longer first exon in transgenic tobacco. Both first exons are about equally represented in the population of large transcripts in maize, but the longer first exon predominates among the large *Spm* transcripts in tobacco cells. The most striking feature of the two alternative first exons is that only the longer form contains translation initiation codons, although both contain translation termination codons. Since the ATG codons for the long ORFs encoded by *tnpA*, *tnpB*, *tnpC* and *tnpD* commence in their respective second exons, the first exon contains none of the protein coding sequence. Thus the longer first exon, but not the shorter one, contains short upstream ORFs, which may affect the translational efficiency of the corresponding mRNAs (Kozak, 1986; Mueller and Hinnebusch, 1986).

## FUNCTIONS OF THE ELEMENT-ENCODED GENE PRODUCTS

The roles of the gene products encoded by the various newly identified transcripts in expression and transposition of the element are as yet poorly understood. The results of preliminary experiments suggest that at least two different element-encoded proteins are required for transposition. Thus, frameshift mutations in ORF 1, ORF2 and ORFA all reduce or eliminate the *Spm* element's ability to *trans*-activate excision of a *dSpm* element in transgenic tobacco (Figure 2; Masson *et al.*, 1989, Masson and Fedoroff, unpublished). A *tnpA* cDNA does not promote *dSpm* excision in tobacco, but will complement the defect in an *Spm* element with a frameshift mutation in ORFA. This suggests that *tnpA* and either *tnpC* or *tnpD* are required for activity in the tobacco excision assay.

The *tnpA*-encoded ORFA protein has been expressed in *E. coli* and shown to bind to the subterminal repetitive regions near element ends, but not to the 13-bp terminal inverted repeats (Gierl *et al.*, 1988). In view of the evidence reviewed below that the *Spm* element encodes a positive autoregulatory gene product, it is possible that the *tnpA* protein is a transcriptional activator and is required for element transcription, but not transposition. It is also possible that the *tnpA* protein participates directly in transposition (or is required for both).

## AN EPIGENETIC MECHANISM AFFECTS EXPRESSION OF THE *Spm* ELEMENT

In her early studies on *Spm*, McClintock noted that the element could be inactivated stably, but impermanently (reviewed in Fedoroff, 1983). An inactive *Spm* element's genetic behavior was indistinguishable from that of a *dSpm* element, except for the occasional appearance among progeny of plants in which the element returned to an active form. In addition, both McClintock and Peterson identified elements that showed heritable differential patterns of expression in development (reviewed in Fedoroff, 1983).

In our recent studies on the element-inactivating epigenetic mechanism, we have established that the element can exist in three genetically distinguishable, interconvertible forms:

STABLY ACTIVE <--> PROGRAMMABLE <--> CRYPTIC

The stably active and cryptic forms of the element are highly heritable. The spontaneous frequency of reactivation of a *cryptic Spm* element has been estimated to be less than 1-2 x $10^{-5}$ (Fedoroff, 1989). Although a comparable estimate has not been made of the frequency with which a stably active element undergoes inactivation *in situ*, it appears to be lower than either the frequency of transposition or intraelement deletion. Programmed *Spm* elements generally undergo a change in phase of activity from inactive to active or vice versa during the developmental cycle. The designation *programmable* derives from the observation that elements in this intermediate, genetically relatively labile state can display a variety of heritable differential patterns or programs of expression during development. For example, plant lines can readily be identified in which the elements are active in tillers, but not in the plant's main stalk (McClintock, 1959; Peterson, 1966) and far more likely to be transmitted in an active form through tiller gametes than through gametes produced on the main stalk (Fedoroff and Banks, 1988). Any given developmental pattern of gene expression is heritable, but subject to reprogramming in a small fraction (0.1-1%) of gametes (Fedoroff, 1989).

There is evidence that determination of the element's phase of activity is genetically distinguishable from that of its program of expression, by which is meant the heritability of the activity phase and the developmental timing and frequency of activity phase reversal during the plant's developmental cycle. An element can, for example, be reprogrammed without undergoing a change in its phase of activity. For example, the probability that an inactive element will return to an active phase is greater if it has been transmitted through tiller than through mainstalk gametes in the previous generation (Fedoroff and Banks, 1988). Moreover, the heritability of the inactive phase is initially low, but increases over several generations of selection for inactivity (Fedoroff and Banks, 1988; Banks and Fedoroff, 1989).

## C-METHYLATION AT THE *Spm* ELEMENT'S 5' END IS CORRELATED WITH ELEMENT INACTIVATION

The results of studies on active, unstably inactive and *cryptic Spm* elements have revealed a correlation between the methylation of C residues near its transcription initiation site and its state of activity (Banks *et al.*, 1988). Changes in element C methylation are confined to the 0.6-kb sequence at the element's 5' end. Other tested methylatable sites within the *Spm* element examined in this study remained methylated, regardless of whether or not the element was genetically active. None of the tested methylatable C residues in sequences flanking the element's site of insertion were found to be methylated (Banks *et al.*, 1988).

Inactive elements were distinguishable from active elements by methylation of C residues upstream of the element's transcription initiation site et nucleotide 209 (Pereira *et al.*, 1986; Banks *et al.*, 1988). This sequence has been designated the upstream control region or UCR (Figure 5). Unstably inactive *programmable Spm* elements were less extensively methylated in the GC-rich sequence immediately downstream from the transcription initiation site than were *cryptic Spm* elements.

The GC-rich internally repetitive sequence, termed the downstream control region (DCR) is not methylated in active elements (Figure 5). Thus an *Spm* element's activity phase is correlated with methylation of the UCR sequence, while the heritability of the inactive state is correlated with the extent of methylation within the DCR.

The increase in heritability of the *Spm* element's inactive state that occurs over several generations is paralleled by an increase in the extent of methylation within the UCR and DCR (Banks and Fedoroff, 1989). UCR sequence methylation appears to precede DCR methylation. After 4 generations of selection for inactivity, neither the heritability of the inactive state nor the extent of methylation had reached the high level observed in a *cryptic Spm* element, suggesting that complete inactivation of an *Spm* element occurs over a number of generations.

The methylation of an element showing a genetic program of differential expression has also been investigated (Banks and Fedoroff, 1989). Element methylation was investigated in different plant parts in a plant family in which the *Spm* element was transmitted in an active form more frequently through tiller than through mainstalk gametes and through female than male gametes. Methylation of the element was found to be less extensive in tillers than in mainstalk leaves. Element methylation was also found to be more extensive in unfertilized ears than in the husks surrounding the ear. The highest levels of element methylation within such plants were detected in pollen and the lowest in embryos from mature kernels. These results suggest that there are at least three distinct developmental points at which element methylation is altered. These appear to coincide with the origin of axillary meristems, the commitment of the apical meristem to floral development, and during development of the zygote.

The close correlation between the *Spm* element's genetic expression program and its 5' terminal methylation pattern suggest that methylation of multiple sites within the element's UCR and DCR sequences determines the somatic and meiotic heritability of its expression phase (Figure 5). The results of these studies provide no insight, however, into the genetic event or events that initiate inactivation of an *Spm* element.

ELEMENT-ENCODED REGULATORY GENE PRODUCTS PROMOTE THE TRANSIENT AND HERITABLE ACTIVATION OF AN INACTIVE *Spm*

McClintock (1958, 1959) reported that an active element can transiently activate an inactive one, suggesting the existence of an element-encoded positive regulatory function. Recent genetic and molecular studies have addressed the ability of an *Spm-w* element to both transiently and heritably activate inactive *Spm* elements (Banks and Fedoroff, 1989; Fedoroff, 1989b). It was found that an unstably inactive *programmable Spm* element is transcriptionally active in the presence of an *Spm-w* element, while a *cryptic Spm* is not. This observation suggests that the *Spm* element's positive regulatory gene product is a transcriptional activator and that it is capable of activating an element that is genetically inactive, methylated in the UCR and partly methylated in the DCR sequences (Figure 5).

As discussed above, it is not yet clear which of the gene products encoded by the *Spm* element's alternatively spliced mRNAs is involved in regulating element expression. However, two kinds of evidence suggest that the *tnpA* protein may either itself be a regulatory gene product or participate in regulation. One indication is that *Spm-w* elements with deletions in *tnpA* intron 1 remain fully capable of *trans*-activating an inactive, partially methylated element. The second indication is that

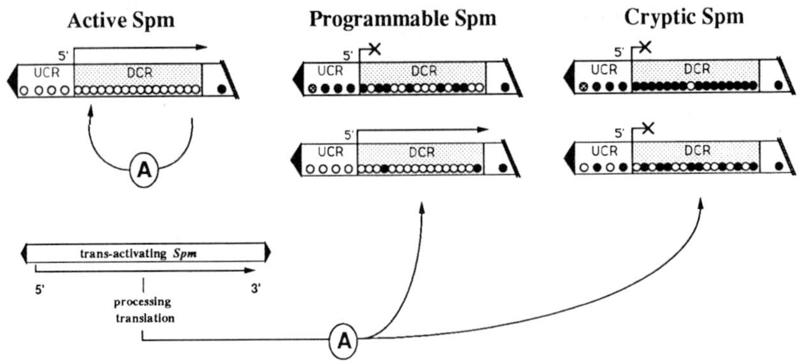


Fig. 5. A diagrammatic representation of *Spm* regulation. The boxes represent the *Spm* element's 5' end. The element's transcription initiation site is marked by the base of the complete arrow (denoting transcriptional activity) or truncated arrow (denoting transcriptional inactivity). The upstream (UCR) and downstream (DCR) control regions are marked and circles within each, as well as downstream from the DCR, represent methylatable C residues. Filled and open circles represent methylated and unmethylated sites, respectively. The hatched circle within the UCR denotes a site which is never fully methylated within a population of elements in a plant. The box at the lower left represents a *trans*-activating *Spm* element and the circled A represents its positive regulatory gene product or products. In this diagram, the active *Spm* element is represented as producing its own regulatory gene product(s) which act to maintain the element in a transcriptionally active and hypomethylated state. The *programmable Spm* is represented as in a transcriptionally inactive form, except in the presence of a *trans*-activating *Spm*, which promotes its transcriptional activation and hypomethylation. The *cryptic Spm* is represented as transcriptionally silent regardless of the presence of a *trans*-activating element. However, the level of methylation of a *cryptic Spm* decreases in the presence of the *trans*-acting element. This, in turn, is correlated with an increase in the probability of genetic reactivation (see text).

*tnpA* protein produced in *E. coli* binds to the element's subterminal sequences, including those immediately upstream of the transcription start site, but does not bind to the element's termini (Gierl *et al.*, 1988).

Although an *Spm-w* element cannot promote the transcriptional activation of a methylated, *cryptic Spm* element, it promotes the cryptic element's genetic reactivation (Fedoroff, 1989). Moreover, extensively methylated *cryptic Spm* elements exhibit a decrease in methylation in the presence of an *Spm-w* element (Figure 5; Banks *et al.*, 1988). Thus, the active element's ability to promote the heritable reactivation of an inactive or cryptic, methylated element appears not to be a passive consequence of its ability to transcriptionally activate an inactive, partially methylated element. Whether the same or different element-encoded proteins are involved in transcriptional *trans*-activation and the heritable reactivation of inactive and cryptic elements is not known.

## SUMMARY

The results of our recent studies on the maize *Spm* transposable element have revealed unexpected complexity in the element's expression and regulation. We have characterized three new, large element-encoded transcripts, designated *tnpB*, *tnpC* and *tnpD*, that are derived from the same primary transcript as the previously identified, more abundant 2.5-kb *tnpA* transcript (Pereira *et al.*, 1986; Masson *et al.*, 1989). The 4 alternatively spliced *Spm* mRNAs code for 4 different proteins with large overlapping domains. The results of preliminary experiments suggest that at least two element-encoded gene products are required for element activity.

The *Spm* element is subject to inactivation by a highly heritable, yet reversible mechanism that can either stably inactivate the element or program it to be differentially expressed during development. Programming and inactivation are accompanied by changes in methylation upstream and downstream of the element's transcription start site. The *Spm* element encodes one or more gene products that can transcriptionally activate an inactive element, as well as promote the heritable reactivation of both unstably inactive programmed and stably inactive *cryptic Spm* elements.

## ADKNOWLEDGMENTS

This work was supported by National Institutes of Health grant 5-R01-GM34296 and by fellowships to J.A.B. from the Center for Agricultural Biotechnology of the University of Maryland at College Park. P.M. is "Chercheur Qualifie" of the Belgian F.N.R.S.

## REFERENCES

Banks, J., Masson, P., and Fedoroff, N., 1988, Molecular mechanisms in the developmental regulation of the maize *Suppressor-mutator* transposable element, Genes Dev. 2:1364.

Banks, J. A., and Fedoroff, N., 1989, Patterns of developmental and heritable change in methylation of the *Suppressor-mutator* transposable element, Develop. Genet. 10:425.

Fedoroff, N. V., 1983, Controlling elements in maize, in "Mobile Genetic Elements", J. Shapiro, ed., Academic Press, New York, pp. 1-63.

Fedoroff, N. V., and Banks, J. A., 1988, Is the *Suppressor-mutator* element controlled by a basic developmental regulatory mechanism?, Genetics 120:559.

Fedoroff, N., 1989a, Maize transposable elements, in "Mobile DNA", M. Howe, and D. Berg, eds., American Society for Microbiology, Washington, pp. 375-411.

Fedoroff, N. V., 1989b, The heritable activation of cryptic *Suppressor-mutator* elements by an active element, Genetics 121:591.

Gierl, A., Lütticke, S., and Saedler, H., 1988, *TnpA* product encoded by the transposable element *En-1* of *Zea mays* is a DNA binding protein, EMBO J. 7:4045.

Gierl, A., Saedler, H., and Peterson, P. A., 1989, Maize transposable elements, Annu. Rev. Genet. 23:71.

Kozak, M., 1986, Point mutations define a sequence flanking the AUG initiator codon that modulates translation by eukaryotic ribosomes, Cell 44:283.

Masson, P., Surosky, R., Kingsbury, J., and Fedoroff, N. V., 1987, Genetic and molecular analysis of the *Spm*-dependent *a-m2* alleles of the maize a locus, Genetics 177:117.

Masson, P., and Fedoroff, N., 1989, Mobility of the maize *Suppressor-mutator* element in transgenic tobacco cells, Proc. Natl. Acad. Sci. USA 86:2219.

Masson, P., Rutherford, G., Banks, J. A., and Fedoroff, J. A. (1989). Essential large transcripts of the maize *Spm* element are generated by alternative splicing. Cell, 58, 755-765.

McClintock, B., 1958, The *Suppressor-mutator* system of control of gene action, Carnegie Inst. Wash. Year Book 57:415.

McClintock, B., 1959, Genetic and cytological studies of maize, Carnegie Inst. Wash. Year Book 58:452.

McClintock, B., 1962, Topographical relations between elements of control systems in maize, Carnegie Inst. Wash. Year Book 61:448.

Mueller, P. P., and Hinnebusch, A. G., 1986, Multiple upstream AUG codons mediate translational control of GCN4, Cell 45:201.

Pereira, A., Cuypers, H., Gierl, A., Schwarz-Sommer, Z., and Saedler, H., 1986, Molecular analysis of the *En/Spm* transposable element system of *Zea mays*, EMBO J. 5:835.

Peterson, P. A., 1966, Phase variation of regulatory elements in maize, Genetics 54:249.

Saiki, R. K., Scharf, S., Faloona, F., Mussil, K. B., Horn, G. T., Erlich, H. A., and Arnheim, N., 1985, Enzymatic amplification of ß-globin genomic sequences and restriction site analysis for diagnosis of sickle cell anemia, Science 230:1350.

# THE *En/Spm* TRANSPOSABLE ELEMENT OF *ZEA MAYS*

Monika Frey, Adriane Menßen, Sarah Grant, Stephanie Lütticke, Julio Reinecke, Stefan Trentmann, Guillermo Cardon, Heinz Saedler and Alfons Gierl

Max-Planck-Institut für Züchtungsforschung
Abteilung Molekulare Pflanzengenetik
5000 Köln 30, FRG

The concept of transposable elements was developed by Barbara McClintock forty years ago. These genetic units move within genomes and frequently can be phenotypically recognized when they integrate into a gene. Characteristically, these insertion mutations are somatically unstable, i.e. excision of the element from the locus restores gene activity in this and all progeny cells. Such variegated phenotypes have appeared in many life forms and were attributed to transposable elements. Soon after the molecular isolation of transposons from *E.coli* (Jordan *et al.*, 1968) it became apparent that transposable elements are natural components of most genomes, because they have been found in representatives of virtually all organisms.

Among the ten transposable element systems which have been genetically identified in maize, the *Enhancer (En)* or *Suppressor-mutator (Spm)* system has been analysed relatively well genetically and molecularly. The current status of the molecular analysis of the *En/Spm* system will we outlined below.

## Structural features of *En/Spm*

The *En* and *Spm* transposable element systems have been identified independently by Peterson (1953) and McClintock (1954). It was shown both genetically (Peterson, 1965) and molecularly (Pereira *et al.*, 1986; Masson *et al.*, 1987) that *En* and *Spm* are virtually identical. The *En/Spm* element is 8.3 kb in length, has a 13 bp perfect terminal inverted repeat (TIR), and causes a 3 bp duplication of the target sequences upon insertion. It encodes at least two products that have been termed tnpA and tnpB (Gierl *et al.*, 1988) (Figure 1). The 69 kDa tnpA protein is encoded by a 2.5 kb transcript and the 132 kDa tnpB protein is encoded by a 6 kb transcript (Figure 1). *TnpA* mRNA is about 100 times more

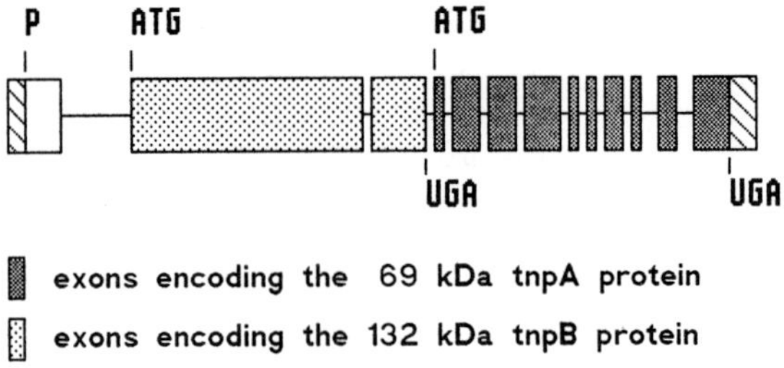


**Figure 1.** Structure of the 8.3 kb *En/Spm* element. Hatched boxes indicate the highly structured termini (see Fig. 2). The unfilled box represents the untranslated first exon of *tnpA* and *tnpB*. The mRNAs of *tnpA* and *tnpB* are derived from pre-mRNA by differential splicing. The 6 kb *tnpB* mRNA spans almost the entire *En/Spm* element and overlaps with *tnpA* mRNA. The exons encoding tnpB protein are contained within the first intron of *tnpA*. The promoter region (P) and translation start and stop codons are indicated.

abundant than *tnpB* mRNA (Pereira *et al.*, 1986). Transcription is initiated at position 209 at the left end of the element and these two transcripts represent alternative splice products of a precursor transcript.

In addition to the structurally intact, transposition competent *En/Spm* element, numerous defective elements have been isolated (Schwarz-Sommer *et al.*, 1985;Masson *et al.*, 1987; Schiefelbein *et al.*, 1988). These elements represent internal deletion derivatives of *En/Spm* and transpose only in the presence of the active intact element. Because of this responsiveness, defective and active elements constitute a two component sytem, consisting of the receptor (defective) element *(dSpm)* and the autonomous (intact) element *(En/Spm)*.

## The *En/Spm* termini serve as substrates for excision

*En/Spm* transposes non-replicatively via excision and reintegration. Two DNA sequence domains at the element's termini are involved in excision. The 13 bp perfect TIR and subterminal regions containing several direct and inverse repetitions of a 12 bp sequence motif (Gierl *et al.*, 1988). These terminal regions comprise 180 bp at the 5' end and 300 bp at the 3' end of *En/Spm* (Figure 2). Deletion of the outermost two nucleotide of the 13 bp TIR drastically reduces the excision frequency of this element (Schiefelbein *et al.*, 1988). Partial deletion of the 12 bp motifs in the subterminal region also correlates with decreased excision rates (Schwarz-Sommer *et al.*, 1985). The extent of the deletions seems to be proportional to the degree of reduction. The entire deletion of the 12 bp motifs at one end of the element completely abolishes excision ability (Gierl, unpublished). In summary, the above observation strongly suggest that the 13 bp TIRs and the subterminal region containing the 12 bp motifs both make up the substrates for excision in the *En/Spm* system.

## *En/Spm*-encoded functions

cDNA of *tnpA* was expressed in *E. coli* and shown to encode a DNA binding protein (Gierl *et al.,* 1988). It recognizes the 12 bp sequence motifs in the subterminal regions of *En/Spm* (Figure 2), which are defined as *cis*-determinants for transposition (see above). Thus a functional role for *tnpA* in transposition is suggested. However, it should be emphasized that tnpA alone is not sufficient to execute excision. This can be concluded from the analysis of mutant *En/Spm* elements like *Spm-w-8011* and *En-2.* These elements still express tnpA, but are defective in promoting excision. In these elements sequences of *tnpB* are deleted.

**Figure 2. Terminal structures of *En/Spm*. About 200 bp of the 5' terminus and 300 bp of the 3' terminus are shown. The filled arrows indicate the 13 bp TIR. The open arrows represent the 12 bp tnpA binding motifs.**

This indicates that a second function is required for transposition. Based on a comparison of *En/Spm* related elements from other plant species, it was speculated that this function could be encoded by the *tnpB* gene (Gierl *et al.,* 1989) and that it might interact with the 13 bp TIR.

In order to test the requirements for excision an assay system was established in transgenic tobacco (M. Frey, unpublished). A defective element that has retained the perfect ends required for excision, but does not encode *En/Spm* functions, was inserted between a promoter and a reporter gene. Preliminary experiments indicate that this element only excises if *tnpA* and *tnpB* are both expressed in the same plant.

In conclusion, two element encoded functions are probably associated with excision of *En/Spm:* Binding of tnpA at both ends could be required to bring the element's termini into contact with each other. The complex formed in this way could then be released from the chromosome by the action of tnpB protein, which may accomplish endonucleolytic cleavage next to the element's ends after interacting with the 13 bp TIR. This complex formation could also be necessary to allow healing of the chromosome ends during excision.

The DNA sequence analysis of excision products (empty donor sites) of *En/Spm* also contributed to the understanding of the excision mechanism. In a systematic analysis of En/Spm excisions from the *waxy* gene Schwarz-Sommer *et al.* (1985) found that only one of ten excision events was precise. The analysis of these altered sequences , "footprints", led to the formulation of a model about plant transposable element excision and integration.

According to the model of Saedler and Nevers (1985), transposition takes place by a "cut and paste" mechanism. The transposase recognizes the ends of the element and introduces staggered nicks at the ends of the target site duplication, in the same fashion as postulated for re-insertion. Footprints result from the action of DNA repair enzymes which act on the protruding single stranded fringes at the excision site and at the element in the transposition complex.

Inaccurate excisions were also reported for other plant transposable elements (Saedler and Nevers, 1985). If these elements are integrated into exon sequences, and if the number of nucleotides added or deleted is a multiple of three, then the consequence will be an allele coding for an altered gene product. In the case that the insertions are in regulatory sequences of genes, new patterns or levels of gene expression may result. Therefore the generation of new variation upon which selection could act, might be accelerated by the action of transposable elements.

## Regulation of *En/Spm* transposition

Transposition is tightly linked to mutation. Therefore it should be regulated to a level which is not deleterious for the cell. There are several levels by which transposition can be regulated and adjusted to compatible levels. First, the *En/Spm* promoter is rather weak. The abundancy of *tnpA* mRNA in poly(A) RNA preparations is about $10^{-6}$. Secondly, the mRNAs of *tnpA* and *tnpB* are derived from one precursor transcript. *TnpB* mRNA is the minor product of this maturation process. TnpB might therefore be rate limiting for transposition in maize.

## *En/Spm* is subject to negative control

An active *En/Spm* can lapse into inactivity (McClintock, 1961). Molecularly, inactivity was correlated with increased levels of C-methylation of the element. A high degree of methylation was also correlated with low levels of transcription (Banks *et al.,* 1988). The consequences of methylation for *En/Spm* activity have been recently reviewed in detail (Fedoroff, 1989).

A completely different type of negative regulation was

detected when the inhibitory effect on *En/Spm* transposition of the *En-I102* element was analysed (Cuypers *et al.*, 1988). This internal deletion derivative of *En* expresses an aberrant polypeptide (tnpR) which shares homology with both tnpA and tnpB. TnpR represses transposition probably by competitive inhibition of tnpA and/or tnpB function. Since there are about 50-100 *En/Spm* homologues (mainly deletion derivatives) distributed throughout the maize genome (Schwarz-Sommer *et al.*, 1984), it is feasible that products encoded by these elements could modulate *En/Spm* transposition.

## *En/Spm* is also subject to positive regulation

A positive regulator for *En/Spm* gene expression has been postulated (Nevers and Saedler, 1977), based on McClintock's observation that inactive elements can be transiently or heritably re-activated by an active element (McClintock, 1971). The finding that *Spm-w-8011*, which probably only encodes *tnpA* (see above), is also capable of *trans*-activating inactive elements (Banks *et al.*, 1988), suggests that tnpA is the activator. The mechanism of activation has been considered in several models, and since it was demonstrated that tnpA binds to the region upstream of the *En/Spm* specific transcription initiation site (Gierl *et al.*, 1988), it is very likely that bound tnpA protein provides a shelter against methylation of the promoter region and thus prevents inactivation of the element. This implies that the presence of tnpA is essential to maintain its own expression.

## Defective *En/Spm* elements can funtion as introns

In several alleles (suppressible alleles), the insertion of defective elements *(dSpm)* reduces, but does not abolish gene expression. In these alleles the insertions are found within the transcribed regions of the genes involved and nearly wildtype sized transcripts are produced, because most of the element sequences are removed by splicing (Schiefelbein *et al.*, 1988). For the the *a2-m1 (state II class)* allele it was shown that this is due to splice sites present at both termini of *En/Spm* (Menßen, unpublished). As a consequence of pre-mRNA processing only 18 bp of *dSpm* sequence remaine within the mRNA of this allele. This slightly altered *A2* product is still functional as shown by the complementation of an *a2* mutant with the corresponding cDNA in a transient assay.

The property to function as intron seems to apply more generally for defective maize transposable elements, because it was also observed for *Ds* elements, the defective members of the *Ac* family of *Z. mays* (for review, see Wessler, 1989). Maybe this feature provides a certain selective advantage by mitigating the impact of transposable element insertion.

## *En/Spm* can control the activity of genes in opposite ways

The interaction of the functions encoded by the autonomous

element may control the activity of genes into which defective elements have integrated. Negative control has been observed with the socalled suppressible alleles. As described above, in these alleles most of the inserted sequences of a defective element are removed by splicing, resulting in an intermediate level of gene expression. However, this is only so in the absence of a functional *En/Spm* element. In its presence, the residual gene expression is blocked by the *Suppressor* function of *En/Spm.* The Suppressor function seems to resemble a negative regulatory circuit in which an *En/Spm* encoded protein acting *in trans* as a repressor. The repressor recognizes and binds a defined *cis*-element located within the defective element. The bound protein is thought to sterically hinder progression of RNA polymerase through the gene, resulting in prematurely terminated transcripts.

The suppressor function was attributed to tnpA protein by reconstruction of the suppressor system in transgenic tobacco (Grant *et al.,* 1990). It was shown that one inverted repeat of the *tnpA* binding motif is sufficient *in cis* for suppression to occur via tnpA protein.

Binding of tnpA can probably also have an opposite effect on gene expression. The expression of the *a1-m2* allele is activated in the presence of *En/Spm.* In this case the insertion of the element disrupts the promotor region of the *A1* gene (Schwarz-Sommer *et al.,* 1987), thereby inhibiting gene expression. However, in the presence of En/Spm encoded functions *A1* expression is partially restored. It was suggested (Gierl *et al.,* 1989) that binding of tnpA to the element in the *A1* promoter activates this promoter, in a similar way as discussed for the positive regulation of the *En/Spm* promotor (see above).

These control systems formally resemble the classical operon model, in which a *cis* regulatory unit (defective element) responds *in trans* to signals (tnpA) emitted by the regulator (autonomous element). Maybe these features have contributed to the generation of new regulatory units during evolution.

## References


Banks, J.A., Masson, P. and Fedoroff, N. 1988. Molecular mechanisms in the developmental regulation of the maize *Suppressor-mutator* transposable element. Genes & Development 2:1364-80

Cuypers, H., Dash, S., Peterson, P.A., Saedler, H. and Gierl, A. 1988. The defective En-I102 element encodes a product reducing the mutability of the En/Spm transposable element system of *Zea mays.* EMBO J. 7:2953-2960

Fedoroff, N. 1989. Maize transposable elements, in Mobile DNA, M. Howe, D. Berg, eds.Dc:ASM Press, 375-411

Gierl, A., Lütticke, S. and Saedler, H., 1988. *TnpA* product encoded by the transposable element En-1 of Zea mays is a DNA binding protein. EMBO J. 7:4045-53

Gierl, A., Saedler, H. and Peterson, P.A. 1989. Maize transposable elements. Annu. Rev. Genet. 23:71-85

Jordan, E., Saedler, H. and Starlinger, P. 1968. 0º and strong polar mutations in the *gal* operon ar insertions. Mol. Gen. Genet. 102:353-63

Masson, P., Surosky, R., Kingsbury, J. and Fedoroff, N.V. 1987. Genetic and molecular analysis of the *Spm*-dependent *a-m2* alleles of the maize *a* locus. Genetics 177:117-37

McClintock, B. 1954. Mutations in maize and chromosomal aberrations in neurospora. Carnegie Inst. Washington Yearb. 53:254-260

McClintock, B. 1961. Further studies on the *Suppressor-mutator* system of control of gene action in maize. Carnegie Inst. Washington Yearb. 60:469-76

McClintock, B. 1971. The contribution of one component of a control system to versatility of gene expression. Carnegie Inst. Washington Yearb. 70:5-17

Nevers, P. and Saedler, H. 1977. Transposable genetic elements as agents of gene instability and chromosome rearrangements. Nature 268:109-15

Pereira, A., Cuypers, H., Gierl, A., Schwarz-Sommer, Zs. and Saedler, H., 1986. Molecular analysis of the *En/Spm* transposable element system of Zea mays. EMBO J. 5:835-841

Peterson, P.A. 1953. A mutable pale green locus in maize. Genetics 38:682-83

Peterson, P.A. 1965. A relationship between the *Spm* and *En* control systems in maize. Am. Nat. 44:391-98

Saedler, H. and Nevers, P. 1985. Transposition in plants: a molecular model. EMBO J. 4:585-590

Schiefelbein, J. W., Raboy, V., Kim, H.Y., and Nelson, O.E. 1988. Molecular characterization of *Suppressor-mutator* (Spm)-induced mutations at the bronze-1 locus in maize: the bz-m13 alleles. In Proc. Internatl.Symp. on Plant Transposable Elements, O. Nelson ed., pp. 261-278, Plenum, New York

Schwarz-Sommer, Zs., Gierl A., Klösgen, R.B., Wienand, U., Peterson, P.A. and Saedler, H., 1984. The Spm (En) transposable element controls the excision of a 2 kb DNA insert at the *wx-m8* locus of *Zea mays.* EMBO J. 3:1021-28

Schwarz-Sommer, Zs., Gierl, A., Cuypers, H., Peterson, P.A. and Saedler, H. 1985. Plant transposable elements generate the sequence diversity needed in evolution. EMBO J. 4:591-597

Schwarz-Sommer, Zs., Gierl, A., Berntgen, R. and Saedler, H. 1985. Sequence comparison of "states" of *a1-m1* suggests a model of *Spm (En)* action. EMBO J. 4:2439-2443

Schwarz-Sommer, Zs., Shepherd, N., Tacke, E., Gierl, A., Rhode, W., et al. 1987. Influence of transposable elements on the structure and function of the *A1* gene of *Zea mays.* EMBO J. 6:287-94

Wessler, S.R. 1989. The splicing of maize transposable elements from pre-mRNA - a minireview. Gene 82:127-133

# THE MECHANISM AND CONTROL OF Tam3 TRANSPOSITION

Rachel Burton[1], Clare Lister[1], Steve Schofield[2], Jonathan Jones[2], and Cathie Martin[1]

1-John Innes Institute, John Innes Centre for Plant Research, Colney Lane, Norwich NR4 7UH, UK
2-Sainsbury Laboratory, John Innes Centre for Plant Research, Colney Lane, Norwich NR4 7UH, UK

## INTRODUCTION

The snapdragon, *Antirrhinum majus*, has a number of transposable elements of the type that move by excision and reintegration. Some of these have been studied genetically and isolated, and the best characterised is Tam3. The activity of this transposon was first recognised in a line where Tam3 is inserted in the *pallida* locus. *Pallida* (*pal*) encodes an enzyme involved in anthocyanin biosynthesis and the phenotype caused by the Tam3 insertion is an attractive variegated flower with an acyanic background (where the transposon is blocking *pal* expression and hence anthocyanin biosynthesis) and full red somatic reversion sites. This line was studied by Darwin and later by Bauer, Mather, Stubbe, Fincham and many other geneticists interested in the nature of its phenotypic instability (Harrison, 1976). Lines have been selected which show high frequencies of somatic and germinal reversion and using this material it has been possible to demonstrate that phenotypic instability associated with Tam3 activity is very susceptible to growing temperature, being 1000-fold greater in plants grown at 15°C compared to those grown at 25°C (Harrison and Fincham, 1964). It has also been shown that instability is dramatically affected by the presence of an unlinked modifier locus, *Stabiliser* (Harrison and Fincham, 1968). When *Stabiliser* (*St*) is homozygous, instability is 400-fold less than when the recessive allele is homozygous. The heterozygous intermediate (*Stst*) supports an intermediate level of instability. Both the environmental and genetic influences on instability have been found to apply to Tam3 activity in other loci such as the *nivea* (*niv*) locus (Carpenter et al., 1987) indicating that they were a property of the causal agent of the instability and not of the host gene.

Tam3 was isolated from an insertion in the *nivea* gene which also affects floral pigmentation in the unstable line JI:98 (Sommer et al., 1985). It was found to be an element of 3.6kb long with terminal inverted repeats that could form an extended stem-loop structure. The terminal 11 bp of the inverted repeats showed strong similarity to the inverted repeat sequences of Ac. The similarity of Tam3 to Ac was emphasised by the discovery that Tam3 usually causes an 8 bp direct duplication of sequence upon insertion, as does Ac, and that the large single open reading frame of Tam3 shows strong homology in its deduced amino acid sequence to the sequence of the transcribed region of Ac (Sommer et al., 1985, 1988; Coen et al., 1986; Martin et al., 1989). Subsequently insertion sequences have been discovered in parsley and pea which also show homology to the ends of Ac in their inverted repeats and which have generated 8 bp direct repeats upon insertion (Herrmann et al., 1988; Bhattacharyya et al., 1990). This suggests that the Ac-type of element may be a mobile unit present in a wide range of species. However, as yet there is no evidence for functional interaction between these Ac element types from different species.

Despite the similarity of Tam3 to Ac, a few striking differences also exist. Whilst the Ac/Ds family in maize consists of many copies of the receptor or non-autonomous Ds element as well as the autonomous Ac, no evidence for non-autonomous or defective Tam3 elements in *Antirrhinum* has been found. While examination of different Tam3 copies from the *Antirrhinum* genome has shown that some may not move it seems likely that these are always inactive, and may not be amenable to trans-activation (Martin et al., 1989). Simlarly no mutation of an autonomous to a non-autonomous form of Tam3 at a fixed locus has ever been observed as it has for Ac. The lack of non-autonomous forms has made the genetic analysis of Tam3 function very difficult. To circumvent some of the problems associated with the genetic analysis of the element, Tam3 has been introduced into a heterologous host (tobacco) to determine whether it is capable of independent transposition. Tam3 appears to be capable of transposition in tobacco (Martin et al., 1989). Empty donor sites have been observed and sequencing reveals that the DNA footprint left by the element conforms to the type found after other Tam3 excisions. Tam3 has also been shown to have inserted into a new position in the tobacco genome, unlinked to the donor site. So either Tam3 contains all the unique sequences required for transposition or it is being efficiently transactivated by the resident transposons of tobacco.

As well as excising at high frequency in *Antirrhinum,* Tam3 also causes rearrangements of long stretches of DNA at high frequency. Such rearrangements involve inversions, deletions and duplications (Martin et al., 1988). Other transposons, including Ac, are known to be able to induce these types of rearrangements. However the frequency at which Tam3 induces them is high enough to allow their development to be followed over single generations and the structural changes to be characerised at the molecular level in the first generation after rearrangement. This analysis of rearrangements, which appear to result from aberrant transposition attempts, has led to further insights into the mechanism of the transposition process.

## Large-scale inversions induced by Tam 3

Three inversions generated by Tam3 have been studied. One, at the *pal* locus, generated in an *Antirrhinum* line JI:42 over twenty years ago, appears to be an inversion of one flanking sequence with a linked reciprocal site, such that only one flanking sequence remains associated with Tam3 (Robbins et al., 1989). The other two inversions, generated recently at the *niv* locus in lines JI:531 and JI:565, involve inversions with a linked reciprocal site, but in each case both sequences flanking Tam3 in the progenitor remain associated with a copy of Tam3 after the inversion (Martin et al., 1988) (Fig. 1). Molecular analysis of one of these inversions has revealed that the linked site defining the other end of the inversion did not contain a copy of Tam3 in the progenitor, showing that the rearrangements did not arise by recombination between the two transposon copies. The sequence data shows that the flanking sequences (y and z) in the inversion carry a direct duplication of 8 bp compared to the progenitor locus (yz). Therefore the inversion arose as a result of a partial transposition. The mechanism whereby this could have occurred is

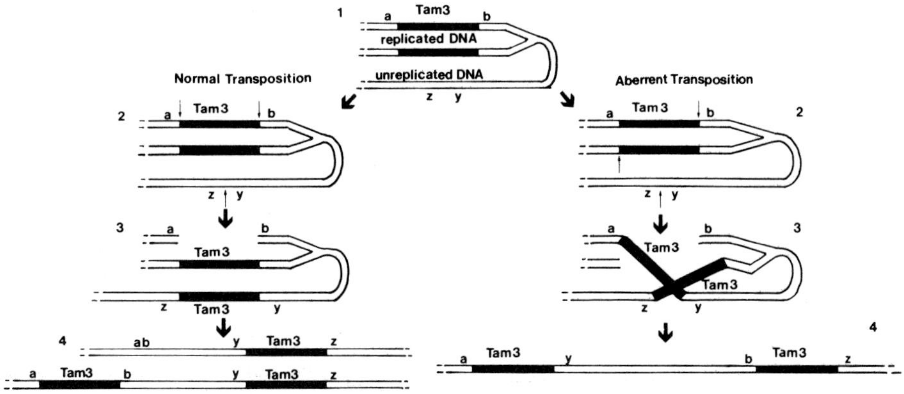


Fig. 1 Diagram to illustrate the formation of inversions producing two copies of Tam3 located at each end of the inversion. 1. We propose that Tam3 preferentially transposes shortly after its replication to a linked but unreplicated site. If transposition continues normally staggered nicks are made in the sequences flanking one copy of Tam3 (2), to leave the empty donor site ab. The new target site yz is nicked. 3. Tam3 integrates at yz. 4. Because yz is at this stage unreplicated, the continuation of replication will leave one replicated chromatid carrying the empty donor site ab and Tam3 at yz. The other chromatid will carry two copies of Tam3 one at ab and the other at yz.

If transposition is aberrant the staggered nicks are made in the DNA flanking recently replicated copies of Tam3, but in this case the ends are from the two different Tam3 copies (2). The other end of each element remains attached to its flanking sequence. The target site yz is also nicked. 3. Integration of the free ends of Tam3 at yz will give some chromosome breakage but one strand should be viable giving rise to an inversion between two Tam3 copies as shown in 4. If the cut ends of the two Tam3 copies integrate at yz in the other orientation such that the Tam3 copy joined to a fuses to z, a deletion of b→y plus one copy of Tam3 will result.

outlined in Fig. 1. If transposition initiates shortly after DNA replication, and has a tendency to select target sites in unreplicated DNA, the transposase has the opportunity for error if it nicks at one end only of each of the replicated Tam3 copies. Transposition may progress with each of the Tam3 ends that is free from flanking DNA (and presumably associated with the transposase) being integrated into the new recipient site. However as the two integrated ends are from different Tam3 molecules the net outcome (in addition to probable chromosomal breakage) on one strand would be an inversion, with the flanking sequences at the fixed end of each Tam3 copy remaining associted with the transposon. The alternative resolution of this event, depending on the orientation of the insertion of the ends of each transposon copy would be a deletion of sequence adjacent to a Tam3 copy, a rearrangement which has also been observed and described at the molecular level (Lister and Martin, 1989).

Inversions and deletions, when generated by aberrant transposition, allow the examinaton of the new recipient site at the same time as the old donor site. In the case of the two insertions that have been examined in detail and the deletion adjacent to Tam3, the element has attempted to move to closely linked sites, suggesting that movement over relatively short distances is preferred. In addition, in each case the new recipient site consists of very low copy DNA suggesting that Tam3 has a preference for integration into low copy sequences. This may be a feature of the structure of such DNA which is often associated with actively-transcribed regions and may be more accessible for integration.

Interaction between Tam3 and flanking inverted repeats

The analysis of a number of footprints left by Tam3 following excision, and in particular the production of short inverted repeats of flanking sequence has led to the suggestion (Coen et al., 1986) that Tam3 excision involves:

1. Nicking by the transposase at the ends of the transposon. The nicks on the two strands of piece of flanking DNA would be staggered by 1 bp.
2. Following release of Tam3 by the staggered nicks the two strands of the flanking sequence could ligate to form a hairpin. The two hairpins formed at the end of each flanking sequence would then fuse to seal the donor site, but slight imprecision in the fusion would lead to deletion or inverted duplications with a cental single base derived from the 1 bp overhang from the staggered nick.

Tam3 has been observed to create larger inverted duplications of flanking sequence without excision of the transposon (Fig. 2A). These longer inverted duplications of flanking sequence may be able to form extended hairpins due to the exact match of their sequences and their length (which may be as great as 4.5kb (Fig. 2B). Fortuitously, the inverted duplications of *niv* sequence seriously disturb the expression of the *niv* gene giving a near acyanic phenotype. However, if half the inverted duplication is removed, in a DNA rearrangement, the expression of the *niv* gene is restored and the production of anthocyanin is observed. These rearrangements can therefore be observed phenotypically as coloured spots or sectors. Excision of Tam3 without flanking DNA leaves a near acyanic phenotype.

The frequency of rearrangement when Tam3 and an adjacent long inverted duplication of *niv* are operating in close proximity is high, both somatically and germinally, when comapred to other lines, using normal excision frequency as a standard. Rearrangements are enhanced (20-fold) at low temperature and suppressed by *Stabiliser*. The influence of

**A**

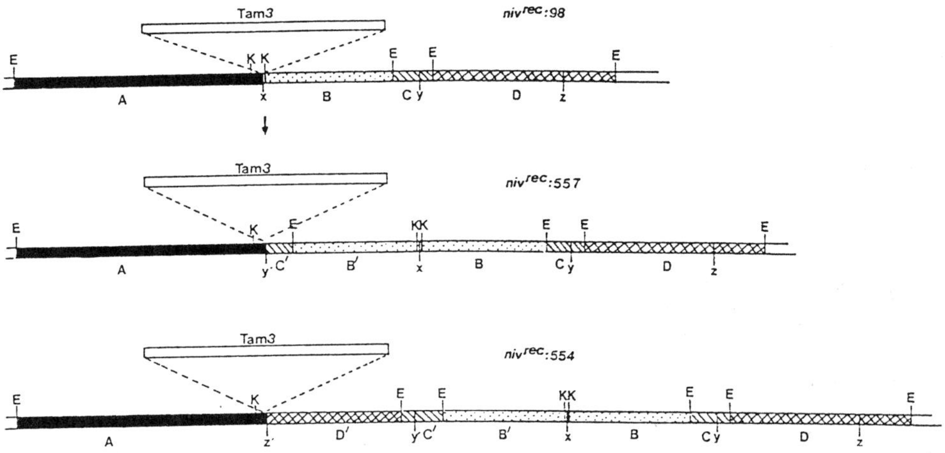


**B**

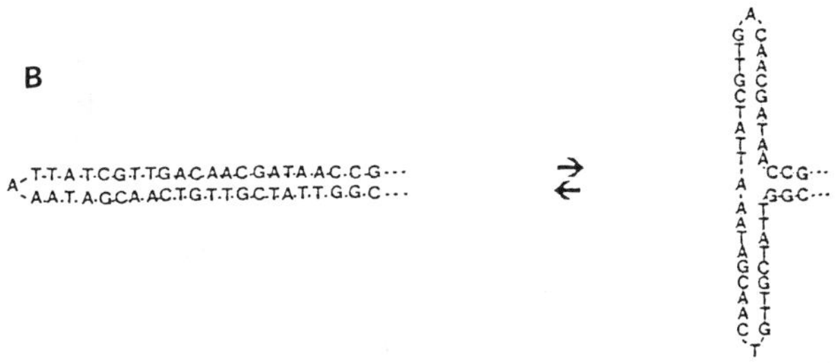

**Fig. 2A.** Restriction maps of *niv* alleles, *niv*$^{rec}$*:554* and *niv*$^{rec}$*:557* in relation to their progenitor allele *niv*$^{rec}$*:98*. The regular *niv* locus structure (of a wild-type) consists of fragments **A** (solid), **B** (stippled), **C** (striped) and **D** (hatched). **A** is a 3.7 kb *Eco*RI-*Kpn*I fragment; **B** is a 1.9 kb *Kpn*I-*Eco*RI fragment, **C** is a 0.6 kb *Eco*RI fragment and **D** is a 2.8 kb *Eco*RI fragment.

In alleles *niv*$^{rec}$*:554* (a 4.3 kb inverted duplication) and *niv*$^{rec}$*:557* (a 2.3 kb inverted duplication), regions of the *niv* locus have been inversely duplicated and inserted between Tam3 and the regular *niv* sequences. The duplicated DNA is indicated as fragments B', C' and D' to relate it to the homologous sequences in the regular *niv* gene structure. The centre of the inverted duplication (point x) is the sequence immediately adjacent to Tam3 in *niv*$^{rec}$*:98*. Point y is the point in the *niv* sequence that forms the end of the inverted duplication (y') in line JI:557; point z is the point in the *niv* sequence that forms the end of the inverted duplication (z') in line JI:554. E = *Eco*RI, K = *Kpn*I, S = *Sma*I.

**Fig. 2B.** Diagram to illustrate how an inverted duplication may form two extended hairpins by a simple change in the bases that cross pair.

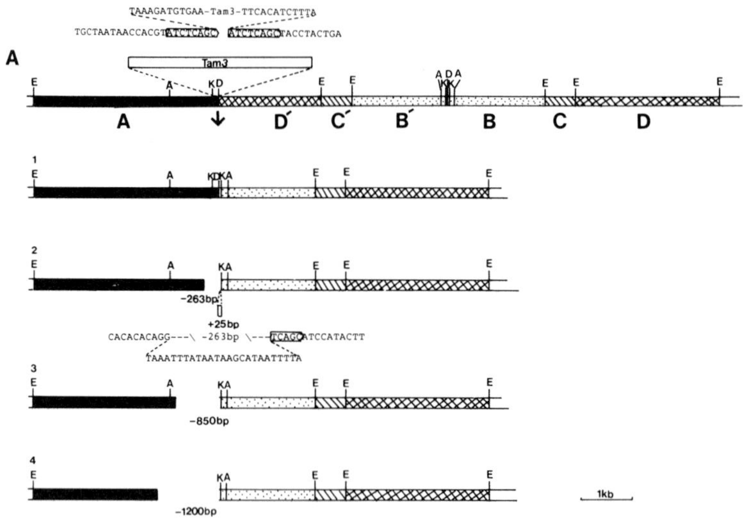


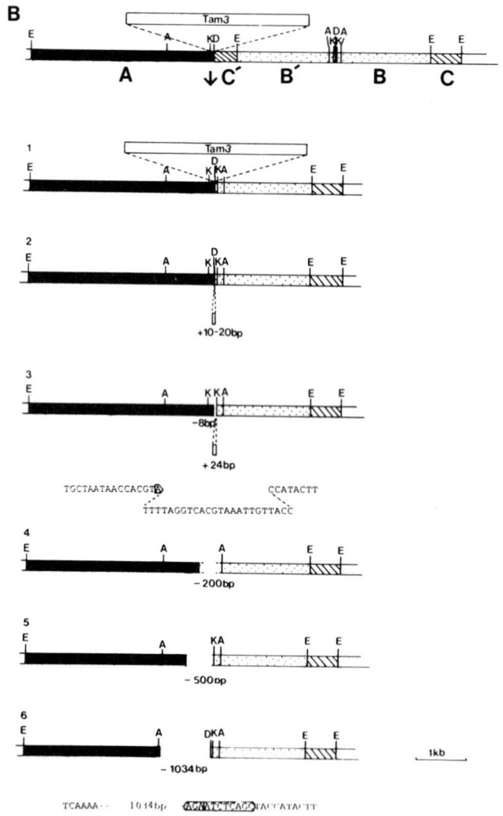


Fig. 3. Restriction maps and sequences of germinal derivatives of two lines carrying inverted duplications of *niv* flanking sequences adjacent to Tam3. The *niv* promoter flanking Tam3 to the left (A) is indicated by solid shading. The inverted duplication is indicated by 3 regions: B and B′ (the inverse of B) by stippled shading, C and C′ by stripped shading and D and D′ by hatched shading. Alterations determined by genomic mapping are indicated as gaps (deletions) or insertions below the line. Where the sequence of alleles has been determined this is shown below the map. The direct duplications flanking Tam3 and formed on its insertion into *niv* in line JI:98 are indicated by boxing with solid lines. Novel sequence of unknown origin is indicated as an insertion below the line of sequence of the *niv* locus. **A.** Progenitor with Tam3 and 4.3 kb inverted duplication that gave rise to derivatives below (1-4). **B.** Progenitor with Tam3 and 2.3 kb inverted duplication that gave rise to derivatives below (1-6). Restriction enzyme sites:- E = *Eco*RI; A = *Ava*II, K = *Kpn*I, D = *Dde*I (for clarity not all *Dde*I sites are illustrated).

temperature and stabiliser indicate that these rearrangements are dependent upon transposition-associated processes. This is confirmed by the phenotype of plants with the inverted duplication of *niv* where Tam3 has excised. In these plants the flowers are acyanic and the frequency of somatic sites is greatly diminished, although the occasional site (1 per 20 flowers) is observed.

We have been able to examine germinal events derived from parents carrying Tam3 plus the *niv* inverted duplication where *niv* expression is restored (Martin and Lister, 1989). We have also examined a number of somatic events involving DNA rearrangements from these lines. The most common event involves loss of Tam3 and the Tam3-proximal half of the inverted duplicaiton, to leave a single copy of the *niv* sequence (Fig. 3). In some cases loss of significant amounts of the sequences flanking Tam3 on the other side have also been lost. In other cases only half the inverted duplication has been lost and Tam3 remains.

The rearrangements therefore represent a category of recurring imprecise excisions or deletions of flanking DNA. The reason why this type of event is relatively common seems to lie in the fact that one end point reproducibly involves the centre of the inverted duplication. The inverted duplication could therefore be forming extended hairpins (without Tam3 excision) which could then actively fuse to other regions of DNA to give the rearrangements. It seems likely that the hairpin preferentially fuses with other hairpins such as those formed in the flanking sequences following Tam3 transposition, hence explaining the high frequency of deletion of Tam3 and half the inverted duplication, or half the inverted duplication alone (Fig. 4B-1,3) and the strong influence of Tam3 on the frequency of rearrangements. However the hairpin may also fuse with other sequences as observed in the examples of rearrangements where deletion has included parts of the sequences flanking Tam3 on the other side to the inverted duplication (e.g. Fig. 4B-2) and in an example where the inverted duplication has folded back to delete all but 700 bp of its tail (Martin and Lister, 1989). The activity of the synthesised hairpin (the inverted duplication) in generating new rearrangements suggests very strongly that hairpins are indeed the intermediates in normal transposition events. The interesting observation that occasional sites may still be observed when Tam3 has excised from *niv* in these lines implies that the inverted duplication alone may be active at a low frequency in generating change. Perhaps its activity is simply determined by the frequency of very closely linked hairpins with which to interact. Some of the rearrangements observed in the progeny of lines carrying Tam3 and the inverted duplication are more complicated that those already described. The most complex so far examined is a triplication of the flanking sequence. This consists of a direct duplication of *niv* gene sequences plus of a 1.6kb piece of DNA (originally lying about 3.5kb downstream of Tam3 in the wild-type unmutagenised lines) between one of the directly duplicated copies and one inversely orientated copy of the *niv* flanking sequence. The complexity of this allele suggests that rearrangement occurred between adjacent copies of the gene and involved several steps. This allele again emphasises that transposition probably occurs very shortly after DNA replication. It also shows the massive changes that can be induced over relatively few generations by this Tam3-based system, without strong deleterious effects on the resultant lines and suggests that this may be a powerful mechanism for generating change in the *Antirrhinum* genome.

<u>Activity of Tam3 in new hosts</u>

We have also wished to study the factors that control Tam3 transposition. The difficulty of studying the function of Tam3 genetically due to the absence of Tam3 mutants in *Antirrhinum* and the difficulties

**Fig. 4.** Models to illustrate **(A)** the formation of inverted duplications flanking Tam3 and **(B)** the subsequent rearrangements resulting from the interaction between inverted repeats and the transposon. **A.** Formation of an inverted duplication of flanking sequence. In the initial stages of transposition staggered nicks are made at the end of Tam3. A hairpin loop forms from ligation of the cut ends of the flanking sequence b-b' to link the two DNA strands. Fusion of one DNA strand of the transposon to one strand of the DNA of the hair pin in an attempted transposition would result in an inverted duplication (or the reciprocal deletion) which would be observed as chromosomal changes following DNA replication or repair. This event could be either (i) intramolecular or (ii) intermolecular between replicated chromatids. B. Formation of derivatives by the interaction of transposon and inverted duplication. The DNA of the inverted duplication may form 2 hairpin loops at one end of the transposon. These hairpins could fuse with other DNA sequences in a similar way to that proposed for the flanking hairpins left after transposon excision (Coen et al., 1986).

B 1. Fusion of the inverted repeat to the flanking hairpin formed during attempted transposition would result in loss of Tam3 and the inverted part of the inverted duplication. 2. Fusion to another sequence upstream of Tam3 would result in loss of part of a, Tam3 and the inverted half of the inverted duplication. 3. Fusion to the downstream end of Tam3 during attempted transposition would result in loss of the inverted half of the inverted duplication. All of these rearrangements could be envisaged involving either intra-molecular fusion of a single chromosome (i) or inter-molecular fusion involving recently replicated chromosomes lying in close proximity (ii). It may be that strand-nicking tends to cause loss of secondary structure as in bacterial plasmids, although the distances involved are relatively large. If so, it is likely that rearrangements tend to result from the interaction of the two recently replicated chromosomes, on one of which the transposon is excising whilst on the other the inverted repeat has adopted the cruciform structure.

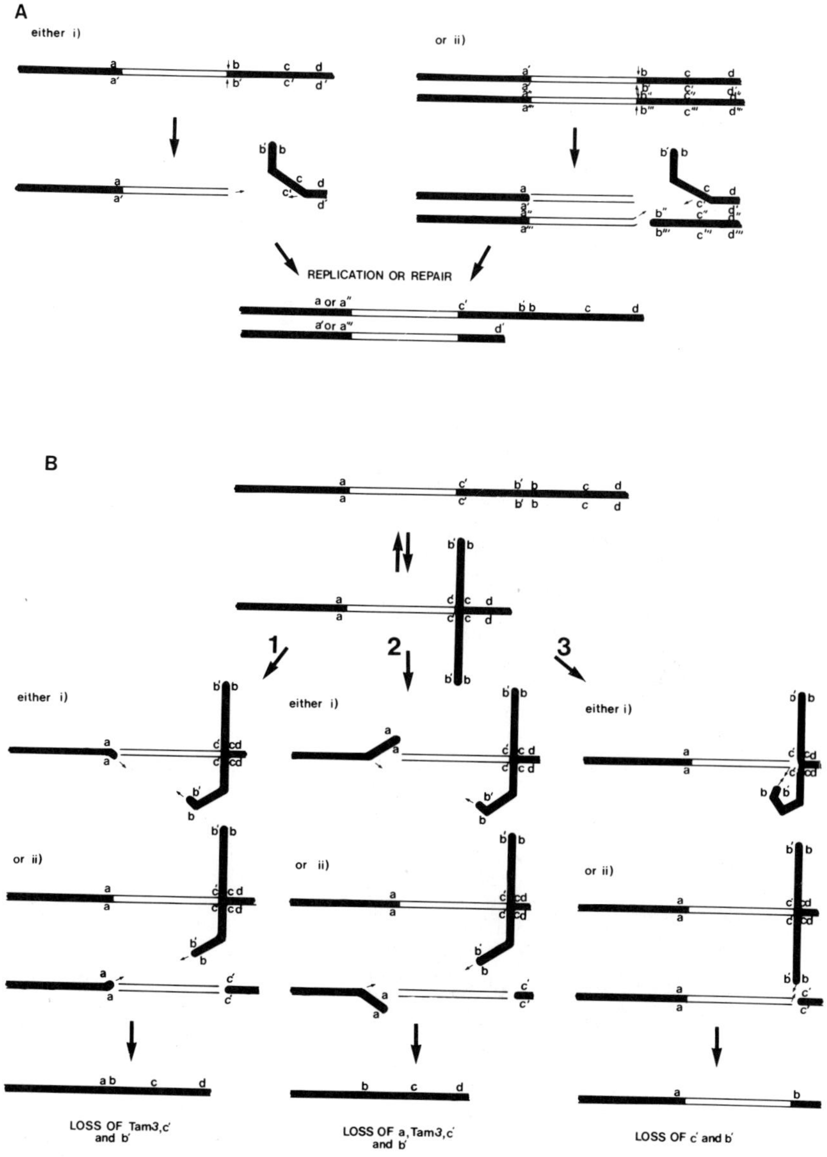

A
either i)
or ii)
REPLICATION OR REPAIR
a or a″
a′or a‴
B
1
2
3
either i)
or ii)
LOSS OF Tam3,c′ and b′
LOSS OF a,Tam3,c′ and b′
LOSS OF c′ and b′

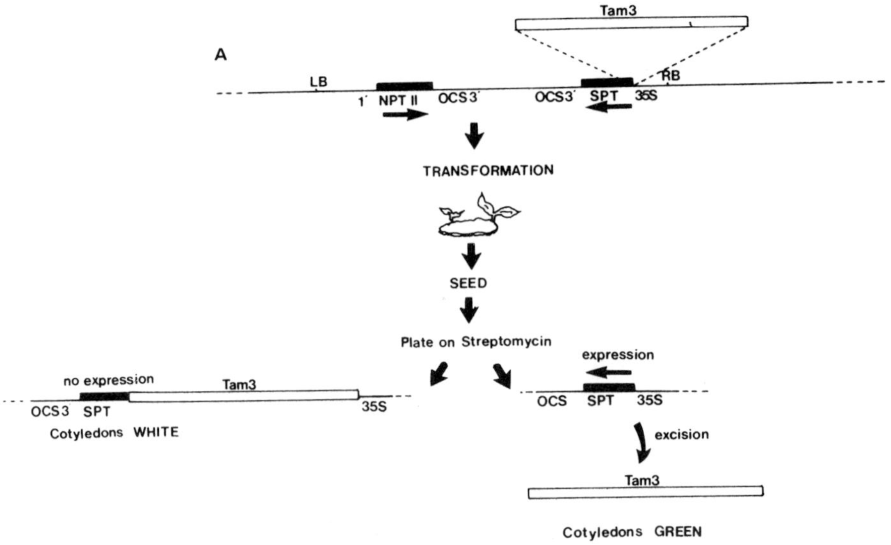


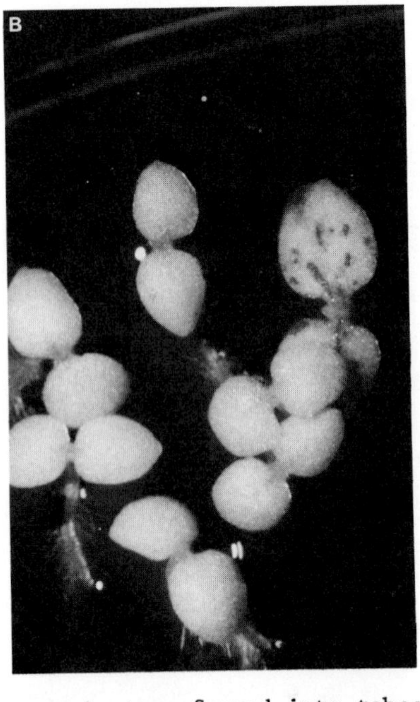


**Fig. 5.** Diagram of the streptomycin-resistance assay for transposon excision. Tam3 is inserted in the leader sequence between the 35S promoter from Cauliflower Mosaic Virus (CaMV) and the streptomycin resistance gene (spt). The spt gene is terminated by the octopine synthase (OCS) terminator from *Agrobacterium tumifaciens*. This construct is transformed into tobacco or *Arabidopsis*. The primary transformants are selfed and the seed collected. If the resultant seedlings are germinated on agar containing streptomycin the cotyledons will either be white where the transposons does not move and the cotyledonary material is streptomycin sensitive, or green where the transposon has jumped out germinally to allow spt expression, or variagated, where the transposon is moving somatically. Constructs have been made with the 5′ end of Tam3 either next to the spt gene or next to the 35S promoter.

NPTII = neomycin phosphotransferase II LB, RB = left border of T-DNA; right border of T-DNA.

B. Varigation observed in tobacco transformed with the Tam3/spt construct.

associated with transforming mutagenised Tam3 back into *Antirrhinum* have led us to attempt to establish a system for studying Tam3 behaviour in new hosts, tobacco and *Arabidopsis*. We have developed the streptomycin assay established by Jones et al (1989) to monitor Tam3 behaviour in these species. (Fig. 5). When germinating seedlings of tobacco (var. petite Havanna) or *Arabidopsis* are exposed to streptomycin they are sensitive to the antibiotic and the cotyledons bleach due to the interference of the drug with chloroplast functioning. When the streptomycin resistance gene (spt) from Tn5 is introduced into the plant driven by a plant promoter, such as the 35S promoter from CaMV, the cotyledons are resistant to the selection and remain green (Fig. 5). The inclusion of a transposable element in the leader sequence between the promoter and spt coding region prevents expression of spt except on excision of the transposon so that somatic excision can be monitored in cotyledons by the observation of green resistant sites on the bleached susceptible background (Fig. 5B). Excision is monitored in the first generation of plants derived by selfing primary transformants. Both susceptible and resistant progeny can be rescued following selection and grown on to maturity. Following insertion of Tam3 into these constructs we have observed a low frequency of varigated individuals amongst the progeny of transformed plants.

<u>SptII instability induced by Tam3 in tobacco</u>

If Tam3 is positioned such that its 5′ end is adjacent to the spt gene, a pale green expression is seen in many of the progeny of primary transformants. As this pale green phenotype segregates with exactly the same ratios as kanamycin resistance in a number of seed capsules from independent transformants we conclude that the element, when inserted in this orientation can promote spt gene expression without excision. As varigation is extremely difficult to score on top of the pale green background we can not score somatic instability in these plants. However, amongst the progeny from selfed primary transformants are a small number (an average between 2 and 5%) of fully green individuals which we suspect to be germinal revertants indicating that Tam3 does excise when inserted in this orientation.

When Tam3 is inserted such that its direction of transcription is the same as the spt gene, the problems of expression of spt prior to excision do not arise. The majority of plants in this case are streptomycin sensitive despite high ratios of kanamycin resistant:kanamycin sensitive seedlings that indicate that the primary transformants have multiple insertions of the Tam3/spt constructs. (These multiple insertions have been confirmed by Southern blot analysis.) On average 17.4% of individuals in the progeny of primary transformants showed some degree of varigation. As the transformed plants have not yet been selected to homozygosity for the Tam3 inserts the minimum varigation one would predict if Tam3 were active in all progeny is 75% (arising from a single insertion segregating 3:1 in the first self generation.) However, in most cases the frequency of varigated individuals fell far below this with just a few seedlings on a plate of about 150 showing varigation despite kanamycin resistant:kanamycin sensitive segregation ratios of 3:1, 15:1 or greater. In these test sowings we have also observed full green seedlings at a frequency of about 1.6% which we presume to represent germinal excision events.

Thus, whilst Tam3 appears to be present in these plants it clearly does not excise in every individual that carries it. We have previously described a rapid and specific methylation of Tam3 in tobacco following transformation which may be responsible for the suppression of Tam3 activity in transformed tissue (Martin et al., 1989). Whilst the varigated phenotype observed in F1 progeny of tobacco suggest that Tam3 movement may

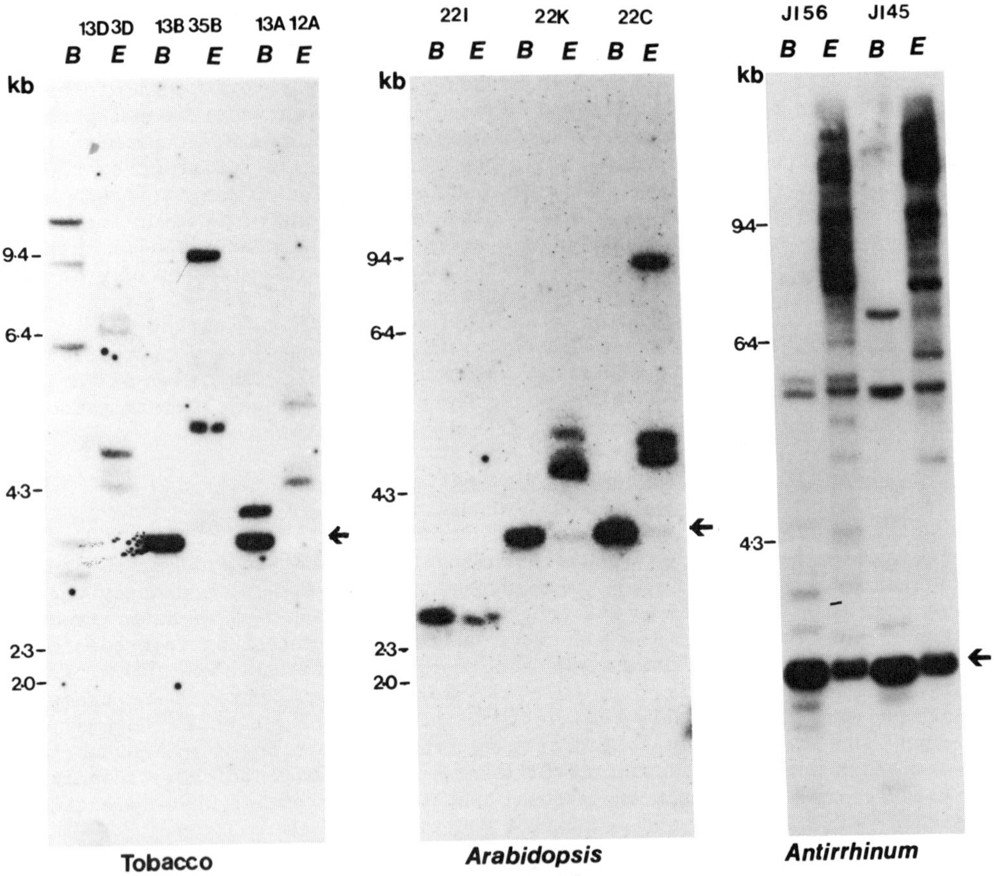


Fig. 6. Southern blots of genomic DNA from tobacco, *Arabidopsis and Antirrhinum* digested with *Eco*RII and *Bst*NI and probed with Tam3. The *Arabidopsis* and tobacco plants had been transformed with Tam3 inserted in the spt construct for the phenotypic assay of transposition. *Bst*NI cuts Tam3 twice at each end to yield an internal 3.3 kb fragment (arrowed). Any other bands produced by this enzyme result from rearrangement of Tam3, usually involving loss of sequences during transformation. *Eco*RII cuts at the same sequence as *Bst*NI but is sensitive to CpNpG methylation. Failure of *Eco*RII to give the same bands as *Bst*NI indicates that these sites are methylated.

In tobacco the results from 3 independent transformants are shown. All the Tam3 copies, whether rearranged or not appear to be highly methylated. In *Arabidopsis* plant 22I which carried a rearrangement of Tam3 was not methylated. Both plants 22K and 22C which carried normal copies of Tam3 (estimated to contain 2 and 3 copies respectively) showed heavy methylation but a small amount of the 3.3kb band was apparent with *Eco*RII (arrowed). This unmethylated DNA is probably visible in *Arabidopsis* but not in tobacco due to the smaller genome size of *Aradidopsis* giving greater sensitivity in the blots. In *Antirrhinum* about 50% of the Tam3 copies appear methylated in a number of lines examined (two are shown here, B = *Bst*NI and E = *Eco*RI). Plant numbers are shown above the gel.

continue in a small proportion of plants derived from an initial transformant, in the majority of seedlings Tam3 movement is suppressed. Occasionally, individuals give rise to higher frequencies of varigation amongst their progeny and at present we are selecting these plants to observe if lines giving high frequency of varigation show significantly lower methylation of Tam3 relative to other transformants. Examination of the DNA of a number of transformants has again illustrated the high degree of methylation of Tam3 that occurs following transformation. To test this genomic DNA has been digested with *Eco*RII (methylation sensitive) and *Bst*NI (methylation insensitive) (Fig. 6). These enzymes cut the same site which occurs twice at each end of Tam3. The release of the 3.3 kb internal Tam3 fragment by *Eco*RII therefore provides a sensitive assay for methylation in the inverted repeats of Tam3.

## Behaviour of Tam3 in Arabidopsis

The phenotypic behaviour of Tam3 in *Arabidopsis* is very similar to that in tobacco. With Tam3 orientated in the same direction as the spt gene we have observed 4.3% of individuals showing a varigated phenotype and 0.2% appear as full green germinal revertants. This low frequency of activity was somewhat unexpected as we had hoped that the lower overall methylation level of *Arabidopsis* DNA would mean that Tam3 would be more active in this species. However, where we have checked genomic DNA from transformed *Arabidopsis* we have again observed a reproducibly high level of methylation of Tam3. Long exposures of the gels do reveal a small proportion of Tam3 DNA that is unmethylated on the test sites, suggesting that in a few cell the element may not be methylated. It therefore seems likely that if such cells can give rise to gametes then the progeny will inherit undermethylated copies of Tam3 which may then represent the small proportion of varigated individuals observed amongst the *Arabidopsis* seedlings scored.

In one *Arabidopsis* plant (22I) we have observed a Tam3 copy that is not methylated on the test sites. This copy was rearranged in that a *Bst*NI digest of genomic DNA revealed a band of 2.7 kb rather than the expected fragment of 3.3 kb (Fig. 6). Mapping of this rearranged copy of Tam3 showed that whilst it maintained the 3′ end of Tam3 adjacent to the spt gene and therefore contained the *Bst*NI sites at one end of the transposon, it had lost the 5′ end of the element. A *Bst*EII site located at position 1258 was absent although an *Nru*I site at 1361 was present. It is likely that this rearrangement resulted from partial insertion of the T-DNA involving loss of the 5′ end of Tam3 during integration. The fact that the *Bst*NI sites at the 3′ end of this copy are unmethylated is of interest. It suggests that the sequences lost in this rearrangement may be required for methylation of the 3′ end of the element. Sequences close to the termini of Tam3 can potentially pair into stem loop structures and such irregularities in DNA structure may be subject to methylation. Removal of one of the inverted repeats of Tam3 may prevent recognition of the other end of the element for methylation. We have previously investigated a similar rearrangement involving loss of the 5′ end of Tam3 in a transgenic tobacco line. In this case methylation of the 3′ end of the element was still observed but the region of 5′ sequences lost was smaller than that observed in *Arabidopsis* plant 22I. This suggests that the target for methylation of the 3′ end requires sequences from the 5′ end of the element which may extend for some distance into Tam3.

## Methylation of Tam3 in *Arabidopsis*

When genomic DNA from *Antirrhinum* is digested with the methylation sensitive/insensitive restriction enzymes *Eco*RII/*Bst*NI a rather different picture emerges to that observed in tobacco or *Arabidopsis*. Despite the

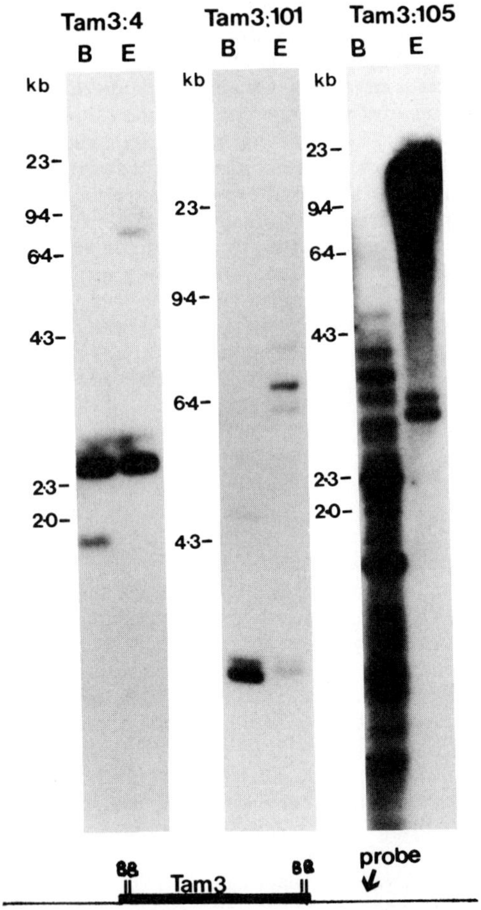


Fig. 7. Investigation of the methylation of Tam3 at three different insertion sites in *Antirrhinum*. Genomic DNA was digested with *Bst*NI and *Eco*RII and probed with the sequences flanking Tam3. Tam3:4 is the copy of Tam3 inserted in the *pal* locus. This is a highly active copy giving rates of germinal excision as high as 30%. The flanking sequence probe shows no difference in cutting between the two enzymes indicating no methylation. The faint band that changes size in the two digests is the somatic excision band, indicating that when Tam3 excises there is a methylated site in the *pal* flanking sequences.

Tam3:101. This transposon copy is though to be active although in *Antirrhinum* lines showing high frequencies of germinal excision of Tam3 from *pal* no excisions of this element copy have been observed for this copy in plants grown at low temperature (15°C). Probing with the flanking sequences indicates that this copy is partially methylated.

Tam3:105. This Tam3 copy lies in highly repeated sequence DNA and has never been observed to transpose germinally or somatically despite examination of plants with high frequencies of Tam3 excision. The flanking sequences, which are highly reitterated are represented by a 1.8kb fragment in *Bst*NI digests. Digestion with *Eco*RII shows this region of the DNA is highly methylated supporting previous data with other enzymes that the ends of this Tam3 copy are methylated (Martin et al., 1989).

large numbers of copies of Tam3 only about 50% are methylated on these sites as shown by the hybridisation to the 3.3 kb internal Tam3 fragment in *Eco*RII-digested DNA as well as DNA digested with *Bst*NI (Fig. 6). When individual copies of Tam3 in *Antirrhinum* are investigated for methylation of their termini, then an inverse correllation between methylation and the degree of activity (as measured by somatic and germinal excision) has been observed for 3 elements (Fig. 7). Tam3:4 at *pal* is highly active and completely unmethylated. Tam3:101 is in single copy DNA but no germinal excisions have been observed. A faint somatic excision band has been seen for this Tam3 copy when probing with the flanking sequences. This copy appears to be partially methylated when the flanking sequences are used to probe DNA cut with *Eco*RII and *Bst*NI. Finally Tam3:105 is known to reside in repeated sequence DNA. Digestion with *Eco*RII and *Bst*NI gives a complex hybridisation pattern but the band representing the flanking sequence fragments of 1.8 kb is not observed in digests with *Eco*RII. This copy has been shown to be capable of transposition in tobacco but no movement, somatic or germinal has ever been observed in *Antirrhinum*. Thus, whilst methylation of Tam3 in *Antirrhinum* does correllate with the observed activity of the particular element, there does not appear to be the degree of specific methylation of the ends of all Tam3 copies observed in *Arabidopsis* and tobacco. This suggests that the mechanism that automatically recognises and methylates Tam3 in foreign hosts may have been lost or suppressed in the natural host. Tam3 copies in *Antirrhinum* may become methylated more as a result of their genomic position generally. This would allow higher frequencies of Tam3 movement in *Antirrhinum*.

REFERENCES


Bhattacharyya, M., Smith,A. M., Ellis, T. H. N., Hedley, C., and Martin, C., 1990, The wrinkled-seed character of pea described by Mendel is caused by a transposon like insertion in a gene encoding starch branching enzyme. *Cell*, 60:115.

Carpenter, R., Martin, C., and Coen, E. S., 1987, Comparison of genetic behaviour of the transposable element Tam3 at two unlinked loci in *Antirrhinum majus*, *Mol. Gen. Genet.*, 207:82.

Coen, E. S., Carpenter, R., and Martin, C, 1986, Transposable elements generate novel spatial patterns of gene expression in *Antirrhinum majus*, *Cell*, 47:285.

Harrison, B. J., 1976, Sixty-Seventh Annual Report of the John Innes Institute, pp. 15-16.

Harrison, B. J., and Fincham, J. R. S., 1964, Instability at the *pal* locus in *Antirrhinum majus*, *Heredity*, 19:237.

Harrison, B. J., and Fincham, J. R. S., 1968, Instability at the *pal* locus in *Antirrhinum majus*. 3. A gene controlling mutation frequency, Heredity, 23:67.

Herrmann, A., Schulz, W., and Hahlbrook, K., 1988, Two alleles of the single copy chalcone synthase gene in parsley differ by a transposon-like element. *Mol. Gen. Genet.*, 212:93.

Jones, J D. G., Carland, F. M., Maliga, P., and Dooner, H. K., 1989, Visual detection of transposition of the maize element Activator (Ac) in tobacco seedlings, *Science*, 244:204.

Lister, C., and Martin, C., 9189, Molecular analysis of a transposon-induced deletion of the *nivea* locus in *Antirrhinum majus*, *Genetics*, 123:417.

Martin, C., Mackay, S., and Carpenter, R., 1988, Large-scale chromosomal restructuring is induced by the transposable element Tam3 at the *nivea* locus of *Antirrhinum majus*. *Genetics*, 119:171.

Martin, C., Prescott, A., Lister, C., and Mackay, S., 1989, Activity of the transposon Tam3 in *Antirrhinum* and tobacco: possible role of DNA methylation, *EMBO J.*, 8:997.

Martin, C., and Lister, C., 1989, Genome juggling by transposons: Tam3-induced rearrangements in *Antirrhinum majus*, Dev. Gen., 10:438.

Robbins, T., Carpenter, R., and Coen, E. S., 1989, A chromosome rearrangement suggests that the donor and recipient sites are associated during Tam3 transposition in *Antirrhinum majus*. *EMBO J.*, 8:5.

Sommer, H., Carpenter, R., Harrison, B. J., and Saedler, H., 1985, The transposable element Tam3 of *Antirrhinum majus* generates a novel type of sequence alterations upon excision. *Mol. Gen. Genet.*, 199:225.

Sommer, H., Hehl, R., Krebbers, E., Piotrowiak, R., Lanning, W-E., and Saedler, H., 1988, Transposable elements in *Antirrhinum majus*, *in*: "Plant Transposable Elements", O. Nelson, ed., Plenum Press, New York.

# CHARACTERIZATION OF MOBILE ENDOGENOUS COPIA-LIKE TRANSPOSABLE ELEMENTS IN THE GENOME OF SOLANACEAE

Marie-Angèle Grandbastien, Albert Spielmann*, Sylvie Pouteau, Eric Huttner, Michèle Longuet, Karl Kunert, Christian Meyer, Pierre Rouzé and Michel Caboche

Laboratoire de Biologie cellulaire, INRA, Centre de Versailles, F-78026 Versailles Cedex, FRANCE
*Present address: ORSAN-Recherche, F-91953 Les Ulis Cedex, FRANCE

## INTRODUCTION

The discovery of transposable elements has been a major step in the evolution of our concepts about heredity. Mobile genetic elements, by their ability to intercalate at different positions in the genome, and to create mutations and genetic rearrangements, appeared then as a challenge to the strict mendelians laws accepted as the basis of eucaryotic genetics. Transposable elements have been now characterized in the widest range of organisms, but the major credit for their discovery must be given to plant geneticists, and especially to Barbara McClintock, for her analysis of maize somatic instabilities in the early 40s. Molecular analysis of plant transposable elements is however quite recent, and the cloning of the first element characterized by McClintock, the maize Ac element, was reported in 1983 (Fedoroff et al). Great progress has been made in this field in the last seven years, mostly sustained by the hopes of considerable applications for genetic engineering, especially for the cloning of genes in higher plants. Gene-tagging techniques allow the cloning of any plant gene, provided that screening for mutants of the corresponding phenotype is possible, and have been successfully used in maize and snapdragon, in which more than a dozen of genes have been cloned this way (Fedoroff et al, 1984; Martin et al, 1985; O'Reilly et al, 1985; Cone et al, 1986; Paz-Ares et al, 1986; Schmidt et al, 1987; Motto et al, 1988; Martienssen et al, 1989). Most of these genes, however, were genes coding for a limited range of functions visible at the plant level, such as flower or seed coloration, or seed starch content, whose cloning required the screening of large population of plants.

However, new types of biochemical mutants can be selected at the cell level, at least for some model species, mostly members of the Solanaceae family, for which cellular genetic systems are quite powerful. It would therefore be useful if the range of genes accessible to gene-tagging could be widened to all the biochemical functions for which cellular selective screens are available. Moreover, it has been shown that in vitro culture can activate transposable elements (Peschke et al, 1987), and a cell culture step could therefore be an efficient way of increasing the probability to obtain insertional mutants. Plant species in which transposable elements have been characterized so far are however not good candidates for cellular genetic experiments. Many studies are engaged in the world to introduce these characterized elements into model species, such as Solanaceae (Baker et al, 1986; Van Sluys et al, 1987; Yoder et al, 1988; Frey et al, 1989; Haring et al, 1989; Pereira and Saedler, 1989;), in order to analyze their behavior in heterologous systems, but above all to benefit from all the technological advantages of these species. But in spite of the fact that these elements usually transpose into heterologous genomes, no success of transposon-tagging in heterologous systems has been reported yet.

*Plant Molecular Biology* 2, Edited by R.G. Herrmann and B. Larkins, Plenum Press, New York, 1991

The work reported here was initiated with two major goals. On one hand, we wanted to demonstrate that the potentials of gene-tagging could be associated in tobacco (Nicotiana tabacum) with the potentials of procedures previously developed in this species for the recovery of biochemical mutants from cell cultures, particularly homogeneous cell cultures derived from mesophyll protoplasts. On the other hand, we wanted to isolate an active transposable element endogenous to tobacco. Such an element would represent a powerful tool, since endogenous transposable elements had not been yet isolated from any member of the Solanaceae family, to which belong some of the best known model species (tobacco, Nicotiana plumbaginifolia, petunia), as well as species of high agronomical importance (tomato, potato).

A good candidate for the isolation of an active tobacco transposable element was the tl mutant line, obtained after seed treatment with the mutagenic agent ethyl methane sulfonate (Dulieu, 1965, unpublished). The nuclear tl mutation causes a chorophyllian deficiency, associated with a somatic instability quite similar to other mutable systems attributed to transposable elements (Deshayes, 1979). Furthermore, a higher mutagenic ability of tl lines has been demonstrated in selections performed on cell cultures (Grandbastien et al, 1989a). The function of the TL nuclear gene involved in the mutation is not known, and the only way to isolate the element presumably active in this line was to search for transposition events into an other gene, for which a molecular probe was available. Insertions into this target gene were searched by selecting spontaneous mutant cells inactivated for this function. As a target gene, we chose the gene coding for the nitrate reductase apoenzyme (NR), because molecular probes for this gene were available (Calza et al, 1987; Vaucheret et al, 1989), and because deficiency in nitrate reductase (NR-) is a good cellular marker, easily selected for by direct screening for chlorate resistance (Grafe et al, 1986).

## SELECTION OF SPONTANEOUS NR- LINES

Tobacco is an amphidiploid species, arising from an ancestral interspecific cross between two Nicotianae species, N. sylvestris and N. tomentosiformis. The NR apoenzyme is encoded by two alloallelic genes, nia1 and nia2, highly homologous, and both functional in wild-type haploid tobacco (Müller, 1983). To induce a NR- phenotype in a haploid tobacco genome, both genes have to be mutated at the same time.

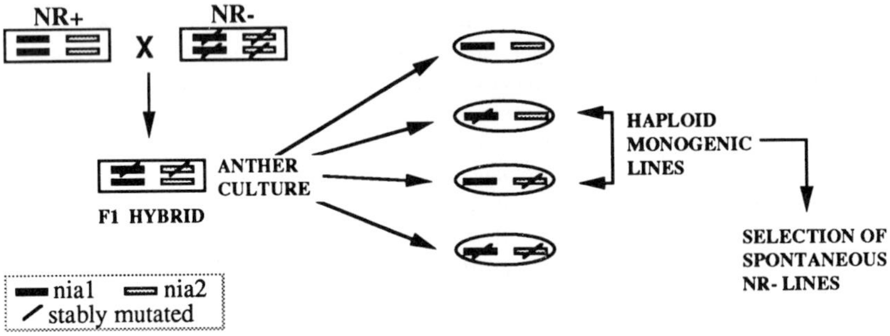


Figure 1. Construction of haploid monogenic lines. The mutant (nia1-;nia2-) nia30 tobacco line (Müller and Grafe, 1978) was used as a source for nia1 and nia2 stably mutated alloalleles, and was crossed to (nia1+;nia2+) tl or TL lines. Anther cultures were performed on F1 heterozygous plants, and the four expected types of haploid lines <(nia1+;nia2+), (nia1+;nia2-), (nia1-;nia2+), (nia1-;nia2-)> were obtained. (nia1-;nia2-) haploids were eliminated by testing for inability to grow on nitrate as the sole nitrogen source. The remaining haploid lines were diploidized in vitro, and monogenic lines detected and classified by subsequent genetic test-crosses with the parental nia30 line or with monogenic lines, directly derived from the nia30 line and kindly provided by Dr A. Müller.

In order to bypass this major drawback, we constructed, by genetic crosses, haploid tl and TL tobacco lines in which either nia1 or nia2 was stably inactivated (see Figure 1). These lines were called haploid monogenic lines, because they behave as if nitrate reductase function is carried by a single gene and a NR- phenotype can be obtained by a single insertional event into the remaining wild type alloallele. Moreover, since a NR- phenotype can also be induced by mutations into other genes than the nia alleles (Müller and Grafe, 1978), and since these genes are also found in two alloallelic functional copies in the haploid tobacco genome, the use of these haploid monogenic lines garanties that the majority of NR-mutants selected from them will be induced by events occuring in nia1 or nia2 only.

Selection of spontaneous chlorate-resistant clones were performed on cell cultures derived from protoplasts isolated from leaves of several tl as well as TL (control) monogenic haploid lines. The NR-phenotype of these clones was confirmed by testing for their inability to grow on nitrate as the sole nitrogen source. Confirmed NR- variants were obtained at frequencies ranging from $10^{-5}$ to $10^{-4}$ in 11 out of 13 independant experiments performed with 7 different tl monogenic lines. Six independant experiments were performed with 3 independant TL control monogenic lines, and spontaneous NR- clones were isolated from only one experiment, at the frequency of $10^{-5}$. Spontaneous NR- clones were regenerated into plants, and grafted in the greenhouse for further molecular and genetic analysis.

## ANALYSIS OF INSERTIONAL NR- MUTANTS

Genomic DNAs from 30 spontaneous NR- tl plants, as well as 3 spontaneous NR- TL control plants, were analysed with different restriction digests, using a NR cDNA probe, corresponding to the central EcoRI fragment of both nia1 and nia2 genes. In 20 of the 30 tl NR- mutants, and in the 3 spontaneous TL control NR- mutants, no modification of the structure of the wild-type nia alloallele was found, and these mutants were assumed to be due to point mutations. But in 9 tl mutants, the NR- phenotype was associated with structural modifications of the wild-type nia alloallele.

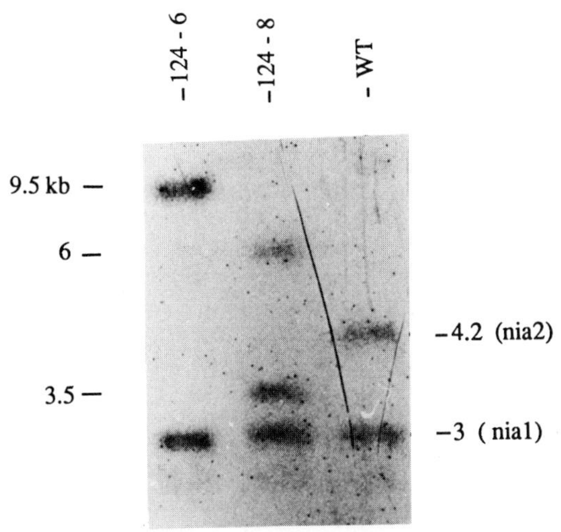


Figure 2. Southern blot analysis of EcoR1 restriction digests of genomic DNA isolated from mutant h124-Nia8, mutant h124-Nia6 and a wild type control tobacco line. The probe is a NR cDNA fragment, revealing, in the wild type genome, a 3 kb band from the nia1 alloallele and a 4.2 kb band from the nia2 alloallele.

Six independant NR- mutants, obtained from 3 different monogenic haploid lines, contained structural modifications typical of an insertion. The three mutants h9-Nia4, h9-Nia5 and h9-Nia6 were obtained from the (nia1-;nia2+) h9 monogenic line (Grandbastien et al,1989b), the two h124-Nia6 and h124-Nia8 mutants were obtained from the (nia1-;nia2+) h124 monogenic line, and the h45-Nia5 mutant was obtained from the (nia1+;nia2-) h45 monogenic line. In mutants h9-Nia4, h9-Nia5, h9-Nia6, and h124-Nia6, the 4.2 kb EcoR1 fragment central to the nia2 gene is replaced by a new band of 9.5 kb, typical of a similar insertion of 5.3 kb. Mutant h124-Nia6 is represented in Figure 2. In the mutant h124-Nia8, the central EcoR1 nia2 band has also disappeared, but is replaced by two new bands of equivalent intensity, one of 6 kb and the other one of 3.5 kb (see Figure 2). The addition of these two new bands gives 9.5kb, which strongly suggests that the insertion found in h124-Nia6 could be closely related to the one present in the other mutants, with a additional internal EcoR1 site. In the mutant h45-Nia5, modifications are found in the nia1 gene, and results are similar to those found in h124-Nia6: the internal 3 kb EcoR1 nia1 band is replaced by two new bands, and results are in accordance with the insertion of a 5.3 kb element containing one internal EcoR1 site (data not shown). More detailed restriction mapping with other genomic probes have revealed that, except for mutant h124-Nia6 (not yet determined), all insertions were located into coding sequences of nia1 or nia2 genes (see diagram in Figure 3).

These results suggest that in 6 out of 9 mutants, isolated from 3 different monogenic haploid lines, the NR- phenotype is associated with insertions of similar, or very closely related elements. This family of elements was called Tnt1 for "transposon of Nicotiana tabacum". In the three remaining NR- mutants, structural modifications of the wild-type nia allolallele were also found, and are presently under investigations.

## Tnt1 IS A RETROTRANSPOSON

EcoR1 restricted genomic DNA from mutants h9-Nia4 and h9-Nia5 were fractionated on sucrose gradients and partial genomic librairies were constructed in phage lambda NM1149. Insertion Tnt1-94, found in mutant h9-Nia4, was cloned, sequenced and characterized by A. Spielmann. Initial description of the Tnt1 element was reported in Grandbastien et al (1989b). The second element, Tnt1-95, present in mutant h9-Nia5 has also been characterized, and is nearly identical to Tnt1-94, but inserted in opposite orientation. Both elements are bounded by short duplications of 5 bp of the nia2 coding sequences, are terminated by the palindromic TG...CA sequence, and contain one single ORF encoding 1328 aminoacids.

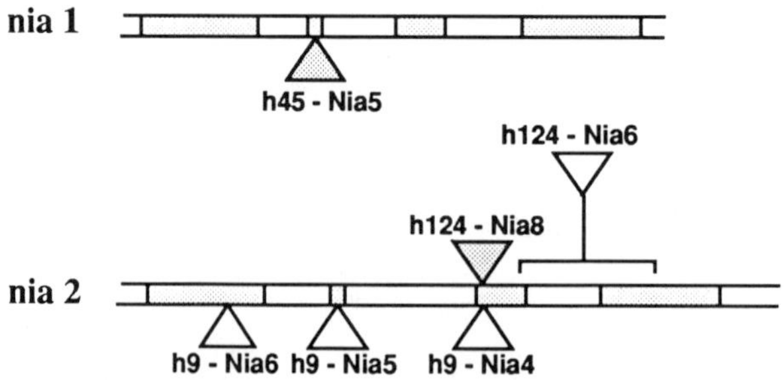


Figure 3. Diagram of Tnt1 and Tnt1-related insertions found into the nia1 or nia2 genes of 6 independant NR- mutant lines. Exons are indicated by shaded boxes and introns by open boxes. Tnt1 insertions are shown by open triangles, and putative modified Tnt1 insertions are shown by shaded triangles.

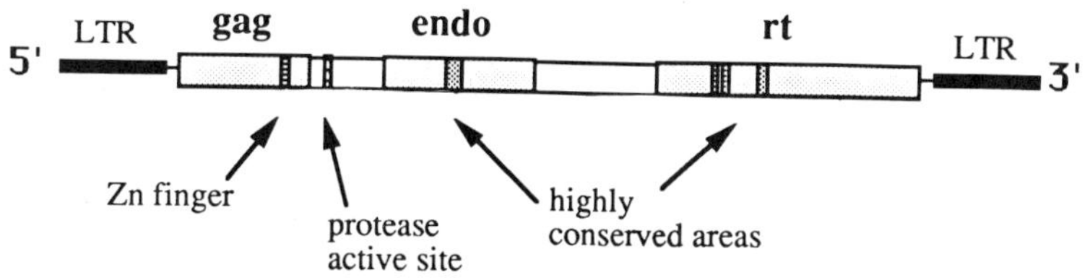


Figure 4. Structure of the Tnt1 retrotransposon. The unique ORF is represented by the large open box located between the 2 LTRs. Shaded boxes indicate regions of extensive aminoacid homologies with functional domains of related retroelements.

The analysis of Tnt1 putative ORF has revealed that it was a member of the retrotransposon family, like the copia element of drosophila. Retrotransposons are very abundant in animal genomes, and have structural and functional analogies with the retroviruses of vertebrates. Retroviruses are RNA viruses whose life cycle requires a phase of integration as a DNA copy, the provirus, into the host genome (for recent review, see Varmus and Brown, 1989). The function of reverse-transcription, which transforms the RNA into double stranded DNA, is encoded by the viral genome.

A typical provirus is bounded by two identical Long Terminal Repeats (LTRs) which contains promoter and regulator sequences, and carries several ORFs encoding for different functional domains:

- the Gag domain codes for the proteins of the viral core found into the virion.
- the Pol domain contains diverse catalytic functions, including the reverse-transcriptase (RT), the endonuclease (Endo) or integrase involved in the provirus integration into the host DNA, the protease (Prot) involved in maturation of viral polyproteins, and the RnaseH function also involved in reverse-transcription. The relative order of these domains is variable.
- the Env domains codes for the virion envelop, involved in the host recognition, that is responsible for the infectious ability.

Retrotransposons have a similar structure and overall organization, but usually do not carry the "env" domain, and are assumed to share common ancestors with retroviruses. Prototypes of retrotransposons are the Ty element of yeast and the copia element of drosophila. It has been demonstrated that Ty transposes via an RNA intermediate (Boeke et al, 1985), and this is generally assumed to be true for all retrotransposons.

Nucleotide and aminoacid sequences of Tnt1 have been compared with other retroviruses and retrotransposons. From N-terminal to C-terminal, the ORF shows extensive aminoacid homologies to Gag, Endo and RT domains (Figure 4). Inside of these regions are found patches of very high homologies with some short domains shown to be conserved in all retroelements, including a DNA binding site (typical Zn finger) in the Gag domain, and a potential protease active site between the Gag and Endo domains. In untranslated areas of Tnt1, sequences corresponding to the primers sites, used by retroelements for the DNA synthesis by reverse transcription, are also found (Figure 5). The first DNA strand is usually primed using a host tRNA, which binds to a region called the Primer Binding Site (PBS), located downstream of the 5' LTR, and showing high homologies to the 3' end of this host tRNA. Two nucleotides downstream the Tnt1 5' LTR, a sequence nearly homologous to 18 bp of the 3' end of the initiator met-tRNA of bean and wheat (Gauss and Sprinzl, 1983) is found, and is most probably the Tnt1 PBS site. A very similar sequence is also found in the Ta1 retrotransposon of Arabidopsis thaliana (Voytas and Ausubel, 1989). It is interesting to note that the met-tRNA is also used as primer by several other retrotransposons, such as the copia (Mount and Rubin, 1985) and 1731 (Fourcade-Peronnet et al, 1988) drosophila elements, and the Lilium henryi del1 element (Smyth et al, 1989). A Polypurine Track (PPT), found upstream of the 3' LTR of retroviruses, is known as the primer site used for synthesis of the second DNA strand. A 11bp PPT is also found in Tnt1, just before the 3' LTR (Figure 5).

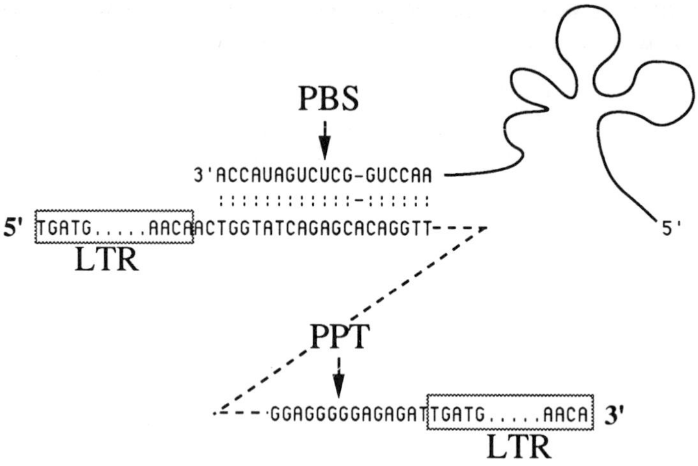


Figure 5. Sequences of the Tnt1 Primer Binding site (PBS) and Polypurine Tract (PPT). Alignment between Tnt1 PBS and the 3' end of bean and wheat initiator met-tRNA is shown.

Tnt1 IS A MEMBER OF THE Ty1-COPIA RETROTRANSPOSON FAMILY

Retrovirus-like retrotransposons have been recently classified in two major subgroups, based on aminoacid homologies, but also on respective order of the coding domains corresponding to the Pol region (Boeke and Corces, 1989). In the first subgroup, defined as the Ty1-copia family, these functional domains are found in the following order: Prot, Endo and RT. Best known members of this family are the yeast Ty1 and the drosophila copia elements. Tnt1 overall organization has been found to be very similar to the organization of copia, and extensive aminoacid homologies have been observed between both elements.

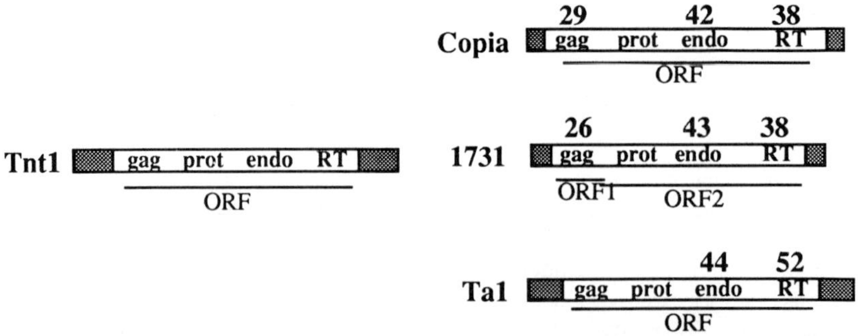


Figure 6. Overall structural similarity and percentage of identical aminoacids between Tnt1 and related retroelements in the three functional domains gag, endonuclease (endo) and reverse-transcriptase (RT).

Tnt1 is for the moment the element most closely related to copia. Tnt1 is also very closely related to another drosophila element, 1731 (Fourcade-Peronnet et al, 1988), and even more closely related to the arabidopsis Ta1 element (Voytas and Ausubel, 198!). In the three cases, homologies are variable depending on the functional domain (Figure 6). It seems therefore that a close relationship is found between elements present in different kingdoms: fungi, animals and plants, which strongly suggests the possibility of horizontal transmission of these elements. Confirmation of such a hypothesis is given by the fact that the second family of retrovirus-like retrotransposons, defined as the gypsy family, also contains related members found in different kingdoms. These elements bear the Pol functional domains in order similar to the retroviral order, that is Prot, RT and Endo, and members of this family include the drosophila gypsy, 17.6 and 412 elements, the yeast Ty3 element, and the del1 element of Lilium henryi (Smyth et al, 1989). Aminoacid alignments have also shown that these elements form a distinct group, and are more closely related to retroviruses than the Ty1-copia group.

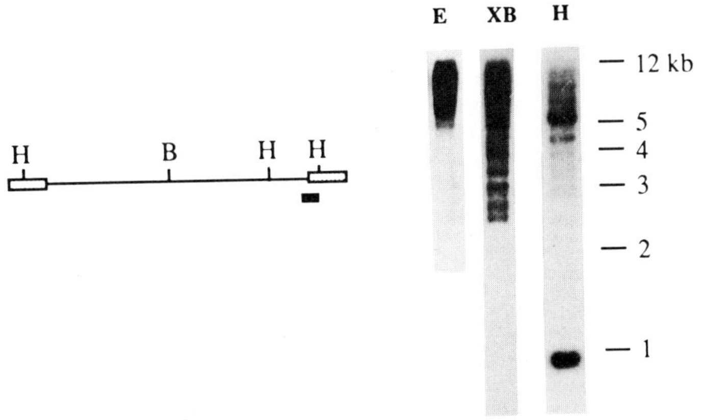


Figure 7. Southern blot analysis of restriction digests of wild type tobacco DNA, hybridized with a small Tnt1 probe, indicated by the dark box on the Tnt1 diagram represented on the left. E = EcoR1; XB = Xho1 + BamH1; H = HindIII.

Tnt1 IS A CONSERVED MULTICOPY FAMILY

Tnt1 has been found in very high copy number (estimated to more than 100 copies) in both tl and wild-type tobacco. Hybridization studies suggest that most family members have a conserved overall structure. In Figure 7 are represented genomic digests of wild-type tobacco, hybridized with a Tnt1 internal probe. The EcoR1 digest suggests that no EcoR1 internal site is found in most elements, and a smear of high molecular weight bands (superior to 5 kb) is found, corresponding to insertions into different genomic locations. The double restriction by Xho1 (no known site in the cloned copy) and BamH1 (one central internal site) creates a smear of fragments superior to 2.5 kb, corresponding to fragments ending in adjacent sequences. Restriction by HindIII reveals the corresponding 0.9 kb internal fragment, and minor hybridizing bands, corresponding probably to minor subclasses of modified members of the Tnt1 family. As described above, among 9 different insertional NR- mutants, 6 are thought to be due to insertions of Tnt1 elements.

Two of them, Tnt1-94 and Tnt1-95, have been sequenced, and both element are 99% identical in the nucleotidic as well as aminoacid sequences of the ORF. Very few restriction site modifications are found. These comparisons show however that both elements probably originate from the reverse transcription of two different Tnt1 parental copies, since the reverse transcriptase mistake rate is estimated to be $10^{-4}$ only (Varmus and Brown, 1989). In addition, it has been observed that in two other insertional mutants, the inserted Tnt1 copy may contain a new EcoR1 restriction site (see above). These results suggest that several members of the multicopy family found in the tobacco genome are active.

No clear relationship between Tnt1 and the somatic instability of the tl mutant has been found yet. Retrotransposons transpose in a replicative way, and do not usually excise at very high frequencies. It is therefore quite possible that Tnt1 is not the element involved in the unstable mutation, and has been fortuitously activated during the cell culture step. It is however important to mention that very few retrotransposons have been isolated in higher plants (only 6 exemples of retroviral-like elements have been reported so far), and that most of them vere found as non-mobile insertion sequences. Only two of them, the maize Bs1 element (Johns et al, 1985) and Tnt1 have been shown to be mobile, and the Adh- mutation induced by insertion of Bs1 into an exon of the Adh gene was originally described as highly unstable (Mottinger et al, 1984). It might be possible that plant retrotransposons have very specific properties regarding their mobility functions, just like classical plant transposons.

## Tnt1-RELATED ELEMENTS ARE FOUND IN OTHER SOLANACEAE SPECIES

The presence of Tnt1 was investigated in several other plant species using probes internal to Tnt1, and stringent hybridization and washing conditions similar to those used in homologous studies.

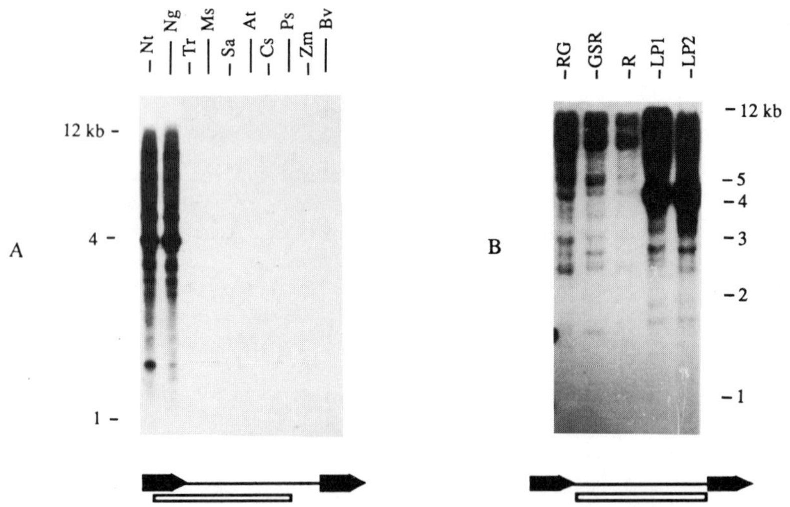


Figure 8. Southern blot analysis of EcoR1 restriction digests of genomic DNA isolated from different plant species, and hybridized with Tnt1 internal probes. The probe used for each blot is indicated by open boxes below the Tnt1 diagram represented under each blot. A) Nt: Nicotiana tabacum, Ng: Nicotiana glauca, Tr: Trifolium repens, Ms: Medicago sativa, Sa: Sinapis alba, At: Antirrhinum majus, Cs: Cucumis sativus, Ps: Pisum sativum, Zm: Zea mays and Bv: Beta vulgaris. B) RG, GSR and R : tomato cultivars Rio Grande, Golden Sunrise and Rimone; LP1 and LP2: 2 varieties of Lycopersicon peruvianum.

Results are presented in Figure 8. No homology was found in several other dicot and monocot species (Figure 8A), but cross hybridization was found with all the Solanaceae species studied. Very strong homology was found in Nicotiana species, such as N. glauca (Figure 8A) and N. plumbaginifolia (not shown). Lower but clear homology was found in potato and petunia (not shown). The most interesting situation has been found in the genus Lycopersicon (Figure 8B), especially in tomato (L. esculentum), where a low copy number family (around 20 copies) is detected. Different tomato cultivars display different patterns of Tnt1-like insertions, suggesting either internal polymorphism or recent transposition of Tnt1-like elements. The situation is however very different in the related wild species L. peruvianum, in which much stronger hybridization is found, with a major EcoR1 band of 5 kb, probably corresponding to an internal fragment (Figure 8B). Stronger hybridization are also found in other wild Lycopersicon species such as L. pennellii, L. hirsutum and L. chilense. It seems from these results that the domestication of tomato has been correlated with a clear loss of the copy number of Tnt1-like sequences. Tnt1 could be a very useful tool for phylogenetic studies in the genus Lycopersicon, and the behavior of such elements in the course of the domestication process could give interesting clues about their natural original role within the host genome.

## CONCLUSION

The isolation of Tnt1 is the first example reported of the use of transposable elements for the selection of biochemical mutants at the somatic cellular level. Tnt1 is also the first transposable element isolated from tobacco, and from Solanaceae in general, and one of the first complete mobile retrotransposon characterized in higher plants. Experiments are presently under way to determine in which conditions Tnt1 transcriptional activity can be precisely detected. Retrotransposons, if they apparently do not induce unstable mutations, can still be advantageous tools for gene-tagging, since they are supposed to transpose independantly from any position effect. Although it does not seem possible to use directly Tnt1 as a gene-tag in tobacco, because of its high copy number, such a possibility remains open in tomato, if it can be confirmed that Tnt1-like elements are still active in this species of high agronomical importance.

But Tnt1 major asset lies in the possibilities of manipulating the element, and of reintroducing it in its own genetic background, where it is known to be functional, in order to follow its response to diverse factors and external stresses. Since tobacco is model species particularly well adaptated to cellular experiments, Tnt1 could be a very useful tool for studying the mechanisms by which transposable elements participate to the somaclonal variation.

## REFERENCES


Baker, B., Schell, J., Lörz, H., and Fedoroff, N., 1986, Transposition of the maize controlling element Ac in tobacco, Proc. Natl. Acad. Sci. USA, 83:4844.

Boeke, J., Garfinkel, D., Styles, C., and Fink, G., 1985, Ty elements transpose through an RNA intermediate, Cell, 40:491.

Boeke, J., and Corces, V., 1989, Transcription and reverse transcription of retrotransposons, Annu. Rev. Microbiol., 43:403.

Calza, R., Huttner, E., Vincentz, M., Rouzé, P., Galangau, F., Vaucheret, H., Chérel, I., Meyer, C., Kronenberger, J., and Caboche, M., 1987, Cloning of DNA fragments complementary of tobacco nitrate reductase mRNA and encoding epitopes common to the nitrate reductases from higher plants. Mol. Gen. Genet., 209:552.

Cone, K., Burr, F., and Burr, B., 1986, Molecular analysis of the maize anthocyanin regulatory locus CA. Proc. Natl. Acad. Sci. USA, 83:285.

Deshayes, A., 1979, Cell frequencies of green somatic variations in the tl chlorophyll mutant of Nicotiana tabacum var. Samsun, Theor. Appl. Genet., 55:145.

Fedoroff, N., Wessler, S., and Shure, M., 1983, Isolation of the transposable maize controlling elements Ac and Ds, Cell, 35:235.

Fedoroff, N. , Furtek, D., and Nelson, O., Jr., 1984, Cloning of the bronze locus in maize by a simple and generalizable procedure using the transposable controlling element Activator (Ac), Proc. Natl. Acad. Sci. USA, 81:3825.

Frey, M., Tavantzis, S., and Saedler, H., 1989, The maize En-1/Spm element transposes in potato, Mol. Gen. Genet., 217:172.

Fourcade-Peronnet, F., d'Auriol, L., Becker, J., Galibert, F., and Best-Belpomme, M., 1988, Primary structure and functional organization of Drosophila 1731 retrotransposon, Nucleic Acids Res., 16:6113.

Gabard, J., Marion-Poll, A., Cherel, I., Meyer, C., Müller, A., and Caboche, M., 1987, M., Isolation and characterization of Nicotiana plumbaginifolia nitrate reductase-deficient mutants: genetic and biochemical analysis of the NIA complementation group, Mol. Gen. Genet., 209:596.

Gauss, D., and Sprinzl, M., 1983, Compilation of tRNA sequences, Nucleic Acids Res., 11:r1.

Grafe, R., Marion-Poll, A., and Caboche, M., 1986, Improved in vitro selection of nitrate reductase-deficient mutants on Nicotiana plumbaginifolia, Theor. Appl. Genet., 73:299.

Grandbastien, M.A., Missonier, C., Goujaud, J., Bourgin, J.P., Deshayes, A., and Caboche, M., 1989a, Cellular genetic study of a somatic instability in a tobacco mutant: in vitro isolation of valine-resistant spontaneous mutants, Theor. Appl. Genet., 77:482.

Grandbastien, M.A., Spielmann, A., and Caboche, M., 1989b, Tnt1, a mobile retroviral-like element of tobacco isolated by plant cell genetics, Nature, 337:376.

Haring, M., Gao, J., Volbeda, T., Rommens, C., Nijkamp, J., and Hille, J., 1989, A comparative study of Tam3 and Ac transposition in transgenic tobacco and petunia plants, Plant Mol. Biol., 13:189.

Johns, M., Mottinger, J., and Freeling, M., 1985, A low copy number, copia-like transposon in maize, EMBO J., 4:1093.

Martienssen, R., Barkan, A., Freeling, M. and Taylor, W., 1989, Molecular cloning of a maize gene involved in photosynthetic membrane organization that is regulated by Robertson's Mutator, EMBO J., 8:1633.

Martin, C., Carpenter, R., Sommer, H., Saedler, H., and Coen, E., 1985, Molecular analysis of instability in flower pigmentation of Antirrhinum majus, following isolation of the pallida locus by transposon tagging, EMBO J., 4:1625.

Mottinger, J., Johns, M., and Freeling, M., 1984, Mutation of the Adh1 gene in maize following infection with barley stripe mosaic virus, Mol. Gen. Genet., 195:367.

Motto, M., Maddaloni, M., Ponziani, G., Brembilla, M., Marotta, R., Di Fonzo, N., Soave, C., Thompson, R., and Salamini, F., 1988, Molecular cloning of the o2-m5 allele of Zea mays using transposon marking, Mol. Gen. Genet., 212:488.

Mount, S., and Rubin, G., 1985, Complete nucleotide sequence of the Drosophila transposable element copia: homology between copia and retroviral proteins, Mol. Cell. Biol., 5:1630.

Müller, A., 1983, Genetic analysis of nitrate reductase-deficient tobacco plants regenerated from mutant cells. Evidence for duplicate structural genes. Mol. Gen. Genet., 192:275.

Müller, A., and Grafe, R., 1978, Isolation and charcterization of cell lines of Nicotiana tabacum lacking nitrate reductase, Mol. Gen. Genet., 161:67.

O'Reilly, C., Shepherd, N., Pereira, A., Schwarz-Sommer, Z., Bertram, I., Robertson, D., Peterson, P., and Saedler, H., 1985, Molecular cloning of the a1 locus of Zea mays using the transposable elements En and Mu1, EMBO J., 4:877.

Paz-Ares, J., Wienand, U., Peterson, P., and Saedler, H., 1886, Molecular cloning of the c locus of Zea mays: a locus regulating the anthocyanin pathway, EMBO J., 5:829.

Pereira, A., and Seadler, H., 1989, Transpositional behavior of the maize En/Spm element in transgenic tobacco, EMBO J., 8:1321.

Peschke, V., Phillips R., and Gengenbach, B., 1987, Discovery of transposable element activity among progeny of tissue culture-derived maize plants. Science, 238: 804.

Schmidt, R., Burr, F., and Burr, B., 1987, Transposon tagging and molecular analysis of the maize regulatory locus opaque-2, Science, 238:960.

Smyth, D., Kalitsis, P., Joseph, J., and Sentry, J., 1989, Plant retrotransposon from Lilium henryi is related to Ty3 of yeast and the gypsy group of Drosophila, Proc. Natl. Acad. Sci. USA, 86:5015.

Van Sluys, M., Tempé, J., and Fedoroff, N., 1987, Studies on the introduction and mobility of the maize Activator element in Arabidopsis thaliana and Daucus carota, EMBO J., 6:3881.

Varmus, H., and Brown, P., 1989, Retroviruses, In " Mobile DNA" Eds Berg, D., and Howe, M., American Society for Microbiology, Washington DC, USA, pp53-108.

Vaucheret, H., Vincentz, M., Kronenberger, J., Caboche, M., and Rouzé, P., 1989, Molecular cloning and characterization of the two homeologous genes coding for nitrate reductase in tobacco, Mol. Gen. Genet., 216:10.

Voytas, D., and Ausubel, F., 1988, A copia-like transposable element family in Arabidopsis thaliana, Nature, 336:242.

Yoder, J., Palys, J., and Lassner, M., 1988, Ac transposition in transgenic tomato plants, Mol. Gen. Genet., 213:291.

# ORGANIZATION AND EVOLUTION OF THE MAIZE MITOCHONDRIAL GENOME

Christiane M.-R. Fauron, Marie Havlik, Mark Casper

Howard Hughes Medical Institute, University of Utah
743 Wintrobe Building
Salt Lake City, UT 84132

## INTRODUCTION

### History

The plant mitochondrial genome (mtDNA) has long presented a puzzle to plant molecular biologists. The research has fallen far behind that of the other organelle DNAs. An important event that increased the interest in the study of the plant mtDNA was the finding that the genetic determinants responsible for the cytoplasmic male sterile (cms) phenotype were located in the mitochondrial genome (Leaver and Gray, 1982; Levings, 1983; Laughnan and Gabay-Laughnan, 1983; Sederoff, 1984). Cms was shown in many cases to segregate along with mtDNA modifications as observed on restriction enzyme profile. This stimulated an extensive effort to study the plant mitochondrial genome. As a result, we have now begun to understand its organization. The early literature reporting the use of restriction enzymes revealed a very complex organization for the plant mitochondrial genome (Quetier and Vedel, 1977). Those complex mtDNA restriction enzyme profiles with submolar and multimolar bands, which led to the idea of genomic molecular heterogeneity has reinvigorated interest in studying plant mtDNA structure by electron microscopy. Unfortunately, a broad size range of mostly linear molecules was visualized, probably representing artefacts of the isolation procedure as it was not possible as yet to isolate intact mtDNA molecules. Closed circular mtDNA molecules represented only a few percent of the total and were of diverse size (Quetier and Vedel, 1977; Levings and Pring, 1978; Levings et al., 1979). For maize, the circles could be graphed into discrete size classes of different abundance (Levings et al., 1979). It is only through the approach of cloning technology during the last six years that the structure of the plant mt genome has been elucidated. It is indeed a complex organization where recombination between repeated sequences plays an important role.

### Cytoplasmic Male Sterility

Cytoplasmic male sterility is very common among higher plants. This trait is characterized by abortion of pollen grain development and is inherited in a non-Mendelian fashion. Three different types of cms have been identified in maize: cmsT, cmsS and cmsC (Laughnan and Gabay Laughnan, 1983). They are distinguished by their interaction to tester

*Plant Molecular Biology* 2, Edited by R.G. Herrmann and
B. Larkins, Plenum Press, New York, 1991

stock carrying nuclear restorer genes, that override the cms phenotype. At the molecular level, mtDNA of the various cms phenotype can be differentiated by restriction fragment length polymorphisms, the presence of different plasmids, changes in these RNAs products and their mt translations products.

In this paper we will concentrate on a single type of cytoplasmic male sterile maize: the cms type T. CmsT has been of importance because it was widely used in the United States as a convenient way to produce hybrid seeds until the late sixties. It was then found to be susceptible to the T toxin produced by the fungal pathogen Helminthosporium maydis and could no longer be used for hybrid corn production. The susceptibility to the T toxin is tightly linked to the male sterile factor. A gene designated urf13-T, present uniquely in the mitochondrial genome of cmsT, appears to be responsible for the cmsT phenotype (Dewey et al., 1986, 1987, 1988). Another advantage to studying cmsT mtDNA is the availability of plant progeny from cmsT tissue culture regenerated plants that have returned to fertility (Gengenbach et al., 1977, 1981; Brettell et al., 1979, 1980) which allowed us to study the mitochondrial DNA rearrangements responsible for the fertility reversion (Fauron et al., 1987; Wise et al., 1987; Rottmann et al., 1987).

## Variation in the Plant Mitochondrial Genome

The most striking aspect of the higher plant mitochondrial genome is the large size variation occurring among the plant species: 200 kb in Brassica (Palmer, 1988) to 2400 kb in muskmelon (Ward et al., 1981). Changes in the size seem to occur quite rapidly since closely related species have quite different mitochondrial genome sizes. Within the curcurbitaceae family the size of the mtDNA can vary by at least 8 fold (Ward et al., 1981).

Various parameters have been hypothesized to explain the variability of the size of the plant mitochondrial genome: the presence of numerous repeated sequences, more mitochondrially encoded functions in the larger genome, presence of larger genes with many introns, or integration of exogenous DNA. We are going to argue that no one of these parameters can alone account for such a wide range in size.

The variation cannot be explained by rapid changes in the amount of repeated sequences because repeated DNA represent at most 10% of the genome and do not seem to vary appreciably between species (Ward et al, 1981). Higher plant mitochondrial genomes are composed principally of single copy DNA.

Larger genome sizes are unlikely to be due to a much larger number of gene as studies from protein synthesis from isolated plant mitochondrial have shown only a slight increase in the number of proteins coding genes compared to the animal mtDNA (Forde et al., 1978; Leaver et al., 1982; Hack and Leaver, 1983) and the number of mitochondrial polypeptides is similar among those species for which the mtDNA shows a great size variation (Stern and Newton, 1985).

The size and structure of known mitochondrial genes can account for only a small part of the extra DNA. So far gene containing introns are cytochrome oxidase subunit II gene with an intron of 794 bp in maize (Fox and Leaver, 1981) also in rice (Kao et al., 1984) and wheat (Bonen et al., 1984). Some genes are larger due to the presence of multiple introns such as in ND1 (Stern et al., 1986; Wahleithner et al., 1990) or ND4 and ND5.

Lastly, plant mitochondrial genomes have integrated foreign DNA. Every plant species examined has shown mtDNA containing chloroplast DNA (ctDNA) sequences as part of their genome (Stern and Palmer, 1984). It was first demonstrated when the maize mtDNA was found to have inserted a 12 kb sequence of the chloroplast inverted repeat (Stern and Lonsdale, 1982). A second major homology to the ctDNA sequence was identified as the ribulose biphosphate large subunit gene (Lonsdale et al., 1983). Other regions of minor homologies to ctDNA have been located on the maize mt genome (Lonsdale, 1984; Sangare et al., 1990). The total amount of ctDNA found in maize mtDNA is less than 5%. However, it has been shown that larger mitochondrial genomes do not contain more chloroplast DNA sequences. Investigation in the curcubit family (Stern et al., 1983) has shown that perhaps more ctDNA homology was found for the 300 kb watermelon than for the 2400 kb muskmelon.

All the parameters described above cannot account for the large size range of the plant mitochondrial genome. Because we are interested in the extent of the variation in size and structure of the plant mitochondrial genome and the mechanisms responsible for this diversity, we have chosen to study those mechanisms of mitochondrial genomic variation within the maize species.

## RECOMBINATION AND GENOME PLASTICITY

Although no direct demonstration of recombination mechanisms in higher plant mitochondrial genome has been possible, all the observations are consistent with the presence of a very active recombinational machinery responsible for the great plasticity of the mtDNA.

1. The multipartite structure is a result of recombination taking place between repeated sequences present in the master chromosome. It can be shown in two ways: First, using repeated elements as hybridization probe, four restriction fragments (if the endonuclease used does not cut within the repeat) can be identified on a total mtDNA digest. Second, clones containing the four pairwise combination of the unique sequences surrounding the repeats have been isolated from the library. In maize the repeats vary in size from as low as a dozen base pairs (unpublished results) to up to 14 kb. Several of them have been sequenced (Houchins et al.; Lonsdale et al., 1988; Fauron et al., 1990). However, no consensus sequence could be identified. It seems, therefore, that the repeat in itself promotes the homologous recombination rather than a consensus sequence within the repeat.

2. Protoplast fusion experiments appear to promote mitochondrial genome rearrangements and recombination (Belliard et al., 1979; Morgan and Maliga, 1987; Ozias-Akins, 1987; Rothenberg et al., 1985; Galum et al., 1982). The somatic hybrids present a mixture of restriction fragments from both parents as well as non-parental fragments. These novel composite genomes are stably transmitted to the progeny and therefore do not result from a slow separation of parental organelles. They are the result of recombination between the two parental types.

3. Recombination can also relate to cms in creating a novel unique open reading frame. One evidence of such an ORF is the Turf13 gene identified by Dewey et al. (1986) that is unique to the cmsT mtDNA and seems to be responsible for the cms phenotype. The sequence analysis of the gene shows that Turf13 is a gene of chimeric origin; it has arisen by recombination within the 26S ribosomal gene and its flanking regions, as well as the atp6 gene flanking region.

4. Sequences highly homologous to chloroplast DNA are present in the mitochondrial genome of higher plants. This was first reported in maize where a 12 kb sequence of the chloroplast inverted repeat is inserted in the mtDNA (Stern and Lonsdale, 1982). Subsequently, many sequences homologous to the chloroplast DNA were inserted in various plant mitochondrial genomes (Stern and Palmer, 1984; Stern et al., 1983). Because the genome size of chloroplast DNA has been highly conserved during evolution, it has been argued that the exchange must have gone from chloroplast to mitochondria. Chloroplast DNA could enter mitochondria during fusion between the two organelles and the chloroplast DNA integrated into mitochondrial DNA through recombination.

5. Recombination with plasmid DNA is another source of variability. Evidence comes from the study of the plasmid-like DNAs in the maize cmsS type. In the presence of the plasmid, the mtDNA can be linearized with the plasmid attached at the ends (Schardl et al., 1984). They are generated through recombination between a 208 bp inverted repeat located at the extremities of the plasmid and also found within the mtDNA (Braun et al., 1986; Escote-Carlson et al., 1988).

## ORGANIZATION OF THE MAIZE MITOCHONDRIAL GENOME

### Presence of Repeated Sequences

The controversy about the molecular heterogeneity was at last resolved by cloning and mapping the mitochondrial DNA of several plants. A model has been adopted that reconciles all the data and accounts for all the original observations: the presence of large size molecules as well as the presence of submolar or multimolar bands in the total mtDNA digest. The model can be understood in terms of recombinational events occurring through sequences repeated in the genome.

The total genetic information can be arranged into a single circular DNA molecule called the master chromosome. This molecule is 208 kb to 242 kb for the Brassica species (Palmer and Herbon, 1988), 305 kb for sunflower (Siculella and Palmer, 1988), 327 kb for spinach (Stern and Palmer, 1986), 386 kb for sugar beet (Brears and Lonsdale, 1988), 430 kb for petunia (Folkers and Hanson, 1989), 440 kb for wheat (Quetier et al., 1985), 540 to 929 for maize (Lonsdale et al., 1984; Fauron and Havlik, 1988; Fauron et al., 1989; Fauron et al., 1990). Alternatively, reciprocal recombination through repeated sequences generate a multipartite organization of the mitochondrial genome. Direct repeats convert the master circle into subgenomic circular molecules, while inverted repeat generate a master chromosome presenting isomeric form with sequence permutations. One exception to the multipartite structure is the unique circular organization of the mtDNA of Brassica hirta where no repeats were found (Palmer and Herbon, 1987).

Depending on the numbers, orientations, and location of the various repeats present on the master circle, the populations of isomeric and subgenomic molecules will be different. An explanatory scheme of some of those conversions is shown on figure 1.

A simple organization of the genome (as illustrated in figure 1B) is found for most of the Brassica mtDNAs. The turnip (Brassica campetris) is organized into three discrete circular chromosomes of 218 kb, 135 kb and 83 kb (Palmer and Shields, 1984), the black mustard (Brassica nigra) has a tripartite structure of 231 kb, 135 kb, 96 kb and radish (Raphanus sativa) with a master chromosome of 243 kb interconvert into subgenomes of 139 kb and 103 kb (Palmer and Herbon, 1986). A

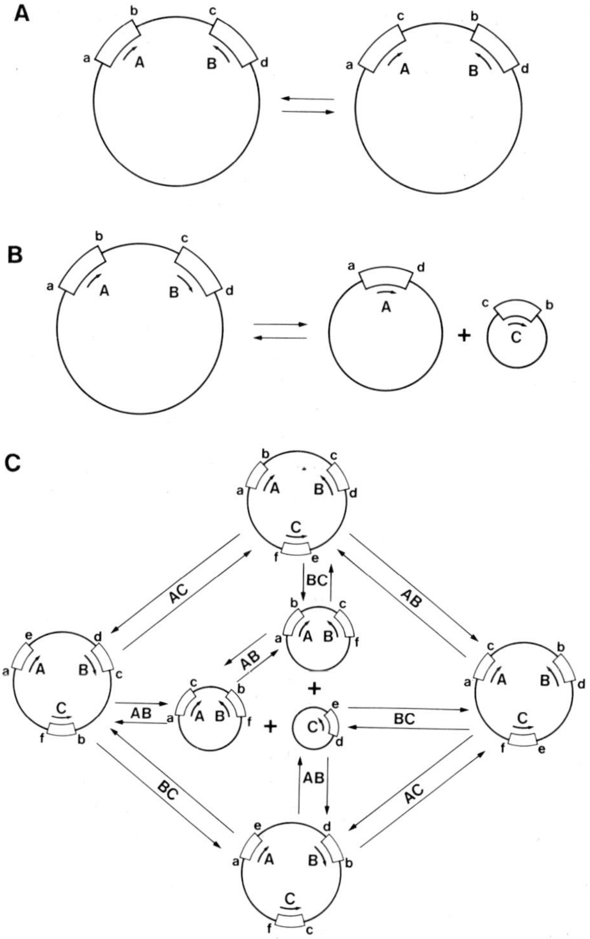


Fig. 1. Illustration of how recombination through repeats can produce various circular forms. (a) Recombination between a single pair of inverted repeat leads an inverse orientation of the sequence located between the two copies of the repeat; (b) recombination between a single pair of direct repeat leads to the formation of two subgenomic circular molecules, each containing one copy of the repeat; (c) recombination between a three copies repeat as found in cmsT mtDNA (two direct copies, one inverted) lead to size different genomic configurations. Four isomeric forms with permuted sequences of the master circle and two sets of subgenomic circular molecules.

tripartite structure has also been identified for spinach (Stern and Palmer, 1986) and sunflower (Siculella and Palmer, 1988). A slightly more complex structure, as illustrated in figure 1C, is represented by the petunia mitochondrial genome and is also found in the maize cmsT mtDNA. In petunia, a repeat of 6.6 kb is represented three times: two copies in direct orientation and the third copy inverted. Recombination gives rise to four permutated isomeric forms of 443 kb as well as 2 subgenomic circles of 244 kb and 199 kb (Folkerts and Hanson, 1989).

A much more complex multipartite organization has been reported for the wheat genome with its ten pairs of repeated sequences (Quetier et al., 1985) and for the maize that we are going to describe in this paper. Maize is the largest mtDNA for which a known physical map has been established.

Multipartite Structure and Plasticity of the Maize mt Genome

The maize mitochondrial genome of a normal fertile (N) cytoplasm and cytoplasmic male sterile type T (cmsT) in a B37 nuclear background can also be represented as a master circular molecule of 570 kb and 540 kb respectively (Lonsdale et al., 1984; Fauron and Havlik, 1988; Fauron et al., 1989). They contain many pairs of repeated sequences, which are mostly different between the two genomes. Many of these repeated elements appear to be involved in homologous recombination and, therefore, the genome organization is a dynamic equilibrium between the master circle present in many isomeric forms and a population of subgenomic circles and perhaps multimeric forms of subgenomic circles as well. Lonsdale et al.,(1984) suggested that the two 5.2 kb repeated sequences among the 6 pairs of large repeats identified in maize N mtDNA were preferred sites of recombination. Therefore, a large proportion of the DNA exists as two circles of 250 kb and 320 kb. A representation of the multicircular organization of the mitochondrial genome from the normal maize can be found in the paper of Lonsdale et al. (1984). The multicircular organization in cmsT mtDNA is even more complex because the primary source of intramolecular recombination is a 1.5 kb sequence repeated three times (Fauron et al., 1989). This gives rise to four isomeric forms of the master chromosomes (I, II, III, IV) that are represented in figure 2 as the framework from which a population of subgenomic circles or new isomeric forms can be observed. Besides the 1.5 kb repeat, three other recombinational repeats: the 6 kb repeat, the 4.6 kb repeat, the 0.12 kb repeat, have been located.

The master chromosome isomer form I with its four sets of repeats can give rise to another two permutated master chromosomes (isomer forms V and VI) by recombination between the two inverted repeats of 6 kb and 0.12 kb respectively. It can also form subgenomic circles of 364 kb / 176 kb and 457 kb / 53 kb by recombination through the direct repeats of 1.5 kb and 4.6 kb respectively. Isomeric form V now contains the two copies of the 6 kb repeat in a direct orientation that has the potential to recombine and give rise to the two subgenomic circles of 247 kb and 293 kb. It also contains the two copies of the 4.6 kb repeats in now inverted orientation leading to a new isomeric master chromosome with new permutation of sequences. Isomeric form VI now contains the two copies of the 0.12 kb repeat in direct orientation where recombination can produce two subgenomic circular molecules 265 kb and 275 kb . The same reasoning about the formation of recombination products can be applied to isomeric forms II, III, IV.

Control of Recombination

Two consequences can be drawn from such a mechanism of recombina-

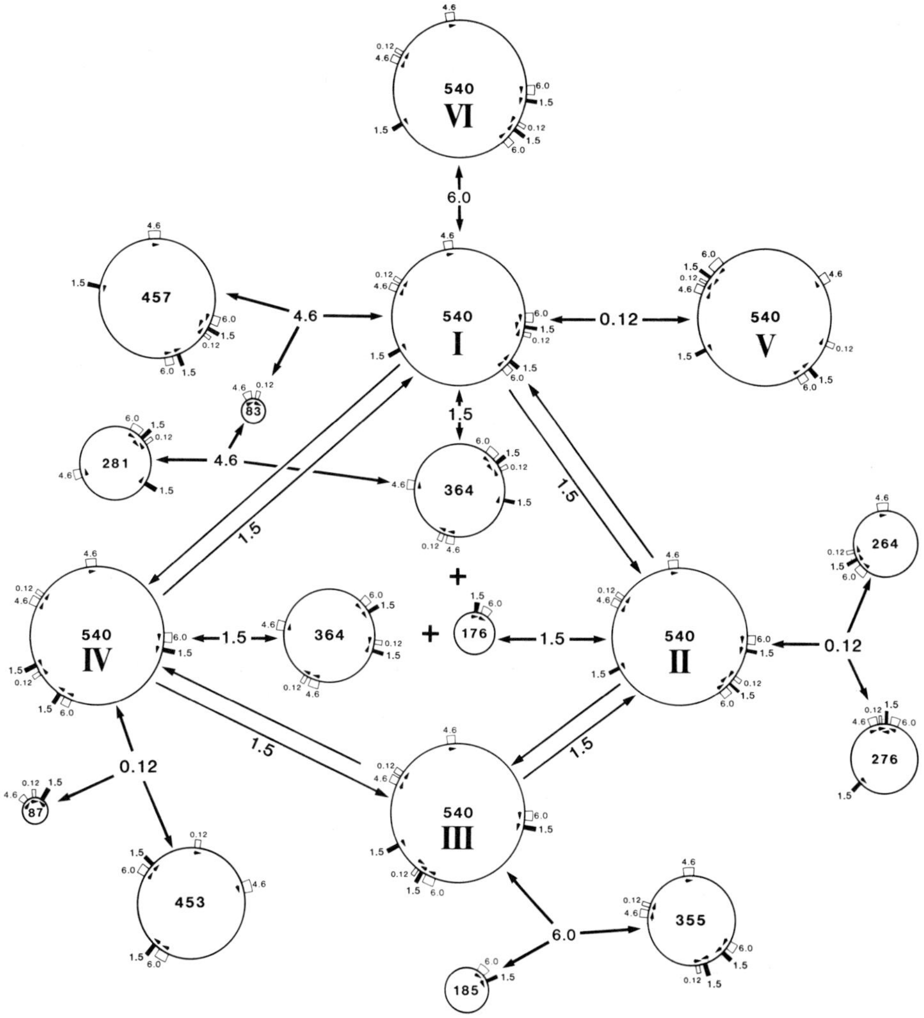


Fig. 2. The multipartite structure of the maize cmsT mitochondrial genome. The four different master chromosomes (I, II, III, IV) and the three sub-genomic circles of 364 kb and 176 kb have been predicted by recombination taking place between pair wise combinations of the three copies of the 1.5 kb repeat. Isomeric form I is the one from which the physical map was drawn (Fauron et al., 1989). Another three sets of repeats have been represented on those various circular forms: repeats of 6 kb, 4.6 kb and 0.12 kb. They will be in direct or inverted orientation depending upon which isomeric form they belong to (see text). Recombination involving those repeats will generate either new isomeric forms of the master chromosome or new subgenomic circles. To simplify the figure, only selected recombinant products have been represented. The recombinant molecules obtained by recombination through the 0.12 kb repeat have been seen only in tissue culture regenerated plants.

tion: First, an elevated number of circularly permuted master chromosomes can be present in the mtDNA populations. These isomeric forms contain the entire sequence complexity of the mitochondrial genome, but the sequence order and orientation are different following recombination between various sets of repeats. Therefore, the concept of direct repeats or inverted repeats needs to be revised as a set of repeated sequences can be either in direct or opposite direction, depending upon which isomeric form they belong to. One example is seen in figure 2: the 6 kb repeat has its two copies directly oriented on isomeric form I while they are inversely oriented on isomeric form III. The second consequence of recombination is that the population of subgenomic circular molecules formed by recombination through direct repeats present on the master chromosome will be variable. Indeed, depending upon which isomeric form of the master chromosome has been selected, the distance separating the two copies of a recombinational direct repeat will vary and so will the size of the subgenomic circles. An example is seen in figure 2. Recombination through the two copies of the direct repeat 0.12 kb from isomeric form II give rise to two circles of 264 kb and 276 kb, while the same event initiated in isomeric form IV give rise to two circles of 453 kb and 87 kb.

It is obvious that a great number of isomeric forms as well as subgenomic circles can be generated. Every time a new repeat is identified, it doubles the number of already known isomeric forms. With more and more small repeats (as small as 12 bp) that are identified, almost an infinite number of isomeric forms of the mtDNA could be present in a cell. Obviously, a control mechanism must be present in determining which one of those recombinant forms are permissible for the cell. If such a variety did exist, the construction of a physical map from overlapping cosmid clones would not have been possible. However, it is certainly true that a discrete subset of cosmid clones identified during mapping experiments represent rare recombinant molecules. An exhaustive search of those true recombinants clones from a cosmid library would give us an idea of the frequency of the rare recombinational events.

Furthermore, we have evidence that a control mechanism is influencing and selecting the recombination events taking place within the mitochondrial genome. Recombination through a 0.12 kb repeat has been represented on figure 2, but this repeat is not functional in the cmsT genome and becomes activated by passages through tissue culture (Fauron et al., 1990).

Variation of the Mitochondrial Genome within the Normal Cytoplasm

The genus Zea Mays is distributed over a large geographical area worldwide. The diverse environments and a continual manipulation by man has generated many regionally dominant groups.

Extensive surveys of the mtDNA restriction fragment length polymorphism have been conducted for maize and related species during the last 15 years (Levings and Pring, 1976; Levings and Pring, 1977; Pring and Levings, 1978; Timothy et al., 1979; Spruill et al, 1980, 1981; Weissinger et al., 1982, 1983; McNay et al., 1983; Kemble et al., 1983).

Weissinger et al. surveyed 93 indigenous maize races of Latin America providing a wide spectrum of the primitive maize cytoplasms. They classified their 93 races of maize into 18 groups according to their EcoRI and BamHI restriction profile (40 bands scored in each digest) and the presence of plasmid-like DNAs. Each group is different from any other group by at most five bands. One important observation we can make from this study is that RU cytoplasms (cytoplasms carrying the plasmid-like DNAs R1 and R2 (Weissinger et al., 1982) were all very similar to the normal maize standard used in this study, the single cross hybrid B73 X Mo17. Kemble et al. (1983) examined 25 accessions of Mexican races of maize. They compared their mtDNA restriction enzyme patterns to the mtDNA of inbred lines of the U.S.A. and found some variants. The observation from this study is that the most common N mtDNAs in Mexican races are similar to the U.S. inbred line N(A188). Several differences could be discerned from the analysis of the BamHI pattern obtained by probing with the plasmid-like R1 (see map from Lonsdale et al., 1984). Homology was found on either a 6.65 kb, a 6.85 kb or a 8.1 kb BamHI fragment. Most of the accessions displayed the 6.65 kb band, characteristic of N(A188); only one race contained the 8.1 kb fragment. A similar study performed by McNay et al. (1983) on the most common U.S. inbred lines revealed the presence of those same three groups regarding R1 homology. However, the most frequent was a 6.9 kb fragment. Therefore, two grouping of normal mtDNA seem to be present within the U.S. inbred lines.

Both of those studies show heterogeneity exists among maize mtDNA. However, the analysis is incomplete. Looking for difference in a restriction enzyme pattern from the EtBr-stained gels will not distinguish a simple mutation in nucleotide sequence from a more complex sequence rearrangement. Furthermore, as the restriction enzyme pattern is very complex, exhibiting submolar and multimolar bands, many differences go undetected and identical patterns do not prove that the mtDNA organization are identical. Similarly, probing the mtDNA with one specific restriction fragment may not reflect the differences encountered within the entire genome.

Complete sets of maize mtDNA clones representing both the entire normal and cmsT mtDNA sequence complexity (in a B37 nuclear background) were used as hybridization probes to survey the changes in the mtDNA of the most common United States inbred lines. Some of the results are shown in figure 3. All the mtDNAs isolated from cmsT in various nuclear backgrounds have been found identical to the already known map (Fauron et al., 1989). However, BamHI, XhoI, and SmaI digests of mtDNA from the normal cytoplasm reveals some heterogeneity depending on the nuclear background. As shown on figure 3, probe N7C9 reveals an extra band in N(A188) mtDNA. Out of the 20 overlapping cosmid clones covering the entire N mtDNA sequence complexity, only two did not reveal variation between N(B37) and N(A188). The great number of rearrangements encountered for the N (A188) mtDNA classifies it in a second group of normal mtDNA that we are calling NA (the group of the N mtDNA already mapped in a B37 nuclear background will be called NB). A sample of the analysis of the restriction fragment length polymorphisms seen between NA, NB and cmsT is illustrated in figure 4. Depending on the clone used as a probe, four kinds of results were obtained:

1. NA and NB are identical but different from cmsT (probe N5B10).
2. NA and NB differ by one fragment or more which is unique to each genome and different from cmsT (probe N5G8).
3. NA and NB differ from each other by one or more fragment that is(are) also present in cmsT (probes N8A1, and Tu4E1).
4. NB is different from NA but identical to cmsT (probe Tu3H11).

Furthermore, preliminary mapping results have shown that NA also contains unique sequences not found in NB or cmsT. The search for other normal cytoplasms fitting within the NA group identified N in the nuclear background W182B and Ky21. From the study McNay et al. (1983) we suspect that F6 strain also belongs to this group although we have not

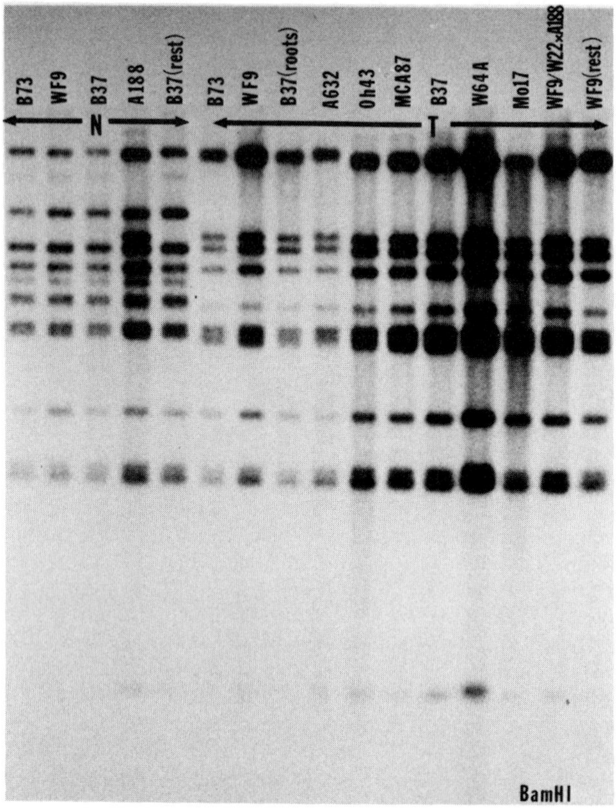


Fig. 3. Hybridization of clone N7C9 to a Southern blot containing BamHI digested mtDNA of N and T in a variety of nuclear backgrounds as indicated on the top of the picture. (Rest) indicates the cytoplasm containing the dominant nuclear genes RF1 and Rf2. All T mtDNAs are identical. NA188 mtDNA is different from the other N mtDNAs.

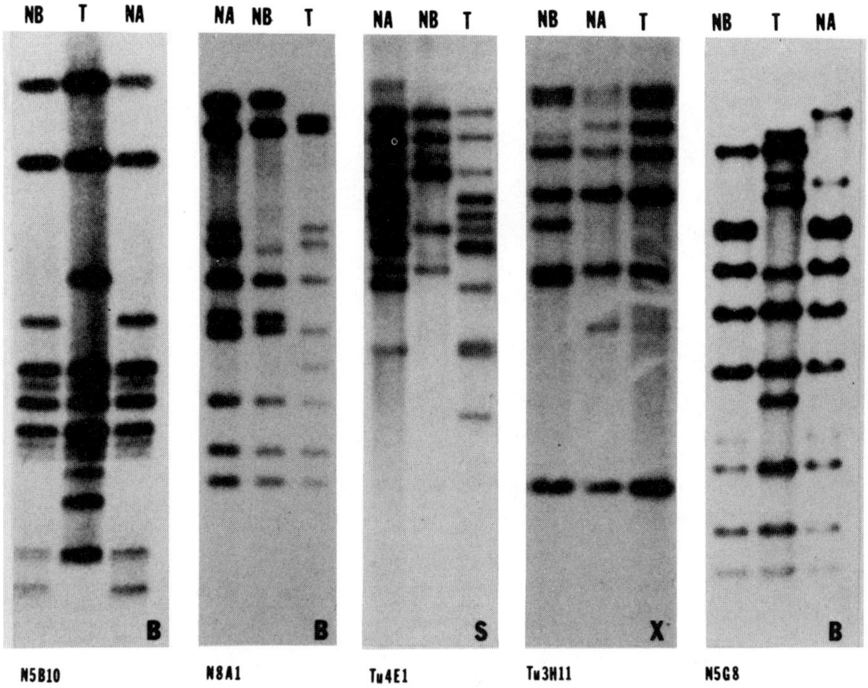


Fig. 4. Extent of variation between the mtDNAs from cytoplasm NB (nuclear background B37), NA (nuclear background A188) and T (B37). The blots used for hybridization contain either BamHI, XhoI or SmaI digests of the various mtDNAs. The probes are cosmid clones from the N or T library whose location is known on the map. NA and NB mt DNAs can be identical but different from cmsT mtDNA (probe N8B11) or rather different (probe N5G8). NA is also related to T sharing one fragment (probe N8A1), many fragments (probe tu4E1) or totally identical to T (probe tu3H11).

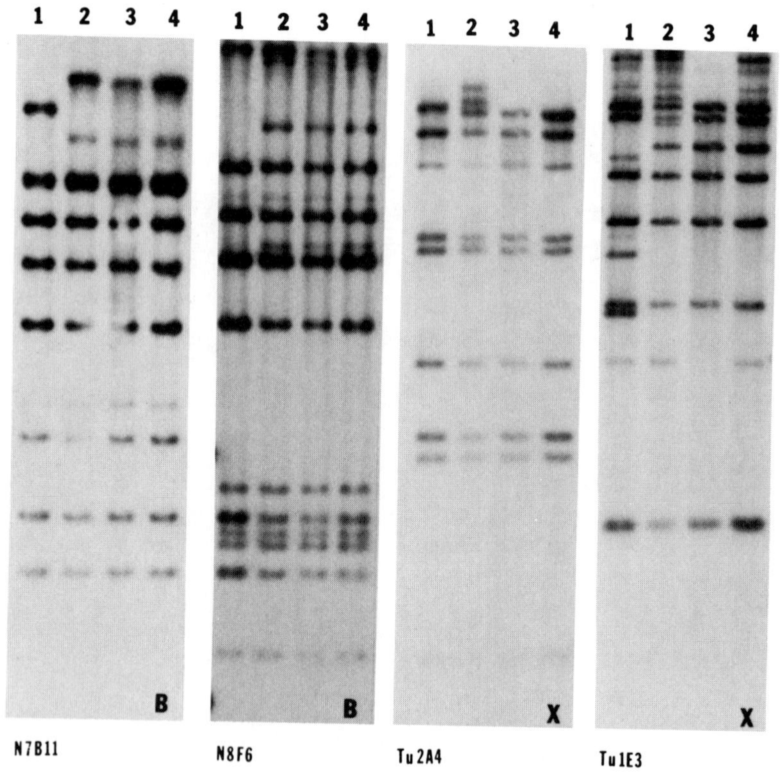


Fig. 5. Restriction length polymorphism of various N mtDNAs digested with BamHI(B) or XhoI(X). NB37 (line 1) represents the NB group while NA188 (line2), NKy21 (line 3), and NW182B (line 4), are members of the group NA. Probes N5BIO and tu4E1 classifies N (A188, Ky21 and W182B) as the same group different from NB37. However, probes tu2A4 and tu3H11 identifies the variations within the NA group.

tested it. Those hybridization studies have shown that the NB group is very homogeneous. The NA group, however, is more heterogeneous, since a few selected probes identify restriction fragment length polymorphism between members of the NA group. It is illustrated in figure 5. Probes N7B11 and N8F6 identify identical patterns between N mtDNAs in the three nuclear backgrounds A188, Ky21 and W182B while probes tu2A4 and tu1E3 reveals unique features in NA188 and NKy21 respectively.

Tissue Culture Regenerated Plants mtDNA

A fair proportion of maize plants regenerated from tissue culture derived from cmsT immature embryos are male fertile (Gengenbach et al., 1977, 1981; Brettell et al., 1978, 1980; Umbeck and Gengenbach, 1983). Studies of the mitochondrial genome from those regenerated plants have shown that the phenotype change was correlated with a change in the mtDNA. Subsequently, it has been shown that the Turf13 gene is responsible for the male sterile phenotype. Those changes have been analyzed in a dozen of those revertants and are the result of recombination events (Fauron et al., 1990; Fauron et al., in preparation). (See figure 6). It is a two-step recombinational event: A part of the multipartite structure, the first step of recombination occurs through two sets of repeats R1/R'1 and R2/R'2 to give rise to four circular subgenomic molecules. The second step is an intermolecular recombination between the subgenomic recombinant products. The recombination is also associated with an unknown mechanism that selectively eliminates or amplifies specific recombination intermediates to reconstruct a novel master chromosome. It presents two new features when compared to its progenitor: a deletion encompassing the Turf13 gene and a duplication whose size is dependent upon the chromosomal location of the repeats involved in recombination.

This mechanism has been described with detail (Fauron et al., 1990a) for the regenerant called V3. The same results were obtained from the study of another ten maize revertants (Fauron et. al., in preparation). However in one instance, the mutant V18, the changes identified in the mitochondrial genome organization were related to different recombinational products (Fauron et al., 1990): different repeats were involved in recombination. The outcome was also the elimination of the Turf13 gene. The study of another regenerant V11 showed an even more complex pattern of mtDNA rearrangements. As in V3 the genomic rearrangements evolved to eliminate the Turf13 locus. However, superimposed on this rearrangement, we found that the same mechanism acting with another two sets of different repeats leads to the elimination of another 6 kb of DNA and duplication of another 21 kb of sequences (Fauron et al., in preparation). A similar kind of mechanism has been described by Small et al. (1989) which seems to take place between small repeats located near the ATPA gene in maize.

## CONCLUDING REMARKS

### The Recombination and mtDNA Size Variation

We have just described the great recombinational capacity and flexibility of the plant mitochondrial genome. Repeats of any size can play a key role in the variation of the mtDNA. As more pairs of repeats or more copies of the same repeat are present in the genome, it gives more chance of generating diverse subgenomic circles, which in turn may recombine through intermolecular recombination.

In principle, much of the paradox about the size variability of

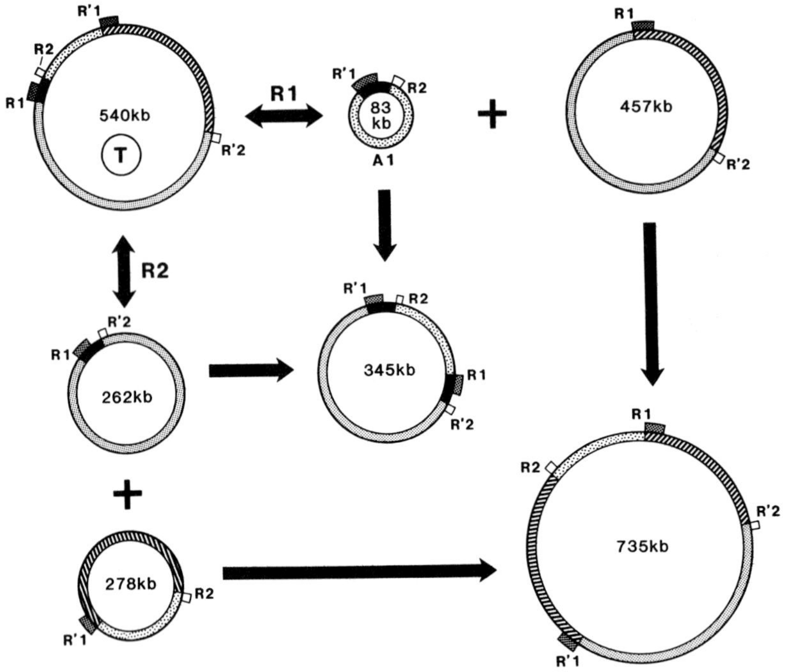


Fig. 6. Model explaining evolution throughout recombination of the mitochondrial genome from maize tissue culture regenerated plants. Two mechanisms are involved for the genomic reorganization: activation of the recombinational repeated R1/R'1 and R2/R'2 and selection of some of the recombinant products. Intramolecular recombination of the two sets of repeats gives rise to four subgenomic circles, two of which are selected throughout intermolecular recombination to produce the new master chromosome of 735 kb. The 345 kb circular recombinant is selectively eliminated.

higher plant mitochondrial genome could be explained by the high frequency of intra- and inter-molecular recombination between repeats. Since small repeats seem to be extremely frequent, and randomly located around the master chromosome, in principle any region of the genome could produce additional repeats through a combined mechanism of intra- and inter-molecular recombinations. If this is the case, why have not more repeated sequences been detected by hybridization experiments, especially in larger genomes. Perhaps repeats accumulate mutation very rapidly, losing rapidly their original identity over time, leaving somehow some prints represented by those small repeat still recognizable.

## Import of Foreign Sequences and mtDNA Size Variation

The mitochondrial genome seems to be a dumping place for foreign sequences that could also play a role in the size diversity. Numerous chloroplast sequences have been imported to the mtDNA. It is very conceivable that nuclear DNA has also been integrated into the mtDNA but no thorough screening has ever been done. A good illustration of the transfer of foreign sequence can be seen with the plasmid R1 from the maize cmsS mitochondria. This plasmid has been found integrated into the nuclear genome (Kemble et al., 1983) and the RI sequence itself contains a fragment of the chloroplast psbA gene (Sederoff et al. 1986). Most of those foreign sequences are probably not functional in their new environment. Consequently, they may evolve rapidly and rapid sequence drift makes ancient importation no longer recognizable by hybridization studies.

## Direction of Evolution

There are differences in the mitochondrial sequence complexity suggesting that the genome in the various groups of maize have been in fluid state since their divergence from a common ancestor. Each contemporary molecule seems to have been established by separate events leading to all kinds of rearrangements. Such diversity of sizes raises the question in which direction did the changes occur since the common ancestor of maize mtDNA? Is it a small genome that was derived from a larger one or vice versa? In other words, has diversity been generated by gain or loss of sequences and what was the most likely size of the ancestral genome? The generalization is that the maize mtDNA has undergone a lot of multiple independent events, both expansionary and reductionary, since the time of the common progenitor.

## Significance of the Diversity

This independent dispersal of sizes raises the question: Will a change in DNA structure alter the phenotype of the plant? In other words, do those variations have any consequences on the organelle gene expression? With the existence of large non-functional spacers regions, plant mitochondrial genomes have more opportunity to let sequence rearrangements, deletions, additions to occur, without interfering with gene expression. However, it is now well documented that recombined sequences have been located within the gene itself or in flanking regions of a gene, possibly affecting the gene activity or changing the transcriptional or translational pattern. Insertion of sequence from the chloroplast genome may alter the coding capacity of a gene. The ct tRNAarg gene provides the last 15 amino acid of ORF25 gene (Dewey et al., 1986). The atp6 gene of cmsC contains an additional 13 amino acid identical to the atp9 gene as well as 147 amino acid from the chloroplast genome (Levings and Dewey, 1988). New regulatory sequences are found at the 5' flanking region of some genes. The coII gene of cmsC is a fusion of the atp6 gene and coII gene of a normal cytoplasm probably contains the promotor regions of the atp6 gene (Levings and Dewey, 1988). Changes in the 5' flanking region of atp9 gene of cmsC provides the gene with a new promoter region which accounts for the variation in transcripts between N cytoplasm and C cytoplasm (Levings and Dewey, 1988). As already seen, new open reading frames have been created as the urf13 gene (Dewey et al., 1986).

REFERENCES


Belliard, G., Vedel, F., Pelletier, G., 1979, Mitochondrial recombination in cytoplasmic hybrids of Nicotiana tabacum by protoplast fusion, Nature, 281:401.

Bonen, L., Boer, P. H., Gray, M. W., 1984, The wheat cytochrome oxidase subunit II gene has an intron insert and three radical amino acid changes relative to maize, EMBO J., 3:2531.

Braun, C. J., Sisco, P. H., Sederoff, R. R., Levings, III, C.S., 1986, Characterization of inverted repeats from plasmid like DNAs and the maize mitochondrial genome, Curr. Genet., 10:625.

Brears, T., Lonsdale, D. M., 1988, The sugar beet mitochondrial genome: a complex organization generated by homologous recombination, Mol. Gen. Genet., 214:514.

Brettell, R. I. S., Goddard, B. V. D., Ingram, D. S., 1979, Selection of TMS-cytoplasm maize tissue cultures resistant to Drechslera maydis T-toxin, Maydica., 24:203.

Brettell, R. I. S., Thomas, E., Ingram, D. S., 1980, Reversion of Texas male sterile cytoplasm maize in culture to give fertile, T-toxin revertant plants, Theor. Appl. Genet., 58:55.

Dewey, R. E., Levings, C. S. III, Timothy, D. H., 1986, Novel recombinations in the maize mitochondrial genome produce a unique transcriptional unit in the Texas male-sterile cytoplasm (cms-T), Cell, 44:439.

Dewey, R. E., Timothy, D. H., Levings III, C. S., 1987, A mitochondrial protein associated with cytoplasmic male sterility in the T cytoplasm of maize, Proc. Natl. Acad. Sci. USA, 84:5374.

Dewey, R.E., Siedon, J. N., Timothy, D. H., Levings III, C. S., 1988, A 13-kilodalton maize mitochondrial protein in E. coli confers sensitivity to Bipolaris maydis toxin, Science, 239:293.

Escote-Carlson, L. J, Gabay-Laughnan, S., Laughnan, J. R., 1988, Reorganization of mitochondrial genomes of cytoplasmic revertants in cms-S inbred line Wf9 in maize. Theor. Appl. Genet., 75:659.

Fauron, C. M.-R., Havlik, M., 1988, The BamHI, XhoI, SmaI, restriction enzyme maps of the normal maize mitochondrial genome genotype B37, Nuc. Acid. Res., 16:10395.

Fauron, C. M.-R., Havlik, M., Lonsdale, D., Nichols, L., 1989, Mitochondrial genome organization of the maize cytoplasmic male sterile type T, Mol. Gen. Gen., 216:395.

Fauron, C. M.-R., Havlik, M., Brettell, R. I. S., 1990, The mitochondrial genome organization of a maize fertile cmsT revertant line is generated through recombination between two sets of repeats, Genetics, 124:423.

Fauron, C. M.-R., Havlik, M., Brettell, R. I. S., Albertsen, M., 1990, Study of two different recombination events in cmsT regenerated plants during reversion to fertility, Theor. Appl. Genet., in press.

Fauron, C. M.-R., Havlik, M. Casper, M., 1990, Evolution through recombination of the maize mitochondrial genome in cmsT regenerated plants, (in preparation).

Folkerts, O., Hanson, M. R., 1989, Three copies of a single recombination repeat occur on the 443 kb master circle of the petunia hybrida 3704 mitochondrial genome, Nuc. Acid. Res., 17:7345.

Forde, B. G., Oliver, R. J. C., Leaver, C. J., 1978, Variation in mitochondrial translation products associated with male-sterile cytoplasms in maize, Proc. Natl. Acad. Sci. U.S.A., 75:3841.
Galun, E., Arzee-Gonen, P., Fluhr, R., Edelman, M., Aviv, D., 1982, Cytoplasmic hybridizaiton in Nicotiana: Mitochondrial DNA analysis in progenies resulting from fusion between protoplasts having different organelle constitution, Mol. Gen. Genet., 186:50.
Gengenbach, B. G., Green, C. E., Donovan, C. M., 1977, Inheritance of selected pathotoxin resistance in maize plants regenerated from cell cultures, Proc. Natl. Acad. Sci., 74:5113.
Gengenbach, B. G., Connelly, J. A., Pring, D. R. Conde, M. F., 1981, Mitochondrial DNA variation in maize plants regenerated during tissue culture selection, Theor. Appl. Genet., 59:161.
Hack, E., Leaver, C. J., 1983, The alpha subunit of the maize F1-ATPase is synthesized in the mitochondrion, EMBO J., 2:1783.
Houchins, J. P., Ginsburg, H., Rohrbaugh, M., Dale, R. M. K, Schardl, C. L., Hodge, T. P., Lonsdale, D. M., 1986, DNA sequence analysis of a 5.27 kb direct repeat occuring adjacent to the regions of episome homology in maize mitochondria, EMBO J., 5:2781.
Kemble, R. J., Mans, R. J., Gabay-Laughnan, S., Laughnan, J. R., 1983, Sequences homologous to episomal mitochondrial DNAs in the maize nuclear genome, Nature, 304:744.
Kemble, R. J., Gunn, R. E., Flavell, R.B., 1983, Mitochondrial DNA variation in races of maize indigenous to Mexico, Theor. Appl. Genet., 65:129.
Kao, T. H., Moon, E., Wu, R., 1984. Cytochrome oxidase subunit II gene of rice has an insertion sequence within the intron, Nuc. Acids Res., 12:7305.
Laughnan, J. R., Gabay-Laughnan, S., 1983, Cytoplasmic male sterility in maize, Annu. Rev. Genet., 17:27.
Leaver, C. J., Dawson, A. J., Isaac, P., Jones, V. P., Hack, E., 1984, Structure and expression of plant mitochondrial genes, in: "Molecular Form and Function of the Plant Genome," Plenum Press, New York.
Levings III, C. S., Pring, D. R., 1976, Restriction endonuclease analysis of mitochondrial DNA from normal and Texas cytoplasmic male-sterile maize, Science, 193:158.
Levings III, C. S., Pring, D. R., 1977, Diversity of mitochondrial genomes among normal cytoplasms of maize, J. of Heredity, 68:350.
Levings III, C. S., Shah, D. M., Hu, W. W. L., Pring, D. R., Timothy, D. H., 1979, Molecular heterogeneity among mitochondrial DNAs from different maize cytoplasms, in: "Extrachromosomal DNA," Cummings, D. J., Borst, P., Dawid, I., Weissman, S. M., Fox, C. F., eds., Academic Press, New York.
Levings III, C. S., Dewey, R. E., 1988, Molecular studies of cytoplasmic male sterility in maize, Phil. Trans. R. Soc., London, B319:177.
Lonsdale, D. M., Hodge, T. P., Howe, C., Stern, B., 1983, Maize mitochondrial DNA contains a sequence homologous to the ribulose 1,5 - hisphosphate carboxylase large subunit gene of chloroplast DNA, Cell, 34:1007.
Lonsdale, D. M., Hodge, T. P., Fauron, C. M.-R., 1984, The physical map and organization of the mitochondrial genome from the fertile cytoplasm of maize, Nuc. Acdis Res., 12:9249.
Lonsdale, D. M., 1984, Chloroplast DNA sequences in the mitochondrial genome of maize, in: "Molecular Form and Function of the Plant Genome," Plenum Press, New York.
McNay, J. W., Pring, D. R., Lonsdale, D. M., 1983, Polymorphism of mitochondrial DNA "S" regions among normal cytoplasms of maize, Plant Mol. Biol., 2:177.

Morgan, A., Maliga, P., 1987, Rapid chloroplast segregation and recombination of mitochondrial DNA in Brassica cybrids, Mol. Gen. Genet., 209:240.
Ozias-Akins, P., Pring, D. R., Vasil, I. K., 1987, Rearrangements in the mitochondrial genome of somatic hybrid cell-lines of Pennisetum americanum (L.) K. Schum. + Panicum maximum, Jacq., Theor. Appl. Genet., 74:15.
Palmer, J. D., Herbon, L., 1986, Tricircular mitochondrial genomes of Brassica and raphanus: reversal of repeat configurations by inversion, Nuc. Acid. Res., 14:9755.
Palmer, J. D., Herbon, L. A., 1987, Unicircular structure of the Brassica hirta mitochondrial genome, Curr. Genet., 11:565.
Palmer, J. D., 1988, Intraspecific variation and multicircularity in Brassica mitochondrial DNAs, Genetics, 118:341.
Palmer, J. D., Herbon, L. A., 1988, Plant mitochondrial DNA evolves rapidly in structure but slowly in sequence, J. Mol. Evol., 28:87.
Pring, D. R., Levings III, C. S., 1978, Heterogeneity of maize cytoplasmic genomes among male sterile cytoplasms, Genetics, 89:121.
Quetier, F., Vedel, F., 1977, Heterogeneous population of mitochondrial DNA molecules in higher plants, Nature, 268:365.
Quetier, F., Lejeune, B., Delorme, S., Falconet, D., Jubier, M. F., 1985, Molecular form and function of the wheat mitochondrial genome, in: "Molecular Form and Function of the Plant Genome," Vol 83, Plenum Press, New York.
Rothenberg, M., Boeshore, M. L., Hanson, M. R., Izhar, S., 1985, Intergenomic recombination of mitochondrial genomes in a somatic hybrid plant, Curr. Genet, 9:615.
Rottmann, W. H., Brears, T., Hodge, T. P., Lonsdale, D. M., 1987, Mitochondrial gene is lost via homologous recombination during reversion of cmsT maize to fertility, EMBO J., 6:1541.
Sangare, A., Weil, J. H., Grienenberger, J. M., Fauron, C., Lonsdale, D., 1990, Localization and organization of tRNA genes on the mitochondrial genomes of fertile and male sterile lines of maize, Mol. Gen. Gen., (submitted).
Schardl, C. L., Lonsdale, D. M., Pring, D. R., Rose, K. R., 1984, Linearization of maize mitochondrial chromosomes by recombinations with linear episomes, Nature, 310:292.
Siculella, L., Palmer, J. D., 1988, Physical and gene organization of mitochondrial DNA in fertile and male sterile sunflower cms-associated alterations in structure and transcription, Nuc. Acid. Res., 16:3787.
Small, I., Suffolk, R., Leaver, C. J., 1989, Evolution of plant mitochondrial genomes via substrichiometric intermediates, Cell, 58:69.
Stern, D. B., Lonsdale, D. M., 1982, Mitochondrial and chloroplast genomes of maize have a 12 kb DNA sequence in common, Nature (London), 299:698.
Stern, D. B., Palmer, J. D., Thompson, W. F., Lonsdale, D. M., 1983, Mitochondrial DNA sequence evolution and homology to chloroplast DNA in angiosperms, Plant Mol. Biol., 1:467.
Stern, D. B., Palmer, J. D., 1984, Extensive and indespread homologies between mitochondrial DNA and chloroplast DNA in plants, Proc. Natl. Acad. Sci. U.S.A., 81:1946.
Stern, D. B., Newton, K. J., 1985, Mitochondrial gene expression in cucirbrtaceae: conserved and vauable features, Curr. Genet., 9:395.
Stern, D. B., Bang, A. G., Thompson, F. W., 1986, The watermelon mitochondrial URF-1 gene: evidence for a complex structure, Curr. Genet., 10:857.
Stern, D. B., Palmer, J. D., 1986, Tripartite mitochondrial genome of spinach: physical structure, mitochondrial gene mapping, and locations of transposed chloroplast DNA sequences, Nuc. Acid. Res., 14:5651.

Sederoff, R. F., Ronald, P., Bedinger, P., Rivin, C., Walbot, V., Bland, M., Levings III, C. S., 1986, Maize mitochondrial plasmid S1 sequences share homology with chloroplast gene psbA, Genetics, 113:469.

Timothy, D. H., Levings II, C. S., Pring, D. R., Conde, M. F., Kernicle, J. L., 1979, Organelle DNA variation and systematic relationships in the genus Zea: teosinte, Proc. Natl. Acad. Sci., 76:4220.

Umbeck, P. F., Gengenbach, B. G., 1983, Reversion of male-sterile T cytoplasm maize to male fertility in tissue culture, Crop Sci., 23:584.

Wahleithner, J. A., Macfarlane, J. L., Wolstenholme, D. R., 1990, A sequence encoding a maturase related protein in a group II intron of a plant mitochondrial nadl gene, Proc. Natl. Acad. Sci., 87:548.

Ward, B. L., Anderson, R. S., Bendich, A. J., 1981, The mitochondrial genome is large and variable in a family of plants (Curcurbitaceae), Cell, 25:792.

Weissinger, A. K., Timothy, D. H., Levings III, C. S., Hu, W. W. L., Goodman, M. M., 1982, Unique plasmide-like mtDNAs from indigenous maize races of latin america, Proc. Natl. Acad. Sci. U.S.A., 79:1.

Weissinger, A. K., Timothy, D. H., Levings III, C. S., Goodman, M. M., 1983, Patterns of mitochondrial DNA variation in indigenous maize races of latin america, Genetics, 104:365.

Wise, R. P., Pring, D. R., Gengenbach, B. G., 1987, Mutation to male fertility and toxin insensitivity in Texas (T) cytoplasm maize is associated with a frameshift in a mitochondrial open reading frame, Proc. Natl. Acad. Sci., U.S.A., 84:2858.

# RNA EDITING IN WHEAT MITOCHONDRIA : A NEW MECHANISM FOR THE MODULATION OF GENE EXPRESSION

Jean-Michel Grienenberger, Lorenzo Lamattina, Jacques Henry Weil, Géraldine Bonnard and José Gualberto

Institut de Biologie Moléculaire des Plantes du CNRS
Université Louis Pasteur, 12 rue du Général Zimmer
67084 Strasbourg, France

RNA editing is a recently discovered mechanism involved in the modulation of gene expression. It is a process which results in the production of mRNAs with a nucleotide sequence differing from that of the template DNA (Simpson and Shaw, 1989). This phenomenom was initially described in the mitochondria (mt) of protozoa (Benne *et al.* 1986). We have recently (Gualberto *et al.*, 1989) shown that RNA editing is also required for the correct expression of plant mt genes. Several wheat mt mRNA sequences were found to differ from the corresponding gene sequences in that a number of cytidine (C) residues are converted into uridine (U) residues. Most C to U conversions analysed cause a change of the amino acid coded by the modified codon. These modifications contribute to the conservation of mitochondrial protein sequences among higher plants. Furthermore, C to U conversions enable the plant mt translation system to use the universal genetic code, since CGG codons, which were postulated to code for tryptophan (Fox and Leaver, 1981), are changed into UGG codons where a tryptophan is to be incorporated in the protein.

## RNA EDITING AT A SPLICING SITE OF *nad*4

The gene coding for subunit IV of NADH dehydrogenase of wheat mt DNA (Lamattina *et al.*, 1989) is interrupted by at least three class II introns, one of which is 3.4 kb long. In order to cross the intron boundaries and to identify the nucleotide sequence corresponding to the junction of the two exons, the sequence of he mature transcript was deduced upon sequencing of the cDNA by the chain termination method, using AMV reverse transcriptase primed by a specific oligonucleotide on total wheat mt RNAs as template.

When determined, the RNA sequence revealed, surprisingly, a difference

*Plant Molecular Biology* 2, Edited by R.G. Herrmann and
B. Larkins, Plenum Press, New York, 1991

with respect to the corresponding DNA sequence (Fig. 1). The transcript contains a uridine at the exon2/exon3 junction, which is part of a non-encoded CUG leucine codon. This experiment has been repeated four times, using different wheat mt RNA extracts, with the same result. As this was the only base change found in the 200 nucleotides sequenced, it is not likely to be the result of a reverse transcriptase error.

The corresponding DNA sequences were determined on both strands using cloned mitochondrial DNA. To eliminate any possible artifacts due to the cloning steps, we have also determined the sequence of the genomic wheat mt DNA after amplification by the polymerase chain reaction (PCR), using *Taq* polymerase. The genomic as well as the cloned wheat mt DNA sequence contain a C residue at this position.

| | | | |
|---|---|---|---|
| E2-I2 Junction | **TGACTATTGGTATGTTTAGTC** | G | GGCGGCGGCCGTTAGG |
| mRNA sequence | **UGACUAUUGGUAUGUUUAGUC** | U | **GAACAUACAGGGAAUU** |
| I2-E3 Junction | TGGGGGAATATTAGGCTCTAT | C | **GAACATACAGGGAATT** |

Figure 1. The sequence of the *nad4* transcript is shown with the DNA sequence corresponding to the exon 2-intron 2 and intron 2-exon 3 boundary regions, aligned above and below it respectively. The non encoded U and the corresponding genomic nucleotides are boxed.

We have also considered the possibility that homologous sequences carrying a T residue could exist in another location in the wheat mt genome. All Southern hybridizations show that each digest gives hybridizing fragments that are unique when they are probed with the corresponding restriction fragments. These experiments show that the comparison which is made between the DNA and RNA sequences is a valid one. The presence of this U residue is necessary to maintain the correct reading frame as shown by its comparaison with already known fungal or mammalian NAD4. After analysis of the 5' and 3' flanking consensus regions of the group II intron, it was deduced that the difference between the DNA and the corresponding RNA is due to a conversion of a C (in the DNA) into a U (in the RNA).

This non-genomically encoded uridine found in the wheat *nad4* transcript is the first demonstration of such a C to U RNA editing in plant mitochondria.

## WHEAT MITOCHONDRIA USE THE UNIVERSEL GENETIC CODE

It has been proposed that the genetic code in plant mitochondria differs from the universal one in that CGG would code for tryptophan instead of arginine (Fox and Leaver, 1981). This hypothesis was based on the fact that CGG was often found in plant mt genes at positions where there are conserved tryptophan in other organisms. However, no CGG-decoding $tRNA^{Trp}$ or $tRNA^{Trp}$ gene has been found in plant mitochondria (Maréchal *et al.*, 1985). Moreover, wheat mt genes coding respectively for *nad3* and *rps12* (Gualberto *et al.*, 1988) and for *cox3* (Gualberto *et al.*, 1990a) contain CGG codons at positions conserved either as tryptophan or as arginine.

The characterization of a C to U conversion at the junction of two exons in wheat *nad4* mRNA showed that RNA editing exists in plant mitochondria. Such C to U conversion, if affecting a CGG codon, would transform it into a UGG coding for tryptophan. To test if this occurs, the mRNA sequences of selected regions flanking CGG codons of wheat mt *nad3*, *rps12* and *cox3* were determined upon sequencing uncloned cDNAs primed by specific oligonucleotides using total mt RNAs as template. The CGG codons of *rps12* and *cox3* which code for conserved arginines remain CGG at the RNA level. However in *cox3,* the CGG located at a position conserved as tryptophan is replaced by a UGG in the deduced mRNA sequence.

In the *nad3* cDNA sequence corresponding to the genomic CGG located where there is a conserved tryptophan, an ambiguity was found (G and A at the same sequence position) that could not be resolved by using different RNA preparations as templates. The RNAs seem therefore to contain two types of molecules, having respectively a CGG or a UGG codon. This could result from the presence of both unmodified and modified transcripts, suggesting that the C to U conversion is a post-transcriptionnal process. The same approach was used to determine the cDNA sequence of regions in the wheat mt *cox*1 (Bonen *et al.*, 1987), *cox*2 (Bonen *et al.*, 1984) and *cob* (Boer *et al.*, 1985) genes where CGG codons are present in the genomic DNA. The results show that the three CGG codons of *cox*2 and the two CGG codons of *cob* (all located at conserved tryptophan positions) are all converted into UGG codons in the mRNAs.

The only CGG of *cox1*, which corresponds to a non-conserved position, is unmodified. Therefore, it is not possible to tell whether a CGG codon found in a wheat mt gene codes for arginine or tryptophan, as this depends whether RNA editing takes place. Our results show that in wheat mitochondria, upon RNA editing, the universal genetic code can be used to specify tryptophan, without the divergence postulated by Fox and Leaver(1981)

## RNA EDITING RESULTS IN THE CONSERVATION OF PROTEIN SEQUENCES

Upon cDNA sequence analyses, other differences were found in addition to the ones involving CGG codons. Out of more than 5000 nucleotides sequenced, 70 differences were found when compared to the genomic sequences (table 1).

Table 1. Differences found in several wheat mt genes between the codons in the genomic sequences and the codons in the mRNAs (as deduced from the cDNA sequences). The positions where a C to U editing has been observed are underlined and the amino acids specified by the corresponding codons are indicated using the one letter code

| DNA mRNA | gene edited | |
|---|---|---|
| ACA - ATA | *nad4* | ACG - AUG |
| ACT - ATT | *nad4* | (T → M) |
| (T → I) | | |
| | | GCN - GUN |
| CAC - UAC | *cob; rps12* | (A → V) |
| CAT - UAU | *cob* | |
| (H → Y) | | |
| | | CAA - UAA |
| CCT - CUU | *nad4; cox3* | (Q → stop) |
| CCG - CUG | *nad3; nad4* | |
| CCA - CUA | *nad4; cox3; orf156* | CAG - UAG |
| CCA - UUA | *nad3* | (Q → stop) |
| (P → L) | | |
| | | CGA - UGA |
| CCC - UCC | *nad4* | (R → stop) |
| CCT - UCU | *nad3* | |
| (P → S) | | no effect |
| CCT - UUU | *nad3* | |
| (P → F) | | ATC - AUU |
| | | (I - I) |
| CGG - UGG | *nad3; nad4; cox2; cox3; cob* | *nad3* |
| (R → W) | | |
| | | CTC - CUU |
| CGT - UGU | *rps12; nad4* | (L - L) |
| (R → C) | | *orf156* |
| CTT - UUU | *nad4; cox3* | |
| CTC - UUC | *cob; orf156* | |
| (L → F) | | |
| TCA - UUA | *cox2; cox3* | |
| TCG - UUG | *rps12; nad3; orf156* | |
| (S → L) | | |
| TCC - UUC | *nad3; nad4; cox3; rps12* | |
| TCT - UUU | *nad4; cob; cox2* | |
| (S → F) | | |

```
NAD3          110     118                     COX3         250      260
Consensus     YeW.q.gLeWte                    Consensus    vDVvWLFLY.siYwWGs
wheat mRNA    -------L-W.-                    wheat mRNA   ----W--L----------
wheat         YEWKRGASDR.E                    wheat        VDVVRLFPFVSIYWWGGI
maize         ----------.-                    maize        -----------------T
petunia       ----------.-                    Oenothera    -------L----------

COB                  100       110       120                   230       240
Consensus     gASfFFiclylHiGRGlYYGSylyp...TWhiG // ipfhPY.t.KDllgflifl
wheat mRNA    -----F--Y---------Y----------W--- // --F-----------W--F-
wheat         GASMFLIVVHLHIFRGLYHASYSSPREFVRCLG // IASYPYFYVKDLVGRVASA
maize         -----------------------------W---    -------------------
Oenothera     ---------------------------------    -SF-----------W--F-

COX2              30        40        50                  90         130
Consensus     lglQDatsPimEeliefHDh.lmivflI.slvsW // EtiWTIL // QWYWsYE
wheat mRNA    --F-----------------------------IW    ---W---    ---W---
wheat         LGSQDAATPMMQGIIDLHHDIFFFLILILVFVSR // EIIRTIF // QWYRTYE
maize         ----------------------------------    -------    -------
Oenothera     ----------------------------------    -------    -------
```

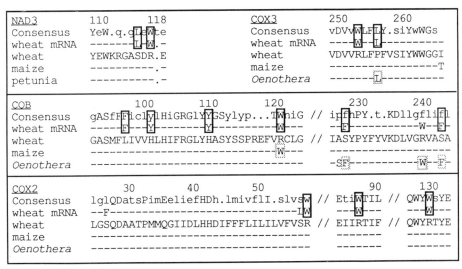

Figure 2. Increase of mt protein sequence conservation due to C to U editing. Selected parts of the amino acid sequence of several wheat mt proteins (deduced from the gene sequences using the universal genetic code) are compared to i) the corresponding amino acid sequence of mt proteins (deduced from the gene sequences) from two other plants (one monocot and one dicot), ii) the aminoacid sequence deduced from the wheat mRNA sequence, iii) the consensus amino acid sequence based on mt protein sequences from non-plant species (found in GenEMBL Data Bank). Dots in the consensus sequence indicate non conserved positions, lower case letters indicate partial conservation, capitals indicate complete conservation. A dash indicate that the amino acid is identical to that deduced from the wheat genomic sequence. Solid line boxes indicate amino acid homology with the consensus sequence, when editing of the wheat mRNA is considered. Dotted line boxes indicate amino acid homology with mt proteins of other plants, when editing of the wheat mRNA is considered (Gualberto *et al.*, 1989)

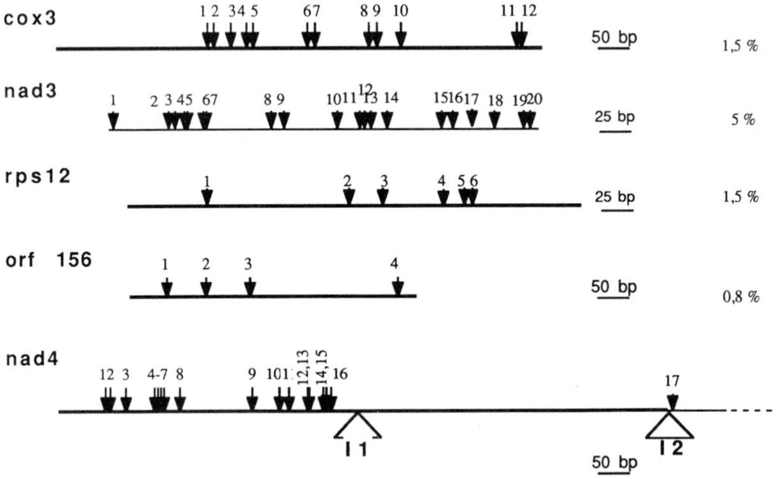


Figure 3. Position of editing sites in five wheat mt mRNAs are indicated by arrows above the line representing the sequence of the transcript and numbered. I1 and I2 indicate the position of two introns of *nad4*. The percentage of nucleotide conversions by RNA editing is indicated

All these differences are C to U conversions which occur in most cases, at the first or second base of the codon, leading to a change of the aminoacid encoded. No modification was found in the non-coding sequences corresponding respectively to 170 and 150 nucleotides upstream *cox3* and *cob* , which suggests that the C to U editing is acting to modulate the coding part of the mRNA.

In agreement with this idea, a comparison of amino acid sequences deduced from the genes and the edited RNAs shows that the latter have a higher homology to the mt proteins of other species (figure 2). Interestingly, in some plants, the mt genome codes for the same amino acid which is specified upon editing in wheat mitochondria (figure 2) and it does so using the same codon as the edited one. This suggests that RNA editing plays a role in the conservation of mt protein sequences during evolution. This C to U RNA editing has also been described to occur in other plants like *Oenothera* (Covello and Gray, 1989. Hiesel *et al.* 1989).

## RNA EDITING INVOLVES A SIGNIFICATIVE NUMBER OF C RESIDUES

In order to estimate the number of C residues which are involved in RNA editing, we have determined the complete sequence of the mRNAs of five wheat mt genes, namely *cox3* (Gualberto *et al.*,1990a), *nad3*, *rps12* (Gualberto *et al.*, 1988), *orf156* (Gualberto *et al.*,1990b) and *nad4* (Lamattina *et al.*, 1989). For this purpose, an oligonucleotide corresponding to the 3' end of each of the known gene sequences was used to prime the synthesis of a first strand cDNA. These cDNAs were then amplified using polymerase chain reaction and cloned.

The nucleotide sequence of a number of the resulting clones was determined and compared with their genomic counterpart. Numerous differences have been found, a number of them being due to misincorporation by the reverse transcriptase or by the *Taq* polymerase. However, these differences appears only once and can then be recognized. After the sequence analysis of more than ten clones for each gene, we found 59 positions in the five genes where RNA editing changes a C into a U (Figure 3). These modifications are consistently found in all clones for *cox3*, *nad4* and *orf156*. For *nad3* and *rps12* which are cotranscribed (Gualberto *et al.*,1988), it was found that some mRNAs are not completely edited, showing that RNA editing is a post transcriptionnal mechanism (manuscript in preparation).

In each of the five genes, there is a definite number of editing sites which account for 0.8% (*orf156*) to 5% (*nad3*) of the total lenght of the coding sequence (Figure 3). All but one of these editing sites modify the nature of the encoded aminoacid leading for exemple to a 15% modification of NAD3 protein as compared with that deduced from the gene sequence. The repartition of the editing sites along the genes appears to be at random, with the possible exception of *nad4* where we found that all edting sites but one are situated in the first exon. We have, at the moment, no explanation for this peculiar distribution.

## CONCLUSION

The C to U editing phenomena, such as described here, raises a number of questions about the transmission of genetic information during gene expression. RNA editing in plant mitochondria implies that the gene sequence is not sufficient to predict the sequence of the corresponding protein. Data on mRNA and/or direct protein sequence data are also required. This raises the question of where is stored the information which is necessary to achiev the specificity of the process. Up to now, there is no evidence that either the primary sequence or potential secondary structures near the editing sites would play a role toward this specificity.

Considering the importance of RNA editing in the modulation of gene expression, studies are being initiated to understand the mechanisms involved and to determine at which stage of mRNA maturation this process takes place. Three different mechanisms could be active during RNA editing. First, deamination of position 4 of the C, as catalyzed by cytidine deaminase, could result in a U. Second, the whole base could be cut off without nicking the sugar-phosphate backbone and replaced. Third, the RNA chain could be cleaved, the C removed by a nuclease and the U added, a model similar to the one proposed for editing in *Trypanosoma*. (Blum *et al.*, 1990). We are currently setting up an *in vitro* test in order to study the mechanism of RNA editing in wheat mitochondria and to characterize the factors which are involved in the specificity of this process.

REFERENCES

Benne, R., Van den Burg, J., Brakenhoff, J., Sloof, P., Van Boom, J. H. and Tromp, M. C., 1986, Major transcript of the frameshift *coxII* from trypanosome mitochondria contains four nucleotides that are not encoded in the DNA. Cell 46, 819-826.

Blum, B., Balakara, N. and Simpson, L., 1990, A model for RNA editing in kinetoplastid mitochondria : small "guide RNA" molecules transcribed from maxicircle DNA provide the edited sequence information. Cell 60, 189-198.

Boer, P. H., McIntosh, J. E., Gray, M. W. and Bonen, L., 1985, The wheat mitochondrial gene for apocytochrome b: absence of a prokaryotic ribosome binding site. Nucleic Acids Res. 13, 2281-2292.

Bonen, L., Boer, P. H. and Gray, M. W., 1984, The wheat cytochrome oxidase subunit II gene has an intron insert and three radical aminoacid changes relative to maize. EMBO J 3, 2531-2536.

Bonen, L., Boer, P. H., McIntosh, J. E. and Gray, M. W., 1987, Nucleotide sequence of the wheat mitochondrial gene for subunit I of cytochrome oxidase. Nucleic Acids Res. 15, 6734.

Covello, P. S. and Gray, M. W., 1989, RNA editing in plant mitochondria. Nature 341, 662-666.

Fox, T. D. and Leaver, C. J., 1981, The *Zea mays* mitochondrial gene coding cytochrome oxydase subunit II has an intervening sequence and does not contain TGA codons. Cell 26, 315-323.

Gualberto, J., Lamattina, L., Bonnard, G., Weil, J. H. and Grienenberger, J. M., 1989, RNA editing in wheat mitochondria results in the conservation of protein sequences. Nature 341, 660-662.

Gualberto, J. M., Domon, C., Weil, J. H. and Grienenberger, J. M., 1990a, Structure and transcription of the gene coding for subunit 3 of cytochrome oxidase in wheat mitochondria. Curr. Genet. 17, 41-47.

Gualberto, J., Weil, J.H. and Grienenberger, J.M., 1990b, Structure and expression of a wheat mitochondrial transcription unit. In "Achievements and Perspectives of mitochondrial research", Quagliariello *et al.*, in press.

Gualberto, J. M., Wintz, H., Weil, J. H. and Grienenberger, J. M., 1988, The genes coding for subunit 3 of NADH dehydrogenase and for ribosomal protein S12 are present in the wheat and maize mitochondrial genomes and are co-transcribed. Mol. Gen. Genet. 215, 118-127.

Hiesel, R., Wissinger, B., Schuster, W. and Brennicke, A., 1989, RNA editing in plant mitochondria. Science 246, 1632-1634.

Lamattina, L., Weil, J. H. and Grienenberger, J. M. ,1989, RNA editing at a splicing site of NADH dehydrogenase subunit IV gene transcript in wheat mitochondria. FEBS Lett. 258, 79-83.

Maréchal, L., Guillemaut, P., Grienenberger, J. M., Jeannin, G. and Weil, J. H., 1985, Sequence and codon recognition of bean mitochondria and chloroplast $tRNAs^{Trp}$: evidence for a high degree of homology. Nucleic Acids Res. 13, 4411-4416.

Simpson, L. and Shaw, J., 1989, RNA editing and the mitochondrial cryptogenes of kinetoplastid protozoa. Cell 57, 355-366.

# TOPOLOGICAL ORIENTATION OF THE MEMBRANE PROTEIN URF13

K. L. Korth, F. Struck, C. I. Kaspi,* J. N. Siedow* and C. S. Levings III

Genetics Department, Box 7614, North Carolina State University
Raleigh, NC 27695-7614, USA

*Department of Botany, Duke University, Durham, NC 27706, USA

## Introduction

Maize carrying the Texas male-sterile cytoplasm (*cms-T*) codes for a mitochondrial gene designated T-*urf13* (Dewey et al., 1986). This unusual gene probably arose by mtDNA rearrangements caused by intramolecular recombinational events. T-*urf13* has been implicated with cytoplasmic male sterility (cms), a cytoplasmically inherited trait causing pollen abortion, and with susceptibility to two fungal pathogens. In maize and most higher plants, the mitochondrial genome is transmitted strictly through the egg. Analysis of *cms-T* revertants provides the most convincing evidence that T-*urf13* is responsible for the two traits (Rottman et al., 1987; Wise et al., 1987a). These studies showed that spontaneous deletions or a frame-shift mutation of the T-*urf13* gene result in reversion to both male-fertile and disease-resistant phenotypes. In addition, these investigations indicate that the two traits are caused by the same gene, T-*urf13*, although the possibility of two closely linked loci cannot be discounted. Moreover, the nuclear restorer gene, *Rf1*, which is one of two genes needed to suppress pollen sterility in *cms-T* maize, reduces the expression of the T-*urf13* encoded protein (Dewey et al., 1987).

T-*urf13* encodes a 13 kDa polypeptide (URF13) that is located in the inner mitochondrial membrane (Dewey et al., 1987; Wise et al., 1987b). Because the T-*urf13* gene is not found in other maize cytoplasms or in other plant species, URF13 must be unnecessary for normal mitochondrial function. It is unclear whether URF13 is associated with a particular electron transfer complex within the inner mitochondrial membrane or the $F_0$-$F_1$ ATP synthase. Earlier results suggested that URF13 could be associated with Complex IV (Ralph Dewey; Christopher Leaver, personal communications) or the ATPase (Wise et al., 1987b). More recently, S. Ferguson-Miller, W. Peiffer and K. Korth obtained results showing that URF13 can associate with several different complexes. In any event, there is no solid evidence indicating that URF13 is specifically bound to a particular inner mitochondrial membrane component.

## Disease Susceptibility and Toxin Sensitivity

In 1969 and 1970 an epidemic of Southern corn leaf blight struck the Southern and Corn Belt regions of the United States, only infesting maize carrying the Texas cytoplasm. At that time the Texas type of cms was widely used for hybrid corn production, and more than 80 percent of U.S. corn acreage carried the Texas cytoplasm. It was immediately evident that the Texas cytoplasm was directly related to the blight epidemic, and the use of *cms-T* for hybrid seed production was quickly abandoned. Southern corn leaf blight is caused by the fungal pathogen, *Bipolaris maydis*, race T, formerly known as *Helminthosporium maydis*, race T. A second fungal pathogen, *Phyllosticta maydis*, also

Plant Molecular Biology 2, Edited by R.G. Herrmann and
B. Larkins, Plenum Press, New York, 1991

uniquely infests maize carrying the Texas cytoplasm. This latter disease, commonly called yellow corn leaf blight, was less serious because the organism is restricted to the northern region of the United States.

The pathogens, *Bipolaris maydis*, race T, and *Phyllosticta maydis*, produce the pathotoxins, *BmT* and *Pm*, respectively. These pathotoxins affect the function of mitochondria from *cms-T* maize but not mitochondria from other maize cytoplasms or other plant species. Effects on mitochondria include inhibition of malate-supported state-3 respiration, stimulation of NADH-mediated state-4 respiration, organelle swelling, leakage of small molecules (e.g., $Ca^{++}$ and $NAD^{+}$) and uncoupling of oxidative phosphorylation (Miller and Koeppe, 1971; Gengenbach et al., 1973; Matthews et al., 1979; Berville et al., 1984; Klein and Koeppe, 1985; Holden and Sze, 1987). *BmT*- and *Pm*-toxin are structurally similar in that they contain repeating linear oxy-oxo or $\beta$-oxydioxo ketol structures, respectively, along a methylene backbone. The *BmT*-toxins range in length from 35 to 45 carbon units, whereas *Pm*-toxins are shorter, ranging from 16 to 24 carbons (Danko et al., 1984; Kono et al., 1985). Both *BmT*- and *Pm*-toxins retain their toxicity for *cms-T* mitochondria, even when their carbonyl groups are reduced with sodium borohydride (Frantzen et al., 1987). Methomyl [S-methyl-N-[(methylcarbamoyl) oxy] thioacetimidate], the active ingredient in Du Pont's systemic insecticide, Lannate, causes equivalent effects on *cms-T* mitochondria, even though it is structurally unlike the pathotoxins (Koeppe et al., 1978).

The fungal pathogen, *Bipolaris maydis*, race T, weakly colonizes normal maize and maize carrying the male-sterile cytoplasms S (*cms-S*) and C (*cms-C*). Colonization is generally limited to boat-shaped lesions on the leaves that do not grow together but remain small and isolated. Thus, *B. maydis*, race T, is normally not a serious pathogen on maize carrying these cytoplasms. In contrast, *B. maydis*, race T, can rapidly and extensively colonize *cms-T* maize. Lesions grow rapidly, coalesce, and spread throughout the plant, causing serious damage and even plant death. The evidence suggests that susceptibility of *cms-T* maize to *B. maydis*, race T, is due to the unique sensitivity of its mitochondria to *BmT* toxin. In contrast, mitochondria of resistant-maize types are insensitive to the *BmT* toxin. Susceptibility to *P. maydis* appears to have a similar basis except that the fungus produces a slightly different pathotoxin, *Pm*-toxin, to which *cms-T* mitochondria are also sensitive.

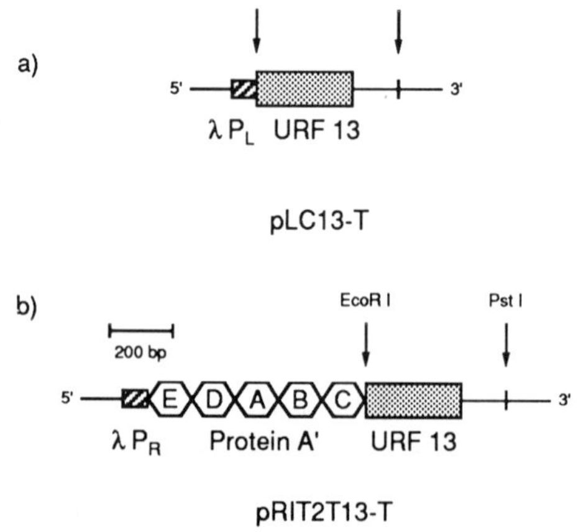


Fig. 1. Constructs for the thermoinducible production of URF13 in *E. coli*.
a) "wild type" URF13 was produced as described (Braun et al.,1989b).
b) Protein A':URF13 fusion product was expressed in the pRIT2T vector (Nilsson et al., 1985). O indicate the IgG binding domains of protein A.

We have demonstrated that T-*urf13* is responsible for pathotoxin sensitivity by showing that expression of URF13 in *Escherichia coli* (Fig. 1a) conveys toxin sensitivity to the bacterium (Dewey et al., 1988). Effects of *BmT*- and *Pm*-toxins and methomyl on *E. coli* are to inhibit glucose-driven respiration and to cause spheroplast swelling and ion leakage (Dewey et al., 1988; Braun et al., 1989a). *E. coli* not expressing URF13 are unaffected by the compounds. The toxic effects in *E. coli* are analogous to those observed in *cms-T* mitochondria. URF13 is a membrane protein in both organisms; it is located in the plasma membrane of *E. coli* and the inner mitochondrial membrane of maize.

We have shown by ion uptake experiments in *E. coli* that the pathotoxin-URF13 interaction causes massive leakage of small ions through the plasma membrane (Braun et al., 1989b). This result demonstrates that the interaction permeabilizes the plasma membrane and that URF13 is a channel-forming protein in the presence of toxin. A similar response to pathotoxin and URF13 is indicated in mitochondria from *cms-T* maize, where incubation of toxin with mitochondria causes rapid leakage of $Ca^{++}$ and $NAD^{+}$ (Holden and Sze, 1984). In *cms-T* mitochondria the interaction permeabilizes the inner mitochondrial membrane, resulting in the dissipation of the membrane potential and loss of mitochondrial function. Methomyl causes effects in *E. coli* and maize mitochondria identical to those produced by the fungal toxins.

The specificity of *BmT*- and *Pm*-toxin toward *cms-T* mitochondria suggests an interaction between URF13 and the toxins. Indeed, we have used reduced, labeled *Pm*-toxin ($^3$H-*Pm*-toxin) to show that toxin specifically binds to URF13 in *E. coli* and *cms-T* mitochondria (Braun et al., 1990). In *E. coli* URF13 binds toxin in a cooperative fashion (Hill coefficient, n = 1.5) with an apparent dissociation constant of 50-70 nM and at a level of 350 pmol of toxin/mg of *E. coli* protein. In *cms-T* maize mitochondria, no cooperativity is detected and much less toxin is bound (≈15 pmol of toxin/mg of mitochondrial protein). Competition and displacement experiments show that toxin binding is reversible and that *BmT*-toxin, *Pm*-toxin and methomyl all bind to either the same site or overlapping sites on URF13.

To learn more about how URF13 confers toxin sensitivity, we have prepared over 100 T-*urf13* mutations by random and site-directed mutational techniques. By testing these mutations for toxin and methomyl sensitivity in *E. coli*, many toxin-insensitive mutants were identified. Several of these mutants were analyzed to determine the specific amino acid changes and why these mutations no longer confer toxin sensitivity (Braun et al., 1990). Toxin-binding studies were used to determine whether toxin insensitivity is associated with changes in the capacity of URF13 to bind labeled toxin. Two insensitive mutants, a change from asp to val at residue 39 and a carboxyl-end deletion of 33 amino acids that produces a truncated URF13 of 82 amino acids, each exhibit a drastic reduction in toxin binding. Although these mutants do bind a little toxin, it is apparently insufficient to cause toxin-URF13-induced membrane permeabilization. It is evident that these mutations affect sites involved in binding because they greatly diminish the binding affinity of URF13 for toxin. A third toxin-insensitive mutation, an internal deletion lacking residues 2 through 11, however, binds substantial toxin (≈300 pmol/mg of protein), almost as much toxin as standard URF13. This result is important because it shows that toxin insensitivity is possible even with toxin binding. Although amino acids 2 through 11 are not required for toxin binding, they must contribute to the changes that bring about membrane permeabilization. Finally, our studies indicate that toxin-URF13 binding is essential for membrane permeabilization.

**Structure of URF13 in Membranes**

We are interested in the structure of URF13 and how the protein is localized in biological membranes because of its relationship to the role of URF13 as a channel-forming protein. URF13 is able to bind toxin and permeabilize membranes in both *cms-T* maize mitochondria and *E. coli*; therefore, these membranes must have common features relative to toxin sensitivity and URF13 function during channel formation. We have initiated studies with *E. coli* to investigate the orientation of URF13 because it is a more practical organism for these experiments, and URF13 is readily altered in *E. coli* to study effects of amino acid changes on URF13 in the membrane and toxin sensitivity.

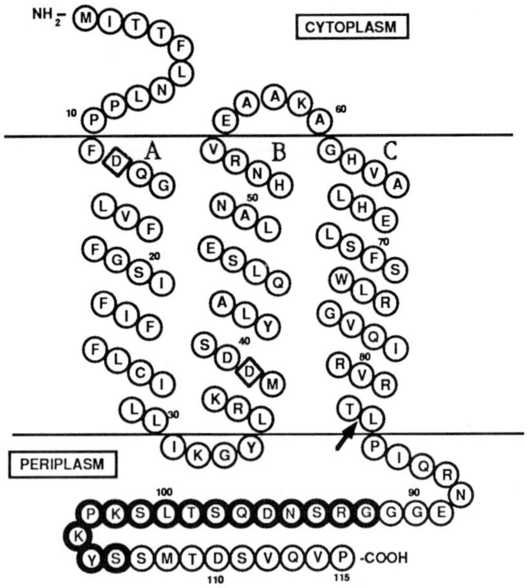


Fig. 2. Proposed arrangement of URF13 within the plasma membrane of *E. coli*. ◇ denote DCCD binding residues. Heavy circles indicate boundaries of the region recognized by antibody (Ab)-C. Three α-helices are designated A, B, and C, respectively.

Our preliminary experiments with *E. coli* indicate that URF13 is oriented as shown in Figure 2. In this model the amino terminus of URF13 is situated in the cytoplasm and the carboxyl end in the periplasmic space. The trans-membraneous localization of the N- and C-termini of URF13 indicates that the protein contains an odd number of membrane-spanning regions; the model shown in Figure 2 suggests that URF13 crosses the membrane three times. The carboxyl end of URF13 contains the binding site for an antibody designated Ab-C. Based on immunogens used in antibody preparation and deletion experiments of the carboxyl terminus of URF13, we know that the Ab-C binding site lies between residues 91 and 107. Therefore, Ab-C specifically binds to the carboxyl end of URF13 and is useful for marking this terminus. When right-side-out *E. coli* membrane vesicles are incubated with Ab-C, washed and subsequently reacted on dot blots with anti-mouse IgG conjugated to either alkaline phosphatase or $[^{125}I]$, heavy staining is detected. Because Ab-C has a specific affinity for the carboxyl terminus of URF13, this result indicates that the carboxyl terminus is available for binding Ab-C and is situated in the periplasmic space (Fig. 3a). When inside-out *E. coli* membrane vesicles are incubated with Ab-C, washed and stained on dot blots, a low level of reactivity, slightly above that of background, is observed. Low level staining is expected if the carboxyl-end of URF13 is located within inside-out vesicles and is unavailable for binding Ab-C. The low level of specific binding of Ab-C detected with inside-out vesicles can be attributed to the vesicle preparation not being 100% inside-out in its orientation. Thus, these experiments suggest that the carboxyl-terminus of URF13 is situated on the periplasmic side of the plasma membrane, as shown in Figure 2.

To determine the location of the amino-end of URF13, we have fused a part of the *Staphylococcus aureus* protein A gene in frame to the amino end of URF13 (Fig. 1b). When this construct is expressed in *E. coli*, it produces a polypeptide with the IgG binding region of protein A fused to URF13. Respiration in the resulting *E. coli* cells is completely inhibited by either toxin or methomyl. Because protein A binds IgG, we have effectively marked the amino-end of this construct for determining its topological orientation in the bacterial plasma membrane. When right-side-out and inside-out

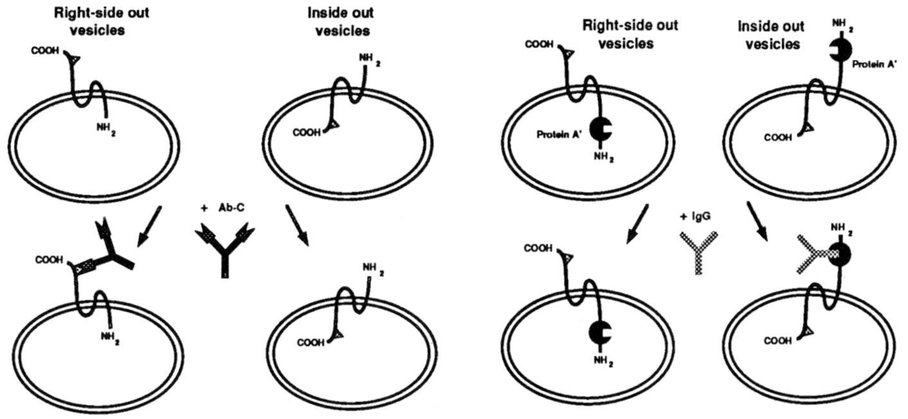


a) Ab-C, specific for the COOH region of URF13, binds to right-side out but not inside-out vesicles containing URF13. ▲ represents the epitope.

b) Non-specific IgG molecules bind at higher levels to inside-out vesicles than to right-side out vesicles containing the ProteinA':URF13 fusion polypeptide.

Fig. 3. Outline of the antibody binding experiments to determine the orientation of URF13 in the membrane. Right-side out vesicles were prepared according to Kaback (1979). Inside-out vesicles were produced as described by Reenstra et al. (1980).

membrane vesicles were incubated with [$^{125}$I]-labeled rabbit anti-mouse IgG, staining revealed that IgG bound more heavily to the protein A-URF13 fusion protein in inside-out vesicles than in right-side-out vesicles (Fig. 3b). This result suggests the amino-terminus is located on the cytoplasmic side of the *E. coli* plasma membrane.

Based upon our observations, we propose a model for the topological orientation of URF13 in the *E. coli* plasma membrane that contains three membrane-spanning segments (Fig. 2). We have no experimental evidence that precisely defines the three membrane-spanning amino acid regions. The actual membrane-spanning regions of URF13 may be longer, shorter or begin and end at different residues than shown in Figure 2. We have proposed that URF13 crosses the membrane three times to accommodate our data showing that the amino terminus of URF13 is located on the cytoplasmic side of the plasma membrane, and the carboxyl terminus is on the periplasmic side. The possibility that URF13 spans the membrane only once or alternatively five times, predicts excessively long or short membrane-spanning domains of URF13, respectively. DCCD binding experiments (Braun et al., 1989b), coupled with mutagenesis studies, reveal that aspartic acid residues at position 12 and 39 covalently bind DCCD. Because DCCD forms a stable adduct only with acidic residues located in a hydrophobic, nonaqueous environment, residues 12 and 39 are predicted to be within the membrane. The model in Figure 2 localizes both asp 12 and 39 within the membrane-spanning regions of helices A and B, respectively, and therefore either residue is expected to be present in a region of high hydrophobicity. Further, when the amino acid residues associated with helices B and C are projected onto an $\alpha$-helical wheel, both helices show considerable amphipathic character, and the only hydrophilic residue associated with the hydrophobic face of helix B is asp-39.

The appearance of cooperativity associated with the binding of toxin to URF13 suggested cross-linking studies to ascertain whether URF13 exists in an oligomeric state within the membrane. Reaction of *E. coli* cells expressing URF13 or isolated *cms-T* mitochondria with lysine-specific cross-linking reagents of varying degrees of hydrophobicity and length gave a distinct band on an immunoblot at a molecular weight corresponding to an URF13 dimer. The longer cross linkers also generated bands on immunoblots consistent with trimeric and tetrameric species. Hydrophobic cross-linkers showed greater reactivity, consistent with the localization of potentially reactive lysine residues within the hydrophobic phase of the membrane. Cross-linking in the presence of added toxin or methomyl did not alter the observed pattern suggesting that channel formation must involve a more subtle change in URF13 structure than the cross-linking reaction is able to resolve. Cross-linking experiments with *E. coli* cells that had been treated with DCCD, to render them insensitive to toxin (Holden and Sze, 1989), showed slightly increased accumulation of the higher molecular weight (trimer/tetramer) products, but the significance of this observation is unclear.

Results of these cross-linking studies indicate that some fraction of the URF13 is present in the membrane as a dimeric species. The cooperative nature of the binding of labeled toxin to URF13 in *E. coli* suggests that either each URF13 molecule binds more than one molecule of toxin or that more than one toxin molecule binds to an oligomeric URF13 structure, and binding of the first toxin molecule facilitates binding of subsequent toxin molecules. The latter explanation is more reasonable given that the product of the URF13-toxin interaction is a membrane channel sufficiently large to accommodate molecules up to 600 $M_r$ (Berville et al., 1984). It is difficult to envision a single URF13 molecule acting to produce such a channel. The current paradigm for the formation of hydrophilic channels within membranes by proteins involves the association of several amphipathic $\alpha$-helices in the membrane such that the polar faces of the helices orient inward to form the lining of the water-filled channel, whereas the hydrophobic faces point outward, interacting with the hydrocarbon phase of the lipid bilayer (Fox and Richards, 1982). Using such a model, we determined that an oligomeric structure consisting of 2-4 individual URF13 molecules may act, in the presence of added toxin, to produce a channel within the membrane. Additional studies are needed to establish the exact structural features associated with URF13 and its interaction with toxin.

**Acknowledgments**

This work was supported by grants from the National Science Foundation to C.S.L. and the U.S. Department of Energy to J.N.S., and by a fellowship from the North Carolina Biotechnology Center to K.L.K.

**REFERENCES**

Berville, A., Ghazi, A., Charbonnier, M., and Bonavent, J.-F., 1984, Effects of methomyl and *Helminthosporium maydis* toxin on matrix volume, proton motive force and NAD accumulation in maize (*Zea mays* L.) mitochondria, Plant Physiol., 76:508-517.

Braun, C. J., Siedow, J. N., and Levings III, C. S., 1989a, The T-*urf13* gene confers fungal toxin sensitivity to maize mitochondria and *E. coli*, in: "The Molecular Basis of Plant Development," vol. 92, R. B. Goldberg, ed., Alan R. Liss, New York, pp. 79-85.

Braun, C. J., Siedow, J. N., Williams, M. E., and Levings III, C. S., 1989b, Mutations in the maize mitochondrial T-*urf13* gene eliminate sensitivity to a fungal pathotoxin, Proc. Natl. Acad. Sci. USA, 86:4435-4439.

Braun, C. J., Siedow, J. N., and Levings III, C. S., 1990, Fungal toxins bind to the URF13 protein in maize mitochondria and *Escherichia coli*, The Plant Cell, 2:153-161.

Danko, S. J., Kono, Y., Daly, J. M., Suzuki, Y., Takeuchi, S., and McCrery, D. A., 1984, Structure and biological activity of a host-specific toxin produced by the fungal corn pathogen *Phyllosticta maydis*, Biochemistry, 23:759-766.

Dewey, R. E., Levings III, C. S., and Timothy, D. H., 1986, Novel recombinations in the maize mitochondrial genome produce a unique transcriptional unit in the Texas male-sterile cytoplasm, Cell, 44:439-449.

Dewey, R. E., Timothy, D. H., and Levings III, C. S., 1987, A mitochondrial protein associated with cytoplasmic male sterility in the T cytoplasm of maize, Proc. Natl. Acad. Sci. USA, 84:5374-5378.

Dewey, R. E., Siedow, J. N., Timothy, D. H., and Levings III, C. S., 1988, A 13-kilodalton maize mitochondrial protein in *E. coli* confers sensitivity to *Bipolaris maydis* toxin, Science, 239:293-295.

Fox, R. O., and Richards, F. M., 1982, A voltage-gated ion channel model inferred from the crystal structure of alamethicin at 1.5-Å resolution, Nature, 300:325-330.

Frantzen, K. A., Daly, J. M., and Knoche, H. W., 1987, The binding of host-selective toxin analogs to mitochondria from normal and "Texas" male sterile cytoplasm maize, Plant Physiol., 83:863-868.

Gengenbach, B. G., Miller, R. J., Koeppe, D. E., and Arntzen, C. J., 1973, The effects of toxin from *Helminthosporium maydis* (race T) on isolated corn mitochondria: swelling, Can. J. Bot., 51, 2119-2125.

Holden, M. J., and Sze, H., 1984, *Helminthosporium maydis* T toxin increased membrane permeability to $Ca^{2+}$ in susceptible corn mitochondria, Plant Physiol., 75:235-237.

Holden, M. J., and Sze, H., 1987, Dissipation of the membrane potential in susceptible corn mitochondria by the toxin of *Helminthosporium maydis*, race T, and toxin analogs, Plant Physiol., 84:670-676.

Holden, M. J., and Sze, H., 1989, Effects of *Helminthosporium maydis* race T toxin on electron transport in susceptible corn mitochondria and prevention of toxin actions by dicyclohexylcarbodiimide, Plant Physiol., 91:1296-1302.

Kaback, H. R., 1979, Bacterial membranes, Methods in Enzymol., 22:99-120.

Klein, R. R., and Koeppe, D. E., 1985, Mode of methomyl and *Bipolaris maydis* (race T) toxin in uncoupling Texas male-sterile cytoplasm corn mitochondria, Plant Physiol., 77:912-916.

Koeppe, D. E., Cox, J. K., and Malone, C. P., 1978, Mitochondrial heredity: a determinant in the toxic response of maize to the insecticide methomyl, Science, 201:1227-1229.

Kono, Y., Takeuchi, S., Kawarada, A., Daly, J. M., and Knoche, H. W., 1980, Structure of the host-specific pathotoxins produced by *Helminthosporium maydis*, race T, Tetra. Lett., 21:1537-1540.

Matthews, D. E., Gregory, P., and Gracen, V. E., 1979, *Helminthosporium maydis* race T toxin induces leakage of $NAD^+$ from T cytoplasm corn mitochondria, Plant Physiol., 63, 1149-1153.

Miller, R. J., and Koeppe, D. E., 1971, Southern corn leaf blight: susceptible and resistant mitochondria, Science, 173:67-69.

Nilsson, B., Abrahmsen, L., and Uhlen, M., 1985, Immobilization and purification of enzymes with staphylococcal protein A fusion vectors, EMBO J., 4:1075-1080.

Reenstra, W. W., Patel, L., Rottenberg, H., and Kaback, H. R., 1980, Electrochemical proton gradient in inverted membrane vesicles from *E. coli*, Biochemistry, 19:1-9.

Rottman, W. H., Brears, T., Hodge, T. P., and Lonsdale, D. M., 1987, A mitochondrial gene is lost via homologous recombination during reversion of CMS T maize to fertility, EMBO J., 6:1541-1546.

Wise, R. P., Pring, D. R., and Gengenbach, B. G., 1987a, Mutation to male fertility and toxin insensitivity in Texas (T)-cytoplasm maize is associated with a frameshift in a mitochondrial open reading frame, Proc. Natl. Acad. Sci. USA, 84:2858-2862.

Wise, R. P., Fliss, A. E., Pring, D. R., and Gengenbach, B. G., 1987b, *urf13-T* of T cytoplasm maize mitochondria encodes a 13 kD polypeptide, Plant Mol. Biol., 9:121-126.

# CYTOPLASMIC MALE STERILITY IN PETUNIA

Maureen R. Hanson*, Marie B. Connett*, Otto Folkerts#, Shamay Izhar+, Susan M. McEvoy*, Helen T. Nivison*, Kim D. Pruitt*

*Section of Genetics and Development, Cornell University, Ithaca, NY 14853 USA; #Dow Chemical Company, 1701 Building, Midland, MI 48674 USA; +ARO Volcani Center, Plant Genetics and Breeding, PO Box 6, Bet Dagan, Israel

A maternally-inherited male sterile phenotype is known in many plant genera, including *Petunia* (Laser and Lersten, 1972; Hanson and Conde, 1985). As all seed on a cytoplasmic male sterile (CMS) plant must result from cross-pollination, the trait has attracted commercial interest for hybrid seed production. Problems with exploitation of the CMS trait, however, include its lack of natural occurrence in certain important crop species as well as unwanted phenotypic "side effects" of the CMS cytoplasm in combination with particular nuclear backgrounds.

The molecular basis of the disruption of pollen development remains unknown. The phenomenon has been under close scrutiny because of several intriguing questions. How can an organelle DNA encode a disruption specifically in pollen development? Why is this trait so commonly found in many plant genera? How do nuclear genes, known as fertility restorers, modify the impact of the organelle DNA to produce normal pollen development?

Substantial progress toward understanding the cause of pollen disruption has been made in *Petunia* and maize, in which abnormal loci that are genetically correlated with the CMS trait have been identified (Boeshore et al., 1985; Dewey et al., 1986; Abbott and Fauron, 1986; Rottman et al., 1987). Information concerning CMS that was derived from *Petunia* will be the focus of this chapter.

In *Petunia*, all natural CMS lines (i.e. those not produced following protoplast fusion) are likely to be derived from a single source. Male sterile plants appeared in a breeding field over forty years ago and were found to exhibit maternal inheritance of the trait (discussed in Izhar, 1984). Unfortunately, because of loss of records it is not clear whether the sterility trait arose spontaneously, was induced by an interspecific cross, or was revealed by interspecific crosses in which nuclear restorer genes segregated from a CMS-encoding cytoplasm present in a wild species (see Hanson and Conde, 1985). The lack of a known progenitor line to existing CMS lines has handicapped genetic analysis of CMS in *Petunia* and in most other genera studied. On the other hand, the ability to produce *Petunia* lines containing recombinant mitochondrial genomes has been extremely valuable for studying CMS. Another valuable genetic resource is a single dominant nuclear gene *Rf* which can confer fertility in plants

*Plant Molecular Biology 2*, Edited by R.G. Herrmann and B. Larkins, Plenum Press, New York, 1991

carrying the CMS cytoplasm. Useful comparisons can be made between fertility-restored and sterile plants carrying the CMS cytoplasm.

While all the natural *Petunia* CMS lines should contain identical organelle DNAs, additional CMS lines that contain novel mitochondrial genomes have been produced following protoplast fusion. Recombination of mitochondrial genomes following protoplast fusion has been well documented, especially in the Solanaceae (reviewed by Hanson, 1984). Some of the existing culture-produced CMS lines are diploid cybrids, apparently containing primarily the nuclear genome of one parental species and recombinant mitochondrial DNAs, while others are tetraploid somatic hybrids in which both parents contributed nuclear genetic information. Izhar's laboratory has produced a collection of *Petunia* cybrids and somatic hybrids containing one or the other parental chloroplast DNAs and novel recombinant mitochondrial DNAs (Izhar et al., 1983; Boeshore et al., 1983; Clark et al., 1985).

## THE SEARCH FOR THE CODING LOCATION OF CMS IN *PETUNIA*

Because both chloroplast and mitochondrial DNAs are maternally inherited in dicots and monocots, the initial search for the coding location of CMS was at the organelle DNA level. Protoplast fusion experiments implicated mitochondrial DNA. The CMS trait segregated independently of the parental chloroplast DNA in somatic hybrids produced from CMS and fertile lines of *Petunia*, *Nicotiana* and *Brassica* (Belliard et al.; 1978; Pelletier et al.; 1983; Clark et al.; 1985).

A region of mitochondrial DNA was correlated with the CMS trait in *Petunia* when a Bgl II restriction fragment length polymorphism (RFLP) was found to segregate with the CMS trait in a population of fertile and CMS somatic hybrid plants produced by protoplast fusion (Boeshore et al., 1985). The CMS parental line and 17 CMS somatic hybrids contained 2 Bgl II restriction fragments lacking from the fertile parent and 24 fertile somatic hybrids.

The mitochondrial DNA region encompassing the CMS-associated RFLP was cloned from the CMS parental line used in the protoplast fusion experiment (Boeshore et al., 1985). Initial sequencing revealed the *Petunia* CMS associated fused (*pcf*) gene (Young and Hanson, 1987) and subsequent sequencing downstream revealed two additional genes (Rasmussen and Hanson, 1989; Hanson et al., 1989): subunit three of NADH dehydrogenase (*nad3*) and small subunit ribosomal protein 12 (*rps12*).

## STRUCTURE OF THE *PETUNIA* CMS-ASSOCIATED REGION

The diagram shown in Fig. 1 presents the genomic DNA sequence of the CMS-associated region, now termed the S-*pcf* locus because of the presence of three different genes. In order to better comprehend the structure of this locus, homologous regions were sequenced from both the CMS and the fertile parental line used in the somatic hybridization experiment.

The 5' transcribed region of the S-*pcf* region and the first 35 codons of the *pcf* gene are identical to a normal *atp*9 gene present in natural CMS lines (Young et al., 1986; Young and Hanson, 1987). Within the non-coding transcribed region is a short region of high similarity to the chloroplast *atp*B gene (K. Wolf, pers. comm.). Such integrations of foreign DNA, especially chloroplast DNA, into higher plant mitochondrial DNA have been well documented (Stern and Lonsdale, 1982; Schuster and Brennicke, 1988). Because the chloroplast DNA sequence is present in both the normal *atp*9 gene and the S-*pcf* locus, this integration event likely preceded the recombination events that produced the *pcf* gene.

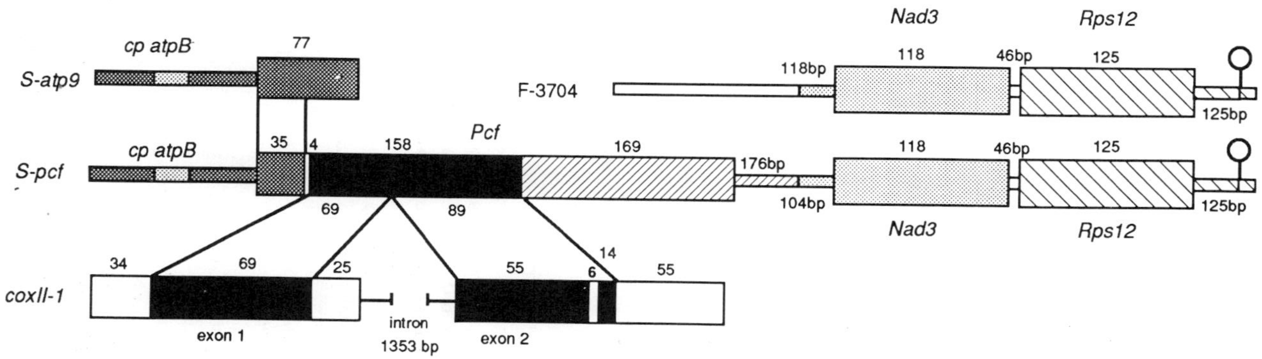


Fig. 1. Diagram of the S-*pcf* locus and homologous mitochondrial genes, according to genomic DNA sequence data. Large boxes indicate coding regions, narrow boxes indicate flanking regions. The number of codons is indicated above the coding region segment; intergenic regions are indicated in base pairs (bp). S-*atp9* is a normal *atp9* gene present in mtDNA of CMS lines. The sequence of the *coxII-1* gene is derived from line 3704, the fertile parent used in the protoplast fusion experiment. Hybridization, mapping, and transcription data indicates a similar gene exists in the mtDNA of the CMS line. The sequences of the *nad3* and *rps12* genes in fertile line 3704 diverge from the S-*pcf* locus 118 bp upstream of the coding region.

Following the region similar to *atp9* is an in-frame fusion of portions of the first and second exon of *coxII* (Young and Hanson, 1987). At the time this sequence was obtained, it was not known that the normal *Petunia coxII* gene carried an intron; the *coxII* sequence of another dicot (*Oenothera*) lacked an intron (Hiesel and Brennicke, 1983). When a *Petunia coxII* gene was sequenced and found to contain a group II intron homologous to the monocot *coxII* introns (Fox and Leaver, 1981; Pruitt and Hanson, 1989), the origin of the *coxII* region of *pcf* was reconsidered.

The absence of the intron in the *coxII* sequences present in *pcf* immediately suggested that these sequences could have been derived from a spliced RNA, which underwent a reverse transcription event into DNA, followed by integration into mitochondrial genomic DNA. However, 75 bp of the first exon are missing as well as the intron. Examining the sequence of the normal *coxII* gene at the location of the deletion in *pcf* revealed a potential cryptic splice site (Pruitt and Hanson, 1989). If this cryptic splice site is sometimes used during RNA processing, the intron as well as about 75 bp could be spliced out of transcripts. Recombination of the reverse-transcribed abnormally-spliced sequence with *atp9* and *urf-S* sequences would produce *pcf* (Fig. 2). A short region of similarity between *atp9* and *coxII*, in the vicinity of the *atp9-coxII* junction in *pcf*, has been detected (Pruitt and Hanson, 1989). Because no progenitor *urf-S* sequence is known, we cannot determine whether or not there is also similarity between *coxII* and an ancestral *urf-S* sequence.

The origin and identity of *urf-S*, the third component of *pcf*, remains unknown, despite numerous database searches. Recently we have discovered a sequencing error in the 3' portion of *urf-S* reported by Young and Hanson (1987). An additional A results in a frame-shift that extends the reading frame of *urf-S* to 169 codons instead of the 157 codons previously described. This correction also changes the predicted molecular weight of the encoded protein from 38 to 39 kD.

The coding region of *nad3* begins 281 bp downstream of *pcf*, followed closely by the coding region of *rps12*. These two genes are also closely linked in wheat and *Oenothera*, though the *Oenothera rps12* gene is not intact (Gualberto et al., 1989; Schuster et al., 1990). For comparison, the *nad3* and *rps12* genes present in a fertile *Petunia* line were sequenced (Rasmussen and Hanson, 1989). The amino acid sequences predicted from the CMS and fertile *nad3* DNA sequences do not differ; the predicted sequences exhibit one conservative amino acid change in *rps12*. Thus, the coding regions of *nad3* and *rps12* in the CMS line appear to be normal. However, the upstream region of *nad3* and *rps12* in the fertile line diverges completely 118 bp 5' to *nad3*; no *pcf* gene is present.

A number of recombination events, certainly at least five, must have been involved in the creation of the S-*pcf* locus. It is important to distinguish the recombination events which created the S-*pcf* locus from the recombination events which produced the novel mitochondrial genomes in the somatic hybrid plants. The S-*pcf* locus was not created by recombination of mitochondrial genomes following protoplast fusion; it was already present in the CMS parental line. Recombination in somatic hybrids merely served to produce many novel mitochondrial genomes, in which the parental S-*pcf* locus was conserved in all somatic hybrids which were male sterile.

The diagrams of the genes shown in Fig. 1 are all based on genomic DNA sequences. With the recent discovery of RNA editing in plant mitochondria, RNAs and therefore protein sequences cannot be predicted from knowledge of the DNA sequence. Edits reported to date alter Cs into Us (Gualberto et al., 1989; Covello and Gray, 1989; Hiesel et al., 1989), except for one report of a U to C edit (Schuster et al., 1990). Conversion of a genomic C

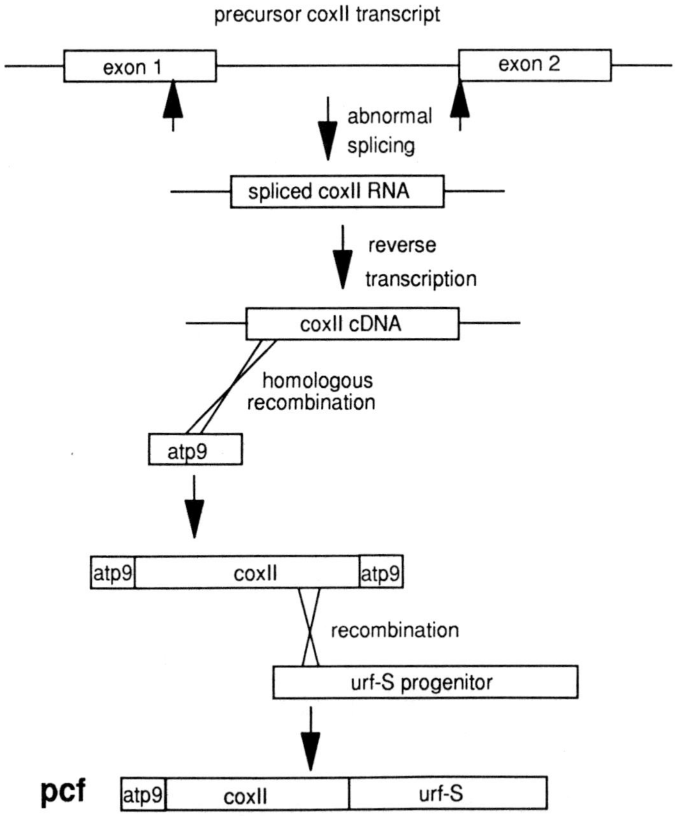


Fig. 2. Model for the origin of the *pcf* gene. Recognition of a cryptic splice site in exon 1 could result in an abnormal spliced RNA. Following reverse transcription, the intronless *coxII* DNA could recombine with a homologous region in an *atp9* gene, producing an *atp9-coxII* recombinant. This recombinant region could in turn recombine with an *urf-S* progenitor gene, to give *pcf*. Although for the purpose of clarity, single cross-overs are shown in the order *atp9* to *coxII* to *urf-S*, a double crossover event could have occurred instead, or *urf-S* could have recombined with *coxII* before recombining with *atp9*.

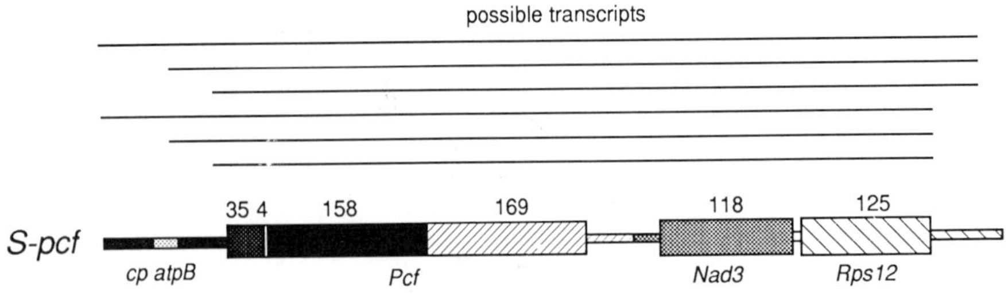


Fig. 3. Possible transcripts of the S-*pcf* locus in suspension culture cells. Three 5' termini and two 3' termini have been identified by S1 nuclease mapping studies.

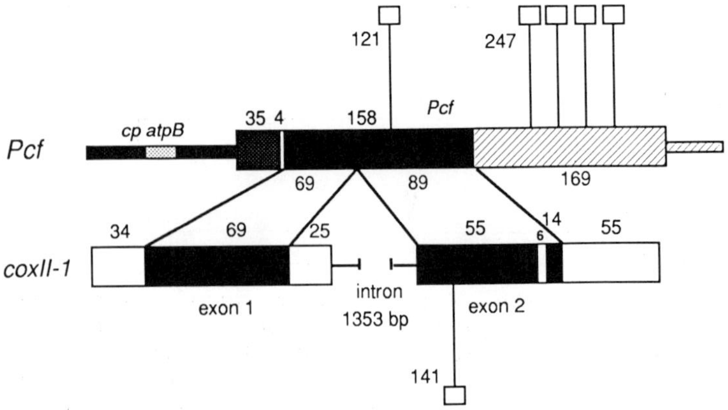


Fig. 4. Location of the sequences used to predict synthetic peptides for antibody production. The COXII synthetic peptide sequence begins at codon 141 of the *coxII-1* gene and corresponds to codon 121 of the *pcf* gene. The URF-S synthetic peptide sequence is tandemly repeated four times in the *pcf* gene; the first repeat begins at codon 247.

to a U in the RNA may alter the predicted amino acid and create start and stop codons. We have detected C to U edits in RNAs of *atp9*, *coxII*, and *pcf* (K. Pruitt, H. Wintz, S. McEvoy, unpublished).

## EXPRESSION OF THE S-*PCF* LOCUS

### Transcripts

Transcript mapping of both *atp9* and S-*pcf* by S1 nuclease protection revealed three 5' transcript termini at identical positions in both genes (Young et al., 1986; Young and Hanson, 1987). In suspension culture RNA of CMS lines, the transcripts terminating closest to the start codon of these genes are the most abundant. S1 protection experiments also revealed that *pcf*, *nad3*, and *rps12* are co-transcribed (Rasmussen and Hanson, 1989; Hanson et al., 1989). More detailed analysis of the 3' transcript termini has revealed the presence of two termini (Pruitt and Hanson, in preparation; Fig. 3). Thus, there could be as many as six transcripts of S-*pcf* differing in length in suspension culture RNA (Fig. 3). However, whether transcripts terminating at all three 5' points also exhibit both 3' termini is not known.

Transcript analysis has also revealed large differences in abundance of S-*pcf* transcripts in plants carrying different nuclear backgrounds (data not shown). Nuclear effects on mitochondrial transcripts have been documented in other plant species (Kennell et al., 1987, 1989). Thus, in order to test whether a nuclear restorer gene may affect an S-*pcf* transcript, it is essential to utilize lines that are isonuclear except for the restorer locus. Unfortunately, there were no isonuclear lines carrying *Rf* in the same nuclear background as the parental lines used in the somatic hybridization experiments. We have undertaken a backcrossing program to introduce *Rf* into the CMS parental line nuclear background. After several backcrosses, the abundance of S-*pcf* transcripts in the *Rf* line is similar to the CMS parent (data not shown). Further analysis of transcripts, comparing different tissues and *Rf* lines, is in progress.

### Polypeptides

In order to produce antibody probes for a protein encoded by the *pcf* gene, two synthetic peptides predicted from the genomic DNA sequence were used to immunize rabbits (Nivison and Hanson, 1987, 1989). The synthetic peptides represented portions of the *coxII* and *urfS* portions of the *pcf* gene (Fig. 4).

Antisera to the COXII-specific synthetic peptide recognized COXII proteins on Western blots of protein from CMS and fertile *Petunia* and other species (Nivison and Hanson, 1989 and unpublished; K. Newton, pers. comm.). However, no signals specific to *Petunia* CMS lines were observed (Nivison and Hanson, 1989).

Unlike the anti-COXII antiserum, antiserum to the URF-S-specific synthetic peptide did detect a protein present in the CMS parental line that was absent from the normal fertile line (Fig. 5). This 25-kd protein was present in CMS somatic hybrids, but missing from fertile somatic hybrids examined (Nivison and Hanson, 1989). The 25-kd protein is much reduced in abundance in lines containing at least one *Rf* allele (Fig. 5). In addition to recognizing the 25-kd CMS-specific protein, the anti-URF-S antiserum recognizes a 20-kd protein which is present in both fertile and CMS lines (Fig. 5).

Cellular and mitochondrial fractionation experiments revealed that the 20 and 25-kd proteins are located in the mitochondria. According to fractionation and protease protection experiments, the 20-kd protein is an integral membrane protein (Fig. 5 and data not shown), while the 25-kd polypeptide exhibits properties characteristic of proteins loosely associated with membranes (Fig. 6). It remains also possible that the 25-kd protein is found in both the matrix and membrane fractions of mitochondria *in vivo*.

The observation that the 25-kd protein is synthesized by isolated mitochondria (Nivison and Hanson, 1989) provided further evidence that the 25-kd protein is encoded by the *pcf* gene. However, a 39-kd protein is predicted from the genomic DNA sequence. Processing of a precursor protein into the 25-kd product would explain both our immunological and molecular weight data. In this model, removal of the N-terminal portion of a full-length pcf-encoded protein by cleavage in or after the COXII synthetic peptide region (Fig. 4) would leave a polypeptide of apparent molecular weight 25-kd recognized by anti-URF-S antibody, but not by anti-COXII antibody.

While we have identified a protein associated with CMS in *Petunia*, we do not yet know its function. Our observation that the nuclear restorer gene affects the abundance of the protein, however, does provide an important second correlation of the *pcf* gene with the sterility trait. Additional fractionation experiments and further characterization of the actual sequence of the 25-kd protein may aid in formulating models. Information from physiological studies, described below, may also produce testable hypotheses.

## RESPIRATION, THE CMS PHENOTYPE, AND THE S-*PCF* LOCUS

Efforts to identify respiratory phenotypes that correlate with CMS in various plant species have met with limited success. At present, there is no respiratory trait known to be common to CMS in more than one species; in other words, there is no physiological aberration known that can explain why pollen developmental disruption is so frequently found as a maternally inherited trait. Either no universal abnormality in mitochondrial function leads to CMS, or the universal trait awaits detection.

We undertook a comparison of respiratory function in *Petunia* CMS and fertile lines because the *pcf* gene included segments similar to an ATP synthase and a cytochrome oxidase subunit. One hypothetical effect of the 25-kd *pcf*-encoded product could be an alteration in cytochrome oxidase complex activity. In higher plants, reducing equivalents may be oxidized either through the cyanide-sensitive pathway or through the cyanide-resistant alternative pathway (Palmer, 1976; Lance et al., 1985; Fig. 7). Therefore, we performed comparative measurements of the cytochrome oxidase and the alternative oxidase pathway.

The activity of the two pathways can be separately measured by inhibiting each one specifically; for example, the cytochrome oxidase is CN-sensitive and SHAM-resistant, while the alternative oxidase is CN-resistant and SHAM-sensitive (Møller et al., 1988). Some residual activity remains after treating cells or mitochondria with both inhibitors.

Initially, we measured oxygen uptake (Estabrook, 1967) of suspension cultures of CMS and fertile lines in the presence of the cytochrome oxidase and alternative oxidase-specific inhibitors. Our results were exactly opposite from the prediction of inhibition of cytochrome oxidase by the PCF polypeptide. Cytochrome oxidase activity was usually higher, not lower, in suspension cultures of CMS than in fertile lines (Connett and Hanson, in press). The frequently lower activity of cytochrome oxidase in fertile

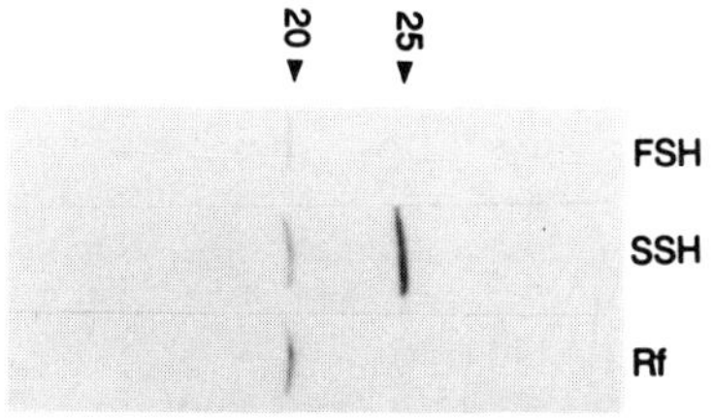


Fig. 5. Immunoblot probed with anti-URF-S peptide antibody. Equal amounts of total mitochondrial protein from fertile (FSH) and sterile (SSH) somatic hybrids and from a fertility-restored *Petunia* line (Rf) were fractionated on an LDS-polyacrylamide gel, transferred to nitocellulose, and probed with the anti-URF-S antibody. The locations of the 25-kD and 20-kD proteins are indicated.

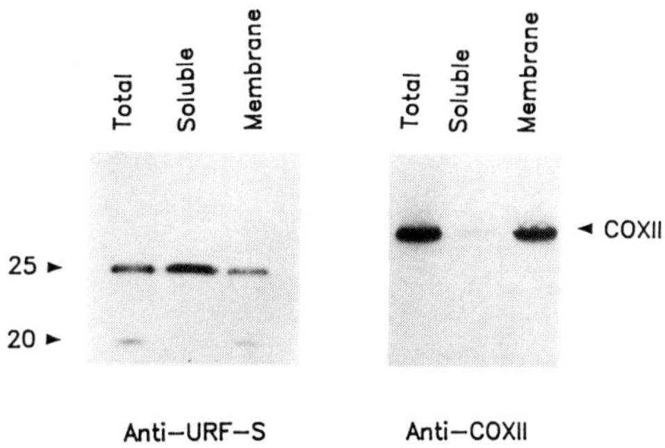


Fig. 6. Fractionation of mitochondrial proteins from a CMS somatic hybrid. Mitochondria were resuspended in hypotonic medium, disrupted in a Dounce homogenizer, and sonicated, and the soluble and membrane fractions separated by centrifugation. Aliquots of each fraction, derived from 100 ug of intact mitochondria, were electrophoresed and transferred to nitrocellulose. Half of the protein blot was probed with the anti-URF-S peptide antibody, and the other half was probed with an anti-yeast COXII monoclonal antibody. (Reprinted from Nivison and Hanson, 1989, *Plant Cell* 1:1121-1130, with permission).

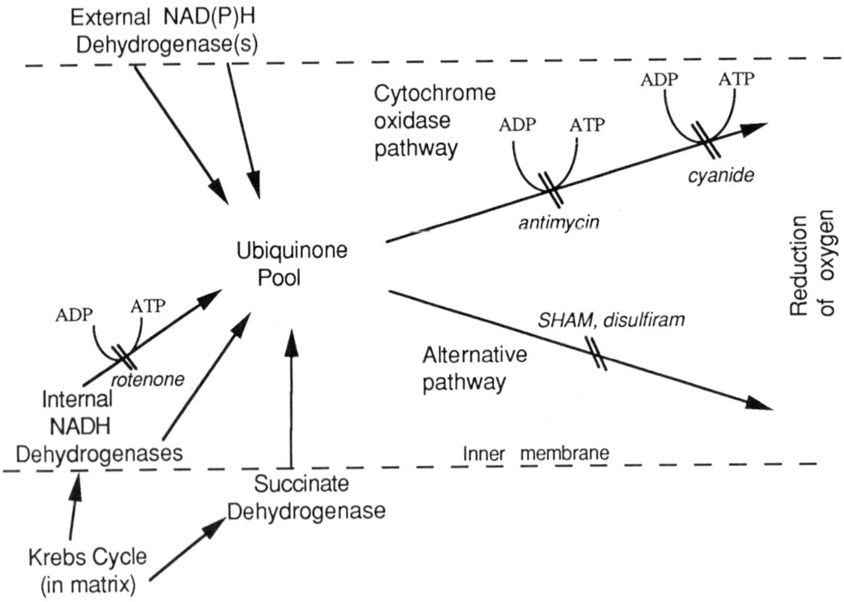


Fig. 7. Mitochondrial electron transport pathways and their inhibitors in higher plants.

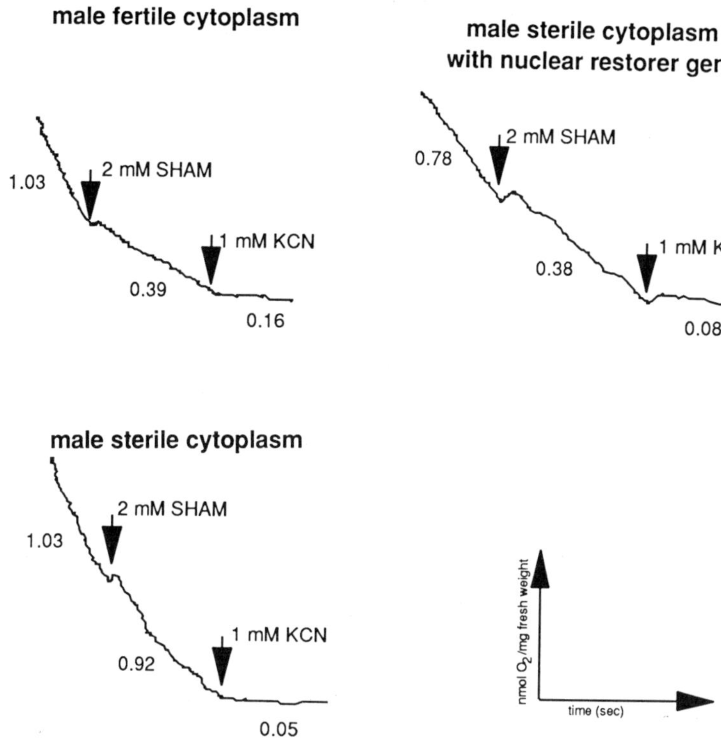


Fig. 8. Representative oxygen electrode traces. These traces show the depletion of oxygen from the medium by the respiration of a suspension of *Petunia* cultured cells. Traces were made using a Clark oxygen electrode, equilibrated at 25°C, and 40 mg suspension cells, measured three to four days after transfer into fresh medium. Units are nmol $O_2$/mg fresh weight/minute. Additions of KCN or SHAM were made as shown.

lines in comparison to CMS lines could not be explained by either substrate saturation or to inhibition by high concentrations of ATP, according to experiments testing these hypotheses (Connett and Hanson, in press). The cytochrome oxidase activity of suspension cultures of different CMS and fertile lines exhibited high variance, and our attention turned instead to the activity of the alternative pathway.

Table 1. Alternative oxidase activity in *Petunia* suspension cultures (% total respiration sensitive to 2mM SHAM minus residual respiration)

| | |
|---|---|
| Isonuclear Pair 1 | |
| CMS | 23.5 ± 18.6 |
| fertile | 88.6 ± 12.5 |
| Isonuclear Pair 2 | |
| CMS | 19.9 ± 12.6 |
| fertile | 60.0 ± 4.8 |
| Isonuclear Pair 3 | |
| CMS | 11.0 ± 1.0 |
| fertile | 51.0 ± 24.0 |
| Restored line 1 | 53.1 ± 8.5 |
| Restored line 2 | 58.0 ± 21.0 |

We analyzed three sets of isonuclear CMS and fertile lines so that any differences in alternative oxidase activity could be ascribed to the cytoplasm rather than to nuclear genes. The activity of the alternative oxidase was estimated as the percentage of total control oxygen uptake sensitive to SHAM minus the percentage of residual oxygen uptake. Suspension cultures of CMS lines measured three days after transfer exhibited consistently lower alternative oxidase activity than fertile lines (Fig. 8 and Table 1). These results were confirmed with mitochondria isolated from suspension cultures, and with other specific inhibitors of the two pathways, such as antimycin and disulfuram (Connett and Hanson, in press). Recently, these results have been extended with suspension cultures of CMS and fertile somatic hybrids (Connett, Hanson, and Izhar, in preparation).

Analysis of suspension cultures of restored lines was particularly intriguing. The nuclear restorer gene confers not only fertility, but also restores normal alternative pathway activity (Table 1). The nuclear restorer lines provide a link between fertility, alternative oxidase activity, and abundance of the 25-kd protein. The simplest interpretation of our data is that both lowered alternative oxidase activity and sterility is encoded by the S-*pcf* locus. We cannot, however, rule out the possibility that *Rf* modifies the expression of both the S-*pcf* locus and a second locus that encodes altered alternative oxidase activity.

Our data indicate that CMS lines are not completely lacking in alternative oxidase activity. Also, low activity is not characteristic of all tissues. When we examined respiration of suspension cultures at time-points immediately after transfer into fresh medium to 7 days after transfer, we found that both CMS and fertile lines initially exhibit high alternative oxidase activity, but that activity in the cells of the CMS line decreases greatly over time, unlike the fertile line (Connett and Hanson, in press). Also, no statistically significant difference in electron transport could be detected between leaf tissues of CMS and fertile lines, but immature anthers exhibited the same lowered activity in CMS vs. fertile lines (Connett and Hanson, in press). Thus the relative activity of the two pathways may be differentially regulated temporally and spatially in CMS and fertile lines.

Could a reduction in alternative oxidase activity result in male sterility in *Petunia*? If this pathway is essential at a critical time during pollen development, then a modification of its activity could be detrimental. In further speculation, we may point out that there is no evidence that can rule out low alternative oxidase activity as a "universal" CMS phenotype present in all species where CMS lines are known. If an aberration in alternative oxidase were normally confined to a small number of cells in a developing anther of a CMS plant, typical physiological measurements would not detect it. On the other hand, the CMS-specific low alternative pathway activity in *Petunia* may be a peculiarity of this genus, and not have any significance with regard to CMS in *Petunia* or in other species. This respiratory trait may simply be correlated with, but not causative of sterility.

## CMS IN *PETUNIA* VS. MAIZE

In order to summarize the present state of knowledge of the molecular mechanism of CMS, the information gained from the *Petunia* and the maize T-CMS systems will be briefly considered below. A more thorough review has been published elsewhere (Hanson et al., 1989). Though several other promising systems are under analysis, the most knowledge concerning CMS to date has been gathered from *Petunia* and the maize T genotypes. In both cases, a mitochondrial DNA region has been genetically correlated with CMS. In maize, the correlation of CMS to the region we will term T-*mcr* could be made by examining changes in mtDNAs that were present in fertile revertants (Fauron et al., 1987; Wise et al., 1987a; Rottman et al., 1987). As in *Petunia*, a protein product of the locus has been identified immunologically and decreases greatly in abundance in fertility-restored lines (Dewey et al., 1987; Wise et al., 1987b).

The components of the *Petunia* and the maize CMS-associated loci do not have any obvious functional similarities (Fig. 9). However, the loci do have several features in common: both contain an abnormal gene comprised of pieces of normally independent genes, and both have an apparently normal mitochondrial gene downstream and co-transcribed with the abnormal gene. In the case of maize, the normal downstream coding region is *orf25*, a gene of unknown function which has been found in the mitochondrial genome of several different species (Stamper et al., 1987; Folkerts and Hanson, 1989). Like the downstream *nad3* and *rps12* genes in *Petunia*, the *orf25* gene downstream of the abnormal fused gene is the only copy of the gene in the mitochondrial genome of the CMS line.

In both maize and petunia, the molecular basis of CMS has not been identified. Although present data implicate the CMS-associated loci, we do not know whether the presence of the fused genes and their expression causes CMS directly through its protein products, or indirectly by adversely affecting the expression of the downstream genes. At the moment, improper expression of the downstream genes is as reasonable a hypothesis for the mechanism of pollen disruption as is production of abnormal proteins encoded by *pcf* and *urf13T*.

One effect of the *urf13T* gene product has, however, been conclusively demonstrated. Maize plants carrying the T cytoplasm have long been known to be sensitive to the toxin produced by a fungal pathogen (Miller and Koeppe, 1971). The *urf13T* gene product binds toxin and confers toxin sensitivity when expressed in *E. coli* and yeast, and therefore presumably in maize (Dewey et al., 1988; Huang et al., 1990; Braun et al., 1990).

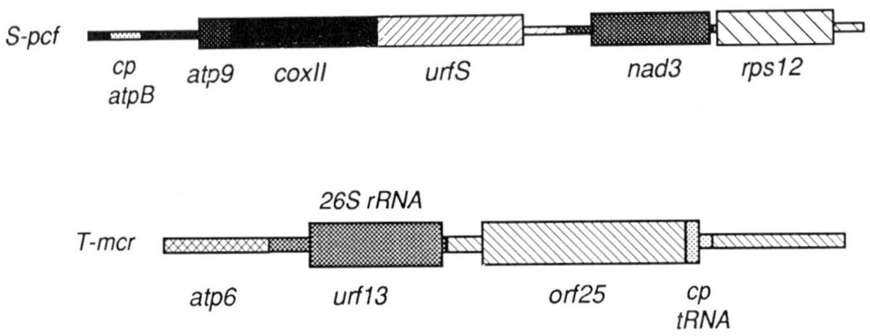


Fig. 9. Diagram of the CMS-associated *Petunia* S-*pcf* locus and the maize T-*mcr* locus.

In addition to pollen developmental disruption, both maize T and *Petunia* CMS have secondary phenotypes, toxin sensitivity and low alternative oxidase, respectively. As discussed above, the maize T toxin-sensitivity has definitely been shown to result from the protein-level expression of the CMS-associated abnormal fused gene. Whether the altered alternative oxidase phenotype in *Petunia* similarly results from expression of the fused gene awaits experimental testing.

ACKNOWLEDGEMENTS

Research in the authors' laboratories was supported by the U.S.D.A. Competitive Grants Program (Genetic Mechanisms), the McKnight Foundation, Hatch Project New York grant DE-FG02-89ER10430-186418, the U.S. Department of Energy Biosciences Program, the U.S.-Israel BARD Program, and the Cornell Biotechnology Program, which is sponsored by the New York State Science and Technology Foundation, a consortium of industries, and the U.S. Army Research Office. We thank Linda Narde for assistance with manuscript preparation.

## BIBLIOGRAPHY


Abbott, A.G., and Fauron, C.M.R., 1986, Structural alterations in a transcribed region of the T type cytoplasmic male sterile maize mitochondrial genome, **Curr. Genet.**, 10:777-783.

Belliard, G., Pelletier, G., Vedel, F., and Quetier, F., 1978, Morphological characteristics and chloroplast DNA distribution in different cytoplasmic parasexual hybrids of *Nicotiana tabacum*, **Molec. Gen. Genet.**, 165:231-237.

Boeshore, M.L., Lifshitz, I., Hanson, M.R., and Izhar, S., 1983, Novel composition of mitochondrial genomes in *Petunia* somatic hybrids derived from cytoplasmic male sterile and fertile plants, **Mol. Gen. Genet.**, 190:459-467.

Boeshore, M.L., Hanson, M.R., and Izhar, S., 1985, A variant mitochondrial DNA arrangement specific to *Petunia* stable sterile somatic hybrids, **Pl. Mol. Biol.**, 4:125-132.

Braun, C.J., Siedow, J.N., and Levings, C.S., III, 1990, Fungal toxins bind to the URF13 protein in maize mitochondria and *Escherichia coli*, **Plant Cell**, 2:153-161.

Clark, E.M., Izhar, S., and Hanson, M.R., 1985, Independent segregation of the plastid genome and cytoplasmic male sterility in *Petunia* somatic hybrids, **Mol. Gen. Genet.**, 199:440-445.

Connett, M.B. and Hanson, M.R., Electron transport through the cyanide-sensitive and cyanide-insensitive pathways in isonuclear lines of cytoplasmic male sterile and male fertile *Petunia*, **Plant Physiol.**, in press.

Covello, P.S., and Gray, M.W., 1989, RNA editing in plant mitochondria, **Nature**, 341:662-666.

Dewey, R.E., Levings, C.S., III, and Timothy, D.H., 1986, Novel recombinations in the maize mitochondrial genome produce a unique transcriptional unit in the Texas male-sterile cytoplasm, **Cell**, 44:439-449.

Dewey, R.E., Timothy, D.H., and Levings, C.S., III, 1987, A mitochondrial protein associated with cytoplasmic male sterility in the T cytoplasm of maize, **Proc. Natl. Acad. Sci. USA**, 84:5374-5378.

Dewey, R.E., Siedow, J.N., Timothy, D.H., and Levings, C.S., III, 1988, A 13-Kilodalton maize mitochondrial protein in *E. coli* confers sensitivity to *Bipolaris maydis* toxin, **Science**, 239:293-295.

Estabrook, R.W., 1967, Mitochondrial respiratory control and the polarographic measurement of ADP:O rations, **Methods Enzymol.**, 10:41-47.

Fauron, C.M.-R., Abbott, A.G., Brettel, R.I.S., and Gesteland, R.F., 1987, Maize mitochondrial DNA rearrangements between the normal type, the Texas male sterile cytoplasm, and a fertile revertant cms-T regenerated plant, **Curr. Genet.**, 11:339-346.

Folkerts, O., and Hanson, M.R., 1989, Three copies of a single recombination repeat occur on the 443 kb mastercircle of the *Petunia hybrida* 3704 mitochondrial genome, **Nucl. Acids Res.**, 17:7345-7357.

Fox, T.D., and Leaver, C.J., 1981, The *Zea mays* mitochondrial gene coding cytochrome oxidase subunit II has an intervening sequence and does not contain TGA codons, **Cell**, 26:315-323.

Gualberto, J.M., Lamattina, L., Bonnard, G., Weil, J.H., and Grienenberger, J.M., 1989, RNA editing in wheat mitochondria results in the conservation of protein sequences, **Nature** 341:660-662.

Hanson, M.R., 1984, Stability, variation and recombination in plant mitochondrial genomes via tissue culture and somatic hybridization, **Oxford Surv. Plant Mol. Cell Biol.**, 1:33-52.

Hanson, M.R., and Conde, M.F., 1985, Functioning and variation of cytoplasmic genomes: lessons from cytoplasmic-nuclear interactions affecting male fertility in plants, **Int. Rev. Cytol.**, 94:213-267.

Hanson, M.R., Pruitt, K.D., and Nivison, H.T., 1989, Male sterility loci in plant mitochondrial genomes, **Oxford Surveys of Plant Molecular & Cell Biology**, 6:61-85.
Hiesel, R., and Brennicke, A., 1983, Cytochrome oxidase subunit II gene in mitochondria of *Oenothera* has no intron, **EMBO J.**, 2:2173-2178.
Hiesel, R., Wissinger, B., Schuster, W., and Brennicke, A., 1989, RNA editing in plant mitochondria, **Science**, 246:1632-1634.
Huang, J., Lee, S.H., Lin, C., Medici, R., Hack, E., and Myers, A.M., 1990, Expression in yeast of the T-URF13 protein from Texas male-sterile maize mitochondria confers sensitivity to methomyl and to Texas-cytoplasm-specific fungal toxins. **EMBO**, 9:339-347.
Izhar, S., Schlicter, M., and Swartzberg, D., 1983, Sorting out of cytoplasmic elements in somatic hybrids of *Petunia* and the prevalence of the heteroplasmon through several meiotic cycles, **Mol. Gen. Genet.**, 190:468-474.
Izhar, S., 1984, Male Sterility in *Petunia*, **in**: "Monographs on Theoretical and Applied Genetics: *Petunia*," K.C. Sink, ed., Springer-Verlag.
Kennell, J.C., and Pring, D.R., 1989, Initiation and processing of *atp6*, T-*urf13*, and *ORF221* transcripts from mitochondria of T cytoplasm maize, **Mol. Gen. Genet.**, 216:16-24.
Kennell, J.C., Wise, R.P., and Pring, D.R., 1987, Influence of nuclear background on transcription of a maize mitochondrial region associated with Texas male sterile cytoplasm, **Mol. Gen. Genet.**, 210:399-406.
Lance, C., Chauveau, M., and Dizengremel, P., 1985, The cyanide-resistant pathway of plant mitochondria, **in**: "Encyclopedia of Plant Physiology 18: Higher Plant Cell Respiration," R. Douce and D.A. Day, eds., Springer-Verlag, Heidelberg, FRG.
Laser, K.D., and Lersten, N.R., 1972, Anatomy and cytology of microsporogenesis in cytoplasmic male sterile angiosperms, **Bot. Rev.**, 38:425-454.
Miller, R.J., and Koeppe, D.E., 1971, Southern corn leaf blight: susceptible and resistant mitochondria, **Science**, 173:67-69.
Møller, I.M., Berczi, A., van der Plas, L.H.W., and Lambers, H., 1988, Measurement of the activity and capacity of the alternative pathway in intact plant tissues: Identification of problems and possible solutions, **Physiol. Plantarum**, 72:642-649.
Nivison, H.T., and Hanson, M.R., 1987, Production and purification of synthetic peptide antibodies, **Plant Mol. Biol. Repr.**, 5:295-309.
Nivison, H.T., and Hanson, M.R., 1989, Identification of a mitochondrial protein associated with cytoplasmic male sterility in *Petunia*, **Plant Cell**, 1:1121-1130.
Palmer, J.M., 1976, The organization and regulation of electron transport in plant mitochondria, **Ann. Rev. Plant Physiol.**, 27:113-157.
Pelletier, G., Primard, C., Vedel, F., Chetrit, P., Remy, R., Rousselle, P., and Renard, M., 1983, Intergeneric cytoplasmic hybridization in *Cruciferae* by protoplast fusion, **Molec. Gene. Genet.**, 191:244-250.
Pruitt, K.D., and Hanson, M.R., 1989, Cytochrome oxidase subunit II sequences in *Petunia* mitochondria: two intron-containing genes and an intron-less pseudogene associated with cytoplasmic male sterility, **Curr. Genet.**, 16:281-291.
Rasmussen, J., and Hanson, M.R., 1989, A NADH dehydrogenase subunit gene is co-transcribed with the abnormal *Petunia* mitochondrial gene associated with cytoplasmic male sterility, **Mol. Gen. Genet.**, 215:332-336.
Rottmann, W.H., Brears, T., Hodge, T.P., and Lonsdale, D.M., 1987, A mitochondrial gene is lost via homologous recombination during reversion of CMS T maize to fertility, **EMBO J.**, 6:1541-1546.
Schuster, W., and Brennicke, A., 1988, Interorganellar sequence transfer: plant mitochondrial DNA is nuclear, is plastid, is mitochondrial, **Plant Science**, 54:1-10.

Schuster, W., Wissinger, B., Unseld, M., and Brennicke, A., 1990b, Transcripts of the NADH-dehydrogenase subunit 3 gene are differentially edited in *Oenothera* mitochondria, **EMBO J.**, 9:263-269.

Schuster, W., Hiesel, R., Wissinger, B., and Brennicke, A., RNA editing in the cytochrome *b* locus of the higher plant *Oenothera* includes a U-to-C transition, **Molecular and Cellular Biology**, in press.

Stern, D.B., and Lonsdale, D.M., 1982, Mitochondrial and chloroplast genomes of maize have a 12-kilobase DNA sequence in common, **Nature**, 229:698-702.

Stamper, S.E., Dewey, R.E., Bland, M.M., and Levings, C.S., III, 1987, Characterization of the gene *urf13-T* and an unidentified reading frame, *ORF25*, in maize and tobacco mitochondria, **Curr. Genet.**, 12:457-463.

Wise, R.P., Fliss, A.E., Pring, D.R., and Gengenbach, B.G., 1987a, *Urf13-T* of T cytoplasm maize mitochondria encodes a 13kD polypeptide, **Plant Mol. Biol.**, 9:121-126.

Wise, R.P., Pring, D.R., and Gengenbach, B.G., 1987b, Mutation to male fertility and toxin insensitivity in Texas (T)-cytoplasm maize is associated with a frameshift in a mitochondrial open reading frame, **Proc. Natl. Acad. Sci. USA**, 84:2858-2862.

Young, E.G., Hanson, M.R., and Dierks, P.M., 1986, Sequence and transcription analysis of the *Petunia* mitochondrial gene for the ATP synthase proteolipid subunit, **Nucleic Acids Research**, 14:7995-8006.

Young, E.G., and Hanson, M.R., 1987, A fused mitochondrial gene associated with cytoplasmic male sterility is developmentally regulated, **Cell**, 50:41-49.

# NUCLEAR AND CHLOROPLAST GENES INVOLVED IN THE EXPRESSION OF SPECIFIC CHLOROPLAST GENES OF CHLAMYDOMONAS REINHARDTII

Jean-David Rochaix, Michel Goldschmidt-Clermont, Yves Choquet, Michael Kuchka and Jacqueline Girard-Bascou[+]

Departments of Molecular Biology and Plant Biology
University of Geneva
Geneva, Switzerland
[+]Institut de Biologie Physico-Chimique, Paris, France

## INTRODUCTION

It is well documented that the photosynthetic multi-molecular complexes of the thylakoid membranes of higher plants and eukaryotic algae consist of subunits encoded by two distinct genetic systems localized in the nucleus and chloroplast. Some subunits of the photosynthetic complexes are encoded by chloroplast DNA and translated on chloroplast 70S ribosomes. Others are encoded by nuclear structural genes, translated on cytoplasmic 80S ribosomes and imported into the chloroplast. Both types of subunits associate with each other to form functional complexes in the thylakoid membrane. During the past years a large number of the nuclear and chloroplast genes of these subunits have been cloned and their structure has been determined. Although the expression of some of these genes has been studied, little is known about their control of expression, in particular about the requirement of nuclear encoded factors for the expression of specific chloroplast genes.

The green unicellular alga Chlamydomonas reinhardtii appears to be a powerful model system for studying this problem. First, its photosynthetic function is dispensable when cells are provided with a reduced carbon source such as acetate. This allows one to easily isolate and maintain photosynthetic mutants. Second C. reinhardtii can be manipulated with ease both at the genetic and biochemical level (cf. Harris, 1989). Third, reliable and efficient transformation methods have recently been established both for the nuclear (Debuchy et al., 1989; Kindle et al., 1989; Kindle, 1990, Mayfield and Kindle, 1990) and chloroplast compartments (Boynton et al. 1988).

We have used C. reinhardtii for studying defined aspects of chloroplast gene expression. Analysis of mutants deficient in either photosystem II (PSII) or photosystem I (PSI) activity has revealed a large number of mutations that affect the expression of specific chloroplast genes either at the level of RNA processing (splicing), RNA stability, translation, protein turn-over or complex assembly. Most of these mutations are nuclear. Here we review some of our recent work on mutants affected in RNA processing and RNA stability.

*Plant Molecular Biology* 2, Edited by R.G. Herrmann and
B. Larkins, Plenum Press, New York, 1991

## UNUSUAL STRUCTURE AND EXPRESSION OF THE psaA GENE OF C. REINHARDTII

The three chloroplast genes psaA, psaB and psaC encode three polypeptides that are part of the PSI reaction center (for review cf. Knaff, 1988). Other chloroplast encoded small molecular-weight PSI polypeptides have recently been identified (Scheller et al. 1989; Ikeuchi et al., 1990). At least 6 nuclear encoded PSI subunits have been found in _C. reinhardtii_ (Girard et al. 1980). With exception of subunit II the genes of the other five subunits have been cloned and sequenced (Franzen et al. 1989a, 1989b). The function of these nuclear encoded subunits is still unknown, especially since no mutant of _C. reinhardtii_ affected in any of these subunits has yet been identified. Fig. 1 presents a scheme of PSI biosynthesis in _C.reinhardtii_.

The psaA gene of _C. reinhardtii_ consists of three exons that are widely separated on the chloroplast genome (fig. 1; Kück et al., 1987). Exon 1 and exon 2 are 50 kb apart while exon 2 and exon 3 are 90 kb apart. Exon 2 is cotranscribed with psbD, the gene which encodes the D2 reaction center polypeptide of photosystem II (fig. 2, Choquet et al. 1988).

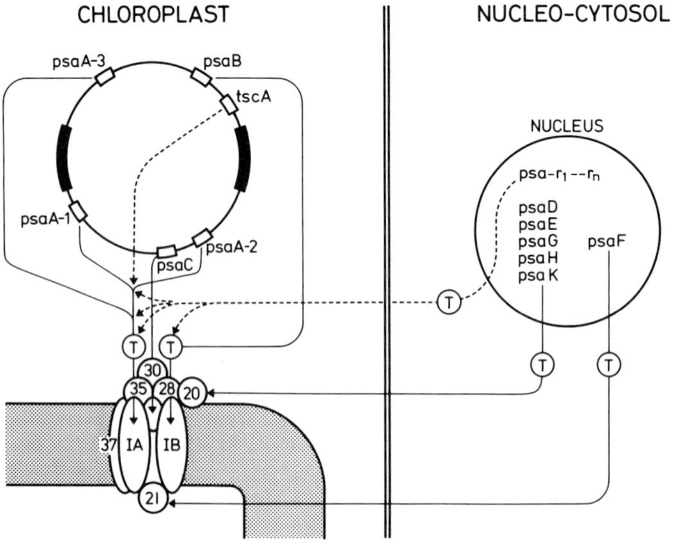


Fig. 1. Schematic view of the biosynthesis of photosystem I in the thylakoids. The chloroplast and nucleo-cytosolic compartments are shown on the left and right, respectively and separated by the chloroplast envelope. Chloroplast DNA is shown as a circle with the two segments of the inverted repeat (black boxes). The locations of the three exons of psaA (psaA-1, -2, 3-), psaB, psaC and tscA are indicated. The subunits of PSI are drawn in the thylakoid membrane (lower left). Subunits marked by arrows are encoded by chloroplast genes and subunits 20, 21, 28, 30, 35 and 37 are encoded by the nuclear genes psaD, psaF, psaH, psaE, psaG and psaK, respectively. T indicates the site of translation on the chloroplast (left) and cytoplasmic (right) ribosomes. psa-$r_1 \ldots r_n$ represent nuclear genes required for the expression of the chloroplast genes of PSI subunits and whose products act at the level of chloroplast RNA processing or translation.

The three psaA exons are flanked by the consensus sequences of group II introns (Kück et al., 1987; Michel and Dujon, 1983) and both psaA introns are split into a 5' and 3' portion. Since exon 1 and exon 2 are oriented in opposite direction on the chloroplast genome, one can exclude a model involving continuous transcription of the chloroplast genome followed by cis-splicing (Choquet et al. 1988). It is therefore clear that transcription of psaA is discontinous and that synthesis of mature message very likely depends on trans-splicing reactions. A fork structure of intron 1 typical for a trans-splicing reaction has indeed been identified recently (cf. fig 2; Choquet et al. unpublished results). In higher plant chloroplasts, psaA is a continuous gene (Fish et al., 1985). However rps12, the chloroplast gene encoding ribosomal protein S12 is disrupted and synthesis of its mature mRNA also appears to involve trans-splicing (Fukusawa et al. 1986; Tonegawa et al. 1986).

To understand the molecular mechanisms that underlie the maturation of the psaA mRNA, we have performed a combined molecular-genetic analysis. A first surprise was that close to one third of the PSI mutants examined are affected in the maturation of psaA mRNA. These mutants can be grouped into three classes A, B and C based on their phenotype with regard to RNA accumulation (fig. 2). Mutants of class A are unable to splice exon 2 and 3 transcripts, but are able to splice exon 1 and 2 transcripts. These mutants accumulate separately an exon 1-exon 2 precursor RNA and an exon 3 transcript. Mutants of class B are unable to perform both splicing reactions and accumulate separate transcripts of exon 1, exon 2 and exon 3. Mutants of class C are unable to splice the transcripts of exons 1 and 2, but are able to perform the second trans-splicing reaction. These mutants accumulate separate transcripts of exon 1 and exon 2-exon 3 (fig. 2). These results indicate that each psaA exon is transcribed independently from the others. The observation that class A mutants are only able to splice the transcripts of exon 1

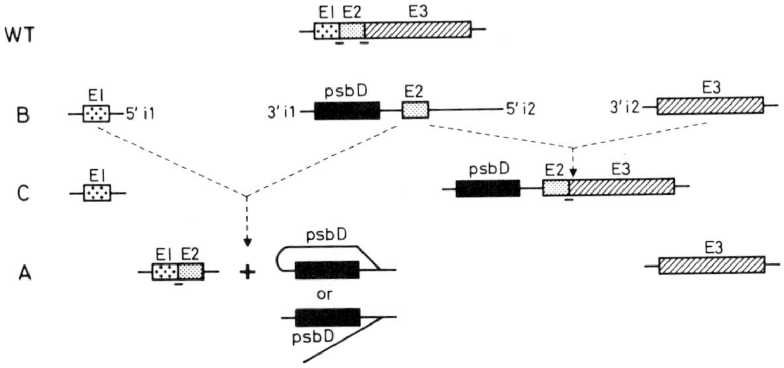


Fig. 2. Three classes of psaA splicing mutants. E1, E2 and E3 represent the transcripts of the three exons of psaA. Each of these exons is transcribed into a precursor RNA (Choquet et al. 1988). Note that psbD is cotranscribed with exon 2. Because of the dispersion of the exons along the chloroplast genome both introns i1 and i2 are disrupted. The 5' and 3' ends of the introns are indicated. Mutants from class B are deficient in both trans-splicing reactions. Class C mutants are unable to splice exons 1 and 2 but can splice exons 2 and 3. Class A mutants are unable to splice exons 2 and 3, but can splice exons 1 and 2. Class A mutants and wild-type cells also accumulate intron 1 (containing psbD).

Table I. Mutants defective in psaA mRNA maturation

| | Splicing defect | location | number of loci |
|---|---|---|---|
| Class A | exons 2-3 | nuclear | 5 |
| Class B | exons 1-2,2-3 | nuclear | 2 |
| Class C | exons 1-2 | nuclear<br>chloroplast | 7<br>1 |

and exon 2 while class C mutants are only able to splice the transcripts of exon 2 and exon 3 suggests that the two splicing reactions can occur in either order and that there are at least two pathways for psaA mRNA maturation.

A second surprise was that these mutations fall into a large number of complementation groups. Classes A, B and C include as least 5,2 and 7 nuclear complementation groups respectively (Table I, Goldschmidt-Clermont et al., 1990). While classes A and B consist at present exclusively of nuclear mutants, class C also includes a group of chloroplast mutants (Table I). The effect of these mutations appears to be specific as the mRNA maturation of other interrupted genes that contain group I introns proceeds normally. However it is not yet known whether other genes with group II introns exist in the C. reinhardtii chloroplast genome. A third surprise was that these chloroplast mutations are deletions that map in a region of the chloroplast genome that is distinct from any of the psaA exons as first shown by Roitgrund and Mets (1990). We have isolated similar chloroplast mutants, Fud3 and H13, with similar deletions (Choquet et al., 1988; Goldschmidt-Clermont et al., 1990). Together these results indicate the presence of a chloroplast trans-acting factor that is required for the first trans-splicing reaction of psaA. Interestingly, all these mutants have a double phenotype: besides being deficient in PSI activity, they are yellow in the dark (Roitgrund and Mets,1990; Choquet et al., unpublished results). In contrast to higher plants, C. reinhardtii is capable of synthesizing chlorophyll in the dark and therefore the wild-type remains green in the dark.

To identify the gene of the chloroplast trans-acting factor (tscA) we have used the newly developed biolistic chloroplast transformation procedure (Boynton et al., 1988). This method uses a particle gun to bombard cells of C. reinhardtii with DNA-coated tungsten particles. Both chloroplast class C mutants, Fud3 and H13, recovered PSI activity upon bombardment with a plasmid containing the 5 kb EcoRI fragment R12 that covers a large portion of the deletion in those mutants. Southern hybridizations confirm that the intact R12 fragment is indeed present in these transformants. Using plasmids containing various subfragments of R12 we have mapped tscA within a region of 700 bp. Sequencing and analysis of this region has not revealed any significant open reading frame. Northern hybridizations with a probe from the tscA region indicates the presence of a small RNA. Experiments are in progress to test whether this RNA is indeed involved in the first trans-splicing reaction of psaA. Possibly, this RNA may basepair with the 5' and 3' parts of the first discontinuous intron of psaA. Whether it also associates with protein factors to form RNP-like structures remains to be seen.

Why does psaA have such an unsual structure ? While there is no obvious answer it may be appropriate to consider how such a gene might have originated. In all higher plants examined and in at least one cyanobacterium (Fish et al., 1985; Kirsch et al., 1986; Lehmbeck et al., 1986; Ohyama et al., 1986; Shinozaki et al., 1986; Cantrell and Bryant, 1987), the psaA gene is continuous and unlike in C. reinhardtii it is closely linked to psaB which encodes the other homologous photosystem I reaction center polypeptide. In higher plants or in C. reinhardtii the similarity between the psaA and the psaB gene products is less (~45%) than that between the psaA or the psaB products from different organisms (>80%) (Kück et al., 1987; Fish et al., 1985). The sequence homology between psaA and psaB could be explained by a gene duplication event which must have occurred before the algal and higher plant branches diverged.

Two possibilities can be considered for the origin of the discontinuous gene structure of psaA in C. reinhardtii. The first is that it reflects an ancient gene structure. In trypanosomes, which are ancient organisms, only trans-splicing but no cis-splicing has been observed (Laird, 1989). In this first scheme duplication of the discontinuous gene might have occurred through some mechanism involving reverse transcription of the messenger and reintegration in the genome, generating a second continuous copy. In the higher plant branch, as well as in cyanobacteria, the split psaA gene might also have been subsequently converted to a continous gene through a similar mechanism involving cDNA intermediates. Loss of introns in mitochondrial genes by this type of mechanism has been demonstrated in yeast (Gargouri et al., 1983).

The second possibility is that the split psaA gene structure of C. reinhardtii occurred later, after the higher plant and algal branches diverged. Since the exons of psaA are bordered by group II intron sequences and since the latter have been implicated in transposition (Michel et Lang, 1985; Dujon et al; 1986), it is possible that the introns mediated the dispersal of the psaA exons. This hypothesis would imply that most factors required for trans-splicing were present when the psaA exons were dispersed.

## NUCLEAR GENES AFFECT THE STABILITY OF CHLOROPLAST TRANSCRIPTS

We have recently described a nuclear mutant of C. reinhardtii nac-2-26 that is deficient in PSII activity because it lacks the D2 polypeptide and that is therefore unable to assemble a stable PSII complex (Kuchka et al., 1989). Analysis of this mutant has revealed that although the psbD gene is transcribed as in wild-type, the psbD transcript is rapidly degraded. This transcript turn-over appears to be specific since the chloroplast transcripts of other PSII genes accumulate normally in this mutant. Since exon 2 of psaA is cotranscribed with psbD (fig. 2) it was of interest to determine the fate of the psaA exon 2 transcript in nac-2-26. In this mutant the psaA mRNA accumulates as in wild-type (Kuchka et al., 1989). This may be due to the fact that the psbA exon 2 precursor transcript is unaffected by the mutation in nac-2-26. Alternatively splicing of the transcripts of exon 1 and exon 2 may proceed significantly faster than the turnover of the psbD transcript in nac-2-26. To test these possibilities a nac-2-26 class B double mutant was constructed. In this mutant the psbD exon 2 precursor does no longer accumulate indicating that the second alternative is correct (Kuchka et al., 1989). This result shows that the increased turnover of the psbD transcript in nac-2-26 also occurs in the absence of a free 3'end of psbD mRNA.

The 3' non translated regions of several higher plant chloroplast transcripts have been shown to contain small inverted repeats that appear to play a role in RNA processing and RNA stability (Stern and Gruissem, 1988). Several proteins have been shown to bind to these regions (Stern et al. 1989). In some cases transcript specific factors have been identified. It therefore seemed of interest to examine whether the fusion of the 3'untranslated region of the psbD mRNA to a reporter RNA could destabilize the chimeric transcript in a nuclear nac-2-26 background. A DNA fragment of 250 bp containing the coding region of the 28 terminal amino acids of D2 and the entire 3' psbD untranslated region was fused a few nucleotides downstrean of the atpB coding region (Fig. 3). This construct was used to transform FuD50, a chloroplast mutant that has a deletion removing the 3' half of atpB (Woessner et al., 1986). Photosynthetic proficient cells were recovered, total RNA was isolated and hybridized with an atpB specific probe. The chimeric transcript was as expected slightly large than the wild-type atpB message and accumulated to the same level (fig. 3). Several transformants (mating type, $mt^+$) were crossed to nac-2-26 ($mt^-$) to determine the level of the chimeric atpB-psbD transcript in a nac-2-26 nuclear background. In such a cross all the offspring in the tetrads will usually inherit the chloroplast genome from the $mt^+$ parent whereas the nuclear mutation will segregate 2:2. As seen in fig. 3 the chimeric RNA accumulated to a similar extent as in wild-type. This experiment shows that the 3'region of psbD is not sufficient to destabilize the RNA in a nac-2-26 nuclear background where the psbD transcript no longer accumulates. Apparently other regions of the psbD transcript are required for promoting degradation of the RNA under these conditions.

The specific destabilization of the psbD mRNA by a single nuclear mutation in nac-2-26 is not a unique case. Similar nuclear mutations affecting the mRNAs of psbB (coding for a PSII polypeptide; Jensen et al., 1986; Monod et al., unpublished results) and of rbcL (coding for the large subunit of ribulose 1,5 bisphosphate carboxylase; Beasley and Mets, unpublished results) have been reported (cf. Table II).

## MULTIPLE NUCLEAR FACTORS ARE REQUIRED FOR THE EXPRESSION OF SPECIFIC CHLOROPLAST GENES

The picture which emerges from these studies is that a surprisingly large number of nuclear encoded factors are involved in the expression of specific chloroplast genes.

Table 2 lists the nuclear mutants affected in the expression of defined chloroplast genes either at the level of RNA processing, RNA turnover or translation. An important point, which has not yet been established is to determine whether these nuclear encoded factors play a regulatory role or whether they are required for the constitutive expression of these chloroplast genes.

The presence of these multiple factors does not appear to be an oddity of <u>C. reinhardtii</u>. Similar findings have been reported for yeast mitochondria (Attardi and Schatz, 1988). As an example at least 14 yeast nuclear complementation groups have been found to be involved in cytochrome oxidase synthesis and assembly (Kloeckener-Gruissem et al. 1987). Recent work in maize also reveals the requirement of nuclear encoded factors for the expression of specific plastid genes (cf. Taylor 1989). There is no obvious answer to explain the requirement of such

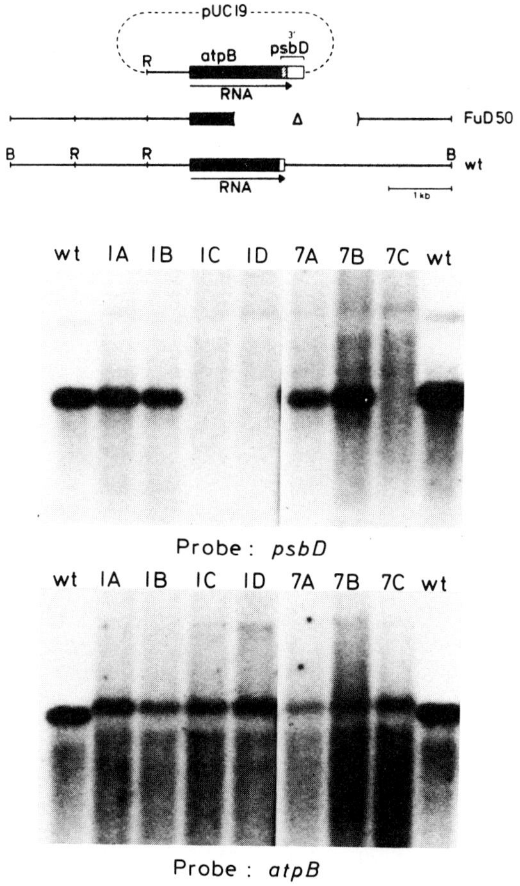


Fig. 3. Expression of a chimeric atpB-psbD transcript in the chloroplast of C. reinhardtii. The atp B deletion mutant Fud50 was transformed with a pUC19 derivative containing the atpB coding region fused to the 3' end of psbD (coding and 3' untranslated regions of psbD are indicated by and respectively).The structure of this plasmid and the chloroplast atpB region of Fud50 and wild-type are shown in the upper part of the figure. The transformants were crossed with nac 2-26. RNA was isolated from a complete tetrad (4A 4B 4C 4D) and from an uncomplete tetrad (7A 7B 7C) and hybridized separately to labelled psbD and atpB probes. Note that the lack of psbD mRNA due to the nuclear nac 2-26 mutation segregates 2:2 whereas the chloroplast encoded chimeric transcript is transmitted to all progeny.

a high number of factors. Perhaps this is a way to tightly integrate the organelle within the cell. It is also possible that some of these factors have other functions in the cell. It is known that some mitochondrial tRNA synthetases in Neurospora and yeast also function as mitochondrial splicing factors (Akins and Lambowitz, 1987; Herbert et al., 1988).

To understand how these factors work, it will be necessary to isolate them. At present these factors are defined only genetically in C. reinhardtii. Some of them might be present in small amounts making their isolation by standard biochemical methods difficult. An obvious

Table II. Nuclear encoded factors required for the expression of specific chloroplast genes in *C. reinhardtii*.

| Chloroplast gene | RNA accumulation | RNA processing | Translation/ protein turn-over | |
|---|---|---|---|---|
| psbA | | | 1 | Girard-Bascou (unpublished) |
| psbB | 1-2 | | | Jensen et al (1986) Monod et al (unpublished) |
| psbC | | | 2 | Bennoun et al (1980) Rochaix et al (1990) |
| psbD | 1 | | 2 | Kuchka et al (1989) Kuchka et al (1988) |
| psaA | | 14 | | Girard-Bascou et al. Goldschmidt-Clermont et al (1990) |
| psaB | | | 2 | Girard-Bascou (unpublished) |
| rbcL | 1 | | | Beasley and Mets (unpublished) |

Numbers refer to the number of nuclear complementation groups affecting expression of a specific chloroplast gene at the level of RNA accumulation, RNA processing or translation / protein turnover.

possibility is to isolate the genes of these factors by complementation of the mutations using the newly developed nuclear transformation system. Alternatively it might be possible to perform gene tagging, with the movable elements Tocl (Day et al., 1988) and Gulliver (Ferris, 1989) of *C. reinhardtii*.

Acknowledgements: We thank O. Jenni and F. Ebener for drawings and photography, J. van Dillewijn and M. Schirmer-Rahire for excellent technical assistance and L. Mets for communicating unpublished results. This work was supported by grant no. 31.26345.89 from the Swiss National Fonds.

## References


Akins, R.A. and Lambowitz, A.M. (1987) Cell 50, 331-345.

Attardi, G. and Schatz, G. (1988) Ann. Rev. Cell Biol. 4, 289-233.

Boynton, J.E., Gillham, N.W., Harris, E.H., Hosler, J.P., Johnson, A.M., Jones, A.R., Randolph-Anderson, B.L., Robertson, D., Klein, T.M., Shark, K.B. and Sanford, J.C. (1988) Science 240, 1534-1538.

Cantrell, A. and Bryant, D.A. (1987) Plant Mol. Biol. 9, 453-468.

Choquet, Y., Goldschmidt-Clermont, M., Girard-Bascou, J.,Kück, U., Bennoun, P. and Rochaix, J.D. (1988) Cell 52, 903-913.

Day, A., Schirmer-Rahire, M., Kuchka, M.R., Mayfield, S.P. and Rochaix, J.D. (1988) EMBO J. 7, 1917-1977.

Debuchy, R. Purton, S. and Rochaix, J.D. (1989) EMBO J. 10, 2803-2809.

Dujon, B., Colleaux, L., Jacquier, A., Michel, F. and Mortheithet, C. (1986) In: Extrachromosomal elements in lower eukaryotes, Wickner, R.B. et al. eds. pp. 5-27, Plenum Publishing Corporation.

Ferris, P.J. (1989) Genetics 122, 363-378.

Fish, L.E., Kück, U. and Bogorad, L. (1985) J. Biol. Chem 260, 1413-1321.

Franzen, L.G., Frank, G., Zuber, H. and Rochaix, J.D. (1988) Plant Mol. Biol. 12, 463-474.

Franzen, L.G., Frank, G., Zuber, H. and Rochaix, J.D. (1989) Mol. Gen. Genet. 219, 137-144.

Fukusawa, H., Kohchi, Z., Shirai, H., Ohyama, K., Umesano, K., Inokuchi, H. and Oseki, H. (1986) FEBS Lett. 198, 11-15.

Gargouri, A., Lazowska, J. and Slonimski, P. (1983) In: Mitochondria Nucleomitochondrial interactions. Schweyen et al. eds. pp. 259-268, De Gruyter, Berlin.

Girard-Bascou, J., Chua, N.H., Bennoun, P., Schmidt, G. and Delosme, M. (1980) Curr. Genet. 2, 215-221.

Goldschmidt-Clermont, M., Girard-Bascou, J., Choquet, Y. and Rochaix, J.D. (1990) Mol. Gen. Genet. in press.

Harris, R.H. (1989) The Chlamydomonas Sourubook, Academic Press.

Herbert, C.J., Labouesse, M., Dujardin, G. and Slonimski, P.P. (1988) EMBO J. 7, 473-483.

Herrin, D.L. and Schmidt, G. (1988) J. Biol. Chem. 263, 14601-14604.

Ikeuchi, M., Hirano, A., Hiyama, T. and Inoue, Y. (1990) FEBS Lett. 263, 274-278.

Jensen, K.H., Herrin, D.L., Plumley, F.G. and Schmidt, G.W. (1986) J. Cell Biol. 103, 1315-1325.

Kindle, K. (1990) Proc. Natl. Acad. Sci. USA 87, 1228-1232.

Kindle, K., Schnell, R.A., Fernandez, E. and Lefebvre, P.A. (1989) J. Cell Biol. 109, 2589-2601.

Kirsch, W., Seyer, P. and Herrmann, R.G. (1986) Curr. Genet.10, 843-855.

Kloeckener-Gruissem, B., McEwen, J.E. and Poyton, R.O. (1987) Curr. Genet. 12, 311-322.

Knaff, D.B. (1988) Trends Biochem. Sci. 13, 460-461.

Kuchka, M., Mayfield, S.P. and Rochaix, J.D. (1988) EMBO J. 7, 319-324.

Kuchka, M., Goldschmidt-Clermont, M., van Dillewijn J. and Rochaix, J.D. (1989) Cell 58, 865-876.
Kück, U., Choquet, Y, Schneider, M. Dron, M. and Bennoun, P. (1987) EMBO J. 6, 2185-2195.
Laird, P.W. (1989) TIBS 5, 2094-208.
Lehmbeck, J., Rasmussen, O.F., Bookjans, G.B., Jepsen, B.R., Stummann, B.M. and Henningson, K.W. (1986) Plant Mol. Biol. 7, 3-10.
Mayfield, S.P. and Kindle, K. (1990) Proc. Natl. Acad. Sci. USA 87, 2087-2091.
Michel, F. and Dujon, B. (1983) EMBO J. 2, 33-38.
Michel, F. and Lang, F. (1986) Nature 316, 641-643.
Oyahama, K., Fukuzawa, H., Kohchi, T., Shirai, H., Sano, S., Umesono, K., Shiki, Y., Takeuchi, M., Chang, Z., Aota, S.I., Inokuchi, H. and Ozeki, H. (1986) Nature 322, 572-574.
Rochaix, J.D. (1987) FEMS Microbiol. Res. 46, 13-34.
Rochaix, J.D., Kuchka, M., Mayfield, S., Schirmer-Rahire, M., Girard-Bascou, J. and Bennoun, P. (1989) EMBO J. 8, 1013-1021.
Roitgrund, C. and Mets, L. (1989) Curr. Genet. 17, 147-153.
Scheller, H.Y., Okkels, J.S., Hoj, P.B., Svendson, I., Roepstorff, P. and Moller, B.L. (1989) J. Biol. Chem. 264, 18402-18406.
Shinozaki, K., Ohme, M., Tanaka, M., Wakasugi, T., Hayashida, N., Matsubayashi, T., Zaita, N., Chungwongse, J., Obokata, J.,Yamagushi-Shinozaki, K., Ohto, C., Torazawa, K., Meng, B.Y., Sugita, M., Deno, H., Kamogashira, T., Yamada, K., Kusada, J., Takaiwa, F., Kato, A., Tohdoh, N., Shimada, H. and Sugiura, M. (1986) EMBO J. 5, 2043-2049.
Stern, D.B. and Gruissem, W. (1987) Cell 51, 1145-1157.
Stern, D.R., Jones, H. and Gruissem, W. (1989) J. Biol. Chem. 264, 18742-18750.
Taylor, W.C. (1989) Ann. Rev. Plant Physiol. Plant Mol. Biol. 40, 211-233.
Torazowa, K., Hayashida, B., Ohokata, J., Shinozaki, K. and Sugiura, M. (1986) Nucl. Acids Res. 14, 3143.
Woessner, J.P., Masson, A., Harris, E.H., Bennoun, P., Gillham, N.W. and Boynton, J.E. (1984) Plant Mol. Biol. 3, 177-190.

# THE THYLAKOID MEMBRANE OF HIGHER PLANTS: GENES, THEIR EXPRESSION AND INTERACTION

Reinhold G. Herrmann, Ralf Oelmüller, Josef Bichler, Alois Schneiderbauer, Johannes Steppuhn, Norbert Wedel, Akilesh K. Tyagi* and Peter Westhoff**

Botanisches Institut der Ludwig-Maximilians-Universität, Menzinger Str. 67, D-8000 München 19, FR Germany

*Department of Plant Molecular Biology, University of Delhi (South Campus), New Delhi-110 021, India

**Institut für Entwicklungs- und Molekularbiologie der Pflanzen, Heinrich-Heine-Universität, Universitätsstraße 1, 4000 Düsseldorf 1, FR Germany

## INTRODUCTION

Chloroplasts are the major sites of energy conversion in the plant cell and play a vital physiological and metabolic role during plant growth and differentiation. Energy, organic matter and oxygen for nearly all biotic processes are provided by photosynthesis and the overwhelming amount of photosynthetic products is formed in the organelle.

The conversion of solar to chemical energy with the associated evolution of molecular oxygen is catalyzed by thylakoid membranes. This spezialized biomembrane system can be resolved into more than 55 polypeptide species most of which are assembled into four membrane-embedded multisubunit protein complexes, the photosystem I and II reaction centers with associated light-collecting chlorophyll *a/b*-containing antennae, the cytochrome *b/f* complex and the ATP synthase. The thylakoid system of higher plants is further characterized by a structural and functional heterogeneity in the lateral plane of the membrane (Anderson and Andersson, 1982). Appressed regions (grana) contain mainly photosystem II, while stroma-exposed lammellae are highly enriched in photosystem I and ATP synthase. The light-trapping and energy-generating complexes, combined with mobile, peripheral extrinsic (plastocyanin, ferredoxines, thioredoxines) and intrinsic (plastoquinon/plastohydroquinon) electron carriers, as well as dynamic changes in the composition and organization of exposed and stacked membrane regions contribute substantially to the enormous physiological flexibility of the photosynthetic process, i.e., to optimize, protect and repair the photosynthetic machinery and its interaction with other principal cellular pathways.

The catalysis of the photosynthetic electron transfer chain is not the only reason why thylakoid membranes have attracted great interest. During the past decade it has been substantiated that the specialized biomembrane is of dual genetic origin, the product of a complex interaction between the nucleo-cytosolic and the organelle's genetic machineries, and that the degree of genetic interaction is enormous. The implications of this interaction, which is specific for eukaryotes, become immediately clear after transfer of the plastids from one species into the nuclear background of another. Such an exchange can severely impair harmoniously balanced growth, specifically of the photosynthetic machinery, since

*Plant Molecular Biology 2*, Edited by R.G. Herrmann and B. Larkins, Plenum Press, New York, 1991

the resulting hybrids or cybrids are often bleached. *Hybrid bleaching*, which is a widely observed phenomenon in nature, reflects an incompatibility between alien genomes and plastomes (plastid genomes) combined in one cell. It is reversible and temporary which distinguishes it fundamentally from plastid or nuclear mutations affecting the organelle, since impaired plastids from disharmonic combinations restore their original morphology and functionality when recombined with their natural genome (Stubbe, 1959).

Yet another kind of complexity has resulted with the advent of multicellular organisms. Higher plants constitute complex cellular societies, each assigned to a limited range of functions essential to the healthy operation of the community. Developmental diversity in a higher plant reflects division of labour between different tissues and organs and, implicitly, regulatory processes operating in space and over long distances. To a large degree, this *additional* information for "three-dimensional" gene expression must be deposited in gene structure.

The enormous complexity of the plant genome and the division of labour in a multicellular organism remain major challenges to contemporary biology. It is obvious that the thylakoid membrane provides an appealing model for the study of both eukaryotism and photosynthesis. In this article, some of our approaches and recent advances in the understanding of this structure will be reviewed. Emphasis will be placed on three aspects, the identification and characterization of polypeptide genes for this membrane, the complexity of membrane biogenesis and on models for compartmental interaction which can accomodate the experimental facts available. The paper follows two previous reviews in this book series that were presented at the NATO-Advanced Study Institutes held in Porto Portese and Renesse, respectively (Herrmann et al., 1982; 1985).

## THYLAKOID PROTEIN GENES

An understanding of the biochemistry and molecular biology of thylakoid membranes depends critically on the availability of gene probes. We have therefore continued our studies on the identification and characterization of genes for this membrane. Fig. 1 and Table 1 summarize the current knowledge of the composition, molecular and genetic organization of the thylakoid membrane. Several generalizations concerning the distribution and organization of genes for this membrane were phrased several years ago (Herrmann et al., 1985) and turned out to be correct.

1. *Genes on plastid chromosomes*: We have previously located and characterized all known genes for components of the four membrane complexes which originate in the plastid chromosome from spinach and noted that plastid chromosomes are principal determinants for thylakoid proteins (summarized in Herrmann et al., 1985). These chromosomes encode all reaction center apoproteins, all cytochromes and catalytic ATP synthase subunits. Completion of an analysis of the *atpA* operon in spinach plastid DNA has led to the discovery of a fourth gene (*atpI*) in this operon encoding a subunit of the membrane-embedded CFo component of the thylakoid ATP synthase which had not previously been suspected. It represents the functional homologue to subunit *a* in the corresponding *E. coli* enzyme (Hennig and Herrmann, 1986, see also Cozens et al., 1986 and below). Sequence analysis of the *psbB* operon, in turn, has uncovered a fourth gene in this transcription unit, *psbH*, coding for a 10 kDa phosphoprotein associated with photosystem II (Westhoff et al., 1986).

   The positions of the thylakoid protein genes and of *rbcL*, the gene for the large subunit of the ribulose bisphosphate carboxylase/oxygenase, on the spinach plastid chromosome map are shown in Fig. 2. Such a map will probably generally apply to all higher plant species studied, with minor variations accounting for a few rearrangements and deletions/insertions (Palmer, 1985). The map and the complete nucleotide sequences available for the plastid chromosomes of *Marchantia*, tobacco and rice (Ohyama et al., 1980; Shinozaki et al., 1986; Hiratsuka et al., 1989) uncover that at least 27 thylakoid proteins are encoded in plastid chromosomes, representing a contribution of approximately half of the structural proteins in the synthesis of the thylakoid membrane. These components with the deduced codon number and predicted membrane topology are listed in Table 1.

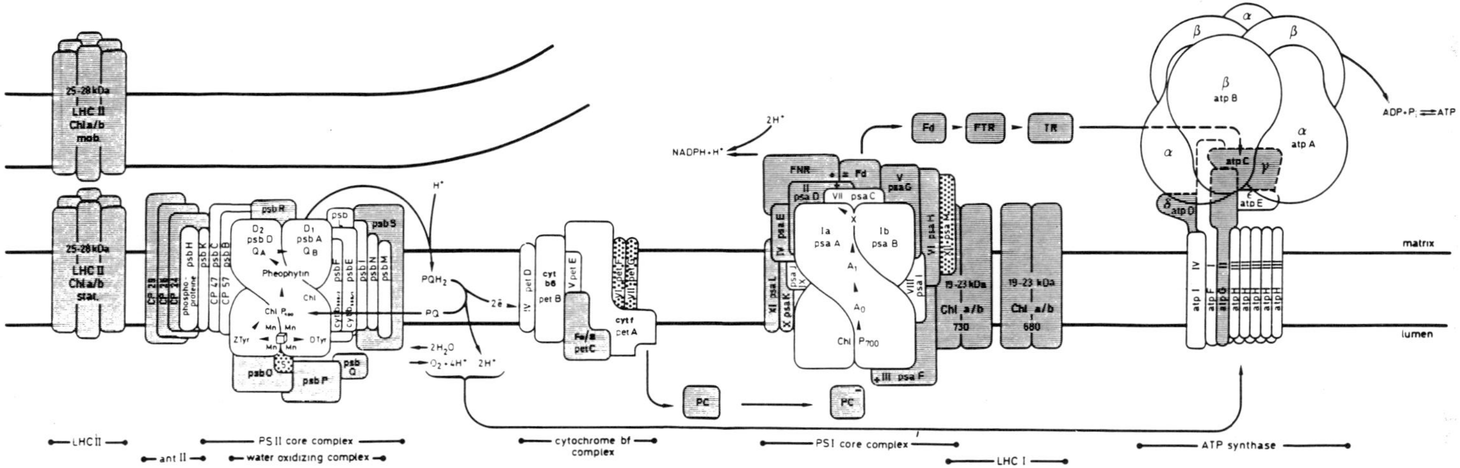


**Fig. 1.** Model for the thylakoid membrane of higher plants illustrating the membrane composition as well as the intergenomic integration of cellular genetic compartments in higher plants. The model is based both on structural data deduced from nucleotide sequences of the corresponding genes studied in this laboratory and on biochemical work of various laboratories. *Plastome*-encoded components are *unmarked, nuclear*-encoded components are *hatched*. The intracellular locations of genes for the dotted polypeptides are unknwon. Photosystem I subunits are numbered with Roman letters I-XII; PC = plastocyanin, Fd = ferredoxin, FNR = ferredoxin oxidoreductase, TR = thioredoxines, FTR = ferredoxin-thioredoxin oxidoreductase. This model is based on a modified, unpublished laboratory-internal version of 1983.

## Table 1. Genes and proteins of the spinach thylakoid membrane.

| protein | gene | coding site | amino acid residues of precursor polypeptide | amino acid residues of mature polypeptide | predicted membrane spans | localization s | localization m | localization l | predicted nuclear gene dosis | ref. |
|---|---|---|---|---|---|---|---|---|---|---|
| **Photosystem II reaction center** | | | | | | | | | | |
| 51 kDa, P680 chl a-apoprotein | psbB | p | | 508 | 5-7 | | x | | | 1 |
| 44 kDa, chl a-apoprotein | psbC | p | | 473 | 5-7 | | x | | | 2 |
| 32 kDa, D1 protein | psbA | p | | 353 | 5 | | x | | | 3 |
| 32 kDa, D2 protein | psbD | p | | 353 | 5 | | x | | | 2 |
| 22 kDa protein | psbS | n | 274 | 205 (?) | 4 | | x | | 2 | 4 |
| 10 kDa phosphoprotein | psbH | p | | 73 | 1 | | x | | | 5 |
| 10 kDa polypeptide | psbR | n | 140 | 99 | 0 | | (x) | | 1-3 | 6 |
| 9 kDa, cytochrome b-559/1 | psbE | p | | 83 | 1 | | x | | | 7 |
| 4.4 kDa, cytochrome b-559/2 | psbF | p | | 39 | 1 | | x | | | 7 |
| 6.5 kDa protein | psbK | p | | 37 | 1 (?) | | x | | | 8 |
| 5.0 kDa protein | psbL | p | | 38 | 1 | | x | | | 9,10 |
| 4.8 kDa protein | psbI | p | | 36 | 1 | | x | | | 9,10 |
| psbM gene product | psbM | p | | 35 | 1 (?) | | | | | 11 |
| psbN gene product | psbN | p | | 44 | 1 (?) | | | | | 11 |
| **Water oxidation complex** | | | | | | | | | | |
| 33 kDa polypeptide | psbO | n | 331 | 247 | 0 | | | x | 1 | 12 |
| 23 kDa polypeptide | psbP | n | 267 | 186 | 0 | | | x | 1 | 13 |
| 16 kDa polypeptide | psbQ | n | 232 | 149 | 0 | | | x | 1 | 13 |
| **Light-harvesting complex of photosystem II** | | | | | | | | | | |
| cabII-1 (lhb-spinA) | | n | 267 | 233 | 3 | | x | | 5 | 14 |
| cabII-2 (lhb-spinB) | | n | 267 | 233 | 3 | | x | | | 15 |
| CP24 | | n | 261 | 210 | 2-3 | | x | | 1 | 16 |
| **Cytochrome b/f complex** | | | | | | | | | | |
| cytochrome f | petA | p | 285 | 250 | 1 | | x | | | 17 |
| cytochrome b6 | petB | p | | 211 | 5 | | x | | | 18 |
| Rieske FeS-protein | petC | n | 247 | 179 | 1 | | x | | 2 | 19 |
| subunit IV | petD | p | | 139 | 3 | | x | | | 18 |
| 3.2 kDa protein, subunit V | petE | p | | 37 | 1 | | x | | | 20 |
| subunit VI | petF | | | | | | | | | 21 |
| subunit VII | petG | | | | | | | | | 21 |
| **Photosystem I reaction center** | | | | | | | | | | |
| P700 chl-a apoprotein 1 | psaA | p | | | 9-11 | | x | | | 22 |
| P700 chl-a apoprotein 2 | psaB | p | | | 9-11 | | x | | | 22 |
| PSI-2, ferredoxin-binding | psaD | n | 212 | 162 | 0 | | (x) | | 1-2 | 23 |
| PSI-3, plastocyanin-binding | psaF | n | 231 | 154 | 0 | | | (x) | 2-3 | 24 |
| PSI-4 | psaE | n | 125 | 91 | 0 | | (x) | | 2 | 23 |
| PSI-5 | psaG | n | 167 | 98 | 0 | | (x) | | 3-5 | 24 |
| PSI-6 | psaH | n | 143 | 95 | 0 | | (x) | | 2-3 | 25 |
| PSI-7 | psaC | p | | 81 | 0 | | (x) | | | 25 |
| PSI-8 | psaI | p | | 36 | 0-1 | | (x) | | | 26 |
| PSI-9 | psaJ | p | | 45 | 1 | | x | | | 27 |
| PSI-10 | psaK | n | 126 | 95 | 2 | | x | | | 28 |
| PSI-11 | psaL | n | 209 | 169 | 2 | | x | | | 29 |
| PSI-12 | psaM | | | | | | | | | 30 |
| **Light-harvesting complex of photosystem I** | | | | | | | | | | |
| cabI (lha-spinA) | | n | 245 | 201 | 3 | | x | | 10 | 31 |
| **ATP synthase** | | | | | | | | | | |
| CF1-alpha | atpA | p | | | 0 | | (x) | | | 32 |
| CF1-beta | atpB | p | | 498 | 0 | | (x) | | | 32 |
| CF1-gamma | atpC | n | 365 | 323 | 0 | | (x) | | 2 | 33 |
| CF1-delta | atpD | n | 257 | 187 | 0 | | (x) | | 1 | 34 |
| CF1-epsilon | atpE | p | | 153 | 0 | | (x) | | | 32 |
| CFo-I | atpF | p | | 201 | 1 | | x | | | 35 |
| CFo-II | atpG | n | 222 | 147 | 1 | | x | | 1 | 36 |
| CFo-III | atpH | p | | 81 | 2 | | x | | | 35,37 |
| CFo-IV | atpI | p | | 247 | 4 | | x | | | 35 |
| **Peripheral proteins** | | | | | | | | | | |
| plastocyanin | | n | 168 | 101 | 0 | | | x | 1 | 38 |
| ferredoxin I | | n | 147 | 97 | 0 | x | | | 2 | 39 |
| ferredoxin-NADP$^+$-oxidoreductase | | n | 369 | 314 | 0 | | (x) | | 1 | 40 |
| thioredoxin m | | n | 181 | 112-114 | 0 | x | | | 2 | 41 |
| thioredoxin f | | n | 190 | 113 | 0 | x | | | 2 | 42 |

p = plastid; m = membrane; s, m, l refer to polypeptide locations in stroma, membrane and lumen, respectively; symbols in brackets indicate peripheral membrane location. Gene nomenclature according to Hallick (1989), Plant Mol. Biol. Rep. 7, 266-275

The thylakoid proteins of plastid origin are all products of single genes per chromosome, which are scattered throughout both single-copy regions and found on both DNA strands. They are generally arranged in operons. However, the genes for components of any given membrane complex are frequently not clustered. Analysis of the RNA species derived from these operons generally shows highly complex patterns. Processing of primary, polycistronic transcripts probably accounts for most of the smaller RNAs (Herrmann et al., 1985; Westhoff and Herrmann, 1988; Barkan, 1988), though there may be transcription initiation sites within gene clusters which could contribute to the overall complexity of the observed patterns (Yao et al., 1989). The implications of the outlined gene arrangement and transcript processing for thylakoid biogenesis will be discussed elsewhere (Herrmann et al., 1991).

2. During recent years our efforts have mainly been directed at characterizing the *nuclear gene complement* for thylakoid proteins. We have isolated cDNAs (and/or genomic clones) for 26 nuclear-encoded components using recombinant DNA technology in the *lambda* gt11 expression system (Table 1). The collection includes the genes/cDNAs for five photosystem I subunits, the Rieske Fe/S protein of the cytochrome *b/f* complex, several peripheral or extrinsic proteins including the thioredoxines *f* and *m*, for three ATP synthase subunits including subunit CFo-II (gene: *atpG*) as well as CP24 and the 22 kDa polypeptide (*psbS*) associated with photosystem II. CFo-II and the 22 kDa protein are phylogenetically and structurally intriguing components. Subunit II was the only component of the CFo assembly for which the gene had not been isolated and this gene was alleged to be of nuclear origin. However, the existence and identity of this protein had not been resolved and remained controversial. Our data show unequivocally that CFo-II is a real (ninth) subunit of thylakoid-located ATP synthases. It is the functional homologue to the cyanobacterial ATP synthase subunit *b'* which, along with subunit *b*, must have arisen

---

*References for Table:* (1) Morris J. & Herrmann R.G. (1984) Nucl. Acids Res. **12**, 2837-2852; (2) Alt J., Morris J., Westhoff P. & Herrmann R.G.(1984) Curr. Genet. 8, 597-606; (3) Zurawski G., Bohnert H., Whitfeld P.R. & Bottomley W. (1982) Proc. Natl. Acad. Sci. USA 79, 7699-7703; (4) Wedel N. & Herrmann R.G. (in preparation); (5) Westhoff P., Farchaus J.W. & Herrmann R.G. (1986) Curr. Genet. 11, 165-169; (6) Lautner A., Klein R., Ljungberg U., Reiländer H., Bartling D., Andersson B., Reinke H., Beyreuther K.& Herrmann R.G. (1988) J. Biol. Chem. 263, 10077-10081; (7) Herrmann R.G., Alt J., Schiller B., Widger W.R. & Cramer W.A. (1984) In: Molecular Form and Function of the Plant Genome. pp 233-256, v. Vloten-Doting L., Groot G.S.P., Hall, T.C. (eds.), Plenum Publishing Coroparation, Amsterdam; (8) Murata N., Miyao M., Hayashida N., Hidaka T. & Sugiura M. (1988) FEBS Lett. 235, 283-288; (9) Webber A.N., Hird S.M., Packman L.C., Dyer T.A. & Gray J.C. (1988) Plant Mol. Biol. 12, 141-161; (10) Ikeuchi M., & Inoue Y. (1988) Plant Cell Physiol. 29, 1233-1239; (11) Ikeuchi M., Takio K. & Inoue Y. (1989) FEBS Lett. 242, 263-269; (12) Tyagi A., Hermans J., Steppuhn J., Vater F. & Herrmann R.G. (1987) Mol. Gen. Genet. 207, 288-293; (13) Jansen T., Rother C., Steppuhn J., Reinke, H., Beyreuther K., Jansson C.,Andersson B. & Herrmann R.G. (1987) FEBS Lett. 216, 234-240; (14) Wedel N. & Herrmann R.G. (unpublished); (15) Schneiderbauer A.& Herrmann R.G. (unpublished); (16) Wedel N. & Herrmann R.G. (in preparation); (17) Alt J. & Herrmann R.G. (1984) Curr. Genet. 8, 51-557; (18) Heinemeyer W., Alt J. & Herrmann R.G. (1984) Curr. Genet. 8, 543-549; (19) Steppuhn J., Rother C., Hermans J., Jansen T., Salnikow J., Hauska G & Herrmann R.G. (1987) Mol. Gen Genet. 210, 171-177; (20) Haley J. & Bogorad L.(1989) Proc. Natl. Acad.Sci. USA 86, 1534-1538 (maize); (21) unpublished data; (22) Kirsch W., Seyer P. & Herrmann R.G. (1986) Curr. Genet. 10, 843-855; (23) Münch S., Ljungberg U., Steppuhn J., Schneiderbauer A., Nechushtai R., Beyreuther K.& Herrmann R.G. (1988) Curr. Genet. 14, 511-518; (24) Steppuhn J., Hermans J., Nechushtai R., Ljungberg U., Thümmler F., Lottspeich F. & Herrmann R.G. (1989) FEBS Lett. 237, 218-224; (25) Steppuhn J., Hermans, J., Nechusthai R., Herrmann G.S. & Herrmann R.G. (1989) Curr. Genet. 16, 99-108; (26) Scheller H.V., Svendsen I., Møller B.L. (1989) Carlsberg Res. Comm. 54, 11-15 (barley); (27) Scheller H.V., Svendsen I., Møller B.L. (1989) J. Biol. Chem. 26, 118402-18406 (barley); (28) Franzen L.G., Frank G., Zuber H. & Rochaix J.D. (1989) Plant Mol. Biol. 12, 463-474 (*Chlamydomonas*); (29) Okkels J.S., Scheller H.V., Jepsen L.B. & Møller B.L. (1989) FEBS Lett. 250, 575-579 (barley); (30) Ikeuchi M.& Inoue Y. (1991) J. Biol. Chem. (in press); (31) Palomares R., Herrmann R.G. & Oelmüller R. (1991) Photochem. Photobiol. (in press); (32) Westhoff P., Nelson N., Bünemann H. & Herrmann R.G. (1981) Curr. Genet. 4, 109-120; (33) Miki J., Maeda M., MukohataY. & Futai M. (1988) FEBS Lett. 232, 221-226; (34) Hermans J., Rother C., Bichler J., Steppuhn J. & Herrmann R.G. (1988) Plant Mol.Biol. 10, 323-330; (35) Henning J. & Herrmann R.G. (1986) Mol. Gen Genet. 203, 117-128; (36) Steppuhn J., Reiländer H., Rother C.& Herrmann R.G. (in preparation); (37) Alt J., Westhoff P., Sears B., Nelson N., Hurt E., Hauska G. & Herrmann R.G.(1983) EMBO J.2, 979-986; (38) Rother C., Jansen T., Tyagi A., Tittgen J. & Herrmann R.G. (1986) Curr. Genet. 11, 171-176; (39) Wedel N., Bartling D. & Herrmann R.G. (1988) Bot. Acta 101, 295-300; (40) Jansen T., Reiländer H., Steppuhn J. & Herrmann R.G. (1988) Curr. Genet. 13, 517-522; (41) Wedel N. & Herrmann R.G. (in preparation); (42) Kamo M., Tsugita A., Wiessner C., Wedel N., Bartling D., Herrmann R.G., Aguilar F., Gardet-Salvi L. & Schürmann P. (1989) Eur. J. Biochem. 182, 315-322

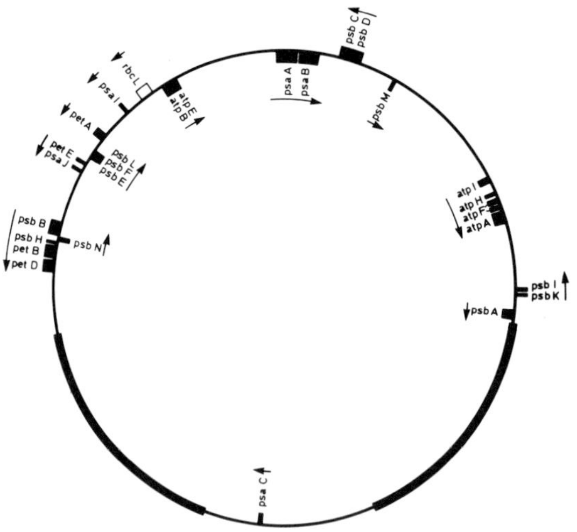


**Fig. 2.** Simplified model of the plastid chromosomes from spinach including the location of genes for thylakoid proteins. The two bold lines mark the inverted repeat. Genes are indicated for photosystem I (*psa*), photosystem II (*psb*), cytochrom *b/f* complex (*pet*), ATP synthase (*atp*) and for ribulose bisphosphate carboxylase/oxygenase (*rbc*). Their transcription polarities are given in parallel arrows. The heteromeric operons are not indicated (see text).

by gene duplication (Cozens and Walker, 1987). Therefore, this subunit is a distinguishing feature between photosynthetic and respiratory ATP synthases. Gene duplication for thylakoid proteins is not unique to CFo-I (the equivalent to the bacterial subunit *b*) and CFo-II. We have previously pointed out (Herrmann et al., 1985) that several plastome-encoded thylakoid proteins must have evolved by a series of duplications of certain primordial DNA segments. However, CFo-I (Hennig and Herrmann, 1986) and CFo-II are the only instance on record for which the genes are found in *different* intracellular compartments. It is worth noting that both proteins integrate into the lipid bilayer with substantially different epitopes (see below and Bartling et al., 1990).

The unique characteristics of the 22 kDa protein, in turn, reside in its remote relationship to antenna apoproteins. Although histidine and methionine residues presumably involved in binding chlorophyll molecules are not conserved and the protein probably does not associate with pigment, secondary structure predictions and conserved domains leave little doubt that this component represents a more distant, yet clearly related member of both the *cab* I/II gene families and CP24. It appears to be the result of a gene-internal duplication. As yet we do not know whether it represents an ancient member to the development of the chlorophycean type of antenna or whether the obvious phylogenetic hierarchy beween *psbS*, CP24 gene and *cab* genes reflects a diverged branch of *cab* genes (in collaboration with B. Andersson, Stockholm).

All components of nuclear origin are decoded from polyadenylated RNA to precursor forms with leader sequences of greatly varying sizes. They are synthesized on free cytosolic ribosomes and imported post-translationally into the organelle for assembly. An interesting side product of our analyses of cDNAs in this context was the accessibility of transit peptides for proteins for which the topographical location is generally well known. Transit peptides carry essential targetting function. Depending on the location of the protein within the organelle, the amino terminal transit peptides are removed in one or two steps by distinct proteases located in the organelle stroma and intrinsically at the lumenal phase of stroma thylakoids (Table 1; Ellis and Robinson, 1987).

Although plastid transit peptides can be clearly distinguished from their mitochondrial counterparts, their primary sequences do not exhibit any remarkable similarity. Several lines of evidence suggest that individual stages of import, targetting, trafficking and assembly rest largely on secondary structure of discrete topogenic sequences (Tyagi et al., 1987; Bartling et al., 1990). Such domains may also be found in the mature polypeptides and exert multiple functions. Codon-correct reciprocal exchanges of transit peptides and mature proteins using a specifically devised cassette system suggest strongly that efficiency and specificity of the import process are frequently determined by a co-evolution of transit peptide and its mature protein (Bartling et al., 1990; Clausmeyer et al., 1991). The presence or absence and the arrangement of these domains determine largely the fate of the protein, and can be used to classify precursor polypeptides into three categories: (i) those that carry composite transit peptides, including predominantly the extrinsic luminal polypeptides such as plastocyanin, the 33, 23 and 16 kDa polypeptides associated with the water oxidation complex, but also subunit III of photosystem I (PSI-III, the plastocyanin docking protein) and CFo-II of the ATP synthase. Composite transit peptides are therefore not confined to extrinsic lumenal components or components attached with hydrophobic domains to the inner membrane surface (probably PSI-III), which have to traverse three membranes. Even CFo-II which is a bitopic transmembrane protein (Fig. 1) contains a bipartite transit peptide. (ii) Polypeptides that carry (hydrophilic) import sequences and are located in the organelle stroma represent a second category, and (iii) those that carry import sequences but are located within the membrane or the lumen and therefore use intrinsic hydrophobic thylakoid-targetting or -translocation signals (uncleaved h-domains) the third. Representatives of the latter class are the Rieske FeS protein, *cab* proteins, CP24 and the 22 kDa polypeptide associated with photosystem II. For further information about transit peptides the reader is referred to Bartling et al. (1990) and to the contributions of v. Heijne, Robinson et al. and de Boer et al. in this volume.

3. *Gene dosage:* One of the unique features of the organization of plastomes is their high degree of reiteration. Individual plastids of higher plants may house up to 100 chromosomes, that is between 500 and 10.000 copies per cell depending on organelle size and developmental stage (Herrmann and Possingham, 1980). Southern analysis based on hybridization of radiolabelled cDNA probes (Table 1) to nuclear DNA restricted with *EcoRI*, *BamHI* or *HindIII*, has yielded the surprising result that, in spinach, approximately one third of the proteins of nuclear origin appear to derive from only a single gene per haploid genome (Table 1). Another 50% appear to be synthesized from two or three gene copies, and only a minor fraction of proteins are the products of small multigene families, comparably to those for the small subunit of ribulose bisphosphate carboxylase/oxygenase (*rbcS* genes) of chlorophyll *a/b* apoproteins (*cab* genes) of light-harvesting antenna (Table 1). These findings have two interesting implications:

    a. The functional interaction between nucleus and plastids has to account for a vast difference in gene dosage between these organelles. This refers to both stoicheiometry of protein production but possibly also to intercompartmental signal intensity (see below). However, reiteration of plastid chromosomes does not necessarily imply that all DNA molecules are active in expression. One of the intriguing questions addressed by these results is therefore whether or not synthesis of thylakoid proteins requires the activity of many copies of plastid chromosomes and, if so, how it is controlled that thylakoid proteins are present in equivalent quantities regardless of whether they are of nuclear or plastid origin.

    b. Sequence comparison of several cDNA (and genomic) clones for a given component suggests that the spinach genes for thylakoid proteins which are present in two (or a few) copies are obviously less diverged than some members of *rbcS* and *cab* gene families and appear to be coding for (very) similar products. This is based on sequence analysis of several cDNAs per component which generally differ by only a limited number of nucleotides; many of them represent silent mutations or are found in untranslated gene regions. This similarity could reflect quite recent duplication events or functional reasons.

Current ideas about gene duplications suggest the existence of stringent mechanisms for keeping to a minimum the number of genes, especially if the genes have a vital role (Ainsworth et al., 1987; Mardsen et al., 1987). Accordingly, duplications would be tolerated only where they are of functional use, for example, to adapt or facilitate the synthesis of a component in response to different environmental or developmental stimuli (Sugita and Gruissem, 1987). It will therefore be of considerable interest to explore whether the nearly identical or "twin" genes are clustered or occupy different positions within the genome, and whether they possess similar regulatory sequences.

## BIOGENESIS, GENE REGULATION

One of the most intriguing questions to be addressed concerning the regulation of eukaryotic genes is how the expression of specific *sets of genes* is controlled *per se*, in a genetically compartmentalized cell and in the multicellular vascular plant with its unique developmental characteristics. The dependence of thylakoid biogenesis on two different sources of protein synthesis implies that in the course of evolution from prokaryotes to eukaryotes intercompartmental controls must have been added to ensure a coordinated delivery of components from either compartment at the proper time and in near stoicheiometric amounts. Understanding the biogenesis of the photosynthetic membrane therefore requires knowledge about the expression of genes (i) in plastids, (ii) in the nucleo/cytosolic compartment, (iii) about the mechanisms of intergenomic integration within cells, (iv) the import of components of nucleo/cytosolic origin into the organelle, (v) processes of subcellular differentiation, and (vi) about mechanisms underlying the response of the membrane system to different environmental conditions.

The differentiation of photosynthetic tissue is organ-specific and accompanied by a massive synthesis of organelle protein. It is a light-controlled process, which can be conveniently studied during greening of etiolated seedlings. Research on plastid biogenesis has concentrated to a large extent on mRNA patterns in recent years or on only a few genes and their products, specifically for nuclear components. However, a better appreciation of the complexity of thylakoid biogenesis requires information upon many components and should include a comparison of RNA and protein data.

Changes in steady-state mRNA amounts for thylakoid proteins in plastids have repeatedly been described and shown to be quite uniform during greening of etiolated material (e. g., Bedbrook et al., 1978); RNA concentrations are (very) low in etiolated material and increase substantially during greening. Changes in the processing patterns, if any, are rare and generally independent of photocontrol. When we analyzed the patterns of nuclear components, we found that they are strikingly similar (Fig. 3A). However, only single major RNA species instead of complex patterns are observed in all instances.

The findings, which rest on approximately 40 components, are substantially different for the corresponding proteins. Western analysis distinguishes two clear biogenetic classes, each of which can be divided into two subgroups (Fig. 3B). From the biochemical standpoint the following aspects are of special interest:

1. The cellular capacity for the synthesis and accumulation of discrete thylakoid proteins appears not to be directly correlated with the amount of mRNA present. This is obvious from the comparison of Northern and Western analysis and has been confirmed by pulse/chase experiments in a few instances.

   Substantial quantities for all components of the cytochrome *b/f* complex, the ATP synthase, for cytochrome *b559* and of peripheral proteins can be produced in the dark from minute RNA quantities (cf. Fig. 3). This synthesis occurs during a relatively short period of etiolated growth implying that its energy requirement must be provided by respiration. The protein concentrations do not undergo substantial variation during greening, i. e., they increase approximately 2- to 3-fold within 48 h of illumination in these instances, while that of the corresponding stationary RNA levels may change as much as 5- to 20-fold. The protein levels remain high for more than 48 h after synthesis has been shut off and appear to be relatively stable, as we have no

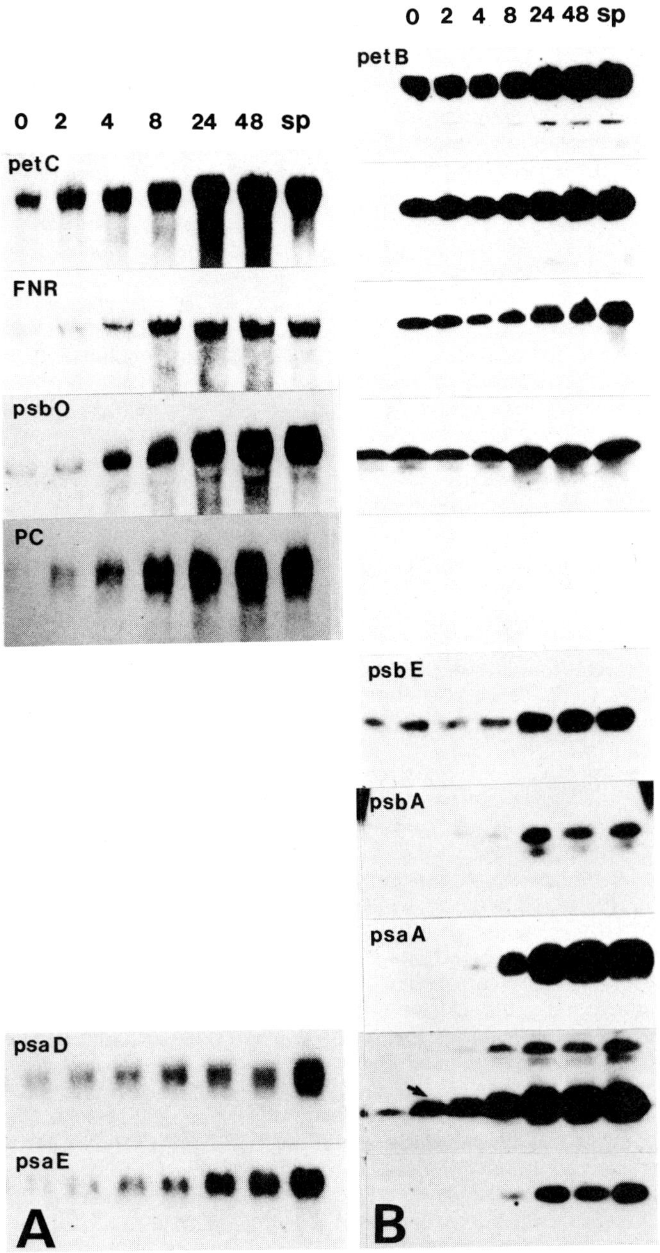


**Fig. 3.** Comparative (A) Northern and (B) Western analysis of mRNA and protein accumulation after transfer of etiolated spinach seedlings to white light of high intensity (appoximately 10,000 Wm $^{-2}$). Seedlings were kept in darkness for seven days before the onset of illumination. Fluorographs; numbers refer to hours in light. The aligned Northern and Western blots correspond to each other. Symbols: *petB* (gene) - cytochrome *b6*, *petC* - Rieske FeS protein, FNR - ferredoxin oxidoreductase, *psbO* - 33 kDa polypeptide of the water oxidation complex, PC - plastocyanin, *psbE* - cytochrome *b559*, *psbA* - D1 protein, *psaA* - P700 chlorophyll *a* apoprotein, *psaD*, *psaE* - subunits II and IV of photosystem I, respectively. Sp = control from developing, green spinach leaves.

indication that they are complicated remarkably by possible turn-over processes, except for the well studied proteins D1 and D2. So, the discordant increase between mRNA and protein levels during that period can therefore not be attributed to protein turnover. It is important to note that this is *independent* of whether a protein is of plastid or nuclear origin.

Some components of the photosystems I and II cannot be detected in the dark although the corresponding (processed) mRNAs are present. A lack of detectable protein could be due to changes in rates of synthesis, rates of degradation, or both. Examination of synthesis rates at different times over a 4 h period by *in vivo* incorporation of amino acids into the protein suggests that significant levels of radiolabel are incorporated into these proteins only during periods immediately following the transition from dark to light (to be published elsewhere). This is indicative of an uncoupling of transcription and translation. No detectable synthesis in the dark is seen, for example, for some of the low molecular units for subunits of photosystem I and subunits of photosystem II. These results strongly suggest that thylakoid protein synthesis may be inhibited for components of both photosystems. However, it is also obvious that this inhibition prevents the synthesis of only some constituents of these membrane assemblies, that this inhibition is not correlated with the position of a polypeptide within a membrane complex (cf. the positions and pattern of D1 and cytochrome *b559* in Figs. 2 and 3), and that inhibition can be released at *different* times during illumination at least for subunits of photosystem I. Thus, the biogenetic classification of thylakoid proteins is not specific for a given membrane complex. Apart from the facts that the synthesis of photosystem I preceds that of photosystem II and that this sequence correlates with the appearance of the corresponding photosynthetic activities (i.e., cyclic electron transport may possibly precede the linear chain under the chosen conditions), no rationale is apparent why the synthesis of components for a given membrane assembly may differ so greatly.

2. The molecular details of the photoregulated signal transduction chain responsible for the expression of thylakoid genes is unknown. However, our studies provide novel aspects relating to the levels at which light controls expression. We have previously shown that light stimulates plastid gene expression at the translational rather than transcriptional level and that (light- regulated?) changes in mRNA stability is a major mechanism of RNA accumulation in plastids during the greening process (Herrmann et al., 1985). These findings have been confirmed (e. g., Kreuz et al., 1986; Laing et al., 1988) and substantially extended by Mullet and coworkers for chlorophyll-binding proteins (Klein and Mullet, 1986; 1987; Klein et al., 1988). The two regulatory switches for plastid gene expression proposed in the literature (Herrmann et al., 1985; Rodermel and Bogorad, 1985) are *a priori* not mutually exclusive, but direct proof for transcriptional *photo*control of plastid genes is still lacking. On the other hand, it is well known that photocontrol on nuclear gene expression operates at the transcriptional level. However, the relatively uniform changes of mRNA levels for *all* thylakoid components studied, the discordant RNA and protein patterns, as well as the uncoupling of transcription and translation in discrete cases demonstrate unequivocally that photocontrol operates at least at *three* levels during thylakoid biogenesis and that posttranscriptional, possibly translational photocontrol is now recognized as being of crucial importance in the synthesis of *both* plastid- *and* nuclear-coded thylakoid proteins. To our knowledge, this level of control has not been considered before for nuclear components. It raises the question for the biological significance of transcriptional photocontrol of nuclear genes.

## ANALYSIS OF NUCLEAR PROMOTORS

Clustering and co-transcription of genes - one strategy widely adopted in prokaryotic organisms - is certainly not the principal strategy for the coordination of the synthesis of components for the thylakoid membrane within and between the organelles. For plastids, this is obvious from the scattered arrangement of genes for subunits of a given membrane complex (Fig. 2). Nuclear genes, in turn, have been mapped to different chromosomes using segregating F2 populations, rare-cutting enzymes and electrophoresis in pulsed fields (unpublished data). It seems more probable therefore that the thylakoid protein genes are coordinately regulated, primarily because they have similar controlling elements which interact with common regulatory factors. Of course, the position of a gene within the nucleus and chromosome may also be an important factor in determining the level of its expression.

The complexity of gene expression during thylakoid biogenesis and the availability of recombinant DNAs for >20 nuclear-coded thylakoid proteins from spinach have therefore prompted us to search for novel aspects, elements and general principles, if any, involved in the photocontrol, temporal and spatial (tissue, organ and developmental specifity) expression patterns of these genes. We have concentrated on single-copy genes (see above). For the practice of studies on gene expression, single-copy genes possess obvious advantages since they represent simpler models than members of gene families (Bichler and Herrmann, 1990). A complicating factor in studying gene families as well as in trying to assess the significance of gene copy number is that the members of such families might be differentially expressed. We have used promotor dissection and transgenic tobacco as well as gel-shift experiments of nuclear proteins that bind sequence-specifically to elements in the 5' flanking gene regions as assay systems. The promotors or promotor segments from (to date) 7 genes, including those of plastocyanin, ferredoxin oxidoreductase (FNR), representatives of the ATP synthase (subunits gamma and delta; *atpC*, *atpD*) and the photosystems I and II, and, as controls, *the rbcS* and CaMV 35 S RNA promotors, were fused to the ß-glucuronidase (GUS) reporter gene and introduced into *Nicotiana tabacum* using an *Agrobacterium*-based binary Ti transformation system (pBIN 19; Bevan, 1984). GUS activity from material grown under defined conditions was monitored both in homogenates of appropriate tissue samples and *in situ* in tissue sections of 10 - 20 randomly selected transgenic tobacco plants per construct (Bichler and Herrmann, 1990). The basic results can be summarized als follows:

1. As judged from steady-state mRNA levels, the 1.4 - 2.0 kb "full-length" promotors confer expression to the reporter gene which appears to be qualitatively and quantitatively comparable to the expression of the homologous spinach system. This expression is light-regulated as well as tissue-, organ- and development-specific.

2. All "photosynthetic" promotors direct high levels of GUS expression in photosynthetically active cells of the pallisade or spongy parenchyma or chloronema cells of midrib and stem. One of the most striking observations of this study has been that promotor activity is apparently not confined to "usual" photosynthetic tissue, nor to the existence of "typical" chloroplasts. For example, the promotors of the genes for plastocyanin or the delta subunit of the ATP synthase are active as well in vascular tissue of anthers, specifically in the phloem regions of leaf and floral stems (Bichler and Herrmann, 1990). This observation has provided the incentive to search for a phloem-located, electron transport chain that would deliver energy for transport processes.

3. *Cis*-elements involved in light-, development- and organ-specific regulation may not be confined to proximal 5' regions. Qualitatively different activities including enhancer- and silencer-like functions both for dark or light conditions appear to be spread over relatively long distances (>1500 nucleotides). For example, the 5' deletion profiles show a positive element between -350 and -380 bp in the PSI-III promotor, involved in supporting high level expression in light (Fig. 4). When basal TATA-box mediated expression is taken into consideration, the extent of light-enhanced expression is more than 10-fold compared to that observed for the -300 bp segment. The level of GUS-mRNA under control of the plastocyanin promotor, in turn, is highly affected by a deletion from -770 to -1180 bp in the corresponding promotor. The data indicate that, in this instance, the repressing action of high mRNA levels in the dark is mediated through a negative sequence element upstream of -770 bp (Bichler and Herrmann, 1990). Comparably, *cis*-elements influencing expression patterns in inner and outer phloem regions respectively, are spread and can be physically separated.

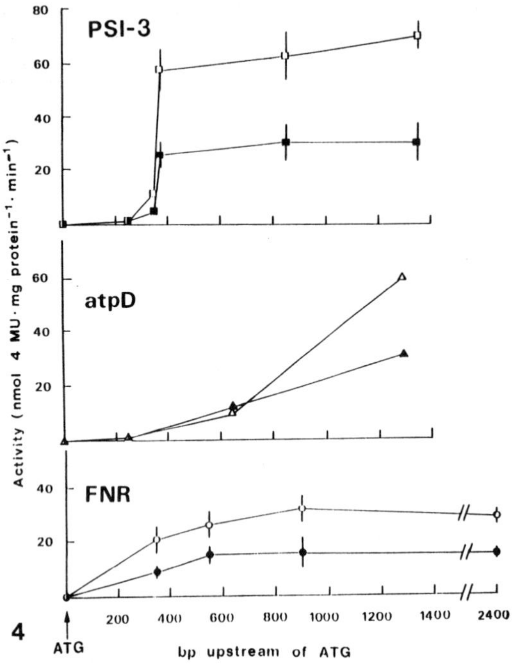

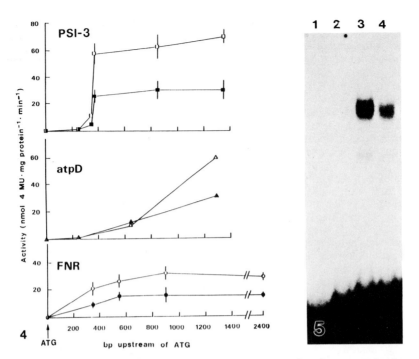


**Fig. 4.** Comparison of GUS gene expression in transgenic tobacco under the control of various 5' promotor fragments of the spinach subunit III of photosystem I (PSI-III), the delta subunit of the ATP synthase (*atpD*) and the ferredoxin-NADP$^+$-oxidoreductase (FNR) genes. The data represent average values of GUS activity of 10 - 20 independently transformed Fo tobacco plants; bars represent S.D. ATG = translation initiation codon. Open symbols: enzyme activity in light-grown plants; close symbols: enzyme activity after 7 days of darkness.

**Fig. 5.** An autoradiogram exemplifying a gel-shift experiment using a 30 bp DNA-fragment from the PSI-III promotor. (1) electrophoretic mobility of the radio-labelled PSI-fragment; (3) profile in the presence of 16 mg protein of a tobacco nuclear extract; (2) as (3), but with a 25-fold molar excess of unlabelled PSI-III fragment; (4) as (3), but with a 200-fold molar excess of linearised pBSC$^+$. The PSI-III region used for these gel shift experiments (-350 to -380 bp upstream from the translation initiation codon) promotes GUS gene expression more than 10-fold in transgenic tobacco (cf. Fig. 4)

4. The DNA segments that constitute plant promotors are complex composites of various *cis*-acting elements which qualitatively and/or quantitatively affect expression. Elements involved in light regulation and tissue-specific expression are probably not congruent although it is possible that *cis*-acting elements are profoundly interrelated and that the recognition features of promotors reside in different *trans*-acting factors (see below). The comparison of various promotors indicates that their architecture is distinctively different in every case, although functionally analogous *cis*-elements appear to operate in all these DNA segments. This refers to both a lack of sequence similarity as well as to the segmental organization. In the promotors of the genes for plastocyanin, the ATP synthase subunit delta and the photosystem I subunit III, *cis*-elements are located over relatively long distances while those in the FNR promotor are predominantly found in the first 500 nucleotides (Fig. 4). A question relevant to the problem of a possible coordinate template function of *cis*-elements is whether short motives known to be involved in expression are found in several promotors or not. Unfortunately, there is no evidence to support this idea in a general sense. Light responsive elements, the best studied cases, are poorly conserved or not obvious in the listed promotors and, if present, they may not necessarily be functional (Bichler and Herrmann, 1990). This fits in well with the results from gel shift assays, since in

many cases no competition between promotors could be observed. Moreover, the available data do not support the idea that spacing of discrete nucleotides rather than the entire conservation of sequence elements is required for function, although this possibility cannot be rigorously excluded presently. This implies that *cis*-elements can only be functionally defined, but also that no ubiquitous concept for light-responsive (or other) elements is yet apparent.

5. With a view to identify *trans*-acting proteins potentially involved in gene regulation, we have performed gel retardation assays using nuclear extracts from spinach, tobacco and tomato. Various regions were found to interact specifically with proteins. The DNA protein complexes are competed by the respective unlabelled homologous DNA fragments (including synthetic oligonocleotides), but are stable to competition by a 25-fold molar excess of other fragments. These results suggest that a number of nuclear factors bind to each of these promotors and that these factors are different in many instances. In a given promotor, a nuclear factor may interact with more than one site; competitive binding is, for example, observed between the CaMV 35S RNA and the *rbcS* promotors. No difference was observed between nuclear extracts from dark-grown and light-exposed material.

   Comparison of the promotor dissection and gel-shift analyses suggests that the correlation between both sets of data is relatively poor; segments identified as functionally active by gene transfer display frequently no interaction with protein factors and vice versa. However, in the case of the promotor of the ferredoxin $NADP^+$ oxidoreductase gene, a 48 bp segment has been identified that binds a *trans*-acting factor in vitro and confers expression to a truncated *cab*-promotor which *per se* is inactive in transgenic tobacco. Similarly, a 30 bp region in the PSI-III promotor which is highly effective in promoting GUS gene expression (Fig. 4) was identified as a *trans*-factor binding site.

Experiments are in progress to refine this analysis in order to obtain a more precise estimate of the complexity of *trans*-acting factors and their interaction with *cis*-elements of different promotors.

## COMPARTMENTAL INTERACTION

Biogenesis of the thylakoid network encompasses regulatory processes between nuclear and plastid genes. To our knowledge no systematic study has been made of the possible nature of the underlying signals. Although the available experimental evidence is as yet very limited, it is obvious that signals must operate in both directions, that intercompartmental control is mutual and occurs at several levels.

1. The import of precursor polypeptides into chloroplasts and their stepwise transport through and into various locations within the organelle are highly regulated and presumably signals and sorting mechanisms including coordinating controls are essential at every step for proper traffic regulation. Thus, cell compartments can certainly cooperate at the level of intracellular fluxes of structural proteins.

2. Although the question of whether RNA processing is obligatory for decoding or not has not been finally settled (cf. Westhoff and Herrmann, 1988; Barkan, 1988), and may, of course, differ from one primary transcript or processing intermediate to another, it is possible that a major step in coordinating organelle gene expression resides in RNA processing (and translation) and that nuclear control is used to make this control more effective. The finding of spliceosome-like, nuclear-coded polypeptides that can be imported into plastids (Li and Sugiura, 1990), the observation that correct RNA processing can occur in pleiotropic plastome mutants that are virtually unable to synthesize any plastid protein at all, and the existence of nuclear mutants affecting RNA modification in the organelle (probably in *trans*), are consistent with this idea. Although this hypothesis provides a reasonable explanation for some observations, it does not readily explain other relevant effects.

3. We have mentioned above that an exchange of plastids and nuclei between species, which can cause severe disturbances of the photosynthetic membrane in the resulting interspecific genome/plastome hybrids, reflects an incompatibility between alien genomes and plastomes in one cell. In principle, hybrid bleaching could either reflect a regulatory or a structural lesion. Arguments in favour of the latter possibility are the detection of variant polypeptides with plastome- (51 kDa chlorophyll *a* apoprotein of photosystem II or ATP synthase subunit alpha) or genome- (subunit II of photosystem I or the 33 kDa polypeptide of the water oxidation complex) associated inheritance (cf. Fig. 2 in Herrmann et al., 1980) and a remarkable correlation between the extent of plastome differences and compatibility relationships. If correct, the underlying principle of plastome/genome cooperation (compatibility/incompatibility) would be compensatory mutation in partner polypeptides originating from different compartments.

4. If plastids are descendants of endosymbionts, it is conceivable that nuclear genes exert an indispensible function to control gene expression of the "invader" while the reverse signal (Oelmüller, 1989; Taylor, 1989) may be metabolic. A plausible working hypothesis is therefore that the information flow of the organelle could be controled by a nuclear protein factor acting in *trans*, and that this signal should be quite abundant, provided all plastid chromosomes of a cell were active (see above). Using specific cDNA libraries, constructed with polyadenylated RNA from polysomal RNA fractions of specifically pretreated greening *Oenothera* seedlings, we have recently isolated a cDNA clone which codes for a Zn-finger protein of approximately 8 kDa (precursor polypeptide: 131 codons) that can be imported into isolated chloroplasts and possesses all attributes for such a signal. This protein binds relatively stable to organelle DNA and can be released after digestion of lysed organelles with DNAase. Another interesting point is that the protein can form dimers probably via its N-terminus and that it contains characteristically spaced aromatic side chains reminiscent to those found in RNA polymerases. Fig. 6 illustrates some of the features of the protein and some ideas how it could interact with DNA. The functional role of this intriguing component is under investigation in our laboratory.

We have reviewed different types of data that bear on thylakoid biogenesis. Their consideration leads us to point out that crucial steps in this biogenesis are not really understood. The synthesis of the membrane begins with the transcriptional activation of genes and ends with the specific location of a protein in its correct position. Such multistep processes are generally subject to elaborate control mechanisms. Although considerable advances have been made, our knowledge how the individual processes of regulation are matched with the biogenesis of the entire membrane system is very limited. The world of

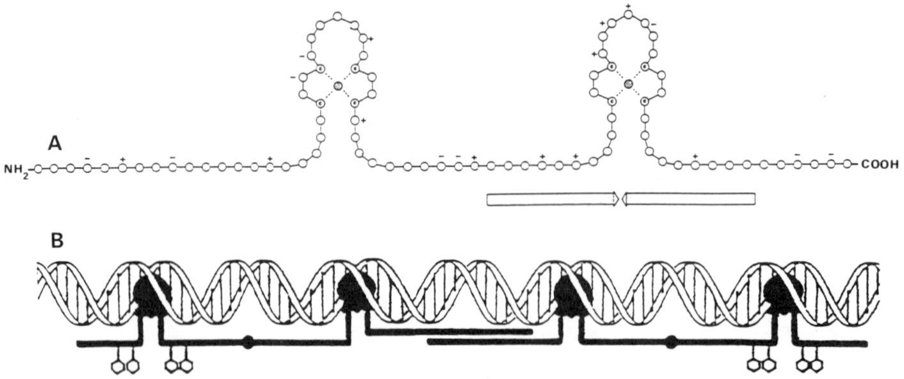


**Fig 6A.** Predicted structure of the organelle-imported DNA-binding Zn-finger protein ICC-1. **B**. Model for a possible interaction with organelle DNA (see text). The arrows in A indicate an inverted amino acid duplication. The bottom symbols in B designate aromatic side chains.

plastid and thylakoid biogenesis still represents a primary challenge in biology. We are convinced that it will have further surprises for us in store.

Acknowledgements: This work was supported in part by the Deutsche Forschungsgemeinschaft (SFB 184, He 693), the Bundesministerium für Forschung und Technologie (0319318A), and the Fonds der Chemischen Industrie.

REFERENCES

Ainsworth CC, Miller TE, Gale MD (1987) a-amylase and ß-amylase homoeoloci in species related to wheat. Gen Res, 49: 93-103.

Anderson JM, Andersson B (1982) The architecture of photosynthetic membranes: Lateral and transverse organization. Trends Biochem Sci, 7: 288-292.

Barkan A (1988) Proteins encoded by a complex chloroplast transcription unit are each translated from both monocistronic and polycistronic mRNAs. EMBO J, 7: 2637-2644.

Bartling D, Clausmeyer S, Oelmüller R, Herrmann RG (1990) Towards epitope models for chloroplast transit sequences. In: "Regulation of Photosynthetic Processes", S. Miyachi, ed., Bot Mag (Tokyo), Spec Issue 2: 119-144.

Bedbrook JR, Link G, Coen DM, Bogorad L (1978) Maize plastid gene expressed during photoregulated development. Proc Natl Acad Sci USA, 75: 3060-3064.

Bevan M (1984) Binary *Agrobacterium* vectors for plant transformation. Nucl Acids Res, 12: 8711-8721.

Bichler J, Herrmann RG (1990) Analysis of the promotors of the single-copy genes for plastocyanin and subunit delta of the chloroplast ATP synthase from spinach. Eur J Biochem, 190: 415-426.

Clausmeyer S, Klösgen RB, Herrmann RG, Targetting efficiency of thylakoid lumen proteins. (submitted)

Cozens AL, Walker JE, Philipps AL, Huttly AK, Gray JC (1986) A sixth subunit of ATP synthase, an Fo component, is encoded in the pea chloroplast genome. EMBO J, 5: 217-222.

Cozens AL, Walker JE (1987) The organization and sequence of the genes for ATP synthase subunits in the cyanobacterium *Synechococcus 6301*. Support for an endosymbiotic origin of chloroplasts. J Mol Biol, 194: 359-383.

Ellis RG, Robinson C (1987) Protein targeting. Adv Bot Res, 14: 1-24.

Henning J, Herrmann RG (1986) Chloroplast ATP synthase of spinach contains nine nonidentical subunit species, six of which are encoded by plastid chromosomes in two operons in a phylogenetically conserved arrangement. Mol Gen Genet, 203: 117-128.

Herrmann RG, Possingham JV (1980) Plastid DNA - the plastome. In: "Results and Problems in Cell Differentiation". vol. Chloroplasts, J. Reinert, ed., Springer: 45-96.

Herrmann RG, Seyer P, Schedel R, Gordon K, Bisanz C, Winter P, Hildebrandt JW, Wlaschek M, Alt J, Driesel AJ, Sears BB (1980) The plastid chromosomes of several dicotyledons. In: "Biological Chemistry of Organelle Function", 31st Colloquium Ges Biol Chem, Th. Bücher, W. Sebald and H. Weiß, eds., Springer: 97-112.

Herrmann RG, Westhoff P, Alt J, Winter P, Tittgen J, Bisanz C, Sears BB, Nelson N, Hurt E, Hauska G, Viebrock A, Sebald W (1982) Identification and characterization of genes for polypeptides of the thylakoid membrane. In: "Structure and Function of Plant Genomes", O. Cifferi and L. Dure III, eds., Plenum Publ Corp, New York: 143-154.

Herrmann RG, Westhoff P, Alt J, Tittgen J, Nelson N (1985) Thylakoid membrane proteins and their genes. In: "Molecular Form and Function of the Plant Genome", L. v. Vloten-Doting, G. Groot, T. Hall, eds., Plenum Publ Corp, New York: 233-256.

Herrmann RG, Westhoff P, Tyagi AK, Link G Biogenesis of plastids in higher plants. In: "Cell Organelles - Plastids, Mitochondria, Glyoxisomes, Peroxisomes" R.G. Herrmann, ed., Spinger Wien, New York, in press.

Hiratsuka J, Shimada H, Whittier R, Ishibashi R, Sakamoto M, Mori M, Kondo C, Honji Y, Sun C-R, Meng B-Y, Li Y-Q, Kanno A, Nishizawa Y, Hirai A, Shinozaki K, Sugiura M (1989) The complete sequence of the rice (*Oryza sativa*) chloroplast genome: Intermolecular recombination between distinct tRNA genes accounts for a major plastid DNA inversion during the evolution of the cereals. Mol Gen Gent, 217: 185-194.

Klein RR, Mullet JE (1986) Regulation of chloroplast-encoded chlorophyll-binding protein translation during higher plant chloroplast biogenesis. J Biol Chem, 261: 11138-11145.

Klein RR, Mullet JE (1987) Control of gene expression during higher plant chloroplast biogenesis: protein synthesis and transcript levels of *psbA*, *psaA-psaB*, and *rbcL* in dark-grown and illuminated barley seedlings. J Biol Chem, 262: 4341-4348.

Klein RR, Mason HS, Mullet JE (1988) Light-regulated translation of chloroplast proteins. I. Transcripts of *psaA-psaB*, *psbA*, and *rbcL* are associated with polysomes in dark-grown and illuminated barley seedlings. J Cell Biol, 196: 289-301.

Kreuz K, Dehesh K, Apel K (1986) The light-dependent accumulation of the P700 chlorophyll-*a* protein of the photosystem I reaction center in barley: evidence for translational control. Eur J. Biochem, 159: 459-467.

Laing W, Kreuz W, Apel K (1988) Light-dependent, but phytochrome-independent, translational control of the accumulation of the P700 chlorophyll-*a* protein of photosystem I in barley (*Hordeum vulgare* L.). Planta, 176: 269-276.

Li Y, Sugiura M (1990) Three distinct ribonucleoproteins from tobacco chloroplasts: Each contains a unique amino terminal acidic domain and two ribonucleoprotein consensus motifs. EMBO J, 9: 3059-3066.

Marsden JE, Schawager SJ, May B (1987) Single-locus inheritance in the tetraploid tree frog *Hyla versicolor* with an analysis of expressed progeny ratios in tetraploid organisms. Genetics, 116: 299-311.

Oelmüller R (1989) Photooxidative destruction of chloroplasts and its effect on nuclear gene expression and extraplastidic enzyme levels. Photochem Photobiol, 49: 229-239.

Ohyama K, Fukuzawa H, Kohchi T, Shirai H, Sano T, Sano S, Umesono K, Shiki Y, Takeuchi M, Chang Z, Aota S, Inokuchi H, Ozeki H (1986) Chloroplast gene organization deduced from complete sequence of liverwort *Marchantia polymorpha* chloroplast DNA. Nature, 322: 572-574.

Palmer JD (1985) Comparative organization of chloroplast genomes. Annu Rev Genet, 19: 325-354.

Rodermel SR, Bogorad L (1985) Maize plastid photogenes: mapping and photoregulation of transcript levels during light-induced development. J Cell Biol, 100: 463-476.

Shinozaki K, Ohme M, Tanaka M, Wakasugi T, Hayashida N, Matsubayashi T, Zaita N, Chunwongse J, Obokata J, Yamaguchi-Shinozaki K, Ohto C, Torazawa K, Meng B-Y, Sugita M, Deno H, Kamogashira T, Yamada K, Kusada J, Takaiwa F, Kato A, Tohdoh N, Shimada H, Sugiura M (1986) The complete nucleotide sequence of the tobacco chloroplast genome: its gene organization and expression. EMBO J, 5: 2043-2049.

Stubbe W (1959) Genetische Analyse des Zusammenwirkens von Genom und Plastom bei *Oenothera*. Z Indukt Abstamm Vererbungsl, 90: 288-298.

Sugita M, Gruissem W (1987) Development, organ-specific and light-dependent expression of the tomato ribulose-1,5-bisphosphate carboxylase small subunit gene family. Proc Natl Aca Sci USA, 84: 7104-7108.

Taylor WC (1989) Regulatory interactions between nuclear and plastid genomes. Ann Rev Plant Phys Plant Mol Biol, 40: 211-233.

Tyagi A, Hermans J, Steppuhn J, Jansson C, Vater J, Herrmann RG (1987) Nucleotide sequence of cDNA clones encoding the complete "33 kd" precursor protein associated with the photosynthetic oxygen-evolving complex from spinach. Mol Gen Genet, 207: 288-293.

Westhoff P, Farchaus JW, Herrmann RG (1986) The gene for the Mr 10.000 phosphoprotein associated with photosystem II is part of the *psbB* operon of the spinach plastid chromosome. Curr Genet, 11: 165-169.

Westhoff P, Herrmann RG (1988) Complex RNA maturation in chloroplasts: the *psbB* operon from spinach. Eur J Biochem, 171: 551-564.

Yao WB, Meng BY, Tanaka M, Sugiura M (1989) An additional promotor within the protein-coding region of the *psbD-psbC* gene cluster in tobacco chloroplast DNA. Nucl Acids Res, 17: 9583-9591.

# GREENING OF ETIOLATED MONOCOTS - THE IMPACT OF LEAF DEVELOPMENT ON PLASTID GENE EXPRESSION

Peter Westhoff and Hans Schrubar

Institut für Entwicklungs- und Molekularbiologie der Pflanzen
Heinrich-Heine-Universität
D-4000 Düsseldorf 1, FRG

## INTRODUCTION

Chloroplast biogenesis consists of a complex series of events which finally leads to the assembly of a functional photosynthetic apparatus. Since genes encoding plastidial proteins are scattered between nuclear and plastid genomes, intergenomic cooperation is an inherent feature of this developmental process. However, neither the signals mediating the exchange of information between the plastid and the nucleo-cytosolic compartments nor master regulatory genes controlling the development of chloroplasts have been identified (Taylor, 1989).

Chloroplast biogenesis in higher plants is remarkably consistent and reproducible (Leech, 1986). Under normal diurnal light-regimes chloroplasts develop from the proplastids of meristematic tissues of the leaf. The photosynthetic capacity of the differentiating plastids increases progressively (Leech, 1984; Wellburn, 1982; Wellburn, 1987). Chloroplast biogenesis under natural conditions is, therefore, a continuous process which leads to the coordinated assembly of the entire photosynthetic machinery. This contrasts with chloroplast development during light-induced greening of etiolated seedlings where it is a step by step process (Leech, 1984).

At first glance photosynthesis may be considered as a cellular property. In higher plants, however, photosynthetic carbon assimilation results from the combined efforts of different cells within a single organ, the leaf. This is particularly obvious in C4 plants of the NADP malic enzyme subgroup. The mesophyll and bundle sheath cells of these species show distinct metabolic capabilities and display a labour-sharing mode of organization (Hatch, 1987; Furbank and Foyer, 1988). The photosynthetic performance of a higher plant, therefore, relies on the smooth integration of cellular and supracellular activities into a common regulatory framework. From this follows that the biogenesis of chloroplasts is an integral part of the developmental programme of the leaf and the expression patterns of photosynthesis genes are likely to depend on the differentiation state of the leaf. Evidence for such an intimate coupling of chloroplast and overall leaf differentiation has recently been obtained from genetic analysis with *Arabidopsis* (Chory et al., 1989).

In the present paper work is summarized which attempts to unravel the influence of leaf development on the expression characteristics of plastid genes. The light-induced greening of etiolated monocotyledonous plants is used as a model system.

## RESULTS AND DISCUSSION

### Plastid gene expression during greening of etiolated *Sorghum* seedlings is controlled by light at the transcriptional level

Greening of etiolated seedlings is widely used to investigate the expression of plastid or nuclear genes that are involved in the biogenesis of chloroplasts. The system is attractive, because a single stimulus, i.e. light, is capable of inducing the sequence of events which transforms a photosynthetically inactive etioplast into a fully functional chloroplast. It has been shown with spinach and barley that all plastid genes analysed to date are transcribed in the dark-grown plants and that the light control of chloroplast biogenesis is mostly due to regulation at the translational or post-translational level (Herrmann et al., 1985; Kreuz et al., 1986; Klein and Mullet, 1987; Klein et al., 1988). From these and other investigations (Deng and Gruissem, 1987; Deng and Gruissem, 1988; Deng et al., 1989) it has been emphasized that post-transcriptional control is the prevalent regulatory principle used by plastids for the expression of their genes (Gruissem et al., 1988).

For an analysis of the impact of leaf differentiation on the expression patterns of plastid genes the monocotyledonous C4 plant *Sorghum bicolor* (Sorghum bicolor cv. Sorghum sudanense hybrid "Golden Harvest"; Rob-Seed Co., Waterloo, Nebraska) was chosen as an experimental system (Westhoff et al., 1988; Westhoff et al., 1990). When grown in the dark for five days *Sorghum* seedlings are characterized by tremendously enlarged meso cotyls. On the other hand, the primary leaf is tiny and enclosed by the coleoptile. Even after prolonged periods of etiolation, i.e. ten to thirteen days of growth in darkness, most of the leaves are still buried within the coleoptile and their growth is retarded. These observations suggest that leaf development in *Sorghum* is blocked in the absence of light. This contrasts with other monocotyledonous plants like barley where leaf development is not arrested during etiolation. Thus by using light as a trigger leaf development can be initiated in etiolated *Sorghum* seedlings (Schrubar et al., 1990).

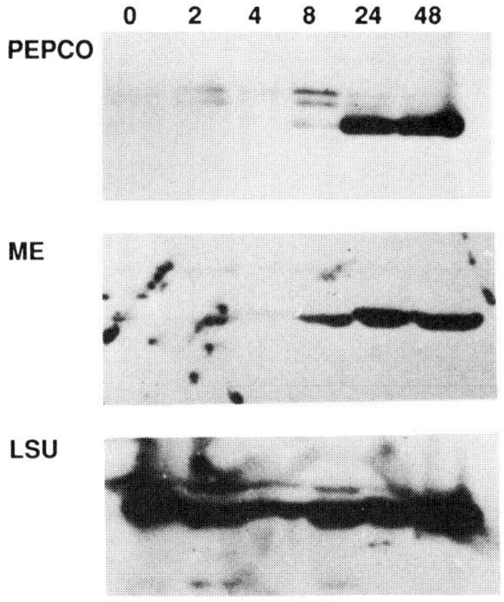


Fig. 1. Accumulation of enzymes of carbon assimilation in greening *Sorghum* seedlings. Seedlings were grown for five days in darkness and illuminated for up to 48 hr. Total soluble proteins were isolated from equal amounts of leaf material at the times given. Levels of phosphoenolpyruvate carboxylase (PEPCO), NADP malic enzyme (ME) and the large subunit of ribulose bisphosphate carboxylase (LSU) were assessed by immunoblotting (Schrubar et al., 1990).

When *Sorghum* seedlings are grown in darkness for five days and are illuminated subsequently, all thylakoid membrane polypeptides appear in a light-dependent fashion. This is true for the chlorophyll apoproteins of both photosystems as well as for the subunits of the cytochrome b/f complex and the accessory polypeptides of the water-splitting unit of photosystem II. In spinach, the latter accumulate at the etioplast stage and their levels do not change significantly during the greening process (Westhoff et al., 1988; 1990). The entire process of thylakoid membrane assembly in *Sorghum* seedlings, therefore, requires the presence of light (Schrubar et al., 1990). A light-dependency is also observed for NADP-malic enzyme and phosphoenolpyruvate carboxylase, two key enzymes of the C4-cycle (Fig. 1). However, the large subunit of ribulose bisphosphate carboxylase does not conform as its levels do not change significantly during the greening (Fig. 1).

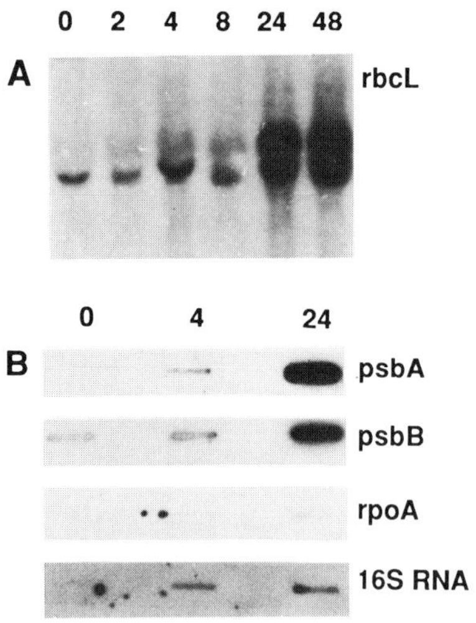


Fig. 2. Titration of plastid RNAs during the greening of *Sorghum* seedlings. Plants were grown as indicated in the legend to Fig. 1. Panel A: Levels of *rbcL* transcripts. Total RNA was analysed by Northern blotting. Two transcripts of 1.8 and 1.9 kb are shown by the fluorograph (Poulsen, 1984). Panel B: Slot blot analysis. RNA was prepared from the same amounts of intact plastids (3 x $10^7$) and transferred to a Biodyne A nylon membrane by use of a slot-blot device. Hybridization was performed by using the corresponding spinach genes as probes. *PsbA* and *psbB* encode the D1 and CP47 polypeptides of photosystem II reaction centre, *rpoA* the $\alpha$ subunit of plastid RNA polymerase, and *rbcL* the large subunit of ribulose bisphosphate carboxylase.

Light affects also the steady state levels of plastid RNAs. Transcripts encoding thylakoid membrane proteins and the large subunit of ribulose bisphosphate carboxylase increase dramatically upon illumination of etiolated seedlings (Schrubar et al., 1990; Fig. 2). Only a slight rise is apparent for transcripts encoding the $\alpha$ subunit of plastid RNA polymerase (*rpoA*) and for plastid ribosomal RNA (Fig. 2). To distinguish whether the changes in transcript levels are due to alterations in the rates of transcription or to alterations in RNA stability, run-on experiments have been performed. It was found that the transcriptional activity of plastids is very low in five day old dark-grown *Sorghum* seedlings and does not change significantly, when growth in darkness is extended to seven days. However, plastids isolated from the illuminated seedlings exhibit up to sevenfold higher rates of transcription as measured by the run-on assay (Schrubar et al., 1990). By using an antiserum raised against a fusion protein of the $\beta$ subunit of plastid RNA polymerase it was shown that the levels of this enzyme increase drastically, when the etiolated seedlings become illuminated. The data demonstrate that in *Sorghum* light is capable of activating the transcription activity of plastids. This is correlated with rising levels of RNA polymerase. The light-induced accumulation of plastid transcripts is accompanied by a light-dependent assembly of all thylakoid membrane complexes. Thus transcriptional control governs the expression of plastid genes during greening of dark-grown *Sorghum* seedlings (Schrubar et al., 1990). However, the non-coordinate accumulation of transcripts and protein in case of the large subunit of ribulose bisphosphate carboxylase indicates additional controlling mechanisms operating at a post-transcriptional level.

A comparative analysis of greening in maize and oat

The question arises why the expression patterns of plastid genes in greening *Sorghum* seedlings apparently differ from that observed in other species. The data reported for *Sorghum* suggest, but does not prove unequivocally that the strict light-dependence of leaf development in this plant accounts for the observed peculiarities in plastid gene expression. For a broader evaluation of these observations greening was analysed in two further monocotyledonous species, i.e. oat and maize, using the same growth conditions as applied for *Sorghum*.

When grown in the dark for five days the primary leaves of maize seedlings (*Zea mays* L., cv. OP Golden Bantum; Quality Seeds, Caldwell, Idaho) are small with respect to the extended mesocotyl and embedded within the coleoptile. The developmental status of five day old etiolated maize seedlings is thus similar to *Sorghum*. Five day old, dark-grown oat seedlings (*Avena sativa*, cv. Pirol; BayWa, München, Bayern), on the other hand, behave differently. Their primary leaves are larger (4-6 cm) than those of *Sorghum* and maize and have penetrated the coleoptile. If relying upon these morphological parameters, leaf development in oat at this stage is more advanced than that of *Sorghum* and maize.

As a first step the patterns of RNA accumulation were established for oat and maize. Fig. 3 shows that levels of plastid transcripts in maize rise in a light-dependent manner (cf. Rodermel and Bogorad, 1985). In contrast to *Sorghum*, however, the increase becomes apparent only after eight hours of illumination. Moreover, RNA levels in the dark-grown maize seedlings are higher as compared to *Sorghum* (Schrubar et al., 1990). On the other hand, almost no changes in the levels of plastid transcripts are observed during the greening of etiolated oat seedlings as has been reported for spinach (Westhoff et al. 1990) and barley (Kreuz et al., 1986; Klein and Mullet, 1987). The light-independent accumulation of plastid transcripts in greening oat is thus strikingly different to *Sorghum* and also to maize which appears to hold an intermediate position.

As far as the assembly of thylakoid membranes is concerned, differences between the three species are more subtle. It is common for angiosperm seedlings that the chlorophyll apoproteins of photosystem I and II reaction centres do not accumulate in the absence of chlorophyll biosynthesis, i.e. in the dark. However, subunits of the cytochrome b/f complex, the ATP synthase and the 34, 23 and 16 kDa proteins of the water-splitting apparatus of photosystem II are present in dark-grown seedlings and their accumulation apparently does not require light

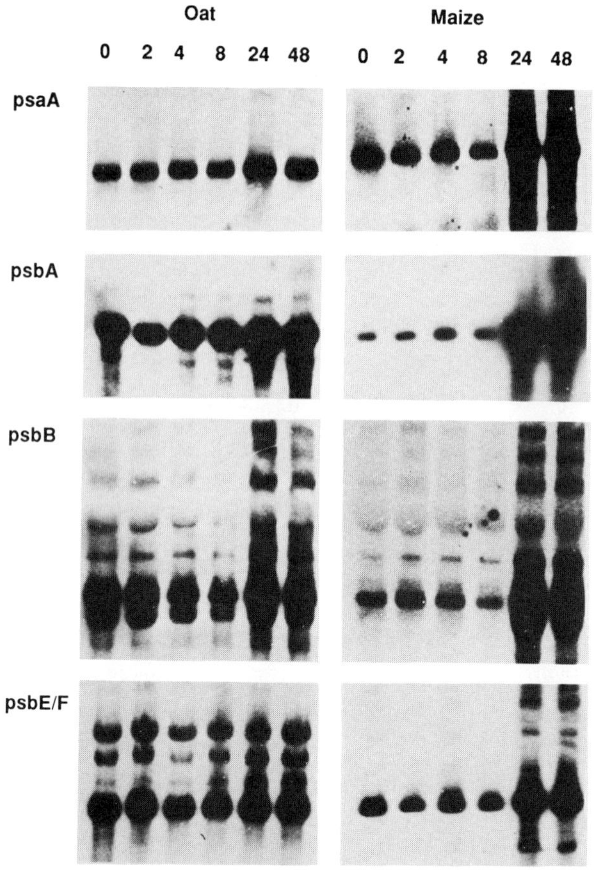


Fig. 3. Plastid transcript levels in greening oat and maize seedlings. Plants were grown in darkness for five days and then illuminated for up to 48 hr. Equal amounts (10 μg) of total cellular RNA were analyzed by Northern blotting (Schrubar et al., 1990) using the corresponding spinach genes as probes for hybridization. *PsaA* encodes the P700 chloropyll apoprotein of photosystem I reaction centre, *psbE/F* the α and β subunits of cytochrome $b_{559}$. For the designation of *psbA* and *psbB* see the legend to Fig. 2

(Nechushtai and Nelson, 1985; Liveanu et al., 1986; Westhoff et al., 1988; 1990). For this reason subunit 4 of the cytochrome b/f complex was selected as a representative for the latter group of polypeptides to compare the mode of thylakoid membrane biogenesis during the greening of five day old dark-grown *Sorghum*, maize and oat seedlings. Fig. 4A shows that the species are almost indistinguishable with respect to the accumulation of this protein. Traces of subunit 4 are detectable in etiolated seedlings, light causes a drastic increase in its steady state levels.

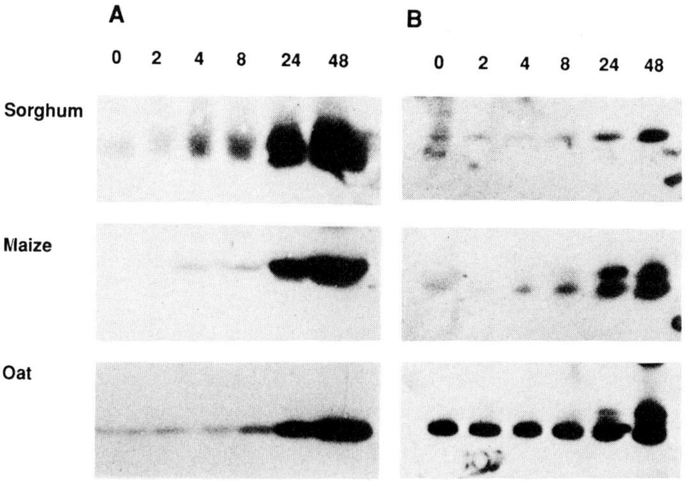


Fig. 4 Accumulation of subunit 4 of the cytochrome b/f complex in greening *Sorghum*, maize and oat seedlings. Seedlings were grown for five (Panel A) or eleven days (Panel B) in darkness and illuminated for up to 48 hr. Total membrane proteins were subjected to immunoblot analysis as described (Schrubar et al., 1990) using an antiserum to the spinach protein (Alt et al., 1983).

To investigate whether the status of leaf development affects the accumulation pattern of subunit 4 during the greening, the etiolation treatment was extended to eleven days. Leaf development in *Sorghum* is still arrested, i.e. leaves are tiny and enclosed in the coleoptile, while the primary and secondary leaves of oat have almost reached their final size. Maize seedlings are intermediate to *Sorghum* and oat with respect to growth characteristics. Their primary and secondary leaves break through the coleoptile only nine days after sowing and are still expanding after eleven days of etiolation. Fig. 4B shows that subunit 4 levels in etiolated *Sorghum* and maize are minute and rise only upon illumination. In contrast, the protein accumulates in etiolated oat seedlings and its level does not change substantially during illumination. While accumulation of subunit 4 in five day old dark-grown oat seedlings is strictly dependent upon light, the light-requirement is abolished when leaf development in the dark has progressed.

## CONCLUSIONS

Plastids are versatile organelles fulfilling quite divergent functions in plant metabolism and development. Thus there may be a need for flexibility in the controlling mechanisms of plastid gene expression. For example, regulation by post-transcriptional means may be one mode plastids can use to alter the expression levels of their genes. However, the importance of the various levels of gene regulation may vary with respect to plastid differentiation. By analysing the light-induced greening of etiolated *Sorghum* seedlings it has been shown that in this plant the photoregulated expression of plastid genes is controlled at the transcriptional level and that the entire process of thylakoid membrane biogenesis requires the presence of light. Since the peculiarities in plastid gene expression observed in *Sorghum* are correlated with an inhibition of leaf development during etiolation, it has been concluded that the status of leaf differentiation controls the expression characteristics of plastid genes (Schrubar et al., 1990). The analysis of protein and RNA accumulation in maize and oat supports this view. The species-specific variations observed in the accumulation patterns can be correlated with differences in the differentiation states of the leaves as assessed by morphological criteria. Work is in progress now to be extend this analysis by following the transcriptional activity of plastids during leaf differentiation. Thus it will be possible to determine whether transcriptional regulation is the prevalent controlling mechanism in the early stages of plastid development.


## ACKNOWLEDGMENTS

We thank Dr. U. Santore (Botanical Institute, University of Düsseldorf) for carefully reviewing this manuscript and M. Streubel for the photographic work. This research is supported by a grant from the Deutsche Forschungsgemeinschaft (SFB 189).


## REFERENCES


Alt, J., Westhoff, P., Sears, B.B., Nelson, N., Hurt, E., Hauska, G. and Herrmann, R.G., 1983, Genes and transcripts for the polypeptides of the cytochrome b6/f complex from spinach thylakoid membranes, EMBO J., 2:979-986.

Chory, J., Peto, C., Feinbaum, R., Pratt, L. and Ausubel, F., 1989, *Arabidopsis thaliana* mutant that develops as a light-grown plant in the absence of light, Cell, 58:991-999.

Deng, X.-W., Tonkyn, J.C., Peter, G.F., Thornber, J.P. and Gruissem, W., 1989, Post-transcriptional control of plastid mRNA accumulation during adaptation of chloroplasts to different light quality environments, Plant Cell, 1:645-654.

Deng, X.W. and Gruissem, W., 1987, Control of plastid gene expression during development: the limited role of transcriptional regulation, Cell, 49:379-387.

Deng, X.W. and Gruissem, W., 1988, Constitutive transcription and regulation of gene expression in non-photosynthetic plastids of higher plants, EMBO J., 7:3301-3308.

Furbank, R.T. and Foyer, C.H., 1988, C4 plants as valuable model experimental systems for the study of photosynthesis, New Phytol., 109:265-277.

Gruissem, W., Barkan, A., Deng, X.W. and Stern, D., 1988, Transcriptional and post-transcriptional control of plastid mRNA levels in higher plants, Trends Genet., 4:258-263.

Hatch, M.D., 1987, C4 photosynthesis: a unique blend of modified biochemistry, anatomy and ultrastructure, Biochim. Biophys. Acta, 895:81-106.

Herrmann, R.G., Westhoff, P., Alt, J., Tittgen, J. and Nelson, N., 1985, Thylakoid membrane proteins and their genes, in: Molecular Form and Function of the Plant Genome, van Vloten-Doting, L., Groot, G.S.P. and Hall, T.C., eds., Plenum Publishing Corporation, New York, pp. 233-256.

Klein, R.R., Mason, H.S. and Mullet, J.E., 1988, Light-regulated translation of chloroplast proteins. I. Transcripts of *psaA-psaB*, *psbA*, and *rbcL* are associated with polysomes in dark-grown and illuminated barley seedlings, J. Cell Biol., 106:289-301.

Klein, R.R. and Mullet, J.E., 1987, Control of gene expression during higher plant chloroplast biogenesis: protein synthesis and transcript levels of *psbA*, *psaA-psaB*, and *rbcL* in dark-grown and illuminated barley seedlings, J. Biol. Chem., 262:4341-4348.

Kreuz, K., Dehesh, K. and Apel, K., 1986, The light-dependent accumulation of the P700 chlorophyll a protein of the photosystem I reaction center in barley: evidence for translational control, Eur. J. Biochem., 159:459-467.

Leech, R.M., 1984, Chloroplast development in angiosperms: current knowledge and future prospects, in: Chloroplast Biogenesis, Baker, N.R. and Barber, J., eds., Elsevier Science Publishers, pp. 1-21.

Leech, R.M., 1986, Stability and plasticity during chloroplast development, in: Symposium of the Society for Experimental Biology: Plasticity in Plants, Vol. 40, Jennings, D.H. and Trewavas, A.J., eds., University of Cambridge, Cambridge, pp. 121-153.

Liveanu, V., Yocum, C.F. and Nelson, N., 1986, Polypeptides of the oxygen-evolving photosystem II complex. Immunological detection and biogenesis, J. Biol. Chem., 261:5296-5300.

Nechushtai, R. and Nelson, N., 1985, Biogenesis of photosystem I reaction center during greening of oat, bean and spinach leaves, Plant Mol. Biol., 4:377-384.

Poulsen, C., 1984, Two mRNA species differing by 258 nucleotides at the 5' end are formed from the barley chloroplast *rbcL* gene, Carlsberg Res. Commun., 49:89-104.

Rodermel, S.R. and Bogorad, L., 1985, Maize plastid photogenes: mapping and photoregulation of transcript levels during light-induced development, J. Cell Biol., 100:463-476.

Schrubar, H., Wanner, G. and Westhoff, P., 1990, Transcriptional control of plastid gene expression in greening *Sorghum* seedlings, Planta, in press.

Taylor, W.C., 1989, Regulatory interactions between nuclear and plastid genomes, Ann. Rev. Plant Physiol. Plant Mol. Biol., 40:211-233.

Wellburn, A.R., 1982, Bioenergetic and ultrastructural changes associated with chloroplast development, Int. Rev. Cytol., 80:133-191.

Wellburn, A.R., 1987, Plastids, Int. Rev. Cytol., 17:149-210.

Westhoff, P., Grüne, H., Schrubar, H., Oswald, A., Streubel, M., Ljungberg, U. and Herrmann, R.G., 1988, Mechanisms of plastid and nuclear gene expression during thylakoid membrane biogenesis in higher plants, in: Photosynthetic Light-Harvesting Systems-Structure and Function, Scheer, H. and Schneider, S., eds., Walter de Gruyter Verlag, Berlin, pp. 261-276.

Westhoff, P., Schrubar, H., Oswald, A., Streubel, M. and Offermann, K., 1990, Biogenesis of photosystem II in C3 and C4 plants - a model system to study developmentally regulated and cell-specific expression of plastid genes, in: Current Research in Photosynthesis, Vol. III, Baltscheffsky, M., ed., Kluwer Academic Publishers, Dordrecht, pp. 483-490.

# REGULATION OF CHLOROPLAST BIOGENESIS IN BARLEY

John E. Mullet, Jeffrey C. Rapp, Brian J. Baumgartner, Tineke Berends-Sexton and David A. Christopher

Department of Biochemistry and Biophysics
Texas A&M University
College Station, Texas 77843-2128

## INTRODUCTION

Higher plant plastids contain multiple copies of a circular genome which encodes between 120 to 140 genes[1,2,3]. Many of these genes encode proteins or RNAs involved in transcription or translation. Among this group of genes are *rpo*A, *rpo*B, *rpo*$C_1$, and *rpo*$C_2$ which encode RNA polymerase subunits, 18 genes encoding ribosomal proteins, 4 rRNA-encoding genes and 30 tRNA genes. Another large group of plastid genes encode proteins involved in photosynthesis. These include *rbc*L, which encodes the large subunit of Rubisco, *psa*A, *psa*B and *psa*C which encode subunits of Photosystem I (PSI), *psbA* through *psb*F and *psb*H through *psb*N, which encode subunits of Photosystem II (PSII), and other genes encoding proteins of the chloroplast ATP synthase and cytochrome $b_6/f$ complexes. While the plastid genome contributes subunits to Rubisco and the electron transport units, these complexes also contain proteins encoded by nuclear genes. As a consequence, biosynthesis of these protein complexes requires coordination of plastid and nuclear gene expression.

Coordination of plastid gene expression and expression of nuclear genes encoding plastid proteins occurs at several levels. If proteins encoded by either genome are synthesized in excess, these subunits are degraded[4,5]. This process maintains the subunit composition of most plastid protein complexes in relatively fixed stoichiometries. A second level of control operates on plastid translation. In this case, inhibition of cytoplasmic translation leads to subsequent decreases in plastid protein synthesis[6]. Expression of genes in the two compartments is also co-regulated by environmental influences such as light. Light acting through the photoreceptor, phytochrome, stimulates transcription of plastid genes[7,8] as well as some nuclear genes which encode plastid proteins (i.e., *cab*, *rbc*S)[9]. Furthermore, plastid genes and nuclear genes encoding plastid proteins are co-induced in a cell specific manner (i.e., leaf mesophyll cells>epidermal

*Plant Molecular Biology 2*, Edited by R.G. Herrmann and B. Larkins, Plenum Press, New York, 1991

or vascular cells), and both gene sets are coordinately induced at a specific stage of leaf development[10]. At present, the molecular basis of cell specific expression and developmental timing is unknown. However, recent studies implicate the existence of a plastid-derived factor which modulates expression of some nuclear genes[11]. This discovery may provide insight into one mechanism which helps coordinate nuclear and plastid gene expression during chloroplast biogenesis.

Chloroplast biogenesis in higher plants occurs during the conversion of meristematic cells of a leaf primordia into leaf mesophyll cells. This developmental process involves an increase in plastid number per cell, activation of plastid DNA synthesis, a build-up of plastid DNA copy number, and coordinated activation of plastid and nuclear genes and biosynthetic pathways (i.e., lipid, tetrapyrrole, terpenoid)[10]. Recently our lab has investigated the order of these events in developing leaves of the monocot barley[12-14]. Monocot leaves are ideal for developmental studies because most leaf cells are derived from a meristematic zone localized in the leaf base. In a growing leaf this results in a gradient of cell and chloroplast development which extends from undifferentiated cells of the leaf base to more mature cells and chloroplasts at the leaf apex. Plastid DNA synthesis and plastid replication are active in the barley leaf basal meristem whereas plastid transcription activity is low[12] (Baumgartner and Mullet, unpublished). This is consistent with a high DNA copy number but low levels of RNA in plastids isolated from this region[12]. When cells of the leaf base stop dividing, they are displaced toward the leaf apex and cell elongation begins. Plastid transcription increases during the period of cell elongation. This activity contributes to a build-up of the plastids translational capacity (i.e., ribosomes, tRNA, rRNA, mRNA) and subsequent accumulation of the photosynthetic apparatus. Once the mature chloroplast population is established, plastid transcription and translation activity decline again[13].

In this paper we have investigated three different aspects of chloroplast biogenesis. First we quantify the build-up of carotenoid and chlorophyll in developing plastids. Second, a procedure for isolation of the plastid RNA polymerase is presented and third the relative stability of *psb*D-*psb*C transcripts is assayed using the chloroplast transcription inhibitor tagetitoxin.

## MATERIALS AND METHODS

Plant Growth - Barley (*Hordeum vulgare* L. var. Morex) seeds were imbibed in distilled water for 1 h with continuous shaking and planted in vermiculite saturated with full strength Hoagland's solution at a depth of 2 cm. All plants were grown in controlled environmental chambers at 23°C. Seedlings were transferred to illuminated chambers ($350 \mu E \cdot m^{-1} \cdot s^{-1}$) for various times or left in darkness throughout the growth period.

Plastid Isolation - Leaves were divided into sections as previously described, prior plastid isolation on Percoll gradients[12]. Chlorophyll was assayed as before[14] and carotenoid content estimated by the method of Bottomley[20].

## Plastid Nucleic Acid Isolation and S1 Nuclease Analysis

Total nucleic acid was isolated from leaves as previously described[15] and resuspended in $H_2O$. Total nucleic acid was treated with RQ1-DNase (Promega Biotech), phenol extracted and EtOH-precipitated. DNA probes (restriction fragments or oligonucleotides) were 5′-end labeled with [$\alpha$-$^{32}$P-ATP]. $S_1$ nuclease and primer extension assays were done as previously described[15].

# RESULTS AND DISCUSSION

## Accumulation of Chlorophyll and Carotenoid During Chloroplast Biogenesis

We previously showed that basal sections of primary barley leaves contain dividing cells and proplastids[12]. Between 1 and 2 cm from the leaf base cell division stops, cell enlargement begins and plastid transcription activity increases. The data in Table I shows changes in the carotenoid and chlorophyll content of plastids as a function of development in barley primary leaves. Carotenoid and chlorophyll levels are low in plastids of the leaf base but the level of these compounds begins to increase once plastids enter the zone of cell elongation. Accumulation of chlorophyll is gradual and parallels accumulation of the chloroplast electron transport system. It should be noted that carotenoid data was from plants grown either 4 days in darkness (basal 4 cm) or 6 days in darkness (apical 10 cm) whereas chlorophyll accumulation is from plants grown 2 days in darkness then illuminated for an additional 2 or 4 days. The data in Table I indicate that activation of tetrapyrrole and terpenoid biosynthetic pathways during chloroplast biogenesis parallels the activation of plastid gene expression. This could indicate that the nuclear genes encoding enzymes in these pathways are modulated by a factor produced by transcriptionally active plastids.

Table I. Accumulation of chlorophyll and carotenoids as a function of plastid development in barley.

| Leaf section (cm from leaf base) | Chlorophyll ($\mu$g/$10^9$ plastids) | Carotenoid ($\mu$g/$10^9$ plastids) |
|---|---|---|
| 0 - 1 | 7 | 1.8 |
| 1 - 3 | 27 ± 2 | 6.4 ± .6 |
| 3 - 4 | 85 ± 2 | 9.4 ± .6 |
| 4 - 6 | 166 ± 17 | ---- |
| 8 - 10 | 519 ± 29 | 16 ± 3 |
| 10 - 12 | 813 ± 94 | 24 ± .7 |
| 12 - 14 | 822 ± 80 | 52 |

## Isolation of the Plastid RNA Polymerase

A key event in chloroplast biogenesis is the activation of plastid transcription. If the increase in plastid transcription which occurs during barley leaf biogenesis is understand the basis of the increase in plastid transcription activity during leaf biogenesis. As a first step in addressing this question we have worked out a procedure to purify the plastid RNA polymerase. We used pea chloroplasts as the starting material for the isolation because large numbers of these organelles were easily obtained. The procedure starts from intact plastids isolated on Percoll density gradients. The plastids were lysed in 50 mM Hepes-KOH (pH 8.0) plus 5 mM $MgCl_2$ and then the thylakoid membranes were collected by centrifugation. Most of the plastid transcription activity remained associated with the thylakoids. RNA polymerase was released from the membrane with 1M NaCl and after removal of stripped membranes by centrifugation (10,000 xg, 10 min) the supernatant was adjusted to 75% ammonium sulfate to precipitate protein. After incubation for 30 min, 4°C the precipitate was collected (10,000 xg, 10 min), resuspended and dialyzed against 50 mM Hepes-KOH (pH 8.0) 50 mM NaCl, 1 mM $MgCl_2$ overnight. The dialyzed protein was next applied to a DE-52 column as previously described[16] and eluted with an NaCl gradient. Fractions stimulating transcription on an exogenous plasmid template containing the *psb*A promoter were combined (fractions eluting between 0.2 and 0.4 M NaCl) and precipitated with 75% ammonium sulfate. The resulting pellet was rinsed, resuspended in 1 ml of DE-52 buffer and applied to a 10 to 30% glycerol gradient. The gradients were centrifuged at 30,000 xg for 12 hrs using an SW-27 swing out rotor at 4°C. After centrifugation, gradients were fractionated, assayed and fractions able to stimulate template dependent transcription were pooled. These fractions were diluted 5-fold in DE-52 buffer and applied to a 5 ml DNA-cellulose column. The DNA cellulose column was washed and then eluted with a gradient of NaCl. Fractions able to stimulate template dependent transcription were pooled, concentrated and a portion of the enriched fraction separated on a 7.5 to 15% SDS polyacrylamide gel (Fig. 1). This analysis revealed a large number of proteins ranging in apparent molecular weight from 150 kd to 14 kd. Although the identity of these polypeptides remains to be conclusively documented, it seems likely that the 150 kd protein is encoded by *rpo*$C_1$, the 120 kd protein by *rpo*B, one of the proteins 70 to 90 kd could be encoded by *rpo*$C_2$ and the 40 kd subunit by *rpo*A (bands marked with arrows). The remaining subunits could be functional subunits of the RNA polymerase or could represent contaminating proteins. The specific activity and yield parameters for this purification procedure are shown in Table II.

Table II. Isolation of RNA-Polymerase from pea chloroplasts

| Fraction | Specific Activity (units/mg protein)[a] | Yield (%) |
|---|---|---|
| Chloroplasts | 0.006 | 100 |
| Thylakoid membranes | 0.02 | 70 |
| 1M NaCl extract | 0.10 | 40 |
| DE-52 | 0.18 | 20 |
| Glycerol gradient | 2.60 | 20 |
| DNA-cellulose | 18.30 | 15 |

[a] Unit - 1 nmole UMP incorporated/30 min/30 C

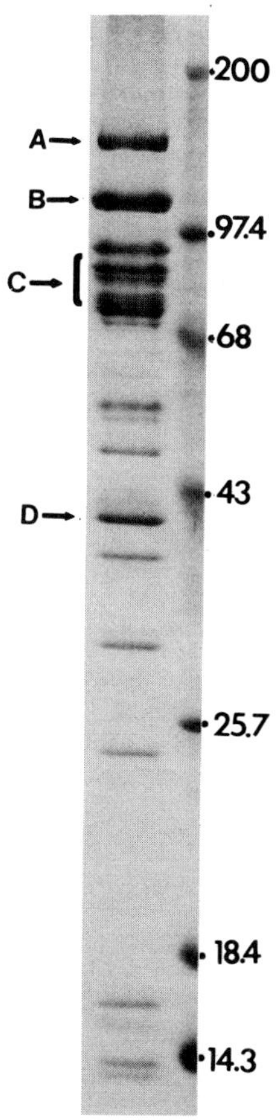


Fig.1. Polypeptide composition of partially purified pea chloroplast RNA polymerase. Molecular weight markers (kd) are shown at the right of the figure. The 150 kd protein marked by arrow A could correspond to the *rpo*$C_1$ gene product. Likewise, arrows B, C and D indicate proteins which may be encoded by *rpo*B, *rpo*$C_2$ and *rpo*A.

## Relative Stability of *psb*D-*psb*C Transcripts

*Psb*D encodes, $D_2$, a PSII reaction center protein and *psb*C encodes, CP43, a chlorophyll apoprotein in PSII. These genes are adjacent to each other in the barley plastid genome and both are part of a complex transcription unit which includes *trn*G(GCC), *trn*S(UGA; GCU), *psb*I, *psb*K and open reading frames of 41, 44 and 62 codons (Fig. 2)[17]. At least 12 different transcripts are produced from the *psb*D-*psb*C transcription unit. The accumulation of most of these RNAs is light independent (Fig. 2, arrows preceded by solid symbols)[18]. However, two RNAs, labeled 4a and 4b in Fig. 2, only accumulate in illuminated plants[18]. Several lines of evidence indicate that the accumulation of RNAs 4a and 4b requires light - induced prevented then accumulation of *rbc*S and *cab* mRNA is inhibited (Rapp and Mullet, unpublished). At present we do not

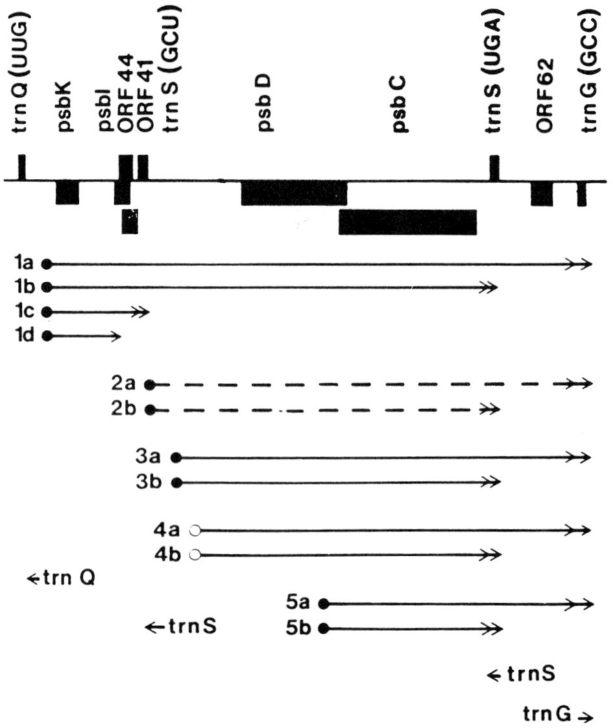


Fig. 2. 6.2 kbp DNA region of the barley plastid genome which contains the *psb*D-*psb*C transcription unit. The solid boxes at the top of the figure represent open reading frames or tRNAs. The arrows below represent RNAs. RNAs which accumulate in dark-grown plants are preceded with solid circles; arrows preceded by open circles correspond to RNAs which accumulate in illuminated plants.

transcription (Sexton, Christopher and Mullet, in preparation). However, part of the differential accumulation of RNAs 4a and 4b in illuminated plants may be due to relatively high stability of these RNAs in illuminated seedlings. To test this possibility barley seedlings were illuminated for 4 hrs to allow accumulation of RNAs 4a and 4b. Barley primary leaves were then excised and placed in water or a solution of 50 $\mu$M tagetitoxin, an inhibitor of plastid RNA polymerase[19]. At various times after initiation of inhibitor uptake, leaf samples were frozen and RNA extracted. $S_1$ nuclease protection analysis was then done on a constant amount of total RNA to determine if the relative ratio of *psb*D-*psb*C RNAs 2a/2b, 3a/3b and 4a/4b changed when plastid transcription was inhibited. The results of this experiment are shown in Fig. 3. $S_1$ nuclease protection analysis of RNA extracted from control non-excised plants is shown at the left of Fig. 3. As expected there is a gradual reduction in the level of RNAs 2a/2b and 3a/3b relative to RNAs 4a/4b as a function of plant illumination. Excised control plants show a similar trend (middle panel, Fig. 3). Tagetitoxin-treated plants, however, show a significant decline in all three *psb*D-

*psb*C RNAs by 24 hrs of inhibitor treatment. RNAs 2a/2b and 3a/3b decline to a greater extent than RNAs 4a/4b. This result indicates that RNAs 4a/4b exhibit greater stabiity than RNAs 2a/2b or 3a/3b under these conditions.

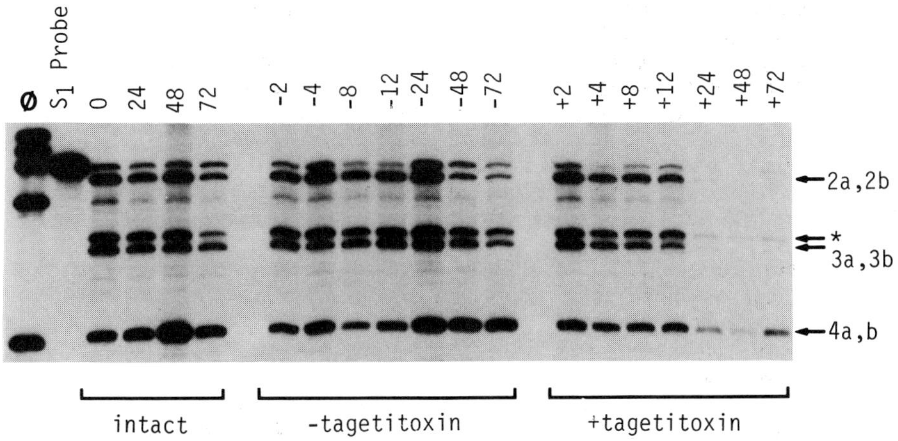


Fig. 3. S1 nuclease assays of *psb*D-*psb*C RNAs 2a/b, 3a/b and 4a/b. Barley seedlings were grown 4.5 days in darkness then illuminated for 4 hrs (time 0). A portion of the seedlings were then excised and their leaf bases placed in water (- tagetitoxin treatment) or in 50 $\mu$M tagetitoxin (+ tagetitoxin). Leaves or plants were then kept in light for up to 72 hrs prior to RNA extraction. $\phi$X174 molecular weight markers are shown at the left and the S1 nuclease probe in the second lane from the left.

**ACKNOWLEDGEMENTS:** We thank Mrs. Sharyll Pressley for her help in preparing this manuscript. This research is supported by NSF grant No. DCB-8916199 to JEM.

**REFERENCES**

1. Shinozaki, K., Ohme, M., Tanaka, M., Wakasugi, T., Hayashida, N., Matsubayashi, T., Zaita, N., Chunwongse, J., Obokata, J., Yamaguchi-Shinozaki, K., Ohto, C., Torazawa, K., Meng, B.Y., Sugita, M., Deno, H., Kamogashira, T., Yamada, K., Kusuda, J., Takaiwa, F., Kato, A., Tohdoh, N., Shimada, H. and Sugiura, M. (1986) The complete nucleotide sequence of the tobacco chloroplast genome: its gene organization and expression. EMBO J. **5**,2043-2049.

2. Ohyama, K., Fukuzawa, H., Kohchi, T., Shirai, H., Sano, T., Sano, S., Umesono, K., Shiki, Y., Takeuchi, M., Chang, Z., Aota, S.-I., Inokuchi, H., and Ozeki, H. (1986) Chloroplast gene organization deducted from complete sequence of liverwort *Marchantia polymorpha* chloroplast DNA. Nature **322**, 572-574.

3. Hiratsuka, J., Shimada, H., Whittier, R., Ishibashi, T., Sakamoto, M., Mori, M., Kondo, C., Honji, Y., Sun, C.-R., Meng, B.-Y., Li, Y.-Q., Kanno, A., Nishizawa, Y., Hirai, A., Shinozaki, K., and Sugiura, M. (1989) The complete sequence of the rice (*Oryza sativa*) chloroplast genome: Intermolecular recombination between distinct tRNA genes accounts for a major plastid DNA inversion during the evolution of the cereals. Mol. Gen. Genet. **217**, 185-194.

4. Schmidt, G.W. and Mishkind, M.D. (1983) Rapid degradation of unassembled ribulose 1,5-bisphosphate carboxylase small subunits in chloroplasts. Proc. Natl. Acad. Sci. USA **80**, 2632-2636.

5. Rodermel, S.R., Abbott, M.S. and Bogorad, L. (1988) Nuclear-organelle interactions: Nuclear antisense gene inhibits ribulose bisphosphate carboxylase enzyme levels in transformed tobacco plants. Cell **55**, 673-681.

6. Nivison, H.T. and Stocking, C.R. (1983) Ribulose bisphosphate carboxylase synthesis in barley leaves. A developmental approach to the question of coordinated subunit synthesis. Plant Physiol. **73**, 906-911.

7. Zhu, Y.S., Kung, S.D. and Bogorad, L. (1985) Phytochrome control of levels of mRNA complementary to plastid and nuclear genes of maize. Plant Physiol. **79**, 371-376.

8. Klein, R.R. and Mullet, J.E. (1990) Light-induced transcription of chloroplast genes. *PsbA* transcription is differentially enhanced in illuminated barley. J. Biol. Chem. **265**, 1895-1902.

9. Nagy, F., Kay, S.A. and Chua, N.-H. (1988) Gene regulation by phytochrome. TIG **4**, 377-42.

10. Mullet, J.E. (1988) Chloroplast development and gene expression. Ann. Rev. Plant Physiol. Plant Mol. Biol. **39**, 475-502.

11. Taylor, W.C. (1989) Regulatory interactions between nuclear and plastid genomes. Annu. Rev. Plant Physiol. Plant Mol. Biol. **40**, 211-233.

12. Baumgartner, B.J., Rapp, J.C. and Mullet, J.E. (1989) Plastid transcription activity and DNA copy number increase early in barley chloroplast development. Plant Physiol. **89**, 1011-1018.

13. Mullet, J.E. and Klein, R.R. (1987) Transcription and RNA stability are important determinants of higher plant chloroplast RNA levels. EMBO J. **6**, 1571-1579.

14. Klein, R.R. and Mullet, J.E. (1987) Control of gene expression during higher plant chloroplast biogenesis. Protein synthesis and transcript levels of *psbA, psaA-psaB*, and *rbcL* in dark-grown and illuminated barley seedlings. J. Biol. Chem. **262**, 4314-4348.

15. Mullet, J.E., Orozco, E.M. and Chua, N.-H. (1985) Multiple transcripts for higher plant *rbc*L and *atpB* genes and localization of the transcription initiation site of the *rbcL* gene. Plant Mol. Biol. **4**, 39-54.

16. Orozco, E.M.Jr., Mullet, J.E. and Chua, N.-H. (1985) An *in vitro* system for accurate transcription initiation of chloroplast protein genes. Nuc. Acids Res. **13**, 1283-1302.

17. Berends, T., Gamble, P.E. and Mullet, J.E. (1987) Characterization of the barley chloroplast transcription units containing *psaA-psaB* and *psbD-psbC*. Nuc. Acids Res. **15**, 5217-5240.
18. Gamble, P.E., Sexton, T.B. and Mullet, J.E. (1988) Light-dependent changes in *psbD* and *psbC* transcripts of barley chloroplasts: Accumulation of two transcripts maintains *psbD* and *psbC* translation capability in mature chloroplasts. EMBO J. **7**, 1289-1297.
19. Mathews, Denis E. and Durbin, J.E. (1990) Tagetitoxin inhibits RNA synthesis directed by RNA polymerases from chloroplasts and *Escherichia coli*. J. Biol. Chem. **265**, 493-498.
20. Bottomley, W. (1970) Deoxyribonucleic acid-dependent ribonucleic acid polymerase activity of nuclei and plastids from etiolated peas and their response to red and far red light *in vivo*. Plant Physiol. **45**, 608-611.

# CHLOROPHYLL BIOSYNTHESIS

Diter von Wettstein

Department of Physiology, Carlsberg Laboratory
Gamle Carlsberg Vej 10
DK-2500 Copenhagen Valby, Denmark

## CHLOROPHYLL SYNTHESIS FROM 5-AMINOLEVULINIC ACID

Chlorophyll is bound to proteins of the photosynthetic membranes. It harvests sunlight and carries out the first reactions in the conversion of light energy to chemical energy, which is conserved in $NADPH_2$ and ATP. The chemical formula of chlorophyll *a* is reproduced in Fig. 1. Four pyrrole rings (I-IV) are bound into a tetrapyrrole ring with a magnesium atom in the center. Ring IV is esterified with a higher alcohol, phytol. For light harvesting higher plants use an additional form of chlorophyll, chlorophyll *b* which contains in position 3 a formyl group (CHO) instead of a methyl group. The ring system with its characteristic conjugated double bonds is assembled in the chloroplast from 8 molecules of 5-aminolevulinic acid (Fig. 2) which contains 5 carbon atoms and as functional groups besides the carboxyl group an amino and a ketogroup. The following experiments have shown that 5-aminolevulinate is the precursor of chlorophyll: If seeds of higher plants are germinated in the dark, the seedlings have yellow leaves due to lack of chlorophyll. If these are placed in a solution of 5-aminolevulinate in darkness they will green in the course of a few hours due to the accumulation of protochlorophyllide, a late precursor of chlorophyll which in higher plants requires light for conversion into chlorophyll. If radioactively labeled 5-aminolevulinate is used with detached leaves or isolated plastids, the label is found in protochlorophyllide. This experiment reveals that all the enzymes building protochlorophyllide from 5-aminolevulinate are present in the plastid of dark-grown leaves and that it is the synthesis of 5-aminolevulinate which is limiting in the dark.

A comparison of the formula of 5-aminolevulinate (Fig. 2) with the heavy emphasized bonds in pyrrole ring IV (Fig. 1) permits the intact skeleton of 5-aminolevulinate to be recognized. The nitrogen coordinated with the magnesium atom originates from the amino group followed by four carbon atoms and the carboxyl-group which is esterified to the phytol. Synthesis of the porphyrin ring is chemically elegant and comprises the following reactions. In the catalytic site of 5-aminolevulinic acid-dehydratase (porphobilinogen synthase) two molecules of the acid are placed staggeredly side by side and by removal of two molecules of water two new carbon bonds are formed to yield a pyrrole ring as shown in Fig. 2. In the subsequent reaction four porphobilinogen molecules are coupled to an open ring or linear tetrapyrrole by removal of the free amino group at left in Fig. 2. In the resulting hydroxymethylbilane molecule, the four pyrrole molecules are connected at the $\alpha$, $\beta$ and $\gamma$ carbons (Fig. 1), while the $\delta$-carbon is linked to the dipyrromethane cofactor of the prophobilinogen deaminase (= hydroxymethylbilane synthase = urogen I synthase). As reviewed in references 1 and 2, this enzyme synthesizes its own cofactor by condensing two molecules of porphobilinogen which are

*Plant Molecular Biology* 2, Edited by R.G. Herrmann and
B. Larkins, Plenum Press, New York, 1991

bound by a thioether linkage to a cystein residue in the apoenzyme. Four porphobilinogen molecules are successively added by the deamination reaction and the tetrapyrrole cleaved off at the δ-carbon of ring I leaving the cofactor on the enzyme for the next round of tetrapyrrole formation. Ring closure of hydroxymethylbilane is catalyzed by uroporphyrinogen III cosynthase and is connected with isomerization of pyrrole ring IV, which is turned around in the process[3,4]. The porphobilinogen in Fig. 2 has at the left side an acetic acid side chain and a propionic acid side chain at the right hand side. In the succeeding step of chlorophyll formation the acetic acid side chains are shortened to methyl groups by the enzyme urogen III decarboxylase. The turning of the pyrrole ring IV with retainment of the methine carbons is seen in Fig. 1 by the fact that the methyl groups of ring IV and I are located in neighbouring positions within the porphyrin ring. Coprogen oxidase trims two of the propionic acid side chains into vinyl groups. The molecule is oxidized by protogen oxidase to establish the conjugated double bond system yielding the red colour of protoporphyrin IX.

Magnesium chelatase which inserts $Mg^{2+}$ into the porphyrin ring has not yet been successfully purified. Methylation of the propionic acid side chain on ring III to yield Mg-protoporphyrin monomethylester is accomplished by an S-adenosylmethionine:Mg-protoporphyrin O-methyltransferase. This is followed by cyclization of the methyl ester side chain to the isocyclic ring V, involving several enzymes[1,2]. A vinyl reductase converts the vinyl group of ring II into an ethyl group giving rise to protochlorophyllide.

In higher plants protochlorophyllide reductase transfers in a photochemical reaction a hydrogen from NADPH to the carbon atom carrying the propionic acid side chain of ring IV and a hydrogen from water (or protein) to the carbon atom carrying the methyl group. Protochlorophyllide is stored in the prolamellar bodies of etioplasts in association with the enzyme and NADPH. Upon illumination the pigment is converted to chlorophyllide and extensive chlorophyll synthesis and chloroplast development initiated.

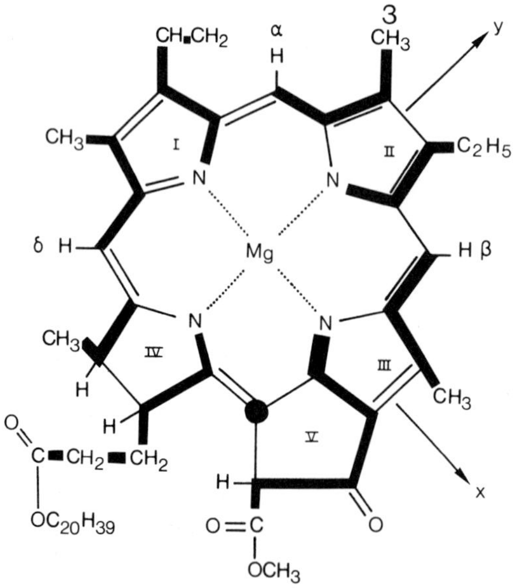

Fig. 1. The chemical formula of chlorophyll *a*. Every chlorophyll molecule is synthesized in the chloroplast from 8 molecules 5-aminolevulinic acid. The eight heavy lines indicate their location in the finished molecule.

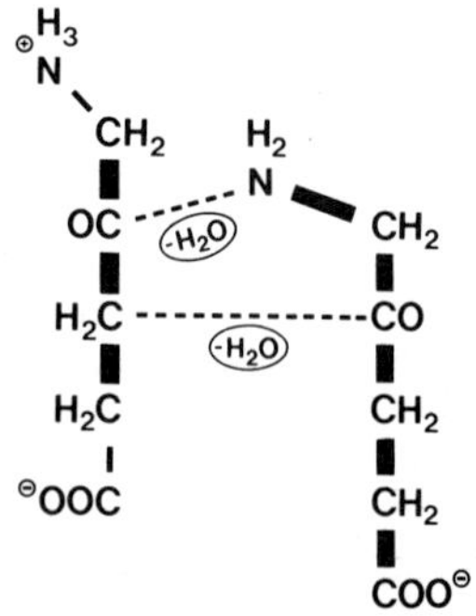

Fig. 2. Formation of the pyrrole ring (porphobilinogen) from two molecules of 5-aminolevulinic acid.

Chlorophyllide *a* is esterified with geranylgeraniol diphosphate and the double bonds in the geranylgeraniol side chain sequentially hydrogenated to give the phytol chain or esterification is directly with phytol diphosphate[2].

Based on the detailed reviews by C.G. Kannangara[1] and W. Rüdiger and S. Schoch[2] the following can be said about the molecular analysis of the enzymes converting 5-aminolevulinic acid into chlorophyll and their structural genes.

1) 5-aminolevulinic dehydratase of spinach[5] and barley is a hexamer with a molecular weight of 250-300 kDa. The detailed catalytic mechanism has been elucidated with the enzyme from *Rhodopseudomonas spheroides*[6]. A structural gene for the *E.coli* enzyme with this function (hem B) has been sequenced[7] as has a cDNA clone of the human enzyme[8].
2) Porphobilinogen deaminase has been purified from *Rhodopseudomonas spheroides*[9], *Chlorella vulgaris*[10], *Euglena gracilis*[11] and *Hordeum vulgare*[12]. The reaction mechanism has been clarified with the enzyme from *E.coli*[13-15], for which the sequence of the structural gene (hem C) is known[16]. A cDNA clone encoding the enzyme which is located in the chloroplast stroma of *Euglena* has been characterized[17].
3) The primary structure of uroporphyrinogen III synthase is known for the corresponding *E.coli* enzyme[18].
4) The amino acid sequence for urogen III decarboxylase from rat and human cells[19,20] has been deduced from corresponding cDNA clones. It is assumed that the same type of enzyme functions in both heme and chlorophyll biosynthesis.
5) Protogen oxidase catalyzing the synthesis of protoporphyrin IX is known at the primary structure level only from the sequence of the yeast Hem B gene[21] which encodes an enzyme function of this type in mitochondrial heme synthesis.
6) Protochlorophyllide reductase is a single polypeptide with a molecular weight of 36 to 38 kDa. A cDNA clone encoding the barley etioplast protochlorophyllide reductase has been sequenced and expressed into an active enzyme in *E.coli*[22].

It is obvious that many interesting functional and regulatory aspects of the enzymes converting 5-aminolevulinate to chlorophyll can be explored with the aid of the few cloned genes encoding enzymes of this pathway. Additional genes of the pathway in higher plants, algae and cyanobacteria can probably be isolated using probes from *E.coli* and mammalian genes with the same functions in tetrapyrrole synthesis. There are additional enzymes which will have to be purified and isolated before their structural genes can be obtained. In *Rhodopseudomonas*[23], in *Chlorella*[23], in *Chlamydomonas reinhardtii*[24], in barley[23,25] and in maize[26] mutants in several genes are known which are unable to metabolize protoporphyrin IX or Mg-protoporphyrin monomethylester. As genetic transformation procedures have been or are being developed for these organisms, it should be possible to isolate the genes affected by complementation using appropriate genomic or cDNA libraries.

## UNIQUE PLANT AND BACTERIAL PATHWAY FOR SYNTHESIS OF 5-AMINOLEVULINIC ACID[27]

The pathway is unique as it uses a transfer RNA in the enzymatic conversion of an amino acid into an amino-aldehyde. In plants glutamate is ligated to the $tRNA^{Glu}$ (UUC), which is then reduced to glutamate 1-semialdehyde and subsequently 5-aminolevulinate is formed by an apparent intramolecular transamination (Fig. 3). In a majority of bacteria this unusual pathway synthesizes heme, vitamine $B_{12}$ or factor F430, a nickel tetrapyrrole cofactor of the methylreductase system in methanogenesis. In human cells, yeast and nitrogen fixing bacteria 5-aminolevulinate is synthesized by 5-aminolevulinate synthase, a polypeptide catalyzing the condensation of succinyl-CoA and glycine ($COO^{-}$-$CH_2$-$CH_2$-CO-S-CoA + $COO^{-}$-$CH_2$-$NH_3^{+}$).

5-Aminolevulinic acid for bacteriochlorophyll is made by 5-aminolevulinate synthase in photosynthetic non-sulfur purple bacteria such as *Rhodopseudomonas spheroides* or *Rhodospirillum rubrum*[28] which oxidize organic acids. In the purple sulfur bacterium *Chromatium*[29] and the green sulfur bacteria *Chlorobium*[30] - which oxidize inorganic sulfur

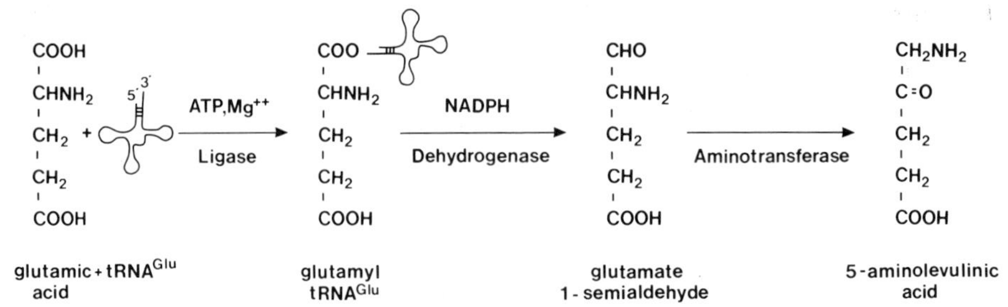


Fig. 3. The pathway from glutamic acid to 5-aminolevulinic acid in higher plants, algae and cyanobacteria requires the activation of glutamate by ligation with a $tRNA^{Glu}$. In higher plants the $tRNA^{Glu}$ is encoded in chloroplast DNA while the enzymes are nuclear coded.

compounds photosynthetically - the 5-aminolevulinate carbon chain for bacteriochlorophyll synthesis is derived from glutamate ($C_5$ pathway).

Barley mutants with defective pigment composition were initially collected and genetically and structurally analyzed in order to understand the development of the photosynthetic apparatus[25]. Two classes of mutants, one representing the lethal *xantha* and *albina* types with characteristics of structural gene mutations blocking the chlorophyll pathway and the other class, comprising *tigrina* mutants which affect the regulation of 5-aminolevulinate synthesis in darkness were found[31]. The discovery of mutants in barley which overproduce protochlorophyllide[32,33] encouraged the investigation of the mechanism of 5-aminolevulinate synthesis in chloroplasts. The findings are in chronological order:

1975 5-Aminolevulinate in plants is made from the intact carbon skeleton of glutamate[34,35]
1976 Synthesis of 5-aminolevulinic acid and chlorophyll is obtained in isolated chloroplasts[36,37]
1977 The components of the pathway are soluble[38]
1978 Discovery of glutamate 1-semialdehyde aminotransferase[39]
1981 Separation of the 5-aminolevulinate synthesizing complex into 3 components[40,41]
1983 Synthesis of the intermediate glutamate 1-semialdehyde[42]
1984 $tRNA^{Glu}$ is a cofactor of the dehydrogenase synthesizing glutamate 1-semialdehyde[43,44]
1985 Inhibition of glutamate 1-semialdehyde aminotransferase by gabaculin causes accumulation of glutamate 1-semialdehyde in leaves[45]
1986 Nucleotide sequence of the $tRNA^{Glu}$ (UUC) required for 5-aminolevulinate synthesis[46]
1987 Purification of the first enzyme in the pathway, barley glutamate-tRNA ligase[47]
1987 Nucleotide sequence of the gene in chloroplast DNA encoding the $tRNA^{Glu}$ (UUC)[48]
1987 Requirement of $tRNA^{Glu}$ for synthesis of 5-aminolevulinate in a methanobacterium[49]
1988 Catalytic mechanism of glutamate 1-semialdehyde aminotransferase studied[50]
1989 *E.coli*, *Salmonella*, *Chlostridium* and *Bacillus subtilis* use the intact five carbon chain of glutamate for synthesis of 5-aminolevulinate[51,52]
1989 Purification of the glutamate 1-semialdehyde aminotransferase from barley and *Synechococcus*[53]. Determination of amino acid sequences.
1989 Simple method of synthesis of glutamate 1-semialdehyde and definitive structure by NMR spectroscopy[54]
1989 Red alga *Cyanidium* uses the $C_5$-pathway[55]
1990 Nucleotide sequence of the genes encoding glutamate 1-semialdehyde aminotransferase from barley, *Synechococcus* and *E.coli*[56,57]
1990 Purification of the glutamyl-tRNA dehydrogenase from *Chlamydomonas*[58]

Fig. 4. Nucleotide sequences of the transfer-RNA molecule from legumes, spinach, *Euglena* and tobacco which participate in the synthesis of chlorophyll.

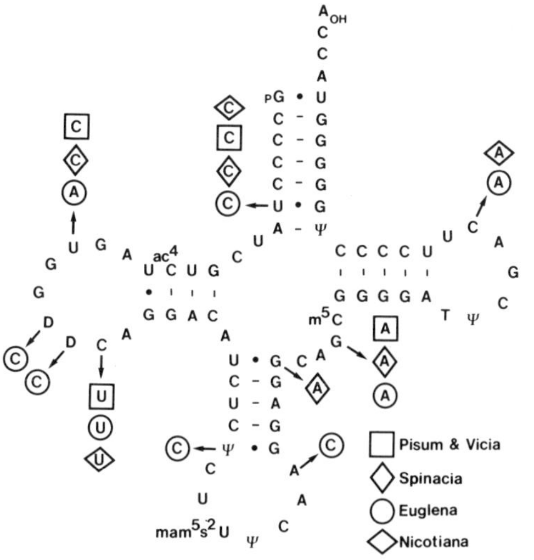


In the first step of the pathway glutamate is activated by ligation to tRNA$^{Glu}$ (Fig. 4) which in higher plants is encoded in chloroplast DNA, while the glutamate tRNA ligase, the glutamyl-tRNA dehydrogenase and the glutamate 1-semialdehyde aminotransferase are encoded in the nucleus, synthesized on cytoplasmic ribosomes and imported into the stroma of the chloroplast. The nucleotide sequences of the tRNA$^{Glu}$ involved have at their 3' ends the characteristic C-C-A sequence, and the U-U-C anticodon sequence[46,59]. Removal of the C-C-A sequence by digestion with snake venom phosphodiesterase leads to loss of the ability of the tRNA to ligate with glutamate and to activate glutamate for 5-aminolevulinate synthesis. Replacing the C-C-A sequence using nucleotidyltransferase restores this ability. The first position of the anticodon of barley chloroplast tRNA$^{Glu}$ is occupied by 5-methyl-aminomethyl-2-thiouridine. This hypermodified nucleotide can be oxidized under mild conditions with iodine and is then unable to function in the ligase reaction but its activity can be restored by reduction with thiosulphate. In the chloroplast genome only one tRNA$^{Glu}$ gene is present and it is therefore inferred that tRNA$^{Glu}$ (UUC) translates both GAA and GAG codons in protein synthesis on chloroplast ribosomes and activates glutamate for chlorophyll synthesis.

Chloroplast glutamic acid tRNA ligase or synthase from barley is a dimer containing two identical subunits of 54,000 mol.weight. The corresponding bacterial synthases function as monomeric proteins of 54,000 to 65,000 mol.weight. The chloroplast glutamyl-tRNA synthase of *Chlamydomonas* is a monomer of 62,000 mol.weight[60]. The chloroplast enzymes ligate glutamate to tRNAs containing glutamate as well as glutamine anticodons[61,62]. The glutamate linked by an ester bond to a tRNA$^{Gln}$ is then converted into glutamine by an amidotransferase. This produces activated glutamine for the incorporation into growing peptide chains on chloroplast ribosomes. Recognition of the tRNA$^{Glu}$ and tRNA$^{Gln}$ most likely involves bases in the acceptor stem and in the anticodon stem[63].

The glutamyl-tRNA dehydrogenase (reductase) reduces the aminoacyl ester linkage by NADPH to generate glutamate 1-semialdehyde. The enzyme has been purified in active form from *Chlamydomonas reinhardtii* as a monomer of 130,000 mol.weight[58]. Several groups are actively working on the purification of this most interesting enzyme in bacteria, algae and higher plants and on the cloning of its structural gene.

The last step in the synthesis of 5-aminolevulinate is catalyzed by glutamate 1-semialdehyde aminotransferase. In Fig. 5 the primary structures of this enzyme from barley, the cyanobacterium *Synechococcus* PCC6301 and *Escherichia coli* are compared[56,57]. The barley

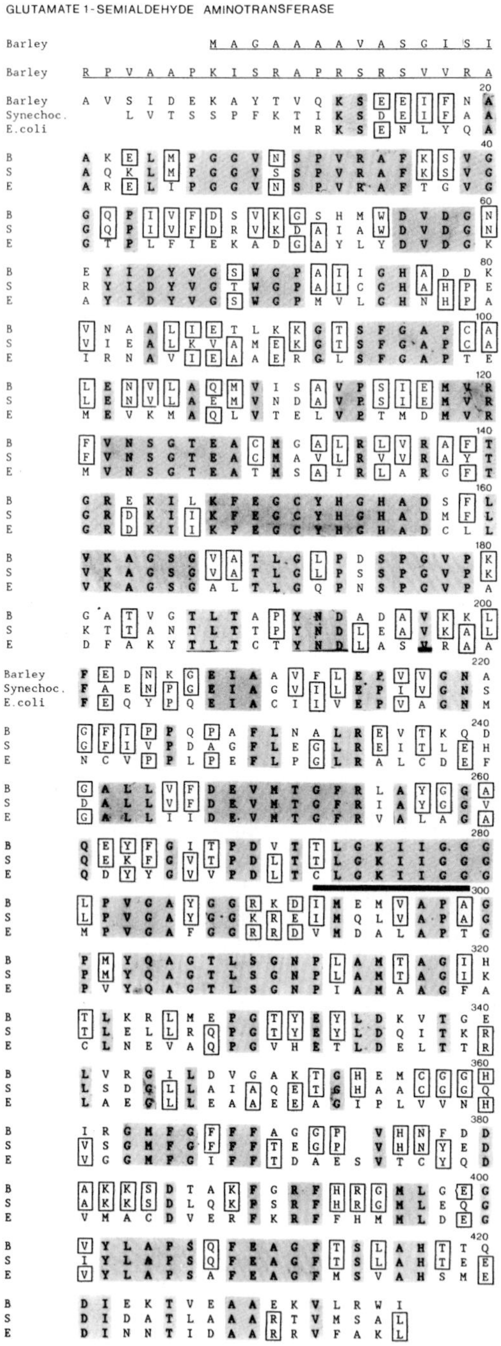


Fig. 5. Conservation of the primary structure of the enzyme glutamate 1-semialdehyde aminotransferase from barley, *Synechococcus* and *E.coli*. Domains with identical amino acid sequences comprise 48% of the primary structure, while the barley and cyanobacterial enzyme share 72% identical residues. The $NH_2$-terminal transit peptide for chloroplast import of the barley enzyme precursor is underlined, while the heavy line marks the putative pyridoxamine phosphate binding lysine and its amino acid neighbours.

amino acid sequence is deduced from a cDNA clone, which has been expressed in a multicopy plasmid in *E.coli* into an active enzyme (46,000 mol.weight) catalyzing the conversion of glutamate 1-semialdehyde into 5-aminolevulinate. The enzyme is synthesized with a transit peptide of 34 amino acids for chloroplast import. The primary structure of the cyanobacterial and *E.coli* enzyme was deduced from their respective genes isolated with the aid of the barley cDNA. In the three enzymes 48% of the primary structure comprise domains with identical amino acid sequences. The cyanobacterial and barley enzyme share 72% identical residues. The peptide containing the putative pyridoxamine phosphate binding lysine is conserved.

The following findings have removed uncertainty regarding glutamate 1-semialdehyde as intermediate in the conversion into 5-aminolevulinate: A new chemical synthesis of glutamate 1-semialdehyde by ozonolysis of 4-vinyl-4 aminobutyric acid has made available unlimited amounts of the compound and permitted its structure to be unequivocally determined by $^{1}$H-NMR and $^{13}$C-NMR spectroscopy[50,54]. Glutamate 1-semialdehyde has an exceptional high affinity for its substrate ($\sim$3 $\mu$M) and is inhibited by gabaculine (3 amino-2,3-dihydrobenzoic acid). Chlorophyll formation in dark-grown seedlings exposed to light is inhibited by gabaculine and the treated leaves accumulate a biosynthetic intermediate which by infrared spectroscopy is found to be identical to the chemical synthesized glutamate 1-semialdehyde[45].

While *E.coli* does not synthesize chlorophyll, it produces 5-aminolevulinate by the $C_5$ pathway and is of interest in this context. Until recently it has been assumed that *E.coli* forms 5-aminolevulinate with the enzyme 5-aminolevulinate synthase and that this enzyme is encoded by the hemA gene at map position 27 min or by the popC gene at map position 4 min of the *E.coli* chromosome. Mutants in both these genes lead to auxotrophy for 5-aminolevulinic acid. Mutant hemA SASX41B has been transformed to 5-aminolevulinate independence by DNA clones from *Rhizobium meliloti*[64], *Bradyrhizobium japonicum*[65], *Rhodobacter sphaeroides*[51,66] and by a cDNA clone encoding 5-aminolevulinic acid synthase from the mouse[67]. The mouse and the mentioned bacteria synthesize 5-aminolevulinate by condensation of succinyl-CoA with glycine in a single enzymatic step and it is therefore not surprising that the *E.coli* hemA mutant can be rescued by providing the synthase gene on a plasmid. When the hemA gene from *E.coli* was sequenced, its deduced amino acid sequence showed no resemblance to authentic 5-aminolevulinate synthase sequences[68,69,70]. Simultaneously precursor studies and heterologous enzyme complementation tests revealed that *E.coli* synthesizes 5-aminolevulinate from glutamate by the $C_5$ pathway[51,52,71]. Extract fractions of *Chlorella* containing the glutamyl-tRNA dehydrogenase complement an extract of auxotroph SASX41B to form 5-aminolevulinate. Further evidence that the hemA gene encodes glutamyl-tRNA dehydrogenase or one component of it was obtained by transforming this auxotroph to prototrophy with a genomic fragment from *Chlorobium vibrioforme* a green sulfur bacterium which synthesizes 5-aminolevulinate from glutamate and contains enzymes using *E.coli* $tRNA^{Glu}$ very efficiently for in vitro substrate synthesis[72]. Sequencing of the popC gene has now revealed it to encode a protein with 52% amino acid sequence identity to barley glutamate 1-semialdehyde aminotransferase[57].

Application of biochemistry and molecular biology to the elucidation of the synthesis of 5-aminolevulinate in plants has uncovered intriguing phenomena. Equally challenging projects lie ahead: Clarification of the catalytic mechanism of glutamate 1-semialdehyde aminotransferase by site directed mutagenesis as well as structural and functional analyses of the dehydrogenase and ligase reactions. We can then approach the questions, if the 5-aminolevulinate overproducing mutants synthesize extra amounts of dehydrogenase and what limits 5-aminolevulinate synthesis in dark-grown seedlings or why chloroplast DNA encoded chlorophyll binding proteins are not translated in the absence of chlorophyll synthesis.

## REFERENCES


1. C. G. Kannangara, Biochemistry and molecular biology of chlorophyll synthesis, *in*: Cell culture and somatic cell genetics of plants. Vol. 7, *The Molecular Biology of Plastids and Mitochondria*, L. Bogorad and I. K. Vasil, eds., Academic Press, New York (1990).

2. W. Rüdiger and S. Schoch, Chlorophylls, *in*: *Plant Pigments*, T. W. Goodwin, ed., Academic press, New York (1988).
3. G. J. Hart and A. R. Battersby, Purification and properties of uroporphyrinogen III synthase (co-synthetase) from *Euglena gracilis*. *Biochem. J.* 232, 151 (1985).
4. A. R. Battersby, Biosynthesis of the pigments of life. *J. Natural Products* 51, 629 (1988).
5. W. Liedgens, C. Lutz and H. A. W. Schneider, Molecular properties of 5-aminolevulinic acid dehydrogenase from *Spinacia oleracea*. *Eur. J. Biochem.* 135, 75 (1983).
6. E. K. Jaffe and G. D. Markham, $^{13}C$ NMR studies of porphobilinogen synthase: observation of intermediates bound to a 280,000 Dalton protein. *Biochemistry* 26, 4258 (1987).
7. Y. Echelard, J. Dymetryszyn, M. Drolet and A. Sasarmann, Nucleotide sequence of the hem B gene of *E.coli* K12. *Mol. Gen. Genet.* 214, 503 (1988).
8. J. G. Wetmur, D. F. Bishop, C. Cantelmo and R. J. Desnick, Human $\delta$-aminolevulinate dehydratase: Nucleotide sequence of a full length cDNA clone. *Proc. Nat. Acad. Sci. USA* 33, 7703 (1986).
9. R. C. Davis and A. Neuberger, Polypyrroles formed from porphobilinogen and amines by uroporphyrinogen synthetase of *Rhodopseudomonas spheroides, Biochem. J.* 133, 471 (1973).
10. Y. Shioi, M. Nagamine, M. Kuraki and T. Sassa, Purification by affinity chromatography and properties of uroporphyrinogen I synthetase from *Chlorella vulgaris. Biochim. Biophys Acta* 616, 303 (1980).
11. D. C. Williams, G. S. Morgan, E. McDonald and A. R. Battersby, Purification of porphobilinogen deaminase from *Euglena gracilis* and studies of its kinetics, *Biochem. J.* 193, 301 (1981).
12. C. G. Kannangara, S. P. Gough and C. Girnth, $\delta$-Aminolevulinate synthesis in greening barley. 2. Purification of enzymes, *in*: *Vth Intern. Congr. Photosynthesis*. V. Chloroplast Development, G. Akoyunoglou, ed., Balaban Int. Science Services, Philadelphia, Pa. (1981).
13. A. D. Miller, G. J. Hart, L. C. Packman and A. R. Battersby, Evidence that the pyrromethane cofactor of hydroxymethylbilane synthase (porphobilinogen deaminase) is bound to the protein through the sulphur atom of cystein-242, *Biochem. J.* 254, 915 (1988).
14. P. M. Jordan, M. J. Warren, H. J. Williams, N. J. Stolowich, C. A. Roessner, S. K. Grant and A. I. Scott, Identification of a cysteine residue as the binding site for the dipyrromethane cofactor at the active site of *Escherichia coli* porphobilinogen deaminase. *FEBS Letters* 235, 189 (1988).
15. A. I. Scott, K. R. Clemens, N. J. Stolowich, P. J. Santander, M. D. Gonzalez and C. A. Roessner, Reconstitution of apo-porphobilinogen deaminase: Structural changes induced by cofactor binding, *FEBS Letters* 242, 319 (1989).
16. S. D. Thomas and P. M. Jordan, Nucleotide sequence of the hem C locus encoding porphobilinogen deaminase of *Escherichia coli* K12, *Nucleic Acids Res.* 14, 6215 (1986).
17. A. L. Sharif, A. G. Smith and C. Abell, Isolation and characterisation of a cDNA clone for a chlorophyll synthesis enzyme from *Euglena gracilis*. The chloroplast enzyme hydroxymethylbilane synthase (porphobilinogen deaminase) is synthesized with a very long transit peptide in *Euglena*. *Eur. J. Biochem.* 184, 353 (1989).
18. A. Sasarman, A. Nepveu, Y. Echelard, J. Dymetryszyn, M. Drolet and C. Goyer, Molecular cloning and sequencing of the hem D gene of *Escherichia coli* K 12 and preliminary data on the uro operon, *J. Bacteriol.* 169, 4257 (1987).
19. P.-H. Romeo, A. Dubart, B. Grandchamp, H. de Verneuil, J. Rosa, Y. Nordmann and M. Goosens, Isolation and identification of a cDNA clone coding for rat uroporphyrinogen decarboxylase. *Proc. Nat. Acad. Sci. USA* 81, 3346 (1984).
20. P.-H. Romeo, N. Raich, A. Dubart, D. Beaupain, M. Pryor, J. Kushner, M. Cohen-Solal and M. Goosens, Molecular cloning and nucleotide sequence of a complete human uroporphyrinogen decarboxylase cDNA. *J. Biol. Chem.* 261, 9825 (1986).
21. M. Zagorec, J.-M. Buhler, I. Treich, T. Keng, L. Guarente and R. Labbe-Bios, Isolation, sequence and regulation by oxygen of the yeast Hem B gene coding for coproporphyrinogen oxidase, *J. Biol. Chem.* 262, 9718 (1988).

22. R. Schulz, K. Steinmüller, M. Klaas, C. Forreiter, S. Rasmussen, C. Hiller and K. Apel, Nucleotide sequence of a cDNA coding for the NADPH-protochlorophyllide oxidoreductase (PCR) of barley (*Hordeum vulgare* L.) and its expression in *Escherichia coli*, *Molec. Gen. Genet.* 217, 355 (1989).
23. S. Gough, Defective synthesis of porphyrins in barley plastids caused by mutation in nuclear genes. *Biochim. Biophys. Acta* 286, 36 (1972).
24. W.-Y. Wang, J. E. Boynton, N. W. Gillham and S. P. Gough, Genetic control of chlorophyll biosynthesis in *Chlamydomonas*: Analysis of a mutant affecting synthesis of $\delta$-aminolevulinic acid, *Cell* 6, 75 (1975).
25. D. von Wettstein, K. W. Henningsen, J. E. Boynton, C. G. Kannangara and O. F. Nielsen, The genic control of chloroplast development in barley, *in*: *Autonomy and biogenesis of mitochondria and chloroplasts*, N. K. Boardman, A. W. Linnane and R. M. Smillie, eds., North Holland, Amsterdam (1971).
26. P. Mascia, An analysis of precursors accumulated by several chlorophyll biosynthetic mutants of maize. *Mol. Gen. Genet.* 161, 237 (1978).
27. C. G. Kannangara, S. P. Gough, P. Bryuant, J. K. Hoober, A. Kahn and D. von Wettstein, $tRNA^{Glu}$ as a cofactor in $\delta$-aminolevulinate biosynthesis: steps that regulate chlorophyll synthesis. *Trends in Biochem. Sci.* 13, 139 (1988).
28. P. M. Jordan and D. Shemin, $\delta$-Aminolevulinic acid synthetase, *in*: *Enzymes*, P. D. Boyer, ed., Vol. 7, Academic Press, New York (1972).
29. T. Oh-hama, H. Seto and S. Miyachi, $^{13}$C-NMR evidence of bacteriochlorophyll *a* formation by the $C_5$ pathway in *Chromatium, Arch. Biochem. Biophys.* 246, 192 (1986).
30. K. M. Smith and M. S. Huster, Bacteriochlorophyll-c formation via glutamate C-5 pathway in *Chlorobium* bacteria. *J. Chem. Soc. Chem. Commun.* 14 (1987).
31. D. von Wettstein, A. Kahn, O. F. Nielsen and S. Gough; Genetic regulation of chlorophyll synthesis analysed with mutants in barley, *Science* 184, 800 (1974).
32. C. G. Kannangara, S. P. Gough and D. von Wettstein, The biosynthesis of $\delta$-aminolevulinate and chlorophyll and its genetic regulation, *in: Chloroplast Development*, G. Akoyunoglou et al., eds., Elsevier/North-Holland Biomedical Press (1978).
33. S. P. Gough and C. G. Kannangara, Biosynthesis of $\delta$-aminolevulinate in greening barley III. The formation of $\delta$-aminolevulinate in tigrina mutants of barley. *Carlsberg Res. Commun.* 44, 403 (1979).
34. S. I. Beale, S. P. Gough and S. Granick, Biosynthesis of $\delta$-aminolevulinic acid from the intact skeleton of glutamic acid in greening barley. *Proc. Nat. Acad. Sci. US* 72, 2719 (1975).
35. E. Meller, S. Belkin and E. Harel, The biosynthesis of $\delta$-aminolevulinic acid in greening maize leaves, *Phytochemistry* 14, 2399 (1975).
36. S. P. Gough and C. G. Kannangara, Synthesis of $\delta$-aminolevulinic acid by isolated plastids, *Carlsberg Res. Commun.* 41, 183 (1976).
37. C. G. Kannangara and S. P. Gough, Synthesis of $\delta$-aminolevulinic acid and chlorophyll by isolated chloroplasts, *Carlsberg Res. Commun.* 42, 441 (1977).
38. S. P. Gough and C. G. Kannangara, Synthesis of $\delta$-aminolevulinate by a chloroplast stroma preparation from greening barley leaves, *Carlsberg Res. Commun.* 42, 459 (1977).
39. C. G. Kannangara and S. P. Gough, Biosynthesis of $\delta$-aminolevulinate in greening barley leaves: glutamate 1-semialdehyde aminotransferase, *Carlsberg Res. Commun.* 43, 185 (1978).
40. W.-Y. Wang, S. P. Gough and C. G. Kannangara, Biosynthesis of $\delta$-aminolevulinate in greening barley leaves IV. Isolation of three soluble enzymes required for the conversion of glutamate to $\delta$-aminolevulinate, *Carlsberg Res. Commun.* 46, 243 (1981).
41. W.-Y. Wang, D.-D. Huang, D. Stachon, S. P. Gough and C. G. Kannangara, Purification, characterization, and fractionation of the $\delta$-aminolevulinic acid synthesizing enzymes from light-grown *Chlamydomonas reinhardtii* cells, *Plant Physiol.* 74, 569 (1984).
42. G. Houen, S. P. Gough and C. G. Kannangara, $\delta$-aminolevulinate synthesis in greening barley V. The structure of glutamate 1-semialdehyde. *Carlsberg Res. Commun.* 48, 567 (1983).
43. C. G. Kannangara, S. P. Gough, R. P. Oliver and S. K. Rasmussen, Biosynthesis of $\delta$-aminolevulinate in greening barley leaves VI. Activation of glutamate by ligation to RNA, *Carlsberg Res. Commun.* 49, 417 (1984).

44. D.-D. Huang, W.-Y. Wang, S. P. Gough and C. G. Kannangara, δ-aminolevulinic acid-synthesizing enzymes need an RNA moiety for activity, *Science* 225, 1482 (1984).
45. C. G. Kannangara and A. Schouboe, Biosynthesis of δ-aminolevulinate in greening barley leaves. VII. Glutamate 1-semialdehyde accumulation in gabaculine treated leaves, *Carlsberg Res. Commun.* 50, 179 (1985).
46. A. Schön, G. Krupp, S. Gough, S. Berry-Lowe, C. G. Kannangara and D. Söll, The RNA required in the first step of chlorophyll biosynthesis is a chloroplast glutamate tRNA, *Nature* 322, 281 (1986).
47. P. Bruyant and C. G. Kannangara, Biosynthesis of δ-aminolevulinate in greening barley leaves. VIII. Purification and characterization of the glutamate-tRNA ligase, *Carlsberg Res. Commun.* 52, 99 (1987).
48. S. Berry-Lowe, The chloroplast glutamate tRNA gene required for δ-aminolevulinate synthesis. *Carlsberg Res. Commun.* 52, 197 (1987).
49. H. C. Friedmann, R. K. Thauer, S. P. Gough and C. G. Kannangara, δ-aminolevulinic acid formation in the archaebacterium *Methanobacterium thermoautotrophicum* requires $tRNA^{Glu}$. *Carlsberg Res. Commun.* 52, 363 (1987).
50. J. K. Hoober, A. Kahn, D. E. Ash, S. Gough and C. G. Kannangara, Biosynthesis of δ-aminolevulinate in greening barley leaves. IX. Structure of the substrate, mode of gabaculine inhibition, and the catalytic mechanism of glutamate 1-semialdehyde aminotransferase, *Carlsberg Res. Commun.* 53, 11 (1988).
51. J.-M. Li, O. Brathwaite, S. D. Cosloy and C. S. Russel, 5-Aminolevulinic acid synthesis in *Escherichia coli*, *J. Bacteriol.* 171, 2547 (1989).
52. Y. J. Avissar and S. I. Beale, Identification of the enzymatic basis for delta-aminolevulinic acid auxotrophy in a hem A mutant of *Escherichia coli. J. Bacteriol.* 171, 2919 (1989).
53. B. Grimm, A. Bull, K. G. Welinder, S. P. Gough and C. G. Kannangara, Purification and partial amino acid sequence of the glutamate 1-semialdehyde aminotransferase of barley and synechococcus, *Carlsberg Res. Commun.* 54, 67 (1989).
54. S. P. Gough, C. G. Kannangara and K. Bock, A new method for the synthesis of glutamate 1-semialdehyde. Characterization of its structure in solution by NMR spectroscopy. *Carlsberg Res. Commun.* 54, 99 (1989).
55. J. D. Houghton, S. B. Brown, S. P. Gough and C. G. Kannangara, Biosynthesis of δ-aminolevulinate in *Cyanidium caldarium*: Characterization of $tRNA^{Glu}$, ligase, dehydrogense and glutamate 1-semialdehyde aminotransferase, *Carlsberg Res. Commun.* 54, 131 (1989).
56. B. Grimm, Primary structure of a key enzyme in plant tetrapyrrole synthesis: Glutamate 1-semialdehyde aminotransferase, *Proc. Nat. Acad. Sci. USA*, in press.
57. B. Grimm, A. Bull and V. Breu, Function of a barley gene identifies the structural genes of glutamate 1-semialdehyde aminotransferase for porphyrin synthesis in *E.coli* and a cyanobacterium. Submitted.
58. M. W. Chen, D. Jahn, G. P. O'Neill and D. Söll, Purification of the glutamyl-tRNA reductase from *Chlamydomonas reinhardtii* involved in δ-aminolevulinic acid formation during chlorophyll biosynthesis, *J. Biol. Chem.* 265, 4058 (1990).
59. G. P. O'Neill, D. M. Peterson, A. Schön, M.-W. Chen and D. Söll, Formation of the chlorophyll precursor δ-aminolevulinic acid in cyanobacteria requires aminoacylation of a $tRNA^{Glu}$ species, *J. Bacteriol.* 170, 3810 (1988).
60. M.-W. Chen, D. Jahn, A. Schön, G. P. O'Neill and D. Söll, Purification and characterization of *Chlamydomonas reinhardtii* chloroplast glutamyl-tRNA synthetase, a natural misacylating enzyme, *J. Biol. Chem.* 265, 4054 (1990).
61. A. Schön, C. G. Kannangara, S. P. Gough and D. Söll, Protein biosynthesis in organelles requires misacylation of transfer RNA, *Nature* 331, 187 (1988).
62. A. Schön and D. Söll, tRNA specificity of a mischarging aminoacyl tRNA synthetase. Glutamyl tRNA synthetase from barley chloroplasts, *FEBS Letters* 228, 241 (1988).
63. M. A. Rould, J. J. Perona, D. Söll and T. A. Steitz, Structure of *E.coli* glutaminyl-tRNA synthetase complexed with $tRNA^{Gln}$ and ATP at 2.8 Å resolution, *Science* 246, 1135 (1989).
64. S. A. Leong, G. S. Ditta and D. R. Helinski, Heme biosynthesis in *Rhizobium*. Identification of a cloned gene coding for δ-aminolevulinic acid synthetase from *Rhizobium meliloti*. *J. Biol. Chem.* 257, 8724 (1982).

65. M. L. Guerinot and B. K. Chelm, Bacterial δ-aminolevulinic acid synthase activity is not essential for leghemoglobin formation in soybean/*Brady-rhizobium japonicum symbiosis*, *Proc. Natl. Acad. Sci. USA* 83, 1837 (1986).
66. T.-N. Tai, M. D. Moore and S. Kaplan, Cloning and characterization of the 5-aminolevulinate synthase gene(s) from *Rhodobacter sphaeroides*. *Gene* 70, 139 (1988).
67. D. S. Schoenhaut and P. J. Curtis, Nucleotide sequence of mouse 5-aminolevulinic acid synthase cDNA and expression of its gene in hepatic and erythroid tissue, *Gene* 48, 55 (1986).
68. M. Drolet, L. Péloquin, Y. Echelard, L. Cousineau and A. Sasarman, Isolation and nucleotide sequence of the *hem A* gene of *Escherichia coli* K 12, *Mol. Gen. Genet.* 216, 347 (1989).
69. J.-M. Li, C. S. Russel and S. D. Cosloy, Cloning and structure of the *hem A* gene of *Escherichia coli* K-12, *Gene* 82, 209 (1989).
70. E. Verkamp and B. K. Chelm, Isolation, nucleotide sequence and preliminary characterization of the *Escherichia coli* K 12 *hem A* gene, *J. Bacteriol.* 171, 4728 (1989).
71. G. P. O'Neill, M.-W. Chen and D. Söll, δ-Aminolevulinic acid biosynthesis in *Escherichia coli* and *Bacillus subtilis* involves formation of glutamyl-tRNA, *FEMS Microbiol. Lett.* 60, 255 (1989).
72. Y. J. Avissar and S. I. Beale, Cloning and expression of a structural gene from *Chlorobium vibrioforme* that complements the *hem A* mutation in *Escherichia coli*, *J. Bacteriol.* 172, 1656 (1990).

# MOLECULAR APPROACHES TO UNDERSTAND SINK-SOURCE RELATIONS IN HIGHER PLANTS

L. Willmitzer, A. Basner, K. Borgmann, W.-B. Frommer, H. Hesse, S. Hummel, J. Koßmann, T. Martin, B. Müller, M. Rocha-Sosa[1], A. v. Schaeven, M. Stitt[2], U. Sonnewald

Institut für Genbiologische Forschung Berlin GmbH
Ihnestr. 63, 1000 Berlin 33
[1]Present address: Centre de Investigation sobre Fijación de Nitrógeno, Cuernavacas, Morelos, Mexico
[2]Institut für Pflanzenphysiologie, Universität Bayreuth
8580 Bayreuth, Germany

## INTRODUCTION

Growth and development of a plant is dependent upon the energy gained by fixing carbon dioxide into carbohydrates during photosynthesis. Primary sites for photosynthesis are leaf and to a much lesser extent stem tissues, whereas other organs such as roots, seeds or tubers do not contribute to carbon assimilation but rather totally depend on photosynthetically active organs. Thus there is a net flow of carbohydrates from photosynthetically active tissues, representing the sources (defined as net exporters of fixed carbon) to photosynthetically inactive parts of the plant, representing the sinks (defined as net importers of fixed carbon). Sink-source relations are constantly adjusted to the developmental stage of the plant (Ho, 1988). Thus all parts of a seedling developing from a seed or a seed tuber represent a sink, whereas the original seed respectively seed tuber represent the source nourishing the developing seedling. Upon further plant development the mature leaves and to a lesser extent the stem tissue will become the major source of a plant whereas young leaves, roots and to some extent stem tissue represent major sinks. Upon seed respectively tuber setting, again a drastic change occurs as now these newly developed organs represent the major sinks in the plant.

It is obvious from this description that one main characteristic of sink-source relations in higher plants is their continuous change throughout plant development and the ongoing competition between different sinks and sources.

Sink-source relations are not only central to plant development but they are also under clear developmental (at least in case of leaves) and environmental control. A young leaf represents a sink which is nearly exclusively dependent upon the import of photoassimilates from mature source leaves. During growth its photosynthetic activity increases and finally it starts exporting carbohydrates. This change follows a strict developmental pattern starting at the tip or rim of the leaves and moving towards the base (Turgeon, 1989). Using albino mutants which are incapable of performing photosynthesis, it has been shown that at least in case of tobacco this sink-source transition is not induced by the ratio between exported and imported photoassimilates but rather under developmental control (Turgeon, 1984).

Sink-source relations have furthermore shown to be under environmental control. Thus shading of some source leaves leads to a rapid establishment of new equilibria between exported and retained photoassimilates in the remaining photosynthetically active non-shaded source leaves which is also reflected by a new equilibrium between the amount of carbohydrate present in the form used for either transport (i.e. sucrose) or present in the form used for transitory storage (i.e. starch). Finally the success of plant breeding during the last 60 years was not primarily due to a large increase in total amount of the biomass formed in various crop plants but rather the preferential partitioning of this biomass into harvestable organs such as seeds or tubers. Thus in Solanum demissum tubers represent only 7 % of the total weight of the plant whereas in case of a modern Solanum tuberosum variety tubers represent more than 80 % of the biomass (Ho, 1989). Similar data are well-established for e.g. soybean or wheat (Gifford et al., 1984). Thus a preferential partitioning of the biomass into harvestable organs has led to an increase in the harvest index and therefore to increased yield.

## MOLECULAR GENETIC TECHNIQUES ALLOW A NEW APPROACH TO UNDERSTAND SINK-SOURCE INTERACTIONS

Despite the fact that sink-source relations are intensively studied on the physiological level, little if any approaches have as yet been described directed at an analysis of this phenomena on the molecular level.

Here we describe a concept aimed at the understanding of these processes using a molecular genetic approach.

This approach can be divided in two parts:

a) Identification of (marker) genes which show a different expression in sink versus source tissue
b) Ectopic expression of genes which might be involved in determining sink-source relations.

### A) IDENTIFICATION OF (MARKER) GENES WHICH SHOW A DIFFERENT EXPRESSION IN SINK VERSUS SOURCE TISSUE

#### A 1) Identification of marker genes

To the best of our knowledge no mutants have been identified which are described as sink or source mutants. As sink-source relations are central to plant development the failure of identifying such mutants could be due to the fact that such mutants would probably display a rather pleiotropic phenotype and not necessarily be recognized as sink or source mutants. In order to overcome this problem, one possible approach is to identify marker genes which show a different expression in sink versus source tissues.

During the last couple of years our group has analyzed the expression pattern of a gene family from potato (Solanum tuberosum) which in green-house grown potato plants shows a tuber-specific expression. This gene family codes for the major glycoprotein species of potato tubers, which has been given the trivial name patatin.

Based on sequence comparisons of the promoter regions, this gene family can be grouped in two different classes, i.e. class I and class II (Mignery et al., 1988). Chimeric genes consisting of the promoter of class I patatin genes fused to a reporter gene such as the ß-glucuronidase from E. coli show a tuber-specific expression in all parenchymatic cells of the tuber (Rocha-Sosa et al., 1989; Wenzler et al., 1989), whereas chimeric genes containing the promoter of a class II gene are expressed in certain regions of the developing tuber such as the pericyclic ring and the phellogen as well as in the rhizodermis just behind the root tip (Köster-Töpfer et al., 1989). Upon transfer of chimeric genes containing class II patatin promoters in other plants such as tobacco or Arabidopsis thaliana, the same specificity of expression is kept, i.e. preferential expression in the vicinity of the root tip in the rhizodermis. This is contrasted by the expression pattern displayed by chimeric genes driven by the class I patatin promoter which in these foreign hosts show a specific expression in all parts of the root except the root tip (W.-B. Frommer, unpublished result) thus being different from the expression of these genes in potato. Class II genes are therefore seemingly under a classical developmental control with signals conserved between different plant species which is not true for class I genes.

Further evidence arguing against class I patatin promoters being under developmental control was obtained from experiments where axenically cultured potato plants where kept on media containing different levels of sucrose. Whereas potato plants kept on low levels of sucrose (i.e. 2 % w/v) showed no expression of patatin genes in their leaves, class I patatin genes were clearly induced in leaves of potato plants cultured on higher levels of sucrose such as 7 % or 10 % (cf. fig. 1).

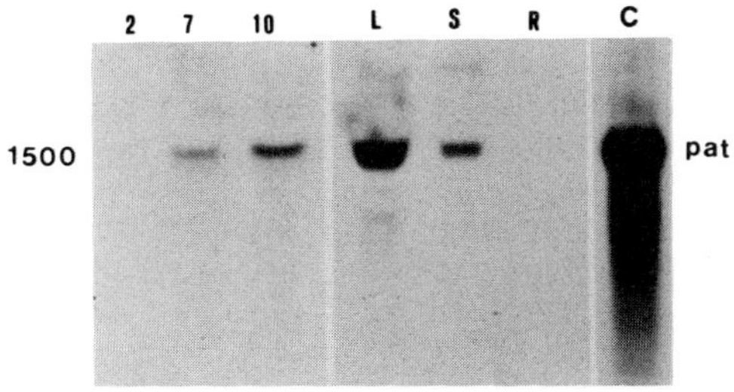


Fig. 1 High levels of sucrose induce the expression of patatin in leaves. Total RNA (50 µg each) isolated from different plants as described below was after electrophoretic separation and transfer on nylon membranes probed for patatin mRNA sequences using the cDNA probe pcT58 (Rosahl et al., 1986) lanes 1-3: RNA was isolated from leaves of potato plantlets axenically cultured for 2 weeks on MS-medium (Murashige and Skoog, 1962) supplemented with 2,7 or 10 % sucrose, lanes 4-6: RNA was isolated from leaf (L), stem (S) and root (R) of a potato plant grown on MS-medium containing 10 % sucrose. Lane 7 shows the hybridization signal obtained with 20 µg total RNA from sink tubers.

The induction of class I patatin accumulation was furthermore as a rule paralleled by an induction of starch accumulation in these leaves.

These data therefore indicate that the expression of class I patatin genes is rather controlled by the metabolic state of the tissue.

Further evidence for the expression of class I patatin genes being related to the state of a tissue being either a sink or a source is based on the following two observations. Firstly upon detaching tubers (and thus the major sinks) from a tuberizing plant, class I patatin gene expression is induced in petiole and stem tissue again in parallel with the accumulation of starch tissue. This can be interpreted as the expression in alternative sinks once the strongest sinks have been removed (Paiva et al., 1983).

Secondly when RNA is isolated from sink (i.e. growing) versus source (i.e. sprouting) tubers, patatin is clearly expressed in sink tubers but little if any RNA is detectable in source tubers (T. Martin, unpublished observation).

Taken together all these data are compatible with a model assuming that class I patatin genes are expressed in sink tissues whereas very little if any expression is detectable in source tissue. If this assumption is true this would allow to set up the following screening for mutants impaired in sink-source relations:

- Chimeric genes consisting of class I patatin promoters fused to the ß-glucuronidase reporter are transferred in Arabidopsis thaliana plants
- Transformants displaying a stable phenotype with respect to the expression of the chimeric gene are mutagenized using either a transposable element such as Ac from maize or EMS as mutagen
- The mutagenized M2 population is screened for mutant phenotypes displaying a different expression pattern of the chimeric class I patatin promoter ß-glucuronidase gene.
- Plants with a mutant phenotype are subsequently analyzed by classical genetics for
  a) the mutation being a transmutation with respect to the chimeric gene
  b) the chromosomal location of the mutation. In case that Ac was the mutagenic agent the mutant and wild-type phenotypes are subsequently cloned using the Ac as a tag; in case the mutant agent was a chemical agent such as EMS, the mutation must be localized using RFLP mapping and subsequently be identified by complementation using YACS libraries.

  It is hoped that in this way mutants impaired in sink-source relationships will be identified, although a large number of other mutants (e.g. impaired in DNA binding proteins) will probably identified in this way, too.

A 2) Identification of genes showing a different expression in sink versus source tissue

In a second approach genes should be identified showing a different expression in sink versus source tissue. To this end, substractive cDNA libraries shall be established and subsequently screened for genes showing either a sink or a source specific expression.

B) ECTOPIC EXPRESSION OF GENES WHICH MIGHT BE INVOLVED IN DETERMINING THE STATE OF A TISSUE WITH RESPECT TO BEING A SINK OR A SOURCE

Plant Molecular Biology has supplied tools allowing:
- Transfer of foreign genes into plants
- Expression of these genes under the control of specific promoters (e.g. limiting the expression to photosynthetically active cells (sources) or seeds respectively tubers (sinks))
- synthesis of alien enzymes in transgenic plants (showing e.g. a different allosteric control)
- direction of a protein into a specific subcellular compartment.

These tools can be applied to genes representing candidates with respect to the sink or source state of a tissue thus allowing to test their assumed role via ectopic expression in transgenic plants.

We are using potato as the model plant to analyze sink-source interactions. In an outgrown potato the main sources are the mature leaves and the strongest sinks are the developing tubers. As in most other plants carbohydrates are transported in the form of sucrose and stored in the form of starch also in potato.

With respect to a source leaf the following characteristics represent potential targets accessible to molecular approaches:
- The photosynthetic activity of the leaf
- The partitioning between exported and retained photoassimilates which is dependent on
  a) partitioning between sucrose and starch
  b) loading of sucrose into the phloem

With respect to a sink tuber the following characteristics represent potential targets accessible to molecular approaches
- The efficiency of unloading of the phloem
- The deposition of photoassimilates in a storage form (i.e. formation of starch)

One central aspect is therefore the biosynthesis of starch and sucrose. Thus enzymes and the respective genes encoding enzymes involved in starch and/or sucrose biosynthesis represent primary targets for a molecular approach as described above. One possibility is e.g. the overexpression of ADP-glucose pyrophosphorylase respectively the expression of ADP-glucosepyrophosphorylase genes showing a different allosteric control in the sink (tuber) and the parallel inhibition of the expression of these genes in the source leaf using the anti-sense approach.

We recently cloned two different ADP-glucose pyrophoshorylase genes from potato (B. Müller et al. submitted) and are presently constructing transgenic potato plants expressing anti-sense RNA's to the two forms of ADP-glucose pyrophoshorylase genes in leaves.

One other central aspect with respect to the efficiency of sources is the photosynthetic activity of the source leaf. Most important with respect to interfering with sink-source relations is the question whether or not the photosynthetic activity of a source leaf is always at its maximum capacity or is rather controlled by e.g. the removal of photoassimilates in sinks. It has been suggested that carbohydrate accumulates in the leaf and photosynthesis is inhibited when the rate of photosynthesis exceeds the demand for sucrose in the sinks (Geiger, 1976). This 'feed-back' or 'sink' regulation of photosynthesis has been studied by inhibiting phloem transport e.g. cold-girdling or detaching leaves (Rufty et al., 1983; Blechschmidt-Schneider et al., 1989) or by altering the sink-source balance in whole plants. The latter can be achieved in several ways e.g. removing potential sinks (Clausen and Biller, 1977; Fondy and Geiger, 1980), inhibiting sink metabolism (Bagnall et al., 1989), increasing the supply of light or $CO_2$ to the source leaf (Cave et al., 1981), or removing other competing source leaves (Sasek et al., 1985).

Many of these studies found that photosynthesis is inhibited when sink demand was decreased. This inhibition of photosynthesis was often accompanied by an accumulation of carbohydrate. However, this was not always the case (Neales and Incoll, 1968; Geiger, 1976; Clausen and Biller, 1976; Carmi and Sholler, 1979; Ntrika and Delrot, 1986), and none of these experiments established an unequivocal causal relation between the accumulation of carbohydrate and the inhibition of photosynthesis. Many of these treatments are also rather unspecific, and would also affect the movement of nitrogen, minerals and hormones.

Sucrose is most likely exported from the leaf by transferring it to the apoplast, and then actively loading it into the phloem (Turgeon, 1989). It should be possible to inhibit sucrose export by introducing and/or overexpressing an invertase into the cell wall. After entering the apoplast, sucrose would be hydrolyzed to glucose and fructose. These free hexoses cannot be taken up into the phloem as well as sucrose (Kallakaral and Komor, 1989). Instead, they will be retrieved by the mesophyll cells (Maynard and Lucas, 1982), rephosphorylated by hexokinase and fructokinase and then reconverted to sucrose. Furthermore inhibiting the export of sucrose from source leaves would represent a rather severe change with respect to sink-source inter- actions and thus these plants should lead to valuable insights into principles governing this interaction.

We have taken advantage of the possibility to construct transgenic plants in order to create such a situation. To this end an invertase gene (suc 2) from Saccharomyces cerevisiae was directed into the apoplastic space (i.e. the cell wall) using signal peptides derived from a vacuolar plant protein, i.e. the proteinase inhibitor II from potato (Keil et al., 1986; Nelson and Ryan, 1980). The rationale for chosing the alien yeast invertase was to minimize the possibility that the enzyme would be inhibited by endogenous invertase inhibitors. Furthermore yeast invertase is known to be active over a broad pH-range (Goldstein and Lampen, 1976) which is important with respect to the fact that the actual pH of the apoplastic space of plants is not precisely known. Transgenic tobacco plants containing this gene varied from wildtype tobacco plants in a number of aspects. Firstly they were much smaller, the smallest tobacco plants reaching a size of only 5 cm. This reduction in size is not due to a reduced number of leaves but rather due to a reduced length of the internodal segments. Secondly as expected for a plant which shows a reduced supply with carbohydrates in its sinks due to reduced export of sucrose from source leaves the invertase expressing plants show a reduced root formation which correlates with the level of expression of the invertase. A third phenotype is the appearance of chlorotic leasons and necrotic regions on the leaves. More significant this phenotype is only visible on mature source leaves, whereas the young sink leaves are devoid of this phenotype. A closer inspection of the leaves shows that the appearance of this phenotype follows a clear developmental pattern starting at the tip and/or rim of the developing leaves and moving towards the base. This, however, is exactly the developmental pattern observed for the sink/source transition in leaves.

Biochemical analysis of the leaves from invertase expressing plants show that they express an about 50 to 500 fold higher invertase level as compared to control plants. The content of reducing sugars, i.e. fructose and glucose is increased by a factor of 4 to 50 over control plants. Another significant difference is the accumulation of large amounts of starch during the day in the invertase expressing leaves which does not get degraded during the dark period.

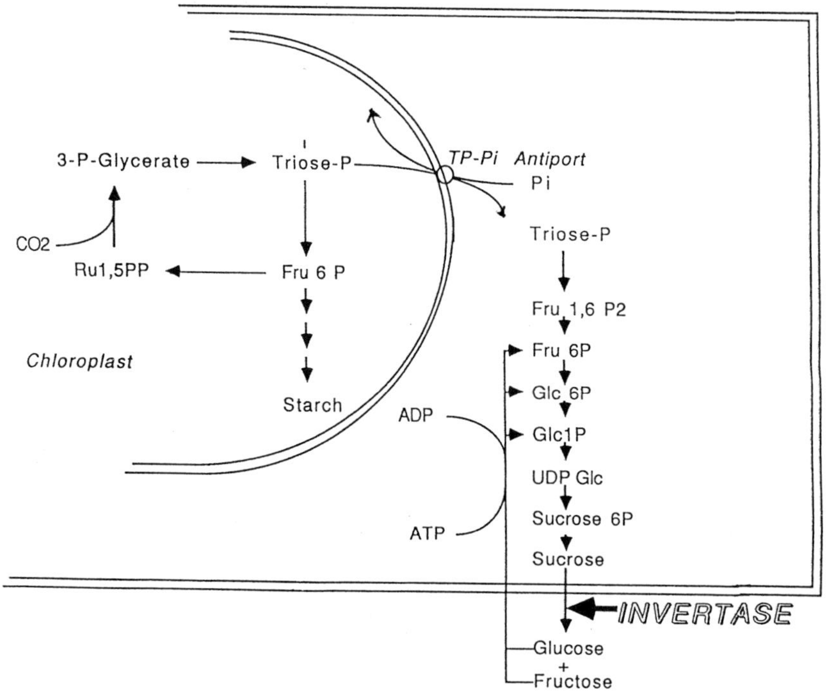


Fig. 2 Schematic representation of the Calvin cycle and the sucrose and starch biosynthetic pathways in a source leaf. The cleavage of the sucrose passing the cellwall by the invertase is indicated.

Most of these phenotypes can be explained by the following model (cf. also fig. 2): In tobacco, like in most other higher plant species sucrose is the preferred if not exclusive compound used in long-distance transport of photoassimilates from source to sink tissues. Young small leaves depend for their development on both photoassimilates formed by the leaf itself and on photoassimilates supplied by older source leaves.

Upon further development the sink develops into a source leaf starting at the leaf tip (cf. Turgeon, 1989). The criterion for becoming a source leaf is defined by a net export of sucrose from the photosynthetically active tissues. In order to export the sucrose the mesophyll cells have to load sucrose into the sieve elements via the companion cells. Whereas it is broadly accepted that the sucrose moves symplastically within the mesophyll cells, the mechanism of loading of the companion cells is a matter of debate. Both symplastic routes as well as routes involving an apoplastic loading step have been proposed (cf. Giaquinta, 1983). The phenotype observed with the transgenic invertase expressing tobacco plants, can, however, only be explained if an apoplastic loading step is involved. Sucrose present in the apoplast is at least partly split by yeast invertase yielding the two hexoses, fructose and glucose. Hexoses are not taken up into the phloem. Thus only two possibilities remain. The hexoses could either accumulate in the apoplast leading to an increase in osmotic potential and thus to plasmolysis of adjacent mesophyll cells which would lead to death of the tissue. A second mechanism would infer that the hexoses are efficiently taken up by the mesophyll cells, where they most likely become phosphorylated yielding hexose phosphates which could re-enter the pathway for sucrose synthesis, resulting in a futile cycle (Huber, 1989). As one consequence these cells would run out of free phosphate probably resulting in a reduction of triosephosphate export from the chloroplast via the triosephosphate / phosphate antiport (Flügge and Heldt, 1984) leading to photoinhibition and finally to pigment bleaching in the leaves due to photodestruction. Both mechanisms are obviously not mutually exclusive and could as well act in a synergistic fashion.

How does the suppression of sucrose export from source leaves influence the photosynthetic activity? Measurements of the $O_2$ evolution show that in source leaves the photosynthetic activity is decreased by a factor of about 5-10, whereas in the young sink leaves it increases by about 50 % when compared to control plants.

These data clearly support the idea of photosynthetic activity being sink-limited. Young leaves are sinks, they retain their photosynthate and use it to support their own growth. Due to the lower supply of these leaves with sucrose from source leaves there is a higher demand for sucrose synthezised within the sink leaf itself thus resulting in a higher rate of photosynthesis. As a leaf matures it turns into a source leaf and exports sucrose to the rest of the plant. In these leaves there is a progressive inhibition of photosynthesis. These plants provide a clear demonstration that photosynthetic metabolism is inhibited when the carbohydrates accumulate in the leaf.

## Conclusions

Due to their central importance for plant development, sink-source interactions have been intensively studied on the physiological and biochemical level. Here we present a new concept which should allow to analyze these interactions on a molecular level. The tools used comprise both ectopic expression of genes in transgenic plants and a purely genetic approach.

The data presented here on the biochemical and physiological changes in a transgenic plant impaired in sucrose export from source leaves indicate that indeed this approach might give new insights in this classical problem.

References


Bagnall, D.J., King, R.W., and Farquhar, G.D., 1988, Planta, 175:348-354.
Blechschmidt-Schneider, S., Ferrar, P., and Osmond, C.B., 1989, Planta, 177:515-525.
Carmi, A., and Sholler, I., 1979, Exp. Bot., 44:479-484.
Cave, G., Tolley, L.C., and Strain, B.R., 1981, Physiol. Plant., 51:171-174.
Claussen, W., and Biller, E., 1976, Z. Pflanzenphysiol., 81:189-198.
Flügge, U.I., and Heldt, H.W., 1984, TIBS, 9:530-533.
Fondy, B.R., and Geiger, D.R., 1980, Plant Physiol., 66:945-949.
Geiger, D.R., 1976, Can. J. Bot., 54:2337-2345.
Giaquinta, R.T., 1983, Ann. Rev. Plant Physiology, 34:347-387.
Gifford, R.M., Thorne, J.H., Hitz, W.O., and Giaquinta, R.T., 1984, Science, 225:801-808.
Goldstein, A., and Campen, J.O., 1978, Methods in Enzymology 42, W.A. Wood, ed., Academic Press, New York.
Ho, L.C., 1989, Ann. Rev. Plant Physiol. Plant Mol. Biol., 39:355-378
Huber, S.C., 1989, Plant Physiology, 91:656-662.
Kallarakal, and J., Komor, E., 1989, Planta, 177:336-341.
Keil, M., Sánchez-Serrano, J., Schell, J., and Willmitzer, L., 1986, Nucl. Acids Res. 14:5641-5650.
Köster-Töpfer, M., Frommer, W.-B., Rocha-Sosa, M., Rosahl, S., Schell, J., and Willmitzer, L., 1989, Mol. Gen. Genetics, 219:390-396.
Maynard, J.W., and Lucas, W.J., 1982, Plant Physiol., 70:1436-1443.
Mignery, G.A., Pikaard, C.S., and Park, W.D., 1988, Gene, 62:27-44.
Murashige, T., and Skoog, F., 1962, Physiol. Plant., 15:473-497.
Neales, T.F., and Incoll, L.D., 1968, The Bot. Rev., 34:107-125.
Nelson, C.E., and Ryan, C.A., 1980, Proc. Natl. Acad. Sci. USA, 77:1975-1979.
Ntrika, G., and Delrot, S., 1986, "Phloem transport", pp. 433-434, Cronshaw, J., Lucas, W.J., and Giaquinta, R.T., eds., Alan Liss Inc., New York.
Pavia, E., Lister, R.M., and Park, W.D., 1983, Plant Physiol., 71:161-168.
Rocha-Sosa, M., Sonnewald, U., Frommer, W.-B., Stratmann, M., Schell, J., and Willmitzer, L., 1989, EMBO J., 8:23-29.
Rosahl, S., Schmidt, R., Schell, J., and Willmitzer, L., 1986, Mol. Gen. Genet., 203:214-220.
Rufty, T.W., and Huber, S.C., 1983, Plant Physiol., 72:474-480.
Sasek, T.W., Delucia, E.M., and Strain, B.R., 1985, Plant Physiol. 78:619-622.
Turgeon, R., 1989, Annu. Rev. Plant Physiol. Plant Mol. Biol., 40:119-138.
Turgeon, R., 1984, Plant Physiol., 76:45-48.
Wenzler, H.-C., Mignery, G.A., Fisher, L.M., and Park, W.D., 1989, Plant Mol. Biol., 12:41-50.

# THE EXPRESSION OF AN OVALBUMIN AND A SEED PROTEIN GENE IN THE LEAVES OF TRANSGENIC PLANTS

Christine Wandelt, Wayne Knibb, Hartmut E. Schroeder, M. Rafiqul I. Khan, Donald Spencer, Stuart Craig, Thomas J.V. Higgins

CSIRO, Divison of Plant Industry, GPO Box 1600, Canberra ACT 2601, Australia

## INTRODUCTION

Many major hurdles in the genetic engineering of plants have been overcome in recent years. These hurdles include the availability of characterized genes for transfer, systems for gene integration into a plant cell nucleus and regeneration of whole plants from cells carrying the integrated DNA. Apart from its value as a tool for unravelling the control of gene expression, genetic engineering has great potential as a means of introducing agronomically useful characters into crop and pasture plants. Although the number of crop and pasture plants that can be transformed and regenerated is still fairly small, the list is increasing rapidly (Gasser and Fraley, 1989).

For this application of genetic engineering the next hurdle is to maximize the level of gene expression in selected organ, tissue or cell types. For instance, we wish to introduce foreign proteins with enhanced nutritional value for animals into the leaves and seeds of pasture and crop plants. This means that these proteins must constitute from 1-10% of the total protein in the leaf or seed. In order to achieve levels of gene expression that would result in such high levels of protein we anticipate that we will need to optimize transcription, translation and protein stability in the transgenic plants.

To this end, we have engineered a chicken ovalbumin cDNA clone and a pea vicilin gene as two test genes for expression in the leaves of tobacco (*Nicotiana tabacum*) and lucerne (*Medicago sativa*). Gene flanking sequences were replaced systematically and the resulting chimeric genes were transferred to the new host plants for expression in leaves. In addition, DNA sequences encoding the vicilin protein were modified or deleted so as to target the protein to new subcellular compartments. The highest levels of protein accumulation were obtained by targetting vicilin to sites other than the vacuole.

## THE EXPRESSION OF CHIMERIC OVALBUMIN cDNAs IN LEAVES OF TRANSGENIC PLANTS

Ovalbumin was detected in the leaves of transgenic tobacco and lucerne (Figure 1) when the ovalbumin cDNA was fused to the CaMV 35S promoter/enhancer. The ovalbumin level was about 0.001% of the total

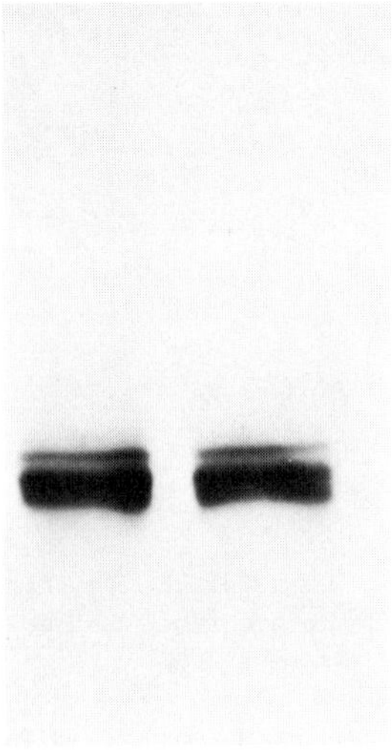


Figure 1. Ovalbumin is Expressed in the Leaves of Transgenic Tobacco and Lucerne.

Tobacco (*Nicotiana tabacum*) and lucerne (*Medicago sativa*) were transformed with an ovalbumin cDNA clone modified for expression in plants. The CaMV 35S promoter enhancer was fused to the 5′ untranslated region and a nopaline synthase 3′-flanking sequence was fused downstream of the 3′- untranslated region of the ovalbumin cDNA. The chimeric gene was transferred to plant tissue pieces using *Agrobacterium tumefaciens* and the binary vector Bin 19 (Bevan,1984). Ovalbumin was detected in protein extracts from leaves of glasshouse-grown lucerne (L) and tobacco (T) plants using a triple antibody technique combined with Western blotting.

extractable protein when this basic construct (No.11 in Figure 2) was used to express ovalbumin in plants. A series of ovalbumin chimeric constructs was then generated (Figure 2) with a view to increasing the level of gene expression.

The level of ovalbumin expression was increased about five-fold when the 3′ flanking sequences of the ovalbumin cDNA was replaced by the corresponding sequence from the plant gene encoding pea albumin 1 (constructs No.7 and 13 in Fig. 2). The addition of a plant 3′ flanking sequence (from nopaline synthase) to the existing avian sequence (construct 55 in Figure 2) was not sufficient to increase expression above the level obtained with the avian sequence alone (Table 1). The replacement of the avian 5′ untranslated sequence with a corresponding sequence from a plant gene encoding the small subunit (SSU) of ribulose bisphosphate carboxylase (Construct No.16 in Fig. 2) resulted in a 2-fold increase in ovalbumin (Table 1) when mRNA levels were low. However, at higher levels of template, no such effect of the SSU 5′ UT was observed (compare Constructs No.7 and 13 in Table 1).

We also measured ovalbumin mRNA levels in the transgenic plants and the results are shown in Table 1. It was clear that ovalbumin mRNA levels were lowest in those plants transformed with genes which contained an avian 3′ flanking sequence (constructs 11, 16 and 55). The addition of a

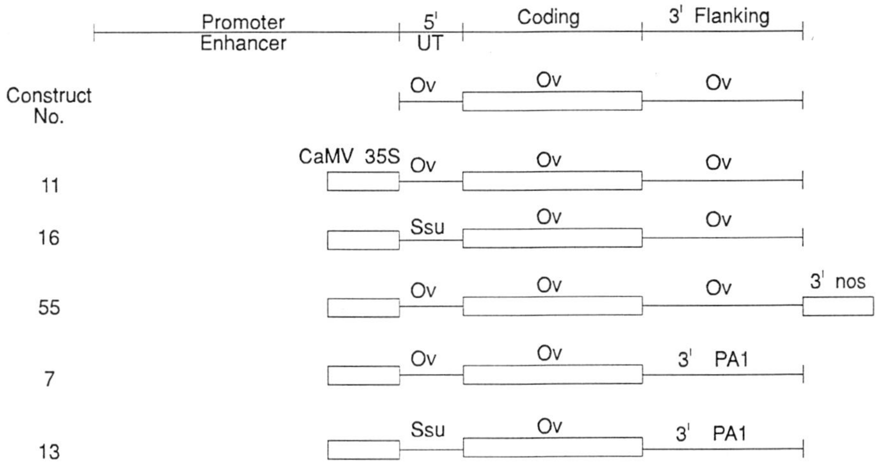


Figure 2. Construction of Chimeric Ovalbumin cDNAs for Maximizing Expression in Plants.

An ovalbumin cDNA was modified by replacement and/or addition of 5' and 3' flanking sequences. The CaMV 35S promoter/enhancer and the 3' nopaline synthase (3' nos) region were as described in Lyon *et al.* (1989) and the 3' PA1 sequence was derived from the pea PA1 gene described in Higgins *et al.* (1986). The SEKDEL sequence was added to the carboxyl terminus of vicilin by *in vitro* mutagenesis and oligonucleotide insertion into the gene.

Table 1. The Level of Ovalbumin mRNA and Protein in the Leaves of Transgenic Tobacco

| | mRNA Level[a] | Protein Level[b] |
|---|---|---|
| Contruct No. | | |
| 11 | 150 | 130 |
| 16 | 96 | 256 |
| 55 | 100 | 212 |
| 7 | 347 | 588 |
| 13 | 823 | 519 |

[a]These are arbitrary values taking the levels obtained from plants transformed with Construct No.55 as a baseline value of 100. Over 20 independently transformed plants were assayed and the averages of all positive plants are presented.

[b]The level of ovalbumin protein is given as ng per 10 mg of total soluble protein extracted from leaves. Over 20 independently transformed plants were assayed and the averages of all positive plants are presented.

nopaline synthase 3' flanking sequence downstream of the avian sequence (construct 55) did not result in high mRNA levels and, it was clear from the size of the ovalbumin mRNA in the 55 series of plants that termination of transcription was occurring in the avian sequence.

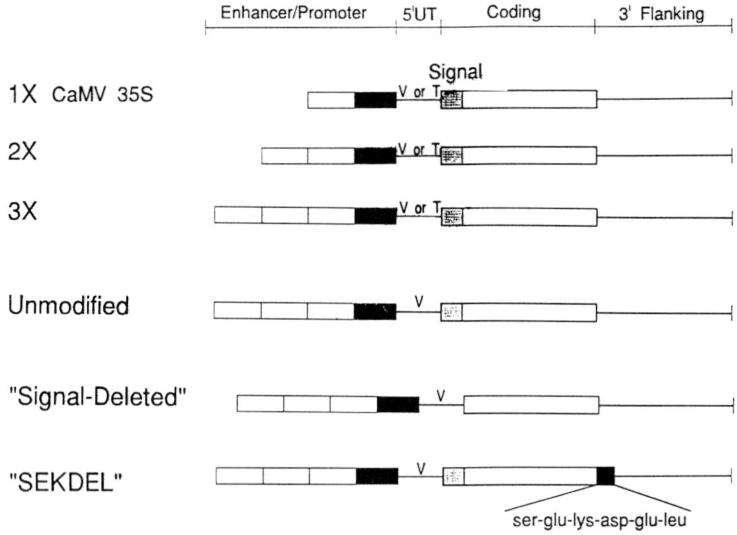


Figure 3. Construction of Chimeric Vicilin Genes for Maximizing Expression in Plants

A vicilin gene (Higgins *et al.*, 1988) was modified by *in vitro* mutagenesis so that unique restriction sites were introduced into the DNA at the cap site, the ATG encoding the initiator methionine, the position encoding the first amino acid of the mature protein and at a site corresponding to the penultimate amino of vicilin. Single (1x) or multiple (2x and 3x) copies of the CaMV 35S promoter/enhancer were inserted upstream of the cap site. In some cases the 5′ untranslated region (V) was replaced with a corresponding sequence (T) from the cowpea strain of tobacco mosaic virus (Meshi *et al.*, 1981).

The DNA encoding the signal-sequence was removed in the "signal-deleted" construct, while the DNA encoding the carboxy-terminal two amino acids was removed and replaced with DNA encoding ser-glu-lys-asp-glu-leu in the SEKDEL construct.

The chimeric genes were transferred to plants using the *Agrobacterium tumefaciens* binary vector pGA492 (An, 1989).

The replacement of the avian 3′ flanking sequence with a corresponding sequence from the gene encoding the pea seed protein, PA1 resulted in a 3-8 fold increase in ovalbumin mRNA levels (constructs 7 and 13, Table 1). This increase in level may be due to more efficient processing during termination and polyadenylation or it may reflect greater stability of the polyadenylated, mature template.

In summary ovalbumin mRNA levels were increased about 10-fold by replacing the avian with a plant 3′ flanking sequence. Replacing the avian with a plant 5′ UT had at most a 2-fold enhancement effect on protein levels. There is also some evidence that there may be a block to ovalbumin mRNA translation, such as in codon usage, since there was not a direct proportional increase in protein level when mRNA increased due to replacement of 3′ flanking sequences. Thus, there appears to be at least two constraints limiting the level of ovalbumin in plants. The first is

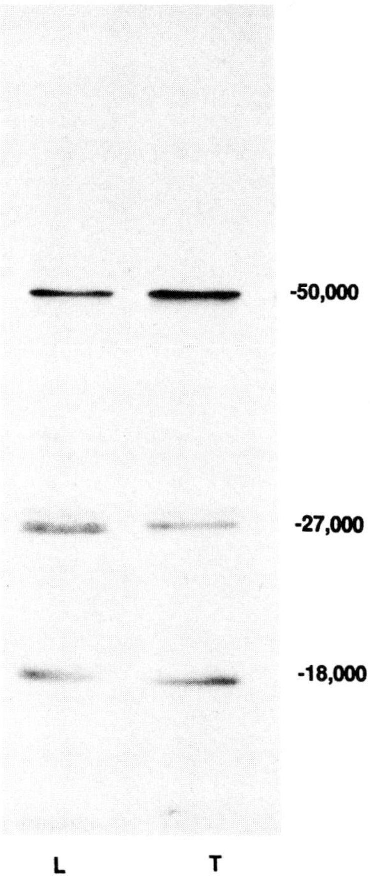


Figure 4. Vicilin is Expressed in the Leaves of Lucerne and Tobacco. A chimeric vicilin gene containing a single copy of the CaMV 35S promoter/enhancer (see figure 3) was used to transform lucerne (L) and tobacco (T). Vicilin was detected in the leaves using a triple antibody technique combined with Western blotting. The Mr 50,000 polypeptide as well as the two major cleavage products of Mr 18,000 and 27,000 are shown.

Table 2. The Level of Vicilin mRNA and Protein in the Leaves of Transgenic Tobacco

| | mRNA Level[a] | | Protein Level[b] | |
|---|---|---|---|---|
| | Vicilin 5′ UT | TMV 5′ UT | Vicilin 5′ UT | TMV 5′ UT |
| 1x CaMV 35S | 98 | 164 | 3020 | 3440 |
| 2x CaMV 35S | 150 | 77 | 2980 | 4800 |
| 3x CaMV 35S | 100 | 200 | 2820 | 2500 |
| 3x CaMV 35S- (Signal Deleted) | 23 | -c | 12500 | -c |
| 3x CaMV 35S- (Vicilin-SEKDEL) | 100 | -c | 250000 | -c |

[a,b] As in Table 1
[c] These constructs were not made.

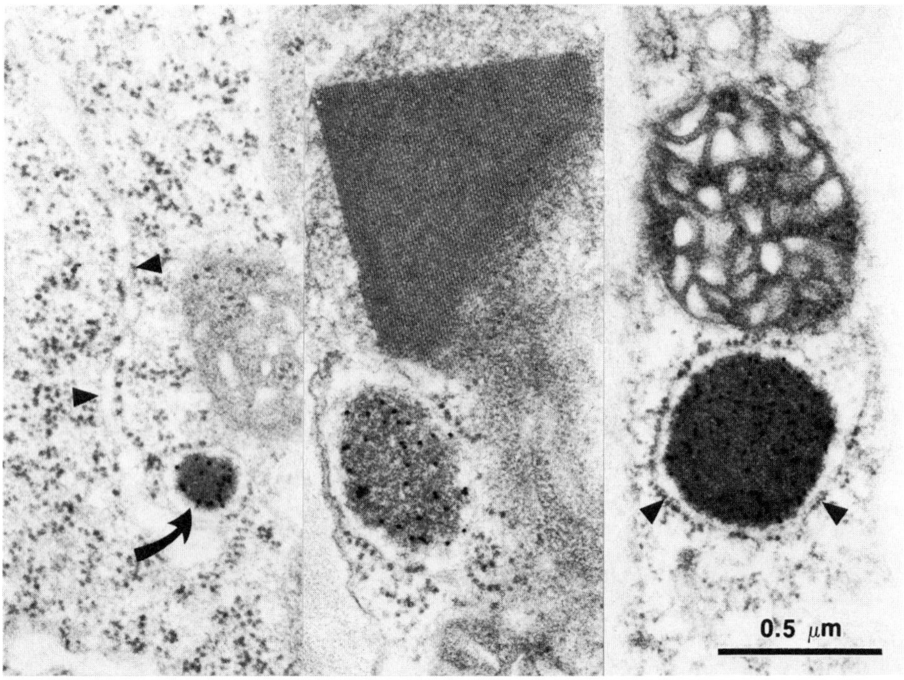


Figure 5. The Vicilin-SEKDEL is Localized in Inclusion Bodies of the Endoplasmic Reticulum.

A set of three electron micrographs of sections of transgenic tobacco leaves in which the vicilin-SEKDEL genes is being expressed. The sections were labelled (after embedding) with immunogold-tagged protein A and vicilin IgG. The arrow indicates a small inclusion surrounded by membrane which appears continuous with rough endoplasmic reticulum (arrowheads).

in mRNA level which may be mediated by transcriptional control and the second may be exerted at the level of translation, for example, there may be rare codons in the ovalbumin mRNA for which there are insufficient amino-acylated tRNAs in the leaf cells.

## THE EXPRESSION OF CHIMERIC VICILIN GENES IN LEAVES OF TRANSGENIC PLANTS

### Multiple Copies of the CaMV 35S Enhancer and Two Different 5′ Untranslated Regions

The gene encoding the pea seed storage protein, vicilin, was modified for expression in leaves of transgenic plants. Mutations were introduced at the start sites of transcription and translation so that precise replacement of the 5′ flanking sequences could be achieved using either the CaMV 35S promoter or multiples thereof, alone, or in combination with the 5′ UT region of the coat protein gene for the cowpea strain of tobacco mosaic virus (CP-TMV) (Fig. 3).

Vicilin was detected in the leaves of transgenic tobacco and lucerne (Figure 4). The level of transgenic protein in the leaves was about 0.03% of total soluble protein when we used a single copy of the CaMV 35S enhancer (Table 2). The addition of one or two further copies of the CaMV 35S enhancer did not increase vicilin mRNA or protein levels (Table 2).

The replacement of the vicilin 5′ UT region with a similar sequence from a highly expressed plant virus gene (CP-TMV) also did not result in enhanced protein levels (Table 2). These results were unexpected in view of the reports that multiple copies of the CaMV 35S enhancer resulted in enhanced gene expression (Kay *et al.*, 1987) and that a viral 5′ UT region enhanced mRNA translation (Sleat *et al.*, 1987). The discrepancy may be explained in a number of ways. One possibility is that vicilin or vicilin gene transcripts are very unstable in the leaves of transgenic hosts. Another possible explanation may be that some of the earlier results were obtained in *in vitro* or transient expression systems whereas our data derive from stable expression in mature plants grown in glasshouse conditions.

Removal of the Vicilin Signal Sequence or the Addition of an SEKDEL Sequence to Vicilin

Vicilin accumulation in the leaves of transgenic tobacco was increased by about 5-fold when the DNA sequence corresponding to the signal peptide sequence was removed (Table 2). It would be predicted that this protein could not enter the endomembrane system and therefore could not be deposited in the leaf equivalent of the seed protein body, the vacuole. Since post-translational processing of vicilin is known to occur in the vacuole, it might be expected that the protein with the deleted signal peptide would not undergo the normal cleavage observed for vicilin. Consistent with these expectations, transgenic vicilin encoded by the modified vicilin gene from which the signal peptide had been deleted resulted in an Mr 50,000 polypeptide without any of the cleavage products normally observed with the unmodified gene when expressed in tobacco leaves (data not shown).

This 5-fold increase in protein level was obtained with a level of vicilin mRNA which was in fact lower than that obtained with the unmodified gene. The simplest explanation of these data is that the protein was targetted to a new compartment with the result that it was now more stable. Pulse-chase labelling studies are now required to confirm this conclusion.

The intact vicilin gene was further modified by removing the DNA encoding the carboxy terminal two amino acids, and replacing it with a DNA sequence encoding "SEKDEL" (Ser-Glu-Lys-Asp-Glu-Leu). When this modified gene was transferred to tobacco, the level of vicilin in leaves was increased over 100-fold compared with the unmodified gene (Table 2) giving a level of 2.5% of the total soluble leaf protein. Again this increase in protein was obtained without any change in vicilin mRNA level.

Based upon studies in mammalian and yeast cells we predict that "Vicilin-SEKDEL" should accumulate in the endoplasmic reticulum and that it would not traverse the entire Golgi apparatus or enter the vacuole (Pelham, 1987). Two pieces of evidence were obtained which are consistent with this prediction. The first was obtained using immunogold labelling of leaf sections with vicilin antibodies (Figure 5). Vicilin was detected in bodies which appear to be distensions of the ER. The second piece of evidence was that the leaf vicilin did not undergo the normal cleavage observed with the product of the unmodified gene. Vicilin-SEKDEL was detected as a single major polypeptide of Mr ~50,000 (data not shown). Together, these data indicate that although vicilin has entered the endomembrane system it has not progressed to the vacuole where it would undergo cleavage. The high level of "vicilin-SEKDEL" accumulation in leaves compared with the unmodified gene may be a direct result of targetting to a new cell compartment in which the protein is far more stable.

CONCLUSION

The level of foreign protein expressed in the leaves of transgenic plants is generally low, although there are notable exceptions (Eckes *et al.*, 1989; Hilder *et al.*, 1987; Hiatt *et al.*, 1989; Johnson *et al.*, 1989). We have designed a range of chimeric genes which were systematically reconstructed so as to examine the relative importance of the promoter/enhancer, 5′ UT, coding region and 3′ flanking sequences in determining the final level of foreign protein. Small responses were obtained to changes in flanking sequences around the coding region while DNA changes designed to target the protein to new cellular compartments resulted in protein levels of over 2% of the soluble leaf protein.

REFERENCES


An, G., 1986, Development of plant promoter expression vectors and their use for analysis of differential activity of nopaline synthase promoter in transformed tobacco cells, Plant Physiol., 81:86.

Bevan, M., 1984, Binary *Agrobacterium* vectors for plant transformation, Nuc. Acids Res., 12:8711.

Eckes, P., Schmitt, P., Daub, W. and Wengenmeyer, F., 1989, Overproduction of alfalfa glutamine synthetase in transgenic tobacco plants, Mol. Gen. Genet., 217:263.

Gasser, C.S. and Fraley, R.T., 1989, Genetically engineering plants for crop improvement, Science, 244:1293.

Hiatt, A., Cafferkey, R. and Bowditch, K., 1989, Production of antibodies of transgenic plants, Nature, 342:76.

Higgins, T.J.V., Chandler, P.M., Randall, P.J., Spencer, D., Beach, L.R., Blagrove, R.J., Kortt, A.A. and Inglis, A.S., 1986, Gene structure, protein structure, and regulation of the synthesis of a sulfur-rich protein in pea seeds, J. Biol. Chem., 261:11124.

Higgins, T.J.V., Newbigin, E.J., Spencer, D., Llewellyn, D.J. and Craig, S., 1988, The sequence of a pea vicilin gene and its expression in transgenic tobacco plants, Plant Mol. Biol., 11:683.

Hilder, V.A., Gatehouse, A.M.R., Sheerman, S.E., Barker, R.F. and Boulter, D., 1987, A novel mechanism of insect resistance engineered into tobacco, Nature, 300:160.

Johnson, R., Narvaez, J., An, G. and Ryan, C., 1989, Expression of proteinase inhibitors I and II in transgenic tobacco plants: Effects on natural defense against *Manduca sexta* larvae, Proc. Natl. Acad. Sci. (USA), 86:9871.

Kay, R., Chan, A., Daly, M. and McPherson, J., 1987, Duplication of CaMV 35S promoter sequences creates a strong enhancer for plant genes, Science, 236:1279.

Lyon, B.R., Llewelllyn, D.J., Huppatz, J.L., Dennis, E.S. and Peacock, W.J., 1989, Expression of a bacterial gene in transgenic tobacco plants confers resistance to the herbicide 2,4-dichlorophenoxyacetic acid, Plant Mol. Biol., 13:533.

Meshi, T., Ohno, T., Iba, H. and Okada, Y., 1981, Nucleotide sequence of a cloned cDNA copy of TMV (Cowpea strain) RNA, including the assembly origin, the coat protein cistron, and the 3′ non-coding region, Mol. Gen. Genet., 184:20.

Pelham, H.R.B., 1988, Evidence that luminal ER proteins are sorted from secreted proteins in a post-ER compartment, EMBO J., 7:913.

Sleat, D.E., Gallie, D.R., Jefferson, R.A., Bevan, M.W., Turner, P.C. and Wilson, T.M.A., 1987, Characterisation of the 5′-leader sequence of tobacco mosaic virus RNA as a general enhancer of translation *in vitro*, Gene, 217:217.

# REGULATION OF PLANT GENE EXPRESSION BY ANTISENSE RNA

Joseph Mol, Alexander van der Krol, Arjen van Tunen, Rik van Blokland, Pieter de Lange and Antoine Stuitje

Department of Genetics, Vrije Universiteit
De Boelelaan 1087
1081 HV Amsterdam, The Netherlands

## INTRODUCTION

The antisense technology is based on blocking the information flow from DNA via RNA to protein by the introduction of an RNA strand complementary to (part of) the sequence of the target mRNA. This so called antisense RNA is thought to basepair to its target mRNA thereby forming double stranded RNA. Duplex formation may impair mRNA maturation and/or translation or alternatively may lead to rapid mRNA degradation. In any event the result mimicks a mutation.

Izant and Weintraub (1984) were the first to demonstrate the effectiveness of antisense gene constructs in eukaryotic cells. Numerous reports have appeared since then on the effective down-regulation of genes either transiently, using cloned DNA, RNA or oligonucleotides or by stable transformation using antisense DNA (for reviews, consult van der Krol et al., 1988a,b; Weintraub, 1990). The antisense approach in eukaryotes has evolved now from a model system to an approach well integrated in the field of molecular and applied genetics.

Plants were the first multicellular organisms in which endogenous genes were successfully down-regulated by antisense counterparts. This was made possible by the unique capacity to regenerate from a single cell. We will review our current knowledge about plant gene regulation by antisense DNA/RNA and speculate on possible mechanisms of action in view of the recent discovery (van der Krol et al., 1990a, Napoli et al., 1990) that "sense" versions of antisense genes sometimes down-regulate homologous gene expression.

## NATURAL ANTISENSE RNA

Antisense regulation was first discovered in naturally occurring bacterial systems (for review, see Simons, 1988). Eukaryotic cells contain RNAs which are complementary to portions of known mRNAs (reviewed in van der Krol et al., 1988a). Recently more examples have appeared in the literature. Kapler and Beverly (1989) report the presence of

*Plant Molecular Biology* 2, Edited by R.G. Herrmann and
B. Larkins, Plenum Press, New York, 1991

RNAs complementary to the dihydrofolate reductase-thymidylate synthase region of the protozoan parasite Leishmania. RNAs complementary to the myelin basic protein gene in mouse have been reported by Tosic et al. (1990). Murine erythroleukemia cells accumulate an antisense RNA involved in the maturation of the transformation associated protein p. 53 (Khochbin and Lawrence, 1989). For plant systems only circumstantial evidence is available for the presence of naturally-occurring antisense RNAs.

## BACTERIAL ANTISENSE AND SENSE GENES IN PLANTS

The first report on artificial antisense regulation of gene expression in plants came from Ecker and Davis (1986). They reported effective transient inhibition of chloramphenicol acetyl transferase (CAT) activity in carrot cells by co-introduction of sense and antisense cat genes in protoplasts. By double transformation of tobacco plants with nos and antisense nos genes, NOS activity can be modulated (Rothstein et al., 1987). More recently other bacterial genes such as bar (Cornelissen and Vanderwiele, 1989) and gus (Robert et al. 1989) have been successfully down-regulated by antisense techniques. Table I gives a summary of the state of the art.

## PLANT [ANTI]SENSE GENES IN PLANTS

The first authentic plant gene successfully down-regulated by antisense technology was the gene for chalcone synthase (chs) encoding the key enzyme of flavonoid biosynthesis (van der Krol et al. 1988c). Since the substrates of CHS are colorless an evenly-reduced pigmentation of the corolla was obtained in independent transformants, as expected. The amount of pigmentation correlates with the level of residual chs mRNA (van der Krol et al. 1990b). Moreover, the effect is specific of the chs mRNA; chi and dfr mRNA levels are unaltered. Unexpectedly, pigmentation patterns were obtained in rings and sectors that showed variation with light and hormone ($GA_3$) regime (van der Krol et al. 1988c, 1990b). Some of the antisense pigmentation phenotypes are shown in Fig. 1.

The different phenotypes observed in independent transformants are attributed to position effects of the antisense chs gene on its own expression. The effects of different promoters, antisense gene fragments and chromosomal location have been described (van der Krol et al. 1990c). An important conclusion is that subgenomic fragments of antisense genes can be ineffective in establishing a phenotype. This may be due to decreased stability of corresponding mRNAs (van der Krol et al. 1990c).

Attempts to enhance coloration by introduction of sense chs or sense dfr genes in Petunia had an inverse effect (Napoli et al. 1990; van der Krol et al. 1990a). Antisense-like effects were observed ranging in phenotype from patterns to fully white (cf. Fig. 1). Transcript analysis showed that transcript levels of both the endogenous and that of the transgene were specifically reduced (co-suppression). The two halves of chs cDNA are equally effective in establishing a

TABLE I Successful Inhibition of Plant Gene Expression by Antisense RNA

| Gene encoding | Reference | Comment |
|---|---|---|
| Chloramphenicol Acetyl Transferase | Ecker & Davis (1986)<br>Delaunay et al. (1988) | Transient expression<br>Double transformation |
| Nopaline synthase | Rothstein et al.(1987)<br>Sandler et al. (1988) | Double transformation<br>Double transformation |
| Phosphinotricin acetyl transferase | Cornelissen & Vandewiele (1989) | Confers resistance to bialaphos |
| ß-glucuronidase | Robert et al. (1989) | |
| Chalcone synthase | van der Krol et al. (1988c) | Role in flower pigmentation |
| Polygalacturonase | Smith et al. (1988,1990)<br>Sheehy et. al. (1988) | Role in fruit softening |
| Ribulose bisphosphate carboxylase (SSU) | Rodermel et al. (1988) | Role in $CO_2$ fixation and photorespiration |
| Peroxidase | Rothstein, unpublished | Role in lignin formation and wound healing |
| Cinnamyl alcohol dehydrogenase | Schuch et al. (1990) | Role in lignin formation |
| Granule-bound starch synthase | Visser, et al. (1990) | |
| Potato virus X coat protein | Hemenway et al.(1988) | |
| Cucumber mosaic virus coat protein | Cuozzo et al. (1988) | |

Fig. 1. Different flower pigmentation phenotypes.
A: VR hybrid; B,C,D: VR hybrid containing antisense chs gene; E: VR hybrid with chs sense gene; F: VR hybrid with dfr sense gene.

sense effect (R. v. Blokland, unpublished data). This indicates that the protein product encoded by the transgene is not a prerequisite. Whether transcription of the transgenes is necessary is under investigation.

Inactivation of chi mRNA should lead to accumulation of yellow chalcones. However, attempts to down-regulate chi gene expression by antisense or sense have failed sofar (van Tunen, unpublished data). Possibly specific sequences or structures are required to get sense or antisense effects.

Polygalacturonase (PG) plays an important role in fruit ripening. Smith et al. (1988, 1990) and Sheehy et al (1988) have observed a dramatic reduction of pg mRNA and PG protein levels after introduction of an antisense pg gene in tomato. This could lead to increased shelf life.

Even the most abundant protein present in plants can be reduced effectively using antisense techniques (Rubisco ss; Rodermel et al. 1988). Amylose-free potatoes were obtained by Visser et al. (1990) by introduction of an antisense granule-bound starch synthase (gbss) gene. Cinnamyl alcohol dehydrogenase (cad) is involved in lignin formation. Schuch et al. (1990) have shown that antisense cad genes completely abolish lignin synthesis.

Attempts to reduce RNA plant virus-specific mRNA levels (e.g. spec. coat protein) by introduction of antisense genes were rather unsuccessful (Hemenway et al. 1988; Cuozzo et al. 1988). Only at relatively low inoculum concentration some degree of protection was observed. The ineffectiveness of this system is probably associated with the cytoplasmic life cycle of these RNA viruses (PVX, CMV). The DNA containing Gemini viruses on the other hand can be efficiently inhibited by an antisense viral gene (Al-1; C. Lichtenstein, unpublished data). Table I gives a complete summary of the data presented.

## ON THE MECHANISM OF ANTISENSE ACTION

Evidence has accumulated over the past years that in prokaryotes antisense RNA forms double-stranded complexes with mRNA, thereby preventing translation. Model experiments in eukaryotes have indicated that duplex RNA can be formed in vivo and that such structures are poorly processed and translated (see van der Krol et al. 1988b). In Xenopus oocytes a double-stranded RNA unwinding activity has been detected, the significance of which is still unclear.

Several lines of evidence suggest that the formation of double-stranded RNA cannot fully account for the phenotypic effects observed in the flower pigmentation system of Petunia. First, duplex chs RNA nor free antisense RNA could be detected in floral tissue of independent transformants (van der Krol et al. 1990b). Second, antisense chs cDNA driven by weak promoters can be very effective. Combined with the observation that antisense chs RNA is less stable than chs mRNA (van der Krol et al. 1990c) we conclude that sub-stoichiometric amounts of antisense RNA are very effective. Third, addition of chs or dfr sense genes leads to antisense-like effects (Napoli et al. 1990; van der Krol et al. 1990a). In this case RNA-RNA interaction is unlikely to occur. At present we cannot discriminate in the latter case, between RNA-DNA and DNA-DNA interaction. It can be envisaged that such an interaction may trigger a mechanism to silence the interacting gene(s) e.g. by base methylation. For an extensive discussion of gene silencing mechanisms, see Napoli et al. 1990 and van der Krol et al. 1990a.

Irrespective the mechanism of action of sense and antisense genes in eukaryotes our experiments with the chi gene and chs gene fragments indicate that specific sequence elements must exist that mediate the silencing. It is encouraging that sense effects are not only seen in the flavonoid pathway, but also in the starch biosynthetic pathway (Visser et al. 1990).

## CONCLUSIONS AND PROSPECTS

Sense and antisense nucleic acids are useful tools to modulate the expression of specific genes. The technique enables one to shut off the expression of entire multigene families and could be very useful to 'probe' kryptic genes e.g. genes that are differentially expressed in a temporal and/or spatial way. The mechanism of action of sense and antisense genes is of interest. Our data suggest that there may be similarities in the way they exert their effect. Future work will concentrate on the possible involvement of RNA-DNA and/or DNA-DNA interactions in the gene silencing.

## REFERENCES


Cornelissen, M. and Vanderwiele, M., 1989, Both RNA level and translation efficiency are reduced by antisense RNA in transgenic tobacco. Nucl. Acids Res., 17: 833.

Cuozzo, M., O'Connel, K.M., Kaniewski, W., Fang, R-X., Chua, N-H. and Turner, N.E., 1988, Viral protection in transgenic tobacco plants expressing the cucumber mosaic virus coat protein or its antisense RNA. Bio/Technology, 6: 549.

Delauney, A.J., Tabaeizadeh, A. and Verma, D.P.S., 1988, A stable bifunctional antisense transcript inhibiting gene expression in transgenic plants. Proc. Natl. Acad. Sci. USA, 85: 4300.

Ecker, J.R. and Davis, R.W., 1986, Inhibition of gene expression in plant cells by expression of antisense RNA. Proc. Natl. Acad. Sci. USA, 83: 5372.

Hemenway, C., Fang, R-X., Kaniewski, W.K., Chua, N-H and Tumer, N.E., 1988, Analysis of the mechanism of protection in transgenic plants expressing the potato virus X coat protein or its antisense RNA. EMBO J., 7: 1273.

Izant, J.G. and Weintraub, H., 1984, Inhibition of thymidine kinase gene expression by antisense RNA: a molecular approach to genetic analysis. Cell, 36: 1007.

Kapler, G.M. and Beverly, S.M., 1989, Transcriptional mapping of the amplified region coding the dihydrofolate reductase-thymidilate synthase of Leishmania major reveals a high density of transcripts, including overlapping and antisense RNAs. Mol Cell. Biol., 9: 3959.

Khochbin, S. and Lawrence, J-J., 1989, An antisense RNA involved in p53 mRNA maturation in murine erythroleukemia cells induced to differentiate. EMBO J.,8: 4107.

Krol,van der, A.R., Mol, J.N.M. and Stuitje, A.R., 1988a, Antisense genes in plants: an overview. Gene, 72: 45.

Krol,van der, A.R., Mol, J.N.M. and Stuitje, A.R., 1988b. Modulation of eukaryotic gene expression by complementary RNA or DNA sequences: an overview. Biotechniques, 6: 958.

Krol,van der, A.R., Lenting, P.E., Veenstra, J., Meer, I.M. van der, Koes, R.E., Gerats, A.G.M., Mol, J.N.M. and Stuitje, A.R., 1988c. Expression of an antisense CHS gene in transgenic plants inhibits flower pigmentation. Nature, 333: 866.

Krol,van der A.R., Mur, L.A., Beld, M., Mol, J.N.M. and Stuitje, A.R., 1990a.Flavonoid genes in Petunia: addition of a limited number of genes copies may lead to suppression of gene expression. Plant Cell, 2: 291.

Krol,van der, A.R., Mur, L.A., de Lange, P., Gerats, A.G.M., Mol, J.N.M. and Stuitje, A.R., 1990b. Antisense chalcone synthase genes in Petunia: visualization of variable transgene expression. Mol. Gen. Genet., 220: 204.

Krol,van der, A.R., Mur, L.A., de Lange, P., Mol, J.N.M. and Stuitje, A.R., 1990c. Inhibition of flower pigmentation by antisense CHS genes: promoter and minimal sequence requirements for the antisense effect. Plant Mol. Biol., 14: 457.

Napoli, C., Lemieux, C. and Jorgensen, R., 1990.Introduction of a chimeric chalcone synthase gene into petunia results in reversible co-suppression of homologous genes in trans. Plant Cell, 2: 279.

Rodermel, S.R., Abbott, M.S. and Bogorad, L., 1988. Nuclear-organelle relationships: nuclear antisense gene inhibits ribulose biphosphate carboxylase enzyme levels in chloroplasts of transformed tobacco plants. Cell, 55: 673.

Robert, L.S., Donaldson, P.A., Ladaigue, C., Altosaar, I., Arnison, P.G. and Fabijanski, S.F., 1989. Antisense RNA inhibition of ß-glucuronidase gene expression in transgenic tobacco plants. Plant Mol. Biol., 13: 399.

Rothstein, S.J., DiMaio, Strand, J.M. and Rice, D., 1987. Stable and heritable inhibition of the expression of nopaline synthase in tobacco expressing antisense RNA. Proc. Natl. Acad. Sci. USA, 84: 8439.

Sandler, S.J., Stayton, M., Townsend, J.A., Ralston, M.L., Bedbrook, J.R. and Dunsmuir, P., 1988. Inhibition of gene expression in transformed plants by antisense RNA. Plant Mol. Biol., 11: 301.

Schuch, W., Knight, M., Bird, A., Frima-Pettenati, J. and Boudet, A., 1990. Modulation of plant gene expression, in press.

Sheehy, R., Kramer, M. and Hiatt, W.R., 1988. Reduction of polygalacturonase activity in tomato fruit by antisense RNA. Proc. Natl. Acad. Sci. USA., 85: 8805.

Simons, R,W., 1988. Naturally occurring antisense RNA control: a brief review. Gene, 72: 35.

Smith, C.J.S., Watson, C.F., Ray, J., Bird,, C.R., Morris, P.C., Schuch, W. and Grierson, D., 1988. Antisense RNA inhibition of polygalacturonase gene expression in transgenic tomatoes. Nature, 334: 724.

Smith, C.J.S., Watson, C.F., Morris, P.C., Bird, C.R., Seymour, G.B., Gray, J.E., Arnold, C., Tucker, G.A., Schuch, W., Harding, S. and Grierson, D., 1990. Inheritance and effect on ripening of antisense polygalacturonase genes in transgenic tomatoes. Plant Mol. Biol., 14: 369.

Tosic, M., Roach, A., de Rivaz, J.C., Dolivo, M. and Matthieu, J.M., 1990. Post transcriptional events are responsible for low expression of myelin basis protein in myelin deficient mice: role of natural antisense RNA. EMBO J., 9: 401.

Visser, R.G.F., Feenstra, W.J. and Jacobsen, E., 1990. Manipulation of granule-bound starch synthase and amylose content in potato by antisense genes. In: Applications of antisense nucleic acids and proteins, J.N.M. Mol and A.R. van der Krol, eds., Marcel Dekker inc. New York, in press.

Weintraub, H.M., 1990. Antisense RNA and DNA. Sci-Am., 1: 34.

# PHOTOCONTROL OF GENE EXPRESSION

E. Schäfer[1], A. Batschauer[1], A.R. Cashmore[2], B. Ehmann[1], H. Frohnmeyer[1,3], K. Hahlbrock[3], T. Kretsch[1], T. Merkle[1], M. Rocholl[1], B. Wehmeyer[1]

[1]Institut für Biologie II, Schänzlestr. 1, D-7800 Freiburg, FRG
[2]Plant Science Institute, University of Pennsylvania, Philadelphia, PA 19104, USA
[3]MPI für Züchtungsforschung, Abt. Biochemie, D-500 Köln 30, FRG

## INTRODUCTION

At least three different photoreceptor classes (phytochromes, blue UV-A- and UV-B receptors) control the transition from skotomorphogenesis to photomorphogenesis (Mohr and Schäfer, 1983). The control of elongation rate, photoperiodism as well as mRNA accumulation after partial reetiolation by phytochromes and blue UV-A receptors is also documented (Schäfer et al., 1984; Kuhlemeier et al., 1987).

In nature plants are exposed to sun light which acts on all possible photoreceptors pigments simultaneously. This may lead to complex interactions among them. Detailed studies of these interactions have been carried out during the last 10 years. However, before describing these apparent interactions it must be established that each photoreceptor system is, in fact, involved in a particular response.

Unfortunately, phytochrome, which is most responsive in red and far-red light, is the only photoreceptor analysed in molecular terms so far. Phytochromes have been purified from several plants and cDNA's and/or genomic clones encoding for phytochrome polypeptides have been sequenced for oat (cf. Sharrok and Quail, 1989), pea, zucchini, rice, corn and Arabidopsis. These data proved that not only two different phytochromes - as predicted from spectroscopical and physiological data (Jabben and Holmes, 1983) - but at least three different phytochromes are present in plants (Sharrock and Quail, 1989) and probably also control photoregulation. Unfortunately, it is extremely cumbersome to obtain physiological evidences for photocontrol by different phytochromes.

The participation of phytochrome in a given situation can be demonstrated under induction (light pulses of 5 min) conditions by testing for repeatable red/far-red reversibility. A reversion of a red light-pulse after a prolonged dark

Plant Molecular Biology 2, Edited by R.G. Herrmann and B. Larkins, Plenum Press, New York, 1991

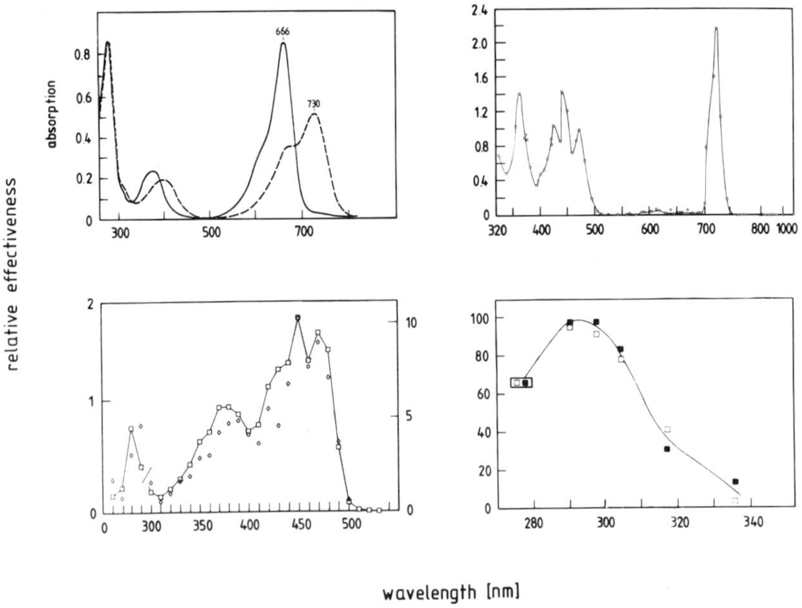


Fig. 1. Comparison of absorption spectra of purified phytochrome ($P_r$ and $P_{fr}$) (a), action spectra for light-controlled inhibition of hypocotyl elongation in lettuce seedlings (HIR, Hartmann, 1967) (b), action spectra for phototropism of oat coleoptiles and of alfalfa hypocotyls (blue light receptor, Briggs and Baskin, 1988) (c) and action spectra for UV induced synthesis of flavonoid glycosides in cell cultures of parsley and anthocyanin in maize coleoptiles (UV-B receptor, Wellmann, 1983).

period by a far-red light pulse is generally taken as indirect evidence for the control by the so-called stable (type II) phytochrome (Jabben and Holmes, 1983).In continuous light (high irradiance response, HIR) dichromatic irradiations are required as a decisive test. Dependence of the HIR on preirradiations with red light pulses can be taken as indirect evidence that this HIR is under the control of type I or labile phytochrome (Holmes and Schäfer, 1981). The role of phytochrome under continuous irradiation in green plants is not understood up to now (Schäfer et al., 1984).

The demonstration that the blue UV-A and/or UV-B photoreceptor systems are active in those systems where also the phytochrome system is active is much more difficult. This is due to the fact that both phytochromes and blue UV-A receptors absorb in blue and UV-A light and that all photoreceptors absorb in UV-B light (Fig. 1). Thus, any response to light may be due to the absorption by a single photoreceptor or to several which may either act independently to each other or may interact.

## IS GENE EXPRESSION UNDER CONTROL OF SEVERAL PHOTORECEPTORS?

### Expression of CAB genes in tomato and tobacco seedlings

Photoregulation mediated by phyotchrome in the dark after light treatment was tested by measuring mRNA accumulation kinetics after red and far-red light pulses. Figure 2 shows that both CAB PSI and CAB PSII mRNA accumulation increases transiently after a red light pulse. The effect of a red light pulse is reversible by a subsequent far-red light pulse almost to the level induced by a far-red light pulse alone. In these systems the inductive level of a far-red light pulse is rather high indicating that this response is very sensitive to even low levels of $P_{fr}$. This phenomenon seems to be observed generally for CAB gene expression (Mösinger et al., 1985; Briggs et al., 1985).

Under continuous irradiation with UV-, blue-, red- and far-red light at 19 $\mu$mol $m^{-2}s^{-1}$ no significant CAB mRNA accumulation was detectable (Wehmeyer et al., 1990). Therefore the question arises whether other photoreceptorsbesides phyotchrome are involved in CAB gene regulation. This could

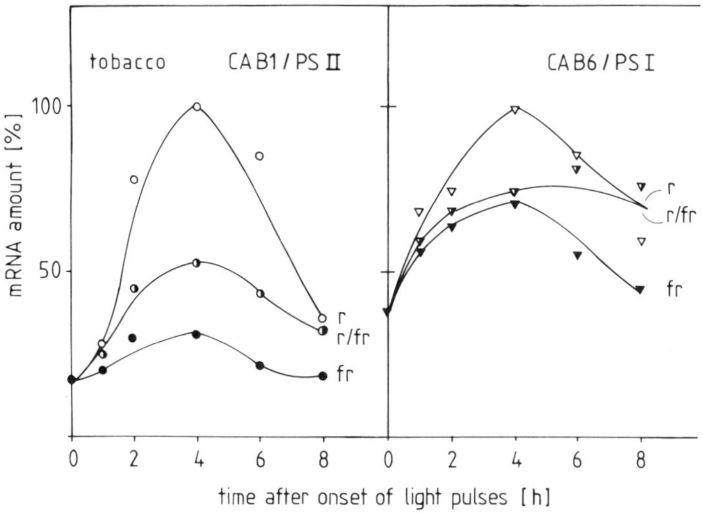


Fig. 2 Time course of photosystem II and I CABmRNA accumulation in 5-day-old etiolated tobacco seedlings. RNA was extracted 1, 2, 4, 6 and 8 h after onset of a 5 min red light pulse (r), 5 min red light pulse followed by 5 min long wavelength far-red light (RG 9 light) pulse (r/fr) and 5 min long wavelength far-red light pulse (fr). Time = 0 represents the mRNA amount in darkness (Wehmeyer et al., 1990)

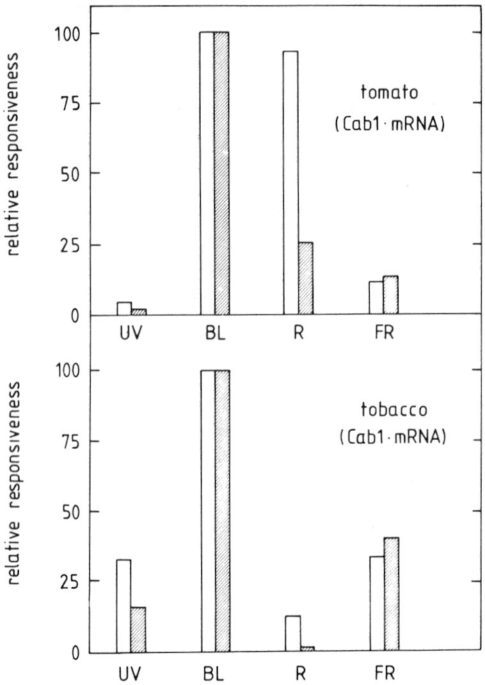


Fig. 3 Action spectra for photosystem II CAB mRNA accumulation in 4-day-old etiolated tomato and 5-day-old etiolated tobacco seedlings after 4 h irradiation. Open bars 50%, hatched bars 70% response level (Wehmeyer et al., 1990).

only be answered after measuring fluence rate response curves. Figure 3 shows the derived action spectra for 50% and 70% maximal response level. These action spectra demonstrate beside $P_{fr}$ action in the dark - i.e. inductive response - a photocontrol by the blue light receptor system, a high irradiance response (response to far-red light) and possibly also a UV-B receptor is involved in CAB mRNA accumulation.

This conclusion can be tested by the analysis of the responsiveness to $P_{fr}$ after a 4 h continuous light pretreatment (Mohr, 1984). The higher effectiveness after UV-B, blue- and far-red compared to red light pretreatment supported the hypothesis that all known photoreceptor systems in higher plants participate in CAB PS II mRNA regulation (Wehmeyer et al., 1990).

Expression of CHS genes in parsley suspension cultures

Based on detailed studies on the photoregulation of flavanoids in parsley cell suspension cultures it is well established that the UV-B-, the blue light receptor systems and phytochrome are involved in regulating this pathway (Duell-Pfaff and Wellmann, 1982). Measurements of CHS mRNA accumulation and run off transcription rates demonstrated that the contributions of the different photoreceptors could also be detected at these levels (Bruns et al., 1986; Ohl et al., 1989). Using a cell culture which are homozygotic with respect to the two allels of the CHS gene the photoregulation by blue and UV-B light was shown to be very similar for both alleles (Ohl et al., 1989).

Expression of CHS genes in mustard

Anthocyanine accumulation - especially in mustard seedlings - is one of the most extensively studied model systems for photoregulation (Lange et al., 1971; Mancinelli, 1983). A cDNA for CHS from mustard (Ehmann and Schäfer, 1989) was used to produce a fusion protein in E. coli to raise antibodies in a rabbit (Kretsch et al., in preparation). Western blot analysis demonstrated phytochrome control and HIR for CHS protein accumulation in hypocotyls and cotyledons of etiolated mustard seedlings (Kretsch et al., in preparation). In 10 day old white light grown plants only strong responsiveness to white light was obtained in primary leaves after 2 days reetiolation in darkness and no responsiveness to red or blue light was detectable. This indicates that under these conditions for strong photocontrol of CHS protein accumulation the UV part of the white light seems to be essential. These analysis has been confirmed at the mRNA accumulation level.

## IS A SINGLE GENE UNDER CONTROL OF SEVERAL PHOTORECEPTORS?

This question can be analysed relative easily if it is a single copy gene - as for CHS in parsley. For multigene families either gene specific probes must be used or promoter studies in transgenic plants should be carried out.

The use of gene specific probes to analyse photocontrol by several photoreceptors

Unfortunately, the number of experiments carried out to address this question is very rare. Developmental and organ-specific expression of the rbcS gene family has been studied in pea (Coruzzi et al., 1984, Fluhr et al., 1986), petunia (Dean et al., 1985, 1987) and tomato (Sugita and Gruissem, 1987), whereby different gene specific expression patterns were observed. Differential light/dark expressions have been described for the members of this gene family in tomato seedlings but the contribution of different photoreceptors has not been analysed.

In dark adapted green potato plants red and blue light induces the same kinetics with the mRNA accumulation for all of the four rbcS genes studied (Fritz et al., 1990). For maintaining high levels of rbcS transcripts in light grown seedlings clearly the contribution of both blue light receptor system and phytochrome could be demonstrated. The results indicate furthermore the different rbcS promotors do not react differentially to the excitations of the different photoreceptors (Fritz et al., 1990).

Developmental, organ-specific and light-dependent expression of the CHS gene family has been described for Petunia hybrida (V30) (Koes et al., 1989). Two of the genes are strongly expressed during flower development and are described to be dependent on red light irradiation in flowers. In contrast, in young seedlings and suspension cultures the same genes can be induced with UV light. In addition under these conditions two other genes of this multi-gene family are also expressed but at a low level. Unfortunately, the photobiolo-

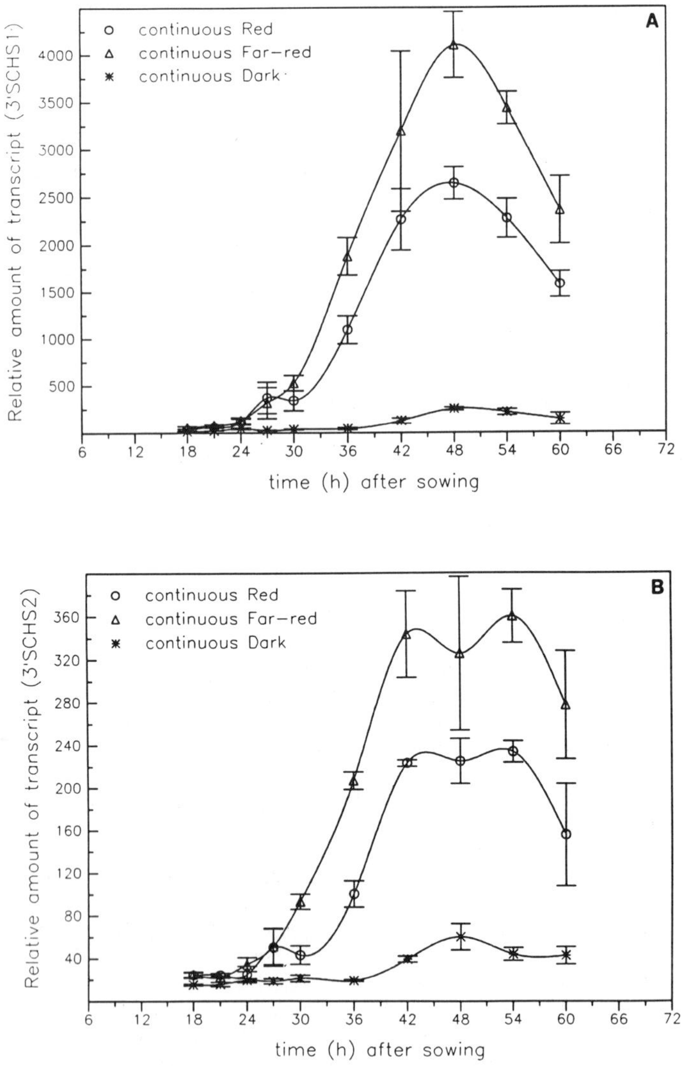


Fig. 4 Transcript accumulation kinetics of chalcone synthase mRNAs in cotyledons of mustard seedlings grown in continuous darkness, continuous red and far-red light from sowing onwards. Cotyledons were harvested at time points indicated. A: mRNA detected with probe 3'SCHS1; B: mRNA detected with probe 3'SCHS2 (Ehmann and Schäfer, unpublished).

gical studies carried out are not detailed enough to conclude that the same genes are in different developmental states under the control of different photoreceptors (Koes et al., 1989). Especially the experiments carried out for dark adapted plants do not exclude that the observed light regulation simply reflects limiting effects of photosynthesis. This is supported by the statement of the authors that photoregulation could not be obtained with repeated 15 min light 45 min dark pulses (Koes et al., 1989).

In etiolated mustard seedlings it could be demonstrated that the two genes for chalcone synthase tested so far are under very similar if not identical light control. A photocontrol by high irradiance response, and induction by type I (labile, fast destructing) and type II (stable, slow destructing) phytochrome was obtained (Fig. 4). Surprisingly in primary leaves the photocontrol is switched mainly to a UV-B receptor. It should be mentioned that this switch is correlated with a change in localisation of the chalcone synthase protein and mRNA from epidermal to mesophyll cells.

Promotor studies using transgenic plants

The question whether a promotor can transfer responsiveness for several photoreceptors has rarely been addressed up to now. Simpson et al (1986) reported that a chimeric gene under control of the 5'-flanking sequences of a RbcS gene from pea - using neomycin phosphotransferase II [NPT(II)] gene from the transposon $T_n5$ as a reporter gene - showed white light but no phytochrome control in etiolated F1 tobacco seedlings. Unfortunately, rigorous tests and further photobiological studies are missing.
Fluhr and Chua (1986) analysed photoregulation of two genes encoding SSU in pea and transgenic petunia plants. It was shown that in young etiolated pea seedlings the 3A and 3C gene is under strong phytochrome control whereas in reetiolated mature green leaves regulation by continuous white light but not by red light pulses was shown. This pattern was also observed in transgenic petunia plants. In addition in mature pea leaves a requirement for blue light for high expression rates was observed. Similar phenomena were observed for the transgenic petunia plants. Up to now detailed photobiological studies in transgenic plants are still missing even for RbcS, CAB and CHS genes.

DEPENDENCE OF THE PATTERN OF PHOTOREGULATION IN DIFFERENT DEVELOPMENTAL STATES AND CELL TYPES

The fundamental question of developmental and cell type specific gene regulation has been rarely addressed with respect to photoregulation. On the other hand the analysis of tissue specific gene expression is wide spread.

Koes et al., 1989, assumed that chalcone synthase genes may be under different light control in young seedlings and floral tissues. Similarly, Fluhr and Chua (1986) conclude differences in photoreceptor dependence for the regulation of rbcS gene from pea in young and mature leaves.

In mustard seeldings chalcone synthase can be found predominantly in epidermal and subepidermal cells from hypocotyls, in the upper and lower epidermal cells from cotyledons, in mesophyll cells of primary leaves, sepals and petals (Kretsch et al., in preparation). The two chalcone synthase genes tested so far are expressed in the same cells (Ehmann, unpublished).

The observed switch of CHS expression in mustard from high irradiance response and phytochrome induction to regulation by a UV-B photoreceptor reflects a developmental cell type dependent regulation pattern.

This cell type specific photoregulation pattern was also observed when chalcon synthase promoter GUS reporter gene fusions with promoters from parsley or mustard genes were tested in transient gene expression in parsley protoplasts. The main photoregulation by UV-B and blue light receptor was obtained for both contructs as it has been described for chalcone synthase in the parsley suspension culture (Ohl et al., 1989, Fig. 5) although the mustard CHS gene is under phytochrome control in cotyledons of young mustard seedlings.

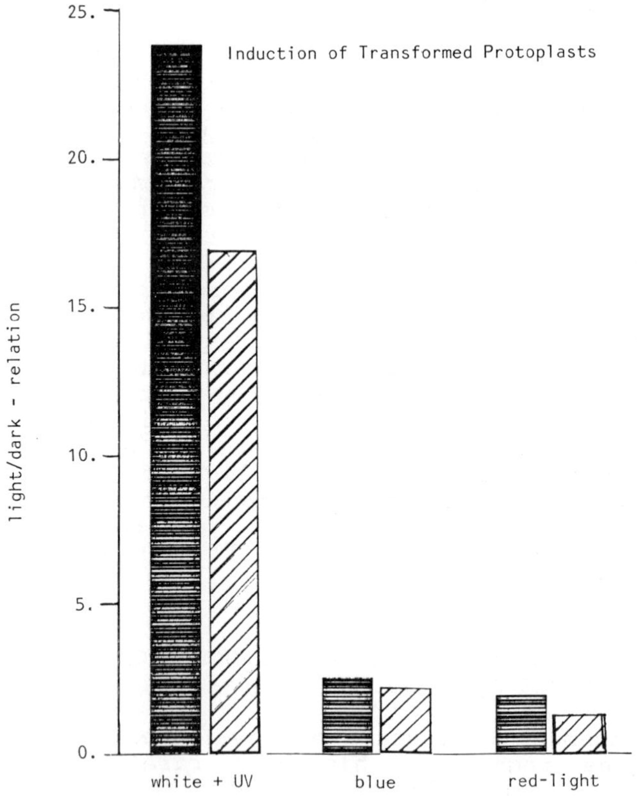


Fig. 5 Photocontrol of parsley and mustard CHS promoter GUS fusion constructs tested in a transient gene expression assay in transformed parsley protoplasts (Schulze-Lefert et al., 1989; Batschauer et al., in press). The transformed protoplasts were irradiated for 9 h and GUS was extracted after the end of the irradiation period. The dark levels were 43 and 37 [pmol MU/mg$P_{tot}$/min] for the parsley and mustard CHS promotor, respectively. The fluence rates were 4.2, 3.7 and 5 $WM^{-2}$ for UV containing white light, blue and red light, respectively.

These observations demonstrate that the photoregulation of gene expression depends strongly on the photoreceptor present and the promotor of the genes but also on the developmental state of the cell in which the expression of the gene will be studied.

REFERENCES

Briggs, W.R., and Baskin, T.I., 1988, Phototropism in higher plants - controversies and caveats, Botanica Acta, 101: 133-139.

Briggs, W.R., Mandoli, D.F., Shinkle, J.R., Kaufman, L.S., Watson, J.C., and Thompson, W.F., 1985, Phytochrome regulation of plant development at the whole plant, physiological and molecular levels, in: "Sensory Perception and Transcution in Aneural Organisms", G. Colombetti, F. Lenci, and P.-S. Song, eds. NATO ASI Series 89, Plenum Press, New York, pp. 265-280.

Bruns, B., Hahlbrock, K., and Schäfer, E., 1986, Fluence dependence of the ultraviolet-light-induced accumulation of chalcone synthase mRNA and effects of blue and far-red light in cultured parsley cells, Planta, 169: 393-398.

Coruzzi, G., Broglie, R., Edwards, C., and Chua, N.-H., 1984, Tissue-specific and light-regulated expression of a pea nuclear gene encoding the small subunit of ribulose-1,5-bisphosphate carboxylase, EMBO J., 3: 1671-1679.

Dean, C., van den Elzen, P., Tamaki, S., Dunsmuir, P., and Bedbrook, J., 1985, Differential expression of the light genes of petunia ribulose bisphophate carboxylase small subunit multi-gene family, EMBO J. 4: 3055-3061.

Dean, C., van den Elzen, P., Tamaki, S., Dunsmuir, P., and Bedbrook, J., 1987, Molecular characterization of the rbcS multi-gene family of Petunia (Mitchell), Mol. Gen. Genet., 206: 465-474.

Duell-Pfaff, N., and Wellmann, E., 1982, Involvement of phytochrome and a blue light photoreceptor in UV-B induced flavonoid synthesis in parsley (Petroselinum hortense Hoffm.) cell suspension cultures, Planta, 156: 213-217.

Ehmann, B., and Schäfer, E., 1988, Nucleotide sequences encoding two different chalcone synthase expressed in cotyledons of SAN 9789 treated mustard (Sinapis alba L.), Plant. Mol. Biol. 11: 869-870.

Fluhr, R., and Chua, N.-H., 1986, Developmental regulation of two genes encoding ribulose-bisphosphate carboxylase small subunit in pea and transgenic petunia plants. Phytochrome response and blue-light induction, Proc. Natl. Acad. Sci. USA, 83: 2358-2362.

Fritz, C.C., Schäfer, E., Schell, J., and Schreier, P.H., 1990, Four different rbcS genes in potato react identically to induction by blue or red light, submitted.

Hartmann, K.M., 1967, Ein Wirkungsspektrum der Photomorphogenese unter Hochenergiebedingungen und seine

Interpretation auf der Basis des Phytochroms (Hypokotylwachstumshemmung bei Lactuca sativa L.), Z. Naturforsch. 22b: 1172-1175.

Holmes, M.G., and Schäfer, E., 1981, Action spectra for changes in the 'high irradiance reaction' in hypocotyls of Sinapis alba L., Planta, 53: 267-272.

Jabben, M., and Holmes, M.G., 1983, Phytochrome and light grown plants, in: "Photomorphogenesis", W. Shropshire, Jr., and H. Mohr, eds., Encycl. Plant Physiol. New Series Vol. 16 B Springer-Verlag, Berlin, Heidelberg, New York, Tokyo, pp. 704-722.

Koes, R.E., Spelt, C.E., and Mol, J.N.M., 1989, The chalcone synthase multigene family of Petunia hybrida (V30): differential, light-regulated expression during flower development and UV light induction, Plant Mol. Biol., 12: 213-225.

Kretsch, T., Ehmann, B., Ocker, B., Speth, V., and Schäfer, E., 1990, in press.

Kuhlemeier, C., Fluhr, R., Green, P.J., and Chua, N.-H., 1987, Regulation of gene expression in higher plants, Annu. Rev. Plant Physiol., 38: 221-257.

Lange, H., Shropshire, W. Jr., and Mohr, H., 1971, An analysis of phytochrome-mediated anthocyanin synthesis, Plant Physiol., 47: 649-655.

Mancinelli, A., 1983, The photoregulation of anthocyanin synthesis, in: "Photomorphogenesis", W. Shropshire, Jr., and H. Mohr, eds., Encycl. Plant Physiol. New Series Vol. 16 B Springer-Verlag, Berlin, Heidelberg, New York, Tokyo, pp. 640-661.

Mohr, H., 1984, Criteria for photoreceptor involvement, in: "Techniques in Photomorphogenesis", H. Smith, and M.G. Holmes, eds., Academic Press, London, pp. 13-42.

Mohr, H., and Schäfer, E., 1983, Photoreception and deetiolation, Phil. Trans. R. Soc. Lond., B303: 489-501.

Mösinger, E., Batschauer, A., Schäfer, E., and Apel, K., 1985, Phytochrome control of in vitro transcription of specific genes in isolated nuclei from barley (Hordem vulgare), Eur. J. Biochem., 147: 137-142.

Ohl, S., Hahlbrock, K., and Schäfer, E., 1989, A stable blue-light-derived signal modulates ultraviolet-light-induced activation of the chalcone-synthase gene in cultured parsley cells, Planta, 177: 228-236.

Schäfer, E., Heim, B., Mösinger, E., and Otto, V., 1984, Action of phytochrome in light-grown plants, in. "Light and the Flowering Process", D. Vince-Prue, B. Thomas, and K.E. Cockshull, eds. Academic Press, London, pp. 17-32.

Sharrock, R.A., and Quail, P.H., 1989, Novel phytochrome sequences in Arabidopsis thaliana: structure, evolution, and differential expression of a plant photoreceptor family, Genes and Development, 3: 1745-1757.

Simpson, J., Van Montagu, M., and Herrera-Estrella, L., 1986, Photosynthesis-associated gene families: differences in response to tissue-specific and environmental factors, Science, 233: 34-38.

Sugita, M., and Gruissem, W., 1987, Developmental, organ-specific, and light-dependent expression of the tomato ribulose-1,5-bisphosphate carboxylase small subunit gene family, Proc. Natl. Acad. Sci. USA, 84: 7104-7108.

Wehmeyer, B., Cashmore, A.R., and Schäfer, E., 1990, Photocontrol of the expression of genes encoding chlorophyll a/b binding proteins and small subunit of ribulose-1,5-bisphosphate carboxylase in etiolated seedlings of Lycopersicon esculentum (L.) and Nicotiana tabacum (L.), Plant Physiol. in press.

Wellmann, E., 1983, UV irradiation in photomorphogenesis, in "Photomorphogenesis", W. Shropshire, Jr., and H. Mohr, eds., Encycl. Plant Physiol. New Series Vol. 16 B Springer-Verlag, Berlin, Heidelberg, New York, Tokyo, pp. 745-756.

# *phyA* GENE PROMOTER ANALYSIS

Peter H. Quail, Wesley B. Bruce*, Katayoon Dehesh, and Jacqueline Dulson†

U.C. Berkeley/USDA Plant Gene Expression Center
800 Buchanan Street; Albany, CA 94710 USA

## Introduction

The autoregulatory control that phytochrome exerts over the transcription of its own *phyA* genes in monocots[1-5] provides a valuable model system for investigating the molecular mechanisms underlying light-regulated gene expression in plants. Conversion of this cytoplasmically localized photoreceptor to its active Pfr form initiates repression of *phyA* transcription within 5 min of light signal perception via a mechanism that proceeds unimpeded in the absence of new protein synthesis[3]. These observations imply a short signal transduction chain from cytoplasm to nucleus with all essential components present in the cell before Pfr formation. Furthermore, Pfr maintains transcriptional repression as long as it is present in the cell above a certain level. Thus, the duration of repression in the dark following pulse-irradiations is dependent on the Pfr-pool size established by the terminal irradiation: red light establishes a large pool (86% Pfr) that sustains repression for > 12 hr; whereas far-red light establishes a small pool (~1% Pfr) that leads to derepression in < 3 hr[3]. Derepression occurs when Pfr is removed from the cell by turnover.

One approach to defining the mechanism by which phytochrome transduces its sensory signal to target genes under its control is to identify sequence elements and DNA-binding factors involved in regulated expression of these genes. We describe here deletion analysis of *phyA* promoters from oats and rice using a microprojectile-mediated gene transfer assay; the detection by gel-shift assay of oat nuclear proteins that bind in sequence-specific fashion to elements in the oat *phyA3* promoter; and the characterization of a cloned factor that binds to a GT-containing sequence element present in a rice *phyA* promoter.

## Functional Analysis of Oat *phyA3* Promoter

We have utilized microprojectile-mediated gene transfer in a transient expression configuration[1,6,7] to begin to define regions of the oat *phyA3* promoter involved in its regulation. Chimeric constructs containing various deletion and sequence substitution

*Present address: Plant Molecular Biology Center, Northern Illinois University, 325 Montgomery Hall, DeKalb, IL 60115-2861.
†Present address: Department of Molecular Biology & Genetics, University of Guelph, Guelph, ONT N1G 2W1, Canada.

mutations of the oat *phyA3* gene fused to a chloramphenicol acetyl transferase reporter (*phyA3*/CAT) have been introduced into etiolated rice seedlings by particle bombardment. CAT activity was determined after 24 hr dark incubation of the tissue that had received terminal red or far-red irradiations at the start of the dark period to establish high or low Pfr levels, respectively.

Figure 1 shows the effects of a series of 5' deletions on *phyA3* promoter activity[8]. As shown previously[1], the "standard" 1-kb promoter supports high levels of expression in response to low Pfr concentrations and is repressed by high Pfr to a level approaching that supported by the TATA box alone in this assay. When this basal TATA-driven expression is subtracted, the extent of phytochrome-controlled repression is 20- to 50-fold, comparable to that observed for endogenous oat *phyA* genes by nuclear run-on transcription assay[3]. The 5' deletion profile shows that a positive element, designated PE1, involved in supporting high level expression in low-Pfr tissue, is located between -381 to -348 bp. In addition, internal deletions indicate the presence of at least two more such positive elements, designated PE2 and PE3, located within regions -635 to -489 and -110 to -76 bp, respectively, upstream of the transcription start site[8]. The data indicate that PE1 and PE2 are functionally redundant, but that PE3 is required in conjunction with either PE1 or PE2 for high level expression.

Figure 2 indicates the locations of PE1 and PE2 in the *phyA3* promoter, as well as various motifs identified by sequence comparison to be conserved in monocot *phyA* promoters[2,4,5,9]. PE3 contains a sequence element, designated here as Box I, which is highly conserved among monocot *phyA* promoters, and is predicted to have a critical role in *phyA3* expression. On the other hand, initial internal deletion and sequence substitution mutations fail to indicate a major role for Box III or GT motifs in the expression or photoresponsiveness of this promoter[8].

The level of high-Pfr-imposed repression of *phyA3* activity is unaffected by any of the 5' terminal deletions (Fig. 1). The data indicate, therefore, that the repressive action of high Pfr levels is not mediated through any negative sequence elements upstream of -381 bp, since deletion of such elements would be expected to lead to high level expression in the

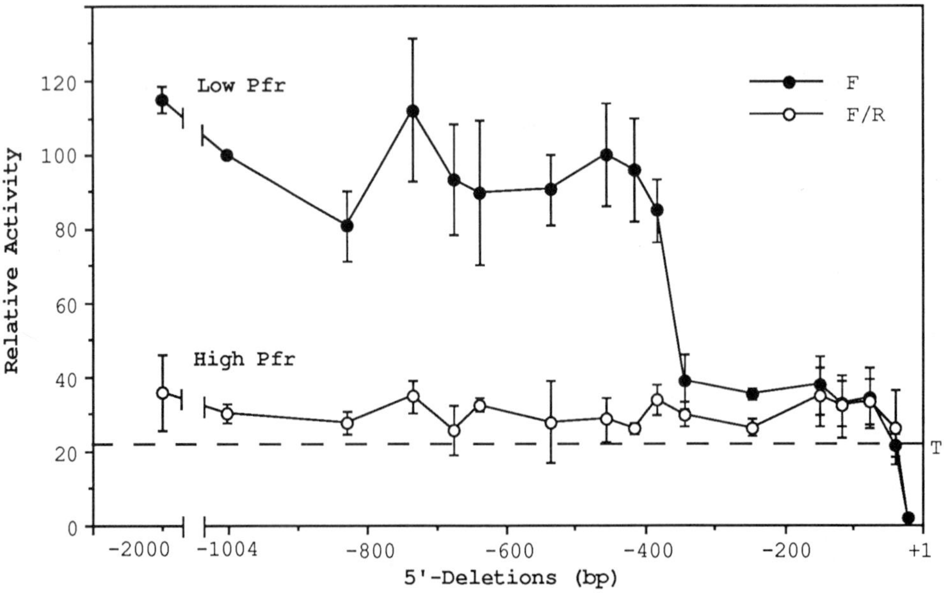


Fig. 1. 5′ deletion analysis of the oat *phyA3* promoter. Various 5′ terminal deletions of the oat *phyA3*/CAT gene were introduced into 2-day old etiolated rice seedlings by particle bombardment[1]. Seedlings were then irradiated with a pulse of either far-red (**F**) light (low Pfr) or **F** followed by red (**R**) light (high Pfr) and returned to the dark for 24 hrs before extraction and measurement of relative CAT activity. The dashed line designated **T** refers to the basal activity driven by the TATA box alone (see text). Adapted from (8).

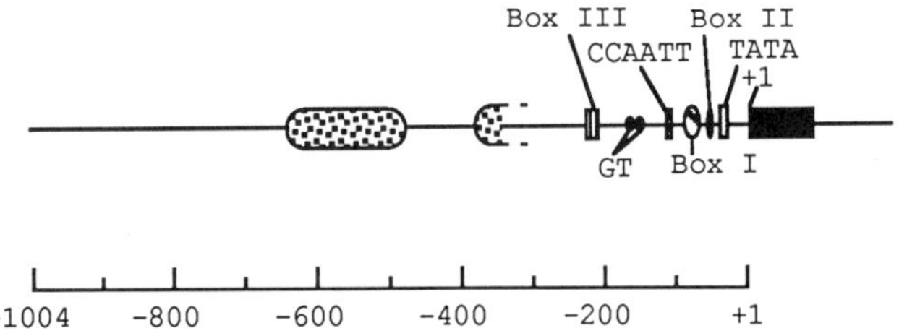


Fig. 2. Schematic of the 5′ flanking region of the oat *phyA3* gene. Box I, II, III, and GT refer to conserved sequences that have been identified by sequence comparison in the 5′ flanking regions of oat, rice, and maize *phyA* genes[2,4,5]. TATA and CCAATT are canonical sequences present in most eukaryotic promoters. PE1 and PE2 are positive promoter elements identified by the gene-transfer experiments described here. A third positive element PE3 detected by gene transfer includes Box I.

presence of high Pfr and this does not occur. Similarly, a series of internal deletions and sequence substitution mutations in the context of the 1-kb promoter also provide no evidence of negatively operating cis-acting elements[8].

## Binding of Nuclear Factors to Oat *phyA3* Promoter

To begin to identify trans-acting factors potentially involved in the regulated expression of the *phyA3* gene, we have performed preliminary gel-shift assays on nuclear extracts from oats. Figure 3 shows the results of an initial survey of the first approximately 1-kb of 5' flanking DNA (-1004 to +88) of the *phyA3* gene. Only two regions display detectable binding: fragment p2 (-396 to -116) and fragment p3 (-492 to -397). No difference in binding pattern was observed between nuclear extracts from dark-grown seedlings and red-light irradiated seedlings (data not shown), indicating the absence of any detectable phytochrome-induced changes in the abundance or binding properties of these factors.

Figure 4 shows that the B2 complex formed with fragment p2 is competed by unlabeled homologous p2 DNA, but is stable to competition by a 200-fold molar excess of a *phyA3* coding-region fragment and to a minimum of 50-fold molar excess of fragment p3. Reciprocal competition experiments with labeled fragment p3 show that complexes B3.1, B3.2, and B3.3 are also stable to competition by 200-fold molar excess of *phyA3* coding-region DNA and to a minimum of 50-fold molar excess of fragment p2 (data not shown). These results indicate that oat nuclear factors bind to the *phyA3* promoter fragments p2 and p3 in sequence-specific fashion and that different factors bind to each of these two regions.

Figure 5 provides evidence that nuclear factors interact with more than one site on fragment p2. The data indicate at least one binding site between -242 and -116 bp (p2A) and one between -396 and -243 bp (p2B). There is a further suggestion from the weak binding to p2D (Fig. 5) and other cross-competition studies (data not shown) that subfragment p2B contains a minimum of two binding sites: one between -396 and -332 bp and one between -331 and -243 bp.

Figure 6A shows that complex B2a formed with subfragment p2A is stable to competition by subfragment p2B, indicating that different factors bind to these two regions of the *phyA3* promoter. Competition analysis with a synthetic oligonucleotide shows that complex B2a results from the binding of a factor to a GT-containing motif between -173 and -151 bp within subfragment p2A (Fig. 6B). The tandem-paired GT-motifs GGTTAAT in the *phyA3* promoter (Fig. 6B) are identical to the core of the *rbcS* promoter sequence designated Box II which has been shown to interact with a factor designated GT-1 in crude nuclear extracts[10-12]. The data indicate, therefore, that oat nuclei contain a GT-1-like factor capable of interacting with the GT elements in the *phyA3* promoter, similar to that reported for rice[5].

Comparison of the gene-transfer and gel-shift assay data indicates a relatively poor correlation between the two approaches to identifying elements important to *phyA3*

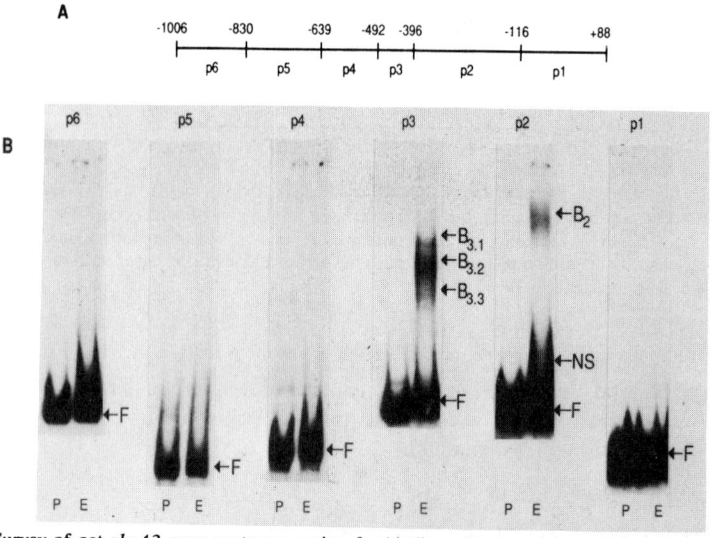


Fig. 3. Survey of oat *phyA3* gene upstream region for binding of oat nuclear proteins. (A) Map of the DNA fragments used as probes (designated p1 through p6). (B) DNA:protein interactions detectable by gel retardation assay. Nuclear proteins were isolated from etiolated oat seedlings[18,19] and assayed according to Fried and Crothers[20]. No difference in pattern was observed between nuclear extracts from irradiated and non-irradiated seedlings. **P**, probe only; **E**, probe + nuclear protein; **F**, free DNA; $\mathbf{B_n}$, specific DNA-protein complexes; **NS**, non-specific binding.

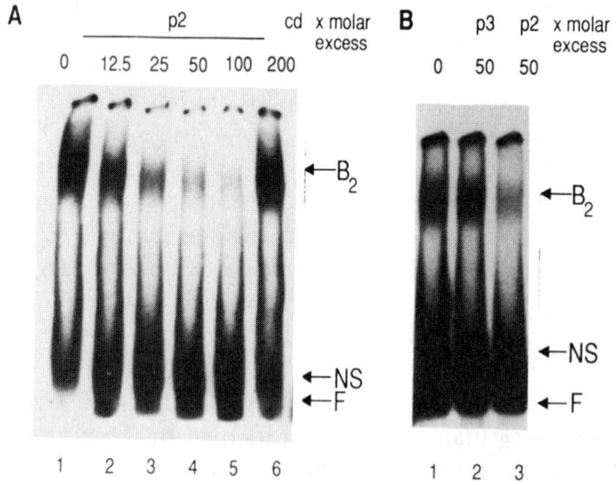


Fig. 4. Characterization of oat nuclear protein binding to oat *phyA3* promoter fragment p2 (-396 to -116 bp). $\mathbf{B_2}$, specific DNA:protein complex; **NS**, non-specific binding; **F**, free probe. (A) Titration of $\mathbf{B_2}$ formation by unlabeled p2, and stability of complex formation to inclusion of the *NcoI-HIII* fragment of *phyA3* coding region (cd)[9]. Either no competitor (lane 1); increasing molar excess of unlabeled p2, as noted above lanes 2-5; or a 200× molar excess of the coding region fragment (lane 6) was included in the binding reaction prior to addition of nuclear protein. (B) Complex $\mathbf{B_2}$ formation is unaffected by inclusion of unlabeled fragment p3 (-492 to -397). Lane 1, no competitor; lane 2, 50× molar excess of unlabeled p3; lane 3, 50× molar excess of unlabeled p2.

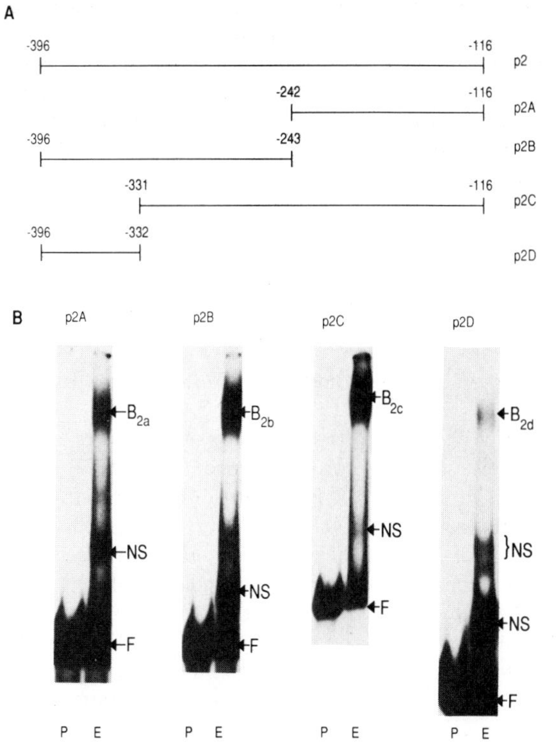


Fig. 5. Binding of oat nuclear protein to subfragments of p2. (A) Map of subfragments generated from p2: p2A, -242 to -116 bp; p2B, -396 to -243 bp; p2C, -331 to -116 bp; p2D, -396 to -332 bp. (B) Gel retardation analysis of oat nuclear protein binding to subfragments of p2. The probes are noted above each panel. **P**, probe only; **E**, probe + nuclear protein; **F**, free probe; $\mathbf{B_{2a}}$, $\mathbf{B_{2b}}$, $\mathbf{B_{2c}}$, $\mathbf{B_{2d}}$ refer to specific complexes formed with probes p2A, p2B, p2C, and p2D, respectively; **NS**, non-specific binding.

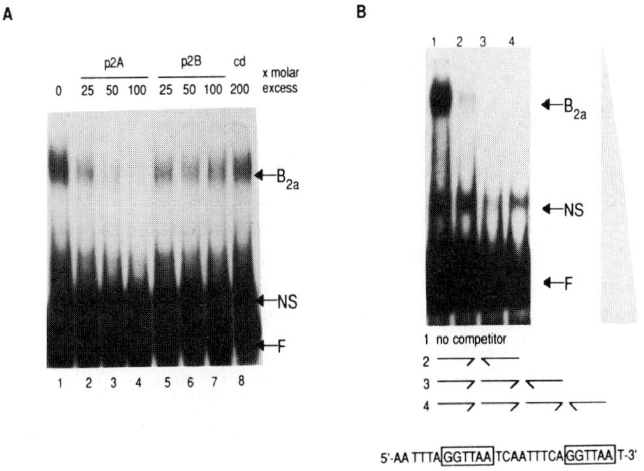


Fig. 6. Competition analysis of complexes formed by oat nuclear protein with subfragment p2A (-242 to -116 bp). $\mathbf{B_{2a}}$, specific DNA:protein complex; **NS**, non-specific binding; **F**, free probe. (A) Sequence in p2B (-396 to -243) fails to reduce complex $\mathbf{B_{2a}}$ formation with subfragment p2A. Either no competitor (lane 1); an increasing molar excess of unlabeled p2A (lanes 2-4), of unlabeled p2B (lanes 5-7); or a 200× molar excess of an unlabeled fragment of the *phyA3* coding region (cd, lane 10) was included in the binding reaction. (B) The tandem GT-element of the *phyA3* gene is involved in formation of complex $\mathbf{B_{2a}}$. Lane 1, no competitor; lanes 2-4, 50× molar excess of synthetic oligonucleotide replicas of the tandem GT-element sequence in the oat *phyA3* gene arranged as a dimer, trimer, and tetramer respectively. The orientation of the elements in each competitor DNA is noted under the lanes. The tandem GT sequence used to generate the multimers is presented at the bottom.

expression. On the one hand, elements PE1 and PE3, identified as functionally active by gene transfer (Figs. 1, 2), are contained respectively within fragments p4 and p1 which exhibit no detectable interaction with factors in oat nuclear extracts by gel-shift assay (Fig. 3). Included in possible reasons for this discrepancy are potentially low abundance of the relevant factors, and *in vitro* instability of these factors. Conversely, the gel-shift assay detects nuclear factors that bind to regions of the promoter that, in the gene-transfer experiments performed thus far, display no evidence of functional elements. These are fragment p3 (-492 to -396); region -332 to -242 in fragment p2B; and the GT-motif in fragment p2A (Figs. 3, 5). The reasons for this discrepancy remain to be resolved. The single instance of a positive correlation occurs with fragment p2D which contains element PE1 detected by gene transfer and which displays weak binding *in vitro* (Fig. 5).

## Cloning of a Trans-Acting Factor that Binds to a GT-Motif in a Rice *phyA* Promoter

Functional analysis of a rice *phyA* promoter using microprojectile-mediated gene transfer indicates that a GT motif with the core sequence GGTAATT is critical to high level expression in low-Pfr cells[13]. This result is in apparent contrast to the behavior of the oat *phyA3* promoter where the resident GT elements were not found to influence expression[8]. However, close inspection of the critical rice motif, which is the 3' member of a pair of tandem GT-containing sequences in the promoter, indicates that it differs from that of the adjacent 5' member of the pair, and from that of both tandem oat GT motifs, all three of which are identical to each other with the sequence GGTTAAT. This latter motif is in turn identical to the core Box II sequence of *rbcS* promoters[10-12]. This difference between the oat and rice *phyA* promoters thus appears to explain why deletion of the GT elements from the oat promoter has little effect on the activity of this promoter[8].

We have cloned a rice nuclear factor, designated GT-2, that binds in a highly sequence specific fashion to the rice GGTAATT motif[13]. We have called this factor GT-2 to distinguish it from the binding activity detected in crude nuclear extracts which interacts with the GGTTAAT motif and has been designated GT-1[5,10,11] (Fig. 6). Figure 7 shows the binding of cloned GT-2 to the rice *phyA* promoter fragment (-364 to -158 bp) that contains the tandem, non-identical GT elements. Footprint analysis indicates that GT-2 binds only to the 3' member of the pair of elements in this fragment[13]. A filter binding assay shows that the GGTTAAT motif of the oat *phyA3* and pea *rbcS* promoters binds GT-2 with an affinity that is two orders of magnitude less than for the rice 3' motif indicating the highly sequence-specific nature of the interaction. Moreover, substitution analysis indicates that all nucleotides within the GGTAATT motif are involved in binding, with the paired G's being most critical[13].

The amino acid sequence of GT-2 shows that it is highly hydrophilic (Fig. 8) but has no striking homology with any proteins currently in the data bases. It does, however, have a number of features that have been shown to be of functional importance to various transcription factors[14-17]. These features include a proline-glutamine rich domain, a basic region, an acidic region, and a region predicted to form a helix-loop-helix structure (Fig. 9). These data, together with the observed sequence-specific binding of GT-2 to the GGTAATT motif, shown by functional analysis to be critical for rice *phyA* expression, suggest that GT-2 may function as a transcriptional activator for the rice promoter. We anticipate finding additional protein sequence at the amino-terminal end of GT-2 since the cDNA insert we have cloned is 1.7 kb whereas the mRNA detected on Northern blots is 3.8 kb[13].

Northern blot analysis shows that the level of GT-2 mRNA declines when dark-grown rice seedlings are exposed to continuous white light for 7 days[13]. However, pulse-irradiations of red or far-red light have no effect on these mRNA levels over a subsequent 3 hr dark period. These results raise the possibility that modulation of GT-2 levels is involved in long-term regulation of the expression of light-controlled genes, but leave open the question of the photoreceptor responsible.

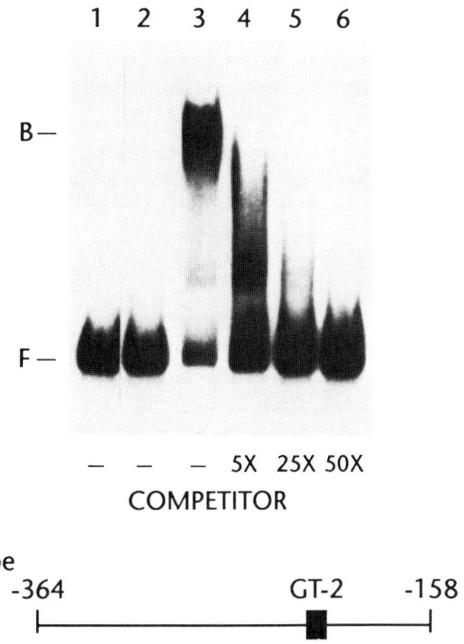


Fig. 7. Gel-retardation analysis of cloned GT-2 binding to the GT-element-containing fragment from the rice *phyA* promoter. GT-2 protein was overexpressed in *E. coli* using a vector containing a T7 promoter[21] into which the GT-2 cDNA had been cloned. A *Bst*Y I-*Hin* PI fragment from the rice *phyA* promoter (-364 to -158 bp) was used as a probe. **B**, bound fragment; **F**, free fragment; lane 1, no protein; lane 2, culture lysate from *E. coli* with GT-2 cDNA in reverse orientation; lanes 3-6, culture lysate with GT-2 cDNA in correct orientation. Lanes 1-3, no competitor; lanes 4-6, increasing molar excess of unlabeled oligonucleotide replica of tandem GT-elements in rice *phyA* promoter: GGTTAATTATTGGCGGTAATT (GT motifs underlined). The indicated molar excess refers to the concentration of tetramers of this oligonucleotide.

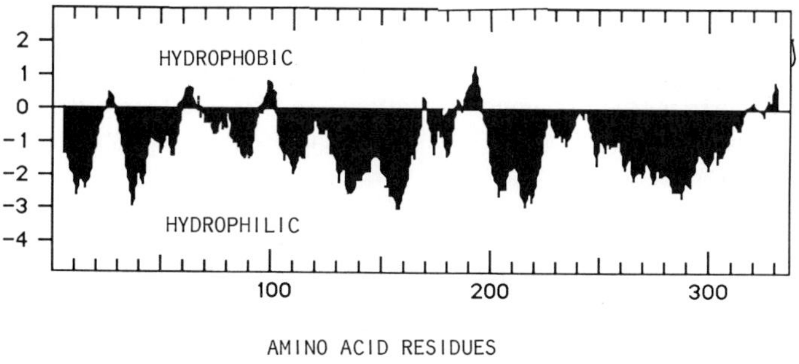


Fig. 8. Hydropathy profile[22] of cloned rice GT-2 protein.

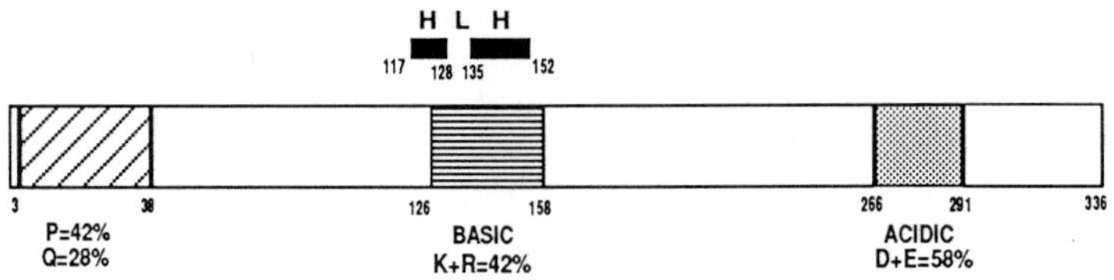


Fig. 9. Schematic representation of rice GT-2 domain structure. Locations of domains rich in proline and glutamine; lysine and arginine (basic); and aspartate and glutamate (acidic) are indicated, together with the percent of these residues within the domain. **HLH** indicates predicted helix-loop-helix motif.

## Conclusions

Comparison of the oat and rice *phyA* promoters provides some initial insights into the regulation of the expression of these genes and at the same time highlights some of the complexities and unresolved questions. On the one hand, the conserved GT motif GGTTAAT, present in both oat and rice promoters[4,9], interacts with factors in nuclear extracts from both species in sequence-specific fashion[5] (Fig. 6), but appears to have no major functional role at least in the expression of the oat *phyA3* gene[8]. These data indicate that, despite the fact that oat and rice contain GT-1-like factors[10,11], which can bind to their target sequences in the *phyA* promoters, these factors do not appear to influence *phyA* expression. The data settle the question in a functional sense, therefore, as to how *phyA* genes which are negatively regulated by light accommodate this core GT-1 target sequence which has been identified as a positive regulatory element involved in light induced expression of *rbcS* genes[10-12]. Still open, however, are the questions as to why this motif would be conserved in the *phyA* promoters, and the mechanism by which it might be rendered nonfunctional in that context.

On the other hand, the contrasting behavior of the oat and rice *phyA* promoters suggests the intriguing possibility that each utilizes one or more different transcriptional activators to drive high level expression in low Pfr cells. The data presented here indicate that the factor GT-2 may function in this capacity in the rice promoter. Conversely, the identical, tandem GT-motifs in the oat *phyA3* promoter neither bind GT-2 nor, as indicated above, appear to have a significant role in oat *phyA3* expression[8]. Instead, the other positive elements PE1 and PE2, which have no obvious sequence relationship to the GT-2 target site, appear to fulfill this role. These results are consistent, therefore, with a modular structure of the *phyA* promoters, with the oat and rice promoters having acquired or retained different regulatory sequence elements which perform the same activator function.

Equally intriguing is the high degree of sequence specificity of GT-2 binding. This factor is clearly able to discriminate with high resolution between GGTTAAT, the core target sequence for GT-1 binding[10,11], and its own closely related target sequence GGTAATT[13]. Whether the converse is true remains to be settled. We have not yet directly tested whether the rice GT-1 factor will bind to the GT-2 target sequence. The question is open, therefore, as to how GT-1 and GT-2 might interact with the tandem target sequences in the rice promoter, and how the factors might interact with each other in occupying these sites. Irrespective, the observations outlined here raise the possibility that plants have a family of functionally distinct trans-acting factors capable of binding in a highly sequence-specific manner to closely related GT-elements.

## Acknowledgments

We thank J. DeWett and V. Walbot for the kind gift of a $\lambda$EMBL3 rice genomic library; A. Christensen for the *ubi*-LUC construct, advice, and encouragement; T. Klein and Biolistics, Inc. for use of a particle gun; S. Goff for help with particle bombardment technology; P.

Oeller for advice with the T7 promoter system; J. Tepperman, D. Apadaca, K. Suslow, H. Hung, and S. Casper for excellent technical assistance; M. Fromm and T. Caspar for helpful discussions; and R. Wells for preparing, editing, and assembling the camera-ready manuscript. Supported by grants 89-37280-4800 from the USDA Competitive Research Grants Program and DCB-8796344 from the National Science Foundation.

**References**


1. W. B. Bruce, A. H. Klein, T. Christensen, M. Fromm and P. H. Quail, Photoregulation of a phytochrome gene promoter from oat transferred into rice by particle bombardment, *Proc. Natl. Acad. Sci. USA*, **86**:9692-9696 (1989).
2. A. H. Christensen and P. H. Quail, Structure and expression of a maize phytochrome-encoding gene, *Gene*, **85**:381-390 (1989).
3. J. L. Lissemore and P. H. Quail, Rapid transcriptional regulation by phytochrome of the genes for phytochrome and chlorophyll a/b binding protein in *Avena sativa, Mol. Cell. Biol.*, **8**:4840-4850 (1988).
4. S. A. Kay, B. Keith, K. Shinozaki and N.-H. Chua, The sequence of the rice phytochrome gene, *Nucl. Acids Res.*, **17**:2865-2866 (1989a).
5. S. A. Kay, B. Keith, K. Shinozaki, M.-L. Chye and N.-H. Chua, The rice phytochrome gene:structure, autoregulated expression, and binding of GT-1 to a conserved site in the 5' upstream region, *Plant Cell*, **1**:351-360 (1989b).
6. T. M. Klein, M. Fromm, A. Weissinger, D. Tomes, S. Schaaf, M. Sletten and J. C. Sanford, Transfer of foreign genes into intact maize cells with high-velocity microprojectiles, *Proc. Natl. Acad. Sci. USA*, **85**:4305-4309 (1988).
7. T. M. Klein, B. A. Roth and M. E. Fromm, Regulation of anthocyanin biosynthetic genes introduced into intact maize tissues by microprojectiles, *Proc. Natl. Acad. Sci. USA*, **86**:6681-6685 (1989).
8. W. B. Bruce and P. H. Quail, Cis-acting elements involved in photoregulation of an oat phytochrome promoter in rice, *EMBO J.*, submitted, (1990).
9. H. P. Hershey, R. F. Barker, K. B. Idler, M. G. Murray and P. H. Quail, Nucleotide sequence and characterization of a gene encoding the phytochrome polypeptide from Avena, *Gene*, **61**:339-348 (1987).
10. P. J. Green, S. A. Kay and N.-H. Chua, Sequence-specific interactions of a pea nuclear factor with light-responsive elements upstream of the *rbcS-3A* gene, *EMBO J.*, **6**:2543-2549 (1987).
11. P. J. Green, M.-H. Yong, M. Cuozzo, Y. Kano-Murakami, P. Silverstein and N.-H. Chua, Binding site requirements for pea nuclear protein factor GT-1 correlate with sequences required for light-dependent transcriptional activation of the *rbcS-3A* gene, *EMBO J.*, **7**:4035-4044 (1988).
12. C. Kuhlemeier, M. Cuozzo, P. J. Green, E. Goyvaerts, K. Ward and N.-H. Chua, Localization and conditional redundancy of regulatory elements in *rbcS-3A*, a pea gene encoding the small subunit of ribulose-bisphosphate carboxylase, *Proc. Natl. Acad. Sci. USA*, **85**:4662-4666 (1988).
13. K. Dehesh, W. B. Bruce and P. H. Quail, Light-regulated expression of a trans-acting factor that binds to a GT-motif in a phytochrome gene promoter, *Science*, submitted, (1990).
14. P. J. Mitchell and R. Tjian, Transcriptional regulation in mammalian cells by sequence-specific DNA binding proteins, *Science*, **245**:371-378 (1989).
15. P. F. Johnson and S. L. McKnight, Eukaryotic transcriptional regulatory proteins, *Ann. Rev. Biochem.*, **58**:799-839 (1989).
16. C. Murre, P. S. McCaw, H. Vaessin, M. Caudy, L. Y. Jan, Y. N. Jan, C. V. Cabrera, J. N. Buskin, S. D. Hauschka, A. B. Lassar, H. Weintraub and D. Baltimore, Interactions between heterologous helix-loop-helix proteins generate complexes that bind specifically to a common DNA sequence, *Cell*, **58**:537-544 (1989).
17. C. Murre, P. S. McCaw and D. Baltimore, A new DNA binding and dimerization motif in immunoglobulin enhancer binding, *daughterless, MyoD*, and *myc* proteins, *Cell*, **56**:777-783 (1989).

18. J. C. Watson and W. F. Thompson, Purification and restriction endonuclease analysis of plant nuclear DNA, *Meth. Enzymol.*, **118**:57-75 (1986).
19. K. D. Jofuku, J. K. Okamuro and R. B. Goldberg, Interaction of an embryo DNA binding protein with a soybean lectin gene upstream region, *Nature*, **328**:734-737 (1987).
20. M. Fried and D. M. Crothers, Equilibria and kinetics of lac repressor-operator interactions by polyacrylamide gel electrophoresis, *Nucl. Acids Res.*, **9**:6505-6525 (1981).
21. F. W. Studier and B. A. Moffat, Use of bacteriophage T7 RNA polymerase to direct selective high-level expression of cloned genes, *J. Mol. Biol.*, **189**, 113-130 (1986).
22. J. Kyte and R. F. Doolittle, A simple method for displaying the hydropathic character of a protein, *J. Mol. Biol.*, **157**:105-132 (1982).

# APPROACHES TO UNDERSTANDING PHYTOCHROME REGULATION OF TRANSCRIPTION IN LEMNA GIBBA AND ARABIDOPSIS THALIANA

E.M. Tobin, J.A. Brusslan, J.A. Buzby, G.A. Karlin-Neumann, D.M. Kehoe, P.A. Okubara, S.A. Rolfe, L. Sun, and S.C. Weatherwax

Biology Department
University of California
Los Angeles, CA 90024, USA

## INTRODUCTION

In order to understand the way in which phytochrome action can affect transcription of specific genes, we have undertaken a study of phytochrome regulated genes in Lemna gibba, an aquatic monocot. Additionally, we have examined phytochrome regulation of cab gene expression in Arabidopsis thaliana, and we have devised selection schemes to isolate Arabidopsis mutants which affect the phytochrome signal transduction pathway leading to altered cab transcription. The ability to grow both these species of plants heterotrophically lends them an advantage compared to many other species in being able to separate the action of light on the phytochrome system from the other effects of light and darkness on plant growth and senescence.

Phytochrome action has been demonstrated to affect the transcription of a number of different genes in many different species (see reviews by Tobin and Silverthorne, 1985, Kuhlemeier et al., 1987). The effect may be either a positive or negative one. Particular short "light responsive elements" which can interact with protein factors have been identified upstream of a number of rbcS and cab genes encoding, respectively, the small subunit (SSU) of ribulose 1,5-bisphosphate carboxylase/oxygenase (Rubisco) and the major apoproteins of the photosystem II light-harvesting chlorophyll a/b-protein complex (LHCII). The evidence suggests these sequences play an important role in the overall response to light/dark conditions (reviewed in Silverthorne and Tobin, 1987; Benfey and Chua, 1989). There is also evidence that phytochrome action can alter RNA levels by effects on additional, post-transcriptional processes (Colbert, 1988; Thompson, 1988; Elliott et al., 1989), as well as influence many other processes, such as membrane permeability, which may not involve altered gene expression (Kendrick and Kronenberg, 1986). Although the phytochrome chromoprotein has itself been the subject of biochemical studies for many years, to date there is no clear understanding of the chain of events by which the phototranformation of phytochrome leads to specific transcriptional changes.

In Lemna gibba we have studied the transcriptional regulation of both rbcS and cab genes in reponse to phytochrome action (Silverthorne and Tobin, 1984). We have also recently identified genes which have a

*Plant Molecular Biology* 2, Edited by R.G. Herrmann and
B. Larkins, Plenum Press, New York, 1991

negative transcriptional response to phytochrome action in this species (Okubara and Tobin, 1990). We are particularly interested in defining the DNA sequences and interacting protein factors important in mediating both types of responses. Because this monocot species cannot be transformed with Agrobacterium tumefaciens-based vectors, we are exploring the possibility of using an homologous transient expression system to investigate the regulatory DNA elements. We have also been able to demonstrate appropriate expression of a reporter gene driven by a Lemna cab promotor in a heterologous system (tobacco).

Analysis of mutant plants has offered some interesting observations and insights into the action of phytochrome (e.g. Adamse et al., 1988; Chory et al., 1989a, 1989b; Koornneef et al., 1980; Parks et al., 1989; Sharrock et al., 1988). Arabidopsis thaliana offers a particularly good opportunity to utilize a genetic approach to understanding the phytochrome signal transduction pathway. The three characterized cab genes of this species (Leutwiler et al, 1986) show phytochrome regulation of their RNA levels, and one of these, cab140, is the most responsive to brief red illumination of etiolated seedlings (Karlin-Neumann et al., 1988). Existing long hypocotyl (hy) mutants, including ones potentially affected in the phytochrome signal transduction pathway (Koornneeff et al., 1980), have been examined and found to have normal phytochrome regulation (Chory et al, 1989; Sun and Tobin, 1990) of their cab RNA levels. Therefore, we are trying to make other additional mutants which will alter the cab transcription increase seen in reponse to phytochrome action, and represent mutations in components of the signal transduction pathway.

## RESULTS AND DISCUSSION

### The rbcS Gene Family of Lemna gibba

The rbcS gene family of Lemna gibba comprises 12-14 genes (Tobin et al, 1985). Figure 1 shows maps of 4 different genomic clones that include six of these genes, and a seventh gene (SSU1) is represented by a cDNA clone (Stiekema et al, 1983). Portions of the 3'untranslated (3'UT) regions of all seven genes have been sequenced and subcloned for use as gene specific probes (Tobin et al, 1985; Silverthorne et al, 1990).

Distinguishing hybridization of two of the 3'UT subclones, SSU1 and SSU13, proved to be complex due to a high degree of sequence identity

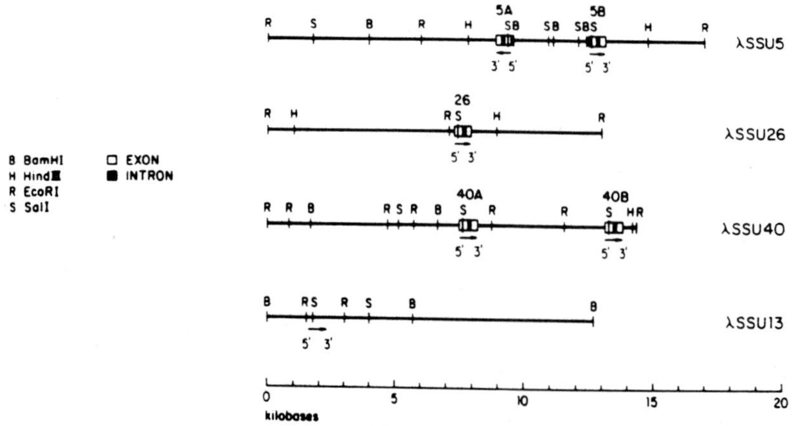


Fig. 1 Restriction maps of Lemna gibba genomic clones containing rbcS genes. The location of HindIII sites was not determined for λSSU13.

(Silverthorne et al., 1990) close to the translation temination codon, as diagramed in Fig. 2. The regions designated as SSU1-S and SSU13-S were derived from regions further downstream and subcloned to provide specific probes for SSU1 and SSU13, respectively. The lower panel of Figure 2 shows genomic Southern analyses using these probes. SSU13-S hybridizes a single band in these digests, whereas the hybridization pattern with SSU1-S suggests that there may be two nearly identical genes corresponding to this probe. Although SSU13-S and SSU1-S hybridize to several bands of the same size, this is coincidental; SSU1-S does not hybridize to genomic clones of SSU13.

Subcloned regions of the 3'untranslated (3'UT) portions of all seven genes have been used to distinguish relative expression levels and phytochrome responsiveness of individual genes (Tobin et al, 1985; Silverthorne et al, 1990). Expression of SSU13 could not be detected,

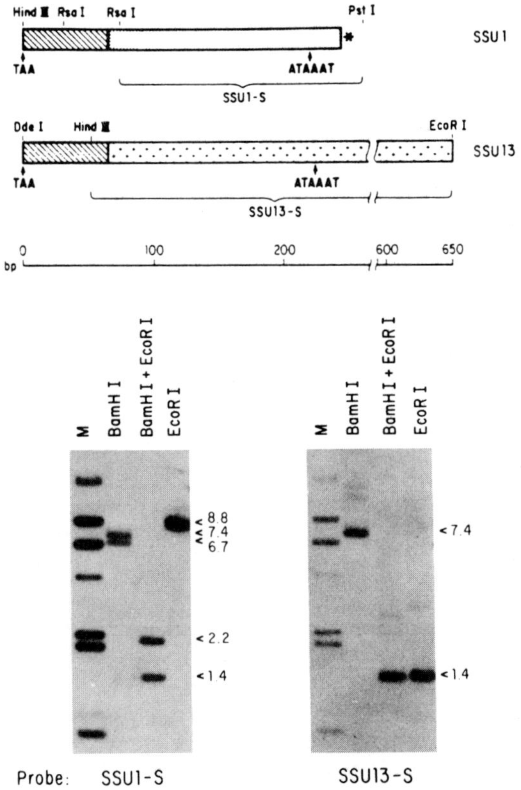


Fig. 2 3'UT specific probes for SSU1 and SSU13. Upper panel: Diagram of subcloned regions of λSSU1 and λSSU13. The regions indicated as SSU1-S and SSU13-S were subcloned to use as gene specific probes. [\\\\\\\\] indicates the region of near identity in the 3'UT region, [        ] is the region that is unique in the SSU1 sequence, and [ + + + + ] is the region unique in the SSU13 sequence. Lower panel: Southern blot analysis of the SSU1-S and SSU13-S sequences. Lemna nuclear DNA (10 μg/lane) was digested to completion with BamHI, EcoRI or a combination of both enzymes. Southern blots were probed first with random-primed DNA of SSU1-S. After washing off all radioactivity, the blot was reprobed with random-primed SSU1-S.

but levels of transcripts corresponding to each of the other clones can be regulated by phytochrome action. Of the transcripts represented by isolated genomic clones, SSU5B shows the highest level of phytochrome regulation and was chosen for further study. Transcriptional activity assayed in isolated nuclei by in vitro "run-on" experiments using a gene-specific probe confirmed that the transcription of the SSU5B gene increases within two hours in response to a single minute of red light (Okubara and Rolfe, unpublished work).

Sequencing of the region between the SSU5A and SSU5B genes (see Fig.1) revealed many stretches of nucleotide identity (T. Yamada and E.M. Tobin, unpublished work). Several short regions within these stretches also show conservation when compared to other rbcS and cab genes from Lemna gibba. Gel retardation assays have been used to demonstrate that a nuclear protein factor can interact with a fragment upstream of the SSU5B gene (-125 to -210) which contains one of these highly conserved sequences. The activity of this factor is considerably lower in nuclei from dark treated plants (1-4 da) than in nuclei from plants grown in the light or returned to light after a 24 h dark treatment (Buzby and Tobin, 1990).

In order to be able to see whether the DNA region that interacts with the protein factor plays an important role in vivo, it will be necessary to test its effect in the plant itself. However, since Lemna gibba, a monocot, cannot be stably transformed, we have tested whether a Lemna promoter-reporter gene construct is normally regulated by phytochrome in transformed tobacco plants. A 1.9 kb fragment upstream of the cab gene AB19 (Karlin-Neumann et al, 1985) was fused to the bacterial gus gene, and transformed into tobacco plants (Rogers et al, 1986; Jefferson et al, 1987). The enzyme activity encoded by this construct can be shown to exhibit phytochrome regulation, but the red induced activity increase is difficult to quantitate over the substantial dark level which is present already in the seeds and increases slightly during germination in complete darkness. For this reason, and also in order to be able to test promoters in an homologous system, we are developing a transient assay system in Lemna fronds.

**Transient Expression in Lemna gibba Fronds**

A transient expression system in which reporter gene fusion constructs can be assayed for phytochrome responsiveness has been reported by Bruce et al. (1989) using an oat phytochrome promoter construct expressed in rice seedlings. We are using the Biolistics Particle Acceleration System (Klein et al, 1988) to deliver DNA to a monolayer of intact Lemna fronds which can then be assayed for the reporter gene activity. As controls, we have used maize ubiquitin promoter-reporter genes which have been shown to be constitutively active in the rice transient expression system and to result in a higher level of activity than CMV 35S promoter constructs (Bruce et al., 1989) We have fused 965 nt of the SSU5B sequence upstream of the transcription start and a short portion of the transcribed region of the gene to a bacterial gene for chloramphenicol acetyl transferase (CAT) with a nos 3'UT. A similar construct was made using a severely truncated SSU5B promoter.

Fig. 3 shows that the 965 nt promoter construct confers substantial CAT activity which continues to increase for 24 hours after the DNA is introduced into the Lemna fronds. Introduction of a truncated promoter construct does not result in any CAT activity accumulating in the fronds (data not shown). Introduction of the SSU5B promoter construct results in substantial CAT activity by 16 hours if the fronds are incubated in the light, but very little activity can be detected if the fronds are maintained (on sucrose) in the dark (Fig. 4). The control construct using the maize ubiquitin promoter fused to CAT was equally

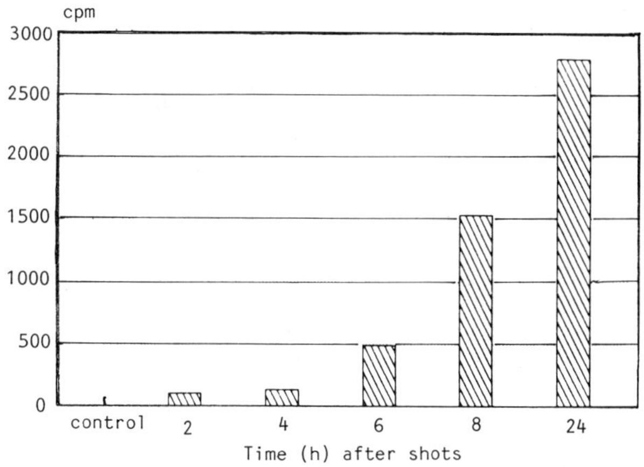


Fig. 3 Time course of increase in CAT activity in constant white light. The average CAT activity of four separate shots is shown. The control value was set at 0 and is an average of two shots with tungsten particles in the absence of any DNA.

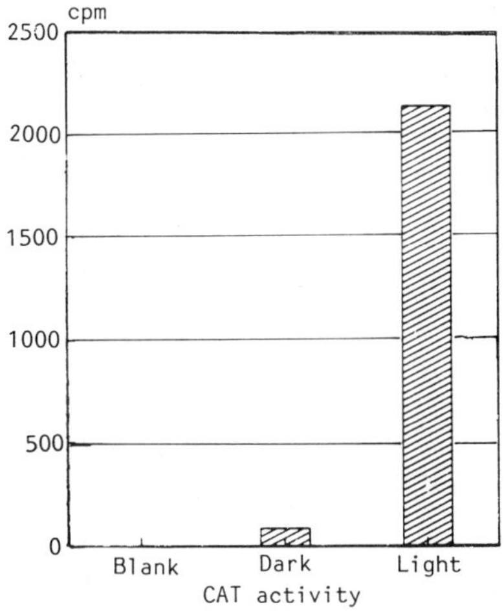


Fig. 4 Incubation in the light or dark affects the expression of the SSU5B:CAT construct in Lemna fronds. The average CAT activity of fronds incubated in white light or darkness for 16 h after three separate shots is shown. The blank value has been set to 0.

expressed in fronds maintained in the light and in darkness. Thus, we hope to use this system to investigate the specific sequences important in light regulation of the SSU5B and other Lemna genes.

**Identification of Genes Negatively Regulated by Phytochrome Action in Lemna**

Differential screening of a Lemna gibba cDNA library (Okubara and Tobin, unpublished) with random-primed probes prepared to RNA isolated from light-grown and dark-treated plants resulted in the isolation of three different cDNAs, designated NR11, NR18, and NR300. These clones represent genes whose transcripts accumulate during dark treatment. The increased RNA levels can be demonstrated to involve increases in the relative levels of transcription of these genes using the in vitro "run-on" isolated nuclei assay (Fig. 5). Furthermore, when nuclei were isolated from plants put into darkness for 4 days (D), and given a 2'red (R) light treatment 4 h before harvest in addition to the dark treatment, a phytochrome induced decrease in transcription of all three NR genes could be demonstrated. This response is reversible by immediate far-red illumination (Okubara and Tobin, 1990). Sequencing of the cDNAs revealed that none of the three are homologous to either phytochrome or protochlorophyllide reductase genes, other genes which have been demonstrated to be negatively regulated by phytochrome action in some species. The function of the encoded proteins is unknown. Genomic clones corresponding to the cDNAs have been isolated from a library made from Lemna DNA partially digested with BamHI and Sau3AI and cloned into Charon 35; these clones are currently being characterized.

## Arabidopsis cab Gene Expression and Mutant Screens

Existing long hypocotyl mutants of Arabidopsis (Koornneef et al, 1980; Chory et al., 1989) fail to exhibit the phytochrome mediated supression of hypocotyl growth, but none of these have substantially altered cab gene responsiveness to phytochrome (Chory et al, 1989; Sun and Tobin, 1990). We are attempting to isolate Arabidopsis mutants in which this transcriptional response is affected by two different strategies. In the first of these we have mutagenized wild type seeds and are screening M2 seeds for those which do not germinate in response to red illumination, but can be rescued by being induced to germinate in the presence of gibberellin.

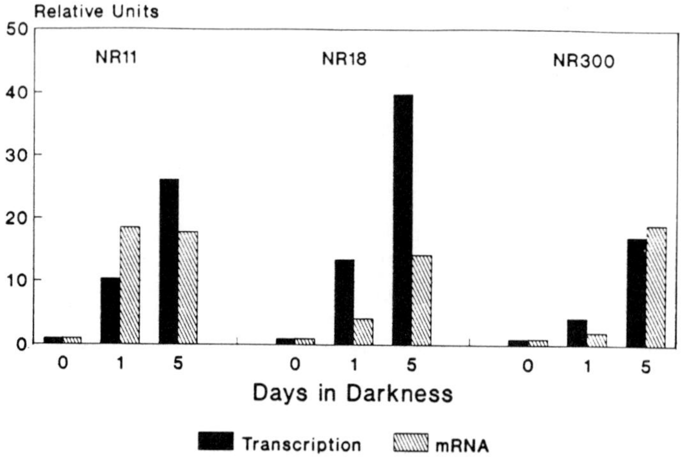


Fig. 5 Relative increases in mRNA levels and relative transcription levels of three negatively regulated genes in Lemna gibba. For the relative mRNA levels, Northern blots were quantitated by densitometry of autoradiograms. For the relative transcription levels, specific transcripts synthesized in vitro were selected by hybridization to the corresponding immobilized genomic clone and quantitated by liquid scintillation counting. The 0 day values were set to 1.0, and the other values were normalized to them for each series.

In a more directed approach we are using a "suicide" gene selection scheme which will allow survival of mutants with disruptions in the phytochrome transduction pathway activiating nuclear gene transcription, while plants with the normal pathway will be severely affected. For this purpose, we have created homozygous lines of Arabidopsis transformed with promoter fusion constructs containing the tms2 gene (Klee et al, 1984) of Agrobacterium tumefaciens. The product of this gene is an auxin amide hydrolase which catalyzes the ultimate step in the formation of auxin in the bacterial pathway (Fig. 6). The plant auxin biosynthetic pathway is different, and the wild type plants lack this activity. Thus, plants expressing the tms2 gene would be expected to be sensitive to otherwise non-toxic levels of auxin amides because they would convert them to toxic levels of auxins. The strikingly different sensitivities to indoleacetamide of wild type and a tms2-containing line of Arabidopsis are illustrated in Fig. 7.

In order to determine the usefulness of this marker for selection of phytochrome response chain mutants, the tms2 gene was placed under the control of either the Arabidopsis cab140 promoter (1 kb upstream of the start of transcription) or the cauliflower mosaic virus 35S promoter and transformed into Arabidopsis (Lloyd et al, 1986). It was important to demonstrate that the expression of the introduced tms2 gene fused to the cab promoter is under phytochrome regulation. Indeed, we have been able to show phytochrome regulation of the tms2 transcript levels, the level of auxin amide hydrolase activity, and the increased growth inhibition in intermittent red light in comparison to complete darkness in such transformed lines; lines transformed with the 35S fusion construct do not show such regulation (Karlin-Neumann et al., 1990). Seeds of the homozygous lines with the cab140:tms2 constructs have been mutagenized, and M2 seeds are being selected for survivors growing under inductive light conditions on medium containing naphthalene acetamide. The M3 progeny which show the auxin amide insensitive phenotype will be tested for alterations in phytochrome regulation of the endogenous cab genes.

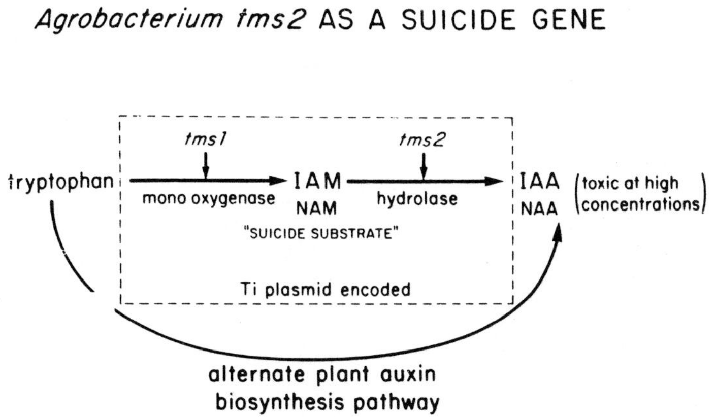


Fig. 6 Bacterial pathway for conversion of tryptophan to auxin.

## ACKNOWLEDGMENT

This research was supported by N.I.H. grant GM-2316 (E.M.T.) and by N.I.H. BRSG grant RR07009-23 to U.C.L.A. We thank Ruth Doxsee and Lu Huang for technical assistance.

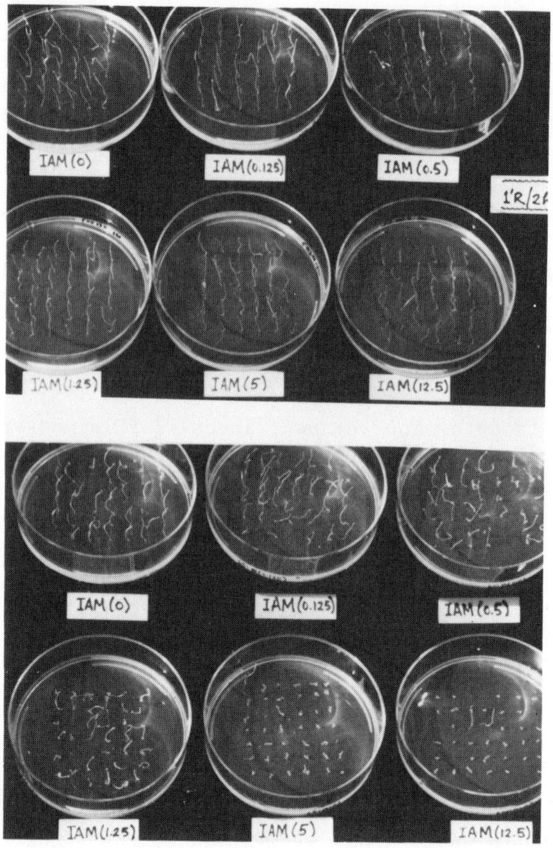


wild type

tms2

Fig. 7 Differential sensitivity of wild-type and tms2 transformed seedlings to indoleacetamide (IAM). Seedlings were germinated and grown for 7 d under intermittent red light (1'/2h) on the concentrations of IAM shown (μM).

## REFERENCES

Adamse, P., Kendrick, R.E. and Koornneef, M. (1988) Photochem. Photobiol. 48:833-841.

Benfey, P.N. and Chua, N.-H. (1989) Science 244:174-181.

Bruce, W.B., Christensen, A.H., Klein, T., Fromm, M. and Quail, P.H. (1989) Proc. Natl. Acad. Sci. USA 86:9692-9696.

Buzby, J. and Tobin, E.M. (1990) Plant Cell 2:805-814.

Chory, J., Peto, C.A., Ashbaugh, M., Saganich, R., Pratt, L. and Ausubel, F. (1 Plant Cell 1:867-880.

Chory, J., Peto, C.A., Feinbaum, R., Pratt, L., and Ausubel, F. (1989b) Plant Cell 58:991-999.

Colbert, J.T. (1988) Plant Cell Environ. 11:305-318.

Elliott, R.C., Dickey, L.F., White, M.J., and Thompson, W.F. (1989) Plant Cell 1:691-698.

Jefferson, R.A., Kavanagh, T.A. and Bevan, M.W. (1987) Embo J. 6:3901-3907.

Karlin-Neumann, G.A., Brusslan, J., and Tobin, E.M. (1991) Plant Cell 3:573-582.

Karlin-Neumann, G.A., Kohorn, B.D., Thornber, J.P., and Tobin, E.M. (1985) J. Mol. Appl. Genet. 3:45-61.

Karlin-Neumann, G.A., Sun L., and Tobin, E.M. (1988) Plant Physiol. 88:1323-1331.

Kendrick, R.E. and Kronenberg, G.H.M., eds. (1986) Photomorphogenesis in Plants (Martinus Nijhoff, Boston) 580 pp.

Klee, H., Montoya, A., Horodyski, F., Lichtenstein, C., Garfinkel, D., Fuller, S., Flores, C., Peschon, J., Nester, E. and Gordon, M. (1984) Proc. Natl. Acad. Sci. USA 81:1728-1732.

Klein, T.M., Fromm, M., Weissinger, A., Tomes, D., Schaaf, S., Sletten, M., and Sanford, J.C. (1988) Proc. Natl. Acad. Sci. 85:4305-4309

Koornneef, M., Rolff, E., and Spruit, C.J.P. (1980) Z. Pflanzenphysiol. 100:147-160.

Kuhlemeier, C., Green, P.J. and Chua, N.-H. (1987) Annu. Rev. Plant Physiol. 38:221-257.

Leutwiler, L.S., Meyerowitz, E.M., and Tobin, E.M. (1986) Nucl. Acids Res. 14:4051-4064.

Lloyd, A.M., Barnason, A.R., Rogers, S.G., Byrne, M.C., Fraley, R.T., and Horsch, R.B. (1986) Science 234:464-466.

Nagy, F., Kay, S.A., and Chua, N.H. (1988) Trends Genet. 4:37-42.

Okubara, P.A. and Tobin, E.M. (1991) Plant Physiol. In press.

Parks, B.M., Shanklin, J., Koornneef, M., Kendrick, R.E. and Quail, P.H. (1989) Plant. Mol. Biol. 12:425-437.

Rogers, S.G., Horsch, R.B., and Fraley, R.T. (1986) Meth. Enz. 11X:627-640.

Sharrock, R.A., Parks, B.M., Koornneef, M., and Quail, P.H. (1988) Mol. Gen. Genet. 213:9-14.

Silverthorne, J. and Tobin, E.M. (1984) Proc. Natl. Acad. Sci. USA. 81:1112-1116.

Silverthorne, J. and Tobin, E.M. (1987) BioEssays 7:18-23.

Silverthorne, J., Wimpee, C.F., Yamada, T., Rolfe, S. and Tobin, E.M. (1990) Plant Mol. Biol. 15:49-58.

Stiekema, W.J., Wimpee, C.F., and Tobin, E.M. (1983) Nucl. Acids Res. 11:8051-8061.

Sun, L. and Tobin, E. M. (1990) Photochem. Photobiol. 52:51-56.

Thompson, W.F. (1988) Plant Cell Environ. 11:319-328.

Tobin, E.M. and Silverthorne, J. (1985) Annu. Rev. Plant Physiol. 36:569-593.

Tobin, E.M., Wimpee, C.F., Karlin-Neumann, G.A., Silverthorne, J., and Kohorn, B.D. (1985). In: The Molecular Biology of the Photosynthetic Apparatus, C.J. Arntzen, L. Bogorad, S. Bonitz, and K.E. Steinback, eds. (Cold Spring Harbor Laboratory, Cold Spring Harbor, N.Y.). pp.373-380.

# CIS-REGULATORY ELEMENTS FOR THE CIRCADIAN CLOCK REGULATED TRANSCRIPTION OF THE WHEAT CAB-1 GENE

A. Pay[1], E. Fejes[1], M. Szell[1], E. Adam[1] and F. Nagy[2]

[1]Institute of Plant Physiology, Biological Research Center, Hungarian Academy of Sciences, Szeged, P.O. Box 521, H-6701 Hungary

[2]Friedrich Miescher-Institute, P.O. Box 2543, CH-4002 Basel, Switzerland

Many genes of higher plants are expressed in a highly regulated fashion. Certain genes are expressed only at specific stages of development and only in certain cell types and/or their expression is regulated by various environmental stimuli[1]. In addition it has also been established that numerous physiological processes of higher plants are regulated by an endogenous circadian rhythm[2]. The mechanism by which an environmental stimulus or an endogenous rhythm regulates gene expression is the subject of considerable interest. Photosynthesis specific genes such as genes encoding the chlorophyll a/b binding protein (Cab) are particularly attractive as model systems to study gene regulation in higher plants. Expression of various Cab genes has been analyzed in both normal and transgenic plants. These experiments revealed that the expression of the Cab genes is induced by light and that the light induction is mediated by at least two photoreceptors, i.e. by the red absorbing phytochrome[3] and by an as yet unidentified blue absorbing photoreceptor[4]. Recent studies also demonstrated that Cab mRNA levels are further modulated by an endogenous rhythm[5,6,7].

Although other physiological phenomena regulated by endogenous rhythms such as flowering in plants, conidium formation in *Neurospora* or eclosion in *Drosophila* have been described, the molecular mechanism by which a circadian clock modulates gene expression has largely remained unknown. In contrast to an assumedly multigene regulated phenomenon such as flowering, the expression of Cab genes provides a relatively simple experimental system to study circadian clock regulated gene expression.

*Plant Molecular Biology 2*, Edited by R.G. Herrmann and B. Larkins, Plenum Press, New York, 1991

To this end we have initiated a series of experiments to characterize the expression pattern of wheat Cab genes under various conditions. We have recently reported that (1) steady state mRNA levels of the wheat Cab genes and particularly that of the Cab-1 gene oscillated with a periodicity of approximately 24 hours in plants grown under 16L/8D photoperiods, (2) transfer of plants to constant light or dark altered the level of the Cab-1 gene specific transcript, but did not eliminate the rhythmic oscillation indicating that the fluctuation (timing) is controlled by a circadian clock, (3) the Cab-1 gene maintained its fluctuating expression in transgenic tobacco plants[8].

<u>The circadian clock regulates the expression of the wheat Cab genes at the level of transcription</u>

The oscillation of the steady state mRNA level could be due to changes of transcription rate or mRNA stability or both. It is generally accepted that *in vitro* nuclear systems from plant cells only elongate transcripts initiated prior to nuclei isolation. As a consequence, it is also assumed that the results of these assays reflect the transcription initiation rate at that time point. To determine the transcriptional status of the Cab genes during a 24 hour period we performed *in vitro* transcription assays. Wheat seedlings were grown for 6 days under a 16 hours light and 8 hours dark photoperiodic cycle (16L/8D). On the 7th day half of the seedlings were transferred to constant light and grown for an additional 2 days. Throughout the 9th day seedlings were harvested at 4 hour intervals, nuclei were isolated and preparative in vitro transcription assays were then carried out using an identical number of nuclei ($1{,}5 \times 10^7$).

Based on three independent experiments we found that the transcription initiation rate of the Cab genes is at least 10-15 times higher in the early morning hours (it peaks around 8 a.m.) than in the late evening hours. In contrast to that of the Cab genes, SSU specific mRNA level did not show significant changes during these 24 hour periods. Transcription initiation rates of the SSU genes were generally 5 fold lower at 8 a.m. than that of the Cab genes and showed a maximum of two fold variation among this different time points; however, this variation lacked a characteristic pattern. Transcription initiation rate of the rRNA genes was nearly identical at different time points (similar to that of the SSU genes) and showed no significant variation under different light regimes. These data clearly show that transcription rates of the wheat Cab genes fluctuate in a very specific manner during a 24 hour period. This specific pattern has been maintained even under constant illumination. These results strongly suggest that the endogenous clock regulates the expression of the wheat Cab genes, at least partially, at the level of transcription.

## Identification of cis-regulatory elements required for regulated expression of the wheat Cab-1 gene in transgenic tobacco plants

In order to define cis-acting elements that mediate the circadian clock controlled transcription of the wheat Cab-1 gene we performed two complementary lines of experiments. In the first we analyzed the expression of a series of 5'-deletion mutants in transgenic plants retaining varying upstream sequences from -1816 to -127. We found, that the -244 mutants still maintain maximum level expression. Further removal of upstream sequences from -244 dramatically decreased the level of expression of the transgenes. Deletion to -230 resulted in a ten fold decrease, additional deletions from -211 to -190 lowered the expression below detection level in the majority of samples. One plant out of 15, however, still expressed the -190 deletion mutant at a low, but detectable level. Moreover, we found that all of the deletion mutants downstream from -1816 to -211 (and the only -190 mutant above detection level) exhibited circadian clock responsive gene expression regardless of the level of transcription. Several conclusions can be drawn from these results: (1) Sequences upstream of -244 are not required for maximum level of transcription in transgenic tobacco plants. (2) a 54 bp sequence located between -244 and -190 contains at least two regulatory elements for maximum level of gene expression. The major cis-acting element for maximum level of gene expression resides between -244 and -230. An auxiliary cis-acting element is located between -211 and -190. Circumstantial evidence indicates, that a third regulatory element is positioned between -190 and -160. The existence of this tentative third element, however, has yet to be proven. We found only one transgenic plant supporting this statement. (3) the 5' boundary of the promoter region that is sufficient to maintain circadian clock responsive Cab-1 gene expression is at -211. We have also attempted to determine the 3'-boundary of the clock responsive cis-acting element. To this end we analyzed the expression of various chimeric genes containing different regions of the Cab-1 promoter. We found that a chimeric gene containing a 1.8 kb region of the Cab-1 gene (-1816 to +31) confers clock responsive expression to the bacterial GUS gene. Removal of upstream sequences from -1816 to -357 did not change the level or pattern of expression. Internal deletion of Cab-1 promoter sequences between -357 to -127, however, completely eliminated the expression of this chimeric gene. We report here that sequences located within -357 and -90 have a distinguishable function in the clock regulated expression of the Cab-1 gene. We found that this 268 bp promoter fragment endows a transcription unit, containing a truncated 35S promoter and the bacterial GUS gene, with fluctuating transcription. These data, together with those obtained from the analysis of the 5'-deletion mutants, indicate (1) the cis-acting element that mediates circadian clock regulated expression is located within an enhancer-like region and (2) the 3'-boundary of this element can be placed at -90.

Does the Cab-1 promoter contains independent cis-regulatory elements for light induced and for circadian clock controlled expression?

We have described above, that the transcription of the wheat Cab-1 gene is regulated by the phytochrome and by an endogenous circadian rhythm. This dual regulation of the Cab-1 gene expression provokes the following questions: (1) how does the phytochrome and the clock interact, (2) what is the relative contribution of the phytochrome and the rhythm to the expression and (3) does the phytochrome act exclusively through the clock or (4) are parallel signal transducing chains involved in fine tuning of the Cab-1 gene expression? To answer at least some of these questions the following experiments were carried out. We prepared several internal deletion mutants of Cab-1 gene, downstream from -357, and analyzed their light induced, circadian regulated expression in transgenic tobacco plants. Up to now, we have completed the analysis of the following three constructs:

1. (-357-179) (-127 Cab/Cab/Cab)
2. (-306-179) (-127 Cab/Cab/Cab)
3. (-267-179) (-127 Cab/Cab/Cab)

We found that the expression of these 5'-internal deletion mutants was clearly regulated by phytochrome in etiolated transgenic seedlings. Transcript levels of these mutants were invariably below detection in dark-grown transgenic seedlings, similar to that of the -127 5'-deletion mutant. We showed that the mRNA level of the transgenes (but not that of the -127 Cab 5'-deletion mutant) was elevated, at least five fold, after a short red light illumination. Moreover the red light dependent induction was clearly reversed by far red light.

We observed, consistent with the low level expression in etiolated seedlings, that mRNA levels were also below detection in dark-adapted (3 days in constant dark) mature transgenic plants. Interestingly, however, the mRNA level of these internal deletion mutants did not show the characteristic fluctuating pattern in transgenic plants grown under a 16L/8D cycle or in constant light. The Cab-1 specific mRNA level was invariably high during a 24 hour period, while the endogenous tobacco Cab genes exhibited the usual oscillatory expression pattern. In contrast to the 5'-internal deletion mutants, the expression of the truncated -127 Cab gene was again below detection level in all of the transgenic plants. It should also be noted that the expression level of these internal deletion mutants was 5-fold lower than that of the -267 Cab 5'-deletion mutant.

We drew three conclusions from these findings: (1) promoter sequences of the Cab-1 gene located between -267 and -179 are responsible for the elevated, light inducible expression of these transgenes,

(2) cis-regulatory elements positioned between -179 and -127 are needed for maximum level expression of the Cab-1 gene and (3) this cis-acting element(s) is absolutely necessary to maintain circadian rhythm responsive expression in transgenic plants. It is tempting, based on these data, to conclude that the wheat Cab-1 promoter contains separated cis-regulatory elements for light-induced and for circadian clock responsive gene expression. There is, however, an alternative interpretation of these data. This would assume the presence of multiple regulatory elements for the optimal (light and rhythm responsive) expression of the Cab-1 gene. Deletion of one or more regulatory element can still result in detectable difference between light and dark expression (in spite of the relatively low level expression of these mutants). The more subtle differences between time points during a 24 hour period are more difficult to observe and may escaped detection at this expression level. In an effort to obtain unambiguous data we are presently undertaking a very detailed analysis of the expression of this mutants.

Based on our preliminary data, however, we suggest that (1) two independent signal transduction chain are involved in controlling the expression of the Cab-1 gene or (2) a light initiated signal transduction chain branches out before reaching the "circadian clock". We are presently mutagenizing this potentially critical region of the Cab-1 promoter and we hope that our data will enable us to distinguish between these alternative models.

## CONCLUSIONS

We have shown, that the short region of the Cab-1 promoter (between -357 and -90) contains multiple cis-regulatory elements that are required for maximum level, regulated expression in transgenic tobacco plants. In addition our data also indicate the presence of separated cis-regulatory elements for light induced and for circadian rhythm responsive gene expression. Based on the analysis of several other Cab promoters from different plant species, the complex structure of the wheat Cab-1 promoter is not unexpected. Multiple regulatory elements were found in the promoter region of Cab genes in tobacco[9], in petunia[10], and in *Arabidopsis*[11]. The contribution of these various cis-acting elements to the regulated gene expression is not fully understood. *In vitro* footprinting experiments combined with analysis of gene expression in transgenic plants, has already yielded valuable information about the fine structure of various plant promoters[12,13,14]. We hope that a similar approach will allow us to determine (1) the contribution of the identified cis-acting elements to the regulated transcription of the wheat Cab-1 gene and (2) to identify the exact position of the cis-acting element that mediates circadian clock responsive gene expression of this wheat gene. The characterization of the clock responsive element could be the first step to elucidating the molecular mechanism by which an endogenous oscillator controls gene expression in higher plants.

REFERENCES


1. C. Kuhlemeier, P.J. Green, and N.-H. Chua, Regulation of gene expression in higher plants, Ann. Rev. Plant. Physiol. 38:221 (1987).
2. D. Vince-Prue, Photomorphogenesis and flowering. In: Encyclopedia of plant physiology, W. Shropshire, and H. Mohr, ed., vol. 16 B, pp. 458 (1983).
3. E. M. Tobin, and J. Silverthorne, Light regulation of gene expression in higher plants, Ann. Rev. Plant. Physiol. 36: 569 (1985).
4. A. K. Marrs, and L. S. Kaufman, Blue-light regulation of gene expression in higher plants, Proc. Natl. Acad. Sci. USA. 86: 4492 (1989).
5. K. Kloppstech, Diurnal and circadian rhythmicity in the expression of light-induced plant nuclear messenger RNAs, Planta 165: 502 (1985).
6. P. Tavladoraki, K. Kloppstech, and J. Argyroudi-Akoyunoglou, Circadian rhythm in the expression of the mRNA coding for the apoprotein of the light-harvesting complex of photosystem II, Plant Physiol. 90: 665 (1989).
7. B. Piechulla, Changes of the diurnal and circadian (endogenous) mRNA oscillations of the chlorophyll a/b binding protein in tomato leaves during altered day/night (light/dark) regimes, Plant. Mol. Biol. 12: 317 (1989).
8. F. Nagy, S. A. Kay, and N.-H. Chua, A circadian clock regulates transcription of the wheat Cab-1 gene, Genes Dev. 2: 376 (1988).
9. C. Castresana, I. Garcia-Luque, E. Alonso, V.S. Malik, and A.R. Cashmore, Both positive and negative regulatory elements mediate expression of a photoregulated Cab gene from *Nicotiana plumbaginifolia*, EMBO J. 7: 1929 (1988).
10. D. Gidoni, P. Brosio, D. Bond-Nutter, J. Bedbrook and P. Dunsmuir, Novel cis-acting element in Petunia Cab gene promoters, Mol. Gen Genet. 215:337 (1988).
11. S. B. Ha, and G. An, Identification of upstream regulatory elements involved in the developmental expression of the *Arabidopsis* Cabl gene, Proc. Natl. Acad. Sci. USA. 85: 8017 (1989).
12. E. Lam, P. Benfey, P. Gilmartin, R. X. Fang and N.-H. Chua, Site-specific mutations alter *in vitro* binding and change promoter

expression pattern in transgenic plants, Proc. Natl. Acad. Sci. USA. 86: 7890 (1989).

13. P. J. Green, A.S. Kay, and N.-H. Chua, Sequence specific interactions of a pea nuclear factor with light-responsive elements upstream of a plant light regulated gene, EMBO J. 6:2543 (1987).

14. G. Giuliano, E. Piechersky, E. Malik, V. S. Timko, P. A. Scolnik and A. R. Cashmore, An evolutionarily conserved protein binding sequence upstream of a plant light regulated gene, Proc. Natl. Acad. Sci. USA. 85: 7089 (1988).

# SELF-INCOMPATIBILITY AS A MODEL FOR CELL-CELL RECOGNITION IN FLOWERING PLANTS

Julie E. Gray, Bruce A. McClure, Volker Haring,
Marilyn A. Anderson, and Adrienne E. Clark

Plant Cell Biology Research Centre
School of Botany
University of Melbourne
Parkville, Victoria 3052, Australia

## INTRODUCTION

Chemical signals which are secreted from one cell type are known to recognise and exert an effect on a specific target cell or cells. In animal systems, recognition of this type is usually achieved by the signal molecule, for example a hormone, binding to a specific receptor on the surface of the target cell. The signal molecule may then be internalised by receptor mediated endocytosis, or alternatively, receptor binding may initiate a second messenger response resulting in an intracellular effect. In plant systems cell-cell interactions are not as clearly understood. Plant cells secrete hormone like substances such as auxins, cytokinins and gibberellins. Receptor mediated endocytosis has recently been demonstrated in cultured soybean cells (Horn et al., 1989) and cyclic AMP, protein kinases, calmodulin and coated pits, which are involved in animal signalling systems, have all been identified in plants. It is therefore possible that in spite of the differences between plant and animal cells, the most important being the presence of a cell wall, analogous cellular recognition systems may operate. Indeed, components of the cell wall, known as "oligosaccharins" have themselves been implicated as hormones (Eberhard et al., 1989).

Self-incompatibility is the inherited ability of a plant to recognise and prevent fertilization by its own pollen. This subject has fascinated scientists since it was first observed in the eighteenth century and the biology and genetics of many self-incompatibility systems are extensively documented (for a comprehensive text see de Nettancourt, 1977). Self-incompatibility is therefore an ideal system to study recognition between plant cells using the modern techniques of molecular biology.

## THE IMPORTANCE OF SELF-INCOMPATIBILITY

Outbreeding within a natural population is important for maintenance of genetic variability. Extensive inbreeding in plants is deleterious to a population, leading to weakened resistance to disease, reduced growth rates and reduced yields. The most effective form of inbreeding is self-fertilization. Self-incompatibility is a feature of many flowering plants which prevents self-fertilization. The effectiveness of self-incompatibility in promoting outbreeding is believed to be one of the most important factors in ensuring the evolutionary success of flowering plants (Whitehouse, 1950).

*Plant Molecular Biology 2*, Edited by R.G. Herrmann and
B. Larkins, Plenum Press, New York, 1991

For fertilization to occur in flowering plants the male gametophyte, or pollen, is transferred from the anther, by wind, insects or another vector, to the female sexual tissues. The pollen grain hydrates on the surface of the stigma and germinates to form a pollen tube which grows down the style and into the ovary. The male gametes are then released and fertilization takes place. Most flowering plants are hermaphroditic and with no barriers to compatibility, self-fertilization would be the most likely outcome. In the animal kingdom, self-fertilization is avoided principally by having the male and female sexes in separate individuals. In contrast, only a small proportion of flowering plant species are dioecious and a variety of alternative mechanisms to prevent inbreeding have evolved.

Other recent reviews of the subject of self-incompatibility include Nasrallah and Nasrallah, 1986; Cornish et al., 1988; Bernatzky et al., 1988; Ebert et al., 1989.

## THE MECHANISMS OF SELF-INCOMPATIBILITY

### Heteromorphic Self-Incompatibility

One mechanism of self-incompatibility in flowering plants involves a difference in the relative height of the stigma and anthers posing a topological barrier to fertilization. This essentially physiological adaptation is known as heteromorphic self-incompatibility although biochemical barriers to selfing may also operate within the same plant. The biology and genetics of this mechanism has been described for a number of species including *Linum grandiflorum* (Lewis, 1943; Ghosh and Shivanna, 1980) and *Primula vulgaris* (Heslop-Harrison and Shivanna, 1981). There has been no research into the molecular biology of heteromorphic self-incompatibility and therefore it will not be further discussed here. For a review of heteromorphic self-incompatibility systems see Gibbs (1986).

### Gametophytic Homomorphic Self-Incompatibility

This is the most common mechanism of self-incompatibility and is found in the greatest number of families of flowering plants. Self-incompatibility is normally encoded by a single, polyallelic locus known as the self-incompatibility or *S*-gene (there are exceptions such as certain grasses in which two polyallelic genes are involved). In gametophytic species, fertilization is prevented if the *S*-allele carried by the haploid pollen matches either of the *S*-alleles present in the diploid style. Recognition is presumed to occur by an interaction between the *S*-gene products of the pollen and the style. If they are products of the same allele, growth of the pollen tube is inhibited thereby preventing self-fertilization (see figure 1). Inhibition of incompatible pollen tubes commonly occurs within the length of the style but may occur anywhere between the stigma surface and the ovule.

In general, the female tissues of gametophytic species have a wet stigma surface, and a transmitting tract or a hollow style down which the pollen tubes grow (see figure 2). The pollen cytoplasm, carrying the nuclei, moves down the growing tube. Callose is deposited in the tube walls and plugs form at regular intervals within the tube, isolating the cytoplasm from the pollen grain. This results in a ladder-like appearance under the microscope. Incompatible tubes are easily distinguished, appearing shorter with characteristically thick, irregular walls. The tip of the tube often appears swollen and may even rupture.

Research on the molecular basis of self-incompatibility has mainly concentrated on the Solanaceae and in particular *Nicotiana alata*.

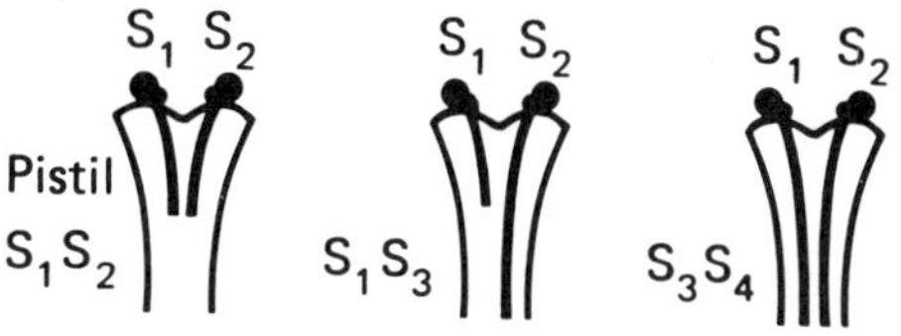


Figure 1. Behaviour of pollen in the gametophytic self-incompatibility system. The pollen parent genotype is $S_1S_2$. When an allele in the individual haploid pollen grain is matched with either allele in the diploid style tissues, growth of the pollen tube is arrested, usually in the style. For example, both $S_1$ and $S_2$ pollen are inhibited in a $S_1S_2$ style and the $S_2$ pollen will grow successfully through the $S_1S_3$ style. Where there is no match of alleles (e.g., $S_1S_2$ pollen grains on a $S_3S_4$ pistil), the pollen tubes of both genotypes grow through the style to the embryo sac. [From Anderson et al., 1983, reproduced with permission from Alan R. Liss, Inc., New York].

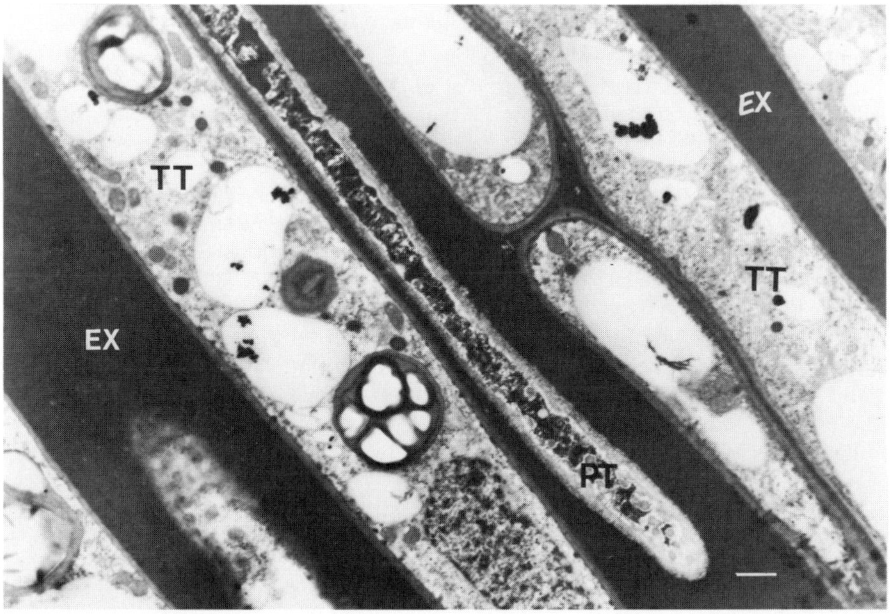


Figure 2. Electron micrograph of a longitudinal section through a *Nicotiana alata* style showing a pollen tube (PT), 12h after pollination, growing through the external matrix (EX) between transmitting tract cells (TT). Bar = 1μm.

### Sporophytic Homomorphic Self-Incompatibility

Sporophytic self-incompatibility is also governed by a single *S*-locus with multiple alleles. In this system, growth or rejection of the pollen is determined by the *S*-alleles of the diploid pollen producing plant, rather than the single *S*-allele carried by the pollen grain. Inhibition of pollen tube growth occurs at an early stage, often before the tube has penetrated the stigma surface. In incompatible pollen tubes, deposits of callose are observed in the tip of the emerging tube and at its point of contact on the stigma.

Sporophytic self-incompatibility has been most extensively studied in *Brassica* species (Nasrallah and Nasrallah, 1986).

## IDENTIFICATION OF STYLE AND STIGMA GLYCOPROTEINS WHICH SEGREGATE WITH *S*-ALLELES

*S*-allele related substances were first detected, in *Oenothera organesis* pollen using the immunological techniques originally developed to characterise blood group antigens (Lewis, 1952). Hinata and Nishio (1978) were the first to convincingly identify an *S*-allele specific protein from a plant stigma. Using isoelectric focussing they demonstrated that a stigma protein associated with a particular allele of *Brassica campestris* segregated with that allele through to the $F_2$ generation. *S*-allele specific proteins have now been characterised from a number of gametophytic (Bredemeijer and Blaas, 1981; Mau et al., 1982; Kamboj and Jackson,1986; Mau et al., 1986; Jahnen et al., 1989; Kirch et al., 1990; Xu et al.,1990; Ai et al., 1990) and sporophytic (Nasrallah and Nasrallah, 1984; Takayama et al., 1987) species. Many of these proteins are glycosylated and hence are referred to as *S*-glycoproteins. The structure of the glycosyl substituents has been partly characterised (Takayama et al., 1986; Woodward et al., 1989), and the possibility that they may play a role in the determination of specificity considered (Woodward et al., 1989).

In general, *S*-glycoproteins are the most abundant proteins in mature styles of self-incompatible species. In most cases they are basic proteins with molecular weights in the range of 27,000 to 33,000 for gametophytic species and 48,000 to 65,000 for the sporophytic species studied.

*S*-glycoproteins purified from *N. alata* inhibit pollen tube growth *in vitro*; however, the allelic specificity of pollen tube inhibition observed *in vivo* was only partially preserved in the *in vitro* assay (Jahnen et al., 1989).

### Cloning of cDNAs Encoding *S*-Glycoproteins

The first cDNA clone to be identified as encoding part of the protein backbone of an *S*-locus specific glycoprotein was that of the $S_6$-allele of *Brassica oleracea*, a species exhibiting sporophytic incompatibility. This clone was isolated from a cDNA library, constructed from stigma mRNA, by differentially screening to select clones which hybridised strongly to mature stigma, but not to leaf or seedling cDNA probes. The identity of the clone was confirmed by Western blotting, (Nasrallah et al., 1985). Since then, the full sequences of cDNAs representing $S_6$-, $S_{13}$- and $S_{14}$-alleles of *B. oleracea* have been published (Nasrallah et al., 1987). Comparison of the derived amino acid sequences from these *B. oleracea* cDNAs and an amino acid sequence determined from *B. campestris* $S_8$-glycoprotein (Takayama et al., 1987) revealed a high degree of homology (approximately 80%).

The first cDNA clone encoding a gametophytic, *S*-allele associated glycoprotein was cloned from N. alata $S_2$-style mRNA. A cDNA library was differentially screened to select clones which hybridised to mature, but not to immature, style cDNA probes. The identity of the clone was confirmed by comparing it's derived amino acid sequence with that of an N-terminal sequence determined from purified $S_2$-glycoprotein (Anderson et al., 1986). Seven cDNAs representing different alleles of the S-gene ($S_2$, $S_3$, $S_6$, $S_1$, $S_z$, $S_a$ and $S_{F11}$) from *N. alata* have now been cloned and their predicted amino acid sequences compared (Anderson et al., 1989; Kheyr-Pour et al., 1990). This revealed a lower level of homology (an average of 54% homology between pairs of sequences, but only 29% overall homology between six alleles compared) between the *N. alata* *S*-glycoproteins relative to the high level of homology found within the Brassica system. Recently three *S*-allele cDNAs from another gametophytic species, *Petunia inflata*, have been sequenced and their derived amino acid sequences shown to have homology to the *N. alata* *S*-alleles (Ai et al., 1990). No homology has been observed between the *S*-glycoprotein sequences of *N. alata* and the two *Brassica* species suggesting functional and/or evolutionary differences between the gametophytic and sporophytic self-incompatibility systems.

Analysis of both the *N. alata* and *Brassica* *S*-glycoprotein systems showed conserved blocks of amino acids between each of their *S*-alleles. It is probable that these conserved regions are important for the function or structure of the *S*-glycoprotein molecules. Furthermore these conserved regions were interspersed by non-conserved, or variable, regions and it has been proposed that allelic specificity must reside within these regions of the molecules (Ebert et al., 1989, Kheyr-Pour et al., 1990).

The *N. alata* $S_2$-cDNA has been used as a probe in *in situ* hybridization studies to show that the *S*-gene is expressed in the transmitting tract cells and a layer of cells surrounding the ovary in the mature pistil. This coincides with the path taken by a pollen tube to effect fertilization of the ovule (Cornish et al., 1987). Similar experiments in *Brassica* showed the *S*-gene to be expressed in cells on the stigma surface (Nasrallah et al., 1988).

Genomic Southern analysis in *N. alata* indicated that the *S*-glycoproteins are probably encoded by a single gene. Restriction fragment length polymorphisms were observed between the different *S*-alleles (Bernatzky et al., 1988).

### Genomic Clones

cDNA probes have been used to isolate genomic clones for *N. alata* $S_2$- and $S_6$-alleles (S-L. Mau, unpublished results). Both were found to contain a single, short intron. Transformation studies are in progress using a transformation system recently developed for *N. alata* (Ebert and Clarke, 1990). Recently, transformation of *Nicotiana tabacum* (a self compatible species) with *B. oleracea* $S_{13}$- and $S_{22}$-genomic constructs has been reported. The *Brassica* gene was expressed in the style transmitting tract (Moore and Nasrallah, 1990) and the *S*-glycoprotein product accumulated in the intercellular matrix of the transmitting tract of the transgenic plants and in specific cells surrounding the ovary (Kandasamy et al., 1990). This pattern of expression is more similar to that of the *N. alata* *S*-gene (Cornish et al., 1987; Anderson et al., 1989) than the normal pattern of expression in *B. oleracea* (Nasrallah et al., 1988).

## SELF-INCOMPATIBILITY GLYCOPROTEINS AS RIBONUCLEASES

### Sequence Homologies with Known RNases

In 1989, analysis of the amino acid sequences derived from *N. alata* $S_2$-, $S_3$- and $S_6$-cDNAs led to the discovery of homology between the *S*-glycoproteins and RNases $T_2$ (Kawata et al., 1988) and Rh (Horiuchi et al., 1988) by McClure and co-workers (McClure et al., 1989). The $T_2$ and Rh RNases are both secreted by fungi. Of the 122 amino acids 30 are perfectly conserved between the three *S*-glycoproteins described by Anderson et al. (1989) and the two fungal RNases. Significant conserved amino acids are the two histidine residues which are implicated in RNase $T_2$ catalysis, and two cysteine residues which are believed to form a disulphide bond in the vicinity of the RNase $T_2$ putative active site. Furthermore these histidine and cysteine residues are also conserved in the sequences of four other recently published *N. alata* alleles (Kheyr-Pour et al., 1990).

**S-Glycoproteins have RNase Activity**

Purified $S_1$-, $S_2$-, $S_3$-, $S_6$- and $S_7$-glycoproteins of *N. alata* were shown, by a perchloric acid precipitation assay to hydrolyse RNA (McClure et al., 1989). RNase activities were found to be in the range of 170 ($S_1$-glycoprotein) to 2200 ($S_7$-glycoprotein) $A_{260}$ units $min^{-1}mg^{-1}$. No DNase activity was detected.

Extracts of styles of a related self-compatible species, *N. tabacum* have at least 100-fold lower RNase activity, which supports the suggestion that RNases play a role in the self-incompatibility response. This result may also have implications for our understanding of interspecific incompatibility. If we accept that RNases are responsible for the inhibition of growth of self-incompatible pollen tubes, it seems reasonable to assume that they could also inhibit growth of pollen from related species. This assumption could explain why *N. tabacum* pollen is rejected by *N alata* styles, which contain high levels of RNase, while *N. alata* pollen grows down styles of *N. tabacum,* which contain much lower levels of RNase.

At the time of writing there is no clear evidence as to whether the RNase activity associated with the *N. alata S*-glycoproteins, is necessarily involved in the self-incompatibility response. It is possible that during the evolution of self-incompatibility an existing RNase molecule was recruited for another function. However, bovine pancreatic RNase A, a structurally unrelated enzyme is able to retard the growth of pollen tubes in vitro (M. Lush, unpublished results). Also, the products of all the *S*-alleles so far tested have substantial RNase activity, and alignment of their amino acid sequences reveals that the active site histidines are highly conserved, even within a region of the sequence which shows allelic variability (McClure et al, 1989). This circumstantial evidence implies, but does not prove, that the RNase activity of the *S*-glycoproteins does have a role in gametophytic self-incompatibility.

There is convincing evidence that RNases are involved in a number of other cell-cell interaction systems. For example, some bacteria produce colicins which are toxic to other strains of bacteria but not to the producing strain. Different strains of bacteria produce different colicins. Self-immunity is due to the specific interaction of the colicin with a plasmid encoded peptide. The cytotoxicity of cloacin DF13 and colicin E3 is due to their ability to cleave a specific bond in 16S ribosomal RNA. A group of highly cytotoxic plant glycoproteins has also been examined in detail. An example is ricin, from the castor bean plant *Ricinus communis*. These proteins inhibit protein synthesis by hydrolysing a specific bond in the eukaryotic 28S ribosomal RNA. Both colicins and ricins require specific receptor molecules for transport into the target cell.

These systems and their relevance to gametophytic self-incompatibility have recently been reviewed by Haring et al., (1990). The observation that extracellular RNases are widespread in animal tissues has led to the proposal that extracellular RNases, RNase inhibitors, and specific RNA molecules may form the basis of a short distance cellular communication system (Benner and Alleman, 1989). In this model, the RNase does not necessarily enter the target cell for recognition to occur.

## POSSIBLE MECHANISMS OF ACTION

The discovery that gametophytic *S*-glycoproteins have RNase activity allows us to speculate on their mode of action. As the products of all the S-alleles tested are RNases, it seems probable that stylar S-allele products inhibit the growth of incompatible pollen tubes by entering the pollen tube cytoplasm and therein disrupting translation. The $S_2$- and $S_6$-glycoproteins of *N. alata* are indeed good inhibitors of a wheatgerm *in vitro* translation system (J. Gray, unpublished results).

On the basis of cytotoxicity being manifested by inhibition of translation, we can envisage how style *S*-RNases might inhibit pollen tube growth, however the question of how the *S*-glycoproteins can distinguish between self and non-self pollen is more difficult to approach. It seems remarkable that compatible pollen tubes are able to grow down a style at a considerable rate, through an exudate containing high levels of *S*-RNase, seemingly unaffected by its cytotoxicity. We now discuss two possible mechanisms by which stylar *S*-RNases could specifically inhibit incompatible pollen tube growth.

### Substrate Specificity of the *S*-RNases

It is possible that style *S*-RNases act by selectively degrading an allele specific pollen component, presumably an RNA containing molecule, which is necessary for continued growth of the pollen tube. This substrate must differ between pollen carrying different *S*-alleles which implies that, in this case, the substrate would be a product of the *S*-locus. The identity of the pollen *S*-allele product, however, still eludes us.

So far experiments with *S*-RNases have revealed only limited substrate preferences (B. McClure, unpublished results).

### Uptake by the Pollen Tube

An alternative mechanism is based on the premise that allelic specificity of the self-incompatibility response lies in the selective uptake, or selective exclusion, of the *S*-RNases by the pollen tube. We are currently testing this hypothesis by investigating the uptake of radioactively labelled *S*-glycoprotein into *in vitro* grown pollen tubes. A selective uptake mechanism must involve recognition of stylar *S*-glycoproteins by the pollen product of the *S*-gene at the pollen tube surface. The pollen *S*-gene product could, therefore in this model, be functioning as a cell surface receptor protein. It is possible that the pollen *S*-gene product is in fact the product of a different but closely linked gene at the *S*-locus. This possibility is supported by the failure to identify a style-like *S*-gene product, or transcript in pollen of *N. alata* which is homologous with the *S*-gene product in styles (B.McClure, unpublished results).

## CONCLUSIONS AND OUTLOOK

There are two theories concerning the evolution of self-incompatibility. One is that it evolved only once in the early history of flowering plants, and that the present variety of incompatibility systems evolved from this common ancestor (Whitehouse, 1950). Alternatively self-incompatibility may have arisen *de novo* on a number of occasions (Bateman, 1952). The acquisition and comparison of more *S*-allele sequence data from different sources, and a knowledge of the biochemical functions of the *S*-glycoproteins associated with the various systems of incompatibility, will help us to resolve this question. However on the basis of the limited data available, it seems likely that the gametophytic and sporophytic systems evolved separately as there are no conserved features between the *S*-glycoproteins characterised from the female tissues of plants exhibiting gametophytic and sporophytic self-incompatibility.

The recent exciting discovery that gametophytic *S*-glycoproteins are RNases has provided an insight into how they may function to retard pollen tube growth. Nevertheless, we still do not understand how allelic specificity is governed. Recognition must involve an interaction between pollen and style components encoded by the *S*-locus. Stylar products of the *S*-locus are now well characterised, but the pollen component remains unidentified, and it now seems probable that it is encoded by a different gene present at the same locus. Further analysis of genomic clones may yield clues as to the identity of the pollen *S*-gene.

Self-incompatibility is an adaptation which runs contrary to most cell-cell recognition systems in both plants and animals in that it is concerned with the recognition and rejection of self rather than non-self. However, a knowledge of how the *S*-RNases function should help us to understand the mechanisms of other cellular recognition processes in plants, for example, response to pathogenic attack or symbiotic interactions.

**Acknowledgements**

Julie Gray and Volker Haring were supported by Fellowships from the Royal Society (London) and the Deutsche Forschungsgemeinschaft (FRG) respectively. We are grateful to Ingrid Bönig who prepared the micrograph and to Joanne Noble who prepared the manuscript for publication. We value discussions with Drs Tony Bacic and Steve Read, Ms Shaio-Lim Mau and our graduate students.

**References**

Ai, Y., Singh, A., Coleman, C.E., Ioerger, T.R., Kheyr-Pour, A. and Kao, T.-H., 1990, Self-incompatibility in *Petunia inflata*: Isolation and characterization of cDNAs encoding three S-allele-associated proteins, 1990, *Sex. Plant Reprod.* , (in press).

Anderson, M.A., Hoggart, R.D., and Clarke, A.E., 1983, The possible role of lectins in mediating plant cell-cell interactions, *in*: "Chemical Taxonomy, Molecular Biology and Function of Plant Lectins," X. Goldstein and X. Edzler, eds, Alan R. Liss, New York, pp 143-161.

Anderson, M.A., McFadden, G.I., Bernatzky, R., Atkinson, A. Orpin, T., Dedman, H., Tregear, G., Fernley, R. and Clarke, A.E., 1989, Sequence variability of three alleles of the self-incompatibility gene of *Nicotiana alata, Plant Cell,* 1: 483-491.

Anderson, M.A., Cornish, E.C., Mau, S-L., Williams, E.G., Hoggart, R., Atkinson, A., Bonig, I., Grego, B., Simpson, R., Roche, P.J., Haley, J.D., Penschow, J.D., Niall, H.D., Tregear, G.W., Coghlan, J.P.,Crawford, R.J. and Clarke, A.E., 1986, Cloning of a cDNA for a stylar glycoprotein associated with expression of self-incompatibility in *Nicotiana alata, Nature,* 321:38-44.

Bateman, A.J., 1952, Self-incompatibility in angiosperms. I. Theory, *Heredity,* 6:285-310.

Benner, S.A. and Alleman, R.K., 1989, The return of pancreatic ribonucleases, *TIBS,* 14: 396-397.

Bernatzky, R., Anderson, M.A. and Clarke, A.E., 1988, Molecular genetics of self-incompatibility in flowering plants, *Dev. Genet.,* 9:1-12.

Bredemeijer, G.M.M. and Blaas, J., 1981, *S*-specific proteins in styles of self-incompatible *Nicotiana alata, Theor. Appl. Genet.,* 59:185-190.

Cornish, E.C., Anderson, M.A. and Clarke, A.E., 1988, Molecular aspects of fertilization in flowering plants, *Ann. Rev. Cell Biol.,* 4: 209-228.

Cornish, E.C., Pettitt, J.M., Bonig, I., and Clarke, A.E., 1987, Developmentally controlled expression of a gene associated with self-incompatibility in *Nicotiana alata, Nature,* 326:99-102.

de Nettancourt, D., 1977, "Incompatibility in angiosperms," Springer-Verlag, Berlin.

Eberhard, S., Doubrava, N., Marfa, V., Mohnen, D., Southwick, A., Darvill, A. and Albersheim, P., 1989, Pectic cell wall fragments regulate tabacco thin-cell-layer explant morphogenesis, *Plant Cell* 1:747-755.

Ebert, P.R., Anderson, M.A., Bernatzky, R., Altschuler, M. and Clarke, A.E., 1989, Genetic polymorphism of genetic self-incompatibility in flowering plants, *Cell* 56:255-262.

Ebert, P.R. and Clarke, A.E., 1990, Transformation and regeneration of the self-incompatible species *Nicotiana alata, Plant Mol Biol,* (in press).

Ghosh, S. and Shivanna, K. R., 1980, Pollen-pistil interaction in *Linum grandiflorum, Planta,* 149:257-261.

Gibbs, P., 1986, Do homomorphic and heteromorphic self-incompatibility systems have the same sporophytic mechanisms? *Plant Syst. Evol.,* 154:285-323.

Haring, V., McClure, B.A. and Clarke, A.E., 1990, Molecular aspects of self-incompatibility in the Solanaceae. *In*: "Advances in Plant Gene Research. Vol 7" (in press).

Heslop-Harrison, J. and Shivanna, K.R., 1981, Heterostyly in *Primula*. 1. Fine structure and cytochemical features of the stigma and style in *Primula vulgaris* HUDS, *Protoplasma,* 107:171-187.

Hinata, K. and Nishio, T., 1978, *S*-allele specificity in stigma proteins of *Brassica oleraceae* and *Brassica campestris, Heredity,* 41:93-100.

Horiuchi, H., Yanai, K., Takagi, M., Yano, K., Wakabayashi, E., Sanda, A., Mine, S., Ohgi, K. and Irie, M., 1988, Primary structure of a base non-specific ribonuclease from *Rhizopus niveus, J. Biochem.,* 103:408-418.

Horn, M.A, Heinstein, P.F. and Low, P.S., 1989, Receptor-mediated endocytosis in plant cells, *Plant Cell,* 1:1003-1009.

Jahnen, W., Batterham, M.P., Clarke, A.E., Moritz, R.L., and Simpson, R.J.,1989, Identification, isolation, and N-terminal sequencing of style glycoproteins associated with self-incompatibility in *Nicotiana alata, Plant Cell,* 1:493-499.

Jahnen, W.,Lush, W.M., and Clarke, A.E., 1989, Inhibition of *in vitro* pollen tube growth by isolated *S*-glycoproteins of *Nicotiana alata, Plant Cell,* 1:501-510.

Kamboj R.K. and Jackson J.F., 1986, Self-incompatibility alleles control a low molecular weight, basic protein in pistils of *Petunia hybrida,Theor. Appl. Genet.,* 71:815-819.

Kandasamy, M.K., Dwyer, K.G., Paolillo, D.J., Doney, R.C., Nasrallah, J.B. and Nasrallah, J.B., 1990, *Brassica S*-proteins accumulate in the intercellular matrix along the path of pollen tubes in transgenic tobacco pistils, *Plant Cell,* 2:39-49.

Kawata, Y., Sakiyama, F. and Tamaoki, H., 1988, Amino acid sequence of ribonuclease $T_2$ from *Aspergillus oryzae, Eur. J. Biochem.,* 176:683-697.

Kheyr-Pour, A., Bintrim, S.B., Ioerger, T., Remy, R., Hammond, S. and Kao, T., 1990, Sequence diversity of pistil *S*-proteins associated with gametophytic self-incompatibility in *Nicotiana alata, Sex. Plant Reprod,*. ( in press).

Kirch, H.H., Uhrig, H, Lottspeich, F., Salamini, F. and Thompson, R.D., 1990, Characterization of the proteins associated with self-incompatibility in *Solanum tuberosum, Theor. Appl. Genet.,* 78:581-588.

Lewis, D., 1943, The physiology of self-incompatibility in plants. II. *Linum grandiflorum, Ann. Bot.,* 7:115-122.

Lewis, D., 1952, Serological reactions of pollen incompatibility substances, *Proc. Roy. Soc. Lond.,* B 140:127-135.

Mau, S-L., Raff J. and Clarke, A.E., 1982, Isolation and partial characterization of components of *Prunus avium* L. styles, including an antigenic glycoprotein associated with self-incompatibility genotype, *Planta,* 156:505-516.

Mau, S-L., Williams, E.G., Atkinson, A., Anderson, M.A., Cornish, E.C., Grego, B., Simpson,R.J., Kheyr-Pour, A. and Clarke, A.E., 1986, Style proteins of a wild tomato (*Lycopersicon peruvianum*) associated with expression of self-incompatibility, *Planta*, 169:184-191.

McClure, B.A., Haring, V., Ebert, P.R., Anderson, M.A., Simpson, R.J., Sakiyama, F. and Clarke, A.E., 1989, Style self-incompatibility gene products of *Nicotiana alata* are ribonucleases, *Nature*, 342:955-957.

Moore, H.M. and Nasrallah, J.B., 1990, A *Brassica* self-incompatibility gene is expressed in the stylar transmitting tissue of transgenic tobacco, *Plant Cell*, 2:29-38.

Nasrallah, J.B., Kao T.H., Chen, C.H., Goldberg, M.L. and Nasrallah M.E., 1987, Amino acid sequence of glycoproteins encoded by three alleles of the *S* locus of *Brassica oleracea*, *Nature*, 326:617-619.

Nasrallah, J.B., Kao T.H., Goldberg, M.L. and Nasrallah M.E., 1985, A cDNA clone encoding an *S* locus-specific protein from *Brassica oleracea*, *Nature*, 318:263-267.

Nasrallah, J.B. and Nasrallah M.E., 1984, Electrophoretic heterogeneity exhibited by the *S*-allele specific glycoproteins of *Brassica*, *Experientia*, 40:279-281.

Nasrallah, M.E. and Nasrallah J.B., 1986, Molecular biology of self-incompatibility in plants, *Trends Genet.*, 2:239-244.

Nasrallah, J.B., Yu, S.M. and Nasrallah M.E., 1988, Self-incompatibility genes of *Brassica oleracea*: Expression, isolation and structure, *Proc. Natl. Acad. Sci. USA*, 85:5551-5555.

Takayama, S., Isogai, A., Tsukamoto, S., Ueda, Y., Hinata, K., Okazaki, K., Koseki, K. and Suzuki, A., 1986, Structure of carbohydrate chains of *S*-glycoproteins in *Brassica campestris* associated with self-incompatibility, *Agric. Biol. Chem.*, 50:1673-1676.

Takayama, S., Isogai, A., Tsukamoto, S., Ueda, Y., Hinata, K., Okazaki, K. and Suzuki, A., 1987, Sequences of *S*-glycoproteins, products of the *Brassica campestris* self-incompatibility locus, *Nature*, 326:102-105.

Whitehouse, H.L.K., 1950, Multiple-alleleomorph incompatibility of pollen and style in the evolution of the angiosperms, *Ann. Botan. New series*, 14:198.

Woodward, J.R., Bacic, A., Jahnen, W. and Clarke, A.E., 1989, N-linked glycan chains on *S*-allele-associated glycoproteins from *Nicotiana alata*, *Plant Cell*, 1:511-514.

Xu, B., Grun, P., Kheyr-Pour, A. and Kao, T.-H., 1990, Identification of pistil-specific proteins associated with three alleles in *Solanum chacoense*,. *Sex. Plant Reprod.*, 3:54-60.

# FLORAL HOMOEOTIC AND PIGMENT MUTATIONS PRODUCED BY TRANSPOSON-MUTAGENESIS IN *Antirrhinum majus*

Rosemary Carpenter, Sandra Doyle, Da Luo, Justin Goodrich, Jóse M. Romero, Robert Elliot, Ruth Magrath and Enrico Coen

Genetics Department, John Innes Institute, John Innes Centre for Plant Research, Colney Lane, Norwich NR4 7UH, U.K.

## INTRODUCTION

Many homoeotic genes affecting flower morphogenesis have been described in diverse species, although there have been few attempts to relate these in a systematic way to the mechanism of floral development (Meyerowitz *et al.*, 1989). In 1742 it was the peloric form of *Linaria vulgaris* that made Linnaeus change his mind about the fixity of species (Linnaeus, 1744) and over a century later Darwin described, and was intrigued by, the behaviour of the peloric *Antirrhinum* (Darwin, 1868). We have used transposon-mutagenesis to generate floral homoeotic mutations in *Antirrhinum* with a view to studying and isolating the genes involved. One advantage of this approach is that transposon integration can be used to identify genes and subsequent excision can be used to prove that the correct gene has been isolated (Martin et al., 1985). In addition, imprecise excision can generate alleles with altered gene expression (Almeida et al., 1989).

*Antirrhinum majus* provides ideal experimental material for this approach. Three different transposons have been isolated from it (Coen et al., 1989) and the technique of transposon-tagging has been successfully used (Martin et al., 1985). Several genes encoding enzymes in the anthocyanin pigment biosynthetic pathway have been isolated and provide good markers for monitoring transposition and the trapping of new transposons. There is also a good genetic map with many well-characterised mutations affecting flower development (Stubbe, 1966). In the wild-type of *Antirrhinum majus* the flowers are borne in a spiral up the stem, forming an inflorescence. Each flower grows in the axil of a bract and is zygomorphic. In transverse section, the flower may be considered as comprising four concentric rings or whorls, each containing several organ members which will be referred to as upper or lower depending on their position relative to the bract which is considered the lowest organ. The first or outermost whorl comprises five sepals, the lowest being alternate to the bract. The second whorl consists of five petals united for part of their length to form the corolla tube with five lobes. Five stamens are initiated and constitute the third whorl; however, the uppermost stamen remains vestigial so that the fully developed flower has only four stamens, the upper two being shorter than the lower two. A bilocular ovary constitutes the fourth whorl. The identity of the whorls will be indicated

**Fig. 1.** Mutations of *cycloidea* isolated by transposon-mutagenesis. The wild-type progenitor is shown at the top.

in sequence starting from the first whorl so that wild-type is sepal, petal, stamen, carpel.

TRANSPOSON MUTAGENESIS

In each case, 10 or 15 plants of lines carrying highly active transposons were grown at a constant 15°C, the temperature at which the greatest frequency of transposition occurs (Harrison and Fincham, 1964; Harrison and Carpenter, 1973). These plants were self-pollinated and seed capsules were collected separately. Seed from these plants gave rise to families of either 15 or 30 plants from each capsule, with a total of 13,000 $M_1$ plants grown in the glasshouse. As most mutations are likely to be recessive many of these plants should have been heterozygotes; therefore these $M_1$ plants were also self-pollinated. For the $M_2$ generation, seed from each family was sown to give either 48 or 96 plants (depending on whether the $M_1$ family comprised 15 or 30 plants) to give a total $M_2$ generation of 40,000 plants grown in the field.

Several flower homoeotic and pigment mutations were obtained and selected for analysis.

Floral homoeotic mutants

The floral homoeotic mutations that were isolated can be divided into four classes. The first class affects the identity of members within the same whorl and includes the *cycloidea* alleles which confer varying degrees of radial symmetry to the flower. Two independent mutations giving this phenotype were obtained (Fig. 1) and genetic analysis of these allowed two groups of *cycloidea* alleles to be defined. Extreme alleles of one group confer a peloric phenotype in which all members of a whorl resemble the lowest member of the wild-type flower. The second group also give flowers which are more symmetrical than wild-type but retain a degree of zygomorphy. Crosses between plants carrying alleles from the groups give $F_1$ plants with an almost wild-type phenotype but with two small notches on the lower flower lip, showing that alleles from the two groups partially complement each other. This suggests that *cycloidea* may be a complex locus composed of two interacting functional components. The second class of homoeotic mutations affect the identity of the whorls. One mutant of this class is *ovulata* which, unlike the other mutations, was detected in the $M_1$ generation, indicating that it was dominant or semi-dominant to wild-type. Several different $M_1$ plants, all derived from the same parent, showed a range of related phenotypes. The most severe of these was self-pollinated and gave an $M_2$ of 3 wild-type, 12 parental phenotypes and 6 mutants with a new extreme phenotype. The first whorl of this extreme phenotype (Fig. 2) comprised five carpels, the uppermost lacking ovules and the four lower being united to give an ovary with four ovule-bearing loculi terminating in a united band of stylar tissue. The second whorl had three lower stamens while the upper two members remained as small vestigial or aborted structures; the third and fourth whorls were wild-type. The near 1:2:1 segregation of the $M_2$ progeny also suggested that the *ovulata* mutation was semi-dominant, the parental phenotype being that of the heterozygote and the extreme phenotype being the *ovulata* homozygote. This was confirmed in the next generation.

Other mutations of this second class were also identified. Mutations at two unlinked loci, *deficiens* and *sepaloidea*, were obtained which, in the most extreme form, resulted in sepals growing in place of petals and carpels instead of stamens to give the phenotype sepal, sepal, carpel. The fourth whorl did not normally develop in the extreme phenotype but comprised the normal bilateral ovary of wild-type in the less severe forms. A third unlinked locus, *globosa*, has also been described which gives this phenotype (Kuckuck and Schick, 1930).

The *deficiens* allele was somatically very unstable. Discrete clonal patches of pigmented petal tissue, sometimes comprising only four cells were observed in the second row of sepals (Fig. 3a and b). Occasionally entire wild-type flowers were produced on the mutant, presumably caused by early somatic reversion events. In addition, small flowers containing much-reduced stamens were sometimes seen and resembled the phenotype of *deficiens*$^{\underline{nicotinoides}}$ (Hertwig, 1926). The phenotype of *sepaloidea* was also very variable with a tendency to produce large areas of petaloid tissue, usually edged with sepal-like areas in the second whorl (Fig. 3 c and d).

The third class of homoeotic genes affect both the identity and number of whorls. This includes the *pleniflora* mutants which give flowers with the first three whorls of the type sepal, petal, petal. The fourth whorl comprises two sepaloid/carpeloid/petaloid structures and within these a proliferation of petaloid whorls occurs.

Two of the mutations were unstable somatically and gave occasional wild-type flowers on otherwise mutant spikes. Seed from these gave both wild-type and mutant progeny. The wild-type progeny were self-pollinated

**Fig. 2.** Inflorescence of the extreme *ovulata* mutation.

and in each case segregated for mutants and wild-types. These results indicated that the mutations were recessive and could revert to wild-type in both somatic and germinal tissue. However, it has not yet been established if these mutations belong to the same or different complementation groups.

The potential for indeterminate growth is further illustrated by the fourth class of mutants where the switch from vegetative growth to that of

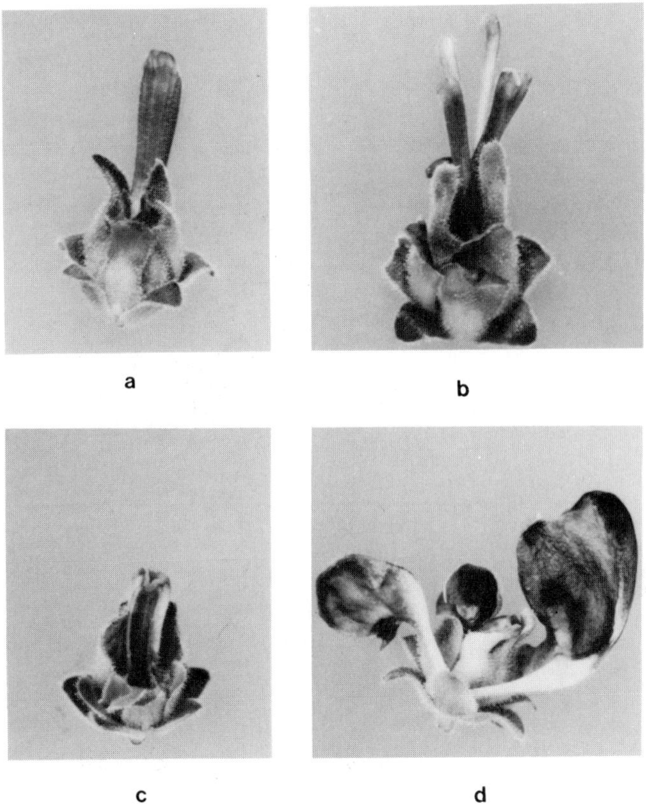

**Fig. 3.** Diversity of flowers produced by *deficiens* and *sepaloidea* mutations. Top row: *deficiens* flowers showing the second whorl of sepals and the united five carpels of the third whorl. Flower **b** shows a clonal patch of petal cells on the second whorl of sepals and a central carpel. Bottom row: flowers from the *sepaloidea* mutation which the most extreme form on the left.

the reproductive spike was as normal. However, in the *floricaula* mutant, instead of a flower being produced in the axil of each bract, a secondary shoot was formed. This shoot in turn produced bracts within the axils of which further shoots were produced. This process could continue indefinitely to form an indeterminate shoot which has lost its ability to flower (Fig. 4). The wild-type *floricaula* product is therefore required for switching indeterminate shoot meristems to floral meristems and presumably activates expression of the other classes of floral homoeotic genes either directly or indirectly.

The mutation was propagated vegetatively and a number of cuttings were grown at 15°C. One of these produced occasional flowers which were self-pollinated and five capsules of viable seed were obtained. The progeny from four of these were all mutant; however one capsule gave progeny which segregated to give 4 mutant and 18 wild-type plants, indicating that germinal reversion to the wild-type allele had occurred.

**Fig. 4.** Spike of the *floricaula* mutation showing indeterminate growth.

The progeny from the wild-types showed that some were heterozygous and others homozygous for the wild-type allele. Further revertant progeny from a number of independent events have now been obtained. Mutant phenotypes have been described in *Arabidopsis* which show many similarities to those of *Antirrhinum* (Haughn and Somerville, 1988). The remarkable similarities between mutants in *Antirrhinum* and *Arabidopsis*, which belong to two taxonomically distant plant subclasses, suggest that the genetic control of whorl identity has been highly conserved in the evolution of dicotyledonous plants.

Cell autonomy of mutants

Many of the homoeotic mutants show instability in somatic or germinal tissue, suggesting that they were transposon-induced. In some cases, the instability may be used to investigate cell-autonomy of the genes affected. The *deficiens-621* allele gives clonal patches of petal tissue on the epidermis of the second whorl of sepals. These patches have very sharp boundaries and may be explained by somatic excision of a transposon from the *deficiens* locus restoring gene function. This suggests that the product of the wild-type *deficiens* gene is not diffusible between cells and acts cell-autonomously, at least in the epidermis. The observation of some very small patches indicates that the *deficiens* product is active throughout development of the petal, from the early stages when petal and sepal primordia become distinct to the final cell divisions of the petal. However, the consequences of *deficiens* expression may be different at each developmental stage so that at early stages it might affect patterns of cell division and expansion and hence the form of the petal, whereas at later stages it may affect cell fate.

Since the clonal patches of the *deficiens*-621 allele occur only in the second and not the first whorl of sepals, there must be other genes that act specifically in the second whorl throughout its development and that are necessary either for *deficiens* expression or for the action of the *deficiens* product. Two such genes may be *globosa* and *sepaloidea*, since mutation of these gives a similar phenotype to the *deficiens* mutant. The *apetala*-3 mutation of *Arabidopsis*, which gives a similar phenotype to *deficiens*, is also thought to act up to a late stage of organ development on the basis of temperature shift experiments (Bowman et al. 1989).

## Transposon-tagging and trapping

Several pigmentation mutants were also obtained and were characterised to determine if they were transposon induced. Five out of six mutants showed somatic and germinal instability, diagnostic of transposon insertions. One mutation, *niv*-600, was in a previously cloned gene and molecular analysis showed that it was caused by insertion in the first exon of a new transposon, Tam4, which is present in about 10 copies in the parental line.

Molecular analysis of *incolorata-601* and its germinal revertants using several Tam probes, showed that it was due to insertion of Taml, whilst an unstable mutation at *delila*, a recessive regulatory gene that blocks pigmentation in the corolla tube, was the result of a Tam2 insertion. The *olive-605* allele, which gives leaves with dark green spots in a yellow background, showed several features typical of Tam3 insertions. The frequency of spots was much greater when the mutant was grown at 15°C compared to 25°C, as has been found for Tam3 insertions at the *pallida* and *nivea* loci (Coen et al., 1989). Furthermore, when *olive-605* was crossed to a line carrying *Stabiliser*, a gene known specifically to inhibit Tam3 excision (Carpenter et al., 1987), the $F_2$ segregated plants showing very few if any dark green spots. These plants failed to grow to maturity, presumably because of photosynthetic deficiency, indicating that the viability of *olive-605* depends on transposon excision. Molecular analysis of *olive-605* and its revertants confirmed that the mutation was caused by Tam3 insertion and has led to isolation of the *olive* locus. This is the first time a yellow leaf pigmentation mutant has been characterised molecularly and this should now allow a new approach to the study of such genes.

The results show that most of the newly produced pigment mutations were caused by transposon insertion and that it is possible to isolate the affected gene using cloned transposons as tags. At least four different transposons are active in the lines used for mutagenesis one of which, Tam4, was newly isolated by trapping it in a previously cloned gene. As further genes and mutations are characterised it should be possible to trap more transposons and so define the full range active in the lines. These could be used to determine which transposons were responsible for the floral homoeotic mutations and so allow the isolation and molecular analysis of the genes involved.

## Acknowledgements

We thank David Hopwood, Cathie Martin and Michal Goldschmidt-Clermont for helpful discussions and Kathryn Blewett, Karen Ingle and Audrey Cooper for summer vacational assistance. We are grateful to Hans Sommer for generously providing us with the sequence and probes of Taml prior to publication, Peter Scott and Andrew Davies for photography, the Glasshouse and Field Services Department for growing the plants and Anne Williams for typing the manuscript.

We acknowledge a generous grant from the Gatsby Foundation which enabled us to carry out the transposon-mutagenesis experiment.

REFERENCES

Almeida, J., Carpenter, R., Robbins, T. P., Martin, C., and Coen, E. S., 1989, Genetic interactions underlying flower color patterns in *Antirrhinum majus*, *Genes and Dev.*, 3:1758.

Bowman, J. L., Smyth, D. R., and Meyerowitz, E. M., 1989, Genes directing flower development in *Arabidopsis*, *The Plant Cell*, 1:37.

Carpenter, R., Martin, C. R., and Coen, E. S., 1987, Comparison of genetic behaviour of the transposable element Tam 3 at two unlinked pigment loci in *Antirrhinum majus*, *Mol. Gen. Genet.*, 207:82.

Darwin, C., 1868, "Variation in plants and animals under domestication", Vol. 2, pp. 59. Murray, London.

Harrison, B. J., and Carpenter, R., 1973, A comparison of the instabilities at the *nivea* and *pallida* loci in *Antirrhinum majus*, *Heredity*, 31:309.

Harrison, B. J., and Fincham, J. R. S., 1964, Instability at the *Pal* locus in *Antirrhinum majus*. 1) Effects of environment on frequencies of somatic and germinal mutation. *Heredity*, 19:237.

Haughn, G. W., and Somerville, C. R., 1988, Genetic control of morphogenesis in *Arabidopsis*, *Dev. Genet.*, 9:73.

Hertwig, P., 1926, Ein neuer Fall von multiplem Allelomorphisms bei *Antirrhinum*, *Z. f. indukt Abst.-u. Vererbungsl.*, 41:42.

Kuckuck, H., von, and Schick, R., 1930, Die Erbfaktoren bei *Antirrhinum majus* und ihre Bezeichung. *Z. f. indukt Abst.-u. Vererbungsl.*, 56:51.

Linnaeus, C., 1744, Peloria (Daniel Rudberg), *Amoenitrates Academicae*, Volumen Primum, Cornelium Haak, Luduri Batavorum, pp.1-19.

Martin, C., Carpenter, R., Sommer, H., Saedler, H., and Coen, E. S., 1985, Molecular analysis of instability in flower pigmentation of *Antirrhinum majus*, following isolation of the *pallida* locus by transposon tagging. *EMBO J.*, 4:1625.

Meyerowitz, E. M., Smyth, D. R., and Bowman, J. L., 1989, Abnormal flowers and pattern formation in floral development, *Development*, 106:209.

Stubbe, H., 1966, "Genetik und Zytologie von *Antirrhinum L.* sect *Antirrhinum*", Veb. Gustav Fischer Verlag, Jena.

# MOLECULAR ANALYSIS OF THE HOMEOTIC FLOWER GENE *deficiens* OF *Antirrhinum majus*

Hans Sommer, Wolfgang Nacken, Peter Huijser, Jose-Pio Beltran*, Peter Flor, Rolf Hansen, Heike Pape, Wolf-Ekkehard Lönnig, Heinz Saedler and Zsuzsanna Schwarz-Sommer

Max-Planck-Institut für Züchtungsforschung
5000 Köln 30, FRG

*Instituto de Agroquimica y Technologia de Alimentos, CSIC, 46019 Valencia, Spain

## INTRODUCTION

Flower formation in higher plants is a very complex process controlled by genetic as well as environmental factors (for review see e.g. Bernier, 1988). Although it is an integrated process, two major phases can be recognized: floral evocation and development. The term "evocation" designates the transition of the vegetative apical meristem to a "floral" meristem - that is a flower primordium generating meristem - following stimulation by internal and/or external "signals". After this initial event the phase of floral development follows, starting with the appearance of floral primordia and ending with the mature flower composed of functionally and structurally distinct organs. During this process the type, number and position of the organs constituting the flower is strictly regulated. How is this achieved and what are the underlying mechanisms? The study of morphogenetic mutants displaying various types of abnormalities could be a first step to answering these questions.

## RESULTS

### Floral morphogenetic mutants in *Antirrhinum majus*

Several different classes of morphogenetic mutants are known in *Antirrhinum* that could be useful for understanding the molecular processes of floral development (Stubbe, 1966). The first class comprises two mutants, *sterilis* (*ste*) and *steriloides* (*sto*). The phenotype of the mutants indicates that the gene products are required for initiation and formation of the floral primordium from the apical meristem after floral evocation, that is after transition of the vegetative meristem to a flower producing meristem. *Sterilis* displays an abnormal inflorescence carrying only bracts but no flowers in the axils of the bracts, in contrast to wild type inflorescences. Unfortunately this mutant is lost. A phenotypically very similar mutant, *steriloides* , was isolated in a transposon mutagenesis program by our group (unpublished). *Steriloides* also has bracts only on the inflorescence, but occasionally produces a few (sometimes deformed) flowers, probably due to leakiness or somatic instability of the recessive mutation.

*Squamata* ( *squam* ) and *squamosa* ( *squa* ) belong to a second class of mutants that bear "shoots" instead of flowers on the inflorescence. While *squamata* is lost, *squamosa* is still available. The phenotype of both mutants suggests that the function of the wild type genes is to establish the identity of the floral primordium; initiation of the primordium seems to be normal in the mutants. If the genes are inactivated, "shoots" will grow in the axils of the bracts. *Squamosa* , like *steriloides* , produces sometimes a few flowers, probably due to leakiness of the recessive mutation.

In the mutant *macho* ( *ma* ), a member of a third class of flower mutants, the floral organs of the first and second whorl are altered homeotically: sepals are transformed to "carpels" and petals to "stamens", similar to *apetala 2* in *Arabidopsis* . Several isolates were recently obtained from a screening program, which are either semi-dominant or dominant and very similar in their phenotypic appearance (our unpublished results). Whether the isolates are alleles or not is currently being tested.

Four different loci have been described (Stubbe 1966) that, when inactivated by mutation, lead to homeotic transformation of the second and third whorl organs: *deficiens* ( *defA* ), *globosa* ( *glo* ), *viridiflora* ( *vir)* and *femina* ( *fem* ). Sepaloid structures are found in the second whorl instead of petals, and the stamina in the third whorl are missing. Instead one finds five ovule-bearing carpels with the upper parts forming a chimney-like structure.

There are two homeotic mutants, *plena* ( *ple* ) and *petaloidea* ( *pet* ), known in which the third and fourth whorl organs are affected. Both mutants are sterile since the reproductive organs are missing. *Plena* displays petals in the third whorl and a repetition of petaloid structures in the center of the flower. *Petaloidea* also has petals in the third whorl like *plena* , but initiates a new peloric (radial) "flower", consisting of sepals and corolla, in the place of the gynoecium. These third class mutants indicate that the wild type gene products may participate in the establishing of whorl and organ identity in the process of flower formation.

A fourth class of mutations affects the symmetry of the flower. The zygomorphic wild type flower displays bilateral symmetry. In the mutant *cycloidea* the flowers have a radial symmetric shape: all organs in the respective whorls are arranged in a radial symmetric fashion. Two independent loci are known that confer this "peloric" phenotype on the flower when mutated (Stubbe 1966; our own observations).

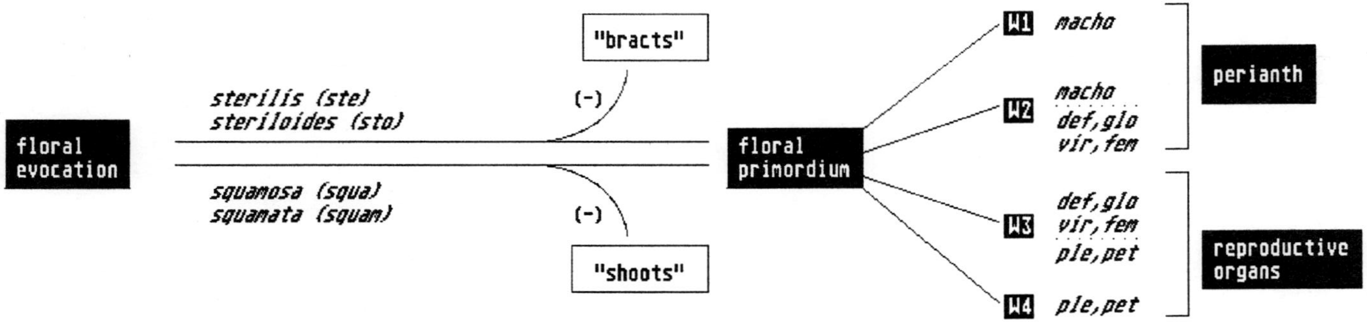


Figure 1 Simplified scheme indicating the "functional" position of various homeotic genes in floral development of _A_. _majus_

A schematic presentation of the possible spatial and temporal activity of the products of the genes identified by the mutants described above is given in Figure 1. Since extensive genetic, morphological and molecular analysis with all of these mutants have not been carried out yet, the scheme can give only an approximate picture of the position and function of these genes in the complex morphogenetic process that starts with floral evocation and ends with the mature flower.

## The *deficiens (defA)* locus and its alleles

In a first attempt to gain insight into structure and function of genes involved in the control of floral development and differentiation we have chosen to study the *deficiens* locus of *A. majus* by genetic, morphological and molecular methods (Sommer et al. 1990).

Three classical mutant alleles of this locus exist that display phenotypic alterations of varying degree. The most extreme allele is *deficiens*$^{globifera}$ (*defA-1*). The five sepals in the first whorl are normal, but the petals (corolla) in the second whorl are replaced by five sepaloid leaves (homeosis). Similarly, the four stamina and the staminoid of the third whorl are converted homeotically to five carpels containing ovules and forming the female organ (gynoecium). The stamina are often fused and form a protruding tubular structure with the stigma on the top. This recessive mutant allele is male sterile.

The second allele, *deficiens*$^{nicotianoides}$ (*defA-2*), is less severely affected in its phenotypic appearance, compared to *globifera*. The petals are reduced in size and greenish, while the stamina look like small leaves with "anthers" on the upper rim, but without pollen. A long style with stigma is protruding from the flower. This allele is also recessive and male sterile; it seems to be dominant over the *globifera* allele in many respects when both are combined in a heterozygote (Klemm 1927; our own observations).

Phenotypically closest to wild type is *deficiens*$^{chlorantha}$ (*defA-3*), the third allele of the deficiens series. The corolla is reduced in size and deformed, with greenish areas on the lower lip. Male fertility is almost normal, although the stamina look slightly different than in wildtype flowers. *Chlorantha* is dominant over the other two alleles in heterozygotes (Klemm 1927).

## Genetic instability of the *globifera* allele

The *globifera* allele shows somatic and germinal instability, as already observed by Baur (1924). Due to somatic reversion late in development, sectors with sharp boundaries of petaloid tissue can be seen on the sepaloid organs in the second whorl. If reversion occurs earlier, then a whole 'sepal' is converted back to a petal; sometimes even a complete corolla was restored, but not the stamina in the third whorl. Somatic reversion restoring only stamen in the third whorl was never observed.

Germinal reversion was observed on the inflorescences of about twenty percent of the globifera plants. One or more wild type like flowers appeared on the inflorescence, which upon selfing, produced segregating progenies of wild type and mutant phenotype in an almost 3:1 ratio, indicating that the reversion event was heritable. The somatic and germinal instability suggests that a transposable element is inserted in the gene, which can frequently excise and thus restore gene function. Furthermore, the sharp boundaries of the somatic sectors on the sepaloid organs in the second whorl of globifera seem to indicate that the gene product of *deficiens* acts cell-autonomously and is not difusible.

**Isolation and characterization of the *deficiens* locus**

For molecular analysis we have chosen the *deficiens* locus for two reasons: first, the existence of revertants isolated from the unstable *globifera* allele, which would be essential to finally prove that the gene cloned is *deficiens* (see below), since *in vitro* complementation is not yet possible in *A. majus* . Secondly, an allelic series of mutants is available (see above) analysis of which could help in understanding the function of the *deficiens* gene.

Our initial attempt was to use the known transposable elements of *A. majus* (for review see Coen and Carpenter, 1986; Sommer et al., 1988) as probes to clone the *defA* gene, since the somatic and germinal instability of the globifera allele seemed to indicate that a transposon was inserted in the gene (see above). This strategy, "transposon tagging", has been successfully used for cloning plant genes (for compilation see e.g. Gierl and Saedler, 1989), but it failed in our case since an unknown transposon was residing at the locus, as became evident later. Therefore, a different strategy was used, based on the assumption that in the *deficiens*$^{globifera}$ mutant the mRNA was either completely abolished or at least strongly reduced, thus allowing a differential screening approach, in order to isolate a cDNA derived from the *defA* gene.

The following strategy was used. PolyA$^+$ RNA was isolated from young wild type inflorescences and converted into first strand cDNA which was then "subtracted" with excess mRNA from young leaves, thus enriching the single strand (ss) cDNA for flower specific sequences. From the enriched sscDNA a cDNA library of 7 x $10^5$ recombinants was prepared by conventional methods (for details see Sommer et al. 1990). This library was then screened differentially with (+) and (-) probes derived from wild type and mutant inflorescences, respectively. In order to increase the sensitivity, these probes were also enriched by subtracting the sscDNAs synthesized from wild type and mutant infores-cence mRNA with excess wild type leaf mRNA. Subsequently the remaining enriched (+) and (-) sscDNAs were labelled by "random priming" and used for differential hybridization. Sixty-eight candidates were obtained by differentially screening 7 x $10^4$ recombinant phages showing a clear difference with the (+) and (-) probes. These candidates belonged to twelve distinct groups comprising between one and ten members, as was shown by cross hybridization and genomic Southerns.

Cosegregation experiments were subsequently carried out with one member from each group of the cDNAs, in order to identify those candidates that were derived from the *defA* gene. For this genomic DNA was isolated from 46 individual F2 plants showing globifera phenotype that were obtained by self-pollinating a wild type heterozygote (defA/def A-1). Of the twelve groups only group VII (consisting of a single member, c23) displayed a restriction polymorphism that cosegregated with the chromosome marked by the defA-1 mutation.

Since cosegregation demonstrates only linkage of a probe with a chromosomal marker, but cannot prove definitely that a clone is derived from the gene in question, additional proof had to be found. Correlation of excision of a transposon with restoration of wild type phenotype is an unequivocal criterion for identity of a probe with the unstable gene. This means that in Southern experiments with genomic DNAs from wild type, mutant and revertant plants a wild type fragment(s) hybridizing with the probe is altered in the mutant, due to the insertion of the transposon, but restored in the revertant(s). Precisely this was observed in Southern hybridizations with wild type, mutant (*globifera*) and revertants (isolated from the *globifera* allele) using the fullsize cDef1 cDNA (isolated with c23) as a probe.

**Structure of the *defA* gene**

Southern hybridization experiments with the fullsize cDNA clone cDef1 showed that the *defA* gene was located on a 6.8 kb EcoRI fragment. This EcoRI fragment was isolated by cloning it into lambda NM 1149; and subsequently its DNA sequence was determined. Comparison of the DNA sequences of both the cDNA and the genomic clone allowed to derive the structure of the *defA* gene which is shown in Figure 2. The coding region, interrupted by five small introns (100 to 250 bp) and one large intron (about 1 kb), spans about three kb. The promoter contains a 250 bp inverted repeat upstream of the putative TATA box, an unusual feature, the function of which is not known at the moment. From the gene a 1.1 kb long mRNA is transcribed, as revealed by Northern analysis. This mRNA has a 681 nucleotides long open reading frame (ORF) that codes for a 227 amino acids long putative protein (DEF A).

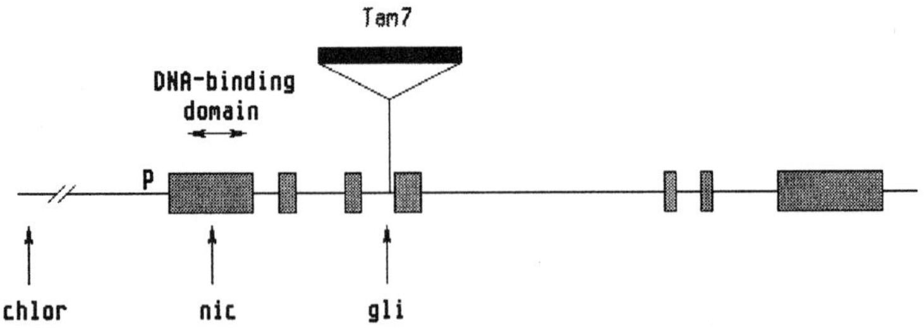


Figure 2 Structure of the *defA* gene. Arrows indicate the position of mutations in the respective mutant alleles

**DEF A has homology to transcription factors**

A search for homology of DEF A to other proteins in the EMBL and Genebank databases disclosed that DEF A has a conserved region in common with two known transcription factors: SRF of mammals (Norman et al., 1988) and MCM1 (=GRM/PRTF), the product fo the *MCM1* gene of yeast (Passmore et al., 1988). SRF (serum response factor) is an essential factor for serum inducible transcriptional activation of the c-*fos* nuclear proto-oncogene which is involved in transcriptional regulation of genes controlling cell growth in response to growth factors. MCM1 is a transcription factor in yeast that, by cooperative interaction with the alpha 2 repressor or with the alpha 1 activator, regulates the expression of cell type specific genes. With both proteins DEF A shares a highly conserved domain of 58 amino acids at its N-terminus: 28 amino acids are identical, while 12 amino acids represent conservative exchanges. The homology between SRF and MCM1 extends further to the C-terminus, in contrast to DEF A.

For SRF and MCM1 it has been shown that the conserved domain of the proteins is involved in DNA binding and protein dimerization. The presence of a similar domain in DEF A, therefore, suggests that DEF A is also a transcription factor regulating the expression of genes determining normal petal and stamen development.

**Expression of the *defA* gene**

The temporal and spatial expression pattern of the *defA* gene was analysed by Northern hybridization with cDef1 as a probe. PolyA$^+$ RNA was prepared from different developmental stages, from 2mm inflorescences to mature flowers. In addition, mRNA was isolated from floral organs, leaves and young shoots. Northern analysis of these mRNAs shows that the *defA* gene is nearly constantly expressed during the whole time of flower development, from very young to very late stages. Expression is flower specific, since no *defA* -specific mRNA is found in young shoots, leaves and bracts (roots were not tested yet). Analysis of the floral organs showed that *defA* is expressed in all organs, although the level of expression is varying quite strongly: very low in sepals, low in carpels, but strong in petals and stamens, the organs that are homeotically transformed in the *deficiens* mutant allele *globifera* . *In situ* hybridization experiments confirmed the strong *defA* expression in petals and stamens but were not able to reveal the weaker expression in sepals and carpels, probably due to lower sensitivity of the method, as compared to Northern analysis.

**Molecular analysis of the *deficiens* mutant alleles**

As mentioned before, three "classical" mutant alleles of the *deficiens* locus have been described: two stable alleles *chlorantha* and *nicotianoides* and the unstable *globifera* allele (Stubbe, 1966). These three alleles were also cloned and molecularly analyzed by Southern and Northern hybridization, and by DNA sequencing.

Southern hybridization experiments with genomic DNA

from the mutant allele *chlorantha* and wildtype using cDef1 as probe showed no difference. Northern blots probed with cDef1 indicated that the size of the mRNA was not altered, but the amount seemed to be only 10% to 15% of that present in wildtype flower buds, indicating a mutation in the promoter region of the mutant allele. This was confirmed by DNA sequencing: there was no difference to wildtype in the structural part of the gene; the only difference, a three bp deletion adjacent to a one bp exchange, was found in the promoter region at about position -1200 (3 kb of the upstream region were sequenced). Thus *chlorantha* represents a "promoter down" mutation of the *defA* gene, thereby generating a homeotically altered morphoallele.

In contrast to *chlorantha* , *nicotianoides* seems to be a "structural" mutant. Southern and Northern blots showed no difference to wild type, but DNA sequencing revealed a one bp exchange in the DNA binding domain leading to a substitution of a glycine by an asparagine in the DEF A protein. The consequence of this subtitution is probably a decrease of the DNA binding affinity and/or specificity of the altered protein thus causing the mutant phenotype.

The unstable *globifera* allele was shown to be an insertional mutant, as revealed by Southern blots and DNA sequencing. A new transposon, not yet described in *A. majus* , Tam7, was found residing at the intron/exon boundary of the third intron and fourth exon of the *defA* gene in the mutant. This 7 kb long transposon carries 20 bp long inverted repeats (IR) at its termini, the outermost 13 bp of which are identical with the terminal IRs of the Tam1 element of *A. majus* ; but no additional homology among the internal parts of Tam7 and Tam1 was discovered. Upon integration Tam7 generates a 3 bp duplication of target sequences, like Tam1. Whether Tam7 is an autonomous element or a defective, non-autonomous one is not known. In Northern blots no *defA* mRNA was detected in *globifera* suggesting that, due to the insertion, an altered mRNA is produced which is rapidly degraded.

**Other regulatory genes homologous to *defA***

When a cDNA library prepared from flower buds was screened with the *defA* DNA binding domain as probe under reduced stringency of hybridization, several additional recombinants displaying varying signal intensities could be isolated at a frequency of about one in 1500 recombinants. These were assigned to distinct groups by their hybridization signal intensity and cross-hybridization with each other. Subsequently one member of each group was characterized by DNA sequencing and Northern analysis. All of these homologues display a high degree of homology in the DNA binding domain (between 65% and 90% of the amino acid sequence) to DEF A and to each other. Of the eight candidates characterized sofar, four are flower specific, while the other four are also expressed in young shoots and leaves. As indicated by Southern hybridization, two of the flower specific homologues can probably be assigned to known homeotic loci, but this still has to be confirmed. Since there are more candidates to be characterized, the total number of DEF A homologous factors present in the flower may well exceed 15. Apparently these proteins represent a family of transcription factors that play an important role in the control of developmental processes in the plant.

## CONCLUSIONS

The results obtained from molecular analysis of the *deficiens* locus of *Antirrhinum majus* demonstrate that the study of morphogenetic mutants can provide some insight into the mechanisms of differentiation and development, but for comprehensive understanding of these complex processes also other approaches are necessary.

We have isolated a regulatory locus by a combination of insertional mutagenesis and differential screening. This method offers an alternative to chromosome walking and transposon tagging for cloning of genes with unknown product and function.

The homeotic *deficiens* locus encodes a putative transcription factor displaying homology to the serum response factor SRF in mammals and the mating type regulating factor MCM1 in yeast. In *Drosophila* many of the homeotic genes also encode transcription factors that play pivotal roles in development. One can probably expect that many of the other known homeotic plant genes also code for such regulatory factors.

The discovery in *Antirrhinum* of other deficiens-homologues containing also the conserved DNA binding domain suggests that these factors constitute a family (like the homeo-box genes) of transcriptional regulators that participate in various developmental processes, not only in the flower but also in other parts of the plant.

## REFERENCES

Baur, E. (1924) Untersuchungen über das Wesen, die Entstehung und die Vererbung von Rassenunterschieden bei *Antirrhinum majus* . Bibliotheca Genetica, 4:1

Bernier, G. (1988) The control of floral evocation and morphogenesis. Ann. Rev. Plant Physiol. Plant Biol., 39:175

Coen, E.S. and Carpenter, R. (1986) Transposable elements in *Antirrhinum majus* : generators of genetic diversity. Trends in Genetics, 2:292

Gierl,A. and Saedler, H. (1989) Transposition in plants. in: "Nucleic Acids and Molecular Biology" 3:251, F. Eckstein and D.M.J. Lilley eds., Springer Verlag Berlin-Heidelberg

Klemm, M. (1927) Vergleichende morphologische und entwicklungsgeschichtliche Untersuchung einer Reihe multipler Allelomophe bei *Antirrhinum majus* . Bot. Archiv., 20:423

Norman, C., Runswick, M., Pollock, R. and Treisman, R. (1988) Isolation and poperties of cDNA clones encoding SRF, a transcription factor that binds to the *c-fos* serum response element. Cell, 55:989

Passmore, S., Elble, R. and Tye B.-K. (1989) A protein involved in yeast binds a transcriptional enhancer conserved in eukaryotes. Genes and Development, 3:921

Sommer, H., Hehl, R., Krebbers, E., Piotrowiak, R. Lönnig, W.-E. and Saedler, H. (1988) Transposable elements of *Antirrhinum majus* . In: "Proc. Internatl. Symp. on Plant Transposable Elements " O. Nelson, ed., Plenum New-York, pp. 227-235

Sommer, H., Beltran, J.-P., Huijser, P. Pape, H. Lönnig, W.-E., Saedler, H. and Schwarz-Sommer, Zs. (1990) *Deficiens* , a homeotic gene involved in the control of flower morhogenesis in *Antirrhinum majus* : the protein shows homology to transcription factors. EMBO J., 9:605

Stubbe, H. (1966) Genetik und Zytologie von *Antirrhinum majus* L. sect. *Antirrhinum* . VEB Gustav Fischer Verlag, Jena

# MUTATIONS OF *KNOTTED* ALTER CELL INTERACTIONS

# IN THE DEVELOPING MAIZE LEAF

Sarah Hake*†, Neelima Sinha†, Bruce Veit*, Erik Vollbrecht* and Richard Walko*

*Plant Gene Expression Center, USDA-ARS; 800 Buchanan St., Albany, CA, 94710; †Dept. of Plant Biology, Univ. of Calif., Berkeley, 94720 USA

## Introduction

### Normal Maize Leaf Development

The mature maize leaf is composed of an encircling sheath and a blade which tilts outward. At the juncture of leaf and blade is the ligule, a fringe-like tissue on the adaxial surface of the leaf (Figure 1). The ligule arises early in leaf development before the sheath and blade have differentiated. Additional features of the maize leaf are parallel longitudinal files of stomata on the adaxial and abaxial surfaces, three types of vascular bundles, and two types of photosynthetic cells. There is the median vein or midrib, 20-30 lateral veins, and intermediate veins, of which 10 or so are found between the lateral veins. The maize leaf exhibits typical Kranz anatomy (Brown, 1975) with the photosynthetic enzymes compartmentalized into two different cell types, the mesophyll cells and the bundle sheath cells. From anatomical studies, Dengler and coworkers (1985) suggested that the lineages of these two cell types were distinct. However, a detailed clonal analysis using genetic instability to create sectors, suggests that the dimorphism into these two cell types is position dependent. Langdale and coworkers (1989) determined that the mesophyll and bundle sheath cells are not clonally separate. In fact, the bundle sheath cells share a lineage with the mesophyll cells in the middle layer of the leaf.

Maize leaf development begins during embryogenesis when 5-6 leaf primordia are formed. After germination, the apical meristem gives rise to the remaining 10-15 leaves in alternate succession. Until the differentiation of the tassel, there are leaf cells at all stages of development in the maize plant. There is a developmental gradient within the plant, with older leaves at the base and younger leaves at the top. The younger leaves have "adult" characteristics, and the lower, older leaves are "juvenile" (Poethig, 1988). The leaf itself is a developmental gradient, cell division and cell elongation begin at the tip and progress downward. Differentiation of most cell types is also basipetal, with the exception of the median and lateral veins which differentiate acropetally (Sharman, 1942). Clonal analysis has been instructive in understanding the relationship of cells within a leaf (Poethig, 1984). Clones of cells extend longitudinally with the length of the leaf indicating that most cell division is restricted to a basal zone in the young leaf primordium. However, there is considerable flexibility in the development of the leaf. Cell invasion is common between the epidermis and inner layers, as well as between longitudinal files of cells (Stewart and Dermen, 1979; Sinha and Hake, in press).

*Plant Molecular Biology* 2, Edited by R.G. Herrmann and
B. Larkins, Plenum Press, New York, 1991

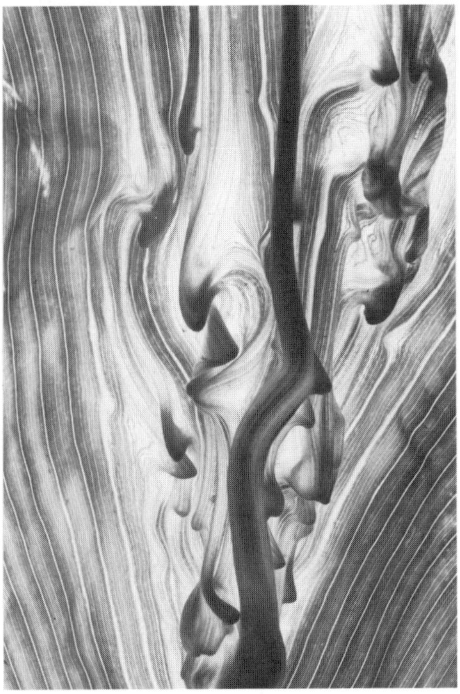

Fig. 1 (*left*). The maize ligule. The sheath wraps around the stalk, the blade tilts out of focus. The translucent fringe at the junction of sheath and blade is the ligule.

Fig. 2 (*right*). Knots along the lateral veins of a *Kn1-O* maize leaf.

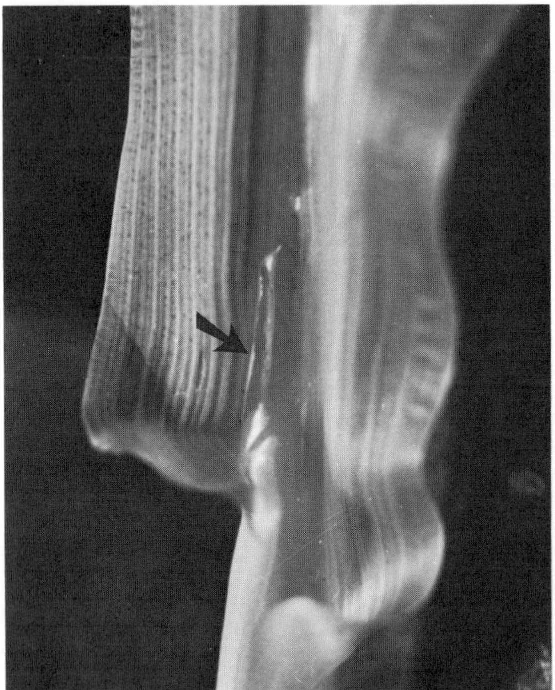

Fig. 3. Displaced patches of ligule in a *Kn1-O* mutant. The continuity of the ligule is broken towards the midrib. Displaced ligule is now parallel with the veins (arrow).

Knotted Leaf Development

*Knotted, Kn1*, is defined by a number of dominant mutations that all map to the same chromosomal region and affect lateral vein development. *Kn1* affects leaf development in a number of profound ways. 1) The lateral veins do not differentiate properly. 2) Regions of extra growth, or knots, form along the lateral veins. These outpocketings involve cell division and cell expansion of all leaf layers (Figure 2). 3) The ligule is displaced from its normal postion into the leaf blade along the lateral veins (Figure 3). Also, ectopic fringes of ligule are found in the leaf blade. Perhaps, as a secondary effect of the knots, is the improper alignment of stomates. In each case, positional signals that regulate cell-cell interactions have been altered. By understanding these attributes of the *Kn1* mutation, we hope to determine how certain components of normal leaf development are regulated. Analysis of *Kn1* will hopefully address the following questions: 1) What signals the formation of the ligule? 2) What cues are involved in the proper differentiation of the mesophyll and bundle sheath cells in the leaf? 3) What signals provide stomatal patterning? 4) How do veins maintain a constant distance between themselves?

## Results

### Phenotypic Analysis

The *Kn1* mutation affects the lateral veins of the leaf blade in a number of ways. The mildest attribute is a "clearing" of the veins. Histological studies first suggested that this "vein clearing" was due to a lack of differentiation in lateral vein cells, particularly the bundle sheath cells (Freeling and Hake, 1985). We have recently reexamined the extent of lateral vein cell differentiation using antibodies to photosynthetic enzymes. In the normal pattern of protein localization, ribulose bisphosphate (RuBP) carboxylase is found in the bundle sheath cells and phosphoenolpyruvate (PEP) carboxylase is localized to the surrounding mesophyll cells. In the "clear" veins, we find that only a few of the bundle sheath cells of the knotted leaf blade express RuBP carboxylase. This localization is analogous to the pattern seen in bundle sheath cells of the leaf sheath. In regions of knots, there is no bundle sheath at all and the veins are devoid of RuBP carboxylase as well as chlorophyll. In the mesophyll cells nearest to a knot, the normal localization of PEP carboxylase is missing, instead, RuBP carboxylase is found. This result is similar to the findings of Langdale and coworkers (1988) in which localization patterns were examined in mesophyll cells quite distant from lateral veins. These mesophyll cells did not contain PEP carboxylase, but instead contained RuBP carboxylase. Langdale and coworkers suggest that the vein provides a signal for the proper repression of RuBP carboxylase and initiation of PEP carboxylase expression. Our data further suggest that the bundle sheath itself is important.

We employed clonal analysis to determine which cell layer was responsible for the knots and ligule displacement (Sinha and Hake, in press). Earlier work showed that knots were not determined by the genetic constitution of the epidermis (Hake and Freeling, 1986). Using a closely-linked albino marker to indicate the absence of the *Kn1* locus, we have shown that it is the inner layer that is required for a knot to be present **and** it is also required for ligule displacement to occur. Thus, the inner cell layer that produces the middle mesophyll and bundle sheath cells (and perhaps the vascular bundle as well) is either the only cell layer that is competent to initiate the signals to make knots and displace the ligule, or the innner layer itself initiates these events.

### Rearrangements and Insertions Within a 10 kb Region Result in Knotted Mutations

Each *Kn1* mutation has a characteristic phenotype, varying in expressivity and timing. We cloned the first *Kn1* mutation by using the *Ds2* element as a transposon tag (Hake, Vollbrecht and Freeling, 1989). This mutation has the mildest phenotype, although it is enhanced in the presence of *Ac*. Using the flanking DNA into which the *Ds2* element had

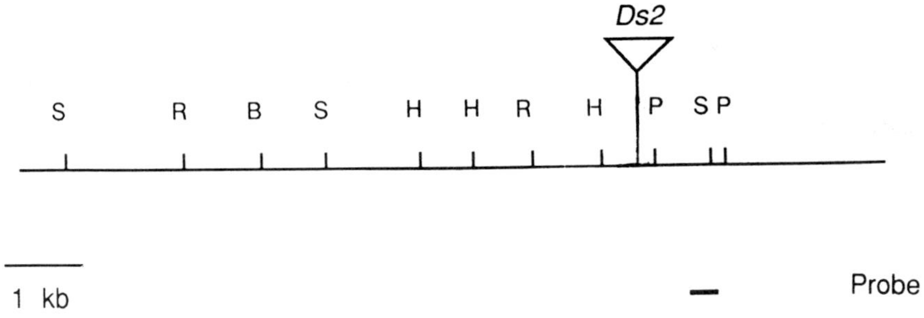


Fig. 4. Restriction map of the *Kn1-2F11* allele that was isolated by transposon tagging with *Ds2*. **S** is *Sac*I, **H** is *Hind*III, **B** is *Bam*HI, **R** is *Eco*RI, **P** is *Pvu*II. The probe used for *in situ*'s and for cDNA cloning is indicated.

inserted (Figure 4), we have further cloned and characterized other *Kn1* mutations. *Kn1-O,* the original *Kn1* mutant, is characterized by a tandem duplication of 17 kb (Veit *et al.*, 1990). In wildtype plants, this region is only represented once. The two repeats are identical at the level of resolution one obtains by digesting the DNA with restriction enzymes that cut genomic DNA frequently. We found a five base pair repeat present at all three junctions; the left border, the tail - head junction, and the right border. Five base pairs is small, but theoretically could have mediated unequal crossing over.

A tight correlation has been established between the number of repeats and the severity of the phenotype. Tassels of *Kn1-O* homozygotes were irradiated and used to pollinate wildtype testers. One exceptionally knotted derivative and four wildtype derivatives were isolated from a screen of 10,000 heterozygotes. Southern analyis demonstrated that the severely knotted individual results from a triplication of the region normally found duplicated in *Kn1-O*. Three of the normal individuals had lost one copy of the tandem duplication and one of them was a deletion for the entire tandem duplication and perhaps flanking DNA. The deletion is not viable in homozygous condition.

Further mutagenesis provides evidence that it is the breakpoint itself and not the duplication *per se* that causes *Kn1-O*. Nine derivatives were found in which the knotted phenotype was absent or reduced from approximately 9000 *Kn1-O* heterozygotes generated in an active Mutator background. Analysis of these derivatives showed that five of them suffered loss of the tandem duplication and the other four contained insertions of foreign DNA. The insertions (three of which are *Mu* elements) are near or at the junction of the tandem duplication (Figure 5). In two derivatives, different *Mu* elements have inserted at the exact same site 310 base pairs to the left of the tandem duplication breakpoint. In another derivative, a third *Mu* element inserted six base pairs to the right of the breakpoint, and the fourth derivative contains a non-*Mu* element approximately 1100 base pairs to the right (Veit *et al.*, 1990). Although the number of insertions is low, the proximity to one another may argue for sequence specificity for Mutator insertions. Alternatively, it may define a narrow window in which insertions can revert the *Kn1-O* phenotype.

The Mutator insertions not only suppress the knotted phenotype, but also increase the instability of the tandem duplication. We have found the spontaneous loss of the tandem duplication to occur at approximately one in 3000 in the absence of mutagens. When a *Mu* element resides at the breakpoint, the loss of the tandem duplication is much more frequent, approximately one in 20. This event is most frequent when the plant is heterozygous for two different *Mu* insertions. In the appropriately marked genetic backgrounds, we have been able to determine that there is no exchange of flanking markers. Therefore, the *Mu* element is mediating intrachromosomal recombination that results in the loss of one of the duplications and the *Mu* insertion.

We have isolated three dominant *Kn1* mutants in non-directed Mutator transposon-tagging experiments. The phenotypes of these mutants, *Kn1-mum1, Kn1-mum2,* and *Kn1-mum3*, are similar to each other. Analysis of genomic DNA suggests that the insertion of Mutator elements is responsible for the mutations. Interestingly, the same *Mu* element, *Mu8*, has inserted in each case in close but not exact proximity. Two other *Kn1* mutants have been analyzed by Southern analysis. These mutations, *Kn1-Z3* and *Kn1-N* were isolated as spontaneous mutations by Dr. Zuber and Dr. Neuffer, respectively. The phenotypes are similar to the *Mu8* mutations, but much less variable. *Kn1-Z3* has a uniform pattern of mild knots on most leaves, while *Kn1-N* expresses knots uniformly, but tends to be more severe, expecially in the first leaves. We have analyzed the DNA by Southern analysis and find that most restriction sites are conserved. In *Kn1-Z3* there is a 5 kb insertion in the same vicinity of the *Mu8* and *Ds2* insertions. We do not have revertant or progenitor data to prove that this insertion causes *Kn1-Z3*, but the data are suggestive. In DNA of *Kn1-N*, the only polymorphism detectable by Southern analysis is a 500 bp insertion approximately 6 kb distal to the other insertions. Again, we do not know that this insertion is responsible for the phenotype.

## Intragenic recombinants between *Kn1-O* and *Kn1-N*

In order to further characterize the *Kn1* locus, we isolated intragenic recombinants of two different *Kn1* alleles. Their phenotypes are quite distinct as well as their DNA polymorphisms. These two *Kn1* alleles are linked to different *Adh1* alleles 1 map unit distal (Figure 6a). Plants that were doubly heterozygous for *Kn1-N Adh1-S* and *Kn1-O Adh1-F6* were crossed to normal plants that were marked with a third *Adh1* allele. The majority of the 2700 plants screened appeared to be either *Kn1-O* or *Kn1-N* heterozygotes. One plant, however, was found that was normal and 2 plants were found that were severely knotted. We determined that recombination had occurred in these derivatives by detecting exchange of the flanking markers, RFLP UMC107 and *Adh1* (Mathern and Hake, 1989). In the normal derivative, unequal crossing appears to have occurred as illustrated in Figure 6b. In the severely knotted derivatives, the tandem duplication was still present as well as the 500 bp insertion that characterizes *Kn1-N*. Therefore, we assume crossing over occurred as in Figure 6c. The results of this recombination delimit the lesion that causes *Kn1-N* to the right of the recombination event in Figure 6c, suggesting that the insertion may be the cause of the *Kn1-N* mutation.

## RNA analysis

RNA transcripts have been analysed from *Kn1-O, Kn1-2F11, Kn1-mum1* and *Kn1-N* seedlings. Genomic subclones of the *Kn1-2F11* allele were used as hybridization probes to RNA blots of knotted and normal siblings. Depending on the subclone, either a complex array of minor transcripts, or a single, non-abundant, transcript of 1.7 kb in length is detected. These same subclones were also used as RNA probes to fixed sections of tissue. Only hybridization of the subclone that detected the single 1.7 kb transcript gives a visible signal in the *in situ*'s.

*In situ* hybridization to leaf blades of normal siblings produces a strong signal at the sclerenchyma cells of the lateral veins. There are two sclerenchyma "caps" that are positioned above and below the bundle sheath cells. In the mature leaf, these cells are

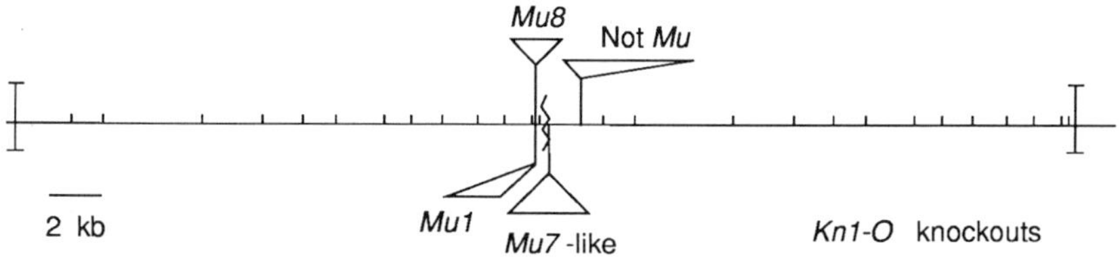


Fig. 5. Map designating the location of the insertions within the 17 kb tandem duplication of *Kn1-O* that suppress the phenotype. The wiggly line in the center is the central junction of the tandem duplication.

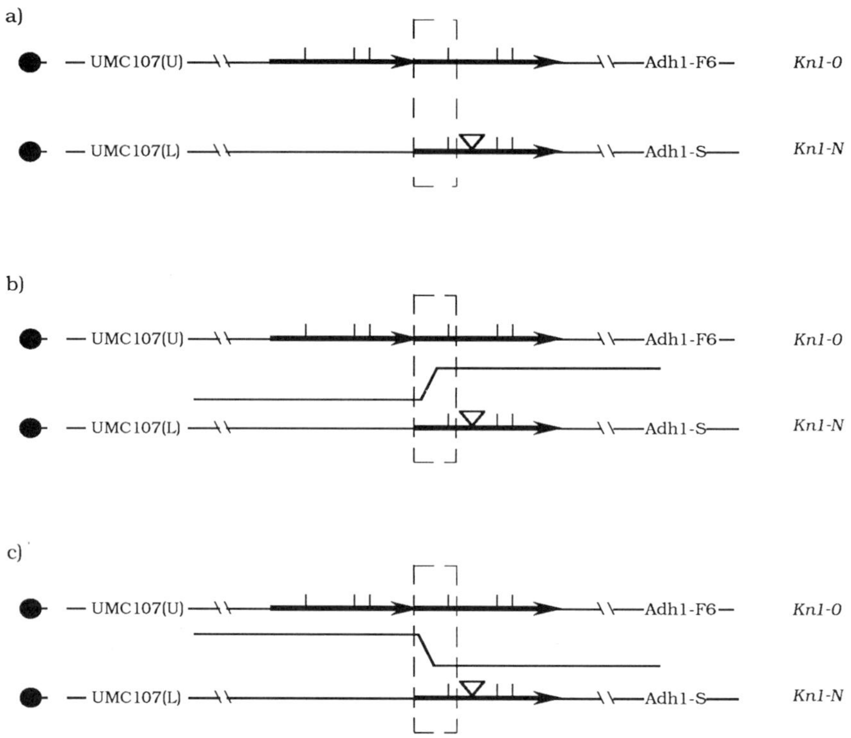


Fig. 6. Intragenic recombination at *Kn1*. Chromosome constitution of a) heterozygous progenitors, b) possible recombination event to produce normal derivatives, c) possible recombination event to produce severe knotted derivatives. Dashed box delimits recombination region.

presumed to play a role in strengthening (Esau, 1953). Signal is also detected along the leaf margin, where a strand of sclerenchyma is found. It is compelling that the signal is primarily localized to the lateral veins which are the focus of the *Kn1* mutation. *In situ* hybridization to knotted leaf blades gives a different pattern that is phenotype dependent, being more obvious when the phenotype is more severe. At the lateral vein, there is no strong signal at the sclerenchyma, only a diffuse signal over the entire vein. The absence of the signal at the sclerenchyma of the lateral veins may be due to these cells not maturing properly in knotted plants. What is very striking, is a new signal over the bundle sheath cells of the intermediate veins.

cDNA structure and sequence analysis

cDNA libraries were prepared from two populations of seedlings that were segregating for wildtype and *Kn1-O* phenotypes. The libraries were screened with the same probe that was successfully used in the *in situ* analysis. The *Kn1-O* and wildtype cDNAs appear identical. The structure of the almost full length cDNA is compared to genomic DNA in Figure 7. The coding region spans approximately 7.7 kb and includes a large intron of 4.9 kb. All three *Mu8* insertions, the *Ds2* insertion and the *Kn1-Z3* polymorphism are localized within a discrete region of this large intron. The structure of the cDNA is also informative in understanding how the tandem duplication of *Kn1-O* conditions the mutant phenotype. The junction of the tandem duplication lies very near to the 5' end of the cDNA, supporting our earlier interpretation of *Kn1-O* (Veit *et al.*, 1990) in which novel sequences brought in by the tandem duplication have altered the regulation of the wildtype gene, *kn1*. Analysis of the polypeptide encoded by the *Kn1* cDNA reveals a region that shares significant homology with homeodomain motifs contained in previously characterized regulatory proteins.

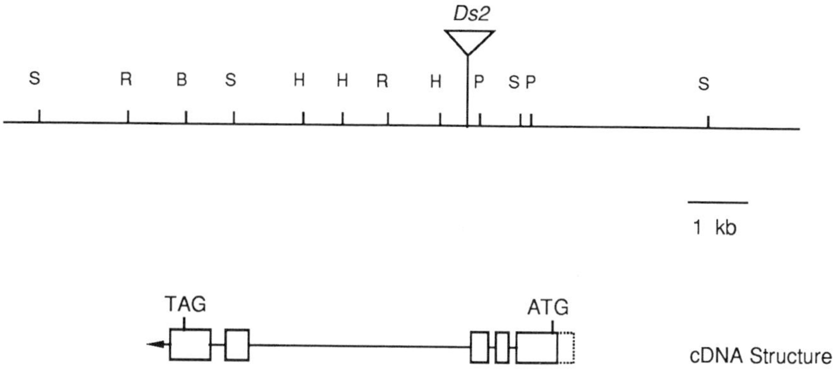


Fig. 7. The structure of the *kn1* cDNA compared to the corresponding genomic sequence of *Kn1-2F11*. Transcription is from right to left. The 5' end is not definite.

The strongest homology is to the yeast DNA binding proteins, such as *MATa2* and *MATPi* (Kelly *et al.*, 1988; Scott *et al.*, 1989). This homology suggests that *kn1* plays a role in gene regulation.

Because the insertions and rearrangements that we have mapped all interrupt non-coding sequence and do not appear to alter RNA size or quantity, we assume that they are altering the regulation of the wildtype transcript. We propose that regulatory sequences exist in this large intron, and similarly, regulatory sequences are present at the 5' end of the gene. The disruption of *kn1* regulation is revealed by altered localization; this in turn, produces a mutant phenotype. One model to explain the phenotype suggests that the wildtype gene, *kn1*, is involved in the maturation of the leaf blade, perhaps signalling cells when they are mature. In the blade, the signal begins at the lateral veins. When the *kn1* gene is expressed in the sclerenchyma cells of the lateral veins, it signals a time to stop dividing and differentiate into appropriate cell types. In the *Kn1* mutations, this signal is reduced in the sclerenchyma cells, and is now found in adjacent intermediate vein cells. Because the *kn1* signal is present in the intermediate veins, these cells are precociously mature. The relocalization of *kn1* signal prevents the proper differentiation of lateral vein cells, instead they divide to make knots or may take on new fates such as becoming ligule. The constraint placed upon the lateral vein cells by the surrounding mature intermediates restricts the lesions to the lateral veins. Thus, it appears to be the interactions of cells that lead to the knotted phenotype. This conclusion is supported by our clonal analysis (Sinha and Hake, in press) in which half a bundle sheath and the adjacent intermediates were the minimal unit required to carry the *Kn1* locus in order to produce knots.

## Acknowledgments

We thank Julie Mathern for help in preparation of the figures and Ron Wells for preparing, editing, and assembling the camera-ready manuscript.

## References

Brown, W. V. 1975. Variations in anatomy, associations and origins of Kranz tissue. *Amer. J. Bot.*, **62**:395-402.

Dengler, N. G., Dengler, R. E. and Hattersley, P. W. 1985. Differing ontogenetic origins of PCR (Kranz) sheaths in leaf blades of C4 grasses (*Poaceae*). *Amer. J. Bot.*, **72**:284-302.

Esau, K. 1953. *Plant Anatomy*, p.195, Wiley, New York.

Freeling, M. and Hake, S. 1985. Developmental genetics of mutants that specify *Knotted* leaves in maize. *Genetics*, **111**:617-634.

Hake, S. and Freeling, M. 1986. Analysis of genetic mosaics shows that the extra epidermal cell divisions in *Knotted* mutant maize plants are induced by adjacent mesophyll cells. *Nature*, **320**:621-623.

Hake, S., Vollbrecht, E. and Freeling, M. 1989. Cloning *Knotted*, the dominant morphological mutant in maize using *Ds2* as a transposon tag. *EMBO J.*, **8**:15-22.

Kelly, M., Burke, J., Smith, M., Klar, A. and Beach, D. 1988. Four mating-type genes control sexual differentiation in the fission yeast. *EMBO J.*, **7**:1537-1547.

Langdale, J. A., Lane, B., Freeling, M. and Nelson, T. 1989. Cell lineage analysis of maize bundle sheath and mesophyll cells. *Develop. Biol.*, 133:128-139.

Langdale, J. A., Zelitch, I., Miller, E. and Nelson, T. 1988. Cell position and light influence C4 versus C3 patterns of photosynthetic gene expression in maize. *EMBO J.*, **7**:3643-3651.

Mathern, J. and Hake, S. 1989. Location of *Adh1* and *Kn1* on the maize linkage map. *Maize Gen. Coop. Newsletter*, **63**:2. (Cited by permission.)

Poethig, R. S. 1984. Cellular parameters of leaf morphogenesis in maize and tobacco. *In: Contemporary Problems in Plant Anatomy*, R. A. White and W. C. Dickison, eds., pp. 235-258. New York: Academic Press.

Poethig, R. S. 1988. Heterochronic mutations affecting shoot development in maize. *Genetics*, **119**:959-973.

Scott, M. P., Tamkun, J. W. and Hartzell III, G. W. 1989. The structure and function of the homeodomain. *Biochim. Biophys. Acta*, **989**:25-48.

Sharman, B. C. 1942. Developmental anatomy of the shoot of *Zea mays* L. *Ann. Bot.*, **6**:245-281.

Sinha, N. and Hake, S. 1990. Mutant characters of *Knotted* maize leaves are determined in the innermost tissue layers. *Develop. Biol.*, in press.

Stewart, R. N. and Dermen, H. 1979. Ontogeny in monocotyledons as revealed by studies of the developmental anatomy of periclinal chloroplast chimeras. *Amer. J. Bot.*, **66**:47-58.

Veit, B., Vollbrecht, E., Mathern, J. and Hake, S. 1990. A tandem duplication causes the *Kn1-O* allele of *Knotted*, a dominant morphological mutant of maize. *Genetics*, in press.

# T-DNA INSERTION MUTAGENESIS IN *ARABIDOPSIS*: A PROCEDURE FOR UNRAVELLING PLANT DEVELOPMENT

Kenneth A. Feldmann, Anna M. Wierzbicki, Robert S. Reiter, and Shirley A. Coomber

E.I. DuPont de Nemours & Co., Plant Sciences
Central Research and Development, PO Box 80402
Wilmington, DE 19880-0402

## INTRODUCTION

With recent advances in molecular biology the study of plant development has gone from a purely descriptive approach to an experimental science. This has been brought on largely by the ability to tag genes with either transposons or T-DNAs and the utility of plant model systems such as *Arabidopsis*. The ability to use restriction fragment length polymorphism (RFLP) maps and "chromosome walking" techniques will, in the very near future, push the study of plant development forward more rapidly. Other techniques such as differential screening and the use of antisense RNA have proven, or will prove, useful when coupled with tagging procedures or known mutants.

In this review we will describe T-DNA tagging, a procedure which we have been employing to generate hundreds of developmental mutants in *Arabidopsis thaliana*. *Arabidopsis* has many features which make it ideal for gene tagging approaches. It is a small, rapid cycling, self-fertilizing, member of the Brassicaceae family with a very small genome (70,000 kb; Leutwiler et al., 1984; Meyerowitz and Pruitt, 1985). Numerous physiological and morphological mutants have been isolated and placed on a genetic map (McKelvie, 1962; Koornneef et al., 1983). In addition, a variety of transformation systems have been developed for this species (Lloyd et al., 1986; Feldmann and Marks, 1987; Valvekens et al., 1988; Damm and Willmitzer, 1989). With these numerous attributes *Arabidopsis* has become the plant model system of choice for plant molecular biologists.

In addition to the T-DNA tagging procedure, we will describe the types of mutants that we have isolated from the transformed population focusing on a number of developmental mutants which we are presently characterizing.

## *ARABIDOPSIS* SEED INFECTION

The seed infection process for transforming *Arabidopsis*, developed several years ago (Feldmann and Marks, 1987), was based on experiments which involved studying the uptake of exogenous DNA into imbibing *Arabidopsis* seeds (Ledoux et al. 1985). We were pursuing a non-tissue culture procedure for transformation with the aim of developing a system for insertion mutagenesis. We hypothesized that it would be necessary to avoid tissue culture because of the high frequency of somaclonal variants that result due to the conditions which are necessary to regenerate whole plants. Data presented at a recent *Arabidopsis* meeting confirmed this hypothesis (Valvekens and Van Montagu, 1990; Van Lisjebettens et al., 1990). Although regeneration procedures for *Arabidopsis* have been improved significantly (Feldmann and Marks, 1986; Valvekens et al., 1988); however, some modifications of the tissue culture process are still required before these rapid regeneration procedures, when coupled with transformation procedures, will decrease the amount of somaclonal variation to a level which makes it practical to generate and screen thousands of transformants for T-DNA induced mutations.

The seed infection procedure has been a very brute force approach and it has taken four years to generate the approximately 7,000 unique transformants we currently possess. A detailed review of all of the observations that have been made utilizing this procedure is in preparation and will be submitted later this year (1990). Briefly, the procedure is as follows: 3,000 wild-type seeds (T1) are imbibed in a nutrient solution for a predetermined length of time, usually around 12-15 hrs (depending on the seed lot) before the addition of 5 x $10^9$ *Agrobacterium* (Feldmann and Marks, 1987; Feldmann et al., 1990). The *Agrobacterium* is a C58C1Rif strain containing the cointegrate plasmid p3850:1003 (Velten and Schell, 1985). The neomycin phosphotransferase II gene, which confers kanamycin resistance ($Kan^R$) to the plant, is located between the T-DNA borders. The seeds and *Agrobacterium* are cocultivated for 24 hrs at which time the seeds are dried on filter paper and sprinkled onto vermiculite presoaked with nutrient solution. The infected T1 plants are grown to maturity and the T2 seeds (~300,000) are collected from each flat of approximately 1,000 T1 plants. The T2 seeds are plated onto agar-solidified medium containing kanamycin and each Kan resistant seedling is transferred separately to vermiculite and allowed to grow to maturity. Seeds (T3) from each selfed $Kan^R$ T2 plant are harvested and numbered consecutively according to the T2 parent.

Genetic and molecular data indicate that 1) 90-95% of the transformants segregate for one to four linked or unlinked functional inserts, with an average of 1.4 inserts per transformant, 2) the inserts are frequently concatamers of T-DNAs in direct and inverted repeats, 3) these inserts are stably inherited, and 4) every $Kan^R$ T2 plant we have isolated and tested is heterozygous for the insert(s) (Feldmann and Marks, 1989; Feldmann et al., 1990; M.D. Marks, unpublished data).

To saturate the genome with inserts (95% probability of an insert every 4 kb) we estimate we would need to generate 37,500 transformants. This assumes 1.4 inserts per transformant, random insertion, 15,000 genes (Leutwiler et al., 1984), and a genome size of 70,000 kb (Leutwiler et al., 1984). If there are 30,000 genes, given that other parameters remain constant, we would need a population of 75,000 transformants. We are in the process of testing the distribution of the T-DNA inserts on the five chromosomes by mapping 50-100 inserts. This will only show whether the inserts are at random on the five chromosomes but may not indicate whether there are localized hotspots in the genome. The work of Koncz et al. (1989) indicates that T-DNAs introduced via a tissue culture procedure preferentially integrate into transcribed regions of the genome; if this is the case for transformants generated by seed infection the number of transformants necessary to saturate the genome should decrease.

We have been unable to ascertain which tissue in the infected plant is being transformed but, from genetic and molecular data, we believe it occurs very late either in the development of the infected sporophytic plant or in the gametophytic generation. Observations that led to this conclusion are: 1) each $Kan^R$ T2 plant tested has been shown to be heterozygous for the insert, 2) screening of all transformants, from a single treatment, generates only one of any particular mutant, and 3) Southern analyses indicate that each transformant tested, from a single treatment, has a unique insertion pattern (in collaboration with M.D. Marks, University of Nebraska). The best hypothesis is that *Agrobacterium* get into interstitial spaces and are carried along as the plant grows. Somewhere during either gametogenesis, sporogenesis, or fertilization a germ line cell is transformed.

## SCREENING THE TRANSFORMED POPULATION FOR ALTERED PHENOTYPES

We are presently screening the transformed population, for visibly altered phenotypes, under two sets of conditions. First, we are screening the epicotyls by planting 30-50 seeds from each segregating line on MetroMix. The seedlings are scored shortly after germination, at the rosette stage, during bolting, late in flowering, at the onset of senescence, and again when the plants are completely senescent. We have observed numerous mutant phenotypes that generally fall into six classes (Table 1). We have screened 1,300 lines in this manner and will be screening thousands of lines in the fall and winter of 1990-91. In the second screen we are plating the lines on agar-solidified medium to screen for transformants with perturbations in root development (e.g., agraviotropy, root hair or lateral root defects), embryonic pattern mutants, seedling-lethals (described later), and photosynthetically incompetent mutants (in collaboration with Phil Benfey, Rockefeller University; Scott Poethig, Univ. of Pennsylvania; and Pablo Scolnik, E.I. Du Pont; respectively).

It is a relatively easy, although very laborious, task to perform the two screens described. Because of the large

Table 1. Types and Frequency of Altered Phenotypes Observed in Screens under Soil Conditions

| Phenotype | Percent of Total Transformants |
|---|---|
| seedling-lethals | 10% |
| size-variants | 7% |
| reduced fertility | 3% |
| embryo-lethals | 2.5%[a] |
| pigment mutants | 1% |
| dramatic mutants | 1%[b] |

[a]Corrected down from 5% following the results of David Meinke (Oklahoma State University) which showed that only half of the transformants that had been scored as embryo-lethals were real mutants.

[b]This class will increase as more careful screening and selection is conducted

number of transformants these screens will generate hundreds of mutants specific to the research interests of a diverse array of scientists. However, when one is interested in screening for a specific phenotype, such as resistance to any selective agent, it would not be practical to screen each line. Aliquoting seeds from each seed container would take many days and would soon deplete the seed lots, so for this type of screening we are pooling the transformed lines. Ten to twelve seeds from each of 100 lines are mixed together and grown to maturity under conditions which allow each of the plants to generate 1,000 seeds on average. These 1.2 million T4 seeds are harvested and used as a seed source for screening on selective agents. Preliminary screens have generated a chlorate resistance mutant, four ethylene affected mutants (Joe Ecker, Univ. of Pennsylvania), and two high florescence mutants. The number of mutants in this physiological class will increase as the population is tested on more agents. In the future the lines will be pooled by harvesting the T4 seeds from T3 plants following the first screening procedure described above. This pooling should generate enough T4 seeds for more than 1,000 different selective screens, of 3,000 seeds each, to be conducted.

## MUTANT DESCRIPTIONS

### Seedling-lethals

In those lines scored as seedling-lethals (Table 1) 25% of the seeds germinate and the cotyledons open but no primary growth is observed. After a few weeks these seedlings turn brown and die. We assume many of these lines are due to perturbations in biosynthetic pathways such that when the cotyledonary reserves are used up the seedlings struggle to survive.

Size-variants

The lines scored as size variants are those in which 25% of the plants are reduced in stature and generally thin and spindly. They produce few seeds and struggle to survive. Again, we predict many of these lines have perturbations in metabolic pathways such that the plant can survive but can not grow at the wild-type level.

Embryo-lethals

Embryo-lethal phenotypes are observed when we collect the T3 seeds from the T2 plant or when we harvest the T4 seeds from the 30 or so T3 plants at the termination of screening. Generally, when we observe that approximately 20% of the seeds are smaller and darker, and therefore aborted, we score the line as an embryo-lethal. Since the plants homozygous for the mutation do not survive the plants scored should segregate 2/3 heterozygous for the mutant phenotype and 1/3 wildtype. David Meinke (Oklahoma State University, Stillwater, OK) has screened through more than 20 of the lines segregating for the embryo-lethal phenotype. Only half of the lines tested showed cosegregation of the T-DNA ($Kan^R$) with the mutant phenotype (David Meinke, personal communication).

Seedling-lethals, size-variants and embryo-lethals account for the bulk of the altered phenotypes that we have observed. Factors that explain this high frequency are the large number of genes involved in these pathways and the observation that these phenotypes can be generated by environmental influences. For example, we had scored 5% of the lines as being embryo-lethals following a screening of the seeds and not the plants. However, David Meinke has shown that half of these lines are not mutants. Large numbers of aborted seeds are generated when the plants are stressed by any number of conditions, e.g., heat, drought, or lodging. As stated above Dr. Meinke has shown that only half of the real mutants are due to a functional insert. Thus the lines scored as embryo-lethals (as is probably true for the lines scored as seedling-lethals and size-variants) are due to a combination of T-DNA insertion mutagenesis, spontaneous mutations, and environmental factors. The last three classes of mutant phenotypes, reduced fertility, pigment, and dramatic, are not produced by growth conditions and are the classes with which we have worked most extensively.

Reduced fertility

The reduced fertility phenotype is easy to score. The mutant plants are not generating as much seed as wild-type plants so they continue to flower later than their sibling wild-type plants. If the plants are grown in a uniform fashion these lines can be picked out rapidly approximately 6 weeks after germination. These phenotypes are due to a failure to make or shed pollen, failure of the filaments to elongate such that the pollen is shed onto the ovary, or defects in transmitting tissue or in the production of viable ovules (in collaboration with Chris Makaroff, Miami University, Oxford, OH).

Pigment mutants

Pigment mutants (albino or yellow-green) are also very easy to score. The phenotype of these mutants, on agar-solidified medium, ranges from seedlings which are transparent to seedlings which are yellow green and fertile. Three of these mutations cosegregate with the $Kan^R$ marker in the T-DNA (Shirley Coomber, E.I. DuPont, and Alejandra Mandel, Irapuato, Mexico). Complementation analysis is in progress.

Dramatic mutants

Dramatic mutants have phenotypes which suggest the perturbation will lead to the dissection of a developmental pathway (Table 2). Three of these mutants, glabrous, agamous and dwarf, have been studied extensively (Marks and Feldmann, 1989 and Herman and Marks, 1989; Yanofsky et al., 1990; Feldmann et al., 1989; and unpublished data, respectively). For all three mutants the plant DNA flanking the border ofthe T-DNA has been isolated and used as a probe to isolate the wild-type sequence. This latter piece of DNA has been transferred to *Agrobacterium* and used to transform the mutant

Table 2. A variety of developmental mutants isolated from a population of transformants generated by the seed infection procedure.

| Mutant name | Inheritance | Reference |
|---|---|---|
| *glabrous* | recessive | a |
| *agamous* | recessive | b |
| *dwarf* | recessive | c |
| *twisted dwarf* | recessive | d |
| *fertile dwarf* | recessive | d |
| *dwarf gigantica* | recessive | d |
| *rapid cycler* | - | d |
| *miniature* | recessive | d |
| *runt*-760 | recessive | d |
| *runt*-1767 | recessive | d |
| *dumpy* | - | d |
| *inverted* | recessive | d |
| rough | recessive | d |
| *late bolter* | recessive | d |
| *stretched* | semidominant | d |
| pistil mutant 1 | recessive | d |
| pistil mutant 2 | recessive | d |
| *degenerate flower* | recessive | d |

[a]Yanofsky, et al., (1990).
[b]Marks and Feldmann, (1989).
[c]Feldmann et al., (1989).
[d]Feldmann, unpublished results.

plant to complement the mutation. *Glabrous* (David Marks, personal communication) and *agamous* each code for a protein homologous to transcription factors (Marks: personal communication; Yanofsky et al., 1990, respectively).

In the remainder of this review we will describe some of the dramatic mutants that have a drastically reduced size and that we have or will be characterizing at the molecular level.

Dwarfs. Dwarfs are distinguished from other size mutants in that they have <25% of the wild-type height, a sturdy stem, and short (but wide) dark green leaves. We have isolated four dwarf mutants from the first 1,700 lines.

- *dwarf (dwf1)* is 5-8 cm in height at maturity (Figure 1A). Every organ we have examined is reduced in length including stem, leaves, roots, hypocotyl, filaments, sepals, anthers, petals, and pistil. This mutant also displays a loss of apical dominance as manifested by prolific branching from the base. The mutants also senesce as long as two months after the wildtype (Feldmann et al., 1989). Microscopic analysis of mutant and wild-type plants has shown that cells in the mutant are smaller (in collaboration with Rick Howard, E.I. DuPont).

Genetic analysis has revealed the T-DNA is tightly linked to the dwarf mutation. In more than 7,000 F2's examined we have not seen any recombination events between the dwarf allele and the $Kan^R$ marker (Feldmann et al., 1989). The plant DNA flanking the T-DNA has been recovered and used as a probe to isolate the wild-type sequence. This latter piece of DNA has been transformed into *Agrobacterium* and the constructed strain has been used to transform dwarf plants. The regeneration of wild-type plants indicates, at least preliminarily, that we have the wild-type piece of DNA which, when disrupted, causes a dwarf phenotype (unpublished results).

- *twisted dwarf* (*dwf2*) is comparable in size to *dwf1* but is distinguished by having a very twisted phenotype. The rosette leaves are more epinastic than *dwf1*, the peduncle twists back and forth in an irregular manner, the pedicels are twisted in the same manner, and the siliques have a spiral phenotype similar to the *angustifolia* phenotype (Koornneef et al., 1983).

- *fertile dwarf* (*dwf3*) is, again, similar in size to dwf1. The leaves of this dwarf are slightly longer and display no curling or bending (Figure 1B). The large seed output is what distinguishes *dwf3* from the other dwarfs. Cosegregation data show that this dwarf is not linked to a functional $Kan^R$ marker.

- *dwarf gigantica* (*dwf4*) does not fit the dwarf phenotype precisely. While it does have a very low stature (7-8 cm) at maturity and a thick stem (Figure 1C), the leaves are large, serrated, thick, and hard. In addition, unlike wildtype, the leaves senesce very rapidly (usually within 24 hrs) once the leaf starts to turn brown. Under our growth conditions wild-type leaves take several days to die completely once the margins start to senesce. The pedicels of *dwf4* are thick, shiny and slightly translucent. *dwf4* is sterile under selfing conditions or when wild-type pollen is used but *dwf4* pollen is viable when outcrossed to a wild-type plant.

All four of the dwarfs described were caused by nuclear recessive mutations. Pairwise crosses have shown that the four mutants fall into different complementation groups.

The remaining mutants described here all have a very small stature. Unlike dwarf mutants they generally have a spindly peduncle with short narrow rosette leaves.

*rapid cycler*. This is also a very small plant (<8cm). The shape of the leaves and the branching pattern is very similar to wildtype. *rapid cylcer* (*rca*) bolts only slightly earlier than the wildtype but what distinguishes it is that it senesces approximately two weeks earlier than the wildtype. We have not compared the seed production of the mutant to the mass of the plant but it appears to be similar to the wildtype.

*miniature*. As the name implies this mutant is a small plant being only 8-10 cm at maturity (Figure 1E). At the seedling stage the small leaves posses high levels of anthocyanins implying the plants are stressed. Little or no branching of the epicotyl is observed. A preliminary examination of the roots on *miniature* plants indicates that they are very short and and the internal structure of the root appears to be disorganized. Genetic analyses show the $Kan^R$ marker and the mutation are tightly linked. No recombination events between the two markers have been observed.

Runts. Two mutants, *runt*-760 and *runt*-1767, have a very small stature with altered root structure (Figure 2A). In these two mutants the roots remain short. In addition the roots are extremely hairy (Figure 2B). It appears that every epidermal cell differentiates into a trichoblast and makes a root hair; however, this has not been confirmed. The cotyledons on runt mutants bend abaxially compared to the horizontal position of wild-type cotyledons. Genetic analyses of the *runt*-760 mutation and the $Kan^R$ marker in the T-DNA indicate that they are tightly linked (unpublished results). We are conducting allelism tests with these two mutants.

Figure 1. A series of dramatic mutants isolated from the transformed population of *Arabidopsis*. A. One month old wild-type and *dwarf* plants (bar = 3 cm). B. Six week old *fertile dwarf* plant (bar = 1 cm). C. Four week old *dwarf gigantica* plant (bar = 1 cm). D. Five and a half week old wild-type and *rapid cycler* plants (bar = 8 cm). E. Four week old wild-type and *miniature* plants (bar = 2.5 cm).

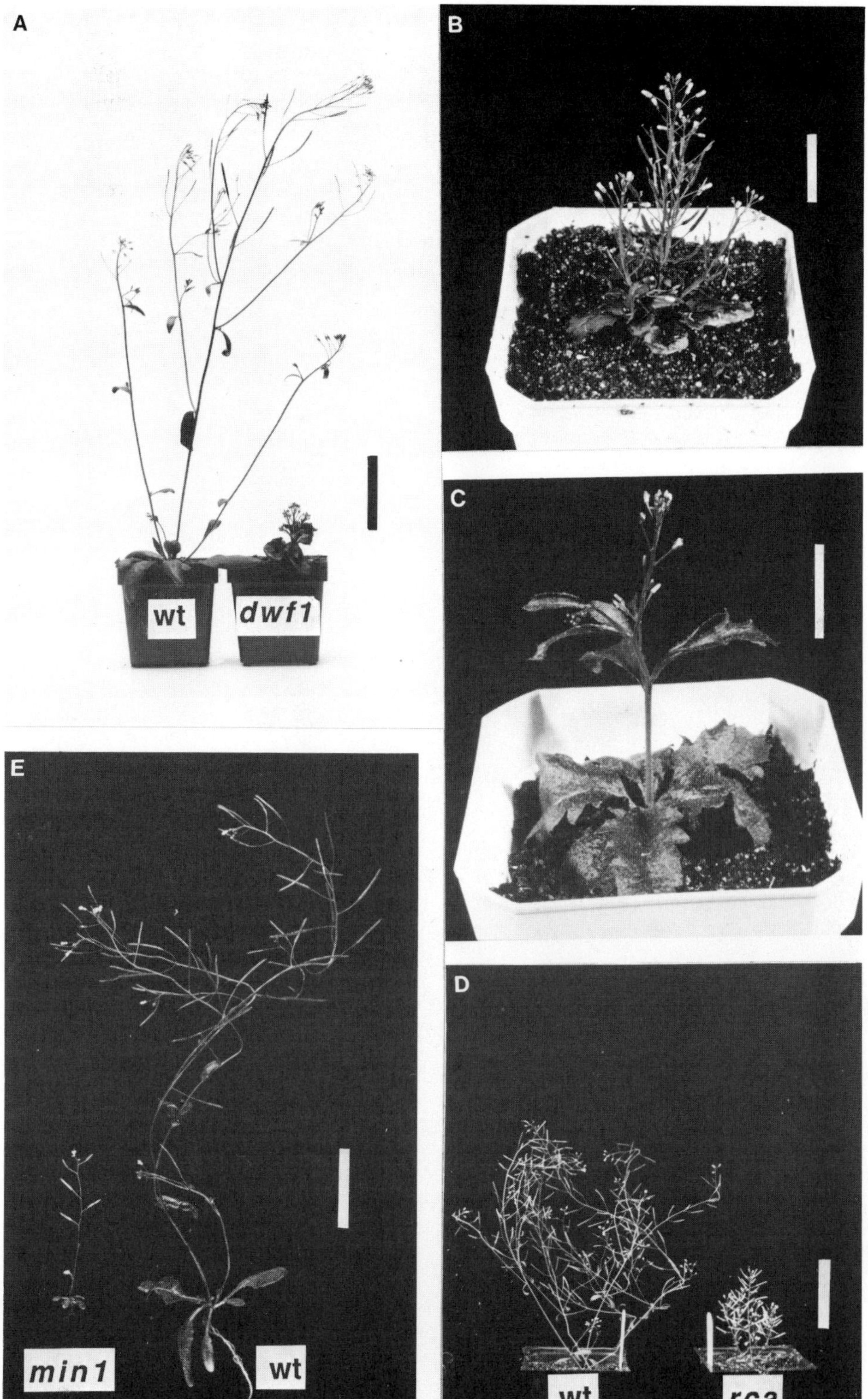
A
wt
dwf1
B
C
D
wt
rca
E
min1
wt

Other mutants. We have isolated numerous other mutants that appear to be affected in a developmental pathway (Table 2). For example, pistil mutants 1 and 2 are clavata-type mutants; the siliques are compsed of multiple carpels. Pistil mutant 1 differs from 2 in that 1 generates pistils possessing 2-5 carpels whereas 2 possesses pistils with two to three carpels. *dumpy* is an interesting mutant in that the cauline leaves, sepals, petals and even the anthers repeat the short round shape of the leaves. *rough* appears to be affected in wax depostion on the surface of the epidermis. The name comes from the roughness of the cauline leaves; they appear to have many more trichomes on these leaves than wildtype.

These dramatic mutants (Table 2) and hundreds of others that are coming out of the various screens that are ongoing in the lab will, if they turn out to be due to the T-DNA insert, be characterized in the near future.

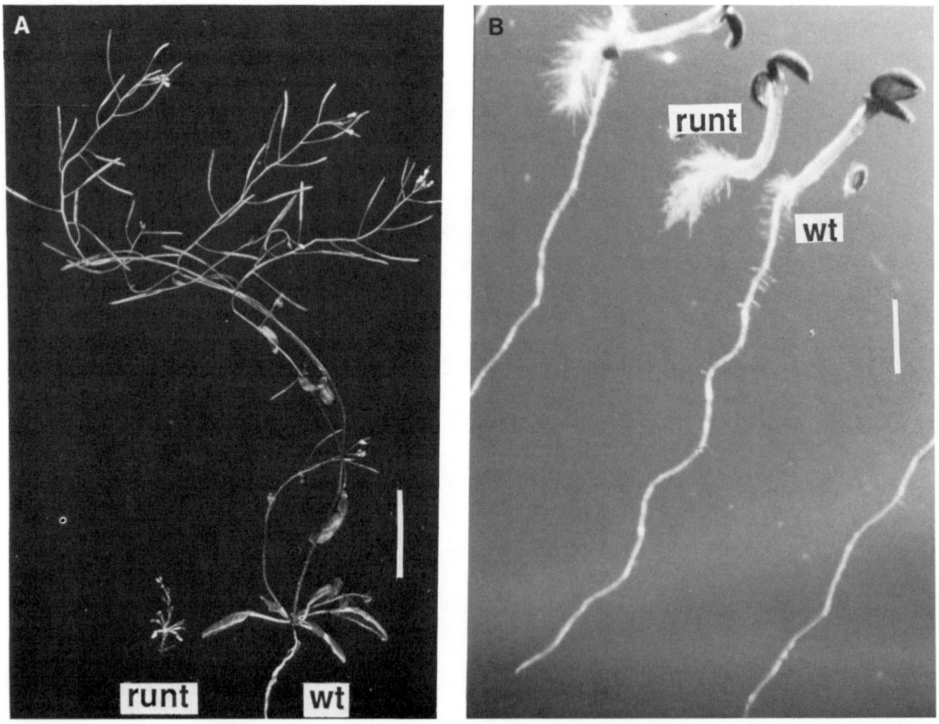


Figure 2. A runt mutant from the T-DNA tagged population of *Arabidopsis* transformants. A. Four week old *runt*-760 and wild-type plant (bar = 2.5 cm). B. Four day old *runt*-760 and wild-type seedlings grown in a vertical position on agar-solidified medium (bar = 0.2 cm).

## CONCLUSIONS

We have generated thousands of transformants using a non-tissue culture approach to transformation. We have also screened part of the population of transformants and isolated numerous developmental mutants. It has been shown genetically that many of the mutants cosegregate with a T-DNA insert. Molecular analyses have shown the T-DNA insertion disrupts a gene resulting in the mutant phenotype. For three of these mutants, *glabrous*, *agamous*, and *dwarf*, the wildtype gene has been cloned and transformed into the mutant plant to show complementation of the mutant phenotype. As we and others continue to screen the expanding population over the next several years hundreds of genes will be isolated in many developmental pathways. These genes and their products will be critical for unravelling various plant developmental pathways.

## REFERENCES


Damn, B., and Willmitzer, L., 1988, Regeneration of fertile plants from protoplasts of different *Arabidopsis thaliana* genotypes, Mol. Gen. Genet., 213:15.

Feldmann, K.A, and Marks, M.D., 1986, Rapid and efficient regeneration of plants from explants of *Arabidopsis thaliana*, Plant Science, 47:63.

Feldmann, K.A., and Marks, M.D., 1987, *Agrobacterium*-mediated transformation of germinating seed of *Arabidopsis thaliana*: A non-tissue culture approach, Mol. Gen. Genet., 108: 1-9 (1987).

Feldmann, K.A., Marks, M.D., Christianson, M.L., and Quatrano, R.S., 1989, A dwarf mutant of *Arabidopsis* generated by T-DNA insertion mutagenesis, Science, 243:1351.

Feldmann, K.A., Carlson, T.J., Coomber, S.A., Farrance, C.E., Mandel, M.A., and Wierzbicki, A.M., 1990, T-DNA insertional mutagenesis in *Arabidopsis thaliana*, in: "Horticultural Biotechnology", Bennett, A.B., and O'Neill, S.D., eds., Wiley-Liss, Inc.

Herman, P.L., and Marks, M.D., 1989, Trichome development in *Arabidopsis thaliana*. II. Isolation and complementation of the *GLABROUSI* gene, The Plant Cell, 1:1051.

Koncz, C., Martini, N., Mayerhofer, R., Koncz-Kalman, Zs., Korber, H., Redei, G.P., and Schell, J., 1989, High frequency T-DNA-mediated gene tagging in plants, Proc. Natl. Acad. Sci. USA, 86:8467.

Koornneef, M., van Eden, J., Hanhart, C.J., Stam, P., Braaksma, F.J., and Feenstra, W.J., 1983, Linkage map of *Arabidopsis thaliana*, J Hered., 74:265.

Ledoux, L., Diels, L., Thiry, M.E., Hooghe, R., Maluszynska, Y., Merckaert, C., Piron, J.M., Ryngaert, A.M., and Remy, J., 1985, Transfer of bacterial and human genes to germinating *Arabidopsis thaliana*, Arab. Inf. Serv., 22:1.

Leutwiler, L.S., Hough-Evans, B.R., and Meyerowitz, E.M., 1984, The DNA of *Arabidopsis thaliana*, Mol. Gen. Genet., 194:15.

Lloyd, A.M., Barnason, A.R., Rogers, S.G., Byrne, M.C., Fraley, R.T., and Horsch, R.B., 1986, Transformation of *Arabidopsis thaliana* with *Agrobacterium tumefaciens*, Science, 234:464.

Marks, M.D., and Feldmann, K.A., 1989, Trichome development in *Arabidopsis thaliana*. I. T-DNA tagging of the *GLABROUSI* gene, The Plant Cell, 1:1043.
McKelvie, A.D., 1962, A list of mutant genes in *Arabidopsis thaliana* (L.) Heynh, Radiation Botany, 1:233.
Meyerowitz E.M., and Pruitt R.E., 1985, *Arabidopsis thaliana* and plant molecular genetics, Science, 229:1214.
Van Lijsebettens, M., Vanderhaeghen, R., Scheres, B., and Van Montagu, M., 1990, Insertional mutagenesis in *Arabidopsis thaliana*: isolation of a T-DNA linked mutation that alters leaf morphology, in: "Fourth International Conference on *Arabidopsis* research", Schweizer, D., Peuker, K., and Loidl. J., eds., University of Vienna, Vienna, Austria.
Valvekens, D., Van Montagu, M., and Van Lusebettens, M., *Agrobacterium tumefaciens*-mediated transformation of *Arabidopsis thaliana* root explants by using kanamycin selection. Proc. Natl. Acad. Sci. USA, 85: 5536 (1988).
Valvekens, D., and Van Montagu, M., 1990, Spontaneous mutagenesis associated with *Arabidopsis* root regeneration. in: "Fourth International Conference on *Arabidopsis* research", Schweizer, D., Peuker, K., and Loidl, J., eds., University of Vienna, Vienna, Austria.
Velten, J., and Schell, J., 1985, Selection-expression plasmid vectors for use in genetic transformation of higher plants, Nucleic Acids Res., 13:6981.
Yanofsky, M.F., Ma, H., Bowman, J.L., Drews, G.N., Feldmann, K.A., and Meyerowitz, E.M., 1990, *Agamous:* an *Arabidopsis* homeotic gene whose product resembles transcription factors Nature, 346:35.

# DEFINING THE VACUOLAR TARGETING SIGNAL OF PHYTOHEMAGGLUTININ

Maarten J. Chrispeels, Craig D. Dickinson, Brian W. Tague, Dale C. Hunt and Antje von Schaewen

Department of Biology
University of California, San Diego
La Jolla, CA 92093-0116

## ABSTRACT

Correct targeting of vacuolar proteins depends on two targeting domains: the signal sequence that allows polypeptides to enter the secretory system and a vacuolar sorting signal that directs proteins to the vacuole. Absence of a vacuolar sorting signal results in secretion by a bulk-flow pathway. The plant vacuolar protein phytohemagglutinin (PHA) is targeted to the vacuoles of yeast cells when the gene for PHA is expressed in yeast. This system was used to demonstrate that a short domain between amino acids 14 and 43 is sufficient to target a passenger protein (invertase) to yeast vacuoles. However, the information in this domain is not necessary to target full length PHA to yeast vacuoles. Experiments with plant cells show that the same domain allows secretion of yeast invertase by *Arabidopsis* protoplasts indicating that in plants the sorting domain is of greater complexity than in yeast and may be a patch rather than a short linear sequence.

## Introduction

The secretory system of plant cells plays an important role in the delivery of proteins to specific cellular compartments and membranes especially the vacuole, the extracellular matrix (cell wall), the plasma membrane and the tonoplast. The biosynthesis of all the proteins located in these compartments starts on the ER when polysomes with nascent polypeptides become attached to ER membranes as a result of the interaction between the signal peptide, the signal recognition particle and the docking protein. The signal peptide functions as the first sorting signal that determines that these proteins will enter the secretory system rather than remain in the cytosol. Other sorting signals, also called targeting signals or domains, specify that proteins will be retained along the transport route (ER, Golgi complex, plasma membrane, tonoplast) or proceed to the vacuoles or be secreted. The identification of these sorting signals is presently the subject of intensive research in mammalian, yeast and plant cells. All living plant cells transport proteins in the secretory system and target some of these proteins to the vacuoles, to the extracellular space and to various membranes. Numerous proteins are delivered to the vacuoles or are secreted (see Table 1). Some cell types devote more than 50% of their protein synthetic capacity to the synthesis of protein that were transported by the secretory system (e.g. storage parenchyma cells in developing seeds or aleurone cells in growing seedlings).

## Secretion Occurs by a Bulk-Flow Mechanism

The available evidence obtained mostly with mammalian and yeast cells indicates that each step along the secretion pathway (ER to Golgi, movement between Golgi cisternae and

Plant Molecular Biology 2, Edited by R.G. Herrmann and
B. Larkins, Plenum Press, New York, 1991

**TABLE 1**

Types of Proteins Sorted in the Secretory System

| | |
|---|---|
| Secreted Proteins | |
| Structural cell wall proteins | |
| | extensin, glycine-rich protein |
| Hydroxyproline-rich glycoproteins | |
| | arabinogalactan protein, potato lectin |
| Hydrolytic enzymes | |
| | α-amylase, ribonuclease, proteinase, phosphatase |
| Plant-defense enzymes and proteins | |
| | ß-glucanase, chitinase, invertase, thionin |
| Cell wall degrading enzymes | |
| | cellulase, polygalacturonase, xylanase |
| Enzymes that modify cell wall components | |
| | peroxidase, laccase |
| Self-incompatibility proteins | |
| | S-locus specific glycoproteins |
| Vacuolar Proteins | |
| Seed storage proteins | |
| | vicilin, legumin, 2S albumins |
| Seed lectins | |
| | phytohemagglutinin, pea lectin, concanavalin A |
| Inhibitors of digestive enzymes | |
| | trypsin inhibitor, chymotrypsin inhibitor, α-amylase inhibitor |
| Vegetative storage proteins | |
| | soybean leaf VSP, sporamin, patatin, bark storage proteins |
| Vegetative lectins | |
| | Sambucus nigra bark lectin, wheat germ agglutinin |
| Plant defense enzymes | |
| | ß-glucanase, chitinase |
| Hydrolytic enzymes | |
| | proteinase, glycosidase, ribonuclease, phosphatase, invertase |

movement from the Golgi to the plasma membrane via secretory vesicles) is a default step; this means that a protein does not require a positive targeting signal to proceed along this pathway or its individual steps. The best evidence that secretion is the default pathway comes from the work of Wieland et al. (1987) who devised a bulk phase marker of the secretory pathway. They observed that when cells are incubated with the tripeptide N-acetyl-Asn-Tyr-Thr-$NH_2$ it enters the ER, is rapidly glycosylated and secreted by the cells. The underlying assumption is that such a tripeptide contains no targeting signals and that its transport therefore defines the default or bulk-flow pathway. Further evidence for the default pathway comes from experiments showing that bacterial or cytosolic proteins are secreted when engineered to contain a hydrophobic signal peptide. Such proteins are presumed to lack positive sorting domains. To demonstrate that in plant cells secretion also occurs by a default mechanism, Denecke et al. (1990) used the signal peptide of a pathogenesis-related protein to target three cytoplasmic enzymes to the lumen of the ER in tobacco cells. The three bacterial enzymes, phosphinotricin acetyltransferase (PAT), neomycin phosphotransferase II (NPTII) and ß-glucuronidase (GUS) were slowly secreted by the cells in which the chimeric genes had been introduced by electroporation. The secretion index (the ratio of enzyme activity outside the cells over activity inside the cells) differed for each

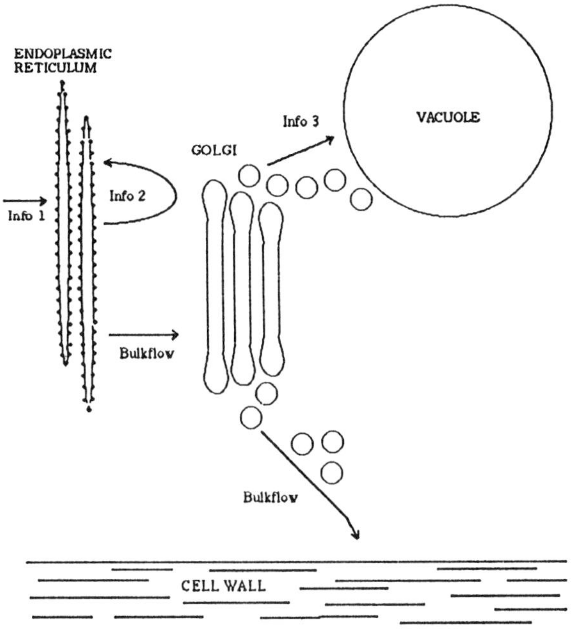


**Fig. 1.** Protein sorting in the secretory system of a plant cell. Info 1 = signal sequence; Info 2 = KDEL/HDEL; Info 3 = vacuolar sorting signal.

enzyme. After 10 h of incubation of the cells subsequent to electroporation, only 5% of the GUS activity, 20% of the NPTII activity and 50% of the PAT activity was outside the cells, indicating that secretion was relatively inefficient and that these proteins were being retained in the secretory system. Usually proteins are secreted within 30-60 min after they are synthesized, and after a 10 h incubation 90-95% of the protein should be outside the cells. We do not yet know why some proteins are efficiently transported through the secretory system whereas others are not. Correct protein folding, formation of oligomers and the presence of glycans, all of which affect the physico-chemical properties of a protein, play a role in efficiency of transport. When protein folding, N-linked glycosylation and oligomerization are blocked by mutations or drug treatment, malfolded proteins may fail to exit the ER.

To demonstrate that secretion is a default pathway while transport to the vacuole requires positive sorting information, we generated a chimeric protein consisting of the signal peptide the first 3 amino acids of phytohemagglutinin-L, and the entire coding sequence of PA2, a cytosolic pea cotyledon storage albumin (Higgins et al., 1987). The chimeric gene, called *phalb*, was expressed in tobacco plants with the seed specific phytohemagglutinin promoter, as well as in tobacco callus with the cauliflower mosaic virus 35 S promoter, and the subcellular location of the PHALB protein ascertained. With the phytohemagglutinin promoter, the gene was expressed only in the seeds of the transformed tobacco plants, but the PHALB protein was not targeted to the prominent protein storage vacuoles present in the seeds (Dorel et al., 1989). We were unable to conclude that PHALB had been secreted because of our inability to determine the subcellular location of the protein with immunocytochemistry. With the CaMV 35 S promoter, the PHALB protein was efficiently secreted by the callus cells. The immunoblot shown in Figure 1 compares the expression of the albumin gene driven by the CaMV 35 S promoter and of the *phalb* gene in NT cells (An, 1985). We compared the abundance of the protein in callus cells and protoplasts. The albumin protein (ALB) is equally abundant in cells and protoplasts (compare lanes 1 and 2) while the PHALB protein is present in callus, but absent from protoplasts (compare lanes 3 and 4). We infer from these results that PHALB is secreted and accumulates in the extracellular matrix/space of the callus. We were able to confirm this by collecting the intercellular washing fluid. This fluid has most of the PHALB associated with callus tissue (Hunt and Chrispeels, unpublished results). PHALB was glycosylated and some of the glycans were modified in the Golgi apparatus. Whether this glycosylation contributes to the efficient secretion of PHALB is not known at present.

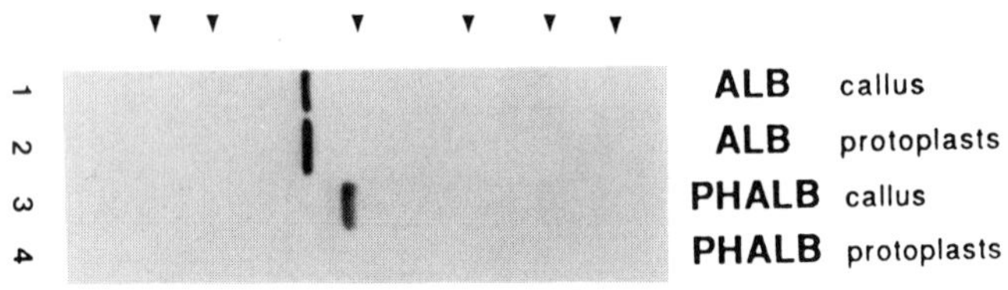


**Fig. 2.** Immunoblot showing the abundance of ALB and PHALB in tobacco callus and protoplasts derived from the callus.

Chimeric genes consisting of the CaMV 35S promoter and the coding sequence of pea albumin or pea albumin with the signal sequence of phytohemagglutinin were expressed in NT (*Nicotiana tabacum*) cells. The proteins, ALB and PHALB respectively were assayed by immunoblot. Equal amounts of protein were loaded in all the lanes.

## Glycans Do Not Contain Vacuolar Targeting Information

Many vacuolar and lysosomal proteins are glycoproteins, and glycans have been shown to play an important role in targeting of lysosomal enzymes in mammalian cells. In mammalian cells, certain high-mannose glycans on the lysosomal enzymes are modified in the Golgi apparatus by enzymes whose action results in the formation of mannose-6-phosphate groups. The targeting of these enzymes to the lysosomes is mediated by membrane-bound receptors that cycle between the trans-Golgi network and a prelysosomal compartment (Brown et al., 1986). The association/dissociation of the enzyme-receptor complex depends on a lowering of the pH along the transport pathway. Could a similar mechanism operate in plant cells? To investigate this question, we eliminated the glycan-attachment sites of PHA by site-directed mutagenesis (Voelker et al., 1989). The high-mannose glycan of PHA is attached to Asn-Glu-Thr$^{14}$, while the second complex glycan is attached to Asn-Thr-Thr$^{62}$. In one mutant, Thr$^{14}$ was changed to Ala$^{14}$, in a second mutant, Asn$^{60}$ was changed to Ser$^{60}$, and a third mutant incorporated both changes. The resulting mutant genes were expressed in tobacco using the ß-phaseolin promoter, and the fate of their protein products was examined in the seeds.

Analysis of the tobacco seeds showed that PHA with only one glycan or without any glycans was correctly targeted to the protein storage vacuoles (Voelker et al., 1989). We also found that the absence of either the complex glycan or the high-mannose glycan did not alter the processing of the other glycan. These results indicate that the targeting signal of this vacuolar protein should be contained in its polypeptide domain. The same approach yielded a similar result for the vacuolar protein wheat germ agglutinin (WGA) (Wilkins et al., 1990). WGA is synthesized as a glycosylated proprotein with a high mannose glycan close to the carboxyterminus. Processing of the protein results in the loss of the glycan and a carboxyterminal peptide. Site-directed mutagenesis resulting in the removal of the glycan attachment site yielded a mutant gene the product of which could not be glycosylated. When the mutant gene was introduced into tobacco, the unglycosylated protein was found in the secretory system and was correctly targeted to the vacuoles. These results confirm earlier work by Bollini et al. (1985) who showed that tunicamycin, an inhibitor of asparagine-linked glycosylation, does not inhibit the transport of unglycosylated PHA to the protein storage vacuoles of bean cotyledons. Furthermore, a number of vacuolar proteins, such as the 11 S globulins, do not have covalently attached glycans, indicating again that the targeting signal must be contained in the polypeptide domain.

## Analyzing a Plant Vacuolar Targeting Signal in Yeast Cells

As a first step in studying the vacuolar targeting signal of a plant protein at the molecular level, we wished to determine whether yeast (*Saccharomyces cerevisiae*) cells would recognize this signal and target PHA to the yeast vacuole. The gene for PHA-L was cloned into the yeast expression vector, pYE7, under control of the acid phosphatase (*PHO5*) promoter. Deletion of phosphate from the culture medium resulted in the expression of the PHA-L gene and the accumulation of PHA-L in the cells up to 0.1% of total yeast protein. PHA-L in yeast was glycosylated, as expected for a protein that enters the secretory system, and all the glycans were of the high-mannose type as expected for a yeast glycoprotein. Cell fractionation studies showed that nearly all the PHA-L accumulated in vacuoles (Tague and Chrispeels, 1987). The next series of experiments was aimed at determining if the vacuolar targeting signal of PHA is contained within a linear domain of the polypeptide. We carried out unidirectional 3' deletions of the PHA-L gene and fused these deletions to a truncated yeast invertase gene that starts at the third amino acid of the mature protein. After transformation of yeast with these chimeric constructs, we selected translational fusions on the basis of their invertase activity and determined the subcellular distribution of invertase as well as its glycosylation status. A similar approach was used by Johnson et al. (1987) to

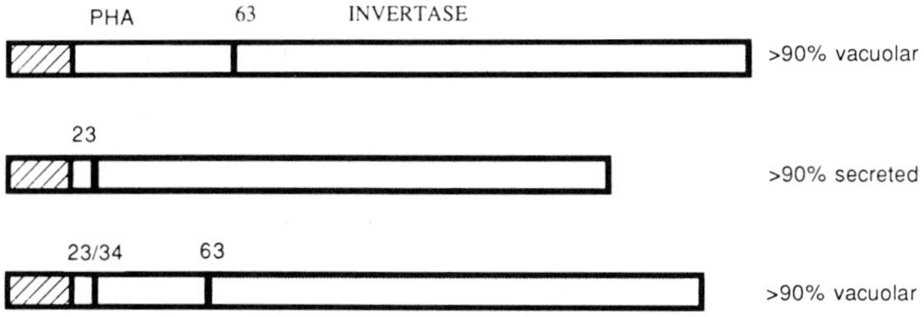


**Fig. 3.** Delineating the vacuolar targeting domain of phytohemagglutinin in yeast. Translational fusions of 3' deletions of the PHA-L gene with the yeast invertase gene were introduced in yeast and the location of invertase ascertained. A vacuolar targeting domain is located between amino acids 14 and 43 of the mature protein. (The signal peptide of PHA-L consists of 20 amino acids.)

| | | | |
|---|---|---|---|
| | | 1 (mature) | |
| **PHA-L** | | SNDIYFNFQRF--NETNLI | **LQRD**ASV-SSSGQ... |
| **PHA-E** | | ASQTSFSFQRF--NETNLI | **LQRD**ATV-SSKGQ... |
| **Soybean agglutinin** | | AETVSFSWNKFVPKQPNMI | **LQGD**AIVTSSG-K... |
| **Concanavalin A** | STHE | TNALHFMFNQFSKDQKDLI | **LQGD**A-TTGTEGN... |
| **Pea lectin** | | TETTSFLITKFSPDQQNLI | **FQGD**GYTT-KE-K... |
| **Favin** | T | DEITSFSIPKFRPDQPNLI | **FQGG**GYTT-KE-K... |
| | | (N-proximal) | |
| **Phaseolin** | | ...SDNSWNTLFKNQYGHIRV | **LQRFD**QQSKRLQNL... |
| | | (C-proximal) | |
| **Pea legumin** | | ...TSSVINDLPLDVVAATFK | **LQRDE**ARQLKSNN... |
| | | (C-proximal) | |
| **Ricin** | | ...LSTAIQESNQGARASPIQ | **LQRDG**SKFSVYDV... |
| **Carboxypeptidase y (yeast)** | | 1 (pro) | |
| | | IS | **LQRPL**GLDKDVLL... |

**Fig. 4.** Comparison of amino acid sequences of plant vacuolar proteins (lectins and storage proteins) and carboxypeptidase Y (yeast). A conserved motif is found in all these proteins. This motif corresponds to the vacuolar sorting signal of carboxypeptidase Y.

define the vacuolar targeting domain of a yeast carboxypeptidase Y. Our data indicate that an amino terminal portion of PHA comprised of a 20 amino acid signal sequence and 43 amino acids of the mature protein is sufficient to target invertase to the yeast vacuole. The vacuolar targeting domain appears to be located between amino acids 14 and 43 of the mature protein. To arrive at this conclusion, we made and tested a large number of constructs, three of which are shown in Figure 3.

A comparison of the amino acid sequences of legume storage proteins and lectins, and of yeast carboxypeptidase Y shows a conserved motif, which in PHA has the sequence LQRD (Fig. 4). The glutamine residue is completely conserved in all legume lectins sequenced to date. The functionality of this sequence has been tested for both carboxypeptidase Y and PHA. A single amino acid change from $L_3QRP$ in normal carboxypeptidase Y (targeted to the yeast vacuole) to $L_3KRP$ resulted in complete mistargeting (secretion) of this vacuolar enzyme in yeast cells (Valls et al., 1987). A three amino acid change from LQRD to LEGN in a portion of PHA that is 43 aa long (in addition to the signal peptide) and used in chimeric constructs with yeast invertase resulted in the secretion of invertase from the cells rather than its transport to the vacuoles (Fig. 5). Thus, the sequence conservation that we observe is in a region that is functionally important for proper targeting in yeast cells. An examination of the sequence of PHA shows that the QRD to EGN mutation actually introduces a new glycosylation site because this tripeptide is followed by alanine-serine causing the conversion of DAS to NAS. A comparison of the electrophoretic mobilities of normal PHA and the EGN mutant shows that the glycosylation site is used in yeast. As a result, a glycan probably masks the targeting domain in the QRD region of the protein. The amino acids in this region are in the form of a loop which is predicted to be at the surface of the folded PHA molecule. This prediction is based on an examination of the predicted crystal structure of other lectins (pea lectin, favin, concanavalin A) that are homologous to PHA.

We tested the importance of the LQRD domain for sorting in another way. The region was mutagenized by random oligonucleotide mutagenesis in the PHA63-invertase fusion and transformed yeast clones were screened for invertase secretion followed by DNA sequencing of the mutated region. High levels (40-60%) of invertase secretion were observed for the following mutants: $L_{38}$ --> P, $L_{38}$ --> R, $Q_{39}$ --> P, $D_{41}$ --> $N_{41}$ (also a glycosylation mutant). All these experiments show that the LQRD motif plays an important role in the peptide domain near the aminoterminus of PHA which is sufficient to target invertase to the yeast vacuole.

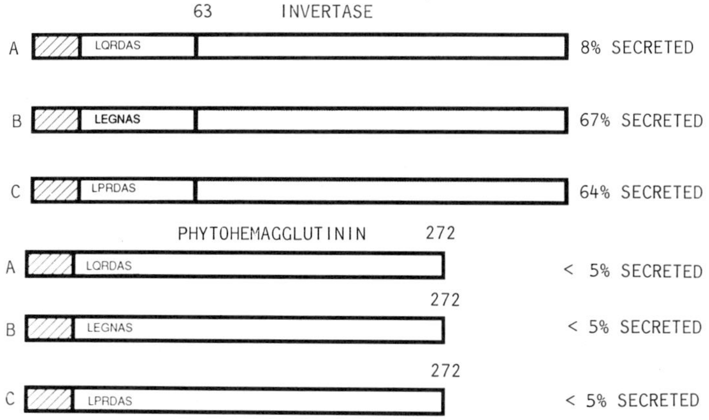


**Fig. 5.** The vacuolar targeting domain at the aminoterminus of PHA is sufficient but not necessary for targeting in yeast. Translational fusions were made between yeast invertase and the first 63 amino acids (including the signal peptide) of PHA with specific mutations. Two mutations (lines B and C) resulted in a high level of invertase secretion. When these same mutations were introduced into full length PHA (lines B' and C'), there was no increase in the secretion of the mutant PHA molecules.

A peptide domain which is sufficient to target a passenger protein to the vacuole (or any other organelle) may not be necessary, if another domain can assume the same function. Thus, a mutation in the first domain that abolishes its function may not be detrimental to correct vacuolar targeting when the same mutation is introduced in the full-length protein. Is the amino-terminal domain of PHA necessary for vacuolar targeting of PHA in yeast? To test this, we introduced the $QRD_{41}$ --> $EGN_{41}$ mutation and the $D_{41}$ --> $N_{41}$ into the sequence of full-length PHA-L and introduced the gene into yeast cells. Most of the PHA synthesized from these two mutant genes remained within the cells and was, presumably, transported to the vacuoles although these same mutations in the PHA63-invertase fusions allowed invertase to be secreted. This means that the LQRD tetrapeptide is not strictly required for proper targeting of PHA, and we infer from these results that another polypeptide domain may be involved in sorting to the yeast vacuole.

## Attempts to Define a Vacuolar Targeting Signal in Plants

To determine the vacuolar targeting signal of phytohemagglutinin in plants, we introduced the same PHA-invertase fusions into *Arabidopsis thaliana* leaf protoplasts and measured the expression of invertase and the extent of its secretion. To do this, we had to switch the 5' and 3' regions and use plant-specific sequences: the CaMV 35S promoter at the 5' end and the octopine synthase poly-A sequences at the 3' end. The constructs were introduced into leaf protoplasts with polyethylene glycol (Negrutiu et al., 1987) and the protoplasts cultivated for 72 h. This period of cultivation is sufficient for regeneration of the cell wall. The amounts of intracellular and extracellular invertase were estimated semiquantitatively after electrophoretic separation of the proteins. The results indicate that an N-terminal domain of 73 amino acids of PHA still allows for the secretion of invertase by the plant cells. The same domain brings about vacuolar targeting in yeast. This means that vacuolar sorting in plants relies on more information than is required for yeast where the information is contained in a relatively short domain near the N-terminus of PHA. Whether this information domain can be defined by a deletion analysis or rather may be dependent on the three-dimensional structure of the protein remains to be demonstrated.

## REFERENCES


An, G., 1985, High efficiency transformation of cultured tobacco cells, Plant Physiol., 79:568-570.

Bollini, R., Ceriotti, A., Daminati, M.G. and Vitale, A., 1985, Glycosylation is not needed for the intracellular transport of phytohemagglutinin in developing *Phaseolus vulgaris* cotyledons and for the maintenance of its biological activities, Physiol. Plant., 65:15-22.

Brown, W.J., Goodhouse, J. and Farquhar, M.G., 1986, Mannose 6-phosphate receptors for lysosomal enzymes cycle between the Golgi complex and endosomes., J. Cell Biol., 103:1235-1247.

Dorel, C., Voelker, T.A., Herman, E.M. and Chrispeels, M.J., 1989, Transport of proteins to the plant vacuole is not by bulk flow through the secretory system, and requires positive sorting information, J. Cell Biol., 108:327-337.

Higgins, T.J.V., Beach, L.R., Spencer, D., Chandler, P., Randall, P.S., Blagrove, R.J., Kortt, A.A. and Guthrie, R.E., 1987, cDNA and protein sequence of a major pea seed albumin (PA2: *Mr* 26,000)., Plant Mol. Biol., 8:37-47.

Johnson, L.M., Bankaitis, V.A. and Emr, S.D., 1987, Distinct sequence determinants direct intracellular sorting and modification of a yeast vacuolar protease, Cell, 48:875-885.

Negrutiu, T., Shillito, R., Potrykus, I., Biasini, G., and Sala, F., 1987, Hybrid genes in the analysis of transformation conditions. I. Setting up a simple method for direct gene transfer in plant protoplasts, Plant Mol. Biol., 8:363-373.

Tague, B.W. and Chrispeels, M.J., 1987, The plant vacuolar protein, phytohemagglutinin, is transported to the vacuole of transgenic yeast, J. Cell Biol., 105:1971-1979.

Voelker, T.A., Herman, E.M. and Chrispeels, M.J., 1989, In vitro mutated phytohemagglutinin genes expressed in tobacco seeds: role of glycans in protein targeting and stability, Plant Cell, 1:95-104.

Wieland, F.T., Gleason, M.L., Serafini, T.A. and Rothman, J.E., 1987, The rate of bulk flow from the endoplasmic reticulum to the cell surface, Cell, 50:289-300.

Wilkins, T.A., Bednarek, S.Y. and Raikhel, N.V., 1990, Role of propeptide glycan in post-translational processing and transport of barley lectin to vacuoles in transgenic tobacco, Plant Cell, 2.

# SIGNALS FOR PROTEIN IMPORT INTO ORGANELLES

Gunnar von Heijne
Department of Molecular Biology
Karolinska Institute Center for Biotechnology
NOVUM
S-141 52 Huddinge
Sweden

## Introduction

The intracellular routing of protein molecules from their site of synthesis to their final intra- or extracellular destinations has been studied intensely for at least two decades. Cell biologists, biochemists, molecular biologists, and even biophysicists have all contributed to our present insights into the mechanisms and signals that have evolved to control protein sorting. Although the focus of this book is on the molecular biology of plant cells, knowledge gleaned form the study of protein sorting in other kinds of cells is highly relevant, since the sorting mechanisms seem to be conserved both in outline and detail throughout the living world.

In this chapter, I will concentrate on the design and physicochemical properties of three classes of targeting peptides (TPs) that are relevant for plant cells: secretory signal peptides (SPs), mitochondrial targeting peptides (mTPs), and chloroplast transit peptides (cTPs). I will also discuss how combinations of these signals serve not only in the primary targeting event to the correct organelle, but how in addition they guide the intra-organellar routing of certain proteins.

## Protein sorting: a biochemical view

Since the biochemistry of protein sorting is dealt with at length in other chapters of this book, I will limit myself to those aspects where TPs have been implied to play a direct role.

There are at least three stages during the transport of a protein from the cytoplasm to its final location where TPs are known to be important: in

*Plant Molecular Biology 2*, Edited by R.G. Herrmann and
B. Larkins, Plenum Press, New York, 1991

keeping the nascent protein in a translocation-competent conformation prior to the membrane-translocation event, in the initial interaction with the components of the translocation machinery, and at the step where the targeting peptide is proteolytically removed after having carried out its job. The effect on protein folding has so far not been correlated with any particular part of a TP, but the two latter roles can in general be mapped to distinct domains in all three classes of TPs discussed here.

A retardation in the folding of a pre-protein has been observed for the periplasmic protein MalE of *Escherichia coli* (Liu, et al., 1988; Liu, et al., 1989; Park, et al., 1988; Randall and Hardy, 1986). In this case, the mature protein refolds much faster than the pre-protein, and it has also been suggested that a partially defective MalE SP can be rendered more efficient by mutations in the mature part of the protein that retard folding, thus leaving more time for the pre-protein in its translocation-competent state to interact productively with the secretory machinery (Cover, et al., 1987). So far, no similar studies showing an effect of mTPs or cTPs on the folding kinetics of precursor proteins have been reported.

The importance of the TP in the initial interactions with the translocation machinery has been well documented in a number of systems. In *E. coli*, an SP, when added together with unfolded, mature protein, has been shown to increase the ATPase activity of the SecA protein in a specific manner (Cunningham and Wickner, 1989; Lill, et al., 1990). In eukaryotes, both the signal recognition particle (SRP) (Wiedmann, et al., 1987a) and the signal sequence receptor protein (Wiedmann, et al., 1989; Wiedmann, et al., 1987b) have been implicated in early interactions with SPs. In the mitochondrial system, a number of putative receptors for preproteins have been identified, but it is not yet clear if they interact directly with the mTP or with other parts of the chain (Pfaller and Neupert, 1987; Pfaller, et al., 1989; Sollner, et al., 1989). Similarly, in chloroplasts candidates for a cTP receptor have been reported (Cornwell and Keegstra, 1987; Pain, et al., 1988), but no definitive proof that they bind productively to cTPs has been forthcoming.

The processing enzymes involved in TP cleavage have been more thoroughly characterized. In *E. coli* two distinct enzymes cleave SPs. One, called signal peptidase I or leader peptidase (Lep) cleaves all the "normal" SPs (Wolfe, et al., 1983a; Wolfe, et al., 1983b), and signal peptidase II cleaves a small number of lipoprotein SPs (Tokunaga, et al., 1985). In eukaryotes, only one signal peptidase (a multi-subunit complex) has been identified is (Greenburg, et al., 1989; YaDeau and Blobel, 1989).

mTPs are cleaved by at least two distinct proteases located in the matrix of the organelle. The first enzyme to be characterized is composed of two subunits. In *Neurospora*, these subunits have been called the matrix processing protease (MPP) and the processing enhancing protein (PEP) (Hawlitschek, et al., 1988); in yeast, the MAS1 protein has been shown to correspond to the MPP (check) (Witte, et al., 1988). Rat mitochondria have been shown to contain two distinct processing activities, with the so-called protease I being equivalent to the MPP+PEP of *Neurospora*. The second activity, protease II, removes a short octa- or nona-peptide from a number of mTPs that have initially been processed by protease I (Kalousek, et al., 1988). It is likely that yeast and *Neurospora* also contain enzymes with protease II-activity, although no such proteins have yet been identified.

In chloroplasts, a stromal enzyme that can cleave cTPs was partially purified already in 1984 (Robinson and Ellis, 1984). In addition, the thylakoid membrane contains a distinct protease that cleaves presequences on proteins imported from the stroma into the thylakoid (Hageman, et al., 1986); the reaction specificity of this enzyme is remarkably similar to that of Lep and of eukaryotic signal peptidase (Halpin, et al., 1989).

**The structure of secretory signal peptides**

Typical signal peptides from prokaryotic as well as eukaryotic proteins have a three-domain structure with a positively charged N-terminal region (n-region), a central hydrophobic region (h-region) and a polar C-terminal region (c-region) that specifies the site of cleavage between the SP and the mature protein, Fig.1. Beyond their overall charged or hydrophobic character, the n- and h-regions display little if any sequence conservation. The c-region is somewhat more constrained in terms of positional amino acid preferences, with positions -3 and -1 being most conserved: here only small, neutral amino acids are allowed - the so-called "(-3,-1)-rule" (von Heijne, 1983; von Heijne, 1986c). In addition, the border between the h- and c-regions is often marked by one or more turn-promoting Pro, Gly, or Ser residues. The low degree of sequence conservation in SPs was dramatically underlined by the observation that some 10-20% of random sequences cloned in front of a normally secreted reporter molecule lacking its own SP could restore measurable levels of secretion; the active sequences selected in this way turned out to be reminiscent of real SPs in terms of charge and hydrophobicity (Kaiser, et al., 1987).

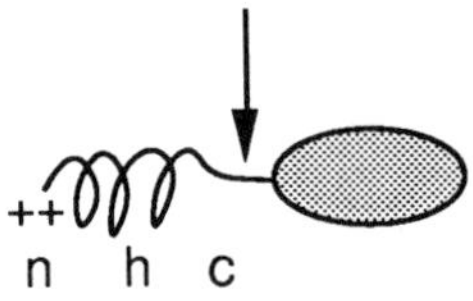


*Fig.1 Signal peptide*

Non-conservative mutations or deletions in the n- and h-regions are generally detrimental to the function of SPs (von Heijne, 1990). This suggests that both the positively charged residues in the n-region as well as the hydrophobic amino acids in the h-region are involved in the initial interactions with the secretory machinery. It has long been hypothesized that these two regions might serve respectively to anchor the SP to the negatively charged phospholipid headgroups on the membrane surface (n-region) and to provide a driving force for the insertion of the SP into the hydrophobic core of the bilayer (h-region). Recent biophysical studies on model peptides in phospholipid mono- and bilayer systems have provided some support for this notion (Batenburg, et al., 1988a; Batenburg, et al., 1988b; Briggs, et al., 1986; Bruch, et al., 1989; Cornell, et al., 1989; Killian, et al., 1990; Mcknight, et al., 1989).

The c-region has been extensively studied by site-specific mutagenesis, and the (-3,-1)-rule has been demonstrated to hold well for both prokaryotic and eukaryotic SPs (Fikes, et al., 1990; Folz, et al., 1988). Thus, cleavage can easily be blocked by the introduction of a residue that violates the (-3,-1)-rule. Conversely, a "cleavage cassette" constructed on the basis of the statistical amino acid preferences in *E. coli* SPs has been shown to serve as a good substrate for Lep and to be cleaved at the intended site (Nilsson and von Heijne, 1990).

Finally, a tendency for a lack of positively charged residues in the region immediately C-terminal to the signal peptide cleavage site (*i.e.* the first 5-10 amino acids of the mature protein) was first detected by statistical studies of known SPs from bacteria (von Heijne, 1986b). More recently, positively charged residues placed in this region have been shown to greatly reduce or even abolish membrane translocation (Li, et al., 1988; Summers, et al., 1989; Summers and Knowles, 1989; Yamane and Mizushima, 1988). As no similar lack of positively charged residues has been observed for eukaryotic SPs, it is not clear if their effect on secretion can be generalized to all SPs.

## Mitochondrial targeting peptides

mTPs are also built from mainly hydrophobic and positively charged amino acids, but their basic design is quite different from that of SPs. Rather than being "linear" amphiphilic structures with one hydrophilic and one hydrophobic end, the polar and apolar residues are strung together in such a way that they can form amphiphilic α-helices with the charged and uncharged residues located on different sides of the helix, Fig.2. This design is suggested both by theoretical (von Heijne, 1986a), genetic (Bedwell, et al., 1989; Lemire, et al., 1989), and biophysical (Endo, et al., 1989; Roise, et al., 1986; Tamm, 1986) studies, although amphiphilic β-strands have also been proposed to be active as import signals (Roise, et al., 1988), but see (Gavel, et al., 1988) for an alternative interpretation.

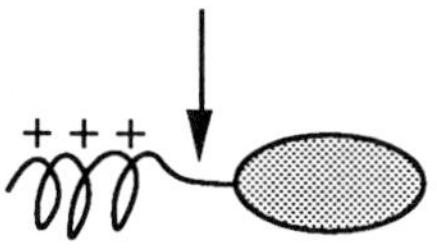


*Fig.2 Mitochondrial targeting peptide*

The one step- *vs.* two-step cleavages discussed above are clearly seen when the amino acid preferences in the region surrounding the cleavage site of known mTPs is analyzed (Hendrick, et al., 1989; von Heijne, et al., 1989). Thus, Arg is frequently found either in position -2 or in position -10 relative to the cleavage site, but only a small number of mTPs have Arg in both of these places. Apparently, something like 50% of all known mTPs are cleaved only once, and typically have an Arg-residue in position -2, whereas the remaining ones are first cleaved according to an $Arg_{-2}$ signal, and subsequently cleaved a second time eight (or in some cases nine) residues further downstream by a protease with a distinct substrate specificity. In addition, we have recently found a distinct subset of the single-step cleaved mTPs that is characterized by a highly conserved Arg-X-Tyr↓(Ser/Ala) pattern around the cleavage site (Gavel and von Heijne, 1990a).

## Chloroplast transit peptides

cTPs from higher plants have a clearly discernible three-domain structure with an uncharged but not very hydrophobic N-terminal region, a central region lacking in negatively charged residues, and a carboxy-terminal region that often has a high potential for forming an amphiphilic β-strand with alternating polar-apolar residues (von Heijne, et al., 1989).

Throughout the length of cTPs, hydroxylated Ser and Thr residues abound. Unfortunately, it is not clear what secondary structure, if any, cTPs are designed to adopt; indeed, their high content of hydroxylated residues and their variability in overall length almost makes one suspect that they may have been selected to *avoid* forming any strong secondary structure!

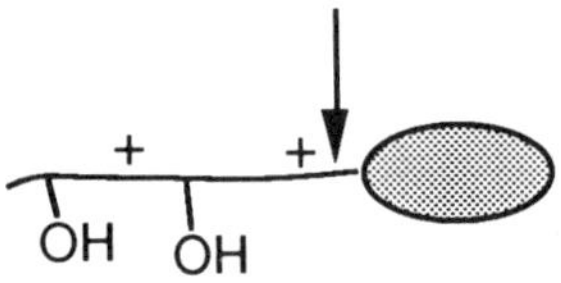


*Fig.3 Chloroplast transit peptide*

The cleavage sites in cTPs are also quite variable, although certain amino acid preferences have been noted lately (Gavel and von Heijne, 1990b). Thus, more or less good matches to a motif (Ile/Val)-X-(Ala/Cys)↓Ala can often be found, but in many cases very different cleavage sites are seen.

An interesting exception to the normal cTP design is observed for *Chlamydomonas*. The cTPs of this alga are much more similar to mTPs than to higher-plant cTPs (Franzen, et al., 1990), with a high content of positively charged arginines and a definite tendency to form highly amphiphilic α-helices. So far, no mTPs are known from *Chlamydomonas*, and it will be most interesting to analyze them once they become available.

**Intra-organellar targeting peptides**

The TPs discussed above serve to effect the initial routing to the correct organelle, and under their guidance the passenger protein is delivered either in the lumen of the ER, in the mitochondrial matrix, or in the stromal compartment of chloroplasts. Many proteins have to be further routed to various intra-organellar locations, such as the intermembrane space of mitochondria, the thylakoids, or to destinations along the secretory pathway. SPs apparently have no influence over sorting processes subsequent to ER-delivery and other signals must be invoked to explain these phenomena (Breitfeld, et al., 1989), but signals for intra-mitochondrial and intra-chloroplastic sorting have been found to be closely related to the TPs discussed above.

For mitochondrial inter-membrane space proteins, a "conservative" sorting mechanism which retains important elements from an ancient, bacterial mode of protein secretion in the organelle progenitor has been demonstrated (Hartl, et al., 1987). Here, a matrix targeting mTP is immediately followed by an SP-like sequence that is exposed once the mTP is removed in the matrix, initiating re-export across the inner membrane, Fig.4. A similar mechanism seems to be responsible for the import of nuclear-encoded proteins into thylakoids: a stroma-targeting cTP fused to an SP leads to an analogous two-step targeting process (Ko and Cashmore, 1989; Smeekens, et al., 1986).

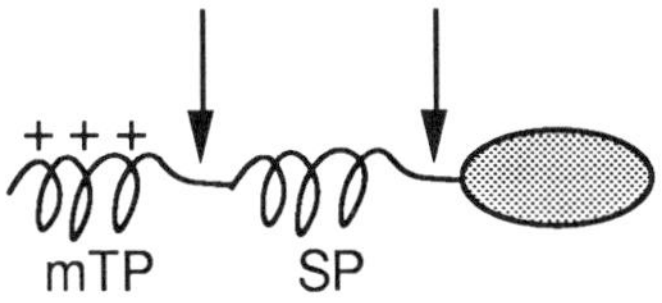


*Fig.4 Composite targeting peptide for intermembrane space routing*

Targeting to other chloroplast compartments is less well understood, although the TP for the phosphate translocator of the inner envelope membrane has been sequenced (Flügge, et al., 1989). It bears no obvious resemblance to stroma-targeting cTPs.

**Conclusion**

The theoretical and experimental study of the targeting peptides responsible for routing proteins to the ER, mitochondria, and chloroplasts has so far allowed the basic design of signal peptides and mitochondrial targeting peptides to be rather precisely defined; chloroplast transit peptides still have many questionmarks surrounding them. All three classes of TPs can be rather easily recognized in newly sequenced genes, and the most likely cleavage site can be reliably identified in signal peptides and (sometimes) in mitochondrial targeting peptides.

Among the major unsolved problems is the details of their functional interactions with other parts of the sorting machinery; indeed, it is not even known if the membrane-active properties of signal peptides and mitochondrial targeting peptides have any relevance to their *in vivo* behavior. A concerted effort from biochemists and biophysicists may help shed some light in this area.

**References**

Batenburg, A. M., Brasseur, R., Ruysschaert, J. M., van, S. G., Slotboom, A. J., Demel, R. A. , and de, K. B., 1988a, Characterization of the interfacial behavior and structure of the signal sequence of

Escherichia coli outer membrane pore protein PhoE, J Biol Chem, 263:4202.
Batenburg, A. M., Demel, R. A., Verkleij, A. J. , and de Kruijff, B., 1988b, Penetration of the signal sequence of Escherichia coli PhoE protein into phospholipid model membranes leads to lipid-specific changes in signal peptide structure and alterations of lipid organization, Biochemistry, 27:5678.
Bedwell, D. M., Strobel, S. A., Yun, K., Jongeward, G. D. , and Emr, S. D., 1989, Sequence and structural requirements of a mitochondrial protein import signal defined by saturation cassette mutagenesis, Mol Cell Biol, 9:1014.
Breitfeld, P. P., Casanova, J. E., Simister, N. E., Ross, S. A., McKinnon, W. C. , and mostov, K. E., 1989, Sorting signals, Curr Opinion Cell Biol, 1:617.
Briggs, M. S., Cornell, D. G., Dluhy, R. A. , and Gierasch, L. M., 1986, Conformations of signal peptides induced by lipids suggest initial steps in protein export, Science, 233:206.
Bruch, M. D., Mcknight, C. J. , and Gierasch, L. M., 1989, Helix Formation and Stability in a Signal Sequence, Biochem, 28:8554.
Cornell, D. G., Dluhy, R. A., Briggs, M. S., McKnight, C. J. , and Gierasch, L. M., 1989, Conformations and orientations of a signal peptide interacting with phospholipid monolayers, Biochemistry, 28:2789.
Cornwell, K. L. , and Keegstra, K., 1987, Evidence that a chloroplast surface protein is associated with a specific binding-site for the precursor to the small subunit of ribulose-1,5-bisphosphate-carboxylase, Plant Physiol, 85:780.
Cover, W. H., Ryan, J. P., Bassford, P. J., Walsh, K. A., Bollinger, J. , and Randall, L. L., 1987, Suppression of a signal sequence mutation by an amino acid substitution in the mature portion of the maltose-binding protein, J Bacteriol, 169:1794.
Cunningham, K. , and Wickner, W., 1989, Specific Recognition of the Leader Region of Precursor Proteins Is Required for the Activation of Translocation ATPase of Escherichia-Coli, P Nas Us, 86:8630.
Endo, T., Shimada, I., Roise, D. , and Inagaki, F., 1989, N-Terminal Half of a Mitochondrial Presequence Peptide Takes a Helical Conformation When Bound to Dodecylphosphocholine Micelles - A Proton Nuclear Magnetic Resonance Study, J Biochem, 106:396.
Fikes, J. D., Barkocy-Gallagher, G. A., Klapper, D. G. , and Bassford, P. J., 1990, Maturation of *Escerichia coli* maltose-binding protein by signal peptiodase I *in vivo*: Sequence requirements for efficient processing and demonstration of an alternate cleavage site, J Biol Chem, 265:3417.
Flügge, U. I., Fischer, K., Gross, A., Sebald, W., Lottspeich, F. , and Eckerskorn, C., 1989, The triose phosphate-3-phosphoglycerate - phosphate translocator from spinach chloroplasts: Nucleotide sequence of a full-length cDNA clone and import of the *in vitro* synthesized precursor protein into chloroplasts, EMBO J., 8:39.
Folz, R. J., Notwehr, S. F. , and Gordon, J. I., 1988, Substrate specificity of eukaryotic signal peptidase, J Biol Chem, 263:2070.
Franzen, L. G., Rochaix, J. D. , and von, H. G., 1990, Chloroplast Transit Peptides from the Green Alga Chlamydomonas-Reinhardtii Share Features with Both Mitochondrial and Higher Plant Chloroplast Presequences, Febs Letter, 260:165.

Gavel, Y., Nilsson, L. , and von, H. G., 1988, Mitochondrial targeting sequences. Why 'non-amphiphilic' peptides may still be amphiphilic, FEBS Lett, 235:173.

Gavel, Y. , and von Heijne, G., 1990a, Cleavage-site motifs in mitochondrial targeting peptides, submitted.

Gavel, Y. , and von Heijne, G., 1990b, A conserved cleavage-site motif in chloroplast transit peptides, FEBS Lett., 261:455.

Greenburg, G., Shelness, G. S. , and Blobel, G., 1989, A Subunit of Mammalian Signal Peptidase Is Homologous to Yeast Sec11 Protein, J Biol Chem, 264:15762.

Hageman, J., Robinson, C., Smeekens, S. , and Weisbeek, P., 1986, A thylakoid processing protease is required for complete maturation of the lumen protein plastocyanin, Nature, 324:567.

Halpin, C., Elderfield, P. D., James, H. E., Zimmermann, R., Dunbar, B. , and Robinson, C., 1989, The Reaction Specificities of the Thylakoidal Processing Peptidase and Escherichia-Coli Leader Peptidase Are Identical, EMBO J, 8:3917.

Hartl, F. U., Ostermann, J., Guiard, B. , and Neupert, W., 1987, Successive translocation into and out of the mitochondrial matrix: targeting of proteins to the intermembrane space by a bipartite signal peptide, Cell, 51:1027.

Hawlitschek, G., Schneider, H., Schmidt, B., Tropschug, M., Hartl, F. U. , and Neupert, W., 1988, Mitochondrial protein import: identification of processing peptidase and of PEP, a processing enhancing protein, Cell, 53:795.

Hendrick, J. P., Hodges, P. E. , and Rosenberg, L. E., 1989, Survey of amino-terminal proteolytic cleavage sites in mitochondrial precursor proteins: Leader peptides cleaved by two matrix proteases share a three-amino acid motif, Proc.Natl.Acad.Sci.USA, 86:4056.

Kaiser, C. A., Preuss, D., Grisafi, P. , and Botstein, D., 1987, Many random sequences functionally replace the secretion signal sequence of yeast invertase, Science, 235:312.

Kalousek, F., Hendrick, J. P. , and Rosenberg, L. E., 1988, Two mitochondrial matrix proteases act sequentially in the processing of mammalian matrix enzymes, Proc Natl Acad Sci USA, 85:7536.

Killian, J. A., de Jong, A. M. P., Bijvelt, J., Verkleij, A. J. , and de Kruijff, B., 1990, Induction of non-bilayer structures by functional signal peptides, EMBO J., 9:815.

Ko, K. , and Cashmore, A. R., 1989, Targeting of Proteins to the Thylakoid Lumen by the Bipartite Transit Peptide of the 33 Kd Oxygen-Evolving Protein, Embo J, 8:3187.

Lemire, B. D., Fankhauser, C., Baker, A. , and Schatz, G., 1989, The Mitochondrial Targeting Function of Randomly Generated Peptide Sequences Correlates with Predicted Helical Amphiphilicity, J Biol Chem, 264:20206.

Li, P., Beckwith, J. , and Inouye, H., 1988, Alteration of the amino terminus of the mature sequence of a periplasmic protein can severely affect protein export in Escherichia coli, Proc Natl Acad Sci U S A, 85:7685.

Lill, R., Dowhan, W. , and Wickner, W., 1990, The ATPase Activity of SecA Is Regulated by Acidic Phospholipids, SecY, and the Leader and Mature Domains of Precursor Proteins, Cell, 60:271.

Liu, G. P., Topping, T. B., Cover, W. H. , and Randall, L. L., 1988, Retardation of folding as a possible means of suppression of a mutation in the leader sequence of an exported protein, J Biol Chem, 263:14790.

Liu, G. P., Topping, T. B. , and Randall, L. L., 1989, Physiological Role During Export for the Retardation of Folding by the Leader Peptide of Maltose-Binding Protein, P Nas Us, 86:9213.

Mcknight, C. J., Briggs, M. S. , and Gierasch, L. M., 1989, Functional and Nonfunctional Lamb Signal Sequences Can Be Distinguished by Their Biophysical Properties, J Biol Chem, 264:17293.

Nilsson, I. , and von Heijne, G., 1990, A signal peptide cleavage cassette, Submitted.

Pain, D., Kanwar, Y. S. , and Blobel, G., 1988, Identification of a receptor for protein import into chloroplasts and its localization to envelope contact zones, Nature, 331:232.

Park, S., Liu, G., Topping, T. B., Cover, W. H. , and Randall, L. L., 1988, Modulation of folding pathways of exported proteins by the leader sequence, Science, 239:1033.

Pfaller, R. , and Neupert, W., 1987, High-affinity binding sites involved in the import of porin into mitochondria, Embo J, 6:2635.

Pfaller, R., Pfanner, N. , and Neupert, W., 1989, Mitochondrial protein import. Bypass of proteinaceous surface receptors can occur with low specificity and efficiency, J Biol Chem, 264:34.

Randall, L. L. , and Hardy, S. J., 1986, Correlation of competence for export with lack of tertiary structure of the mature species: a study in vivo of maltose-binding protein in E. coli, Cell, 46:921.

Robinson, C. , and Ellis, R. J., 1984, Transport of proteins into chloroplasts. Partial purification of a chloroplast protease involved in the processing of imported precursor polypeptides, Eur J Biochem, 142:337.

Roise, D., Horvath, S. J., Tomich, J. M., Richards, J. H. , and Schatz, G., 1986, A chemically synthesized pre-sequence of an imported mitochondrial protein can form an amphiphilic helix and perturb natural and artificial phospholipid bilayers, Embo J, 5:1327.

Roise, D., Theiler, F., Horvath, S. J., Tomich, J. M., Richards, J. H., Allison, D. S. , and Schatz, G., 1988, Amphiphilicity is essential for mitochondrial presequence function, Embo J, 7:649.

Smeekens, S., Bauerle, C., Hageman, J., Keegstra, K. , and Weisbeek, P., 1986, The role of the transit peptide in the routing of precursors toward different chloroplast compartments, Cell, 46:365.

Sollner, T., Griffiths, G., Pfaller, R., Pfanner, N. , and Neupert, W., 1989, Mom19, an Import Receptor for Mitochondrial Precursor Proteins, Cell, 59:1061.

Summers, R. G., Harris, C. R. , and Knowles, J. R., 1989, A Conservative Amino Acid Substitution, Arginine for Lysine, Abolishes Export of a Hybrid Protein in Escherichia-Coli - Implications for the Mechanism of Protein Secretion, J Biol Chem, 264:20082.

Summers, R. G. , and Knowles, J. R., 1989, Illicit Secretion of a Cytoplasmic Protein into the Periplasm of Escherichia-Coli Requires a Signal Peptide Plus a Portion of the Cognate Secreted Protein - Demarcation of the Critical Region of the Mature Protein, J Biol Chem, 264:20074.

Tamm, L. K., 1986, Incorporation of a synthetic mitochondrial signal peptide into charged and uncharged phospholipid monolayers, Biochemistry, 25:7470.

Tokunaga, M., Loranger, J. M., Chang, S. Y., Regue, M., Chang, S. , and Wu, H. C., 1985, Identification of prolipoprotein signal peptidase and

genomic organization of the lsp gene in Escherichia coli, J Biol Chem, 260:5610.
von Heijne, G., 1983, Patterns of amino acids near signal-sequence cleavage sites, Eur J Biochem, 133:17.
von Heijne, G., 1986a, Mitochondrial targeting sequences may form amphiphilic helices, Embo J, 5:1335.
von Heijne, G., 1986b, Net N-C charge imbalance may be important for signal sequence function in bacteria, J Mol Biol, 192:287.
von Heijne, G., 1986c, A new method for predicting signal sequence cleavage sites, Nucleic Acids Res, 14:4683.
von Heijne, G., 1990, The signal peptide, J Membr Biol, in press.
von Heijne, G., Steppuhn, J. , and Herrmann, R. G., 1989, Domain structure of mitochondrial and chloroplast targeting peptides, Eur.J.Biochem., 180:535.
Wiedmann, M., Goerlich, D., Hartmann, E., Kurzchalia, T. V. , and Rapoport, T. A., 1989, Photocrosslinking Demonstrates Proximity of a 34-kDa Membrane Protein to Different Portions of Preprolactin During Translocation Through the Endoplasmic Reticulum, Febs Letter, 257:263.
Wiedmann, M., Kurzchalia, T. V., Bielka, H. , and Rapoport, T. A., 1987a, Direct probing of the interaction between the signal sequence of nascent preprolactin and the signal recognition particle by specific cross-linking, J Cell Biol, 104:201.
Wiedmann, M., Kurzchalia, T. V., Hartmann, E. , and Rapoport, T. A., 1987b, A signal sequence receptor in the endoplasmic reticulum membrane, Nature, 328:830.
Witte, C., Jensen, R. E., Yaffe, M. P. , and Schatz, G., 1988, MAS1, a gene essential for yeast mitochondrial assembly, encodes a subunit of the mitochondrial processing protease, Embo J, 7:1439.
Wolfe, P. B., Wickner, W. , and Goodman, J. M., 1983a, Sequence of the leader peptidase gene of Escherichia coli and the orientation of leader peptidase in the bacterial envelope, J Biol Chem, 258:12073.
Wolfe, P. B., Zwizinski, C. , and Wickner, W., 1983b, Purification and characterization of leader peptidase from Escherichia coli, Methods Enzymol, 97:40.
YaDeau, J. T. , and Blobel, G., 1989, Solubilization and characterization of yeast signal peptidase, J Biol Chem, 264:2928.
Yamane, K. , and Mizushima, S., 1988, Introduction of basic amino acids residues after the signal peptide inhibits protein translocation across the cytoplasmic membrane of *Escherichia coli*, J.Biol.Chem., 263:19690.

# PROTEIN TRANSLOCATION ACROSS THE CHLOROPLAST ENVELOPE MEMBRANE

Douwe de Boer, Johan Hageman, Rien Pilon,
Twan America and Peter Weisbeek

Department of Molecular Cell Biology and Institute of Molecular Biology, University of Utrecht, P.O. Box 80.056, 3508 TB Utrecht, The Netherlands

## INTRODUCTION

During the last few years the amount of articles dedicated to chloroplast import has increased steadily. Many aspects have been elucidated, especially the energetics of the import reaction, the processing of the precursor during translocation and the organization of the transit peptide in functional domains have gained much attention (Keegstra et al., 1989 and Smeekens et al., 1990). The ATP requirement for the import process is already known for quite some time (Grossman et al., 1980), however the site were this ATP is used is still a subject of discussion (see Theg et al., 1989). Proteins destined to the chloroplast stroma contain an N-terminal transit peptide that is cleaved off by a stromal processing peptidase during or shortly after translocation across the envelope membrane. Proteins that reside in the thylakoid lumen also contain an N-terminal transit sequence but this one is larger and is composed of two distinct domains. The first part is functionally equivalent to the transit peptide of stromal proteins and is called chloroplast import domain. The second part has a lot of similarities to prokaryotic and endoplasmic reticulum signal sequences and is called the thylakoid transfer domain (Smeekens et al., 1986 and Hageman et al., 1990). The chloroplast import domain is removed by the stromal processing peptidase and the second domain by the thylakoidal processing peptidase (Hageman et al., 1987).

However little is known about the protein conformation during import and whether cytosolic factors are required. It is also poorly documented whether proteins in the envelope membrane are necessary for import. It is shown that protease treatment abolishes chloroplast import (Chua et al., 1979 and Cline et al., 1985) and one group reports the identification of a receptor (Pain et al., 1988). But details about the mechanism of translocation across the envelope are not yet revealed.

*Plant Molecular Biology* 2, Edited by R.G. Herrmann and
B. Larkins, Plenum Press, New York, 1991

To investigate the conformation of precursor proteins during translocation we constructed a chimeric fusion protein containing the complete transit peptide of plastocyanin fused to the dihydrofolate reductase (PCDHFR) protein (Hageman et al., 1990). The conformation of DHFR can be stabilized by complexing with methotrexate (MTX). To be able to study protein import with pure precursor protein in the absence of cytosolic factors we expressed both the precursor to ferredoxin and the precursor to plastocyanin in E.coli cells. After purification we used these proteins in in vitro chloroplast import studies. With these precursor proteins we are also able to perform binding and competition studies to obtain more information about the amount and diversity of the putative receptor present in the envelope membrane.

## PROTEIN CONFORMATION DURING IMPORT

The chimeric protein PCDHFR was constructed by fusing the coding sequence for the transit peptide of plastocyanin (PC) in front of the mature coding sequence of mouse cytosolic DHFR (fig. 1).

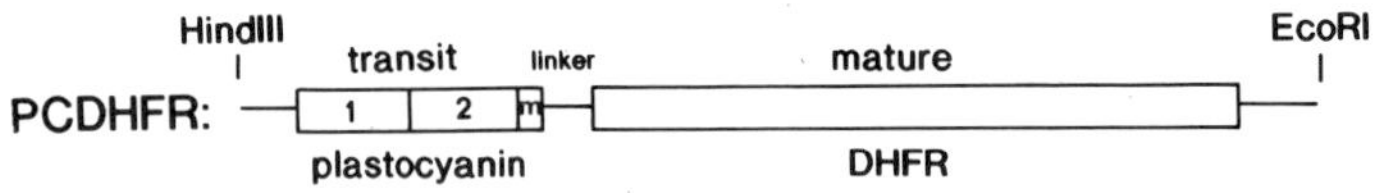


Fig. 1. Structure of the PCDHFR fusion protein. The coding sequence between the HindIII and EcoRI site in the transcription vector pSP64 (Melton et al., 1984) is shown. 1= first domain transit peptide, 2= second domain transit peptide, m= part of mature plastocyanin.

The conformation of PCDHFR is stabilized considerably when the substrate analog MTX is present. This can be monitored by protease treatment, because the folded protein has a much higher resistance to degradation than its unfolded counterpart (Eilers and Schatz, 1986). Fig. 2 shows that the PCDHFR construct is only reduced in size, but not degraded when 1 $\mu$M MTX is present, although up to 625 $\mu$g/ml of thermolysine was added. This protection is not an effect of MTX on the protease itself, because the control plastocyanin precursor protein is already completely degraded at a concentration of 2.5 $\mu$g/ml thermolysine whether or not MTX is present. The reduction in size of PCDHFR is probably caused by the degradation of the PC transit peptide, that is supposed to be external to the folded protein core.

The PCDHFR construct imports efficiently into isolated chloroplasts and most of the imported protein is present in the stroma as an intermediate sized protein together with degradation products (fig. 3A). The plastocyanin transit sequence is not able to efficiently direct the foreign DHFR sequence to the thylakoid lumen, only a limited amount of mature sized protein is found. This phenomenon is also found with other passenger proteins attached to the plastocyanin transit sequence (Smeekens et al. 1986 and De Boer et al., in preparation). The degradation products are not observed in vitro after incubation with stroma (not shown) suggesting that they are associated with the import process. We expect that this degradation is due to partial or complete unfolding of the precursor during import since a folded DHFR molecule is supposed to be resistant against protease (see fig. 2).

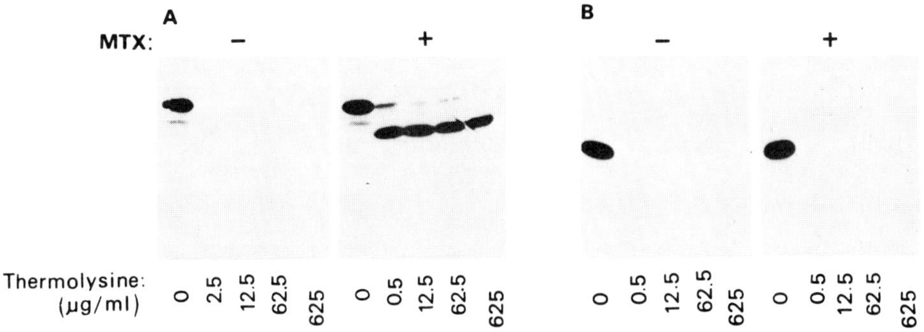


Fig. 2. Protection against thermolysine. Resistance of PCDHFR (**A**) and plastocyanin (**B**) precursor proteins to increasing concentrations of thermolysine in the presence (+) or absence (-) of 1 μm MTX. p= precursor, m= mature.

When MTX is present during import (fig. 3B) most of the degradation products in the stroma disappear, but PCDHFR can still be found imported into the chloroplast. Apart from the intermediate sized band an additional larger band becomes visible and a mature sized band appears, both localized in the stroma. The origin of the larger band is not clear, it might be the result of an aberrant processing. The mature sized band probably is the protease resistant core of PCDHFR present in the stroma. None of the above products was present in the envelope membrane fraction (not shown). The inhibition of the degradation suggests that the fusion protein is sufficiently

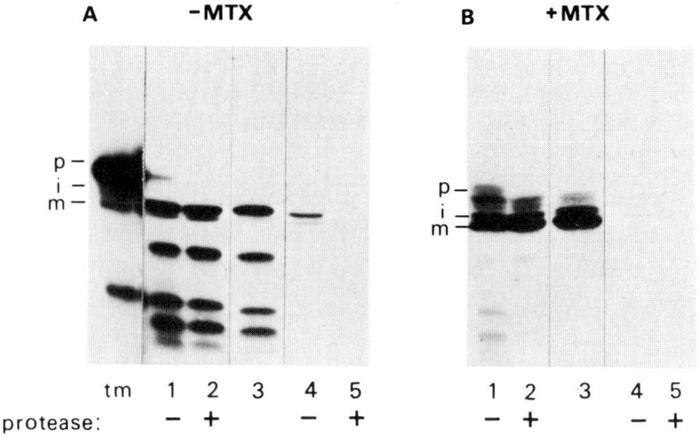


Fig. 3. The effect of MTX on import of PCDHFR. **A.** Chloroplast fractionation after PCDHFR import without MTX, TM= translation mixture, 1= total chloroplasts, 2= total chloroplasts protease treated, 3= stroma, 4= thylakoids, 5= thylakoids protease treated, p= precursor, i= intermediate, m= mature. **B.** As A only PCDHFR import with 1$\mu$ M MTX present.

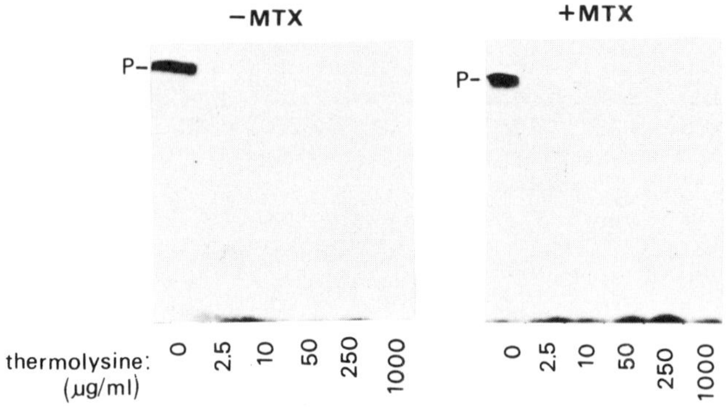


Fig. 4. Protection against thermolysine after binding to chloroplasts. Increasing concentrations of thermolysine were added in the presence (+) or absence (-) of 1$\mu$m MTX. p= precursor.

stabilized when MTX is present. Quantification of the import data shows that the ratio of imported MTX molecules to PCDHFR molecules is one. This 1:1 ratio does not exist when pPC is imported in the presence of MTX, indicating that the imported PCDHFR molecules are present in the stroma complexed with MTX.

Two different models can account for these data. Either PCDHFR is imported into the chloroplast complexed with MTX and therefore has to be translocated in a folded state or the complex unfolds during translocation and MTX is bound to PCDHFR in the stroma after diffusion into the chloroplast. An observation during the import studies with MTX was that bound PCDHFR precursor protein could be removed from the chloroplast by protease treatment although it was not clear whether it was completely degraded. To discriminate between the two models we therefore repeated the protease protection experiments of fig. 2 but this time with proteins bound to the chloroplast. Fig. 4 shows that binding of PCDHFR to the chloroplasts abolishes the stabilizing effect of MTX. This experiment shows that chloroplast are able to unfold PCDHFR despite the complexation with MTX. Either a strong membrane bound unfoldase or the lipids of the envelope are able to unfold the complex. This result supports the second model in which the complex is first unfolded and MTX is complexed again in the stroma after independent diffusion over the envelope membranes. To test this we first imported PCDHFR into isolated chloroplasts and subsequently incubated the chloroplasts with MTX. Quantification of these data (not shown) shows that again a 1:1 ratio of MTX to PCDHFR is present inside the chloroplasts, whereas only background levels of MTX are observed inside the chloroplast when other proteins are imported. We therefore favor the second model.

## CHLOROPLAST IMPORT OF THE FERREDOXIN AND PLASTOCYANIN PRECURSOR AFTER EXPRESSION IN E.COLI

The ferredoxin and plastocyanin precursor sequences were cloned between the NcoI and PstI sites in the inducable expression plasmid pKK233.2 (Amann and Brosius, 1985), to be able to obtain large amounts of pure protein. The constructs were expressed in the E.coli strain PC2495 (Van der Plas et al., 1989). The lacI$^q$ gene, necessary for repression of the trc promoter in pKK233.2, is present on the F plasmid in this strain. An NcoI site had been introduced by mutagenesis at the start of the coding sequence for the precursor of both proteins. The ferredoxin protein was found to be unstable in PC2495 (Pilon et al. 1990) and therefore the SstI fragment from the ferredoxin construct was recloned into the high copy number plasmid pUC18 (Yanisch-Perron et al., 1985) together with the lacI$^q$ gene from plasmid pMMB22 (Bagdasarian et al., 1983). The lacI$^q$ gene was necessary for regulated expression in the E.coli lon$^-$ strain AB1899 (Howard-Flanders et al., 1964). Fig. 5 shows the resulting ferredoxin and plastocyanin constructs, called pPAFI$^q$20 and pKPAP14 respectively. The trc-promoter was induced with IPTG, a concentration of 0.5-1 mM was enough to repress the lacI$^q$ gene product and to give maximum expression (fig. 6). Both proteins have the same molecular weight as the corresponding precursor proteins made in a wheat germ system.

They were localized in the cytoplasm as part of insoluble aggregates, that could be resolved with 6 M urea or guanidine/HCl.

Both proteins were purified to homogeneity (not shown) and the isolated precursor proteins were subjected to N-terminal sequencing. The N-terminal sequences were exactly the same as the sequences deduced from the c-DNA clones, except for the first methionine that was removed in both cases. To test whether these proteins are active in protein import into isolated chloroplasts, E.coli cells were labelled with [35S]-sulphate and the precursor proteins were isolated and resuspended in 6 M urea. The ferredoxin precursor protein was radiochemically pure, but the plastocyanin precursor was only partially purified. The ferredoxin precursor was diluted in the

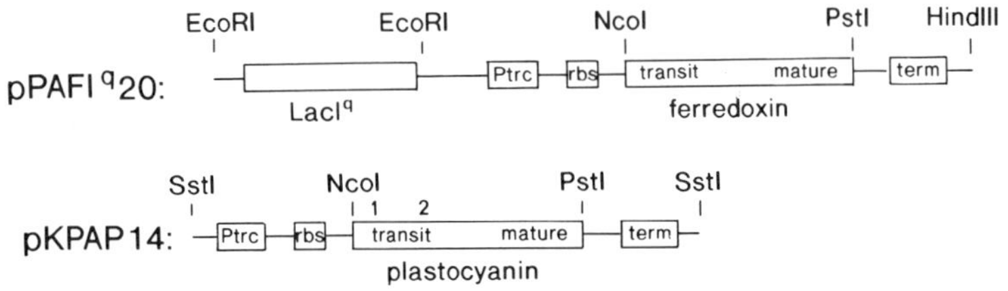


Fig. 5. Schematic representation of the inserts in the expression plasmids pPAFI$^q$ and pKPAP14. **A.** The ferredoxin expression plasmid shown between the EcoRI and HindIII sites in pUC18. **B.** The plastocyanin expression plasmid shown between the SstI sites in pKK233.2.

chloroplast mixture to a final concentration of 0.2 M urea and the plastocyanin precursor to a final concentration of 0.5 M urea. These high urea concentrations did not affect protein import of wheat germ synthesized precursors proteins. The chloroplast mixtures only contained pea chloroplasts resuspended in import buffer (50 mM Hepes/KOH pH= 8 and 330 mM sorbitol) and 2 mM Mg-ATP. Fig. 7 shows that the two precursor proteins are processed to their mature sizes and that they are localized in their appropriate compartments. Apart from intact chloroplasts no other proteins were added to the import reaction (although some contaminating proteins were present in the case of plastocyanin), therefore we conclude that cytosolic factors are not absolutely necessary. Presently we are using the purified precursor proteins for binding studies with isolated chloroplasts and competition studies with precursor proteins synthesized in a wheat germ system.

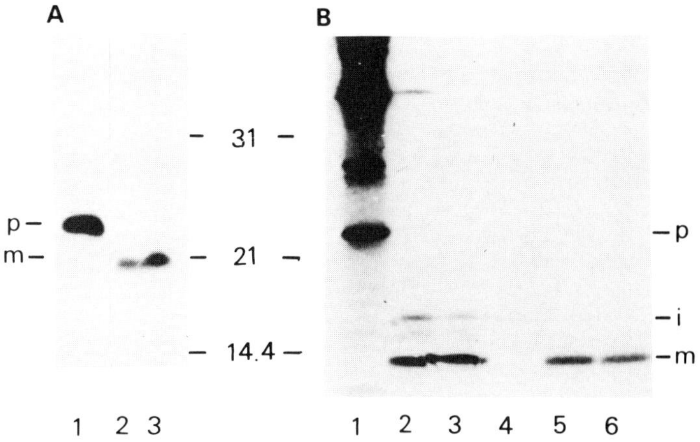


Fig. 6. Expression in E.coli. **A.** Western blot of total protein isolated from E.coli cells harbouring plasmid pPAFI$^{q}$20. Lane 1, before induction; lane 2, after induction with IPTG. Antibodies against spinach ferredoxin were used. Molecular weight markers are in kDal. **B.** As A only E.coli cells harbouring pKPAP14 and antibodies against spinach plastocyanin were used. p=precursor.

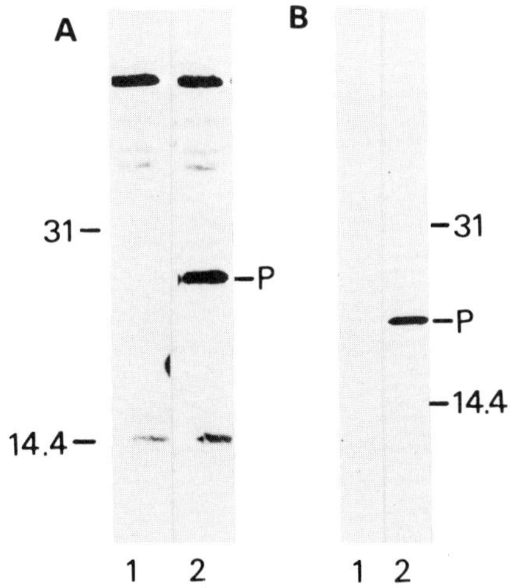


Fig. 7. Import into isolated chloroplasts. **A.** Import of the ferredoxin precursor. Lane 1, isolated precursor; lane 2, total reisolated chloroplasts after import; lane 3 reisolated chloroplasts after protease treatment. **B.** Import of the plastocyanin precursor. Lane 1, isolated precursor; lane 2, total reisolated chloroplasts after import; lane 3 chloroplasts after protease treatment; lane 4, chloroplast stroma; lane 5, thylakoid fraction; lane 6, thylakoid fraction after protease treatment. p= precursor, i= intermediate, m= mature.

## CONCLUSIONS

Import experiments with the chimeric protein PCDHFR show that proteins become at least partly unfolded during translocation across the envelope. This conclusion was also reached for mitochondrial import when DHFR was used in combination with a mitochondrial transit peptide (Eilers and Schatz, 1986). The experimental data however conflict between mitochondrial and chloroplast import with this protein. When MTX was used in mitochondrial import studies it completely blocked import of DHFR. No protein entered the mitochondrium. When MTX was used in chloroplast import studies, normal import rates were found. This suggests that chloroplasts have the capacity to unfold the DHFR/MTX complex more strongly than mitochondria. The absence of e.g. a membrane potential in chloroplasts and the presence of this strong unfolding activity emphasize the differences in the mechanism of import between these two organelles. Whether this unfolding is mediated by the recently detected hsp-70 protein in the envelope membranes of the chloroplast (Marshall et al., 1989) or by another protein or whether lipids are involved remains to be elucidated.

In this article we also show that it is possible to use precursor proteins that are synthesized in E.coli cells to study protein import into isolated chloroplasts. Both chloroplast import and routing occurs as determined with proteins synthesized in the wheat germ system. The experiments revealed that cytosolic factors are not strictly needed for the translocation across the envelope membrane, in vitro translocation can occur without them.

## REFERENCES


Amann, E. and Brosius, J., 1985, 'ATG vectors' for regulated high-level expression of cloned genes in E.coli, Gene, 40:183.

Bagdasarian, M. M., Amann, E., Lurz, R., Rückert, B. and Bagdagarian, M., 1983, Activity of the hybrid trp-lac (tac) promoter of Escherichia coli in Pseudomonas putida. Construction of broad-host-range, controlled-expression vectors, Gene, 26:273.

Chua, N. -H. and Schmidt, G. W., 1979, Transport of proteins into mitochondria and chloroplasts, J.Cell Biol., 81:461.

Cline, K., Werner-Washburne, M., Lubben, T. H. and Keegstra, K., 1985, Precursors to two nuclear-encoded chloroplast proteins bind to the outer envelope membrane before being imported into chloroplasts, J.Biol.Chem., 260:3691.

De Boer, D., Cremers, F., Teertstra, R., Smits, L., Hille, J., Smeekens, S. and Weisbeek, P., 1988, In-vivo import of plastocyanin and a fusion protein into developmentally different plastids of transgenic plants, EMBO J., 7:2631.

Eilers, M. and Schatz, G., 1986, Binding of a specific ligand inhibits import of a purified precursor protein into mitochondria, Nature, 322:228.

Grossman, A., Bartlett, S. and Chua, N. -H., 1980, Energy-dependent uptake of cytoplasmically synthesized polypeptides by chloroplasts, Nature, 285:625.

Hageman, J., Robinson, C., Smeekens, S. and Weisbeek, P., 1986, A thylakoid processing protease is required for complete maturation of the lumen protein plastocyanin, Nature, 324:567.
Hageman, J., Baecke, C., Ebskamp, M., Pilon, R., Smeekens, S. and Weisbeek, P., 1990, Protein import into and sorting inside the chloroplast are independent processes, Plant Cell, in press.
Howard-Flanders, P., Simons, E. and Thericot, L., 1964, A locus that controls filament formation and sensitivity to radiation in E. coli K12, Genetics, 49:237.
Keegstra, K., Olsen, L. J. and Theg, S. M., 1989, Chloroplastic precursors and their transport across the envelope membranes, in: Annual Review of Plant Physiology and Plant Molecular Biology, Briggs, W. R. ed., Annual Reviews Inc., Palo Alto, California USA.
Marshall, J. S., DeRocher, A. E., Keegstra, K. and Vierling, E., 1990, Identification of heat shock protein hsp70 homologues in chloroplasts, Proc. Natl. Acad. Sci. USA, 87:374.
Melton, D. A., Krieg, P. A., Rebagliati, M. R., Maniatis, T., Zinn, K. and Green, M. R., 1984, Efficient in vitro synthesis of biologically active RNA and RNA hybridisation probes from plasmids containing a bacterial SP6 promoter, Nucl. Acids Res., 12:7035.
Pain, D., Kanwar, Y. S. and Blobel, G., 1988, Identification of a receptor for protein import into chloroplasts and its localization to envelope contact zones, Nature, 331:232.
Pilon, M., De Boer, A. D., Knols, S. L., Koppelman, M. H. G. M., Van der Graaf, R. M., De Kruijff, B. and Weisbeek, P. J., 1990, Expression in E. coli and purification of a translocation-competent precursor of the chloroplast protein ferredoxin, J. Biol. Chem., 265:3358.
Smeekens, S., Bauerle, C., Hageman, J., Keegstra, K. and Weisbeek, P., 1986, The role of the transit peptide in the routing of precursors toward different chloroplast compartments, Cell, 46:365.
Smeekens, S., Weisbeek, P. and Robinson, C., 1990, Protein transport into and within chloroplasts, TIBS, 15:73.
Theg, S. M., Bauerle, C., Olsen, L. J., Selman, B. R. and Keegstra, K., 1989, Internal ATP is the only energy requirement for the translocation of precursor proteins across chloroplastic membranes, J.Biol.Chem., 264:6730.
Van der Plas, J., Bovy, A., Kruyt, F., De Vrieze, G., Dassen, E., Klein, B. and Weisbeek, P., 1989, The gene for the precursor of plastocyanin from the cyanobacterium Anabaena sp. PCC 7937: isolation, sequence and regulation, Molec. Microbiol., 3:275.
Yanisch-Perron, C., Vieira, J. and Messing, J., 1985, Improved M13 phage cloning vectors and host strains: nucleotide sequences of the M13mp18 and pUC19 vectors, Gene, 33:103.

# PROTEIN TRANSPORT ACROSS THE THYLAKOID MEMBRANE

Colin Robinson, Ruth Mould and Jamie Shackleton

Department of Biological Sciences
University of Warwick
Coventry
CV4 7AL

## INTRODUCTION

The biogenesis of thylakoid proteins is a complex issue, since these proteins are encoded by two distinct genomes. Approximately half are encoded by chloroplast DNA and synthesised within the organelle, whereas the remainder are synthesised in the cytoplasm and post-translationally imported into the chloroplast. Imported proteins are initially synthesised as larger precursors containing aminoterminal pre-sequences; in some cases, such pre-sequences have been shown to contain all of the information specifying transport into the chloroplast (reviewed in Keegstra, 1989; Smeekens, Weisbeek and Robinson, 1990).

The import of nuclear-encoded thylakoid lumen proteins is particularly interesting, because these proteins have to cross all three chloroplast membranes to reach their sites of function. Such proteins include plastocyanin, which is essentially soluble in the lumen, and the 33 KD, 23 kD and 16 kD proteins of the photosynthetic oxygen-evolving complex, which are loosely bound to the lumenal face of the thylakoid membrane. These proteins appear to be transported into the thylakoid lumen by a two-step mechanism, which is illustrated in Fig. 1. Initially, precursors are imported into the stroma and cleaved to intermediate forms by a stromal processing peptidase (SPP), after which the intermediates are transferred across the thylakoid membrane and processed to the mature sizes by a thylakoid processing peptidase, TPP (Smeekens et al., 1986; Hageman et al., 1986; James et al., 1989). In keeping with this proposed import mechanism, the pre-sequences of lumenal proteins have been shown to contain two distinct targeting signals; a postively-charged, hydrophilic "envelope transfer domain" followed by a more hydrophobic "thylakoid transfer domain" (Smeekens et al., 1985; Tyagi et al., 1987).

In order to analyse in greater detail the mechanism of protein transport across the thylakoid membrane, we recently developed an <u>in vitro</u> assay for the import of the 33 kD protein (33 K) by isolated thylakoids (Kirwin et al., 1989). In this assay, pre-33 K was incubated with thylakoids, stromal extract and ATP. Under these conditions, pre-33 K is processed to the intermediate form (int-33 K) by SPP in the stromal extract, and mature-size 33 K appears inside the thylakoid

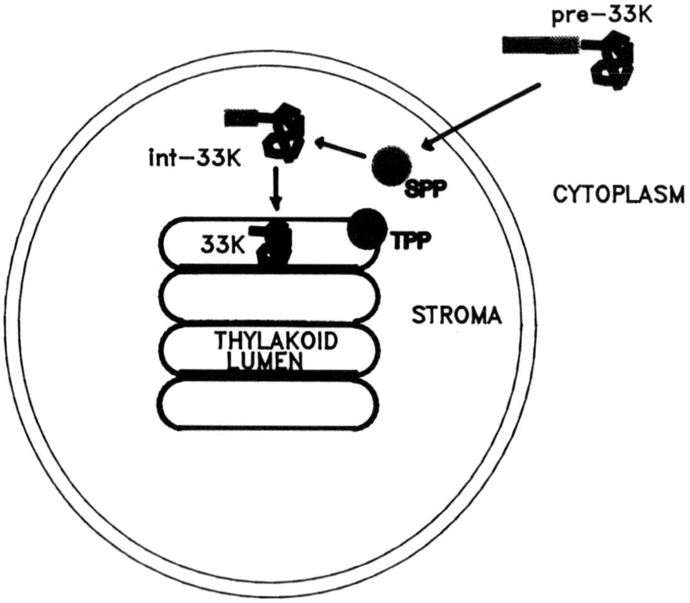


Fig. 1. Two-step model for the import of a thylakoid lumen protein. The precursor of the 33 kD oxygen-evolving protein (pre-33 K) is imported into the stroma and cleaved to an intermediate form by SPP, a stromal processing peptidase. The intermediate is then transported across the thylakoid membrane and processed to a mature size by TPP, a thylakoidal processing peptidase.

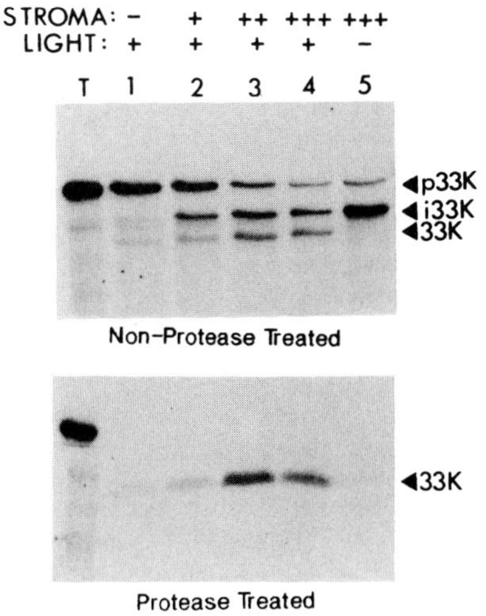


Fig. 2. Light driven import of 33 K by isolated thylakoids. Pre-33 K (denoted p33 K) lane T) was incubated with thylakoids in the absence of stroma (lanes 1) or with increasing levels of stroma (lanes 2-4) under illumination. Lane 5: incubation as in lane 4 but in the dark. Samples were analyzed directly (upper panel) or after protease treatment (lower panel).

vesicles. Approximately 20% of the available int-33 K (the presumed substrate for the thylakoid import system) was imported in this assay.

In this article we briefly describe more recent data on the thylakoidal protein import system, and we show how this assay can be used to probe the requirements for the transport of two lumenal proteins across the thylakoid membrane.

## LIGHT-DRIVEN PROTEIN IMPORT BY ISOLATED THYLAKOIDS

Although import of 33 K into isolated thylakoids can be driven with reasonable efficiency by ATP, we have since found that more efficient import is observed when driven by light. Fig. 2 shows the effects of varying the light regime and concentration of stromal extract on the import of 33 K. In the presence of light and reasonably high amounts of stroma (lanes 3 and 4) pre-33 K is converted to int-33 K (by SPP) and also to the mature size. The mature size 33 K is resistant to added protease, indicating that import into the lumenal space has occurred. When incubation takes place in the presence of stroma, but in the dark (lane 5), pre-33 K is efficiently processed to int-33 K, but no import is observed. Clearly, efficient import of 33 K by isolated thylakoid requires both light and stromal extract. It should be emphasised that exogenous ATP was omitted from these incubations. However, the transport system may utilise ATP which is contributed by the pre-33 K translation mixture.

The requirement for stromal extract may reflect a requirement for SPP if, as the import model suggests, it is the intermediate form of 33 K which is imported by thylakoids. However, it is also possible that the stromal extract supplies soluble factors which assist in the translocation process. For example, the integration of the light-harvesting chlorophyll a/b binding protein into isolated thylakoid membranes requires at least one stromal protein factor (Cline, 1986; Chitnis et al., 1987).

The light-driven import assay has also been used to study the biogenesis of a second lumenal protein : the 23 kD polypeptide of the oxygen-evolving complex (23 K). As with 33 K, this protein is initially synthesised in the cytoplasm with a bipartite pre-sequence (Tyagi et al., 1987) and in vitro pre-23 K is processed to intermediate or mature sizes, respectively, by partially purified SPP or TPP (James et al., 1989). Fig. 3 shows the effects of stromal extract or light on the import of 23 K by isolated thylakoids. The results show that light drives efficient import in both the presence or absence of stromal extract (lanes 1-3). The efficiency of import is in fact higher than that observed using 33 K as a substrate; greater than 50% of available 23 K can be imported in this type of assay. In the dark (lanes 4) little or no import is observed, emphasising that light is a critical factor in this assay system for the import of both 33 K and 23 K. However it is notable that, whereas 33 K import requires stromal extract in the incubation mixture, 23 K is efficiently imported by thylakoids even in the absence of stroma. The clear implication is that thylakoids can import the full precursor of 23 K in the absence of cleavage by SPP. This has been confirmed by other data; we have been able to block cleavage of pre-23 K by SPP by chemically modifying the precursor, but the modified pre-23 K is nevertheless imported into the thylakoid lumen of isolated chloroplasts.

These data indicate that the second thylakoid transfer domain of pre-23 K is capable of acting as an internal targeting signal. This raises the obvious question : why does the 23 K pre-sequence contain an SPP cleavage site? The answer to this question is presently unclear, but it is possible that 23 K is imported into thylakoids more efficiently as the intermediate form than as the full precursor.

## TRANSPORT OF 23 K ACROSS THE THYLAKOID REQUIRES A PROTON GRADIENT

Further experiments were carried out to determine the basis for the stimulation of protein import by light. Thylakoid import studies were carried out in the presence of several inhibitors : a mixture of DCMU and methyl viologen (to block electron transport in the thylakoid membrane), valinomycin/$K^+$ (to dissipate the $\Delta\psi$ component of the proton motive force) and nigericin/$K^+$ (to dissipate the $\Delta pH$ component). Figure 4 shows that valionmycin/$K^+$ inhibits 23 K import to a small extent, but that both DCMU/methyl viologen and nigericin/$K^+$ almost totally inhibit import. We conclude from these results that efficient transport of 23 K into isolated thylakoids requires an energised membrane, and that the $\Delta pH$ component of the proton motive force is the critical factor rather than $\Delta\psi$.

This conclusion has been reinforced by results from experiments using intact chloroplasts rather than thylakoids. We have found that pre-23 K is imported into the chloroplast stroma in the presence of any of the above inhibitors, when ATP is supplied to drive import. However, in the presence of DCMU/methyl viologen or nigericin ($K^+$, imported 23 K accumulates in the stroma as the intermediate form, and transport across the thylakoid membrane is almost completely inhibited.

Work is in progress to analyse how the transport of 23 K is harnessed to the thylakoidal $\Delta pH$, and to determine whether the transport of other proteins across the thylakoid membrane requires an energised membrane. Other work is aimed at understanding the thylakoidal import

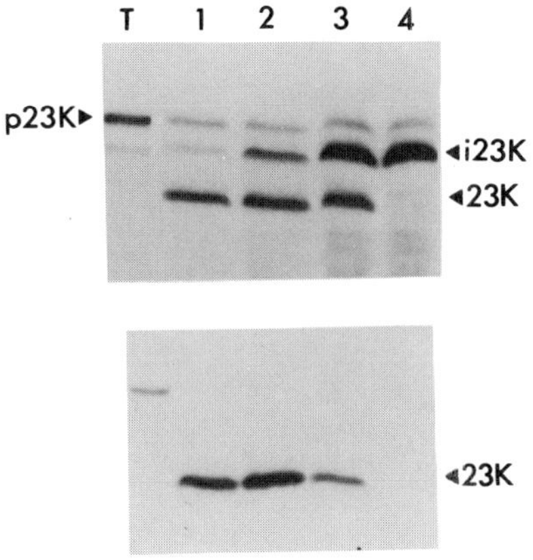


Fig. 3. Light-driven import of 23 K by thylakoids. Pre-23 K (lane T) was incubated with thylakoids in the absence of stroma (lane 1) or with increasing amounts of stroma (lanes 2 and 3) in the light. Lane 4 as lane 3 but in the dark. After incubation, samples were analyzed directly (upper panel) or after protease treatment (lower panel).

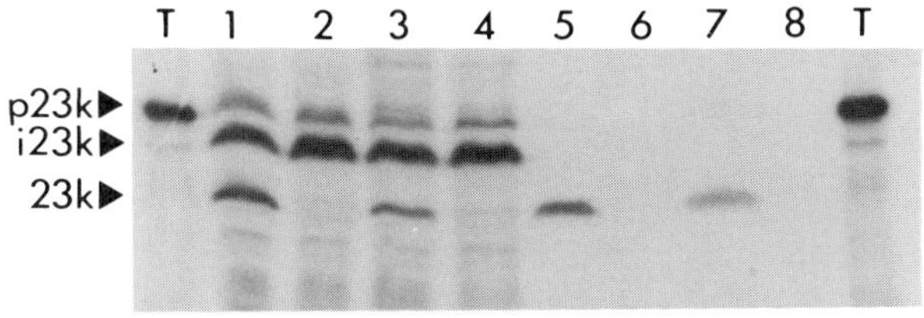


Fig. 4. Import of 23 K by thylakoids requires a proton gradient. Pre-23 K (lanes T) was incubated with thylakoids and stroma in the light (lane 1). Similar incubations also contained DCMV/methyl viologen (lane 2) valinomycin/$K^+$ (lane 3) or nigeicin/$K^+$ (lane 4). Lanes 5-8 represent protease-treated samples of lanes 1-4.

mechanism, about which little is currently known. It is likely that the transport system has features in common with bacterial export systems; bacterial leader sequences have several features in common with thylakoid transfer domains (von Heijne et al., 1989), and E.coli leader peptidase and TPP have remarkably similar reaction specificities (Halpin et al., 1989). However, further work is required in order to determine the composition and mechanism of action of the thylakoidal system, since no soluble or membrane-bound components have been isolated to date.

## REFERENCES


Chitnis, P.R., Nechushtai, R. and Thornber, J.P. (1987). Plant Mol. Biol. 10, 3-11.

Cline, K. (1986). J. Biol. Chem. 261, 14804-14810.

Hageman, J., Robinson, C., Smeekens, S. and Weisbeek, P. (1986). Nature 324, 567-569.

Halpin, C., Elderfield, P.D., James, H.E., Zimmermann, R., Dunbar, B. and Robinson, C. (1989). EMBO J. 8, 3917-3921.

James, H.E., Bartling, D., Musgrove, J.E., Kirwin, P.M., Hermmann, R.G. and Robinson, C. (1989). J. Biol. Chem. 264, 19573-19576.

Keegstra, K. (1989). Cell 56, 247-253.

Kirwin, P.M., Meadows, J.W., Shackleton, J.B., Musgrove, J.E., Elderfield, P.D., Mould, R., Hay, N. and Robinson, C. (1989). EMBO J. 8, 2251-2255.

Smeekens, S., De Groot, M., van Binsbergen, J. and Weisbeek, P. (1985). Nature 317, 456-458.

Smeekens, S., Bauerle, C., Hageman, J., Keegstra, K. and Weisbeek, P. (1986). Cell 46, 365-375.

Smeekens, S., Weisbeek, P. and Robinson, C. (1990). Trends in Biochem. Sci. 15, 73-76.

Tyagi, A., Hermans, J., Steppuhn, J., Jansson, C., Vater, F. and Herrmann, R.G. (1987). Mol. Gen. Genet. 207, 288-293.

Von Heijne, G., Steppuhn, J. and Herrmann, R.G. (1989). Eur. J. Biochem. 180, 535-545.

# PROTEIN IMPORT INTO PLANT MITOCHONDRIA

François Chaumont and Marc Boutry

Unité de Biochimie physiologique
Université Catholique de Louvain
Place Croix du Sud 2
1348 Louvain-la-Neuve
Belgium

## INTRODUCTION

Plant cells are divided into distinct compartments surrounded by a single or a double membrane. Two of these subcellular structures, mitochondria and chloroplast, fulfill various bioenergetic functions. In addition to the oxidative phosphorylation providing energy in non photosynthetic tissues or during the dark period, mitochondria contribute to important metabolite biosynthesis (amino acids, phospholipids, nucleotides...). For their part, chloroplasts are the location not only of photosynthesis, but also of many metabolic processes (fatty acid synthesis, starch metabolism, ammonium assimilation...).

Several hundred proteins are necessary to achieve these metabolic functions. Mitochondrial and chloroplast DNA encode only a limited set of proteins. Consequently, most proteins are encoded by nuclear genes, synthesized as precursors in the cytosol and transported into the organelle. The information for specific targeting is mostly located in the amino-terminal extension (presequence or targeting signal) of the precursor. During or after import into mitochondria, the presequence is removed to give a mature protein.

Protein import into fungal or mammalian mitochondria has been well characterized (for review, see Rosenberg *et al.*, 1987 ; Attardi and Schatz, 1988 ; Hartl *et al.*, 1989 ; Hartl and Neupert, 1990). With regard to plants, most available information concerns protein transport into chloroplasts (for review , see Keegstra *et al.*, 1989 ; Weisbeek *et al.*, 1989 ; Smeekens *et al.*, 1990). Despite the few mitochondrial precursors which have been characterized, some data has been recently obtained by *in vitro* and *in vivo* experiments. In this paper, we will review this information and relate it to models drawn up for other organisms.

*Plant Molecular Biology 2*, Edited by R.G. Herrmann and
B. Larkins, Plenum Press, New York, 1991

```
                    10        20        30        40        50
                     |         |         |         |         |

sw. po.
b. bean.
               ++      +    +      +       +   +   +   +     +
N. plu.     MASRRLLASLLRQSAQRGGGLISRSLGNSIPKSASRASSRASPKGFLLNR
                                               +           ++
human                                    MLGFVGRVAAAPASGALRRLT
                                               +          +
bovine                                   MLGLVGRWAASPASGALRGLS
                                                ++      ++
S. pom.                                       MLKKQALSGIRRFSLA
S. cer.                                                   MVLP
                                                       +
N. cra.                                              MFKSGISAF


                    60       70        80        90
                     |        |         |         |
               v        +   +    -
sw. po.         XXAAAPAXKPAAKPAXNE/
               v
b. bean         ATSAAAAXTPXVPXP/
                               +   -   +  --        +
N. plu.     AVQYATSAAAPASQPSTPPKSGSEPSGKITDEFTGAGSIGKVCQVIGA/
                        +  +++ +  +-   v     +      +
human       PSASLPPAQLLLRAVRRRSHPVREYAAQTSPSPKAGAATGRIVAVIGA/
                        +         +-   v     +      +
bovine      PSAPLPQAQLLLRAAPAALQPARDYAAQASPSPKAGATTGRIVAVIGA/
             +    +   ++      +    + v  -  +  +   +
S. pom.     TKQSFVKTSYKLPRKSWLNTAKFNTIRYASTEAAKHNKGSIKQAVIGA/
            +      +   +  +          ++  v          +
S. cer.     RLYTATSRAAFKAAKQSAPLLSTSWKRCMASAAQSTPITGKVTAVIGA/
             +  +       ++  +      +     +v       - +
N. cra.     ARTARPSFAAASRRAVRPAALNLRAPALSRFASSAGVGDGK/
```

**Figure 1**. Comparison of amino-terminal sequences of the mitochondrial ATPase ß subunit from different organisms. The presequences of ß subunit precursor from *N. plumbaginifolia* (*N.* plu.) (Boutry and Chua, 1985), human (Otha *et al.*, 1988), bovine (Breen *et al.*, 1988), *S. pombe* (*S. pom*) (Falson and Boutry, in preparation), *S. cerevisiae* (*S. cer.*) (Vassarotti *et al.*, 1987) *N. crassa* (*N.cra*) (Hartl *et al.*, 1989) and the sequences of mature ß subunit from sweet potato (sw. po.) (Kobayashi *et al.*, 1986) and broad bean (b. bean.) (Leterme, personnal communication) are depicted. The beginning of the conserved sequences of the mature part is underlined. Arrows indicate the cleavage sites of the mature proteins. Charged residues are indicated (+ or -).

PRESEQUENCE STRUCTURE AND FUNCTION

The $NH_2$-terminal extensions of a number of fungal and mammalian precursors have been analyzed (Hartl *et al* ., 1989). They are different in length and primary structure but they share several common features : they are rich in positively charged (mostly Arg), hydroxylated (more Ser than Thr) and unpolar residues like Ala and Leu and lack acidic amino acids. However, the most important common characteristic may be their ability to form an amphiphilic alpha helix where the positively charged and hydrophobic residues are on opposite sides (von Heijne, 1986 ; Roise *et al.*, 1986 ; von Heijne *et al.*, 1989).

```
                  +    ++          +        +v       -
SOD-m        :    MALRTLASKKVLSFPFGGAGRPLAAAASARGVTTVTLPDL/
                  +    ++          +        +v       -
SOD-N.p.     :    MALRTLVSRRTLAT--GLGFRQQ-----LRGLQTFSLPDL/
                  +          +     + +        +  v   -
GDC          :    MALRMWASSTANALKLSSSSRLHLSPTPSISRCPSNVLDG/
                  ++      +    +      +       +   +   +
ß            :    MASRRLLASLLRQSAQRGGGLISRSLGNSIPKSASRASSR
                  +     +    v
                  ASPKGFLLNRAVQYATSAAAPASQP/
```

**Figure 2.** Comparison of plant mitochondrial presequences. Presequences of manganese superoxide dismutase from maize (SOD-m) (White and Scandalios, 1989) and *N. plumbaginifolia* (SOD-*N.p.*) (Bowler *et al.*, 1989a), of glycine decarboxylase (GDC) (Kim and Oliver, 1990) and of ATPase ß subunit from *N. plumbaginifolia* (ß) (Boutry et Chua, 1985), are depicted. Arrows indicate the cleavage sites. Charged residues are indicated (+ or -).

To our knowledge, the sequence of four plant mitochondrial precursors have been obtained. The first one is the $F_1$-ATPase ß subunit from *Nicotiana plumbaginifolia*. (Boutry and Chua, 1985). By comparison with the sequence of the mature ß subunit from sweet potato (Kobayashi *et al.*, 1986) or broad bean (Leterme, personnal commmunication), the presequence should be 55 or 54 residue-long (Figure 1).

The other plant mitochondrial precursors which have been identified are those of manganese superoxide dismutase from *N. plumbaginifolia* (Bowler *et al.*, 1989a) and maize (White and Scandalios, 1989) and of glycin decarboxylase (Kim and Oliver, 1990) (Figure 2). The presequences are respectively 24, 31 and 34 residue-long. The plant presequences share the common features of fungal and mammalian mitochondrial presequences: they contain arginines or lysines interspersed with uncharged amino acids and lack acidic residues. Interestingly, the four first residues of the presequences are conserved ($MA^{L}/_{S}R$) with a positive charge at the fourth position (Figure 2). Residues 7 to 24 of the ß subunit presequence have been shown to form an alpha amphiphilic helix (von Heijne, 1986). Such a structure is hardly found for the other presequences.

The involvement of the plant F1-ATPase ß presequence in targeting has been demonstrated by chimeric gene studies (Boutry *at al.*, 1987). The sequence coding the 90 $NH_2$-terminal residues of the ß subunit precursor was fused to the bacterial chloramphenicol acetyltransferase (cat) gene. After introduction of the chimeric gene into tobacco, the CAT protein was retrieved exclusively in mitochondria showing that the first 90 residues contain information for specific targeting. To determine the essential region of the ß presequence, 3' deletions of the latter were fused to the cat gene. Analysis of transgenic plants has shown that the 23 first residues are capable of specifically targeting the CAT protein into plant mitochondria (Chaumont and Boutry, in preparation). Suprisingly, the fusion residues located between the ß presequence and the CAT protein contain information for a second cryptic cleavage site. In order to confirm that the 23 $NH_2$-terminal residues of the presequence are sufficient and that the passenger protein is not involved in the targeting, the deleted presequences were fused to the glucuronidase gene. The transgenic plants containing the gus gene flanked of 89 or 59 residues of the ß precursor displayed high GUS activity in mitochondria. Lower activity was found in mitochondria of transgenic plant with the 23 amino acid construct (Chaumont and Boutry, in preparation). However its resistance to external proteinase K indicated that the enzyme was located inside the organelle.

Deletion analysis of the maize superoxide dismutase presequence has shown that the central portion is necessary for efficient transport but single residue mutation has not been obtained yet and essential residues may be evenly distributed (White and Scandalios, 1989).

The targeting information does not necessarily reside in the $NH_2$-terminal region of precursors. The mitochondrial ADP/ATP carrier from fungi is imported into the organelle inner membrane without any apparent proteolytic processing (Adrian *et al.*, 1986). The targeting signal was found to be contained in an internal segment (residues 72 to 111) (Smagula and Douglas, 1988). In fact, the ADP/ATP carrier consists of three homologous domains of about 100 residues which evolved by triplication of an ancestral gene (Saraste and Walker, 1982). The carboxy terminal of each domain is predicted to form an alpha helical structure (Aquila *et al.*, 1985) wich has features of mitochondrial presequences. The maize ATP/ADP carrier was found to be 74 % homologous to the *Neurospora* enzyme (Baker and Leaver, 1985) and is probably targeted in the same way.

## PRECURSOR RECEPTORS

An early step of protein import is the recognition of precursors by mitochondria. Two outer membrane proteins of *Neurospora* (MOM 19 and MOM 72) have been identified as specific receptors for precursors (Söllner *et al.*, 1989 ; Hartl and Neupert, 1990). In addition, a component of the outer membrane, called the general insertion protein, was identified as an intermediate step to the import pathways of different precursors (Pfaller *et al.*, 1988). This protein could be MOM 38 and ISP 42, two outer membrane protein from *Neurospora* (Hartl and Neupert, 1990) and yeast (Vestweber *et al.*,1989) respectively. No receptors have been identified yet for plant mitochondria. Proteinaceous component(s) of the mitochondrial surface are necessary since trypsin pretreatment of plant mitochondria inhibited the import of $F_1$-ATPase ß subunit precursor from *N. plumbaginifolia* (Chaumont, O'Riordan and Boutry, submitted).

## ENERGY REQUIREMENT

Transport of *N. plumbaginifolia* and *Neurospora* $F_1$-ATPase ß subunits and maize Mn-superoxide dismutase precursors into plant mitochondria requires a membrane potential accross the inner membrane (White and Scandalios, 1987 ; Whelan *et al.*, 1988 ; Chaumont, O'Riordan and Boutry, submitted). This electrical potential may have an electrophoretic effect on the positive charges spread throughout the targeting sequences.

In addition to a membrane potential, nucleoside triphosphates (ATP or analogues) are required to complete import of plant precursors into mitochondria (White and Scandalios, 1989 ; Chaumont, O'Riordan and Boutry, unpublished results). This requirement was previously observed for other organisms (for review, see Hartl *et al.*, 1989). A possible function of nucleoside triphosphates is to keep the precursor in an unfolded conformation (Pfanner *et al.*, 1987 ; Verner and Schatz, 1987), possibly involving a heat shock protein (Deshaies *et al.*, 1988). But ATP or analogues seem also to be required in the matrix (Hwang and Schatz, 1989), where their role is not known exactly. However, a heat shock protein (Hsp 60) localized in the mitochondrial matrix, seems to be involved in an ATP dependent conformational change of imported proteins (Ostermann *et al.*, 1989).

## A CONSERVED MECHANISM

Although the sequence homology between the yeasts *Saccharomyces cerevisiae* and *Schizosaccharomyces pombe*, *Neurospora*, bovine, human and *N. plumbaginifolia* F1-ATPase ß subunit is very high all along the mature protein (over 65 %), the presequences are different in length and primary structure (Figure 1). We have tested whether this dissimilarity was reflecting distinct import mechanisms. We found that the yeast precursors were correctly translocated and processed into plant mitochondria (submitted). The *Neurospora* ATPase ß precursor was also correctly addressed to plant mitochondria (Whelan *et al.*, 1988). Conversely, the plant ß precursor was well imported and cleaved in *S. cerevisiae* mitochondria. Interestingly, trypsin pretreatment of plant mitochondria also inhibited import of *S. pombe* ß precursor (Chaumont and Boutry, submitted). This observation indicates that the heterologous import is also using protein factors of the outer membrane.

A conserved import mechanism is confirmed by *in vivo* experiments. The presequence of the mitochondrial tryptophanyl tRNA synthetase from yeast targets the bacterial ß-glucuronidase into plant mitochondria (Schmitz and Lonsdale, 1989). Conversely, the plant Mn-superoxide dismutase is efficiently imported and processed by yeast mitochondria (Bowler *et al.*, 1989b). Thus import of both yeast and plant precursors into mitochondria of both organisms seem to use a highly conserved pathway.

## ORGANELLE SPECIFICITY

*In vitro* and *in vivo* experiments have shown that the chloroplast presequence of ribulose-1,5-biphosphate carboxylase from *Chlamydomonas reinhardtii* could target other proteins into yeast mitochondria, but with low efficiency (Hurt *et al.*, 1986). Since this import was not affected by protease pre-treatment of mitochondria, this import seemed to bypass proteinaceous receptors (Pfaller *et al.*, 1989). In another set of experiments, protein transport into chloroplast and mitochondria was found to be specific. A same reporter protein (CAT) was specifically targeted to the chloroplast or the mitochondrion according to the fused presequence (Boutry *et al.*, 1987). Other chloroplast presequences did not import proteins into yeast mitochondria (Smeekens *et al.*, 1987). In the same way, the chlorphyl a/b binding protein from *N. plumbaginifolia* is not imported in mitochondria isolated from plant and yeast (Chaumont, O'Riordan and Boutry, submitted).

Thus, in spite of similar properties between mitochondrial and chloroplast presequences, the information for specific targeting resides in this part of the protein which probably interacts with receptors on the outer membrane of each organelle.

## PROSPECTIVES

Characterization of protein import into plant mitochondria is only beginning. Even if the import mechanism seems to be conserved in many aspects between plants and other organisms, more presequences need to be characterized and import processes have to be investigated in detail. In particular, the four presequences discussed in this review are all involved in targeting to the matrix. It will be of interest to identify presequences directing proteins to other compartments of the plant mitochondria. To this respect, we have initiated the purification of the four plant NADH dehydrogenases (Leterme, unpublished results) which were reported to be located in different mitochondrial compartments (Moller and Lin, 1986). On the other hand, the identification, purification (and cristallization) of receptors of the outer membrane of mitochondria and chloroplast should allow to resolve the interesting problem of organelle specificity.

Understanding the functions of mitochondria during plant growth and development would greatly benefit from genetic transformation of the mitochondrial genome. Unfortunately, this requires to solve several problems such as the introduction and stability of the foreign DNA and predicting RNA editing. Meanwhile, an alternative is to transform the nucleus and to import proteins into the organelle using a targeting sequence. In addition, this system provides the possibility to modulate the expression of the genes introduced with appropriate transcription promoters.

ACKNOWLEDGEMENTS

F.C. is recipient of a fellowship from the *Institut pour l'Encouragement de la Recherche Scientifique dans l'Industrie et l'Agriculture.* M.B. is Research Associate of the *Fonds National de la Recherche Scientifique.* This work was supported by grants from the *Service de Programmation de la Politique Scientifique* (86/91-87) and the *Commission of the European Communities* (BAP-0019-B).

REFERENCES

Adrian, G.S., McGammon, M.T., Montgomery, D.L. and Douglas, M.G. (1986) Mol. Cell. Biol. 6, 626-634.
Aquila, H., Link, T.A. and Klingenberg, M. (1985) EMBO J. 4, 2369-2376.
Attardi, G. and Schatz, G. (1988) Ann. Rev. Cell. Biol. 4, 289-333.
Baker, A. and Leaver, C.J. (1985) Nucl. Acid Res. 13, 5857-5867.
Boutry, M. and Chua, N.H. (1985) EMBO J. 4, 2159-2165.
Boutry, M., Nagy, F., Poulsen, C., Aoyagi, K. and Chua, N.H. (1987) Nature 6128, 340-342.
Bowler, C., Alliotte, T., De Loose, M., Van Montagu, M. and Inzé, D. (1989a) EMBO J. 8, 31-38.
Bowler, C., Alliotte, T., Van den Bulcke, M., Bauw, G., Vandekerckhove, J., Van Montagu, M. and Inzé, D. (1989b) Proc. Natl. Acad. Sci. USA 86, 3237-3241.
Breen, G.A.M., Holmans, P.L. and Garnett, K.E. (1988) Biochemistry 27, 3955-3961.
Deshaies, R.J., Koch, B.D., Werner-Washburne, M., Craig, E.A. and Schekman, R. (1988) Nature 332, 800-805.
Hartl, F.U., Pfanner, N., Nicholson, D.W. and Neupert, W. (1989) Biochim. Biophys. Acta 988, 1-45.
Hartl, F.U. and Neupert, W. (1990) Science 247, 930-938.
Hurt, E.C., Soltanifar, N., Goldschmidt-Clermont, M., Rochaix, J.D. and Schatz, G. (1986) EMBO J. 5, 1343-1350.
Hwang, S.T. and Schatz, G. (1989) Proc. Natl. Acad. Sci. USA 86, 8432-8436.
Keegstra, K., Olsen, L.J. and Theg, S.M. (1989) Ann. Rev. Plant Physiol. Plant Mol. Biol. 40, 471-501.
Kim, Y. and Oliver, D.J. (1990) J. Biol. Chem. 265, 848-853.
Kobayashi, K., Iwasaki, Y., Sasaki, T., Nakamura, K. and Asahi, T. (1986) FEBS Letters 203, 144-148.
Moller, I.M. and Lin, W. (1986) Ann. Rev. Plant Physiol. 37, 309-334.
Ohta, S., Tomura, H., Matsuda, K. and Kagawa, Y. (1988) J. Biol. Chem. 263, 11257-11262.
Ostermann, J., Horwich, A.L., Neupert, W. and Hartl, F.U. (1989) Nature 341, 125-130.
Pfaller, R., Steger, H.F., Rassow, J., Pfanner, N. and Neupert, W. (1988) J. Cell. Biol. 107, 2483-2490.
Pfaller, R., Pfanner, N. and Neupert, W. (1989) J. Biol. Chem. 264, 34-39.

Pfanner, N., Tropschug, M. and Neupert, W. (1987) Cell 49, 815-823.
Roise, D., Horvath, S.J., Tomich, J.M., Richards, J.H. and Schatz, G. (1986) EMBO J. 5, 1327-1334.
Rosenberg, L.E., Fenton, W.A., Horwich, A.L., Kalousek, F. and Kraus, J.P. (1987) Ann. New York Acad. Sci. 488, 99-108.
Saraste, M. and Walker, J.E. (1982) FEBS Letters 144, 250-254.
Schmitz, U.K. and Lonsdale, D.M. (1989) Plant Cell 1, 783-791.
Smagula, C. and Douglas, M.G. (1988) J. Biol. Chem. 14, 6783-6790.
Smeekens, S., van Steeg, H., Bauerle, C., Bettenbroek, H., Keegstra, K. and Weisbeek, P. (1987) Plant Mol. Biol. 9, 377-388.
Smeekens, S., Weisbeek, P. and Robinson, C. (1990) Trends in Biochem. Sci. 15, 73-76.
Söllner, T., Griffiths, G., Pfaller, R., Pfanner, N. and Neupert, W. (1989) Cell 59, 1061-1070.
Vassarotti, A., Chen, W.J., Smagula, C. and Douglas, M.G. (1987) J. Biol. Chem. 262, 411-418.
Verner, K. and Schatz, G. (1987) EMBO J. 6, 2449-2456.
Vestweber, D., Brunner, J., Baker, A. and Schatz, G. (1989) Nature 341, 205-209.
von Heijne, G. (1986) EMBO J. 5, 1335-1342.
von Heijne, G., Steppuhn, J. and Herrmann, R.G. (1989) Eur. J. Biochem. 180, 535-545.
Weisbeek, P., Hageman, J., De Boer, D., Pilon, R. and Smeekens, S. (1989) J.Cell Sci. Suppl. 11, 199-223.
Whelan, J., Dolan, L. and Harmey, M.A. (1988) FEBS Letters 236, 217-220.
White, J.A. and Scandalios, J.G. (1987) Biochim. Biophys. Acta 926, 16-25.
White, J.A. and Scandalios, J.G. (1989) Proc. Natl. Acad. Sci. USA 86, 3534-3538.

# ASSEMBLY OF MAIZE STORAGE PROTEINS INTO PROTEIN BODIES IN DEVELOPING ENDOSPERM

Brian A. Larkins, Craig R. Lending, and
Everaldo de Barros

Department of Plant Sciences
University of Arizona
Tucson, Arizona 85721 USA

## INTRODUCTION

Storage proteins are synthesized in developing seeds and stored for use during germination. In legumes and most other dicots, the principle storage proteins are globulins that are distinguished as 7S and 11S based on their sedimentation coefficients.[1] The 7S globulin is composed of three spherical subunits and approximates a flattened, six-pointed spheroid. The 11S globulin is composed of six subunits that are arranged in two sets of trimers stacked one atop the other and offset by 60°. The two types of proteins are randomly distributed within membrane-bound aggregates (protein bodies) that form after deposition of the proteins into the vacuole.

The embryos of cereals contain small amounts of 7S-type globulins, but most of the storage protein is synthesized and deposited in the endosperm.[1] Oat and rice endosperms contain globulins that are structurally related to the 11S globulins in dicot seeds.[2,3] These proteins also form insoluble aggregates within membranes derived from the vacuole. However, the most abundant storage proteins in other cereals are the alcohol-soluble prolamines.

Prolamines have quite varied structures and several pathways of synthesis and deposition.[1,4] In some cereals they aggregate into protein bodies directly within the rough endoplasmic reticulum (RER). In others, they are transported from the RER, via Golgi, to vacuole-derived membranes. Because prolamines are insoluble in aqueous solvents, it has been difficult to study their conformation and structural interactions. A number of questions remain to be answered regarding the evolutionary and genetic origin of these proteins and whether they have roles other than simply storing nitrogen, sulfur, and carbon skeletons for the germinating embryo.

To further characterize the synthesis and deposition of maize prolamines, the zeins, we have used antibodies against several of these structurally distinct proteins to investigate their organization within individual protein bodies and their distribution throughout the endosperm. These analyses reveal an unexpected pattern of association of zeins into protein bodies. The ordered distribution of these proteins in protein bodies may be involved in maintaining the structural integrity of protein bodies and endosperm tissue during seed desiccation.

*Plant Molecular Biology* 2, Edited by R.G. Herrmann and
B. Larkins, Plenum Press, New York, 1991

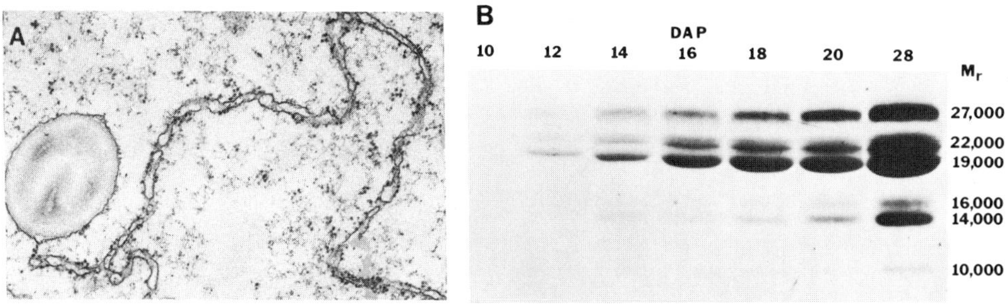


Fig. 1. Analysis of zein proteins from maize endosperm. (A) Electron micrograph of a protein body from a maize endosperm cell at 19 days after pollination. The RER is discontinuous in cross-section but continuous in surface section (×33,500); (B) SDS-polyacrylamide gel analysis of zein proteins from developing endosperm of the maize inbred W64A. The proteins were extracted with 70% ethanol containing 2% 2-mercaptoethanol. DAP, days after pollination; $M_r$, apparent molecular mass.

## ZEIN SYNTHESIS AND PROTEIN BODY FORMATION

Zeins are synthesized by membrane-bound polyribosomes in developing maize endosperm.[5] The proteins are co-translationally transported into the lumen of the RER, where they aggregate into insoluble masses called protein bodies (Fig. 1A). In normal maize genotypes, protein bodies reach a diameter of 1 to 2 $\mu$m. The RER in maize endosperm occurs in long ribbon-like sheets, and continuities between the membranes surrounding the protein body and the RER are not always visible. It is generally thought that protein bodies bud off from the RER, but there is no experimental evidence to confirm this.

Zein synthesis occurs during the period between free nuclear division and seed desiccation. At maturity these proteins account for around 60% of the total endosperm protein. Traditional procedures for extracting zeins involve solubilization in alcoholic solutions in the absence or presence of reducing agents, such as 2-mercaptoethanol. However, we recently described a simpler procedure in which all the zein proteins are extracted in a single fraction.[6] With this procedure it is possible to use an enzyme-linked immunosorbant assay (ELISA) to make quantitative comparisons of zein proteins between different genotypes.

SDS-polyacrylamide gel electrophoretic analysis of zeins from developing seeds reveals a mixture of polypeptides that separate into bands of $M_r$ 27,000, 22,000, 19,000, 16,000, 14,000, and 10,000 (Fig. 1B). Since mRNAs encoding each of these proteins have been cloned and sequenced, some detail is known of the structure of the proteins.[1,7] We designate the proteins of $M_r$ 22,000 and 19,000 as alpha-zeins, the protein of $M_r$ 14,000 as the beta-zein, and the proteins of $M_r$ 27,000 and 16,000 as gamma-zeins.[7] The protein of $M_r$ 10,000 is designated the delta zein. The beta-, gamma-, and delta-zeins are all sulfur-rich proteins, while the alpha-zeins contain only low percentages of cysteine and methionine.

## IMMUNOLOCALIZATION OF ZEINS IN PROTEIN BODIES

Each of these zeins differs sufficiently in structure so that non-crossreacting antisera can be produced.[8] With these antisera it was possible to investigate the location of the alpha-, beta-, and gamma-zeins within protein bodies, as well as to study their distribution in different parts of the endosperm. Figure 2 shows an analysis of protein bodies from endosperm cells at 14 days after pollination. For this study we compared protein bodies in the subaleurone layer (Fig. 2A-D), the first starchy en-

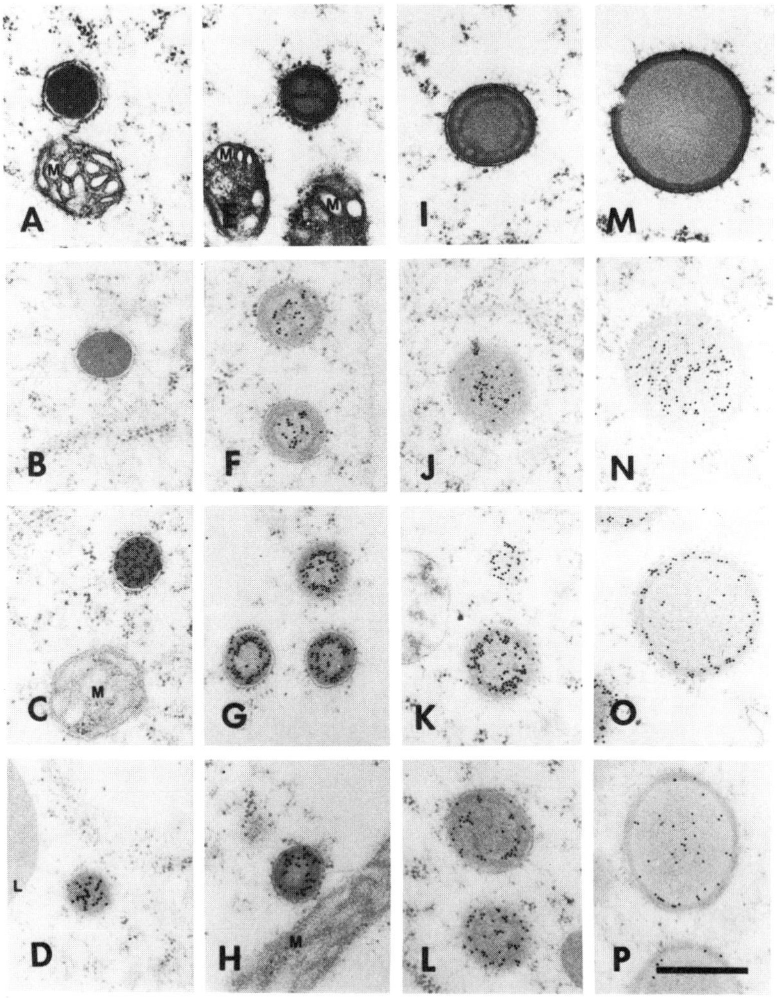


Fig. 2. Electron micrographs showing representative protein bodies from different cell layers of the endosperm at 14 days after pollination. The protein bodies are representative of the morphology and size of protein bodies within the cell layer described and represent a developmental series (from left to right). In general, alpha-zeins are located in the light-staining regions, and beta- and gamma-zeins are located in the dark-staining regions. For immunostaining, sections were incubated in a rabbit antibody and then in goat anti-rabbit/colloidal gold (10 nm). Bar = 0.5 $\mu$m. M, mitochondria; L, lipid body. (A-D) Protein bodies from the subaleurone layer: (A) stained with uranyl acetate and lead citrate; (B) section immunostained for alpha-zein; (C) section stained for beta-zein; (D) section stained for gamma-zein. (E-H) Protein bodies from the first starchy endosperm layer: (G) the labeling for beta-zein is preferentially distributed over the dark-staining region near the interface with the light-staining material. (I-L) Protein bodies from the second starchy endosperm layer. (M-P) Protein bodies from the fifth starchy endosperm layer. The sequence of stainings in (E-H), (I-L), and (M-P) is the same as in (A-D). (From Lending and Larkins[9])

dosperm layer (Fig. E-F), the second starchy endosperm layer (Fig. I-L), and the third through the fifth starchy endosperm layers (Fig. 2M-P).

Protein bodies in the first subaleurone layer stain darkly and uniformly with uranyl acetate and lead citrate (Fig. 2A). However, in the next cell layers the dark-staining material is infiltrated with locules of light-staining material (Fig. 2E). By the fifth starchy endosperm cell layer, the light-staining locules are coalesced (Fig. 2M), and the dark-staining material is mostly peripherally distributed.

Reaction of the protein bodies with antibodies against the alpha-, beta-, and gamma-zeins reveals that the dark- and light-staining regions correspond to different proteins. The small dark-staining protein bodies in the subaleurone layer show no reaction with alpha-zein antibodies (Fig. 2B), whereas the beta- and gamma-zein antibodies react strongly with these dark-staining regions (Fig. 2C and 2D).

The larger protein bodies containing light-staining inclusions show some reaction with the alpha-zein antibodies (Fig. 2F). In these, the staining with beta- and gamma-zein antibodies is more peripheral (Fig. 2G and 2H). In protein bodies from the fourth and fifth endosperm cell layers, the central part of the protein body stains intensely with the alpha-zein antibodies (Fig. 2J and 2N). In these protein bodies, most of the staining with beta- and gamma-zein antibodies is peripheral (Fig. 2K and 2L), although there is some staining in the central part of the protein body (Fig. 2O and 2P).

The differences in size and zein composition of protein bodies in the subaleurone cells correspond with stages of cell maturity. The cells just beneath the aleurone are the youngest and are meristematic, while the deeper, underlying cells are more mature.[10,11] Thus, the small protein bodies containing only beta- and gamma-zeins just beneath the aleurone represent the earliest developmental stage of a protein body. The larger protein bodies containing alpha-zein that are found in deeper cell layers represent more fully developed protein bodies.

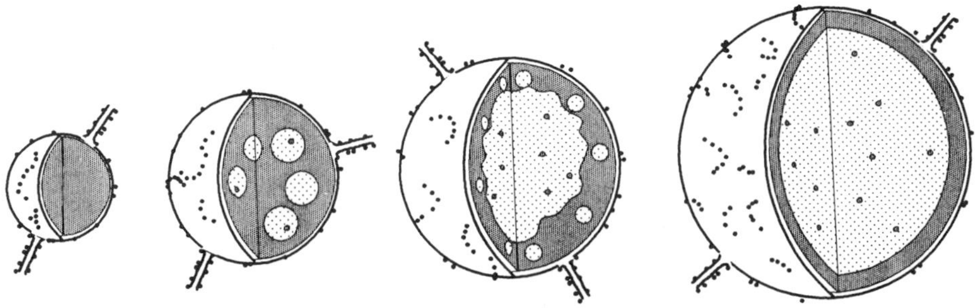

Fig. 3. Development of protein bodies in maize endosperm (not drawn to scale). The heavily stippled regions correspond to regions that are rich in beta- and gamma-zeins, and the lightly stippled regions correspond to regions rich in alpha-zeins. The protein body is surrounded by RER (the dark dots represent ribosomes). Some beta- and gamma-zeins are found within the regions that consist primarily of alpha-zein (heavily stippled inclusions). We do not know whether these are strands of protein that run through the protein body or small, individual aggregates. (From Lending and Larkins[9])

## A STRUCTURAL MODEL FOR PROTEIN BODIES

Based on the differences in the distribution of zeins in protein bodies, we proposed a model for the pattern of zein deposition during protein body formation.[9] As illustrated in Figure 3, initial accretions within the RER consist of dark-staining deposits of both beta- and gamma-zeins and contain no alpha-zein proteins. Subsequently, alpha-zeins begin to accumulate as discrete, light-staining locules within the matrix of beta- and gamma-zeins. Eventually, the locules of alpha-zeins fuse and fill the center of the protein body. The beta- and gamma-zeins form a more or less continuous layer at the periphery of the protein body, but small dark-staining patches of beta-zein and, more commonly, gamma-zeins may remain within the interior.

We do not have sufficient data to explain the significance of the developmental appearance of the different zeins in protein bodies. Little is understood about the secondary and tertiary structure of the sulfur-rich zeins, and we do not know the nature of the structural interactions between them. We know that these proteins are linked by disulfide bonds, but it is not clear whether they are crosslinked to one another. We recently determined that the delta-zein is principally found within the interior rather than the surface of the protein body (data not shown), so not all of the sulfur-rich zeins are on the surface.

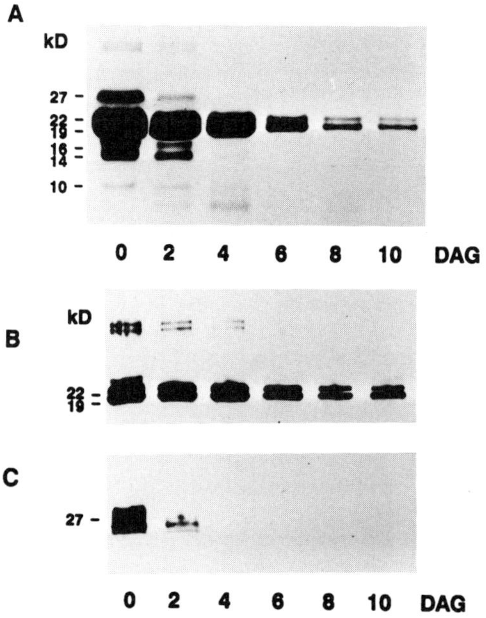


Fig. 4. SDS-polyacrylamide gel electrophoretic separation of zein proteins from germinating maize seeds. Total zein was extracted from seeds at 0 to 10 DAG with 70% ethanol containing 1% 2-mercaptoethanol and 0.1% Triton X-100. (A) Samples of 2.5 $\mu$l from each stage; gel stained with Coomassie blue; (B) samples were diluted 1:100; following electrophoretic separation, the proteins were transferred to nitrocellulose membrane and reacted with anti-alpha zein serum; (C) samples were diluted 1:20; following electrophoretic separation the proteins were transferred to nitrocellulose membrane and reacted with anti-gamma serum. kD, kilodalton; DAG, days after germination. (From de Barros and Larkins[12])

Evidence supporting this model of zein deposition within protein bodies also comes from analyses of the proteolysis of zeins during germination.[12] In germinating maize seed, as in other cereals, proteases are synthesized and secreted into the endosperm to degrade the storage proteins, so that their nitrogen, sulfur, and carbon skeletons can be utilized in the developing embryo.[13] The initial enzymes, identified as proteinase A,[14] are sulfhydryl proteases that cleave internal peptide bonds and make the molecules susceptible to other proteases, including carboxypeptidases, and amino- and dipeptidases.

Figure 4 shows an SDS-polyacrylamide gel analysis of zein proteins recovered from the endosperm of germinating maize seed. By two days after germination (DAG), there was a noticeable reduction in the amount of the gamma-zeins ($M_r$ 27,000 and 16,000) and beta-zein ($M_r$ 14,000) (Fig. 4A); none of these proteins were detectable by four DAG. This rapid disappearance of gamma-zein was further confirmed by a sensitive immunoblot analysis (Fig. 4C). In contrast to the sulfur-rich zein proteins, the alpha-zeins appear to be hydrolyzed more slowly in germinating seeds. In Coomassie blue-stained gels (Fig. 4A) and immunoblots (Fig. 4B), these proteins are observed as late as ten DAG.

Proteases degrading zeins are released from the aleurone into the starchy endosperm during the early stages of germination.[12] It is likely, therefore, that proteins on the surface of protein bodies are among the first to become hydrolyzed. Furthermore, since protein bodies near the surface of the grain (close to the aleurone) contain a greater proportion of the beta- and gamma-zeins, one might expect to detect a rapid and preferential disappearance of these proteins. Alternatively, the rapid disappearance of the gamma-zeins may be associated with their solubility in aqueous solvents following reduction and proteolytic cleavage. The alpha-zeins, in addition to being more centrally located, are much more insoluble, and this may also contribute to their slower hydrolysis during germination.

Although we do not know the structural basis for the distribution of various zeins within protein bodies, it is possible that mutations leading to variation in the amounts and distribution of these proteins may result in phenotypic differences in mature kernels. We recently reported that modified *opaque-2* mutants, which have a normal rather than a starchy endosperm, contain three to four times as much gamma-zein as typical normal and *opaque-2* genotypes.[6] In addition, *floury-2* mutants, which also have a soft, floury endosperm,[15] contain protein bodies with an atypical, asymmetric organization of zein proteins. Whether or not these phenotypes result from qualitative or quantitative differences in zeins and their organization into protein bodies remains to be determined.

## REFERENCES

1. M.A. Shotwell, and B.A. Larkins, The biochemistry and molecular biology of seed storage proteins, *in*: "The Biochemistry of Plants, A Comprehensive Treatise," A. Marcus, ed., Academic Press, New York (1989).
2. M.A. Shotwell, C. Afonso, E. Davies, R.S. Chesnut, and B.A. Larkins, Molecular characterization of oat seed globulins, *Plant Physiol.* 87:698 (1988).
3. F. Takaiwa, S. Kikuchi, and K. Oono, *Mol. Gen. Genet.* 208:15 (1987).
4. P.R. Shewry, and A.S. Tatham, The prolamin storage proteins of cereal seeds: structure and evolution, *Biochem. J.* 267:1 (1990).
5. B.A. Larkins, and W.J. Hurkman, Synthesis and deposition of zein in protein bodies of maize endosperm, *Plant Physiol.* 62:256 (1978).
6. J.C. Wallace, M.A. Lopes, E. Paiva, and B.A. Larkins, New methods for extraction and quantitation of zeins reveal a high content of $\gamma$-zein in modified *opaque-2* maize, *Plant Physiol.* 92:191 (1990).

7. B.A. Larkins, J.C. Wallace, G. Galili, C.R. Lending, and E.E. Kawata, Structural analysis and modification of maize storage proteins, *Develop. Indust. Microbiol.* 30:203 (1989).
8. C.R. Lending, A.L. Kriz, B.A. Larkins, and C.E. Bracker, Structure of maize protein bodies and immunocytochemical localization of zeins, *Protoplasma* 143:51 (1988).
9. C.R. Lending, and B.A. Larkins, Changes in the zein composition of protein bodies during maize endosperm development, *Plant Cell* 1:1011 (1989).
10. E.L. Fisk, The chromosomes of *Zea mays*, *Am. J. Bot.* 14:53 (1927).
11. L.F. Randolph, Developmemtal morphology of the caryopsis in maize, *J. Agric. Res.* 53:881 (1936).
12. E.G. de Barros, B.A. and Larkins, Purification and characterization of zein-degrading proteinases from germinating maize endosperm, *Plant Physiol.* (submitted).
13. G.B. Fincher, Molecular and cellular biology associated with endosperm mobilization in germinating cereal grains, *Ann. Rev. Plant Physiol. Plant Mol. Biol.* 40:305 (1989).
14. A.D. Shutov, and I.A. Vaintraub, Degradation of storage proteins in germinating seeds, *Phytochem.* 26:1557 (1987).
15. D.D. Christianson, U. Khoo, H.C. Nielsen, and J.S. Wall, Influence of *opaque-2* and *floury-2* genes on formation of proteins in particulates of corn endosperm, *Plant Physiol.* 53:851 (1974).

# GENETIC AND MOLECULAR STUDIES ON ENDOSPERM

# STORAGE PROTEINS IN MAIZE

Natale Di Fonzo, Hans Hartings, Massimo Maddaloni, Stefan Lohmer*, Richard Thompson*, Francesco Salamini*, Mario Motto

Istituto Sperimentale per la Cerealicoltura, Via Stezzano 24, 24100 Bergamo, Italy
* Max-Planck-Institut für Züchtungsforschung - Köln (FRG)

## INTRODUCTION

In maize (Zea mays), as in other cereals, the endosperm is specialized in the accumulation of starch and proteins as the major food reserves for germinating plant embryo. The composition of these reserves is responsable for determining the nutritional quality of the grain with respect to both human and livestock consumption. Because of the relevance of zein synthesis in determining the nutritional value of maize kernel proteins understanding of the genetic organization and regulation of genes involved in storage protein accumulation may result in novel approaches to improve nutritional value of maize grain.

In this communication we summarize recent progress toward understanding the regulation of zein gene expression. Brief reviews of the mutations that affect zein gene expression and the molecular characterization of the b-32 and Opaque-2 regulatory loci are also presented.

## MOLECULAR FEATURES OF MAIZE STORAGE PROTEINS

The storage proteins can be separated into four distinct groups by SDS-PAGE having apparent mol wts of 23,000 to 21,000 (collectively termed the 22 kDa class), 20,000 to 18,000 (20 kDa class), 16,000 to 14,000 (15 kDa class) and 12,000 to 10,000 (10 kDa class)[1]. When separated by 2-dimensional gel electrophoresis, up to 30 individual polypeptides can be identified[2,3,4]. The charge heterogeneity observed in isoelectric focusing (IEF) polyacrylamide gels was found to be the result of amino acid differences among the IEF zein fractions[5]. Furthermore, the charge heterogeneity is independent of environmental interaction and developmental stage of the endosperm[5].

The 22 kDa and 20 kDa zein classes, the major storage proteins, constitute about 70-80% of this fraction. The second largest fraction of storage proteins constituents are the 28,000 and 15,000 polypeptides, also called reduced soluble protein (RSPs) or glutelin-2, which account for about 15% of total endosperm protein[6,7,8] and differ in amino acid composition and sequence from zeins. The two lower mol wt prolamines (15 and 10 kDa classes), even though having an amino acid composition similar to the other zeins, contain a higher level of methionine and cysteine. Based on immunocytochemical analysis, the 22 and 20 kDa components are distributed throughout the matrix of the protein body, whereas the 28, 15, and 10 kDa proteins are localized near its surface[9].

*Plant Molecular Biology* 2, Edited by R.G. Herrmann and
B. Larkins, Plenum Press, New York, 1991

At the molecular level, the 22 kDa and 20 kDa zeins are encoded by a multigene family (80-100 genes) with a pool of active and inactive genes [10,11,12,13]. Each subfamily encodes for structurally distinct groups of polypeptides belonging predominantly, but not exclusively, to single size classes [14]. Sequence analysis of zein cDNA and genomic clones indicates that zein proteins contain a central domain of 8 to 10 repeats of a block of 20 amino acids[15,16,17]. The basic overall similarity of all 22- and 20-kDa zeins sequenced to date suggests that the genes originated by duplication of an ancestral gene followed by subsequent mutations[17].

The comparison among nucleotide sequences of 5 zein genomic clones encoding the 20-kDa zeins reveals 93% to 98% homology on their coding regions; this homology persists also in the 5' and 3' DNA sequences flanking the gene[18].

Zein genomic clones contain consensus CAAT and TATA sequences in their 5' noncoding regions and one or more consensus polyadenylation sequences (AATAAA/G) in their 3' noncoding regions[16, 19]. The relative positions of these putative regulatory sequences are similar to those reported in other eukaryotic genes. In a comparison between the 5' flanking sequences of zein genes coding for the two size classes of zeins (20- and 22-kDa) the largest single stretch of homology was identified as a 15-bp region at -336 relative to the translation starts, which is highly conserved in all zein genes[20]. The 15-bp sequence, CACATGTGTAAAGGT, can bind nuclear proteins[21] and may play a role in the regulation of zein gene expression. Additional work[22] indicated that an upstream sequence in the 5' noncoding sequence of a gene encoding a Mr 19,000 zein, delimited by nucleotides -337 and -125 with respect to the mRNA cap site, is required for maximal transcription of the gene.

## CHROMOSOMAL LOCATION OF ZEIN GENES

Taking advantage of the great genetic variability, existing among maize inbreds, in IEF and SDS polyacrilamide gel electrophoresis banding patterns, 20 structural loci, encoding for different zeins, have been mapped. The preferential location of these genes is into two regions: the short arms of chromosome 7 and chromosome 4. Interestingly in these regions are also located some of the genes, i.e. Opaque-2 (O2), Defective endosperm*-B30 (De*-B30), Floury-2 (Fl2), controlling the rate of storage protein accumulation.

## REGULATION OF STORAGE PROTEIN ACCUMULATION

The storage proteins of maize represent a striking example of coordinated gene expression during development. The individual polypeptides are synthesized at a constant rate relative to each other as indicated by the qualitative and quantitative invariance of the IEF and SDS patterns. Even the amount of nutrients, especially the relative content of carbohydrates and amino acids, transported from the plant to the seeds, affects the rate of storage protein accumulation.

This complex type of control implies the existence of several regulatory functions acting on the transcription of the structural genes, on the translational efficiency and stability of their products and on the ability to sense the state of development of the cells. It is not surprising, therefore, that the expression of few genes is controlled by several loci acting on the onset and on the rate of zein accumulation[23] and that several mutant alleles of these loci have been discovered which disrupt the coordinated accumulation of maize storage proteins. The mutant alleles at these loci reduce the zein level to a different extent and affect the deposition of all the zein polypeptides or of some group preferentially[24].

The first mutants studied related to the zein regulatory loci were opaque-2 (o2) and floury-2 (fl2). Subsequently, the opaque-7 (o7), opaque-6 (o6), floury-

3 (fl3), Mucronate (Mc), and De*-B30 mutants were identified. The o2 and De*-B30 mutants preferentially reduce the level of the 22 kDa zeins, o7 preferentially reduces the 20 kDa zeins, while fl2, Mc, and o6 reduce the synthesis of all the zein classes to the same extent. These regulatory genes have been mapped to regions nearby those occupied by some zein genes; O2 is on the short arm of chromosome 7, Fl2 is on the short arm of chromosome 4, O7 is on the long arm of chromosome 10 and O6 is on chromosome 8 (long arm) linked or corresponding to the locus coding for the b-32 protein[25,26]. The mechanism by which these mutations influence zein synthesis is complex and not fully understood. The genes o2, o7, and fl2 decrease zein mRNA populations while o6, Mc, and De*-B30 lower the level of translatable zein mRNA[11,27,28] although this later set has not yet been thoroughly probed at the mRNA level to determine if mRNA population sizes are affected.

In o2, a recessive mutation, the onset of zein synthesis is delayed and total accumulation of zeins in the mature kernel is significantly reduced[29]. The O2 locus must act through a diffusable regulatory molecule, because it affects genes not linked to it suggesting that the O2 gene is a trans-acting regulator of zein expression. It has been found that a 32 kDa protein (b-32 protein) is missing in o2 and o6 endosperm mutants[30]. The b-32 protein is present in the soluble cytoplasm of the endosperm of normal kernels but absent in seven recessive alleles at the O2 as well as in o6 mutant[30]. However, the o2 gene does not encode the b-32 protein. Instead it appears to be possibly encoded by, or tightly linked, to the O6 locus. A hierarchy between the O2 and O6 loci has been proposed: O2 activates O6, which positively controls the expression of zein genes[30]. In such a mechanism, zein genes would be placed in the terminal position of a cascade of a regulatory pathway. However, in o6 mutants, both major zein classes are reduced whereas in o2 only the 22 kDa class is reduced. Thus O2 must also act via proteins other than the O6 product to control zein synthesis. Further evidence indicates that o2 and o6 mutants have decreased levels of lysine-ketoglutarate reductase, an enzyme involved in lysine catabolism[31]. The o7 mutation causes a reduction in the synthesis of the 20 kDa zein class. The fact that the two regulatory loci affect zein polypeptides suggests that multiple regulatory pathways may be active in zein synthesis[24]. Both probably function via diffusable regulatory molecules because they can affect either linked and unlinked zein genes.

The fl2 mutant appears to act by a different mechanism than the opaque-2 and opaque-7 mutants; it reduces overall zein synthesis in the kernel and acts in a dosage-dependent manner[32,33]. The mutant fl3 shows a similar dosage effect because, as the dose increases, zein production is decreased. De*-B30 mutation is dominant and is closely linked to O2. This mutant again preferentially reduces expression in the 22 kDa zein class.

An association has been found between dominant or semidominant zein regulatory loci (fl2 Mc, and De*-B30) and the overproduction of a 70 kDa protein[24,34]. It is possible that this b-70 protein is involved in the zein synthetic-secretory system.

The interactions between some of the mutant alleles previously described have also been investigated analyzing the level of zeins in double mutants with pairs of alleles in all possible combinations. The recessive alleles o2 and o7 are epistatic over the gene action conditioned by Fl2, in the sense that Fl2-fl2 differences are no more evident in o2 or o7 recessive situations. In similar experiments concerning the effects of the o2 allele on O7 and Mc, epistatic relationships among the cited loci were not found[28,35,36]. Not all of the zein structural genes are equally affected by a particular regulatory locus, and therefore, genes belonging to the two pathways of regulation may possess regulatory functions used in different combinations: the O2 locus, for instance, affects O6 which is reported to positively control the synthesis of 20 kDa and 22 kDa proteins; by itself O2, however, plays a positive role only on the 22 kDa class.

## CLONING OF REGULATORY GENES

Further progress to increase understanding of the normal control of zein genes is expected from the molecular cloning and analysis of the regulatory loci affecting the accumulation of zein polypeptides in developing kernels. In order to study at the molecular level the b-32 gene,[37,38] cDNA clones containing sequences related to b-32 were isolated by screening a lambda gt11 library for the expression of the b-32 protein using antibodies raised against the purified protein. By Northern analysis the size of the b-32 mRNA was estimated to be 1.2 kb. Hybrid-selected translation assay showed that the message codes for a protein with an apparent mol wt of 30-35 kDa. The nucleotide sequence of a full-length cDNA clone contains an open reading frame of 909 nucleotides which codes for a polypeptide of mol wt 32,430 (and shows the following features: no signal peptide is observable; it contains seven tryptophan residues, an amino acid absent in maize storage proteins; polar and hydrophobic residues are spread along the sequence; several pairs of basic residues are present in the N-terminal region). From the structural analysis of the b-32 protein it was possible to predict the existence of an acidic central domain separated by two compact domains which may accomplish a regulatory role by interacting with specific chromatin structures located nearby the transcriptional machinery at the 5' flanking region of zein genes. Three b-32 genomic clones were selected from two genomic libraries obtained from the maize inbred lines W64A and A69Y. The nucleotide sequences of the complete coding region of each b-32 gene, as well as long stretches of their 5' and 3' flanking regions, were determined suggesting that introns are not present in the b-32 genomic sequences. Minor variations among the three genes and the reported b-32 cDNA sequence indicate that they constitute a gene family showing a characteristic polymorphism. Such a polymorphism is highly evident in large segments of the upstream regulatory sequences. Interestingly, when compared with cDNA (W64A) or with gene b-32.120 (W64A), the genes b-32.129 (W64A) and b-32.152 (A69Y) show three 1-nucleotide insertions in the central part of the coding region (positions 324, 441, 537), resulting in a completely different sequence of the b-32 protein central domain. In all cases, variations in the N- and C- terminal domains account only for microheterogeneity. The two b-32 genes isolated from the inbred line W64A are very similar with regard to the flanking sequences and they possess the same motifs that apparently are relevant for gene expression.

It would be of great interest to determine if both b-32 protein messages are present and active (giving rise to two differentiated albumin components) during maize endosperm development. Potential on-and-off regulation of each message at different times during development could result in a clue to the role of this protein in the expression of storage protein genes and the control that the Opaque-2 locus exerts on them. We have carried out preliminary experiments (chemical cleavage of Tyr and Met residues) on b-32 protein from mature endosperm (40 days after pollination). Peptides were recovered, electrophoresed, blotted on nitrocellulose paper and detected by an antibody reaction. Our results indicate that the b-32 component present in mature endosperm cells derives most, if not all, from clone b-32.120. As far as the reported possibility that the b-32 protein is coded by the O6 locus[30], recent linkage and biochemical data[25,26,39] have provided positive circumstantial evidence. Further work is needed in this area to assess the existence and nature of b-32 mutations in o6 endosperms.

As far as the O2 locus is concerned,[40,41] genetic analyses suggested that this locus codes for a positive, trans-acting, transcriptional activator of the zein protein genes. Because the molecular cloning of the O2 locus and the analysis of its structure and expression, could contribute to an understanding of the nature of its regulatory effects, transposon tagging with the transposable element Ac (Activator) was carried out. Several transposon-induced mutable alleles were obtained and genetic data suggested that in the case of the o2-m5 a functional Ac element is responsable for the observed somatic instability. The isolation of genomic clones containing flanking sequences corresponding to the O2 gene was possible by screening an o2-m5 genomic library with a probe corresponding to internal Ac sequences usually absent in the defective element Ds. A sequence representing a XhoI fragment of 0.9 kB lying, in the 6IP clone, adjacent to the Ac element, was subcloned and utilized to prove that it corresponded to a part of the O2 gene. To obtain this information we made use of: 1) DNAs from several reversions originating from the unstable o2-m(r) allele, which, when digested with SstI, showed a correct 3.4 kb fragment typical of non-inserted alleles of the O2 locus; and 2) recessive alleles of the O2 locus which were devoid of a 2.0 kb mRNA, present on the contrary in the wild type and in other zein regulating mutants different from O2. The structure of this gene was determined by sequence analysis of both genomic and cDNA clones.

The size of O2 mRNA is 1751 bp |poly(A) tail not included| of 1380 bp preceeded by three short ORFs of 3, 21 and 20 amino acid residues. The main ORF comprises 1362 bp and is composed of six exons ranging in size from 465 to 61 bp and five introns of 678 bp to 83 bp. As can be seen in the cDNA clone, a poly(A+) addition site is present 40 bp after the TGA-stop codon. A putative site for the transcription initiation (TGACAT) can be found 300 bp upstream of the start-codon, a sequence (CTATTG) present 341 bp upstream of the start-codon is the best fit for a putative TATA box. A putative protein 454 amino acids long was derived by the theoretical translation of the genomic sequences corresponding to exons. The length of the polypeptide is in agreement with the estimated length of about 58 kDa of the in vitro-translation product of the O2 transcript.

The protein translation of the nucleotide sequence of the main ORF present in O2 reveals further interesting features. A striking homology of this amino acid sequence was found with proteins known to be involved in transcriptional activation, such as the gene products of GCN4, jun, fos, myc and C/EBP[42]. The region of homology is restricted to O2 protein domains encoded by exons 4 and 5. In the first of these two domains, 11 residues out of 30 are positively charged. This amino acid sequence is particularly well conserved compared, for instance, with the one present in jun, fos and GCN4 (for instance, at the nucleotide level, from amino acid 253 to 260 the homology between GCN4 and O2 is 86%). That this region is required for DNA binding is confirmed by in vitro studies with fos where mutations in this region abolished binding[43]. This DNA binding, basic domain is common to other DNA binding proteins possessing leucine repeats[44]. In the o2 protein basic domain, the presence of lysine and arginine residues which favor alpha-helices, and the absence of proline which is not present in this type of secondary structure, reinforced at a functional level the cited homology with the transcription regulatory proteins of yeast and mammals.

The second protein domain of Opaque-2 that, with the exception of a first critical leucine, is encoded by exon 5, is characterized by a periodic repetition of leucine conforming to the Leu-$X_6$-Leu motif reported for the so-called leucine zipper present in proteins encoded by several transcriptional activators[42]. A significant homology with GCN4 is moreover evident at intermediate positions between two leucines, particularly in the region coding the amino acids from 260 to 276. In the same protein domain of O2 containing the described leucine zipper, a second stretch of amino acids conforms to the leucine periodicity Leu-$X_6$-Leu and extends for two further groups of heptamers.

The very C-terminal portion of the major o2 protein may have a third domain with interesting binding properties. In exon 6 two cysteines are present which are followed by two other residues and then by histidine. It is known that such sequences may constitute a metal binding core[45]. The sequence reported, for instance, is a component of the binding site of plastocyanin, a Cu metalloprotein[46]. A similar Cys-$X_2$-His motif participates in DNA fingers capable of nucleic acid binding[47], or flanks a Cys-$X_5$-Cys loop responsible for Zn binding[48]. The presence of the major O2-encoded product of the two Cys-$X_2$-His repeats spaced by eight amino acids suggests a folding or at least a binding capacity of the molecule which may be of interest for gene regulation. The association of finger structures with transcriptional activation is in fact so well accepted that the presence of the corresponding motifs may be considered diagnostic of new transcription factors[49].

## Molecular interaction between the O2 and b-32 loci

To obtain experimental evidence that the Opaque-2 locus encodes a trans acting factor, we have started protein-DNA binding experiments. For this purpose the O2 gene was expressed as a protein in E. coli (fusion protein to glutathione-S-transferase, gst).

Band shift assays were performed in the presence of purified gst-o2 fusion protein and a 250 bp long b-32 promoter fragment (-350 to -100). Two complexes with different electrophoretic mobilities appeared indicating clearly that the gst-o2 fusion protein binds to the b-32 promoter. To demonstrate that the complexes formed are not the result of a potential DNA binding activity of gst, the labeled b-32 promoter fragment was incubated in the presence of gst alone. Under this condition no complex formation was seen. It is concluded that the retarded complexes are the results of DNA binding activity of the o2 protein.

Is is also inferred that the control of the O2 locus on the b-32 gene (possibly O6)[30] is efficiently mediated by a protein-promoter interaction. A similar approach has been used by Schmidt and collaborators[50] to show that O2 protein bound to two specific regions of the 5' side of the coding sequence in a zein genomic clone of the 22 kDa class.

## REFERENCES


1. E. Gianazza, P. G. Righetti, F. Pioli, E. Galante, and C. Soave, Size and charge heterogeneity of zein in normal and opaque-2 maize endosperms, Maydica 21:1 (1976).
2. G. Hagen and I. Rubenstein, Two-dimensional gel analysis of the zein proteins in maize, Plant Sci. Lett. 19:217 (1980).
3. A. Vitale, C. Soave, and E. Galante, Peptide mapping of IEF zein components from maize, Plant Sci. Lett. 18:57 (1980).
4. F. A. Burr and B. Burr, In vitro uptake and processing of prezein and other maize preproteins by maize membranes, J. Cell Biol. 90:427 (1981).
5. P. G. Righetti, E. Gianazza, A. Viotti, and C. Soave, Heterogeneity of storage proteins in maize, Planta 136:115 (1977).
6. C. M. Wilson, P. R. Shewry, and B. J. Miflin, Maize endosperm proteins compared by sodium dodecylsulphate polyacrylamide gel electrophoresis and isoelectric focusing, Cereal Chem. 58:275 (1981).
7. A. Vitale, E. Smaniotto, R. Longhi, and E. Galante, Reduced soluble proteins associated with maize endosperm protein bodies, J. Exp. Bot. 33:439 (1982).
8. M. D. Ludevid, M. Torrent, J. A. Martinez-Izquierdo, P. Puigdomenech, and J. Palau, Subcellular localization of gluteline-2 (Zea mays L.) endosperm, Plant Mol. Biol. 3:227 (1984).
9. C. R. Lending, A. L. Kriz, B. A. Larkins, and C. E. Bracker, Structure of maize protein bodies and immunocytochemical localization of zeins, Protoplasma 143:51 (1988).
10. G. Hagen and I. Rubenstein, Complex organization of zein genes in maize, Gene 13:239 (1981).
11. F. A. Burr and B. Burr, Three mutations in Zea mays affecting zein accumulation, J. Cell Biol. 94:201 (1982).
12. A. Viotti, D. Abildsten, N. Pogna, E. Sala, and V. Pirrotta, Multiplicity and diversity of cloned zein cDNA sequences and their chromosomal location, EMBO J. 1:53 (1982).

13. G. Heidecker and J. Messing, Structural analysis of plant genes, Ann. Rev. Plant Physiol. 37:439 (1986).
14. A. Viotti, G. Cairo, A. Vitale, and E. Sala, Each zein gene class produce polypeptides of different sizes, EMBO J. 4:1103 (1985).
15. D. E. Geraghty, J. Messing, and J. Rubenstein, Sequence analysis and comparison of cDNAs of the zein multigene family, EMBO J. 1:1329 (1982).
16. K. Pedersen, J. Devereux, D. R. Wilson, and B. A. Larkins, Cloning and sequence analysis reveal structural variation among related genes in maize, Cell 29:1015 (1982).
17. A. Spena, A. Viotti, and V. Pirrotta, A homologous repetitive zein genes, EMBO J. 1:1589 (1982).
18. A. L. Kriz, R. S. Boston, and B. A. Larkins, Structural and transcriptional analysis of DNA sequences flanking genes that encode 19 kilodalton zeins, Mol. Gen. Genet. 207:90 (1987).
19. J. Messing, D. Geraghty, G. Heidecker, N. T. Hu, J. Kridl, and I. Rubenstein, Genetic engineering of plant genes, in: "Genetic engineering of plants", A. Hollaender, T. Kosuge, and C.A. Meredith, eds., Plenum Press, N.Y. (1983).
20. J. W. S. Brown, C. Wandelt, C. Feix, G. Neuhaus, and H. G. Schweiger, The upstream regions of zein genes. Sequence analysis and expression in the unicellular alga Acetabularia, Eur. J. Cell Biol. 42:161 (1986).
21. U. G. Maier, J. W. S. Brown, C. Toloczyki, and G. Feix, Binding of a nuclear factor to a consensus sequence in the 5' flanking region of zein genes from maize, EMBO J. 6:17 (1987).
22. D. L. Roussell, R. S. Boston, P. B. Goldsbrough, and B. A. Larkins, Deletion of DNA sequences flanking an Mr 19,000 zein gene reduces its transcriptional activity in heterologous plant tissue. Mol. Gen. Genet. 211:202 (1988).
23. L. A. Manzocchi, M. G. Daminati, and E. Gentinetta, Viable defective endosperm mutants in maize, Maydica 25:199 (1980).
24. C. Soave and F. Salamini, Organization and regulation of zein genes in maize endosperm, Phil. Trans. R. Soc. London 304:341 (1984).
25. H. Hartings, S. Lanzini, P. Ajmone Marsan, L. Pirovano, J. Palau, N. Di Fonzo, Regolazione della sintesi zeinica in mais: caratterizzazione della famiglia genica b-32, Atti XXXIII SIGA, p. 115 (1989).
26. H. W. Bass, P. H. Sisco, D. L. Murray, and R. S. Boston, Probes for the b-32 protein hybridize to loci on 7L and 8L, Maize News Lett. 64:97 (1990).
27. P. Langridge, J. A. Pintor-Toro, and G. Feix, Transcriptional effects of the opaque-2 mutation of Zea mays, Planta 156:166 (1982).
28. F. Salamini, N. Di Fonzo, E. Fornasari, E. Gentinetta, R. Reggiani, and C. Soave, Mucronate, Mc, a dominant gene of maize which interact with opaque-2 to suppress zein synthesis, Theor. Appl. Genet. 65:123 (1983).
29. R. A. Jones, B. A. Larkins, and C. Y. Tsai, Storage protein synthesis in maize. II. Reduced synthesis of a major zein component by the opaque-2 mutant of maize, Plant Physiol. 59:525 (1977).
30. C. Soave, I. Tardani, N. Di Fonzo, and F. Salamini, Regulation of zein level in maize endosperm by a protein under control of the opaque-2 and opaque-6 loci, Cell 27:403 (1981).
31. P. Arruda, J. M. G. Tocozzilli, A. Vieira, and W. J. Da Silva, Degradation of lysine in maize: possible pathway and genetic control, Maize Gen. Coop. News Lett. 58:50 (1984).
32. K. H. Lee, R. A. Jones, A. Dalby, and C. Y. Tsai, Genetic regulation of storage protein synthesis in maize endosperm, Biochim. Gen. 14:641 (1976).
33. C. Soave, S. Dossena, C. Lorenzoni, N. Di Fonzo, and F. Salamini, Expressivity of the floury-2 allele at the level of zein molecular components, Maydica 23:145 (1978).
34. E. Galante, A. Vitale, L. Manzocchi, C. Soave, and F. Salamini, Genetic control of a membrane component and zein deposition in maize endosperm, Mol. Gen. Genet. 192: 316 (1983).

35. N. Di Fonzo, E. Fornasari, F. Salamini, R. Reggiani, and C. Soave, Interaction of the mutants floury-2, opaque-7 with opaque-2 in the synthesis of endosperm proteins, J. Hered. 71:397 (1979).
36. E. Fornasari, N. Di Fonzo, F. Salamini, R. Reggiani, and C. Soave, Floury-2 and opaque-2 interaction in the synthesis of zein polypeptides, Maydica 27:185 (1982).
37. N. Di Fonzo, H. Hartings, M. Brembilla, M. Motto, C. Soave, E. Navarro, J. Palau, W. Rohde, and F. Salamini, The b-32 protein from maize endosperm, an albumin regulated by the O2 locus: nucleic acid (cDNA) and amino acid sequences, Mol. Gen. Genet. 212:481 (1988).
38. H. Hartings, N. Lazzaroni, P. Ajmone Marsan, A. Aragay, R. Thompson, F. Salamini, N. Di Fonzo, J. Palau, and M. Motto, The b-32 protein from maize endosperm: characterization of genomic sequences encoding two alternative central domains, Plant Mol. Biol.: in press (1990).
39. L. Manzocchi, C. Tonelli, G. Gavazzi, N. Di Fonzo, and C. Soave, Genetic relationship between o6 and pro-1 mutants in maize, Theor. Appl. Genet. 72:778 (1986).
40. M. Motto, M. Maddaloni, G. Ponziani, M. Brembilla, R. Marotta, N. Di Fonzo, C. Soave, R. Thompson, and F. Salamini, Molecular cloning of the o2-m5 allele of Zea mays using transposon marking, Mol. Gen. Genet. 212:488 (1988).
41. H. Hartings, M. Maddaloni, N. Lazzaroni, N. Di Fonzo, M. Motto, F. Salamini, and R. Thompson, The O2 gene which regulates zein deposition in maize endosperm encodes a protein with structural homologies to transcriptional activators, EMBO J. 8:2795 (1989).
42. W. H. Landschulz, P. F. Johnson, S. L. McKnight, The leucine zipper: a hypothetical structure common to a new class of DNA binding proteins, Science 240: 1759 (1988).
43. M. Neuberg, M. Schuermann, J. B. Hunter, and R. Müller, Two functionally different regions in Fos are required for the sequence-specific DNA interaction of the Fos/jun protein complex, Nature 338:589 (1989).
44. T. Kouzarides and E. Ziff, The role of the leucine zipper in the fos-jun interaction, Nature 336:646 (1988).
45. J. Miller, A. D. McLachlan, and A. Klug, Repetive zinc-binding domains in the protein transcription factor IIIA from Xenopus oocytes, EMBO J. 4:1609 (1985).
46. W. Haehnel, Plastocyanin, in: "Encyclopedia of Plant Physiology", A. Pirson, M. H. Zimmermann, eds., Springer, NY (1986).
47. J. M. Berg, Potential metal-binding domains in nucleic acid binding proteins, Science 232:485 (1986).
48. J. M. Berg, Proposed structure for the zinc-binding domains from transcription factor IIIA and related-proteins, Proc. Natl., Acad. Sci. USA 85:99 (1988).
49. R. M. Evans and S. M. Hollenberg, Zinc fingers: gilt by association, Cell 52:1 (1988).
50. R. J. Schmidt, F. A. Burr, M. J. Aukerman, and B. Burr, Maize regulatory gene opaque-2 encodes a protein with a "leucine-zipper" motif that binds to zein DNA, Proc. Natl. Acad. Sci. USA 87:46 (1990).

# ASSEMBLY PROPERTIES OF MODIFIED SUBUNITS IN THE GLYCININ SUBUNIT FAMILY

William J.P. Lago, M. Paul Scott and Niels C. Nielsen

Crop Production and Pathology Research, United States Department of Agriculture, Agricultural Research Service, Agronomy Department, Purdue University, West Lafayette, IN 47907

Soybeans provide an interesting experimental system to study for a variety of reasons. They are an extremely rich but underutilized source of dietary protein that could be used directly for human nutrition. In order to make soybean protein more acceptable, efforts will need to be directed toward correcting an imbalance in the essential amino acids, improving the digestibility of the major seed storage proteins, and reducing the tendency of soybean proteins to elicit serious antigenic responses in animals that consume them. The soybean seed storage proteins are also an interesting biological system to study. Two types of proteins, glycinin and $\beta$-conglycinin, account for about 70% of the total seed protein, and these are synthesized during an intensive burst of metabolic activity in a carefully orchestrated developmental sequence. This developmental process is characterized not only by temporal and spatial control of gene expression, but also by a complicated assembly process that involves at least three subcellular compartments and several proteolytic cleavages. These latter events occur as the storage proteins make their way through the endomembrane system of the cell to their final site of deposition in the seed. The purpose of this communication is to review some of the key elements in the pathway that leads to the deposition of the glycinin in the protein bodies of soybean cotyledons.

Glycinin is isolated from seeds as a 350,000 dalton complex that consists of six subunits. The subunits contained in the complex are the products of a small multi-gene family. At least five genes are involved that are designated $Gy_1$ to $Gy_5$, and the corresponding proteins that are produced from them are identified as G1 to G5 (1). The five subunits are grouped according to sequence similarity, with Group 1 consisting of G1, G2, and G3 and Group 2 containing G4 and G5. The pathway followed during synthesis and assembly of glycinin subunits has been described earlier (2,3). In summary, pre-proglycinin monomers are synthesized into the endoplasmic reticulum like secreted proteins in animal cells, and during this process a signal peptide is removed cotranslationally. Once in the endoplasmic reticulum, the proglycinin monomers assemble into trimers, and then the trimers move from the endoplasmic reticulum, through the golgi apparatus, and into vacuolar protein bodies where the proglycinin subunits are cleaved. After post-translational cleavage, each mature subunit consists of an acidic polypeptide chain that is linked to a basic chain by a single disulfide bond (4). The trimers subsequently become assembled into a higher order complex, although the features of their surface of the subunit, the majority of the alterations we have evaluated have been directed to this region. Using the *in vitro* assembly systems described in detail elsewhere (3,5,6), it has been possible to demonstrate that deletion of all DNA encoding the HVR from a G4 cDNA results in the production of proglycinin monomers (G4D25) that still assembled into both trimers and hexamers (6, see also Table 1). To extend these observations, an attempt was made to insert additional methionine into the HVR using one of two approaches. In the first, multiple copies of a symmetrical hexanucleotide (CGCATG) were inserted into a unique Accl site in the HVR of a cDNA that encoded G4 proglycinin. These multiple insertions resulted in an alternating ARG-MET sequence in a region of the HVR that is predicted to be in a turn secondary structure conformation. Up to five methionines could be inserted in this fashion with no measurable effect on either the extent or rate of proglycinin self-assembly. However, the introduction of more than 5 copies of the six base pair oligonucleotide into the region encoding the HVR resulted in an unstable plasmid, apparently because of recombination at the site where the multiple copies of the oligomer were inserted (6). In the second approach, we determined whether there were examples in nature where multiple methionine or cysteine residues were contained in exposed turn or helical regions of proteins whose three dimensional structure was known. Upon searching the EMBL data base for such motifs, regions in citrate synthase and ferridoxin were found to contain such structures. To test the effect of introducing these structures into G4 proglycinin, oligonucleotides that encoded these high methionine structures were synthesized and inserted into the HVR of pSP65/248 (6). When these constructions were tested *in vitro*, the proglycinin monomers whose synthesis they directed were able to assemble at the same rate and to the same extent as unmodified G4 controls.

The results summarized thus far in this communication concern an evaluation of proglycinin assembly *in vitro*. The logical extension of these experiments is to ask whether results obtained *in vitro* bear any relationship to what transpires *in vivo*. To address this question, plasmid constructs encoding the subunits referred to in Table 1 were ligated to an authentic $Gy_4$ promoter region that contained a signal sequence and an authentic 3' flanking region, and these were used to transform tobacco. Tobacco was used for these experiments for several reasons. First, it would be a problem to distinguish between engineered and native subunits of glycinin if soybean was used. Also, it is relatively easy to transform and regenerate tobacco, whereas this is not straightforward in soybean. Finally, seeds of tobacco contain a homologous 11S globulin and the regulatory sequences of soybean seed storage protein genes are known to be recognized by the tobacco developmental programs (8). We suspected that the proglycinin monomers might interact and co-assemble with the homologous tobacco legumins, and thereby would not be detrimental to normal seed development in the transformed plants.

For transformation, standard triparental mating procedures involving *Agrobacterium tumifaciens* were used. The binary vectors that were employed for this purpose contained either an engineered or an unaltered cDNA derived glycinin gene, as well as a gene for kanamycin resistance. Southern analyses were performed on the regenerated tobacco plants to determine that the genes were transferred, and that the glycinin gene had not undergone gross rearrangement during transformation. To detect and quantitate glycinin in the seeds from the transformed tobacco, ELISA was used in conjunction with polyclonal anti-glycinin antibodies. Because we were concerned that the anti-glycinin antibodies might not recognize the modified subunits as efficiently as the unmodified ones, all of the subunits described in Table 1 were tested for their ability to interact with the antibodies. While small decreases in the ability of the

Table 1. Assembly Assay Results of Insertion and Deletion Modified Proglycinins and Ranges of Protein Accumulation in Transgenic Seeds.

| Protein | Modification | Self-Assembly | Relative Amount ($\mu$g/mg seed) |
|---|---|---|---|
| G4 | None | + | 3.5 - 33.5 |
| G4GMR5 | HVR insertion HGMRGMKHASFLSS | + | 6.0 - 13.0 |
| G4GMR6 | HVR insertion HASFLSSHGMRGMK | + | 1.0 - 2.5 |
| G4GMT2 | HVR insertion HGPMTEMNGPHGPMTEMNGP | + | 1.0 - 1.6 |
| G4RM5 | HVR insertion RMRMRMRMRM | + | 1.0 - 9.0 |
| G4Δ25 | HVR deletion | + | 18.5 |
| G4HX | Basic polypeptide deletion | - | 0.0 |

Insertion site in each case was immediately after position 330, in the $G_4$ cDNA, for the RM repeat and immediately after position 304 for the other insertions (1). Relative amounts are based on the highest accumulation found in a plant in the series containing the insert noted using G4 as the standard.

antibodies to recognize certain of the modified subunits were observed, statistical evaluation revealed that these decreases were not significant ($P > 90\%$, data not shown).

As a control, tobacco was transformed with a cDNA derived gene that encoded unmodified G4. Upon quantitating G4 in seeds from the transformed and regenerated tobacco plants, individual transformants were found to accumulate glycinin in amounts that ranged from 3.5 to 33.5 μg of protein per mg seed. The 10-fold difference in accumulation could be explained in at least two ways. Random insertion of the modified gene into different regions of the tobacco genome could in result different levels of gene expression. Alternatively, the differences could be due to copy number differences among individual transformants. Experiments are underway to distinguish among these various possibilities.

Table 1 also summarizes data about accumulation of several modified glycinin subunits in seeds from transformed tobacco plants. Plants with the G4GMR5 gene insertion accumulated a modified glycinin subunit that varied in concentration among transformants from 6.0 to 13.0 μg per mg seed. Plants containing the gene encoding G4D25, a modified G4 subunit in which 90% of the hypervariable region was deleted, accumulated the subunit at a concentration of 18.5 μg per mg seed in the single transformed plant that has been evaluated. Plants transformed with modified glycinin genes that encoded G4RM5 and G4GMR5 also accumulated sizable amounts of modified glycinin subunits.

While our evaluations of transformed tobacco that contain modified glycinin subunits are still in progress, not all constructions seem to accumulate effectively. Of the five transformed plants that have been tested which contain G4GMT2, only two contained detectable amounts of the modified subunits. They accumulated at 1.0 and 1.6 μg per mg seed, amounts far below that accumulated by the controls. A similar situation occurred in the case of transformants that contained the glycinin gene for the G4GMR6 subunit. The low levels of accumulation occurred despite the fact that these particular modified subunits appeared to assemble to a normal extent and at a normal rate compared to G4 controls *in vitro* (6). The reason for this discrepancy in not clear. It may be that certain types of modifications are discriminated against by the plant. Among the possibilities being considered is that these modifications affect targeting of the precursors to the protein body. In this regard, deletions in the basic polypeptide also are of interest. Those modifications, such as in G4HX, that eliminate the capacity for assembly *in vitro* are not accumulated at a detectable level in any of the eight independently transformed tobacco plants that were tested.

Together, our results demonstrate that it is technically feasible to modify storage protein genes to improve nutritional quality, and then have the modified subunits accumulate to high levels in the seed. Obviously, if such modifications are to have a maximum beneficial effect, they should be introduced into an agronomically superior genetic background that is devoid of glycinin subunits with low nutritional quality. In the case of glycinin, the two group 2 subunits, G4 and G5, are the likely candidates for elimination because they each contain only 3 methionine residues (1). With this in mind, the soybean germplasm collection has been screened to identify varieties that lacked individual glycinin subunits. Null alleles for the glycinin subunits G3, G4 and G5 exist. The *gy3* null allele is the result of a chromosomal rearrangement and can be detected using its unique RFLP (9). The *gy4* null allele results from a point mutation in the initiator ATG codon (10), and its absence form the seed can be detected using SDS-PAGE. The nature of the *gy5* null allele has not established yet, but it can be detected using SDS-PAGE in the presence of urea (11). By combining "null alleles" of genes for several of the glycinin subunits, a genetic background has been constructed in which modified proteins can be introduced and have a maximal effect.

The recent development of ballistic transformation (12) will make it possible to transform our soybean lines which lack several of the glycinin subunits with genes modified to improve nutritional quality. By bombarding soybean embryos with gold particles coated with DNA, chimeric plants in which some of the cells have been stably transformed can be produced. If any of the originally transformed cells give rise to germ cells, then the S1 seeds from these cells will produce a transformed plant that will be useful for the production of varieties with improved quality. Thus, the technology is currently in place for all of the steps in this plan, making engineered soy protein a likely component of the markets of the future.

## REFERENCES


1. Nielsen, N.C., Dickinson, C.D., Cho, T.-J., Thanh, V.H., Scallon, B.J., Fischer, R.L., Sims, R.L., Drews, G.N. and Goldberg, R.B., 1989, Characterization of the glycinin gene family in soybean, *Plant Cell* 1:313-328.

2. Chrispeels, M.J., Higgins, T.J.V., and Spencer, D., 1982, Assembly of storage protein oligomers in the endoplasmic reticulum and processing of the polypeptides in the protein bodies of developing pea cotyledons, *J. Cell Biol.* 93:306-313.

3. Dickinson, C.D., Hussein, E.A. and Nielsen, N.C., 1989, Role of post-translational cleavage in glycinin assembly, *Plant Cell* 1:459-469.

4. Staswick, P.E., Hermodson, M.A., and Nielsen, N.C., 1984, Identification of the cystines which link the acidic and basic components of the glycinin subunits, *J. Biol. Chem.* 259:13431-13435.

5. Dickinson, C.D., Floener, L.A., Lilley, G.G. and Nielsen, N.C., 1987, Self-assembly of proglycinin and hybrid proglycinin synthesized *in vitro* from cDNA, *Proc. Natl. Acad. Sci. USA* 84:5525-5529.

6. Dickenson, C.D., Scott, M.P., Hussein, H.A., Argos, P. and Nielsen, N.C., 1990, Effect of structural modifications on the assembly of a glycinin subunit. *Plant Cell* (in press).

7. Argos, P., Narayona, S.V.L. and Nielsen, N.C., 1985, Structural similarity between legumin and vicilin storage proteins from legumes. *EMBO J.* 4:1111-1118.

8. Chen, Z.-L., Pan, N.-S., and Beachy, R.N., 1988, A DNA sequence element that confers seed-specific enhancement to a constitutive promoter, *EMBO J.* 7:297-302.

9. Cho, T.J., Davies, C.S., Fischer, R.L., Tumer, N.E. Goldberg R.B., and Nielsen, N.C., 1989, Molecular characterization of an aberrant allele for the *Gy3* glycinin gene, *Plant Cell* 1:339-350.

10. Scallon, B.J., Dickinson, C.D., and Nielsen, N.C. (1986) Characterization of a null-allele for the Gy4 glycinin gene from soybean. *Molecular and General Genetics* 208:107-113.

11. Fontes, E.P.B., Moreira, M.A., Davies, C.S., and Nielsen, N.C., 1984, Urea-elicited changes in relative electrophoretic mobility of certain glycinin and β-conglycinin subunits, *Plant Physiology* 76:840-842.

12. McCabe, D.E., Swain, W.F., Martinell, G.J., Christou, P., 1988, Stable transformation of soybean (*Glycine max*) by particle acceleration, *Biotechnology* 6:923-926.

# THE PROLAMINS OF THE TRITICEAE (BARLEY, WHEAT AND RYE) : STRUCTURE, SYNTHESIS AND DEPOSITION

Peter R. Shewry[a], Arthur S. Tatham[a], Gillian Hull[b], Nigel G. Halford[b], Janey Henderson[c], Nick Harris[c] and Martin Kreis[d]

[a] Department of Agricultural Sciences, University of Bristol, AFRC Institute of Arable Crops Research, Long Ashton Research Station, Long Ashton, Bristol, BS18 9AF, UK
[b] AFRC Institute of Arable Crops Research, Rothamsted Experimental Station, Harpenden, Herts, AL5 2JQ, UK
[c] Department of Biological Sciences, University of Durham, Durham, DH1 3LE, UK
[d] Université de Paris-Sud, F-91400 Orsay Cedex, France

## INTRODUCTION

Prolamins are present only in the seeds of grasses, and form the major storage protein fraction in all cultivated cereals except oats and rice (where they are minor components). They are characterised by solubility, either in the native state or after reduction of inter-chain disulphide bonds, in alcohol/water mixtures but insolubility in water or dilute saline, and contain high proportions of glutamine and proline.

Wheat, barley and rye are closely related species which are classified together in the Triticeae, and this relationship is reflected in the structure and genetical control of their prolamins. In the present paper we briefly summarise the structures, distribution and evolutionary relationships of these proteins, and discuss structural features in relation to their deposition and packaging within protein bodies.

### Protein Groups and Families

Two-dimensional analyses of prolamin fractions from members of the Triticeae show many individual components, usually about 20 to 30 in diploid barley and rye but 50 or more in hexaploid bread wheat. These vary in their $M_r$'s (about 30 to 100,000) pI's and amounts. There is also considerable variation in the individual components present in different genotypes of the same species. Despite this high degree of polymorphism the prolamins of all three species can be classified into

Table 1. Families of S-rich, S-poor and HMW prolamins present in barley, wheat and rye

| Group | Family | Species | Specific Name | Characteristics |
|---|---|---|---|---|
| S-rich | γ-type | Wheat | γ-gliadin | [a]Mr 30-45,000 |
| | | Rye | 40Kγ-secalin | (75Kγ-secalins |
| | | | 75Kγ-secalin | Mr 75,000) |
| | | Barley | γ-hordein | |
| | α-type | Wheat | α-gliadin | 30-40 mol % Gln |
| | | Rye | absent | 15-20 mol % Pro |
| | | Barley | absent | 2-3 mol % Cys. |
| | aggregated type | Wheat | LMW glutenin subunits | |
| | | Rye | absent | |
| | | Barley | B hordein | |
| S-poor | | Wheat | ω-gliadin | [a]Mr 40-70,000 |
| | | Rye | ω-secalin | 40-50 mol % Gln |
| | | Barley | C hordein | 20-30 mol % Pro |
| | | | | 0 mol % Cys |
| | | | | 0-0.2 mol % Met |
| | | | | 8-9 mol % Phe |
| HMW | | Wheat | HMW glutenin subunits | [b]Mr 65-90,000 |
| | | Rye | HMW secalin | 30-35 mol % Gln |
| | | Barley | D Hordein | 10-16 mol % Pro. |
| | | | | 15-20 mol % Gly |
| | | | | 0.5-1.5 mol % Cys |

[a] Mr by SDS-PAGE

[b] Absolute Mr

only three groups, and into families within these. The properties of these groups and the names of the constituent polypeptides in the three species are summarised in Table 1.

Amino Acid Sequences

Complete amino acid sequences are now available for representative members of all three groups, derived in most cases from cloned cDNA's and genes. The structures of an S-poor ω-secalin, an HMW glutenin subunit and an S-rich B1 hordein are summarised in Table 2. They all contain extensive repeated sequences, based on one or more short peptide motifs, which account for about 30% (B1 hordein) to almost 100% (ω-secalin) of the total protein. These are flanked by

Table 2. Structural Domains of Typical S-rich, S-poor and HMW Prolamins

| Protein Group | S-rich | S-poor | HMW |
|---|---|---|---|
| Species | Barley | Rye | Wheat |
| Protein | B1 hordein | ω-secalin | Subunit 1By9 |
| N-terminal Domain | | | |
| No of residues | absent | 12 | 104 |
| No. of cysteines | - | 0 | 5 |
| Repetitive Domain | | | |
| No of residues | 79 | 322 | 538 |
| No. of cysteines | 0 | 0 | 1 |
| Consensus motifs | PQQPX(X)(X) | PQQPFPQQ | GYYPTSLQQ PGQGQQ |
| C-terminal Domain | | | |
| No. of residues | 195 | 4 | 42 |
| No. of cysteines | 7 | 0 | 1 |
| Whole Protein | | | |
| No. of residues | 274 | 338 | 684 |
| No. of cysteines | 7 | 0 | 7 |

Based on sequences reported by Halford *et al.* (1987), Forde *et al.* (1985) and unpublished results of G.A. Hull, M. Kreis and P.R. Shewry.

short non-repetitive sequences in the ω-secalin and HMW subunit. Short (up to 15 residue) non-repetitive sequences are also present at the N-termini of most S-rich prolamins (see Kreis *et al.*, 1985), but are absent from B1 hordein.

Protein Conformations

The secondary structures of prolamins have been studied by computer prediction, and by spectroscopic (circular dichroism, nuclear magnetic resonance and fourier-transform infra-red) analyses of purified proteins and peptides and synthetic peptides based on unique sequences and repeat motifs (Tatham *et al.*, 1984, 1985a,b,c, 1987, 1989, 1990a,b,c; Tatham & Shewry, 1985; Field *et al.*, 1986, 1987). The non-repetitive domains are generally rich in α-helix, with smaller proportions of β-sheet, β-turn and unordered structure. They therefore resemble many globular metabolic proteins. In contrast the repetitive domains contain little or no α-helix or β-sheet, but are rich in β-turns. Structural prediction indicates that these are regularly arranged, within individual repeat motifs and spanning the junctions between adjacent motifs. Their regularity reflects the primary structure and is most marked in the S-poor and HMW prolamins which have highly conserved repeat structures (Fig. 1). They are less regularly distributed in the S-rich prolamins where the repeat structure is more degenerate (see Tatham *et al.*, 1990c).

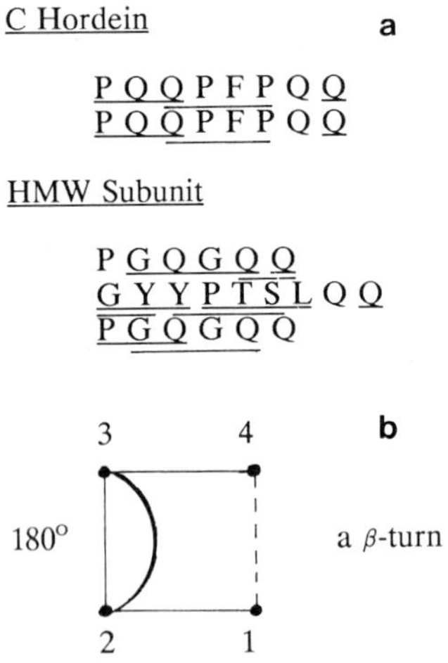


Fig. 1a Predicted β-turns in consensus repeat motifs of C hordein and a wheat HMW subunit (see Tatham *et al.*, 1985a,b).
b Diagrammatical representation of a β-turn. 1,2,3 and 4 represent amino acid residues. The turn is often (but not always) stabilised by a hydrogen bond (broken line) between residues 1 and 4.

Hydrodynamic studies of C hordein and an HMW subunit show that both are semi-rigid rods, with diameters of about 18 Å and lengths varying (depending on the solvent) from 280 to 360 Å for C hordein and about 500 to 600 Å for the HMW subunit (Table 3). Field *et al.* (1986, 1987) proposed that these rods resulted from the formation of a novel spiral supersecondary structure, based on repetitive β-turns. A structure of this type (called a β-spiral) is formed by a synthetic polypentapeptide based on a repeat motif of elastin (Venkatachalam & Urry, 1981), a protein of mammalian connective tissue.

Assuming that the repetitive domain of C hordein and the HMW subunits form β-spirals of similar dimensions to that formed by the polypentapeptide of elastin it is possible to calculate their overall dimensions. The values obtained (Table 3) are broadly consistent with the experimentally determined dimensions of the whole proteins.

More recently Miles *et al.* (1990) have obtained direct evidence for the formation of a spiral structure by HMW subunits, using scanning tunnelling microscopy. This showed the presence of aligned rods, with diameters of about 19.5 Å. In additional diagonal striations with a pitch of about 15 Å were also visible, presumably corresponding to turns of the spiral.

The observed alignment of the rods could have resulted from protein-protein interactions, indicating a relevance to protein packing within the protein body. However, an alternative explanation is that it resulted from interaction with the graphite substrate used for sample preparation.

Table 3. Comparison of the determined molecular dimensions of C hordein and HMW subunit 2 with the calculated dimensions of β-spirals formed by their repetitive domains (from Field *et al.*, 1986,1987)

| | | **Observed Dimensions in Å** | | **Calculated Dimensions of ß-spiral** |
|---|---|---|---|---|
| | | dilute* HAc | aqueous* alcohols | |
| C hordein | L | 282 | 363 | 294 |
| | D | 19.1 | 17.0 | 18 |
| HMW subunit | L | 504 | 492 | 440** |
| | D | 17.5 | 17.9 | 18 |

*at 30°C
**repetitive domain only

Tatham *et al.* (1990b) suggested that the repetitive domains of γ-type gliadins could also form a loose spiral structure but this appears unlikely for the α-type and aggregated families (Tatham *et al.*, 1990c).

## Prolamin Synthesis and Deposition

Prolamins are synthesised on the rough ER with a signal peptide which directs the newly-synthesised protein into the lumen of the ER. There is some uncertainty about the subsequent pathway of deposition and the origin of the protein bodies. Miflin *et al.* (1981) prepared protein bodies from developing endosperms by sucrose density ultracentrifugation and showed the presence of marker enzymes associated with the ER (NADH cytochrome c reductase) but not with the vacuole (RNA-ase, phosphodiesterase and N-acetylglucosaminidase). They proposed that the protein bodies were formed by direct deposition and accummulation within the ER, as in maize (Larkins & Hurkman, 1978). However, studies using electron microscopy and immunogold labelling have provided convincing evidence that at least some protein bodies in wheat and barley are formed by transport via the Golgi apparatus into the vacuole (Kim *et al.*, 1988; Cameron Mills & von Wettstein, 1980; unpublished results of J. Henderson, N. Harris, P.R. Shewry and B.J. Miflin, and of D. Bechtel and P.R. Shewry). It is possible to reconcile these two conflicting lines of evidence by suggesting that two alternative pathways may operate within the same tissue. The protein bodies formed in the early stages of development (when it is easier to fix and section tissue) may be of vacuolar origin, whereas direct accummulation within the lumen of the ER may occur later in development when the cells become distended with starch obstructing protein transport.

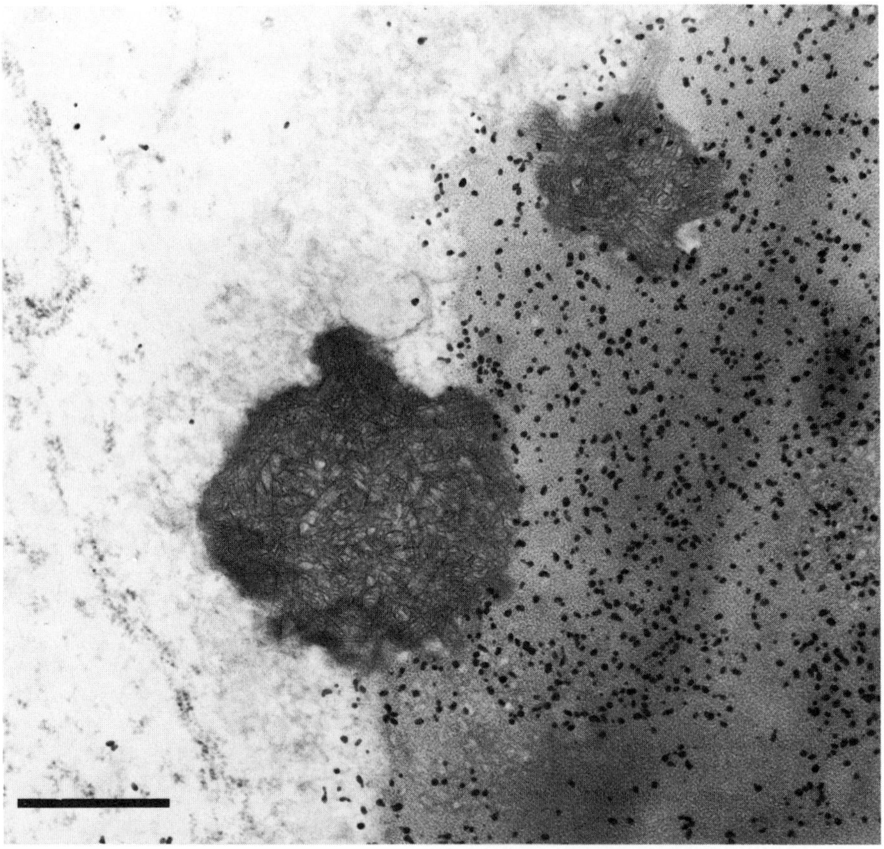

Fig. 2. Immunogold labelling of a protein body present in an endosperm cell of developing barley, using an antibody raised against C hordein. Note that only the homogeneous phase is labelled. (Unpublished result of J. Henderson, B.J. Miflin, P.R. Shewry and N. Harris). The bar represents 0.5 μm.

Transmission electron microscopy of developing endosperms of barley and wheat shows protein bodies consisting predominantly of a homogeneous matrix with more densly stained granular inclusions (see Miflin & Shewry, 1979; Cameron-Mills & von Wettstein, 1980). The latter do not react with antisera raised against prolamins (Fig. 2), but do react with antisera to the triticins (Bechtel *et al.*, 1989) a group of globulin storage proteins which are homologous with the 11S legumins (Singh *et al.*, 1988). The antiserum also gave a less intense reaction with the homogeneous phase, possibly due to antigenic cross-reaction with prolamins. Thus, in the Triticeae, as in oats (Lending *et al.*, 1989), the protein bodies appear to consist of prolamins and globulin in separate phases. It is not known whether these phases have separate origins, as proposed for oats by Lending *et al.* (1989).

## Prolamin Processing

Apart from signal peptide cleavage, the only known post-translational processing event is disulphide bond formation. Individual prolamins may be monomers with intra-chain disulphide bonds (α-type and γ-type S-rich prolamins) or present in polymers stabilised by inter-chain disulphide bonds (aggregated S-rich and HMW prolamins). The latter many have Mr's in excess of 1 x $10^6$ (Field *et al.*, 1983a,b), and comprise the glutenin fraction which has a key role in determining the technological properties of wheat doughs. Although protein disulphide isomerase has been demonstrated to be present in the ER of developing wheat grains (Roden *et al.*, 1982) and catalyses disulphide bond formation in prolamins of wheat synthesised *in vitro* (Bulleid & Freidman, 1988a,b), its role *in vivo* remains to be established conclusively. In addition it is not known whether any rearrangement or synthesis of disulphide bonds occurs in the Golgi apparatus or the protein bodies.

Thus, the precise pathway of prolamin synthesis processing and deposition in members of the Triticeae remains unclear and is an important target for future research.

## REFERENCES


Bechtel, D.B., Wilson, J.D. and Shewry, P.R. 1989. Identification of legumin-like proteins in thin sections of developing wheat endosperms by immunocytochemical procedures. Cereal Foods World, 34: 784.

Bulleid, N.J. and Freedman, R.B. 1988a. The transcription and translation *in vitro* of individual cereal storage protein genes from wheat (*Triticum aestivum* cv. Chinese Spring). Biochem. J., 805-810.

Bulleid, N.J. and Freedman, R.B. 1988b. Defective co-translational formation of disulphide bonds in protein disulphide-isomerase-deficient microsomes. Nature 335: 649-651.

Cameron-Mills, V. and von Wettstein, D. 1980. Endosperm morphology and protein body formation in developing wheat grain. Carls. Res. Commun., 45: 577-594.

Field, J.M., Shewry, P.R., Burgess, S.R., Forde, J., Parmar, S and Miflin, B.J. 1983a. The presence of high molecular weight aggregates in the protein bodies of developing endosperms of wheat and other cereals. J. Cereal Sci. 1: 33-41.

Field, J.M., Shewry, P.R., and Miflin, B.J. 1983b. Aggregation states of alcohol-soluble storage proteins of barley, rye, wheat and maize. J.Sci. Food Agric. 34: 362-369.

Field, J.M., Tatham, A.S., Baker, A. and Shewry, P.R. 1986. The structure of C hordein. FEBS Lett., 200: 76-80.

Field, J.M. Tatham, A.S., and Shewry, P.R. 1987. The structure of a high molecular weight subunit of wheat gluten. Biochem. J., 247: 215-221.

Forde, B.G., Heyworth, A., Pywell, J. and Kreis, M. 1985. Nucleotide sequence of a B1 hordein gene and the identification of possible upstream regulatory elements in endosperm storage protein genes from barley, wheat and maize. Nucleic Acids Res., 13: 7327-7339.

Halford, N.G., Forde, J., Anderson, O.D., Greene, F.C. and Shewry, P.R. 1987. The nucleotide and deduced amino acid sequences of an HMW glutenin subunit gene from chromosome 1B of bread wheat (*Triticum aestivum* L.), and comparison with those of genes from chromosomes 1A and 1D. Theor. Appl. Genet., 75:117-126.

Kim, W.T., Franceschi, V.R. Krishnan, H. and Okita, T.W. 1988. Formation of wheat protein bodies: involvement of the Golgi apparatus in gliadin transport. Planta, 176: 173-182.

Kreis, M., Shewry, P.R., Forde, B.G., Forde, J. and Miflin, B.J. 1985. Structure and evolution of seed storage proteins and their genes, with particular reference to those of wheat, barley and rye. in: "Oxford surveys of plant cell and molecular biology", Vol. 2, B.J. Miflin, ed., pp 253-317, Oxford.

Larkins, B.A. and Hurkman, W.J. 1978. Synthesis and deposition of zein in protein bodies of maize endosperm. Plant Physiol. Lancaster, 62: 256-263.

Lending, C.R., Chesnut, R.S., Shaw, K.L. and Larkins, B.A. 1989. Immunolocalisation of avenin and globulin storage proteins in developing endosperm of *Avena sativa* L. Planta, 178: 315-324.

Miflin, B.J. and Shewry, P.R. 1979. The synthesis of proteins in normal and high lysine barley seeds, in: "Recent advances in the biochemistry of cereals", D. Laidman and R.G. Wyn Jones, eds, pp. 239-273, Academic Press; London.

Miflin, B.J., Burgess, S.R. and Shewry, P.R. 1981. The development of protein bodies in the storage tissues of seeds. J. exp. Bot., 32: 119-219.

Miles, M.J., Carr, H.J., McMaster, T., Belton, P.S., Morris, V.J., Field, J.M., Shewry, P.R. and Tatham, A.S. 1990. Scanning tunnelling microscopy of a wheat gluten protein shows a novel supersecondary structure. Submitted.

Roden, L.T., Miflin, B.J. and Freedman, R.B. 1982. Protein disulphide-isomerase is located in the endoplasmic reticulum of developing wheat endosperm. FEBS Lett., 138: 121-124.

Singh, N.K., Shepherd, K.W., Langridge, P., Gruen, L.C., Skerrit, J.H. and Wrigley, C.W. 1988. Identification of legumin-like proteins in wheat. Plant Mol. Biol., 11: 633-639.

Tatham, A.S., and Shewry, P.R. 1985. The conformation of wheat gluten proteins. The secondary structures and thermal stabilities of α, β, γ, and ω-gliadins. J. Cereal Sci. 3: 103-113.

Tatham, A.S., Shewry, P.R. and Miflin, B.J. 1984.Wheat gluten elasticity: a similar molecular basis to elastin? FEBS Lett. 177: 205-208.

Tatham, A.S., Drake, A.F. and Shewry, P.R. 1985a. A conformational study of 'C' hordein, a glutamine and proline-rich cereal seed protein. Biochem. J. 226, 557-562.

Tatham, A.S., Miflin, B.J., and Shewry, P.R. 1985b. The β-turn conformation in wheat gluten proteins: relationship to gluten elasticity. Cereal Chem. 62: 405-412.

Tatham, A.S., Shewry, P.R. and Belton, P.S. 1985c. Carbon-13 NMR study of C hordein. Biochem. J., 232: 617-620.

Tatham, A.S., Field, J.M., Smith, S.J. and Shewry, P.R. 1987. The conformation of wheat gluten proteins 2. Aggregated gliadins and low molecular weight subunits of glutenin. J. Cereal Sci. 5: 203-214.

Tatham, A.S., Drake, A. F. and Shewry, P.R. 1989. Conformational studies of a synthetic peptide corresponding to the repeat motif of C hordein. Biochem. J. 259, 471-476.

Tatham, A.S., Marsh, M.N., Wieser, H. and Shewry, P.R. 1990a. Conformational analysis of peptides corresponding to the coeliac toxic regions of wheat α-gliadin. Submitted.

Tatham, A.S., Masson, P. and Popineau, Y. 1990b. Conformational studies of peptides derived by the enzymic hydrolysis of a gamma-type gliadin. J. Cereal Sci., 11:1-13.

Tatham, A.S., Shewry, P.R. and Belton, P.S. 1990c. Structural studies cereal prolamins, including wheat gluten. Advances in Cereal Science and Technology. Vol. 10. Ed. Y. Pomeranz, AACC St. Paul, Minnesota. In press.

Venkatachalam, C.M. and Urry, D.W. 1981. Development of a linear helical conformation from its cyclic correlate. β-spiral model of the elastin polypentapeptide (VPGVG)n. Macromol., 14: 1225-1232.

# GENES INDUCED BY ABSCISIC ACID AND WATER STRESS IN MAIZE

Montserrat Pagès, Dolors Ludevid, Josep Vilardell, M.Angel Freire, Maria Pla, Margarita Torrent and Adela Goday

Departamento Genética Molecular. C.I.D. (C.S.I.C.) Jorge Girona 18-26. 08034 Barcelona. Spain

SUMMARY

The plant hormone abscisic acid (ABA) appears to modulate the responses of plants under conditions of water deficit (Davies and Mansfield, 1983). In general, there are two basic types of response to ABA which have been correlated with regulation of gene expression. First, a slow response during angiosperm embryo development (Quatrano, 1986), with ABA levels increasing (Jones and Brenner, 1987) during the embryogenic period prior to the desiccation of the embryo, and inducing the synthesis of specific proteins and mRNAs (Galau et al. 1987, Sanchez-Martinez et al. 1986). Second, a rapid response in water-stressed plant tissues where the level of ABA increases (Wright and Hiron, 1969) and alters the level of specific gene expression (Heikkila et al. 1984).

In earlier studies on the regulation of gene expression during embryogenesis in Zea mays L we have described the expression of a set of specific polypeptides which were rapidly induced by ABA in young embryos upon hormone treatment (Sanchez-Martinez et al. 1986). These polypeptides also appear during normal embryogenesis when the development of the embryo is progressing to the maturation stage, coinciding with the period when the endogenous level of ABA attains a maximum. After accumulating in dry embryos these polypeptides disappear during the first hours of germination (Goday et al. 1988)

cDNAs and genomic clones corresponding to two proteins, RAB-17 and MAH9, were obtained (Vilardell et al. 1990, Gomez et al. 1988). Northern blot hybridization showed that these genes are involved in generalized plant ABA responses, as their products accumulate in leaves under conditions of water-deficit (Pla et al. 1989, Vilardell et al. 1990).

The mode of action of the hormone remains unclear. Functional analyses using transient systems and transgenic plants have identified ABA-responsive DNA elements in the promoters of different ABA-regulated genes (Marcotte et al. 1989, J. Vilardell, unpublished). However, current evidence suggests that more than one mechanism may determine the level of expression of various ABA-responsive genes, and some of the ABA-regulated proteins seem to be functionally unrelated.

REGULATION OF GENE EXPRESSION IN DEVELOPING MAIZE EMBRYOS

The characterization of gene expression at specific stages of embryogenesis and germination is important for a proper undertanding of the control mechanisms of plant cell development. Developmental morphology of the caryopsis in maize has been the subject of many studies (Randolph 1936): organogenesis in the maize embryo includes the formation of three lobes (posterior, distal, and anterior) which are evident 10 days after pollination (d.a.p.). The development of the embryo is completed by about 30 to 40 d.a.p. and maturation and drying take place during the next 10 to 20 days . However, little information is available on the molecular mechanisms involved in this process.
An important part of this study is the characterization of proteins synthesized at the various stages of embryo maturation and germination and the corresponding expression of their mRNAs.

Changes in relative abundance of mRNA species during development have been described in different plant systems (Galau and Dure, 1981).
A previous study in maize reported the patterns of protein synthesis in embryos at several stages of embryogenesis and at very early stages of germination (Sanchez-Martinez et al 1986).

We used maize embryos isolated from the endosperm in order to enhance the uptake of labeled aminoacids and to distinguish clearly between embryo and endosperm mRNA (Ludevid et al. 1981). Throughout the period of maize ontogeny studied (from 10 d.a.p. to embryo germination) a number of labeled polypeptides were clearly resolved by two-dimensionl gel electrophoresis. In this group of spots we identified three sets of expressed polypeptides, characterized as follows:

Embryonic set. This group with 7 members is expressed both in young and in mature embryos. Three members of this group were also found in the mRNA stored in dry embryos but they disappear by 8 hours of germination.

Maturation set. This group has 44 members appearing in the embryo maturation period (31-50 days after pollination). Eight of them were not present in the mRNA of dry embryo, 16 disappeared 8 hours after imbibition, and 20 of them were still present in embryos after 8 hours of imbibition. Some polypeptides of this set can be induced prematurely in young embryos by incubation with ABA.

Germination set. This group has 13 members appearing during the first hours of germination, only 6 of them were apparent in 8 hours imbibed embryos, and expressed after 2 hours imbibition. Presumably, the mRNAs for 3 of this group are already present in the stored mRNA of dry embryos.

The synthesis of most of the mature embryo polypeptides during the first 2 hours of imbibition of dry embryos suggests a significant role for stored mRNA shortly after the onset of imbibition. On the other hand some spots are detected in the in vitro translation products from young embryos, but are missing in the in vivo pattern at the same stage. These polypeptides are clearly identified in the mature embryo, indicating the existence of a time lag between the appearance of the mRNA in the cytoplasm of the cells and the detection of the final product. These results suggest the existence of various different mechanisms of control at the level of gene expression during embryogenesis and early germination.

ABSCISIC ACID REGULATION OF PROTEIN SYNTHESIS IN MAIZE EMBRYO

It has been reported that when the maize embryo progresses from the cell division phase to the maturation stage, the level of ABA in the seed rises, preventing the precocious germination of the embryo (King 1976). A number of studies have reported that ABA inhibits germination of young embryos in culture, (Quatrano 1986). This hormone also enhances the synthesis of a group of embryo-specific polypeptides and increases steady-state levels of their associated mRNA (Galau and Dure 1981, Williamson et al. 1985). In maize, there are 22 polypeptides present in the maturation set that can be induced prematurely in excised young embryos when incubated in the presence of ABA. These polypeptides may be examples of the response to this hormone in maize development (Sanchez-Martinez et al. 1986).

Regulation by ABA of a highly phosphorylated 23 kD protein

A subset of polypeptides of special interest was identified in mature embryos as a group of spots in the 23 kD region by means of two-dimensional electrophoresis from "in vivo" labeled or "in vitro" translated products. These polypeptides appear during normal embryogenesis when the development of the embryo is progressing to the maturation stage which coincides with the period where endogenous ABA attains the maximum (Jones and Brenner, 1987). After accumulating in mature and dry embryos (Figure 1), they disappear during the first hours of germination. These polypeptides can be induced prematurely in immature embryos by ABA treatment.

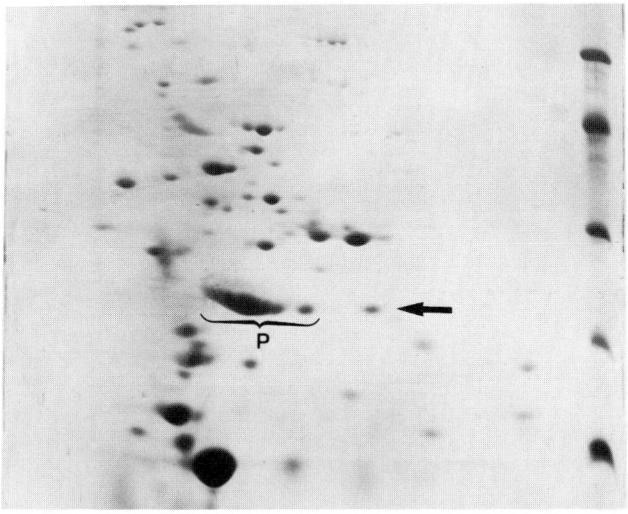


Figure 1. Coomassie blue staining of total protein extracts from dry embryos, resolved by two-dimensional electrophoresis. The procedure followed is essentially described in Sanchez-Martinez et al 1986. Arrow indicates the 23 kD proteins. P localizes the position of the protein phosphorylated forms

While all these polypeptides were detected in experiments "in vivo" by incubating excised embryos in the presence of the amino acid precursor $^{35}$S-methionine, only some of them were detected in the corresponding "in vitro" experiments, where the more acidic proteins were lacking (Goday et al 1988).

This is due to a post-translational phosphorylation of the more basic protein forms, as assesed by treatment of protein extracts with alkaline phosphatase before two dimensional electrophoresis. After alkaline phosphatase treatment there is a simultaneous disappearance of the more acidic polypeptides and a significant increase in amount of basic polypeptides. More evidence for the existence in mature embryos of phosphoproteins in the 23 kD range was obtained from double labeling experiments. The two dimensional electrophoretic pattern of proteins isolated after incubation of mature embryos both with $^{35}$S-methionine and $^{32}$P-phosphate allowed us to identify the proteins phosphorylated "in vivo" as being the same phosphorylated proteins which disappear after alkaline phosphatase treatment (Goday et al, 1988). From these data we can conclude that the 23 kD proteins, are regulated during embryogenesis, precociously induced in young embryos upon ABA treatment, and post-translationally modified by phosphorylation.

Protein phosphorylation and dephosphorylation are considered to be important regulatory mechanisms by which the activity of key enzymes and receptor molecules is altered in response to a wide variety of external stimuli. The next step, to determine whether the restricted pattern of expression and phosphorylation of this particular protein set actually reflects any developmental or functional role, was to obtain the sequences of the cDNA clones coding for these ABA regulated proteins, in order to achieve a better understanding of the relationship between ABA and embryogenesis.

## Isolation of cDNA and genomic clones for the 23 kD protein

A cDNA library from mature embryos was plated on filters in duplicate and each replicate was hybridized with radioactive single stranded cDNA made from young embryo or mature embryo polyA$^+$ RNA and forty clones were detected which had mRNA complementary sequences abundant in mature embryo.

In order to identify which clones had cDNAs encoding the 23kD proteins, a cDNA was made from size selected mRNAs enriched in this sequence. PolyA$^+$ RNA from mature embryos was first electrophoresed and transferred to an mAP filter which was cut according to RNA size. RNA was eluted from consecutive fractions, in which the RNA length diminishes stepwise in 150 nucleotides per fraction, then translated "in vitro" and products analyzed by one-dimensional electrophoresis to identify the fraction encoding the 23 kD protein. A cDNA probe was made from the mRNA fraction of about 1000 nucleotides and hybridized to the forty colonies which displayed differential expression in embryos. Six positive clones were obtained. These clones were included in a larger set classified as ABA inducible, which correpond to mRNAs expressed when young embryos are treated with the hormone.

From these clones the plasmid MA12 containing the longest insert was chosen for characterization. Hybrid selection experiments demonstrated that MA12 cDNA specifically hybridizes with mRNAs encoding the 23 kD protein (Goday et al 1988). This cDNA was used to screen a maize genomic library and one positive clone (gRAB-17) was sequenced.

```
maize  M E Y G Q Q G Q R G H G R T G H V D Q Y G N P V G G V E H G T G G M R H G T G T   40
rice   M E - - H Q G Q H G H V - T S R V D E Y G N P V G T - - - G A G - - - H G Q - M

maize  T G G M G Q L G E H G G A G M G G G Q F Q P A R E E H K T G G I L H R S G S S S   80
rice   - G T A G - M G T H G T A G T G R - Q F Q P M R E E H K T G G V L Q R S G S S S

maize  S S S S E D D G M G G R R K K G I K E K I K E K L P G G H K - D D Q H A T A T T   119
rice   S S S S E D D G M G G R R K K G I K E K I K E K L P G G N K G E Q Q H A M G G T

maize  G G A Y G Q Q G H T G S A Y G Q Q G H - T G - - - G A - - - Y A T G T E G T G E   152
rice   G T G T G T G T G T G G A Y G Q Q G H G T G M T T G T T G A H G T T T T D T G E

maize  K K G I M D K I K E K L P G Q H                                                   168
rice   K K G I M D K I K E K L P G Q H
```

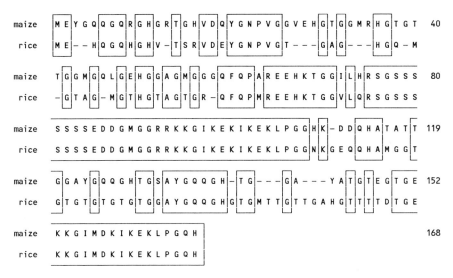

Figure 2. Comparison of the amino acid sequences of the RAB-17 protein from maize and RAB-21 from rice. Amino acids homologous in both sequences are boxed. The sequences have been aligned using the "microgenie" computer program.

The deduced amino acid sequence of the 23 kD protein in maize (Vilardell et al 1990) indicates conserved tracts of amino acid sequence with the late embryogenesis abundant (Lea) protein named D11 from cotton (Baker et al. 1988). Both proteins are rich in glycine and contain a central region of high homology of about 30 amino acids beginning with the cluster of contiguous Ser residues. Virtual identity is also present in the center and in the C terminal domains of RAB-21 from rice (Mundy and Chua 1988) and 23kD protein from maize, and the conserved sequence of repeat KKGI-KIKEKLPG which appears twice in both proteins is also the same. Due to this close similarity we decided to call the gene RAB-17 (Responsive to ABA) the number indicating the mol. wt. predicted from the sequence.

The molecular weight of the RAB proteins deduced from their sequences are however, smaller than that predicted from the mobility of these proteins in SDS-PAGE. Phosphoamino acid analysis showed that only the serine residues of RAB-17 are phosphorylated. It is of interest that serine residues are particularly well-suited to altering the structure of a protein by phosphorylation (Weller, 1979). Phosphorylation of the serine hydroxyl residue will prevent the formation of hydrogen bonds and can thus cause profound alterations in the structure of the protein, apart from the added effect of the negative charge of the phosphate group.

Current molecular studies aim to delineate the transduction pathway mediating ABA-responsive gene expression. ABA induces a rapid increase in intracellular $Ca^{2+}$ in guard cells (MC. Ainsh et al. 1990) and in roots following water stress (Lynch et al. 1989). The amino acid environment around the serines RSG(S)7EDD and HTGSAYG do not contain the consensus pseudosubstrate site of protein kinase C, (RFARKGSLRQ) (House and Kemp, 1987). This indicates that models implicating $Ca^{2+}$ as "second messenger", in which the action of the hormone causes the phosphorylation of specific sets of proteins by the activation of a protein kinase C, are not operating in RAB-17 protein phosphorylation. However, the acidic residues following the serine cluster may be a substrate for other kinases such as casein kinase II. The recognition sequence for this enzyme requires two acidic residues following the phosphorylatable residue and has been identified as S(T)-E(S-P)-D(E) (Hathaway and Traugh 1983). Protein kinase activities have been detected in a number of higher plants (Trewavas, 1975) and casein-type protein kinases have been isolated from maize and other plants with properties similar to casein kinases found in animals. Preliminary results show that RAB-17 isolated protein serves as a substrate for casein kinase II obtained from either rat liver cytosol or from maize embryos ( Martinez, unpublished).

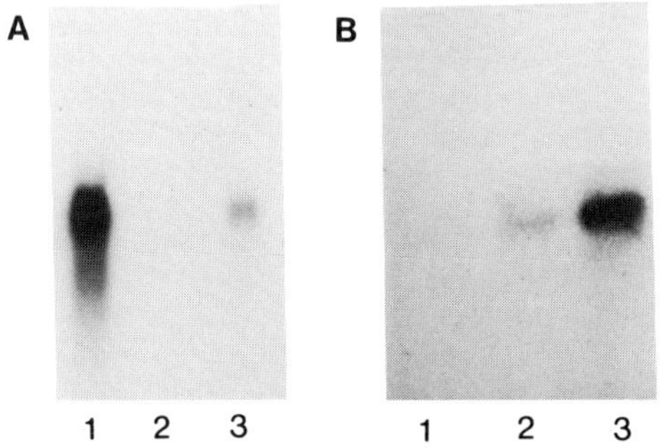


Figure 3. RAB-17 mRNA levels in 7 day-old leaves. (A) Blot hybridization of total RNA from non-treated leaves (lane 1), leaves from seedlings subjected to water stress (lane 2) and leaves from seedlings incubated for 24 hours with 100 uM ABA, with additional spraying with the hormone solution at regular intervals (lane 3). (B) Total RNA from leaves dehydrated for one hour (lane 1), two hours (lane 2) and three hours (lane 3)

REGULATION OF RAB-17 BY WATER STRESS

The plant hormone ABA appears to modulate the responses of plants under conditions of water deficit (Davies and Mandsfield 1983). Current evidence indicates that plants subjected to osmotic stress respond with a series of physiological, morphological and biochemical changes which appear to be elicited by the reduction of cell turgor. Among them, closure of stomata, changes in ionic composition, and accumulation of organic osmolytes have been reported. ABA levels increase in water-stressed vegetative tissues (Wright et al 1969) and this alters the expression of specific genes encoding proteins of largely unknown function.

To determine the regulation of RAB-17 by water stress in vegetative tissues, Northern blots were used to analyze the steady state levels in young leaves of well-watered or desiccated 7 day-old plants. Figure 3,A shows that no significant levels of the RAB-17 mRNA is found in well-watered leaves, however, upon water-deficit or exogenous ABA treatment, the mRNA accumulates.

In maize leaves subjected to water stress the level of ABA shows a ten fold increase in a three-hour period (Gomez et al. 1988). The results obtained from Northern blot hybridization using total RNA from seven-day-old leaves dehydrated for one, two or three hours, shows (Figure 3,B) that RAB-17 transcript accumulates on dehydration and the maxima correlate with the maxima of ABA abundance.

Regulation of RAB-17 by water stress in vegetative tissues of viviparous mutants

The ABA-deficient viviparous mutants of maize (Robertson, 1955) seem to have great potential for providing information concerning the regulation of the expression of genes involved in ABA responses, both in embryo and vegetative tissues. Seeds of viviparous mutants germinate while still attached to the ear, before full maturity has been reached, and without having first undergone the normal period of desiccation. They are deficient in ABA (Neill et al. 1986). However, when young embryos are explanted into culture media containing ABA, their growth is arrested as in the wild type embryos. Seedlings rescued from viviparous kernels contain less ABA than do wild seedlings and the ABA concentration does not increase in response to water deficit.

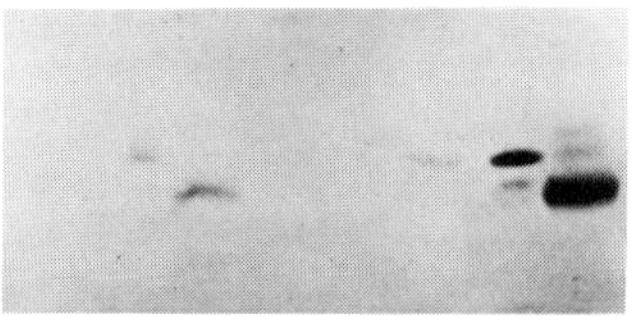

1 2 3 4 5 6

Figure 4. Immunoblotting of total protein extracts from vp2 embryos at 25 d.a.p. (lane 1) 36 d.a.p. (lane 2) and 45 d.a.p. (lane 3), and corresponding wild type embryos at 30 d.a.p. (lane 4), 42 d.a.p. (lane 5) and dry (lane 6).

Northern analysis of RNA from leaves of viviparous mutants subjected to water stress showed that the up-regulation of mRNA was not detected in any of the ABA deficient mutants tested (Pla et al 1989). These mutants do not produce ABA in response to water-deficit stress. However, when young seedlings of viviparous mutants were treated with ABA, both proteins and mRNA are expressed.

These results indicate that RAB-17 gene is inducible by ABA in embryos, and in vegetative organs challenged by hydric stress. This gene and the previously described ABA-regulated gene encoding a glycine rich protein (Gomez et al 1988)), may be part of a larger group of genes which are involved in generalized maize ABA responses.

Expression of RAB-17 protein in viviparous embryos

We investigated whether this difference in response between the wild-type and the *viviparous* mutants would also be found in the embryo.

Considering that these proteins are normally expressed in late embryogenesis when desiccation starts and since *viviparous* mutants are devoid of desiccation and have low levels of endogenous ABA, we examined whether the 23 kD polypeptides would also accumulate naturally in the mutants.

Total protein extracts of *vp2* embryos were blotted to nitrocellulose filters and the polypeptides were detected using the specific antiserum. Surprisingly, the 23 kD proteins are detected in *vp2* (Figure 4) and in the *viviparous* embryos of all the mutants tested (Pla et al 1989). The only differences that could be appreciated, were the lower amounts accumulated by the mutants and a faster disappearence, which was probably due to the precocious germination of the mutants.

This result was further confirmed by protein labeling "in vivo" of embryos which have been treated or not with ABA for 24 hours. Two-dimensional electrophoresis showed that the pattern and ABA effect in the 23 kD protein synthesis is similar in both the mutant and the wild-type.

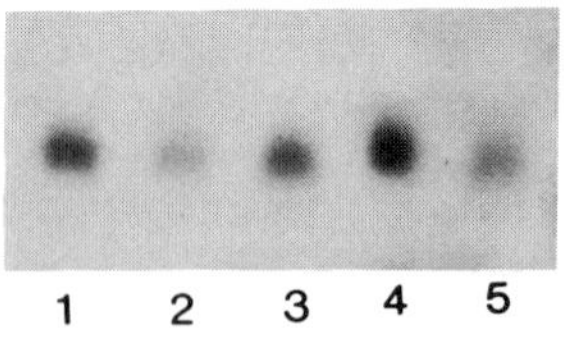


Figure 5. Northern blot analysis of MAH9 mRNA in ABA-treated calli. Total RNAs were analyzed using the MAH9 cDNA probe. Embryonic calli (lane 1). Embryonic calli incubated for one day (lane 2) and six days (lane 3) with 10 uM ABA. Meristematic calli (lane 4) incubated for one day (lane 5) with the hormone.

DISCUSSION

The differential regulation of the RAB-17 gene in embryo and vegetative tissues of the ABA-deficient mutants of maize, raises several questions on the mode of action of ABA in the modulation of gene expression during embryogenesis and water-stress.

The molecular mechanism of the stimulatory effects of ABA on gene expression is not clear. However, direct evidence of ABA regulating gene transcription has recently been provided. A DNA promoter fragment responding specifically to ABA treatment in ABA-resposive genes was reported (Marcotte et al. 1988, Mundy et al 1990). Identification of specific regulatory sequences involved in the hormone action may allow us to check whether the same region of the gene, or different regions in the gene are responsive to the different signals (ABA and water stress) in embryo and vegetative tissues.

As a first step to determine how RAB-17 expression is regulated by ABA, chimaeric genes containing the CAT reporter gene, linked to the 5´ flanking fragments from RAB genomic clone were constructed and introduced into rice protoplasts. After uptake of the constructs rice protoplasts were incubated for 24 hours in the presence or absence of ABA and then assayed for CAT activity. Preliminary results (Vilardell unpublished) indicate that the RAB-17 promoter region of -229 is sufficient for the ABA induction of the reporter gene CAT. This region contains the motif ACGTGCGCC, which has been identified as a functional fragment in the 5´region of other ABA-regulated genes (Marcotte et al 1989, Mundy et al 1990).

These data clearly indicate that at least part of the ABA control in gene expression is at the transcriptional level. However, current evidence suggests that more than one mechanism determines the level of expression of various ABA-responsive genes (Peña-Cortes et al 1990). In fact the gene MAH9 in maize which is developmentally regulated during embryogenesis and up regulated by ABA in embryos and by water-stress in leaves, is not responsive to exogenous ABA in cultured cells where a high level of expression of MAH9 mRNA is found (Figure 5).

There is much that still needs to be known. Further studies are needed to define functions for the proteins encoded by ABA-responsive genes. Hypothetical functions in desiccation survival have been suggested for some Lea and RAB proteins based on sequence features (Dure et al 1989). However, recent evidence suggests that some ABA-regulated proteins are functionally unrelated.

Knowledge already available indicates that the MAH9 protein is an RNA binding protein. The sequence of the predicted MAH9 protein contains a ribonucleoprotein consensus sequence-type RNA binding domain (CS-RBD) of 90 amino acids with two highly conserved motifs, the octapeptide RGFGFVTF and the hexapeptide CFVGGL which conform perfectly to the consensus sequences RNP1 and RNP2. Like several RNA binding proteins, it also has a glycine-rich terminal domain. By means of ribohomopolymer binding assays and using specific antibodies against the protein we demonstrated that MAH9 protein is a "bona fide" RNA-binding protein (Ludevid et al. submitted) and that it is located in the cytoplasm.

Many RNPs have been identified in the cytoplasm which have been involved in a wide range of cellular processess including mRNA translation stability (PABP), mRNA repression and degradation (prosomes). On the other hand accumulation of ABA regulated mRNAs has been ascribed to both transcriptional and post-transcriptional regulation (Marcotte et al. 1989) and a temporal separation between transcription and translation has been described for some germination specific genes

Further elucidation of the functional role of RAB proteins in embryo and seedlings, and the regulatory role played by the RNA-binding protein in the hormone action, may provide us with a better understanding of the mechanisms of desiccation tolerance in plants.

Acknowledgements

This work is supported by grants BIO88-0162 from CICYT, and 151/87 from CIRIT.

## REFERENCES


Baker JC, Steele C, Dure III L: Sequence and characterization of 6 Lea proteins and their genes from cotton. Plant Molec. Biol 11: 277-291 (1988).

Davies MJ, Mansfield TA: The role of abscisic acid in drought avoidance. In Addicott, F.T. (ed.) Abscisic Acid Praeger New York pp 237-268 (1983).

Dure L, Crouch M, Harada J, Ho THD, Mundy J, Quatrano RS, Thomas T, Sung ZR: Common amino acid sequence domains among the LEA proteins of higher plants. Plant Mol. Biol. 12: 475-486 (1989).

Galau GA, Dure III LS: Developmental biochemistry of cottonseed embryogenesis and germination. Changing mRNA populations as shown by reciprocal heterologous complementary DNA/mRNA hybridization. Biochemistry 20:4169-4185 (1981).

Galau GA, Bijaisoradat N, Hughes DW: Accumulation kinetics of cotton late embryogenesis-abundant mRNAs and storage protein mRNAs coordinate regulation during embryogenesis and the role of abscisic acid. Dev. Biol. 123: 198-212 (1987).

Goday A, Sanchez-Martinez D, Gomez J, Puigdomenech P, Pages M: Gene expression in developing Zea mays embryos: Regulation by abscisic acid of a highly phosphorylated 23- to 25- kD group of proteins. Plant Physiol. 88: 564-569 (1988).

Gomez J, Sanchez-Martinez D, Stiefel V, Rigau J, Puigdomenech P, Pages M : A gene induced by the plant hormone abscisic acid in response to water stress encodes a glycine-rich protein. Nature 334: 262-264 (1988).

Hathaway GM, Traugh JA: Casein Kinase II. Meth. Enzymol. 99: 317-331 (1983).

Heikkila JJ, Papp JET, Schultz GA, Bewley JD: Induction of heat shock protein messenger RNA in maize mesocotyls by water stress, abscisic acid and wounding. Plant Physiol. 76:270-274 (1984).

House C, Kemp BE: Protein kinase C contains a pseudosubstrate prototope in its regulatory domain. Science 248: 1726-1728 (1987).

Jones RJ, Brenner ML: Distribution of abscisic acid in maize kernel during grain filling. Plant Physiol. 83: 905-909 (1987).

King RW: Abscisic acid in developing wheat grains and its relationship to grain growth and maturation. Planta 132: 43-51 (1976).

Ludevid MD, Torrent M, Martinez-Izquierdo JA, Puigdomenech P, Palau J: Subcellular localization of glutelin-2 in maize (Zea mays L.) endosperm. Plant Mol.Biol. 3: 227-234 (1984).

Lynch J, Polito VS, Lauchli A: Salinity stress increases cytoplasmic $Ca^{2+}$ activity in maize root protoplasts. Plant Physiol. 90: 1271-1274 (1989).

Marcotte Jr WR, Bayley ChC, Quatrano RS: Regulation of a wheat promoter by abscisic acid in rice protoplasts. Nature 335: 454-457 (1988).

Marcotte WR, Russel SH, Quatrano RS: Abscisic acid-responsive sequences from the Em gene of wheat. Plant Cell 1: 969-979 (1989).

McAinsh MR, Brownlee C, and Hetherington AM: Abscisic acid-induced elevation of guard cell cytosolic $Ca^{2+}$ precedes stomatal closure. Nature 343: 186-188 (1990).

Mundy J, Chua NH: Abscisic acid and water-stress induce the expression of a novel rice gene. EMBO J. 7: 2279-2286 (1988).

Mundy J, Yamaguchi-Shinozaki K, Chua NH: Nuclear proteins bind conserved elements in the abscisic acid-responsive promoter of a rice rab gene. Proc. Nat. Acad. Sci. USA in press (1990).

Neill SJ, Horgan R, Parry AD: The carotenoid and abscisic acid content of viviparous kernels and seedlings of Zea mays L. Planta 169: 87-96 (1986).
Pla M, Goday A, Vilardell J, Gomez J, Pages M: Differential regulation of the ABA induced 23-25 kD proteins in embryos and vegetative tissues of the viviparous mutants of maize Plant Mol. Biol. 13: 385-394 (1989).
Peña-Cortes H, Sanchez-Serrano JJ, Willmitzer L, Prat S: Abscisic acid is involved in the wound-induced expression of the proteinase inhibitor II gene in potato and tomato. Proc. Nat. Acad. Sci. USA 86: 9851-9855 (1989).
Quatrano RS: Regulation of gene expression by abscisic acid during angiosperm embryo development. Oxford Syreys Plant Mol. Cell Biol. 3:467-477 (1986).
Randolph LF: Developmental morphology of the caryopsis in maize. J. Agric. Res. 53:881-916 (1936).
Robertson DS: The genetics of vivipary in maize. Genetics 40: 745-760 (1955).
Sanchez-Martinez D, Puigdomenech P, Pages M: Regulation of gene expression in developing Zea mays embryos Protein synthesis during embryogenesis and early germination of maize. Plant Physiol. 82: 543-549 (1986).
Trewavas AJ: Post-translational modifications of proteins by phosphorylation. Annu. Rev. Plant Physiol. 27: 349-374 (1975).
Vilardell J, Goday A, Freire MA, Torrent M, Martinez MC, Torne JM, Pages M: Plant Mol. Biol. 14: 423-432 (1990).
Weller M: General aspects and functions of proteins which contain covalently bound phosphorus. In Protein phosphorylation. Ed Pion Ltd., London 1-10 (1979).
Wright ST, Hiron RWP: (+)- Abscisic acid the growth inhibitor in detached wheat leaves following a period of wilting. Nature 224: 719-720 (1969).

# MOLECULAR ANALYSIS OF DESICCATION TOLERANCE IN THE RESURRECTION PLANT CRATEROSTIGMA PLANTAGINEUM

D. Bartels, K. Schneider, D. Piatkowski,
R. Elster, G. Iturriaga, G. Terstappen,
Le Tran Binh, F. Salamini

Max-Planck-Institut für Züchtungsforschung
Carl-von-Linne-Weg 10, 5000 Köln 30, FRG

## INTRODUCTION

Understanding the molecular mechanisms of desiccation tolerance in higher plants is an important goal of applied biology. Many efforts in plant breeding have been made to select for drought resistance. The selection has always been difficult because complex traits are involved. Using suitable experimental systems molecular biology offers the possibility to investigate drought resistance on the level of single genes. Two model systems (developing cereal embryos and resurrection plants) have been chosen to understand mechanisms of desiccation tolerance at the molecular level. Common to both these experimental systems is that they can experience protoplastic dehydration. This mechanism should be distinguished from biochemical or morphological adaptations which allow plants to withstand dry environments.

In most higher plants tolerance to protoplastic dehydration is restricted to the seed. The embryo tolerates water potentials as low as 50% relative humidity (Gaff, 1980)), a property it acquires during seed maturation (Bartels et al., 1988). After the seed germinates the ability to survive desiccation is lost. Therefore, tolerance to dehydration is developmentally programmed, and with the exceptions of the resurrection plants (Gaff, 1971) restricted to the embryo.

A remarkable situation occurs in a small group of angiosperms termed poikilohydric or resurrection plants (Walter, 1955; Gaff, 1977). The natural habitat of these plants are arid areas and most of them are native to South Africa or Australia where they serve as pioneer plants in the colonization of rocks. Resurrection species are found among several monocotyledonous and dicotyledonous families.

*Plant Molecular Biology 2*, Edited by R.G. Herrmann and
B. Larkins, Plenum Press, New York, 1991

The unique feature of the resurrection plants is that they possess mature foliage that withstands virtually complete dryness. They can survive in a dehydrated state for long periods, but after rehydration dried plants resume full physiological activity within several hours.

EXPERIMENTAL SYSTEM

For our studies we have chosen a representative of the poikilohydric plants, the resurrection plant Craterostigma plantagineum (Fam. Scrophulariaceae) (Volk and Leippert, 1971). As completely desiccated Craterostigma plants recover and are viable after rehydration, it was inferred that it should be possible to isolate molecular components essential for desiccation tolerance. The plant offered another advantage: undifferentiated desiccation intolerant callus tissues could be 'resurrected' after drying if pretreated with the plant hormone abscisic acid (ABA). The correlation of the ABA treatment and the viability of callus tissue after desiccation is shown in Fig. 1. The Craterostigma plant is an excellent system to screen for gene products relevant to desiccation tolerance, because two metabolically different types of tissue which both have the ability to 'resurrect' after severe desiccation can be analysed.

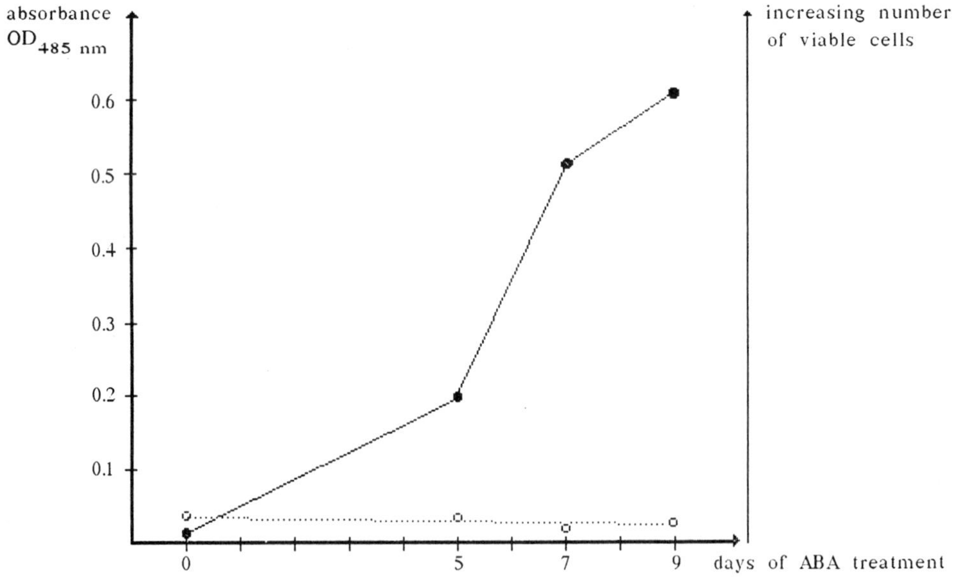


Fig. 1. Callus tissue was cultured on MS medium containing 5 mg $l^{-1}$ ABA, then it was desiccated and put again on MS medium. Three weeks later the callus was tested for viability by the reduction of 2,3,5-triphenyltetra zolium chloride (TTC) to water insoluble red formazan which was measured via absorbance at 485 nm according to Towill and Mazur (1975).

## EFFECTS OF DRYING ON PROTEINS

In order to investigate the effect of a drying or ABA treatment on gene expression in *Craterostigma* total soluble proteins were analysed by two dimensional electrophoresis and visualized by silver staining from the following tissues: untreated leaves, desiccated leaves, untreated callus, ABA-treated callus, ABA-treated dried callus, dried callus. Several differences in the protein pattern were found between fresh and dried leaves as well as between untreated and ABA-treated callus. In ABA-treated callus several new proteins appeared with molecular weights ranging between 16 and 130 kD (Fig. 2). A very similar set of desiccation stress proteins was also identified in dried leaves in comparison to untreated leaves. To summarize these experiments three major groups of proteins can be distinguished: (i) those predominantly present in dried leaves and ABA-treated callus but not in untreated tissues (ii) those which disappeared during drying and (iii) those present in more or less equal amounts in untreated and dried leaves or ABA-treated callus. It was shown that the desiccation or ABA-induced proteins disappear when the leaves are rehydrated or when the ABA is withdrawn from the callus culture medium.

These changes in the expression pattern of the proteins were confirmed when mRNAs were extracted and the *in vitro* synthesized proteins were analysed by electrophoresis. These results pointed to differential gene activation induced by desiccation or ABA (Bartels et al., 1990).

## CHARACTERIZATION OF DESICCATION-RELATED cDNA CLONES

In order to study the desiccation-related proteins and their genes a cDNA clone bank was established using poly(A) RNA from dried *Craterostigma plantagineum* leaves. This library was screened with $^{32}$P-labelled RNA probes from dried leaves and ABA-treated callus as well as from untreated tissues. Gene products which could play a role in osmoprotection should be expressed in dried leaves and in ABA-treated callus and only cDNA clones hybridizing to RNA from both tissues were selected. Through this differential screening procedure a number of desiccation-related clones were obtained. For further analyses the cDNA clones were classified into ten groups based on close sequence homologies. These hybridization groups are depicted in a Venn diagram (Fig. 3). Around 30% of the cDNA clones did not fall into any of these groups and represent rarely expressed genes.

The abundantly expressed genes were further characterized in hybrid-selected translation experiments. The hybrid-release translation products derived from six different cDNA clones are shown in Fig. 4. The results confirmed that the selected clones code for genes abundantly expressed during dehydration. The *in vitro* synthesized proteins were also separated in a two-dimensional electrophoresis and compared to the desiccation-induced *in vivo* proteins (Bartels et al. 1990). Proteins selected for

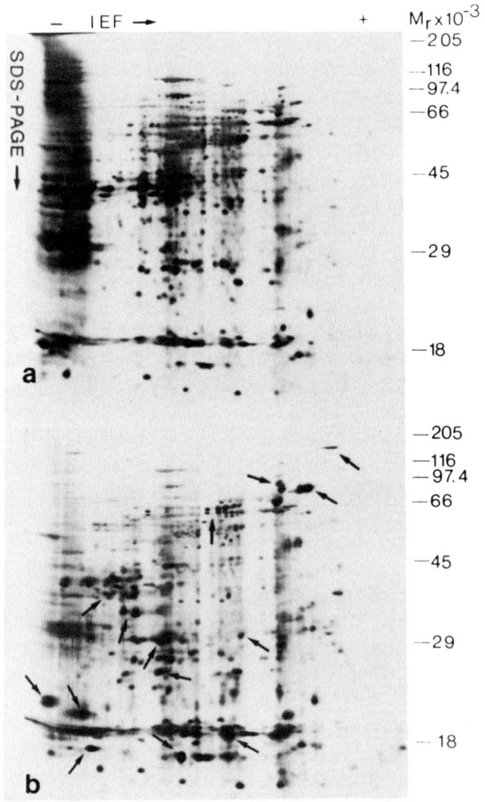


Fig. 2. Two-dimensional separaton (IEF/SDS-PAGE) of total proteins extracted from a) dried callus b) ABA-treated, dried callus. The arrows point to the proteins which are induced by ABA.

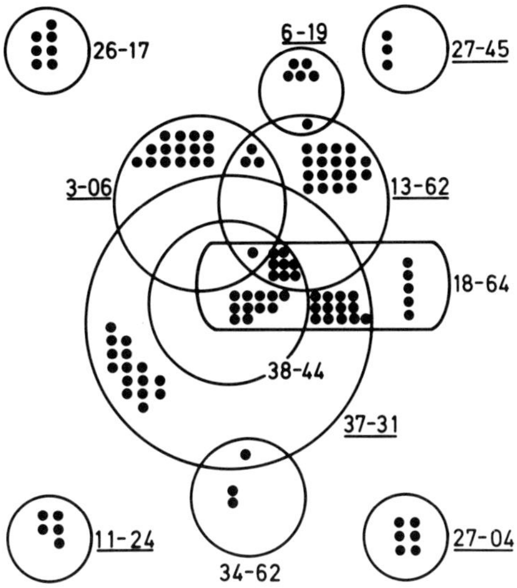


Fig. 3. Venn diagram illustrating the cross-hybridization between the most abundant cDNA clones and ten sets of related sequences were identified. All selected cDNA clones strongly hybridized to mRNA of desiccated leaves and ABA-treated callus. - (This figure is reproduced from Bartels et al. 1990 with the kind permission of the Springer Verlag).

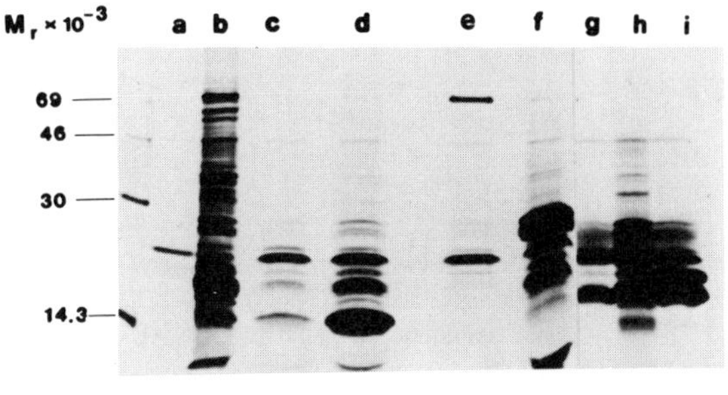


Fig. 4. Separation of <u>in vitro</u> translation products on an SDS-polyacrylamide gel. The $^{35}$-S-labelled translation products were visualized after fluorography. The <u>in vitro</u> synthesized proteins were derived from the following samples a) control, no RNA added b) total poly(A) RNA from desiccated leaves c) vector only, d) hybrid release translation (Hrt) of pcC 27-45 e) Hrt of pcC 11-24 f) Hrt of pcC 3-06 g) Hrt of pcC 6-19, h) Hrt of pcC 13-62 i) pcC 27-04; on the left hand side $^{14}$C-labelled proteins are shown as molecular weight standards.

by the cDNA clones pcC 27-45, pcC 3-06, pcC 6-19, pcC 27-04 and pcC 11-24 could be correlated to in vivo proteins. This observation needs to be confirmed by immunoblots but it suggests that the investigated desiccation-related proteins do not undergo major postranscriptional modifications.

RNA hybridization experiments demonstrated that only a minimal amount of water loss is necessary to activate transcription of these genes. The signal transduction from the environmental stress factor (water stress) to drastically increased transcript levels is very fast. Already after 30 minutes of a drying treatment enhanced transcript levels were measured (Bartels et al. 1990). The effect of the desiccation stress is as rapid as it has been described for heat shock regimes (for a review see Neumann et al., 1990).

Not only in callus but also in leaves the expression of the water stress related transcripts can be mediated by exogenous ABA. This fact suggests that ABA may be a major component in the pathway from water stress to gene activation. Progress in unravelling this transduction pathway comes from promoter studies of two ABA responsive genes isolated from embryos of wheat and rice. In transient expression studies Mundy et al. (1990) and Marcotte et al. (1989) have identified sequences which may act as ABA-responsive elements. Promoter analyses of the Craterostigma gene pcC 27-45 have been initiated to isolate regulatory sequences responding to ABA and to water stress. Two interesting questions should be answered with these experiments: (i) are the ABA responsive elements the same in embryos as in leaves of resurrection plants and (ii) how efficient are these different promoters with respect to rapid transcript accumulation.

CRATEROSTIGMA cDNA CLONES AND THEIR RELATIONSHIP TO OTHER WATER STRESS GENES

To study the structure and function of the desiccation-induced genes from Craterostigma the nucleotide sequence of five genes were determined and the deduced proteins were analysed for their features. Determined were the sequences of the following cDNA clones, pcC 3-06, pcC 6-19, pcC 13-62, pcC 27-04, pcC 27-45. The sequences are presented in another manuscript (Bartels et al., in preparation). The deduced proteins revealed some interesting features; there is in general a preference for hydrophilic amino acids, four proteins are nearly devoid of cysteine and tryptophan residues, but contain a high number of lysine residues. Noticeable in pcC 27-04 and pcC 6-19 is the high glycine content. Although the sequences of pcC 27-04 and pcC 6-19 did not crosshybridize under stringent conditions (0.3 M $Na^+$, 65$^oC$), they display conserved sequence tracts: a cluster of serines and lysine-rich motifs, but a diverged amino-terminal part of the gene. This peculiar gene structure has been found in a number of ABA responsive genes isolated from dry embryos (Baker et al., 1988; Mundy et al., 1990) as well as from dehydrated barley and corn seedlings (Close et al., 1989). The basic structure of this type of gene is depicted in the cartoon in Fig. 5. It shows a mosaic pattern of tracts with very conserved hydrophilic regions, a

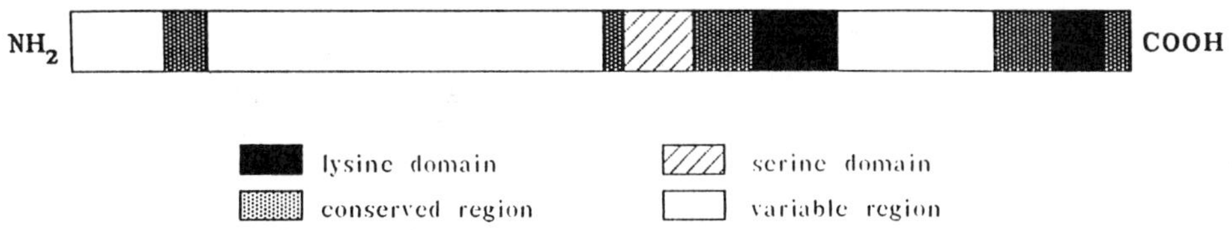


Fig. 5. Conserved sequence features in dehydration-induced proteins: characteristic for this type of gene family are a cluster of serines and a lysine rich domain. The variable regions differ in lengths and in composition between plant species and within Craterostigma.

highly conserved carboxy terminal part, a region with medium conservation in the amino terminus part and in between regions with no conservation. A conservation between such different species (monocotyledonous and dicotyledonous) suggests that these motifs must be of functional relevance for these proteins. The finding that desiccation leads to the expression of the type of genes in embryos and leaves of resurrection plants is of importance as it implies that at least in part the same gene products are involved in the response to desiccation in both tissues.

Besides the ones mentioned several other ABA-responsive genes which are abundant in desiccated seeds have been cloned and characterized (Gomez et al., 1988; Harada et al., 1989; Leah and Mundy, 1989; Litts et al., 1987; Raikhel and Wilkins, 1987). No major regions of homologies could be found between these genes and the Craterostigma genes analysed up to now. But the characterization of several other abundantly expressed desiccation-related genes from Craterostigma has not been done, and further nucleotide sequencing will reveal whether there is a set of genes unique to Craterostigma or whether more regions with significant sequence homologies will be discovered. An interesting observation is made by Dure et al. (1989) where they describe common amino acid domains among several LEA (late embryogenesis abundant) proteins of higher plants. Although the function of these proteins is unknown, structural analyses suggest that they could play a role in osmoprotection (Close et al., 1989; Dure et al., 1989). Protein analyses suggest that similar structural elements (amphiphilic alpha helices) could also exist in the desiccation-induced proteins from Craterostigma.

CONCLUSIONS

A number of genes have been cloned from desiccation tolerant tissues: dry embryos as well as leaves of resurrection plants. Common to all these genes is that they are abundantly expressed upon water stress or can be induced by ABA. This suggests that in both systems ABA may be a mediator in the signal transduction pathway. Sequence homologies indicate that among higher plants important sequence domains are conserved and that resurrection plants are not an exception. As in resurrection plants the desiccation stress proteins can be expressed during the whole life cycle of the plant mainly in leaves but also in other organs like roots (Bartels et al., in preparation) it becomes a realistic goal that these genes could be functionally expressed in desiccation-intolerant plants. In future experiments it needs to be investigated whether there are gene products specific to resurrection plants essential for survival after dehydration or whether the quantities of expressed desiccation stress proteins could be an important factor.

## REFERENCES


Baker JC, Steele C, Dure III L: Sequence and characterization of 6 Lea proteins and their genes from cotton. Plant Mol Biol 11: 277-291 (1988)

Bartels D, Schneider K, Terstappen G, Piatkowski D, Salamini F: Molecular cloning of abscisic acid-modulated genes which are induced during desiccation of the resurrection plant Craterostigma plantagineum. Planta 181: 27-34 (1990)

Bartels D, Singh M, Salamini F: Onset of desiccation tolerance during development of the barley embryo. Planta 175: 485-492 (1988)

Close TJ, Kortt AA, Chandler PM: A cDNA based comparison of dehydration-induced proteins (dehydrins) in barley and corn. Plant Mol Biol 13: 95-108 (1989)

Dure III L, Crouch M, Harada J, Ho T-H, Mundy J, Quatrano R, Thomas T, Sung ZR: Common amino acid sequence domains among the LEA proteins of higher plants. Plant Mol Biol 12: 475-486 (1989)

Gaff DF: Desiccation-tolerant flowering plants in Southern Africa. Science 174: 1033-1034 (1971)

Gaff DF: Desiccation tolerant vascular plants of Southern Africa. Oecologia 31: 95-109 (1977)

Gaff DF: Protoplasmic tolerance of extreme water stress. In: Adaptation of plants to water and high temperature stress (Turner NC, Kramer PJ, eds). Wiley-Interscience, New York: 207-229 (1980)

Gomez J, Sanchez-Martinez D, Stiefel V, Rigau J, Puigdomenech P, Pages M: A gene induced by the plant hormone abscisic acid in response to water stress encodes a glycine-rich protein. Nature 334: 262-264 (1988)

Harada JJ, Deliste AJ, Baden CS, Crouch ML: Unusual sequence of an abscisic acid-inducible mRNA which accumulates late in Brassica napus seed development. Plant Mol Biol 12: 395-401 (1989)

Leah R, Mundy J: The barley amylase/subtilisin inhibitor: nucleotide sequence and patterns of seed-specific expression. Plant Mol Biol 12: 673-682 (1989)

Little JC, Colwell GW, Chakerian RL, Quatrano RS: The nucleotide sequence of a cDNA clone encoding the wheat Em protein. Nucl Acids Res 15: 3607-3618 (1987)

Marcotte WR Jr, Russell SH, Quatrano RS: Abscisic acid - responsive sequences from the Em gene of wheat. The Plant Cell 1: 969-976 (1989)

Mundy J, Yamaguchi-Shinozaki K, Chua N-H: Nuclear proteins bind conserved elements in the abscisic acid-responsive promoter of a rice rab gene. Proc Natl Acad Sci USA 87: 1406-1410 (1990)
Mundy J, Chua N-H: Abscisic acid and water stress induce the expression of a novel rice gene. EMBO J 7: 2279-2286 (1988)
Neumann D, Nover L, Parthier B, Rieger R, Scharf KD, Wollgiehn R, Zur Nieden U: Heat shock and other stress response systems of plants. Biol Zentralblatt 108: 1-146 (1989)
Raikhel NV, Wilkins TA: Isolation and characterization of a cDNA encoding wheat germ agglutinin. Proc Natl Acad Sci USA 84: 6745-6749 (1987)
Towill LE, Mazur P: Studies on the reduction of 2,3,5-triphenyltetrazolium chloride as a viability assay for plant tissue cultures. Can J Bot 53: 1097-1102 (1975)
Volk OH, Leippert H: Vegetationsverhältnisse im Windhoeker Bergland, Südwest Afrika. In: Journal XXV-S.W.A., pp 5-44, Wissenschaftl Gesellschaft, Windhoek (1971)
Walter H: The water economy and the hydrature of plants. Ann Rev Plant Physiol 6: 239-252 (1955)
Yamaguchi-Shinozaki K, Mundy J, Chua N-H: Four tightly linked rab genes are differentially expressed in rice. Plant Mol Biol 14: 29-39 (1989)

# THE ANAEROBIC RESPONSIVE ELEMENT

Mark R. Olive, John C. Walker*, Karambir Singh, Jeff G. Ellis, Danny Llewellyn, W. James Peacock and Elizabeth S. Dennis

CSIRO Division of Plant Industry, Canberra ACT, Australia

## INTRODUCTION

The response to anaerobic stress such as occurs during flooding, is one of the best-characterised stress responses of plants, at the physiological and biochemical levels. Recent advances in molecular biology have facilitated the analysis of some of the molecular events triggered by anaerobic stress and the mechanisms regulating this physiologically important stress response. Within 2 min of the transfer of maize roots to an anaerobic environment oxidative phosphorylation is inhibited, as indicated by an increase in cytoplasmic NADH levels (Roberts *et al.*, 1984). In anaerobic cells of higher plants the $NAD^+$ required for continued glycolysis may be regenerated via the lactate dehydrogenase (*Ldh*) and alcohol dehydrogenase (*Adh*) reactions.

### The Anaerobic Polypeptides

The metabolic shift from oxidative respiration to lactic and/or ethanolic fermentation is achieved by the suppression of normal protein synthesis and the induced synthesis of a specific set of polypeptides, the anaerobic polypeptides (ANPS, Sachs *et al.*, 1980). When maize seedlings are labelled with [$^{35}S$]-methionine following 5 hr of anaerobiosis and the polypeptides analysed on 2-dimensional polyacrylamide gels, approximately twenty ANPs are resolved, accounting for greater than 70% of total proteins synthesised. Six anaerobically inducible enzymes have currently been attributed to nine specific molecular weight ANPs (Kelley, 1987), although a total of twelve enzymes have been shown to be anaerobically regulated in maize. These enzymes may be classified into four functional groups:

1) Enzymes mobilising glucose - sucrose synthase I (ANP 87: Springer *et al.* 1986)
2) Stem glycolytic enzymes - phosphoglucomutase (Bailey -Serres *et al.*, 1987), phosphoglucoisomerase (ANP33A, 35.5; Kelley and Freeling, 1984a), fructose-1,6-diphosphate aldolase (ANP 33A, 35.5; Kelley and Freeling, 1984b; Hake *et al.*, 1985),glyceraldehyde-3-phosphate dehydrogenase III (Bailey-Serres *et al.*, 1987; Russell and Sachs, 1989), phosphoglycerate

*Division of Biological Sciences, University of Missouri-Columbia, Columbia MO 65211 USA

*Plant Molecular Biology 2*, Edited by R.G. Herrmann and B. Larkins, Plenum Press, New York, 1991

mutase (Bailey-Serres *et al.*, 1987) and enolase (Bailey-Serres *et al.*, 1987).
3) Fermentative enzymes - pyruvate decarboxylase (ANP 65A; Laszlo and St Lawrence, 1983), *Adh1* (ANP 40A,40B; Hagemann and Flesher, 1960; Sachs *et al.*, 1980; Dennis *et al.*, 1985), *Adh2* (ANP 42A,42B) and lactate dehydrogenase (Roberts *et al.*, 1985; Hoffmann *et al.*, 1986).
4) Amino acid biosynthetic enzymes - alanine aminotransferase (Good and Crosby, 1989)

Thus, with the exception of the alanine aminotransferase enzyme, all of the enzymes induced in response to anaerobic stress are enzymes of anaerobic carbohydrate metabolism. In particular, the induction of pyruvate decarboxylase and *Adh* is significant, as the regeneration of $NAD^+$ via the ethanolic fermentation pathway enables anaerobic glycolysis to continue in submerged roots for approximately 70 hr. Transient lactic fermentation does occur at the onset of anaerobic stress in maize seedlings, however metabolism generally switches to predominantly ethanolic fermentation after 30 min of anaerobic stress. Maize seedlings that are homozygous for a non-functional *Adh1* gene continue to reduce pyruvate to lactate for at least 80 min, resulting in unrestrained cytoplasmic acidification, irreversible cell damage and death of the plants (Roberts *et al.*, 1984).

### Transcriptional Regulation of Genes Encoding the Anaerobic Polypeptides

Maize genes and/or cDNAs encoding several anaerobically-inducible enzymes have been cloned and Northern analyses demonstrate that the steady-state levels of mRNAs for sucrose synthase I (Werr *et al.*, 1985; Springer *et al.*, 1986), glyceraldehyde-3-phosphate dehydrogenase III (Russell and Sachs, 1989), aldolase (Hake *et al.*, 1985), *Adh1* (Gerlach *et al.*, 1982; Dennis *et al.*, 1984; Hake *et al.*, 1985) and *Adh2* (Gerlach *et al.*, 1982; Dennis *et al.*, 1985) increase in anaerobically-stressed maize roots. It is likely that the increase in mRNA levels for these enzymes is the result of increased transcription of the corresponding genes in response to anoxia. For the maize *Adh1* gene at least, nuclear run-off experiments support this conclusion, although there is also some evidence for greater stability of *Adh1* transcripts in anaerobic compared to aerobic roots (Rowland and Strommer,1986). In addition to the transcriptional regulation of genes encoding anaerobic polypeptides the repression of normal aerobic protein synthesis at the onset of anaerobiosis may be regulated at the translational level, as the *in vitro* translation products of mRNAs isolated from anaerobic roots contain the normal aerobic complement of polypeptides in addition to the anaerobic polypeptides (Sachs *et al.*, 1987). Thus, mRNAs encoding anaerobic polypeptides are preferentially translated *in vivo*, during periods of anaerobic stress. This phenomenon is restricted to the roots of seedlings treated anaerobically for only 5 hr, since after 24 hr of anaerobiosis, only mRNA encoding anaerobic polypeptides is present (Sachs *et al.*, 1980), suggesting that the aerobic mRNA has been turned over by this time.

## THE ANAEROBIC RESPONSIVE ELEMENT REGULATES ANAEROBIC INDUCTION OF THE MAIZE *Adh1* GENE

We have sought to understand the molecular mechanisms regulating the increased transcription of the maize *Adh1* gene in response to anaerobic stress. The system we have used to identify the *cis*-acting regulatory sequence of the *Adh1* promoter essential for anaerobic responsiveness (the Anaerobic Responsive Element, or ARE) is a transient expression assay originally described by Howard *et al.* (1987). When the full-length *Adh1* promoter is linked to a structural reporter gene, either bacterial chloramphenicolacetyl-transferase (CAT) or $\beta$-glucuronidase (GUS) and the

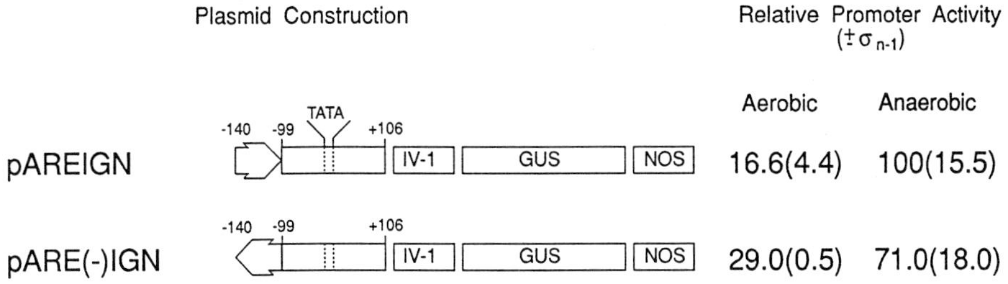


Figure 1. The ARE functions in both orientations. The ARE sequence from position -140 to -99 was cloned in both orientations upstream of a truncated maize *Adh1* promoter (positions -99 to +106), the intron 1 sequence of the *Adh1* gene (IV-1), $\beta$-glucuronidase (GUS) reporter gene and the nopaline synthase (NOS) 3'-termination signal, as indicated in the figure. The ARE in construct pARE IGN is present in the same orientation as in the maize *Adh1* promoter. In construct pARE(-) IGN, this sequence is the inverse orientation. Maize protoplasts were electroporated with constructs and incubated for 20 hr in either aerobic (20%$O_2$/80%$N_2$) or anaerobic (5%$O_2$/95%$N_2$) conditions, after which time GUS enzyme activity was determined. Promoter activity in maize protoplasts is indicated relative to the expression of pARE IGN in anaerobic incubation conditions.

resulting construct is electroporated into maize suspension cell protoplasts, reporter gene expression may be anaerobically induced by incubation in 5% $O_2$/95% $N_2$. Transcriptional initiation from the chimaeric gene is at the same site as in the endogenous *Adh1* gene and anaerobic expression of the chimaeric gene qualitatively parallels expression of the endogenous *Adh1* gene (Howard *et al.*, 1987), indicating that the same cellular signal transduction pathway and transcription machinery is being used. Promoter deletions downstream of position -140 or upstream of position -99 in the *Adh1* promoter eliminate anaerobically-regulated CAT gene expression in electroporated maize protoplasts (Walker *et al.*, 1987), indicating that the ARE sequence is located in this region of the promoter.

We have further demonstrated that the ARE sequence shares some properties with enhancer sequences. For example, inversion of the entire ARE sequence between positions -140 and -99 does not significantly affect the element's function in maize cells, suggesting that the productive interaction between the ARE and other promoter-specific elements such as the TATA box, is undisturbed when their relative orientation is changed (Figure 1). In addition, the ARE sequence is able to function independent of its distance from the TATA box in maize protoplasts, at least up to a total distance of 300 bp (Olive, unpublished). Interestingly, the ARE may be able to function over distances as great as 1.85 kb from the TATA box element *in vivo*, as insertion of Robertson's mutator element *Mu*3 between the ARE and TATA box element of the maize *Adh1-3F* allele does not eliminate anaerobic responsiveness of this gene in maize roots (Bailey-Serres *et al.*, 1987).

When additional copies of the ARE sequence are juxtaposed in the 5'-flanking sequences of the *Adh1* promoter upstream of position -140, the level of reporter gene expression increases almost linearly up to six copies in anaerobic conditions only (Figure 2). In aerobic conditions

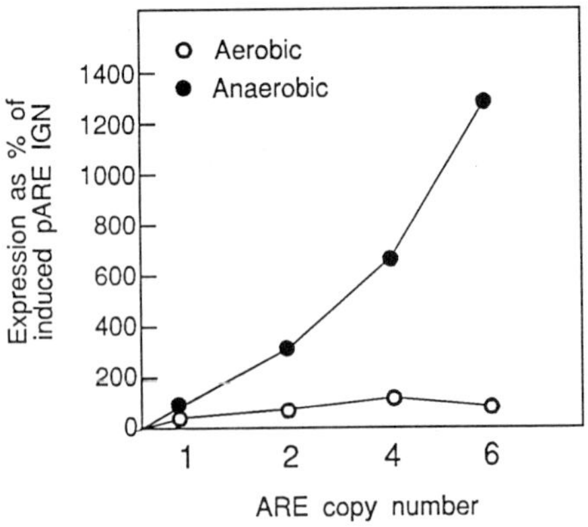


Figure 2. Promoter activity in anaerobic conditions increases in proportion to ARE copy number. Plasmids containing 1,2,4, or 6 copies of the ARE sequence upstream of a truncated *Adh1* promoter (positions -99 to +106) driving GUS gene expression were electroporated and promoter activity assayed as described in the legend to figure 1. Relative promoter activity is as in figure 1.

expression increases at a slower rate and is saturated when four ARE copies are present in the promoter. This observation is consistent with the inducible enhancer activity of single ARE copies in the *Adh1* promoter. The interaction between nuclear proteins and functionally significant DNA enhancer sequences is an important transcriptional control mechanism in eukaryotes (Voss *et al.*, 1986; Ptashne, 1986; Mitchell and Tijan, 1989; Struhl, 1989). The binding of *trans*-acting cellular factors to the ARE sequence was initially demonstrated by *in vivo* DNA methylation interference (Ferl and Nick, 1987). We have confirmed the presence of nuclear proteins in maize nuclei that specifically interact with the ARE sequence *in vitro*, using gel bandshift assays (Olive, unpublished). Given the inducible activity of the ARE, nuclear protein factors binding to the ARE sequence maybe limiting, or present in a substantially inactive form, in aerobic maize cells. The results of Ferl and Nick (1987) indicated that the ARE sequence is protected from DNA methylation in both aerobic and anaerobic conditions, suggesting that these factors may be post-translationally modified in anaerobic conditions to activate ARE-mediated transcription. Ferl and Nick (1987) also identified other regions of the *Adh1* promoter, between positions -181 and -177 and between positions -99 and -91, which appeared to be protected from DNA methylation only in anaerobic conditions. Functional analyses of the *Adh1* promoter in maize protoplasts do not support a role for these sequences in mediating the anaerobic responsiveness of gene expression (Walker *et al.*, 1987; Olive, unpublished results).

Using linker-scanning mutagenesis of the *Adh1* promoter, Walker *et al.*, (1987) identified two sub-region domains within the ARE that interact co-operatively to generate anaerobic responsiveness. These sub-regions are between positions -133 and -124 (sub-region I) and between positions -113 and -99 (sub-region II) and map to regions binding nuclear proteins *in vivo* (Ferl and Nick, 1987). Mutation within sub-region I or II domains eliminates inducible promoter activity and reduces the level of expression

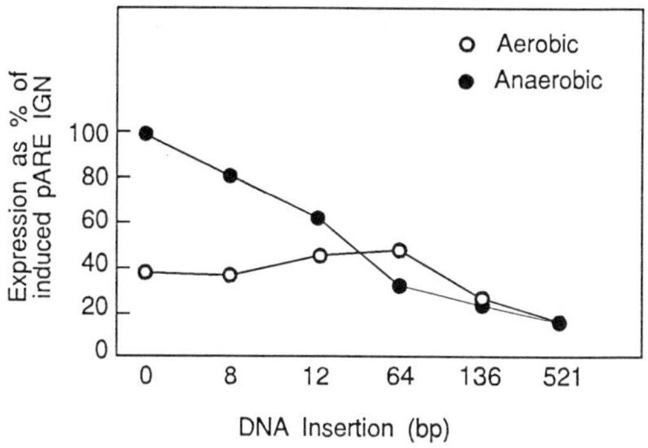


Figure 3. The effect of increased spacing between sub-regions I and II on promoter activity in aerobic and anaerobic conditions. DNA fragments of the sizes indicated were cloned between sub-regions I and II in pARE IGN (for details of this plasmid refer to figure 1). The resultant plasmids were electroporated into maize protoplasts and promoter activity assayed as described in the legend to figure 1. Relative promoter activity is as in figure 1.

in aerobic conditions to 20% or 4% respectively, of the wild type expression (Walker *et al.*, 1987). We have studied the co-operativity between sub-regions I and II further, by cloning stretches of pUC19 DNA between these sequences at position -125 and assaying for promoter activity in maize protoplasts. Promoter activity in aerobic conditions is independent of the distance between sub-regions up to 64 bp but is reduced to background levels when an additional 136 bp of DNA is inserted between sub-regions I and II (Figure 3), a pattern similar to that observed for the domains of the SV40 enhancer (Zenke *et al.*, 1986). In contrast, promoter activity in anaerobic maize protoplasts declines steadily as the distance separating sub-regions I and II is increased (Figure 3). These data argue that anaerobically-regulated expression of the *Adh1* gene in maize requires the close physical association of the ARE sub-region sequences on the DNA helix. It is likely that the close proximity of sub-region I and II facilitates interaction between *trans*-acting protein factors bound to the DNA at these positions in the formation of a transcriptionally active complex. Expression data indicate it is unlikely that an interaction between ARE-bound protein factors occurs to any appreciable degree in aerobic maize cells. Protein factors may be bound to one or both sub-region domains in aerobic cells and anaerobic stress may promote their increased binding or dimerisation, a scenario entirely consistent with the *in vivo* footprint data obtained by Ferl and Nick (1987).

Both sub-regions of the ARE possess a conserved core sequence motif GGTTT, suggesting that these sub-region domains may control similar activities. To address this question we replaced the wild type ARE sequence of a truncated *Adh1* promoter linked to a GUS reporter gene with either a duplicated sub-region I domain or a duplicated sub-region II domain and assayed for GUS activity in maize protoplasts. Both of these mutations appear to result in elevated aerobic levels of expression, a phenomenon for which we have no explanation at present (Figure 4).

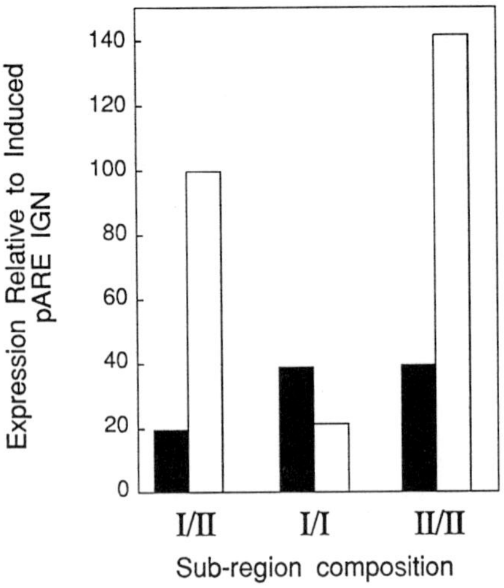


Figure 4. Anaerobic responsiveness of the *Adh1* promoter in maize requires at least one sub-region II domain. The ARE sequence of a truncated *Adh1* promoter driving GUS gene expression (I/II) was replaced with either a duplicated sub-region I domain (I/I) or a duplicated sub-region II domain (II/II). Electroporations of maize protoplasts and promoter activity assays were as described in the legend to figure 1.

However, only the wild-type ARE and the sub-region II dimer mutant are capable of responding to anaerobic stress (Figure 4). Thus, while two sub-region domains are essential for anaerobic responsiveness of gene expression (Walker *et al.*, 1987), our data indicate clearly that at least one of these domains must be a sub-region II sequence. *In vitro* analysis of the direct interaction of nuclear proteins with sub-region dimer mutants indicates that both sub-regions bind the same nuclear protein factor (Olive,unpublished). This finding is not surprising in view of the conserved core sequence in sub-regions I and II. The ability of a second sub-region II sequence to functionally replace the distal sub-region I motif of the *Adh1* promoter also supports this conclusion. Thus, the apparently different activities of sub-regions I and II in anaerobically-stressed maize protoplasts may suggest that each domain has a different affinity for the same nuclear protein. Nucleotide sequences flanking the conserved core sequence motif present in each sub-region differ considerably and it is not difficult to envisage how such differences might affect the formation of DNA/protein complexes.

The ARE sequence is able to regulate expression from a heterologous promoter, a property of many enhancer sequences (Voss *et al.*, 1986). As shown in Figure 5, the CaMV 35S promoter directs GUS gene expression in maize protoplasts independent of hypoxia, and truncation of this gene to position -90 reduces the level of gene expression to only 20-30% of the level observed for the full-length 35S promoter. Addition of the ARE sequence from position -140 to position -99 upstream of the truncated 35S promoter results in 2- to -3 fold greater expression in aerobic conditions and 4- to 6-fold greater expression in anaerobic conditions (Figure 5) and therefore the ARE is a positive inducible enhancer in maize. More

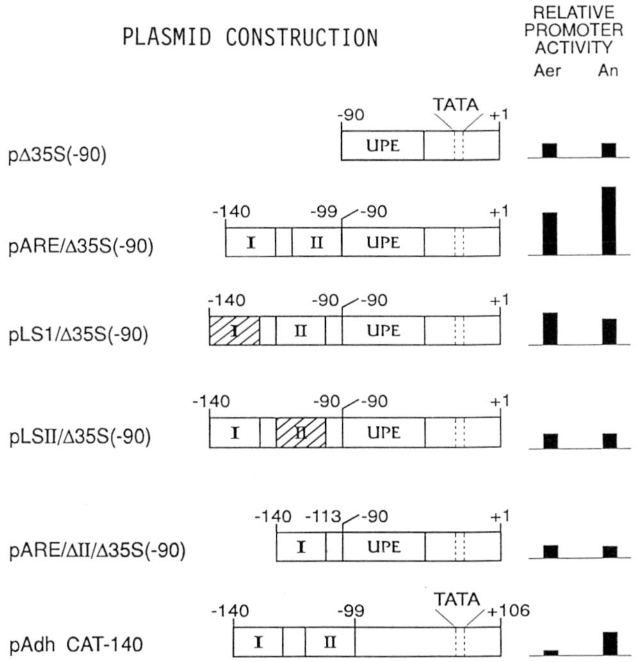


Figure 5. The ARE regulates expression in maize from the heterologous CaMV 35S promoter. ARE sequences were cloned in front of a truncated CaMV 35S promoter (-90 to +1; stippled areas) driving CAT gene expression and promoter activity assayed in maize protoplasts. The ARE sequences present in constructs pLSI/Δ35S(-90) and pLSII/Δ35S(-90) carried mutations in sub-regions I and II, respectively (hatched areas) while the sub-region II sequence was deleted from construct pAREΔII/Δ35S(-90). Construct pAdhCAT-140 contains the *Adh1* promoter from position -140 to +106, driving CAT gene expression. Promoter activity in $20\%O_2/80\%N_2$ (Aer) and $5\%O_2/95\%N_2$ (An) is shown relative to the activity of the full-length 35S promoter, which is hypoxia-independent.

importantly, the ARE possesses a general enhancer function which is expressed at all oxygen concentrations, in addition to the well-characterised inducible enhancer function, when assayed in the context of the CaMV 35S promoter. Interestingly, mutations within sub-region II of the ARE alone reduce the level of aerobic expression from the hybrid ARE/35S promoter (Figure 5), suggesting that the sub-region II but not the sub-region I sequence is transcriptionally active at this oxygen concentration. The 35S promoter contains a functional enhancer at position -75 which has 70% identity to the octopine synthase (*ocs*) enhancer element (Bouchez *et al.*, 1989; Singh *et al.*, 1989) and interacts with the maize nuclear protein factor OCSTF (Bouchez *et al.*, 1989). Assuming that regulated gene expression is the sum of productive interactions between all *trans*-acting factors bound to a promoter, then this general enhancer activity of the ARE may arise from interaction between factors bound to the sub-region II sequence and the 35S enhancer sequence. In addition, the activity of sub-region II in aerobic conditions supports the conclusion that nuclear factors recognising the ARE are present in aerobic conditions and able to bind DNA. Consistent with earlier studies examining activity of the ARE sequence in the context of the *Adh1* promoter (Walker *et al.*, 1987), mutation of either sub-region of the ARE destroys anaerobic responsiveness of the hybrid ARE/35S promoter (Figure 5), indicating that both sub-regions are also required for full activity of the hybrid promoter in anaerobic conditions.

CONSERVATION OF *CIS* AND *TRANS*-ACTING FACTORS IN THE ANAEROBIC RESPONSE

The induced expression of genes encoding the anaerobic polypeptides *Adh1*, *Adh2*, aldolase, sucrose synthase-1 and glyceraldehyde-3-phosphate dehydrogenase II occurs within 1 to 3 hrs of anaerobiosis (Hake *et al.*, 1985; Russell and Sachs, 1989). The coordinate induction of these genes suggests that similar molecular mechanisms are controlling the expression of these genes, for example common recognition sequences within the promoters of these genes and common *trans*-acting protein factors. In addition, the ability to respond to anaerobic stress is widespread in the plant kingdom. The anaerobic induction of the *Adh1* gene has been observed in maize, barley, wheat, pea, petunia, *Arabidopsis thaliana*, potato, cotton, tobacco and carrot, suggesting that the metabolic shift from oxidative phosphorylation to ethanolic fermentation is the preferred adaptive mechanism utilised in this stress response. We have examined the nucleotide sequences of the maize *Adh2*, maize sucrose synthase-1, maize aldolase, pea *Adh1* and *Arabidopsis thaliana Adh1* genes for the presence of the ARE core sequence motif. As shown in Figure 6, this sequence is present in the promoters of these genes. In the case of the maize aldolase gene, functional analysis of the promoter in maize protoplasts indicates that the ARE consensus sequence is located in a region of the promoter necessary for anaerobic expression (Dennis *et al.*, 1988) Similarly, the ARE core sequence present in the pea *Adh1* gene is located in the region critical for anaerobic induction of that gene. Functional analysis of the pea *Adh1* promoter in tobacco indicates that there are three sub-region domains responsible for anaerobic induction of this gene, clustered between positions -145 and -98 and expression requires the presence of at least two intact,adjacent sub-regions (Llewellyn, unpublished). The region of the pea *Adh1* promoter which possesses the strongest anaerobic activity contains the ARE core consensus sequence between positions -112 and -101, in the inverse orientation to that found in maize (Figure 6). Interestingly, all of the sequences with the exception of the ARE sub-region I domain from the maize *Adh1* gene, possess a thymidine residue at nucleotide position 3 in the consensus sequence (Figure 6). Perhaps the presence of a cytosine at this position in the

| | |
|---|---|
| ARE sub-region I, maize *Adh1-1S* gene (Dennis *et al.*, 1984) | $^{-133}$CCCGGTTTCG$^{-124}$ |
| ARE sub-region II, maize *Adh1-1S* gene (Dennis *et al.*, 1984) | $^{-113}$CGTGGTTTGC$^{-104}$ |
| maize *Adh2-N* gene (Dennis *et al.*, 1985) | $^{-143}$CGTGGTTTCT$^{-134}$ |
| maize sucrose synthase gene (Werr *et al.*, 1985) | $^{+431}$CGTGGTTTGC$^{+440}$ |
| | $^{-171}$TCTGGTTTTG$^{-162}$ |
| maize aldolase (Dennis *et al.*, 1987) | $^{-68}$GCTGGTTTCT$_{-59}$ |
| pea *Adh* (reverse orientation) (Llewellyn *et al.*, 1987) | $^{-101}$TTTGGTTTGT$^{-112}$ |
| Arabidopsis *Adh* (reverse orientation) (Chang and Meyerowitz, 1986) | $^{-153}$TTTGGTTTTG$^{-162}$ |

**CONSENSUS SEQUENCE**

**TC        TC**
**CGTGGTTTGT**
**GT        CG**

Figure 6. Sequences conserved in the promoters of anaerobically induced genes. The positions of bases in the gene sequences are given, relative to the transcription start site. The ARE core sequence is underlined in each case. The overall consensus sequence is shown in the lower part of the figure.

sub-region I sequence accounts for the relative weakness of this domain in expression studies (Figure 4, Figure 5).

The apparent ubiquity of the core ARE sequence in the promoter of anaerobically inducible genes suggests that a similar *trans*-acting protein factor might mediate the increased transcription of several genes in response to anaerobic stress. Ellis *et al.* (1987) demonstrated that transcription from the maize *Adh1* promoter is correctly initiated in tobacco protoplasts, although the level of transcription observed is low. A region of the *ocs* gene (-292 to -116) containing the 16 bp palindromic *ocs*-enhancer, when placed upstream of position -140 of the maize *Adh1* promoter driving CAT gene expression allows detection of anaerobically-inducible expression in tobacco (Ellis *et al.*, 1987). Replacement of the ARE sequence in this hybrid promoter with pBR322 DNA eliminates expression in tobacco tumour-tissue (Ellis, unpublished) consistent with the conclusion that the ARE is responsible for the anaerobic induction of reporter gene expression. When the *ocs* gene sequences and the ARE sequence are placed upstream of a crippled nopaline synthase (NOS) promoter driving CAT gene expression, the construct is also anaerobically-inducible in the leaves of stably-transformed tobacco plants (Ellis, unpublished). Thus, the nuclear protein factors recognising the ARE in tobacco and maize are sufficiently conserved to bind the same DNA recognition sequence. However, the requirement for the *ocs*-enhancer sequences to boost expression in tobacco suggests that the interaction between tobacco nuclear factors and the maize ARE is imperfect, perhaps requiring stabilisation- by ASF-1, the tobacco nuclear factor recognising the *ocs*-enhancer (Lam *et al.*, 1989).

## CONCLUSION

We have identified a functionally significant inducible enhancer sequence which is necessary for increased transcription of the maize *Adh1* gene in response to anaerobic stress. This element consists of two sub-region domains, each of which contains a conserved core sequence and binds an identical *trans*-acting nuclear protein. In addition there is evidence suggesting that these *cis*- and *trans*-acting factors are conserved in mediating the response to anaerobic stress, between different anaerobically-regulated genes in maize and also to some degree across species barriers.

The least understood aspect of cellular regulation in plants is the signal transduction-pathway from the cell wall to the nucleus. It is possible that plant hormones may initiate a cascade of events leading to increased activity of the ARE-binding factor and increased transcription of the *Adh1* gene. The activity of this transcription factor may be regulated at the transcriptional or the translational level, however, the presence of ARE-binding factor in aerobic maize cells suggests that post-translational modifications such as phosphorylation/dephosphorylation may be an important control of anaerobically-regulated gene expression. The cloning and purification of genes encoding ARE-binding factor and an analysis of how this protein is regulated will assist in elucidating further the molecular mechanisms involved in the response of plants to environmental stress.

## REFERENCES


Bailey-Serres, J., Kloeckener-Gruissem, B., and Freeling, M., 1987, Genetic and molecular approaches to the study of the anaerobic response and tissue specific gene expression in maize, *Plant, Cell and Environ.*, 11:351.

Bouchez, D., Tokuhisa, J.G., Llewellyn, D.J., Dennis, E.S., and Ellis, J.G., 1989, The ocs-element is a component of the promoters of several T-DNA and plant viral genes, *EMBO J.*, 8:4197.

Chang, C., and Meyerowitz, E.M., 1986, Molecular cloning and DNA sequence of the *Arabidopsis thaliana* alcohol dehydrogenase gene, *Proc. Natl Acad. Sci. (USA)*, 83:1408.

Dennis, E.S., Gerlach, W.L., Pryor, A.J., Bennetzen, J.L., Inglis, A., Llewellyn, D., Sachs, M.M., Ferl, R.J., and Peacock, W.J., 1984, Molecular analysis of the alcohol dehydrogenase (*Adh1*) gene of maize, *Nucl. Acid Research*, 12:3983.

Dennis, E.S., Sachs, M.M., Gerlach, W.L., Finnegan, E.J., and Peacock, W.J., 1985, Molecular analysis of the alcohol dehydrogenase 2 (*Adh2*) gene of maize, *Nucl. Acid Research*, 13:727.

Dennis, E.S., Gerlach, W.L., Walker, J.C., Lavin, M., and Peacock, W.J., 1988, Anaerobically regulated aldolase gene of maize : A chimaeric origin?, *J. Mol. Biol.*,202:759.

Ellis, J.G., Llewellyn, D.J., Dennis, E.S., and Peacock, W.J., 1987, Maize *Adh-1* promoter sequences control anaerobic regulation: addition of

upstream promoter elements from constitutive genes is necessary for expression in tobacco, *EMBO J.*, 6:11.
Ferl, R.J., and Nick, H.S., 1987, *In vivo* detection of regulatory factor binding sites in the 5' flanking region of maize *Adhl*, *J. Biol. Chem.*, 262:7947.
Gerlach, W.L., Pryor, A.J., Dennis, E.S., Ferl, R.J., Sachs, M.M., and Peacock, W.J., 1982, cDNA cloning and induction of the alcohol dehydrogenase gene of maize, *Proc. Natl Acad. Sci. (USA)*, 79:2981.
Good, A.G., and Crosby, W.L., 1989, Anaerobic induction of alanine aminotransferase in barley root tissue, *Plant Physiol.*, 90:1305.
Hagemann, R.H., and Flesher, D., 1960, The effect of anaerobic environment on the activity of alcohol dehydrogenase and other enzymes in corn seeds, *Arch.Biochem. Biophys.*, 87:203.
Hake, S., Kelley, P., Taylor, W.C., and Freeling, M., 1985, Co-ordinate induction of alcohol dehydrogenase 1, aldolase and other anaerobic RNAs in maize, *J. Biol. Chem.*, 260:5050.
Hoffmann, N.E., Bent, A.F., and Hanson, A.D., 1986, Induction of lactate dehydrogenase isozymes by oxygen deficit in barley root tissue, *Plant Physiol.*, 82:658.
Howard, E.A., Walker, J.C., Dennis, E.S., and Peacock, W.J., 1987, Regulated expression of an alcohol dehydrogenase chimaeric gene introduced into maize protoplasts, *Planta*, 170:535.
Kelley, P.M., 1987, Identifying the anaerobic proteins of maize, *in*: "Environmental Stress in Plants," J.H. Cherry, ed., Springer-Verlag,Berlin.
Kelley, P.M., and Freeling, M., 1984a, Anaerobic expression of maize glucose phosphate isomerase I, *J. Biol. Chem.*, 259:673.
Kelley, P.M., and Freeling, MN., 1984b, Anaerobic expression of maize fructase-1,6-diphosphate aldolase, *J. Biol. Chem.*, 259:14180.
Laszlo, A., and St Lawrence, P., 1983, Parallel induction and synthesis of PDC and ADH in anoxic maize roots, *Mol. Gen. Genetics*, 192:110.
Llewellyn, D.J., Finnegan, E.J., Ellis, J.G., Dennis, E.S., and Peacock, W.J., 1987, Structure and expression of an alcohol dehydrogenase 1 gene from *Pisum sativum* (cv. Greenfeast), *J. Mol. Biol.*, 195:115.
Mitchell, P.J., and Tijan, R., 1989, Transcriptional regulation in mammalian cells by sequence-specific DNA binding proteins, *Science*, 245:371.
Ptashne, M., 1986, Gene regulation by proteins acting nearly and at a distance, *Nature*, 322:697.
Roberts, J.K., Callis, J., Wemmer, D., Walbot, V., and Jardetzky, O., 1984, Mechanism of cytoplasmic pH regulation in hypoxic maize root tips and its role in survival under hypoxia, *Proc. Natl Acad. Sci. (USA)*, 81:3379.
Rowland, L.J., and Strommer, J.N., 1986, Anaerobic treatment of maize roots affects transcription of *Adhl* and transcript stability, *Mol. Cell. Biol.* 6:3368.
Russell, D.A., and Sachs, M.M., 1989, Differential expression and sequence analysis of the maize glyceraldehyde-3-phosphate dehydrogenase gene family, *The Plant Cell*, 1:793.
Sachs, M.M., Freeling, M., and Okimoto, R., 1980, The anaerobic proteins of maize, *Cell*, 20:761.
Springer, B., Werr, W., Starling, P., Clark Bennett, D., Zokolica, M., and Freeling, M., 1986, The *Shrunken* gene on chromosome 9 of *Zea mays* L. is expressed in various plant tissues and encodes an anaerobic proteins, *Mol. Gen. Genet.*, 205:461.
Struhl, K., 1989, Helix-turn-helix, zinc finger, and leucine zipper motifs for eukaryotic transcriptional regulatory proteins, *TIBS*, 14:137.
Voss, S.D., Schlokat, U., and Gruss, P., 1986, The role of enhancers in the regulation of cell-type-specific transcriptional control, *TIBS*, 11:287.

Walker, J.C., Howard, E.A., Dennis, E.S., and Peacock, W.J., DNA sequences required for anaerobic expression of the maize alcohol dehydrogenase 1 gene, *Proc. Natl Acad. Sci. (USA)*, 84:6624.

Werr, W., Frommer, W.B., Maas, C., and Starlinger, P., 1985, Structure of the sucrose synthase gene on chromosome 9 of *Zea mays* L., *EMBO J.*, 4:1373.

Zenke, M., Grundström, T., Matthes, H., Winzerith, M., Schatz, C., Wilderman, A., and Chambon,P., 1986, Multiple sequence motifs are involved in SV40 function, *EMBO J*, 5:387.

# THE INDUCTION OF THE HEAT SHOCK RESPONSE: ACTIVATION AND EXPRESSION OF CHIMAERIC HEAT SHOCK GENES IN TRANSGENIC PLANTS

Fritz Schöffl, Mechthild Rieping, Klaus Severin

Universität Bielefeld
Biologie (Genetik)
D-4800 Bielefeld
F.R.G.

present address:

Universität Tübingen
Biologie (Allgemeine Genetik)
D-7400 Tübingen
F.R.G.

## INTRODUCTION

The heat shock (hs) response is a highly conserved, almost universal genetic system in living organisms. This response appears to have a protective function for many types of cells and tissues which have been exposed to heat stress, but also certain other chemical and physico-chemical stressors have the capacity to elicit the hs response (for reviews see Neumann et al., 1989; Lindquist and Craig, 1988; Schöffl et al., 1988). Hence, stressed cells seem to seek protection from the detrimental effects of environmental stress, at least to some extent, by the hs response. In physical terms, the stressor is an external force causing an internal strain. The adoption of these terms to biological systems has changed the meaning in a way that the term stress is used for "any environmental factor potentially unfavourable to living organism" and "strain is any physical or chemical change produced by a stress" (Levitt, 1980). The term hs response is commonly used for the reprogramming of cellular activities which is rapidly induced by heat stress. One important feature of this response is the de novo synthesis of a number of hs proteins (hsps). Following severe but sublethal heat stress, cells are able to recover and they may even tolerate a subsequent, higher dosage of the stressor.

Although the precise function of the different classes of hsps is still unknown, one property common to all of them is their hs-dependent interaction with other cellular proteins, which is typical for the high molecular mass (hmm) hsp (> 60KD), or with other hsp, shown in particular for the small hsp (< 30KD) (for review see Neumann et al., 1989; Schöffl et al., 1988, 1990). Hmm-hsps and related but constitutively expressed proteins may be considered as molecular chaparoninins which are involved in transport, maturation, and assembly of the temporarily associated proteins. The small hsps (17-30 KD), the most complex and prevalent groups

*Plant Molecular Biology 2*, Edited by R.G. Herrmann and
B. Larkins, Plenum Press, New York, 1991

of hsps in plants, aggregate with each other in a temperature dependent fashion and the formed granular structures seem to be involved in the preservation of non-hs mRNAs (Nover et al., 1989).

Transcriptional regulation of gene expression has been demonstrated for soybean hs genes (Schöffl et al., 1987). This property is functionally related to conserved hs elements (HSE) present in multiple copies upstream from the TATA box sequences of these and also most of the other hs genes in different organisms. HSEs are cis-active binding sites for the trans-active hs factor HSF. HSF has been isolated from yeast, Drosophila and human cells (Sorger and Pelham, 1987; Wu et al., 1987; Goldenberg et al., 1988) but not yet from plants. The most highly conserved motif within HSE sequences is nGAAnnTTCn, more generally defined as blocks of alternating GAA with a spacing of 2 bp between them. The functional analysis of native and synthetic HSEs showed that two copies of interlocked HSEs, equivalent to one-and-a-half copies of the 10 bp consensus sequence or three blocks of alternating nGAAn elements, are minimally required for hs induced transcription (Amin et al., 1988; Xiao and Lis, 1988).

This tripartite structure of the binding site is consistent with the association of HSF to trimers which have been isolated from Drosophila (Perisic et al., 1989), and yeast (Sorger and Nelson, 1989), but it is still a matter of discussion whether the binding site for a trimeric HSF is a degenerate palindromic or, alternatively, a directly repeated sequence (Raibaud, 1990). The structural features of a soybean HSE consensus sequence, which proved to be functional in chimaeric gene constructructions in transgenic tobacco, would favour an imperfect palindromic sequence (Schöffl et al., 1989, 1990).

The binding of HSF to conserved promoter elements is not a sufficient criterion for transcriptional activation. In yeast, HSF binds HSE sequences already prior to the transcriptional activation by hs (Sorger et al., 1987). In contrast, Drosophila and mammalian HSF seem to bind to HSE only after a hs (Sorger et al., 1987; Zimarino and Wu, 1987; Kingston et al., 1987). Thus, binding and transcriptional activation are two separate processes, the latter may possibly require phosphorylation of HSF (Sorger et al., 1987; Larson et al., 1988). The yeast HSF is an 833 amino acid protein encoded by a single copy gene that is essential for growth at all temperature (Sorger and Pelham, 1988; Wiederrecht et al., 1988). The region of HSF between residues 167 and 284 has been shown to be involved in DNA-binding (Wiederrecht et al., 1988), sequences between residues 327 and 424 are important for oligomerization of HSF (Sorger and Nelson, 1989). The mechanism by which HSF is activated in response to hs is the key for the understanding of the signal transduction pathway. First steps toward this goal have been the selections of mutations affecting the hs response in Drosophila (Parker-Thornburg and Bonner, 1987), and in yeast (Findly et al., 1988). High efficiency of stimulated transcription may additionally require other cis-acting elements and their corresponding trans-acting factors. CCAAT sequences, as shown in mammalian cells (Morgan et al., 1987) and AT-rich sequences as suggested by the analyses of plant hs genes (Baumann et al., 1987; Czarnecka et al., 1989) may be involved in processes regulating the amplitude of transcription.

The functional analyses of plant hs promoters in transgenic plant cells revealed the minimal requirements of cis-active sequences for the regulated transcription of native soybean hs genes (for reviews see Schöffl et al., 1988, 1990; Neumann et al., 1989; Nover, 1990). It has not yet been demonstrated that these hs-mRNAs are translated into polypeptides in transgenic plants. However, using chimaeric hs promoter: reporter gene constructions, we and others (Schöffl et al., 1989; Strittmatter and Chua, 1987) have shown that native and synthetic consensus promoter elements can be used in modular constructions to express functional gene products of the linked genes in a temperature-dependent fashion. Here we summarize the analyses of chimaeric hs genes in transgenic plants and discuss gene regulatory phenomena of both, transcriptional and posttranscriptional mechanisms, in the context of new experimental results of our laboratory. It will be demonstrated that combinations of HSEs and constitutive promoter elements are involved in transcriptional activation, the amplitude of transcription, and translational competence of mRNA chimeras during hs, and that selectable marker genes driven by a soybean hs promoter can be used for the creation of a suitable genetic background in transgenic plants which may allow selection of mutations that act in trans to alter the hs response.

## RESULTS AND DISCUSSION

### Cis-Acting Sequences Involved in the Regulation of Transcription and Translation

The functional significance of multiple HSE sequences of the soybean hs gene Gmhsp 17.3-B was investigated by 3'-promoter deletions linked to the chloramphenicol acetyltransferase (CAT) reporter gene (Schöffl et al., 1989). Constructs containing the native TATA box sequence, multiple HSEs located upstream from the TATA box within an approximately 150 bp region, downstream sequences including the mRNA start site and almost the entire non-translated mRNA leader region, showed high levels of heat-inducible CAT activity in transgenic tobacco plants. In these plants the CAT activities correlated with the level of CAT-mRNAs. Similar high levels of mRNA and reporter enzyme activity were present in plants containing translational fusion constructs driven by the same promoter. These results suggest that most of the cis-regulatory sequences, which are important for transcription and translation of linked reporter genes during hs, are included within approximately 320 bp upstream from the translational start site (+1) of the native hsp protein coding region. However, the question arose whether additional upstream sequences, located beyond position -321, would have an effect on the amplitude of gene expression. The analysis of 5'-promoter deletions of the native Gmhsp 17.3-B gene showed that heat-induced transcript levels were significantly enhanced (by a factor of 10) by contructs containing upstream sequences (up to position -439) in comparison with contructs terminated at -298 (Baumann et al., 1987). These sequences, delimited by the deletions -439 and -298, lack HSEs; however, structural features including an imperfect palindromic sequence with its centre at position -407 and a run of adenine bases between positions -357 and -371 are present. The importance of the AT-rich "simple sequences" was further underscored by gel retardation analyses (Severin et al., 1989; Schöffl et al., 1990) suggesting that nuclear proteins bind to this region. Nuclear protein binding AT-rich upstream sequences of Gmhsp 17.5-E seem to be also involved in modulation of the quantity of transcription (Czarnecka et al., 1989).

In order to test the effect of distal promoter upstream sequences on the amplitude of gene expression, chimaeric gene constructions were generated and gene expression was studied in transgenic tobacco. These constructs (Fig. 1) differed in two aspects from the previously described CAT gene constructions: (i) the CAT reporter gene was replaced by the glucuronidase (GUS) gene, terminated by the 3'-region (NOSter) of the nopaline synthase gene (Jefferson, 1987); (ii) the addition of a DNA-segment, spanning the sequences between positions -321 and -593 upstream from the hs gene (Fig.1). Seven to 23 independently transformed "expressor" plants were examined for heat-inducible GUS activity for each of the six different constructions (Table 1). Surprisingly, no significant differences in the maximum levels of GUS activities, induced after 2 hours heat shock, 40°C, are detectable for -12 and -19 constructs, irrespective of an addition of the upstream sequences. The apparently higher levels in plants containing constructs with the upstream region seem to result from an elevated basal expression of GUS under non-hs conditions. The basal levels are 2-3 fold higher than in -12, -19 plants. The calculated hs-induction ratio is unchanged for -12 plants and even lower for -19 plants containing the upstream region. Hence, no enhancer function can be assigned to the additional upstream sequences.

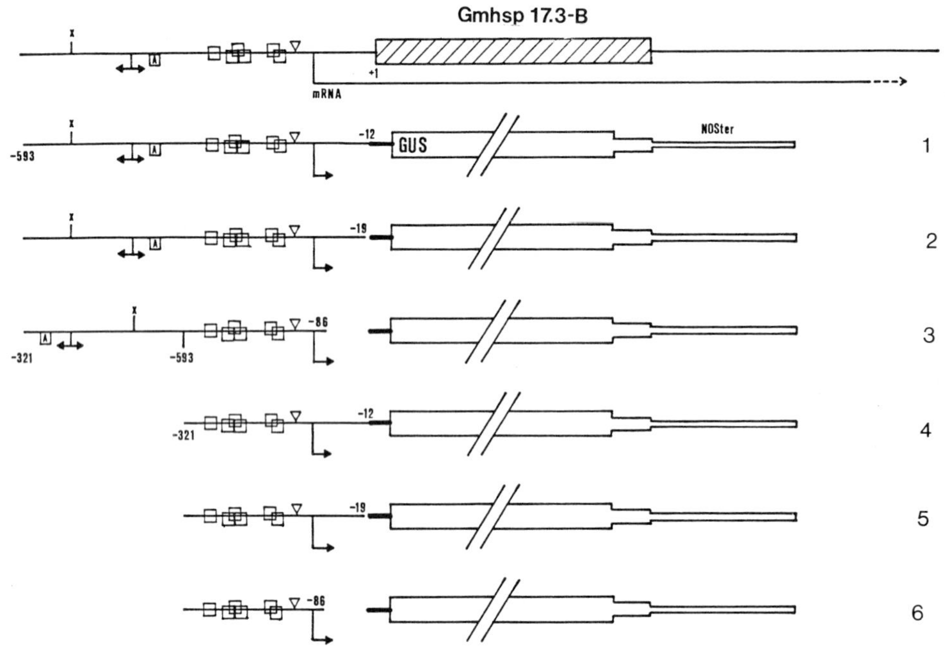


Fig. 1. Schematic diagrams of hs promoter:GUS fusions. (1) -593/-12/NOSter, (2) -593/-12/NOSter, (3) -321/-593/-86/NOSter, (4) -321/-12/NOSter, (5) -321/-19/NOSter, (6) -321/86/NOSter. The numbers within constructs refer to the translational start site of the Gmhsp 17.3-B protein coding region. Constructs number 4-6 contain the hs promoter up to the 5'-position -321, construct number 1 and 2 contain additional upstream sequences up to position -593 in the native orientation, number 3 contains an inverted upstream region. Symbols are used for: TATA-box (triangle), HSE (square), A/T-run (A), imperfect palindrom (bipointed arrow), XbaI site (X), transcriptional start site (arrow). The gaps indicate the loss of hs mRNA 5'-leader sequences (Schöffl et al., 1989).

Table 1. Heat shock induced GUS activities in transgenic tobacco plants

| Nr[a] | Construct | Number of plants tested | Average GUS activity[b] (pMol MU/min/ mg protein) 25°C | 40°C | Heat shock induction ratio (fold) |
|---|---|---|---|---|---|
| 1 | -593/-12/NOSter | 21 | 12.5 | 36 | 2.9 |
| 2 | -593/-19/NOSter | 7 | 17.5 | 30 | 1.7 |
| 3 | -321/-593/-86/NOSter | 7 | 5.5 | 14 | 2.5 |
| 4 | -321/-12/NOSter | 23 | 5.5 | 16 | 2.9 |
| 5 | -321/-19/NOSter | 7 | 5.5 | 23 | 4.2 |
| 6 | -321/-86/NOSter | 15 | 5.5 | 9 | 1.6 |

[a]) numbers refer to constructions shown in Fig. 1

[b]) GUS assays were performed according to Jefferson (1987) using 20μg total leaf protein extracted after 2 h incubation at the respective temperature

Basal levels of GUS activity can be explained by a leakyness of the hs promoter, resulting in a low level of mRNA at normal temperatures (Schöffl and Baumann, 1985). The high stability of GUS enzyme, respectively its slow turnover in plant tissue may lead to the accumulation of this enzyme to significant levels in non-hs leaves. The higher levels of basal GUS activity in plants containing additional upstream sequences cannot be explained conclusively. It seems possible that the putative enhancer affects, in contrast to the native hs gene, only the basal transcription of chimaeric gens. It has to be pointed out that an inversion of the upstream region, as shown for the -86 construct (Fig. 1), does not increase the basal level of GUS activity in these plants.

CAT-fusions driven by -86 deletion constructs are not efficiently translated during hs, but translation is resumed during subsequent recovery periods (Schöffl et al., 1989). Plants containing the -86/GUS fusions show a low but significant level of heat-inducible GUS activity without requireing recovery from heat treatment (Table 1). This may be due to the higher sensitivity of the GUS assay compared to the CAT assay. It cannot be excluded that the efficiency of translation is higher for chimaeric GUS mRNA than for CAT mRNA.

The modular use of hs promoters and promoter elements for transcriptional regulation of chimaeric genes rises also a number of questions which cannot be answered to satisfaction at the present time. The basic units of cis-acting sequences are multiple copies of HSE-like sequences located upstream from the TATA box sequence. Synthetic HSE2 sequences and the otherwise silent viral CaMV 35S promoter represent a compatible combination of binding sites for hs-specific and, respectively, basic _trans_-acting factors (Schöffl et al., 1989). Strittmatter and Chua (1987) have shown that overlapping HSEs were sufficient to confer heat-inducibility to a light regulated small subunit rubisco gene promoter, but only in leaves. Heat induction of this gene is repressed in roots, probably via interference by a silencer region. The TATA box of the human hsp70 gene can be replaced by the TATA boxes derived from the adenovirus EIIA and SV40 earlier promoters without loss of function (Green and Kingston, 1990). A replacement of the proximal HSEs by the upstream enhancer-like sequences of the _Gmhsp 17.3-B_ gene is ineffective in transcriptional activation. Neither constitutive nor heat-inducible activity of chimaeric CaMV:CAT gene constructs was detected in tobacco leaves. However, a 3'-extension of this region by a sequence that includes one single HSE of the native hs promoter yields high levels of heat-inducible CAT activity in leaf tissue of transgenic plants (Rieping and Schöffl, unpublished). These results suggest a cooperation between different distal and proximal elements. It is conceivable that the CCAAT box sequences, located upstream from the most distal HSE, may become crucial for the hs activation if HSEs are limited, and they may be involved in an elevation of transcription of the native promoter. The lack of a noticable effect on reporter gene expression by the extension of 5'-sequences (5' to position -321, see Fig. 1, Table 1) is in agreement with this interpretation; the CCAAT box sequences are already present in the non-extended constructions. Further experimental evidence is required to prove unambigiously the function of CCAAT box sequences in transcriptional regulation of plant hs genes.

## Heat-Inducible Hygromycin Resistance in Transgenic Tobacco

The modular use of hs promoters as temperature-sensitive switches for the expression of heterologous genes becomes increasingly important for applied aspects, e.g. agricultural crop improvement and biotechnology by genetic engineering, and basic plant science. Due to the conservation of _cis_- and _trans_-acting regulatory components, hs promoters of plants are recognized in heterologous species and can be used for the regulated expression of chimaeric genes (see previous chapter). One important application is the creation of a suitable genetic background in transgenic plants allowing selection of second site mutants. Mutation that interfere with the signal transduction that elecits the hs response would be highly valuable for the identification of components of the pathway, the elucidation of its molecular mechanism, and the investigation of thermotolerance in plants with altered hs-response. The activation of HSF appears to be the key to the signal transfer pathway. Mutations that act in _trans_ to alter the transcription of hs genes have been selected in _Drosophila_ (Parker-Thornburg and Bonner, 1987) and in yeast (Findly et al., 1988) by the use of hs promoter-driven marker genes. Similar selection schemes may be also used for plants (Schöffl, 1988).

We have constructed a chimaeric gene consisting of the promoter of the soybean hs gene _Gmhsp 17.6-L_, the coding region of a hygromycin phosphotransferase (_hpt_) gene, and the nos-termination sequence (Severin and Schöffl, 1990). The structural features of this hs promoter are very similar to those of the _Gmhsp 17.3-B_ gene described above (for review see Schöffl et al., 1990); the point of fusion with the _hpt_ gene is within the 5'-nontranslated leader sequence. A very large part of the hs-mRNA leader has been preserved in this construct. The _hpt_ gene has been chosen because it is one of the few resistance markers for which sufficient high levels of resistance to drug can be established in plants if the gene is driven by a constitutive promoter (Van den Elzen et al.,

1985). Transgenic tobacco plants containing the $hpt^{hs}$ gene are unable to grow on hygromycin at normal temperature. However, one hour/day heat treatment (40°C), applied over several weeks, was sufficient to express the resistant phenotype in these plants and in about 75% of the seeds in the F1-generation from selfed plants. These data suggest that the hygromycin resistance gene is functional, and its expression is faithfully controlled by the soybean hs promoter. Since uninduced and hs-induced hsL-hpt-nos plants can be easily discriminated on hygromycin containing media, F1 seeds represent a suitable material for mutagenesis and selection of trans-dominant mutations selected by growth on hygromycin under normal temperature conditions (25°C). Such mutations should affect the hs-response in a way that hs genes are constitutively expressed, probably by an alteration of components involved at any step of the signal transfer pathway which consequently results in constitutive activation of HSF. Alternative selection strategies could be used for the identification of mutations with adverse effect on the activation of HSF (Schöffl et al., 1988).

## GENERAL CONCLUSIONS

The use of hs promoters in chimaeric gene constructions offer a number of possibilities to alter gene expression of specific genes and genetic traits, and for scrutiny of the basic molecular mechanisms which are effective in intramolecular signalling that leads to the hs response. Prerequisite to applications, more knowledge is required to understand how cis- and trans-regulatory elements, respectively factors interact and mediate transcriptional and posttranscriptional regulation of gene expression. The assignment of structural features and functional properties of the soybean hs gene Gmhsp 17.3-B (Fig.2) is typical for the plant hs genes studied to date. Several of these features appear conserved in hs genes from higher eucaryotes. Conserved HSE sequences, favourably in an overlapping configuration are the dicisive cis-acting promoter sequences for heat-inducibility of transcription. It seems clear, that at least two copies of HSEs are necessary. However, evidence from research in animal cells, and recent experiments in plants, suggest that one HSE may be sufficient if a second copy is replaced by other proximal elements, e.g. CCAAT box sequences. CCAAT box sequences are involved in the regulation of hsp70 gene expression in Xenopus and human cells (Wu et al., 1986; 1986; Bienz and Pelham, 1986; Bienz, 1986). A complex interaction of CTF

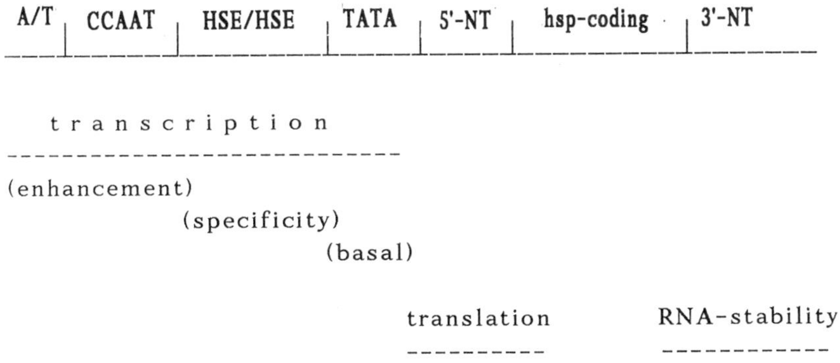


Fig. 2. Model of the structural and functional organization of cis-acting sequences of the soybean hs gene Gmhsp 17.3-B. The particular sequence elements are not drawn to scale.

and HSF, bound to mutually replaced CCAAT and HSE sequences, seems to be involved in tissue specificity and quantity of hsp70 expression (Bienz and Pelham, 1986; Bienz, 1986). The molecular mechanism of the cooperation between HSE-bound HSFs, respectively between HSF and CCAAT box binding factors CTF, which may stimulate transcription synergistically, are still unknown. It is speculated that these complexes may facilitate binding, isomerization or initiation of RNA polymerase II. Despite the high conservation of HSEs from various organisms, some properties of HSF are seemingly different. The activation of HSF binding to HSF is hs-dependent Drosophila, human, and plant cells (Scharf and Nover, personal communication), but in yeast HSF binds DNA constitutively. The activation of DNA binding and the activation of transcription appear to be two separate processes, the latter seems to be mediated by components involved in posttranslational modifications of HSF and possibly other secondary factors. The isolation of genes encoding HSF is of utmost importance for studying the processes of binding and activation and for the manipulation of the hs response in transgenic plants.

Posttranscriptional regulation of hs gene expression is attributed to 5'- and 3'-nontranslated sequences of the mRNA. The 5'-NT sequences seem to have dual effects, on both transcription and translation of Gmhsp 17.3-B (Schöffl et al., 1989, 1990). The preferential translation of hs-mRNAs might result from a low potential for secondary structure formation intrinsic to the native mRNA leader sequences. It is speculated that the low secondary structure allows a Cap-binding-protein-independent translation during hs. A partial preservation of 5'-leader sequences in transcriptional fusions with reporter genes may not be sufficient for efficient translation of the mRNA chimaeras during hs. Reporter gene dependent differences in expression of otherwise identical hs gene constructions cause it difficult to predict the precise type of posttranscriptional regulation. It must be taken into account that novel transcriptional fusions between hs and reporter genes may create changes in secondary structure at the 5'-terminus of the chimaeric mRNAs which would either favour or adversely affect translation during hs. An example for a favoured translation of a transcriptional fusion is the linkage of the 5'-end of the CaMV 35S RNA to the CAT gene (Schöffl et al., 1989). A second type of posttranscriptional regulation of hsp expression seems to involve changes of hs mRNA stability. Hsp70-mRNAs in Drosophila and human cells are selectively degraded during recovery periods at low temperatures following heat treatments (Theodorakis and Morimoto, 1987; Petersen and Lindquist, 1988). The shorter half-life of hsp mRNAs may limit the competition with the preserved normal mRNAs during recovery from hs, thus a preferential synthesis of non-hs-proteins can take place. It is not yet known what parts of the hsp mRNA are recognized by the cellular machinery which controls RNA levels during hs and recovery. Preliminary studies on chimaeric Gmhsp 17.3-B:GUS genes in transgenic tobacco suggest an involvation of 3'-NT-sequences (Rieping and Schöffl, unpublished). Further investigations are required to delimit cis-active sequences and to identify trans-acting factors for the posttranscriptional regulations of hs-RNAs. The modular use of sequences for posttranscriptional regulation of gene expression may greatly influence the spectrum of genetic manipulation involving chimaeric hs genes.

ACKNOWLEDGEMENTS

We thank Drs. L. Nover and D. Scharf (Akademie der Wissenschaften, Halle) for sharing unpublished information with us, Heike Behrens and Ute Lutterschmid for their help in preparing the manuscript. The research in our laboratory has been supported by grants of the Deutsche Forschungsgemeinschaft (Scho 242/4-4 and 5-1).

## REFERENCES

Amin, J., Anathan, J., and Voellmy, R., 1988, Key features of heat shock regulatory elements, Mol. Cell.,8: 3761.

Baumann, G., Raschke, E., Bevan, M., and Schöffl, F., 1987, Functional analysis of sequences required for transcriptional activation of a soybean heat shock gene in transgenic tobacco plants, EMBO J., 6: 1161.

Bienz, M., 1986, A CCAAT box confers cell-type-specific regulation on the Xenopus hsp70 gene in oocytes, Cell, 46: 1037.

Bienz, M., and Pelham, H. R. B., 1986, Heat shock regulatory elements function as an inducible enhancer in the Xenopus hsp70 gene and when linked to a heterologous promoter, Cell, 45: 753.

Czarnecka, E., Key, J. L., and Gurley, W. B., 1989, Regulatory domains of the Gmhsp17.5-E heat shock gene promoter of soybean: a mutational analysis, Mol. Cell. Biol., 9: 3457.

Findly, R. C., Alavi, H., and Platt, T., 1988, Isolation of mutations that act in trans to alter expression from a yeast hsp70 promoter, Mol. Cell. Biol., 8: 3423.

Goldenberg, C. J., Luo, Y., Fenna, M., Baler, R., Weinmann, R., and Voellmy, R., 1988, Purified human factor acivates heat-shock promoter in a HeLa cell-free transcription system, J. Biol. Chem., 263: 19734.

Green, J. M., and Kingston, R. E., 1990, TATA-dependent and TATA-independent function of the basal and heat shock elements of a human hsp70 promoter, Mol. Cell. Biol., 10: 1319.

Jefferson, R. A., 1987, Assaying chimeric genes in plants: the GUS gene fusion system, Plant Mol. Biol. Rep., 5: 387.

Kingston, R. E., Schütz, T. J., and Larin, Z., 1987, Heat-inducible human factor that binds to a human hsp70 promoter, Mol. Cell. Biol., 7: 1530.

Larson, J. S., Shuetz, T. J., and Kingston, R. E., 1988, Activation in vitro of sequence-specific DNA binding by a human regulatory factor, Nature, 355: 372.

Levitt, J., 1980, Responses of plants to environmental stresses, Vol. I, Academic Press, New York.

Lindquist, S., and Craig, E. A., 1988, The heat shock proteins, Annu. Rev. Genet., 22: 631.

Morgan, W. D., Williams, G. T., Morimoto, R. J., Greene, J., Kingston, R. E., and Tjian, R., 1987, Two transcriptional activators, CCAAT-box-binding transcription factor and heat shock transcription factor, interact with a human hsp70 gene promoter, Mol. Cell. Biol., 7: 1129.

Neumann, D., Nover, L., Parthier, B., Rieger, R., Scharf, K.-D., Wollgiehn, R. and zur Nieden, U., 1989, Heat shock and other stress response systems of plants, Biol. Zent.bl., 108: 1.

Nover, L., 1990, Heat shock response, CRC Press, Boca Raton.

Nover, L., Scharf, K.-D., and Neumann, D., 1989, Cytoplasmic heat shock granules are formed from precursor particles and contain a specific set of mRNAs, Mol. Cell. Biol. 9: 1298.

Parker-Thornburg, J., Bonner, J. J., 1987, Mutations that induce the heat shock response of Drosophila, Cell, 51: 763.

Perisic, O., Xiao, H., and Lis, J. T., 1989, Stable binding of Drosophila heat shock factor to head-to-head and tail-to-tail repeats of a conserved 5 bp recognition unit, Cell, 59: 797.

Petersen, R., and Lindquist, S., 1988, The Drosophila hsp70 message is

rapidly degraded at normal temperatures and stabilized by heat shock, Gene, 72: 161.
Raibaud, O., 1990, Direct repeats in HSF binding sites, Nature, 344: 204.
Schöffl, F., 1988, Genetic engineering strategies for manipulation of the heat shock response, Plant Cell Envir., 11: 339.
Schöffl, F., and Baumann, G., 1985, Thermo-induced transcripts of a soybean heat shock gene after transfer into sunflower using a Ti plasmid vector, EMBO J., 4: 1119.
Schöffl, F., Baumann, G., and Raschke, E., 1988, The expression of heat shock genes. A model for environmental stress response, in: "Plant Gene Research - Temporal and Spacial Regulation of Plant Genes," Verma, D. P. S., Goldberg, R. B., eds., Springer Verlag, Wien, New York.
Schöffl, F., Rieping, M., Baumann, G., Bevan, M. W., and Angermüller, S., 1989, The function of plant heat shock promoter elements in the regulated expression of chimaeric genes in transgenic tobacco, Mol. Gen. Genet., 217: 246.
Schöffl, F., Rieping, M., and Raschke, E., 1990, Functional analysis of sequences regulating the expression of heat shock genes in transgenic plants, in: "Genetic Engineering of Crop Plants," Lycett, G. W., Grierson, D., eds., Butterworth Ltd., London.
Schöffl, F., Rossol, I., and Angermüller, S., 1987, Regulation of the transcription of heat shock genes in nuclei from soybean (Glycine max) seedlings, Plant Cell Envir., 10: 113.
Severin, K., Kliem, M., and Schöffl, F., 1989, Binding of nuclear proteins to the promoter upstream regions of soybean heat shock genes, in: "Proceedings of the Braunschweig Symposium on Applied Plant Molecular Biology," Galling, G., Technische Universität Braunschweig.
Severin, K., and Schöffl, F., 1990, Heat-inducible hygromycin resistance in transgenic tobacco, (submitted).
Sorger, P. K., Lewis, M. J., and Pelham, H. R. B., 1987, Heat shock factor is regulated differently in yeast and HeLa cells, Nature, 329: 81.
Sorger, P. K., and Nelson, H. C. M., 1989, Trimerization of a yeast transcriptional activator via a coiled-coil motif, Cell, 59: 807.
Sorger, P. K., and Pelham, H. R. B., 1988, Yeast heat shock factor is an essential DNA-binding protein that exhibits temperature-dependent phosphorylation, Cell, 54: 855.
Strittmatter, G., and Chua, N.-H., 1987, Artificial combinatin of two cis-regulatory elements generates a unique pattern of expression in transgenic plants, Proc. Natl. Acad. Sci. USA, 84: 8986.
Theodorakis, N. G., and Morimoto, R. I., 1987, Posttranscriptional regulation of hsp70 expression in human cells: effects of heat shock, inhibition of protein synthesis, and adenovirus infection on translation and mRNA stability, Mol. Cell. Biol., 7: 4357.
Van den Elzen, P., Townsend, J., Lee, K. Y., Bedbrook, J. R., 1985, A chimaeric hygromycin resistance gene as a selectable marker in plant cells, Plant Mol. Biol., 5: 299.
Wiederrecht, G., Seto, D., and Parker, C. S., 1988, Isolation of the gene encoding the S. cerevisiae heat shock transcription factor, Cell, 54: 841.
Wu, B. J., Kingston, R. E., Morimoto, R. J., 1986, Human hsp70 promoter contains at least two distinct regulatory domains, Proc. Natl. Acad. Sci. USA, 83: 629.
Wu, C., Wilson, S., Walker, B., Dawid, I., Paisley, T., Zimarino, V., and Ueda, H., 1987, Purification and properties of Drosophila heat shock activator protein, Science, 238: 1247.
Xiao, H., and Lis, J. T., 1988, Germline transformation used to define key features of heat shock response elements, Science, 239: 1139.
Zimarino, V., and Wu, C., 1987, Induction of sequence-specific binding of Drosophila heat shock activator protein without protein synthesis, Nature, 327: 727.

# OXIDATIVE STRESS IN PLANTS

Chris Bowler, Luit Slooten[1], Ed W.T. Tsang[2], Wim Van Camp, Marc Van Montagu, and Dirk Inzé

Laboratorium voor Genetica, Rijksuniversiteit Gent, B-9000 Gent (Belgium); [1]Laboratorium voor Biofysica, Vrije Universiteit Brussel, B-1050 Brussel (Belgium); [2]National Research Council of Canada, Plant Biotechnology Institute, Saskatoon, Saskatchewan S7N 0W9 (Canada)

## INTRODUCTION

A consequence of aerobic life is the formation of reactive forms of oxygen such as superoxide radicals ($O_2^{\overline{\cdot}}$), hydrogen peroxide ($H_2O_2$), and hydroxyl radicals (OH•). In particular, hydroxyl radicals are one of the most reactive species known to chemistry, being able to react with DNA, proteins, lipids, and almost any other constituent of living cells (Halliwell, 1984). Its primary route of formation is thought to be an iron-catalyzed reaction of superoxide radicals with hydrogen peroxide as follows:

$$H_2O_2 + O_2^{\overline{\cdot}} \xrightarrow{Fe^{2+},\ Fe^{3+}} OH^- + O_2 + OH\cdot$$

Not only are these reduced oxygen species generated as by-products of indigenous biological oxidations, but their formation is greatly increased during environmental stress, as caused for example by herbicides (Harbour and Bolton, 1975; Orr and Hogan, 1983), air pollutants, such as ozone (MacKay et al. 1987), redox-active compounds (Hassan and Fridovich, 1977), heat shock (Lee et al., 1983), and chilling (Clare et al., 1984).

Organisms have evolved a battery of mechanisms which prevent the formation of such deleterious oxygen species, one example being the superoxide dismutases (SOD; EC 1.15.1.1), a class of metalloproteins which catalyze the dismutation of superoxide radicals to oxygen and hydrogen peroxide. Their importance has been demonstrated in SOD-deficient mutants of *Escherichia coli* (Carlioz and Touati, 1986) and yeast (Biliński et al., 1985; van Loon et al., 1986) which are hypersensitive to oxygen.

Three classes of SOD can be distinguished according to their metal cofactor: the copper/zinc, the manganese, and the iron forms (for review, see Bannister et al., 1987). The iron enzyme (FeSOD) is present in prokaryotes and in some plants. The manganese SOD (MnSOD) is widely

*Plant Molecular Biology* 2, Edited by R.G. Herrmann and B. Larkins, Plenum Press, New York, 1991

distributed among prokaryotic and eukaryotic organisms; in eukaryotes it is most often found in the mitochondrial matrix. The iron and manganese SODs are very similar in their primary, secondary, and tertiary structure (see for example, Parker and Blake, 1988). The copper/zinc enzyme (Cu/ZnSOD) represents a distinct class, found almost exclusively in eukaryotic species, where it is often present as several isoforms. One of these is always present in the cytosol and a chloroplastic isoform appears commonplace in plants (Halliwell, 1984).

Plant superoxide dismutases have been implicated as a component in stress tolerance. For example, resistance to air pollutants, such as ozone and $SO_2$, has been shown to be correlated with increased levels of enzymes involved in superoxide detoxification in poplar (Tanaka and Sugahara, 1980) and *Phaseolus vulgaris* (Lee and Bennett, 1982). Increased resistance to chilling injury in *Chlorella ellipsoidea* and to photooxidative damage in ripening tomato fruits has also been correlated with SOD activity (Rabinowitch et al., 1982; Clare et al., 1984). Our ultimate goal is to understand the role which superoxide dismutases play during environmental adversity, since it is conceivable that a plant which has been genetically engineered to produce elevated levels of superoxide dismutase will be more tolerant to stress.

## RESULTS AND DISCUSSION

### Analysis of SOD enzymes in *Nicotiana plumbaginifolia*

As a first step in this study, we have characterized the SOD enzymes present in *Nicotiana plumbaginifolia*. By using the *in situ* staining technique of Beauchamp and Fridovich (1971) directly on non-denaturing polyacrylamide protein gels, we have demonstrated the presence of five major bands of SOD activity in protein extracts derived from leaves of *N. plumbaginifolia* (Figure 1). Inhibitor studies with $H_2O_2$ and KCN, commonly used to distinguish between the different classes of SOD, identified one of these as a MnSOD (resistant to both $H_2O_2$ and KCN), three as Cu/ZnSODs (sensitive to both $H_2O_2$ and KCN), and the other as an FeSOD (resistant to KCN, but sensitive to $H_2O_2$). Subcellular fractionations reveal the MnSOD to be mitochondrial and the FeSOD to be chloroplastic.

### Cloning SOD

cDNA clones encoding MnSOD, Cu/ZnSOD, and FeSOD have now been isolated. To clone the MnSOD, an N-terminal sequence was first obtained from the protein (Bauw et al., 1987), which allowed the design of an oligonucleotide probe for screening a *N. plumbaginifolia* cDNA library. A full-length clone encoding the MnSOD was subsequently isolated and is shown in Figure 2 (Bowler et al. 1989a). The sequence starting from amino acid 25 is identical to the previously determined N-terminal amino acid sequence of the mature protein. The amino acids in front of this (numbered 1–24) have characteristic features of a transit peptide for translocation to the mitochondrial matrix, as would be predicted from subcellular fractionations (Figure 1). This leader sequence has also been shown to target the MnSOD to yeast mitochondria (Bowler et al., 1989b).

Cu/ZnSOD is perhaps the best characterized of the SOD enzymes. Several amino acid sequences are available (see, for example, Getzoff et al., 1989) and conserved regions have been identified. Consequently, it was straightforward to design degenerate oligonucleotide probes based upon these homologous regions for isolating Cu/ZnSOD cDNAs from *N. plum-*

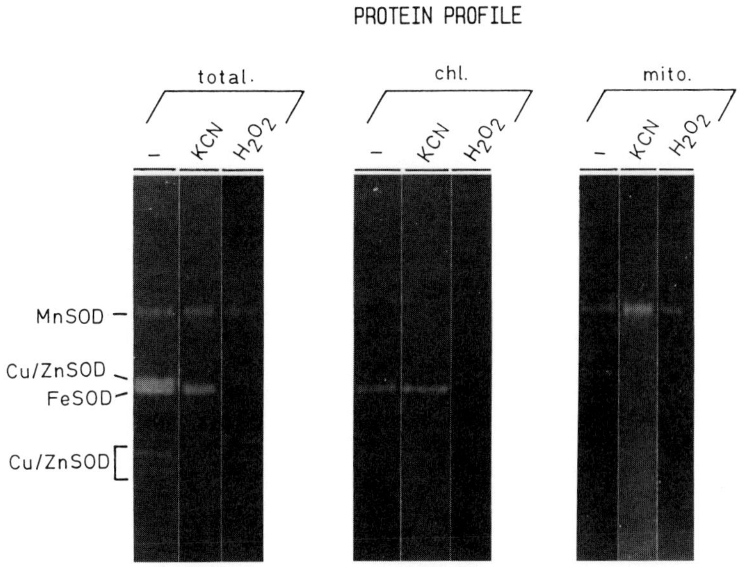


Fig. 1. Profile of SOD enzymes in *N. plumbaginifolia*. Protein samples (150 µg total protein or 75 µg organellar protein) were separated on native protein gels and stained for SOD activity immediately (-) or after incubation in 2 mM KCN (lanes KCN) or 5 mM $H_2O_2$ (lanes $H_2O_2$). The subcellular localization of the SOD isozymes was determined by analyzing chloroplastic (chl.) and mitochondrial (mito.) fractions in addition to total extracts. Based upon this analysis, *N. plumbaginifolia* contains one mitochondrial MnSOD, one chloroplastic FeSOD, and three cytosolic Cu/ZnSODs as indicated.

*baginifolia*. This strategy allowed the isolation of full-length Cu/ZnSOD cDNAs (Tsang et al., in preparation).

FeSODs have traditionally been considered as being present only in prokaryotes and it is only recently that they have been found in some plant species. Amino acid composition analysis of an FeSOD from tomato (Kwiatowski et al., 1985) suggests that these plant FeSODs are sufficiently different from bacterial FeSODs to make it impossible to isolate clones by conventional hybridization techniques using a bacterial FeSOD as probe. Instead, we attempted to isolate FeSOD cDNAs by using a genetic selection technique based upon the complementation of a SOD-deficient strain of *Escherichia coli*, which is unable to grow aerobically on minimal medium (Carlioz and Touati, 1986). A plasmid-based cDNA expression library was constructed and introduced into the mutant by electroporation. Some of the complemented *E. coli* clones isolated in this manner were indeed found to express a plant-derived FeSOD, although they lack the chloroplast targeting sequence which we presumed to be present at the N-terminus. We have subsequently isolated full-length cDNA clones which contain sequences upstream from the mature protein which resemble chloroplast translocation sequences, thus confirming our subcellular fractionation data (Van Camp et al., in preparation).

The availability of cDNAs encoding all the different types of SOD in *N. plumbaginifolia* will allow a detailed analysis of their role in

```
1     GGGGGGGGGG GGGGGGCTGG CCTCTCTGGG CATGACCTGC AACTATAAAA GGACACCATA GAGTTAACAG
                                                                        HpaI

                                             M   A   L   R   T   L   V   S   R   R   T   L
71    CTAGAAAGCA TTTAGGAATA TCTCAAAA ATG GCA CTA CGA ACC CTA GTG AGC AGA CGG ACC TTA
                                                                             10
                                                      ↓
       A   T   G   L   G   F   R   Q   Q   L   R   G   L   Q   T   F   S   L   P
135   GCA ACA GGG CTA GGG TTC CGC CAG CAA CTC CGC GGC TTG CAG ACC TTT TCG CTC CCC
                                   20                                      30

       D   L   P   Y   D   Y   G   A   L   E   P   A   I   S   G   D   I   M   Q
193   GAT CTC CCC TAC GAC TAT GGA GCA CTG GAG CCG GCA ATT AGC GGT GAC ATA ATG CAG
                                       40                                      50

       L   H   H   Q   N   H   H   Q   T   Y   V   T   N   Y   N   K   A   L   E
249   CTC CAC CAC CAG AAT CAC CAT CAG ACT TAC GTC ACC AAT TAC AAT AAA GCC CTT GAA
                                           60

       Q   L   H   D   A   I   S   K   G   D   A   P   T   V   A   K   L   H   S
306   CAG CTA CAT GAC GCC ATT TCC AAA GGA GAT GCT CCT ACC GTC GCC AAA TTG CAT AGC
       70                                      80

       A   I   K   F   N   G   G   G   H   I   N   H   S   I   F   W   K   N   L
363   GCT ATC AAA TTC AAC GGC GGA GGT CAC ATT AAC CAC TCG ATT TTC TGG AAG AAT CTT
           90                                      100

       A   P   V   R   E   G   G   G   E   P   P   K   G   S   L   G   W   A   I
420   GCC CCT GTC CGC GAG GGT GGT GGT GAG CCT CCA AAG GGT TCT CTT GGT TGG GCT ATC
               110                                     120

       D   T   N   F   G   S   L   E   A   L   V   Q   K   M   N   A   E   G   A
477   GAC ACT AAC TTT GGC TCC CTA GAA GCT TTA GTT CAA AAG ATG AAT GCA GAA GGT GCT
                   130                HindIII                140

       A   L   Q   G   S   G   W   V   W   L   G   V   D   K   E   L   K   R   L
534   GCT TTA CAG GGC TCT GGC TGG GTG TGG CTT GGT GTG GAC AAA GAG CTT AAG CGC CTG
                   150                                     160

       V   I   E   T   T   A   N   Q   D   P   L   V   S   K   G   A   N   L   V
591   GTG ATT GAA ACC ACT GCT AAT CAG GAC CCT TTG GTT TCT AAA GGA GCA AAT TTG GTT
                           170                                     180

       P   L   L   G   I   D   V   W   E   H   A   Y   Y   L   Q   Y   K   N   V
648   CCT CTT CTG GGA ATA GAC GTT TGG GAA CAT GCA TAC TAC TTG CAG TAC AAA AAT GTA
                               190                                     200

       R   P   D   Y   L   K   N   I   W   K   V   M   N   W   K   Y   A   N   E
705   AGA CCT GAT TAT CTG AAG AAC ATA TGG AAA GTT ATG AAC TGG AAA TAT GCA AAT GAA
                                   210                                     220

       V   Y   E   K   E   C   P   *
762   GTT TAT GAG AAA GAA TGT CCT TGAACAGGGA TATTTGATGT TGTTTTGAGG ACGTCTGTAA

823   AACTTTTTGA TGGGAAATAA GGCTGAGTGA CATGAGCAGG TGTCCTGTTT TTCTTGCATG TAGTCGCTGG

893   CTGATGTACT TGATGTATTT CTGGAAAAGG TTGATGTATG TACTTGATAT ATGGAGCCTA [AATAAA]ACTA

963   CTCTATCGTT TGAGCGCAAA CCCCCCCCCC CCCC
```

Fig. 2. Sequence of a manganese superoxide dismutase (MnSOD) of *N. plumbaginifolia*. Complete nucleotide sequence of the MnSOD cDNA insert of 996 base pairs with its flanking G/C homopolymer tails added during the cloning procedure. The sequence homologous to the oligonucleotide probe is underlined. The potential polyadenylation site AATAAA (position 953) is indicated by boxing. The predicted coding sequence starting at position 99 (start codon ATG) to position 783 (stop codon TGA marked with an asterisk) has been translated into the corresponding amino acid sequence, written above the nucleotide sequence in the one-letter code. The arrow on amino acid 25 points to the N-terminus of the mature protein (Bauw et al. 1987).

the plant's response to oxidative stress. Moreover, since the enzymes are each present in different cellular compartments, it will be possible to study their role in protecting their respective subcellular environments from superoxide-mediated damage. As an example of this, data concerning the expression of MnSOD is given in the following sections.

## Expression analysis of MnSOD

MnSOD has previously been shown to be a highly abundant protein in *N. plumbaginifolia* cell suspension cultures (Bauw et al., 1987). To characterize the expression at the mRNA level, Northern analysis on total RNA with a probe synthesized from the MnSOD cDNA insert was carried out. This revealed considerable variations in the expression of MnSOD in plant cells (Figure 3a). Expression is very weak in leaves from intact plants, 2-3-fold higher in roots, and approximately 50 times higher in dark-grown, heterotrophic cell suspension cultures.

This dramatic difference between the expression of MnSOD in leaves and cell suspensions could be due to the differences in the cell types

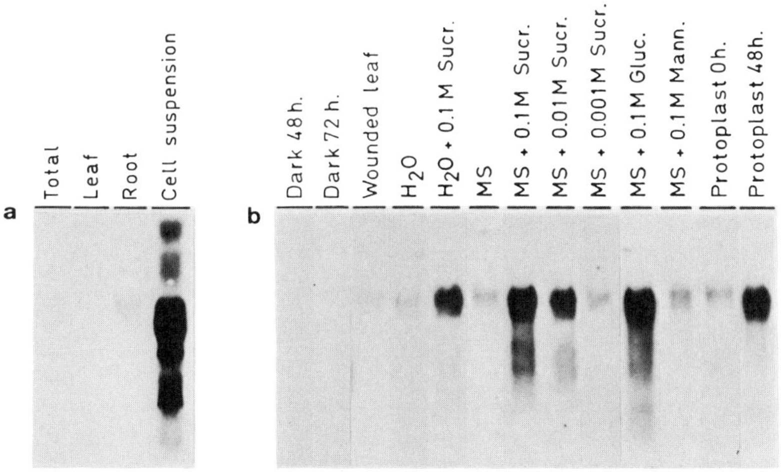


Fig. 3. Northern analysis of MnSOD mRNA in *N. plumbaginifolia*. (a) Hybridization of a MnSOD cDNA-derived probe to total RNA from whole plants, leaves, roots, and dark-grown cell suspension cultures. The size of the hybridizing RNA is 1.25 kb. (b) Effects of sugars on the accumulation of MnSOD mRNA. Lanes 1 and 2 represent RNA extracted from whole plants incubated in the dark for 48 hours and 72 hours, respectively. The control for wounding (lane 3) involved making several cuts on leaves of intact plants and extracting RNA 24 hours later. Two samples of protoplasts are shown (lanes 12 and 13): immediately after isolation (time 0) and after a further 48-hour incubation at low-light intensity in K3 medium (Nagy and Maliga, 1976), supplemented with 0.4 M glucose, 0.1 mg/l naphthalene acetic acid (NAA), and 0.2 mg/l 6-benzylaminopurine (BAP). All other samples were extracted from leaf discs incubated for 48 hours under dark conditions in the indicated liquid media. Abbreviations: gluc., glucose; mann., mannitol; sucr., sucrose.

or to differences in culture conditions. To resolve this question, *N. plumbaginifolia* leaf discs were incubated for 48 hours in the dark in Murashige and Skoog (MS) liquid medium (Murashige and Skoog, 1962) and assayed for MnSOD mRNA. Figure 3b shows that leaf tissue maintained in growth medium contained levels of MnSOD mRNA comparable to those seen in cell suspension cultures (Figure 3b). This response was not due to a switch-off of photosynthesis, as a 72-hour exposure of whole plants to the dark did not result in increased expression of MnSOD. Also, incubation of leaf discs in the dark in pure water caused only a weak (2–3-fold) induction, similar to the effect of wounding whole plants. Thus, the massive induction of MnSOD in tissue culture appears to be an intrinsic effect of the medium itself. This induction was also seen when freshly isolated protoplasts were incubated in regenerating medium (Figure 3b). To analyze the effect of the medium, sucrose and salts were tested separately on leaf discs. We found sucrose to be the crucial factor for induction and its effect was greatest in the presence of salts. A linear dose-response could be observed at different concentrations of sucrose (0.001 M–0.1 M), with the highest level of expression being reached after a 48-hour incubation. Since the response was amplified by salts, we tested combinations of iron, manganese, copper, and zinc ions in the presence of sucrose, and we concluded that the enhancement in induction was not due to these salts in particular (data not shown). The massive increase in MnSOD expression was also produced by glucose, but not by mannitol (Figure 3b). Since mannitol is used as a non-catabolizable osmoticum in plant tissue culture (Street, 1959), our observation suggested that the induction of MnSOD by sugars was due to a trophic rather than an osmotic effect.

<u>Analysis of SOD and cytochrome oxidase activities</u>

From SOD assays performed on non-denaturing protein gels we estimated the abundance of the MnSOD protein in cell suspension cultures to be approximately 20 times higher than that found in leaves of intact plants (Figure 4). When leaf discs were incubated with sucrose, the MnSOD protein was shown to follow a similar, albeit delayed, induction profile in relation to its mRNA (Figure 4a). These experiments confirmed that sucrose caused an accumulation of MnSOD in leaf discs similar to that seen in cell suspension culture.

In contrast to the MnSOD, the other SODs showed no significant alterations in expression (data not shown). As the MnSOD proved to be the mitochondrial enzyme, its induction could be considered as a response against superoxide radicals generated in mitochondria by enhanced respiratory oxidation of sugars. It has been established that superoxide radicals are produced from the mitochondrial respiratory chain of eukaryotes (Loschen et al., 1974), so we attempted to measure the activity of this pathway by measuring cytochrome C oxidase activity. This is the terminal enzyme, catalyzing the reduction of oxygen to water. In Figure 4b, we show a clear correlation under the conditions tested, between the activities of MnSOD and cytochrome oxidase.

A requirement for SOD activity within the mitochondria in times of high respiratory activity has also been shown in yeast, since a yeast mutant lacking a mitochondrial MnSOD was found to be unable to grow on a non-fermentable carbon source such as ethanol (Van Loon et al., 1986). Although a cytosolic SOD remains functional in this mutant, compartmentalization evidently precludes it from substituting for the mitochondrial SOD. Correspondingly, the specific and dramatic induction of MnSOD in *N. plumbaginifolia* could indicate that it alone can protect against the severe oxidative stress created in mitochondria by tissue culture.

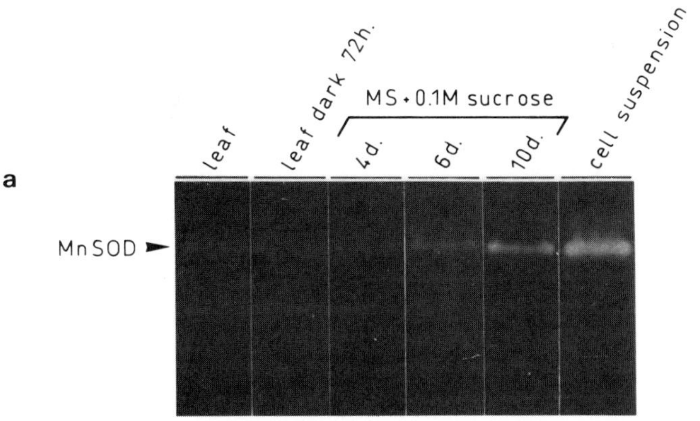


**b**

| | | | | | | |
|---|---|---|---|---|---|---|
| MnSOD activity | 1 | 1 | 1.5 | 3 | 6 | 20 |
| Cytochrome oxidase activity | 1 | 1.2 | 2.9 | 5.2 | 14.5 | 18.2 |

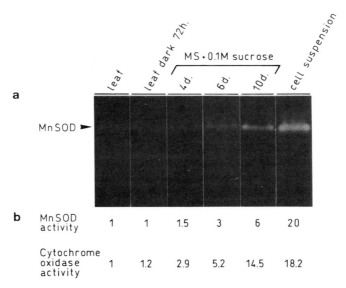

Fig. 4. Analysis of superoxide dismutase and cytochrome oxidase activities. (a) Induction of MnSOD activity in leaf pieces incubated in MS medium with sucrose and in cell suspensions, as revealed by SOD staining on non-denaturing gels. Changes in MnSOD activity were best visualized in samples containing 50 $\mu$g protein. (b) SOD and cytochrome oxidase activity measurements shown as fold increases over the activity in an untreated leaf. Experimental conditions have been described in Figure 3. Leaf discs were incubated for the periods indicated in MS + 0.1 M sucrose supplemented with 0.5 mg/l NAA and 0.1 mg/l BAP.

## The involvement of SOD in the plant defense response

Some reports demonstrate that changes in SOD activity occur when a plant is infected with pathogens (Arrigoni et al., 1981; Zacheo et al., 1982; Zacheo and Bleve-Zacheo, 1988), and evidence is accumulating that the generation of superoxide radicals is responsible for the occurrence of necrotic lesions upon infection (Sekizawa et al., 1987; Doke and Ohashi, 1988). To study the involvement of SOD in the defense response of *N. plumbaginifolia*, we chose to examine the effect of ethylene (and ethephon), salicylic acid, and infection with *Pseudomonas syringae*. Ethylene is known to induce several proteins involved in plant defense responses (Ecker and Davis, 1987) and endogenous levels of ethylene control many processes within a plant (Yang and Hoffman, 1984). Ethephon, which is hydrolyzed to ethylene in the plant, can mediate similar responses (van Loon, 1985). Several benzoic acid derivatives, such as salicylic acid, can also induce stress responses; for example, it has been shown to induce several of the pathogenesis-related (PR) proteins (Hooft van Huijsduijnen et al., 1986). We used a strain of *P. syringae* that is non-pathogenic to *N. plumbaginifolia*, but which does elicit the hypersensitive response. Infection is characterized by the rapid development (within 6–12 hours) of localized necrosis around the infected area which prevents further spread to other parts of the leaf.

Figure 5 shows the MnSOD mRNA to increase considerably following treatment for 48 hours with ethylene, ethephon, and salicylic acid. MnSOD induction in response to *P. syringae* appeared most apparent after 24 hours and was followed by a slow decay. Subsequent analyses of Cu/ZnSOD and FeSOD expression in response to *P. syringae* reveal similar inductions, although with differing kinetics (Tsang et al., in preparation), which distinguishes the response from that mediated by sucrose, in which only the MnSOD was appreciably induced. However, cytochrome oxidase activity was again found to be stimulated by the infection (data not shown). Salicylic acid and ethylene are also known to enhance plant respiration (Laties, 1982; Raskin et al., 1987), thus reinforcing the hypothesis that MnSOD must be induced whenever mitochondrial activity increases, regardless of the cause.

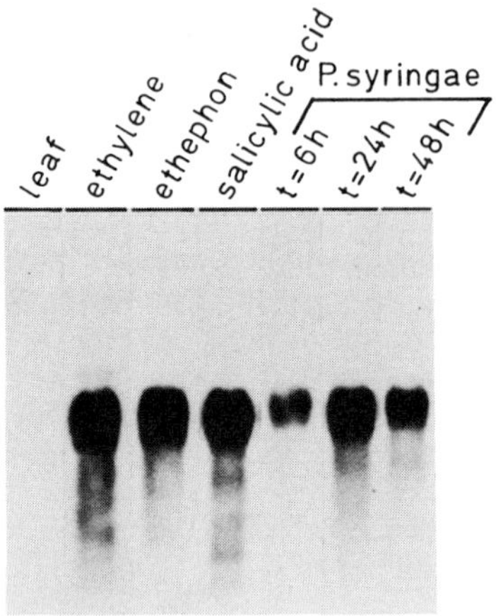


Fig. 5. Induction of MnSOD mRNA during the pathogenesis response. Whole plants were incubated under an atmosphere of ethylene (10 ppm in purified air), or sprayed with ethephon (1 mg/ml) or salicylic acid (10 mM) and left for 48 hours. *P. syringae* infections were performed on non-flowering plants grown in soil under greenhouse conditions. An end-log culture grown in LPG medium (0.3% yeast extract, 0.5% bactopeptone, 0.5% glucose) was diluted to 1 x $10^7$ cells/ml and injected into the leaves with a hypodermic syringe.

Possible mechanisms of MnSOD regulation

Taken collectively, these results reveal the connection between cytochrome oxidase and MnSOD expression. An obvious hypothesis is that this close correlation reflects a common regulatory mechanism which allows a stoichiometric "switching on" of all the genes involved in mitochondrial respiration. In yeast, it has been shown that the expression of MnSOD and cytochrome oxidase are coupled, and a gene which controls oxygen-regulated functions has been identified (Lowry and Zitomer, 1984). It is possible that further characterization of the response in plants may reveal a similar mechanism, although the presence of two respiratory pathways (cyanide-sensitive, the other cyanide-resistant) would presumably make the situation somewhat more complex.

However, the function of the enzyme might suggest that its induction would somehow be mediated by superoxide radicals. The *oxyR* gene product, which regulates several oxidative stress-inducible genes in

*Salmonella typhimurium* and *E. coli* is an example of such a mechanism. Essentially, the oxidation/reduction state of the protein determines whether it potentiates or represses transcription of these genes (Storz et al., 1990). If superoxide radicals were the triggering factor, it should be possible to block the induction of the endogenous enzyme by constitutively overexpressing SOD in that particular compartment, since the inducing molecule would then be removed. We are currently in the process of testing this hypothesis.

## PERSPECTIVES

The above example of SOD regulation serves to demonstrate the efficacy of using a molecular approach to clarify the role of SOD enzymes in plants during stress conditions. Work currently in progress in the Laboratorium voor Genetica (Gent, Belgium) aims to exploit this approach further by using probes to each of the SOD enzymes to study their expression in response to environmental stresses such as heat shock, chilling, and pollutants.

We are also interested to determine what effects may result from overexpressing SOD in plants. Since superoxide radicals are likely to be generated in different compartments of the cell during different stress conditions, an analysis of the importance of SOD in providing better protection can only be carried out by increasing the SOD activity in different subcellular locations. To do this, we have constructed expression vectors designed to overexpress MnSOD in mitochondria, cytosol, and chloroplasts. These have been introduced into tobacco and the transgenic plants obtained were found to target the MnSOD enzyme very efficiently to the desired subcellular locations. A preliminary analysis of these plants has been carried out by studying the effect of methyl viologen on transgenic material. This compound is known to generate superoxide radicals primarily within the chloroplasts. Our results show that superoxide radical-mediated disruption of membranes and photosynthetic pigments is significantly reduced in plants which overexpress MnSOD and interestingly, plants which overexpress MnSOD in the chloroplasts appear to show the highest level of protection. We are currently examining the behaviour of these plants to other more natural stress conditions.

### Acknowledgments

We are grateful to Martine De Cock, Jeroen Coppieters, Karel Spruyt, Stefaan Van Gijsegem, and Vera Vermaercke for their help in preparing this article. This work was carried out with support from the "Algemene Spaar- en Lijfrentekas-Kankerfonds", the Services of the Prime Minister (U.I.A.P. 12120C0187), and the International Atomic Energy Agency (no. 5285). WVC and CB are indebted to the "Instituut ter aanmoediging van het Wetenschappelijk Onderzoek in Nijverheid en Landbouw" and SERC (NATO), respectively for predoctoral fellowships; DI is a Senior Research Assistant of the National Fund for Scientific Research (Belgium).

## REFERENCES

Arrigoni, O., Zacheo, G., Bleve-Zacheo, T., Arrigoni-Liso, R., and Lamberti, F., 1981, Changes of superoxide dismutase and peroxidase activities in pea roots infested by *Heterodera goettingiana*, Nematol. Medit. 9:189.

Bannister, J.V., Bannister, W.H., and Rotilio, G., 1987, Aspects of the structure, function and applications of superoxide dismutase, CRC Crit. Rev. Biochem. 22:111.

Bauw, G., De Loose, M., Inzé, D., Van Montagu, M. & Vandekerckhove, J., 1987) Alterations in the phenotype of plant cells studied by $NH_2$-terminal amino acid sequence analysis of proteins electroblotted from two-dimensional gel-separated total extracts, Proc. Natl. Acad. Sci. USA 84:4806.

Beauchamp, C.O., and Fridovich, I., 1971, Superoxide dismutase: improved assays and an assay applicable to acrylamide gels, Anal. Biochem. 44:276.

Biliński, T., Krawiec, Z., Liczmański, A., and Litwińska, H., 1985, Is hydroxyl radical generated by the Fenton reaction in vivo? Biochem. Biophys. Res. Comm. 130:533.

Bowler, C., Alliotte, T., De Loose, M., Van Montagu, M., and Inzé, D., 1989a, The induction of manganese superoxide dismutase in response to stress in *Nicotiana plumbaginifolia*, EMBO J. 8:31.

Bowler, C., Alliotte, T., Van den Bulcke, M., Bauw, G., Vandekerckhove, J., Van Montagu, M., and Inzé, D., 1989b, A plant mitochondrial preprotein is efficiently imported and correctly processed by yeast mitochondria, Proc. Natl. Acad. Sci. USA 86:3237.

Carlioz, A., and Touati, D., 1986, Isolation of superoxide dismutase mutants in *Escherichia coli*: is superoxide dismutase necessary for aerobic life? EMBO J. 5:623.

Clare, D.A., Rabinowitch, H.D., and Fridovich, I., 1984, Superoxide dismutase and chilling injury in *Chlorella ellipsoidea*, Arch. Biochem. Biophys. 231:158.

Doke, N., and Ohashi, Y., 1988, Involvement of an $O_2^-$ generating system in the induction of necrotic lesions on tobacco leaves infected with tobacco mosaic virus, Physiol. Mol. Plant Pathol. 32:163.

Ecker, J.R., and Davis, R.W., 1987, Plant defense genes are regulated by ethylene, Proc. Natl. Acad. Sci. USA 84:5202.

Getzoff, E.D., Tainer, J.A., Stempien, M.M., Bell, G.L., and Hallewell, R.A., 1989, Evolution of CuZn superoxide dismutase and the Greek key ß-barrel structural motif, Proteins 5:322.

Halliwell, B., 1984, "Choroplast metabolism - The structure and function of chloroplasts in green leaf cells", Clarendon Press, Oxford.

Harbour, J.R., and Bolton, J.R., 1975, Superoxide dismutase in spinach chloroplasts: electron spin resonance detection by spin trapping, Biochem. Biophys. Res. Commun. 64:803.

Hassan, H.M., and Fridovich, I., 1977, Enzymatic defenses against the toxicity of oxygen and of streptonigrin in *Escherichia coli*, J. Bacteriol. 129:1574.

Hooft van Huijsduijnen, R.A.M., Alblas, S.W., De Rijk, R.H., and Bol, J.F., 1986, Induction by salicylic acid of pathogenesis-related proteins and resistance to alfalfa mosaic virus infection in various plant species, J. Gen. Virol. 67:2135.

Kwiatowski, J., Safianowska, A., and Kaniuga, Z., 1985, Isolation and characterization of an iron-containing superoxide dismutase from tomato leaves, *Lycopersicon esculentum*, Eur. J. Biochem. 146:459.

Laties, G.G., 1982, The cyanide-resistant alternative path in higher plant respiration, Ann. Rev. Plant Physiol. 33:519.

Lee, A.H., and Bennett, J.H., 1982, Superoxide dismutase. A possible protective enzyme against ozone injury in snap beans (*Phaseolus vulgaris* L.), Plant Physiol. 69:1444.

Lee, P.C., Bochner, B.R., and Ames, B.N., 1983, AppppA, heat-shock stress, and cell oxidation, Proc. Natl. Acad. Sci. USA 80:7496.

Loschen, G., Azzi, A., Richter, C., and Flohé, L., 1974, Superoxide radicals as precursors of mitochondrial hydrogen peroxide, FEBS Lett. 42:68.

Lowry, C.V., and Zitomer, R.S., 1984, Oxygen regulation of anaerobic and aerobic genes mediated by a common factor in yeast, Proc. Natl. Acad. Sci. USA 81:6129.

Mackay, C.E., Senaratna, T., McKersie, B.D., and Fletcher, R.A., 1987, Ozone induced injury to cellular membranes in *Triticum aestivum* L. and protection by the triazole S-3307, Plant Cell Physiol. 28:1271.

Murashige, T. and Skoog, F., 1962, A revised medium for rapid growth and bio assays with tobacco tissue cultures, Physiol. Plant. 15:473.

Nagy, J.I., and Maliga, P., 1976, Callus induction and plant regeneration from mesophyll protoplasts of *Nicotiana sylvestris*, Z. Pflanzenphysiol. 78:453.

Orr, G.L., and Hogan, M.E., 1983, Enhancement of superoxide production in vitro by the di phenyl ether herbicide nitrofen, Pestic. Biochem. Physiol. 20:311.

Parker, M.W., and Blake, C.C.F., 1988, Iron- and manganese-containing superoxide dismutases can be distinguished by analysis of their primary structures, FEBS Lett. 229:377.

Rabinowitch, H.D., Sklan, D., and Budowski, P., 1982, Photo-oxidative damage in the ripening tomato fruit: protective role of superoxide dismutase, Physiol. Plant. 54:369.

Raskin, I., Ehmann, A., Melander, W.R., and Meeuse, B.J.D., 1987, Salicylic acid: a natural inducer of heat production in *Arum* lilies, Science 237:1601.

Sekizawa, Y., Haga, M., Hirabayashi, E., Takeuchi, N., and Takino, Y., 1987, Dynamic behavior of superoxide generation in rice leaf tissue infected with blast fungus and its regulation by some substances, Agric. Biol. Chem. 51:763.

Storz, G., Tartaglia, L.A., and Ames, B.N., 1990, Transcriptional regulator of oxidative stress-inducible genes: direct activation by oxidation, Science 248:189.

Street, H.E., 1959, Special problems raised by organ and tissue culture. Correlation between organs of higher plants as a consequence of specific metabolic requirements, in "Heterotrophy" (Encyclopedia of Plant Physiology, Vol. XI), W. Ruhland, ed., Springer-Verlag, Berlin.

Tanaka, K., and Sugahara, K., 1980, Role of superoxide dismutase in defense against $SO_2$ toxicity and an increase in superoxide dismutase activity with $SO_2$ fumigation, Plant Cell Physiol. 21:601.

van Loon, L.C., 1985) Pathogenesis-related proteins, Plant Mol. Biol. 4:111.

van Loon, A.P.G.M., Pesold-Hurt, B., and Schatz, G., 1986, A yeast mutant lacking mitochondrial manganese-superoxide dismutase is hypersensitive to oxygen, Proc. Natl. Acad. Sci. USA 83:3820.

Yang, S.F., and Hoffman, N.E., 1984, Ethylene biosynthesis and its regulation in higher plants, Ann. Rev. Plant Physiol. 35:155.

Zacheo, G., and Bleve-Zacheo, T., 1988, Involvement of superoxide dismutases and superoxide radicals in the susceptibility and resistance of tomato plants to *Meloidogyne incognita* attack. Physiol. Mol. Plant Pathol. 32, 313-322.

Zacheo, G., Bleve-Zacheo, T., and Lamberti, F., 1982, Role of peroxidase and superoxide dismutase activity in resistant and susceptible tomato cultivars infected by *Meloidogyne incognita*, Nematol. Medit. 10:75.

# ELUCIDATING LIPID METABOLISM USING MUTANTS OF *ARABIDOPSIS*

John Browse[1], Suzanne Hugly[2], Martine Miquel[1]
and Chris Somerville[2]

[1]Institute of Biological Chemistry
Washington State University
Pullman, WA 99164 USA

[2]MSU-DOE Plant Research Laboratory
Michigan State University
East Lansing, MI 48824 USA

## INTRODUCTION

The biochemistry of membrane glycerolipid synthesis is receiving increasing attention. This is due in part to a realization that lipid structure has a very important role in maintaining the integrity of membranes and therefore of the cell and its organelles. In addition, there is an increasing appreciation that the problems of defining lipid synthesis and of understanding the significance of particular lipid compositions are tractable.

In this article, we review work on a series of *Arabidopsis* mutants with defects in the synthesis of membrane lipids. These mutants have contributed to our understanding in three main areas. First, they have provided an alternative approach to elucidating the mechanisms regulating synthesis and desaturation of membrane glycerolipids.

---

[1]Supported in part by grants from the U.S. Department of Energy (DE-ACO2-76ER01338) and the National Science Foundation (DCB-8803855).

[2]Abbreviations: ACP, acyl carrier protein; X:Y fatty acyl group containing X carbons with Y *cis*-double bonds; DAG, diacylglycerol; DGD diacyldigalactosylglycerol (digalactosyldiglyceride); MGD diacylgalactosylglycerol (monogalactosyldiglyceride); PA, phosphatidic acid; PC, phosphatidylcholine; PE, phosphatidylethanolamine; PG, phosphatidylglycerol; PI, phosphatidylinositol; SL, sulfoquinovosyldiacylglycerol (sulfolipid).

*Plant Molecular Biology* 2, Edited by R.G. Herrmann and
B. Larkins, Plenum Press, New York, 1991

Second, the availability of a series of mutants with specific alterations in leaf lipid composition offers a novel method for studying the role of lipids in the structure and function of plant membranes. Finally, because of the advantages of *Arabidopsis* as a model for molecular biology (Pang and Meyerowitz, 1987) the mutations can be used as markers to facilitate the cloning of the desaturase genes by chromosome walking or gene tagging.

## TWO PATHWAYS OF LIPID SYNTHESIS

It is now generally accepted that there are two distinct pathways in plant cells for the biosynthesis of glycerolipids and the associated production of polyunsaturated fatty acids. The evidence for the two-pathway model has been summarized in a review by Roughan and Slack (1982). In brief, the model proposes that fatty acids synthesized *de novo* in the chloroplast may either be used directly for production of chloroplast lipids by a pathway in the chloroplast (the prokaryotic pathway), or may be exported to the cytoplasm as CoA esters where they are incorporated into lipids in the endoplasmic reticulum by an independent set of acyltransferases (the eukaryotic pathway). The essential features of this model are depicted in Figure 1. Both pathways are initiated by the synthesis of 16:0-ACP by the fatty acid synthase in the chloroplast. 16:0-ACP may be elongated to 18:0-ACP and then desaturated by a soluble desaturase so that 16:0-ACP and 18:1-ACP are the primary products of chloroplast fatty acid synthesis. These thioesters may be used within the chloroplast for the synthesis of phosphatidic acid (PA) by acylation of glycerol-3-phosphate, or they may be hydrolyzed to free fatty acids which move through the chloroplast envelope to be converted to CoA thioesters in the outer envelope membrane by acyl-CoA synthetase.

Because of the substrate specificities of the plastid acyltransferases (Frentzen *et al.*, 1983) the PA made by the prokaryotic pathway has a 16-carbon fatty acid at the *sn*-2 position and an 18-carbon fatty acid at the *sn*-1 position. This PA is used for the synthesis of phosphatidylglycerol (PG) or is converted to diacylglycerol (DAG) by a PA-phosphatase located in the inner chloroplast envelope (Block *et al.*, 1983; Andrews *et al.*, 1985). This DAG pool can act as a precursor for the synthesis of the other major thylakoid lipids monogalactosyldiacylglycerol (MGD), digalactosyldiacylglycerol (DGD) and sulfolipid (SL) (Coves *et al.*, 1986; Heemskerk *et al.*, 1985; Kleppinger-Sparace *et al.*, 1985).

Acyl groups exported from the chloroplast as CoA esters are used for the synthesis of PA, mainly in the endoplasmic reticulum. In contrast to the plastid isozymes, the acyltransferases of the endoplasmic reticulum only produce PA with an 18-carbon fatty acid at the *sn*-2 position; 16:0, when present, is confined to the *sn*-1 position (Frentzen *et al.*, 1984). This PA gives rise to the phospholipids such as phosphatidylcholine (PC), phosphatidylethanolamine (PE) and phosphatidylinositol (PI) which are characteristic of the various extrachloroplast membranes (Moore, 1982). In addition, however, the diacylglycerol moiety of PC is returned to the chloroplast envelope where it enters the DAG pool and contributes to the synthesis of thylakoid lipids (Fig. 1) (Roughan and Slack, 1982; Heinz and Roughan, 1983).

In many species of higher plants, PG is the only product of the prokaryotic pathway and the remaining chloroplast lipids are synthesized entirely by the eukaryotic pathway. In those species, such as *Arabidopsis*, in which both pathways contribute to the synthesis of MGD, DGD and SL, the leaf lipids characteristically contain substantial amounts of hexadecatrienoic acid (16:3) which is found only in MGD and DGD molecules produced by the prokaryotic pathway. These plants have been termed 16:3 plants to distinguish them from the other angiosperms (18:3 plants) whose galactolipids contain predominantly α-linolenate (Jamieson and Reid, 1971).

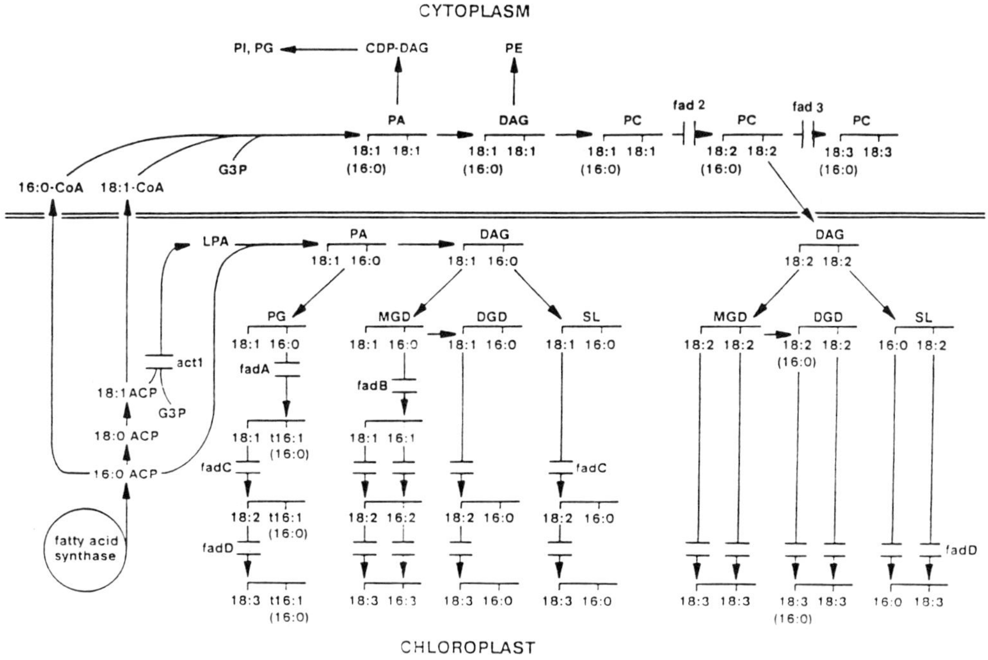


**Figure 1.** An abbreviated scheme showing the two pathways for glycerolipid synthesis in *Arabidopsis* leaves showing locations for enzyme deficiencies in six mutants listed in Table 1. See text for details.

A detailed quantitative analysis of the relative contributions of the two pathways was carried out in *Arabidopsis* (Browse *et al.*, 1986b). These studies indicated that approximately 38% of newly synthesized fatty acids enter the prokaryotic pathway of lipid biosynthesis. Of the 62% which is exported as acyl-CoA species to enter the eukaryotic pathway, 56% (34% of the total) is ultimately reimported into the chloroplast. Thus, chloroplast lipids of *Arabidopsis* are about equally derived from the two pathways. The results of these studies are schematically presented in Figure 2a.

## FATTY ACID DESATURATION

The membrane lipids of higher plants contain primarily 16-carbon and 18-carbon acyl groups with 0 to 3 *cis* double bonds (Frentzen, 1986). As a rule, the first unsaturation is at the ω-9 position, the second is at the ω-6, and the third is at the ω-3 position (i.e., 3 carbons from the methyl end of the acyl chain). The major exception to this is the presence of a *trans* unsaturation which occurs uniquely at the ω-12 (i.e., Δ3) position of the 16-carbon acyl group on the *sn*-2 position of chloroplast PG. The only plant desaturase that has been characterized in any detail is the soluble stearoyl-ACP desaturase which inserts a double bond at the ω-9 position in 18:0-ACP. This enzyme has been highly purified from developing safflower seeds and has been shown to require oxygen and reduced ferrodoxin (McKeon and Stumpf, 1982); it is essentially inactive with 16:0-ACP as substrate.

In the last few years, it has become clear that 16:0 and 18:1 fatty acids are only desaturated to their final products (predominantly 16:3 and 18:3) after incorporation into glycerolipids. The desaturases are assumed to be membrane bound enzymes, but to date none has been solubilized. For this reason, there is little information about the details of the desaturation reactions in plants. Isolated spinach chloroplasts supplied with [$^{14}$C]-acetate synthesize 1-18:1,2-16:0-MGD which is sequentially desaturated to 18:3,16:3-MGD (Roughan *et al.*, 1979). In the same way $^{14}$C-18:1 in PG or SL was desaturated to 18:2 and 18:3 (Sparace and Mudd, 1982; Joyard *et al.*, 1986). These observations imply the existence of a family of fatty acid desaturases which are located in the chloroplast and which use as substrates glycerolipids produced by the prokaryotic pathway. These desaturases can also act on 18:1- and 18:2-containing chloroplast lipids derived from the eukaryotic pathway (Browse *et al.*, 1986). However, in the eukaryotic pathway, the predominant conversion of 18:1 to 18:2 and at least some conversion of 18:2 to 18:3 takes place with PC of the endoplasmic reticulum as substrate. It is possible to measure NAD(P)H-dependent 18:1-PC desaturase activity in microsomal preparations from pea leaves (Slack *et al.*, 1876) and developing safflower seeds (Stymne and Appelqvist, 1978). An 18:2 desaturase has also been demonstrated in microsomes from developing linseed cotyledons (Browse and Slack, 1981), but this activity is not stable.

## THE PHYSIOLOGICAL ROLE OF LIPID UNSATURATION

Chloroplast membranes have a characteristic and unusual fatty acid composition. Typically, 18:3 or a combination of 18:3 and 16:3 fatty acids account for approximately two-thirds of all the thylakoid membrane fatty acids and over 90% of the fatty acids of MGD, the most abundant chloroplast lipid (Gounaris and Barber, 1983). The atypical fatty acid Δ3,*trans*-hexadecenoate (*trans*-16:1) is present as a component of the major thylakoid phospholipid, PG. The fact that these and other characteristics of chloroplast lipids are common to most or all higher plant species, suggests that the lipid fatty acid composition is important for maintaining photosynthetic function. Furthermore, the thylakoid membrane is the site of light absorption and oxygen production. The free radicals which are by-products of these reactions will stimulate oxidation of the polyunsaturated fatty acids. Since this might be expected to mediate against a high degree of unsaturation, it has been inferred that there is a strong selective advantage to having such high levels of trienoic fatty acids in thylakoid membranes.

Many different approaches have been used in attempts to elucidate the significance of membrane fatty acid composition to photosynthetic function. These include the correlation of events during chloroplast development (Van

**Table 1.** Fatty Acid Composition of Total Leaf Lipids From Some Mutants of *Arabidopsis*.

| Mutant Line[1] | 16:0 | 16:1 | 16:1t | 16:2 | 16:3 | 18:0 | 18:1 | 18:2 | 18:3 |
|---|---|---|---|---|---|---|---|---|---|
| | Fatty Acid Composition | | | | | | | | |
| Wild type | 16 | tr | 2 | tr | 12 | 1 | 4 | 18 | 47 |
| *fad*A | 18 | tr | **0** | tr | 12 | 1 | 3 | 19 | 47 |
| *fad*B | **24** | - | 2 | - | - | 1 | 2 | 14 | 56 |
| *fad*C | 15 | **8** | 2 | - | - | 1 | 21 | 18 | 34 |
| *fad*D | 17 | 3 | 3 | **6** | **2** | 1 | 9 | **39** | **19** |
| *fad*2 | 15 | tr | 4 | tr | 10 | 1 | **27** | 4 | **36** |
| *fad*3 | 17 | tr | 2 | tr | 15 | 1 | 5 | **26** | **33** |
| *act*1 | 10 | tr | 2 | tr | **tr** | 1 | 8 | 23 | 54 |

[1]The gene symbols correspond to the putative defects in lipid metabolism shown in Figure 1.

Walraven *et al.*, 1984; Leech *et al.*, 1973), physical studies of model membrane systems (Galey *et al.*, 1980; Sen *et al.*, 1981; Quinn and Williams, 1983), reconstitution of photosynthetic components with lipid mixtures (Gounaris *et al.*, 1983), alteration of lipids *in situ* by heat stress (Gounaris *et al.*, 1984), lipase treatment (Rawyler and Siegenthaler, 1981), chemical inhibitors (Leech *et al.*, 1985) or hydrogenation of unsaturated fatty acids (Vigh *et al.*, 1985; Thomas *et al.*, 1986). To date, however, these approaches have not been successful in establishing any unequivocal relationships between membrane form and function. The importance of lipid diversity in other membrane systems of the cell is equally ill-defined. In microbial systems, the role of lipid unsaturation in thermal adaptation of the cell is quite well understood (Cronan and Rock, 1987; McElhaney, 1989), but it is not yet clear how these simple models apply to eukaryotic organisms which contain more highly unsaturated membrane lipids (Lyons *et al.*, 1979).

## MUTANT ISOLATION

Since there was no obvious way to identify mutants with altered lipid composition on the basis of gross phenotype, we screened for mutants by direct assay of fatty acid composition of leaf tissue by gas chromatography (Browse *et al.*, 1985b). Using this method, it is possible to obtain quantitative information on the fatty acyl composition of the lipids from as little as 5 mg of leaf tissue within several hours. We employed this procedure to measure the fatty acyl composition of total lipids from single leaves from randomly chosen plants in an ethylmethane sulfonate mutagenized population of *Arabidopsis*. From among the first 2,000 M2 plants examined in this way, we identified 7 mutants with major changes in fatty acyl composition (Browse *et al.*, 1985a), and in subsequent searches we have identified several additional mutant lines. The fatty acid composition from 7 mutants are shown in Table 1. In all of these mutants the alteration in fatty acid composition has been found to be due to a single recessive nuclear mutation. The putative sites of the enzymatic lesions in these mutants are illustrated in Figure 1.

## A MUTANT DEFICIENT IN THE CHLOROPLAST 18:1/16:1 DESATURASE

The *fad*C mutant was identified because gas chromatographic analysis revealed substantial increases in both 18:1 and 16:1 fatty acids, the virtual absence of 16:3, and a significant reduction in the level of 18:3 compared with the wild type (Table 1). The fatty acid composition of individual leaf lipids and data from labelling studies on mutant and wild type plants (Browse *et al.*, 1989) both support the conclusion that the *fad*C gene controls the activity of a chloroplast ω-6 glycerolipid desaturase. Analysis of the *fad*C mutant and a *fad*D mutant which controls the activity of the corresponding ω-3 desaturase (Browse *et al.*, 1986a, 1989) have provided insights into the substrate specificity of the desaturases. Both the ω-3 and ω-6 desaturases act on 16-carbon, as well as 18-carbon fatty acids and, therefore, appear to determine the site of desaturation relative to the methyl end of the chain. Both enzymes will act on acyl groups at either the *sn*-1 or *sn*-2 positions of the glycerol backbone. Finally, both enzymes appear to effect desaturation of all the major glycerolipids of the chloroplast membranes.

A major difference between the ω-6 and ω-3 desaturases is the extent to which they contribute to the desaturation of lipids of the extrachloroplast membranes. In the *fad*D mutant, the ratio of 18:3 in the mutant to that in the wild type was similar for all the major leaf polar lipids (Browse *et al.*, 1986a). This observation is consistent with the existence of at least two 18:2 desaturases with the *fad*D gene product contributing equally to the desaturation of chloroplast and extrachloroplast membrane lipids. In contrast, the mutation at *fad*C has only a small effect on desaturation of extrachloroplast membranes.

The accumulation of 18:1 and 16:1 in the *fad*C mutant (rather than ω-9, ω-3 dienoic fatty acids) indicates that monoenoic fatty acids are not suitable substrates for the ω-3 desaturase. The simplest explanation is that, while the ω-3 desaturase determines the position for insertion of the double bond relative to the methyl end of the molecule (Browse *et al.*, 1986a), it has an additional requirement for existing double bond at ω-6. Similarly, the fact that the ω-6 desaturase does not act on 16:0 at *sn*-2 of SL, and PG indicates that this enzyme requires there to be an existing double bond at ω-9. These suggestions are consistent with the findings of Howling *et al.*, 1972) for the desaturase of *Chlorella* presented with various monoenoic fatty acid substrates. Thus, our results are consistent with a model for the ω-6 and ω-3 chloroplast desaturases as enzymes which reside mainly in the hydrophobic region of the thylakoid (or envelope) bilayer, so that they have little or no interaction with the hydrophilic headgroups of the membrane lipids and which desaturate any fatty acids which meet the requirement of having a specific distance between an existing double bond and the methyl end of the acyl chain. In contrast, the desaturases controlled by the *fad*A and *fad*B genes appear to be substrate specific. The first acts only on 16:0 esterified to position *sn*-2 of chloroplast PG to produce *trans* (16:1 (Browse *et al.* 1985a) while the second inserts a *cis* double bond at the ω-9 position of 16:0 at the *sn*-2 position of MGD (Kunst *et al.* 1989b).

The mutation at the *fad*C locus has a very large effect on the degree of unsaturation of the photosynthetic membranes of the LK3 line. The average number of double bonds per glycerolipid molecule was 4:57 in chloroplasts from the wild type, but only 3:132 in chloroplasts from LK3. In MGD, the major thylakoid lipid, the number of double bonds per molecule fell from 5:78

in the wild type to 3:69 in the mutant. This change in the degree of thylakoid membrane unsaturation has no pronounced effects on the growth rate or vigor of the mutant under our standard growth conditions. There is, however, one indication from the biochemical studies reported here that lipid biosynthesis in the mutant may be regulated to ameliorate the effect of the *fad*C mutation. Compared with the wild type, the LK3 line shows a 30 to 35% decline in synthesis of MGD via the prokaryotic pathway and a corresponding increase in MGD synthesis by the eukaryotic pathway. The net effect is that a greater proportion of the acyl groups on MGD is desaturated by the endoplasmic reticulum 18:1 desaturase so that more MGD containing 18:3 is produced. In the absence of such a shift in metabolism, MGD from leaves of the mutant would contain an average of less than 2:8 double bonds per molecule.

Since we do not yet understand what factors mediate the partitioning of lipid synthesis between the two pathways, we are not able to definitively confirm that this altered balance between the two pathways of MGD synthesis is a regulated response to the loss of chloroplast ω-6 desaturase activity. However, these observations are consistent with our characterization of the *act*1 mutants (see below) from which it is apparent that demand for a balanced complement of different lipids required for the correct assembly of chloroplast membranes can indeed strongly regulate lipid metabolism in leaf cells.

## MUTANTS DEFICIENT IN THE SYNTHESIS OF PROKARYOTIC LIPIDS

Three independent, but allelic, *act*1 mutants were identified, because they lack 16:3 without showing any increase in the precursors 16:0, 16:1 or 16:2 (Table 1) (Kunst *et al.*, 1988). Enzyme assays showed that these mutants contain less than 5% of the wild type activity of the chloroplast 18:1-ACP:glycerol-3-P acyltransferase - the first enzyme in the prokaryotic pathway of lipid synthesis (Fig. 1).

In wild type *Arabidopsis*, approximately half of the chloroplast glycerolipids are derived from the prokaryotic pathway (Browse *et al.*, 1986b) (Fig. 2a). However, labeling experiments and fatty acid analyses of leaf lipids (Kunst *et al.*, 1988) indicate that loss of the acyltransferase activity does not result in the accumulation of precursors (18:1- and 16:0-ACP) upstream of the enzyme defect but instead causes a redirection of lipid metabolism so that the eukaryotic pathway predominates in the mutants (Fig. 2b). Somewhat unexpectedly, this redirection has little effect on the amount of each lipid that accumulates or the total glycerolipid content of the tissue. The enhanced flux through the eukaryotic pathway in the mutant is accompanied by a change in the proportion of individual lipids synthesized by this pathway. Thus, the amount of lipids, such as PE and PI, in the extrachloroplast membranes of the mutant is similar to that in the wild type. However, to compensate for the loss of the prokaryotic pathway, the amount of PC synthesized must be increased about two-fold to provide the DAG moieties required for normal levels of glycerolipid synthesis by the chloroplast. The net amount of PC that accumulates is apparently only that required for the synthesis of the extrachloroplast membranes, since the level of this lipid in the mutant is only slightly higher than in the wild type. The partitioning of eukaryotic DAG to MGD, DGD and SL synthesis is also altered in the *act*1 mutants (Fig. 2). Clearly, the synthetic reactions of the eukaryotic pathway, including synthesis of PC and export of lipid from the endoplasmic reticulum membranes are regulated in concert to meet the requirements for correct thylakoid membrane biogenesis.

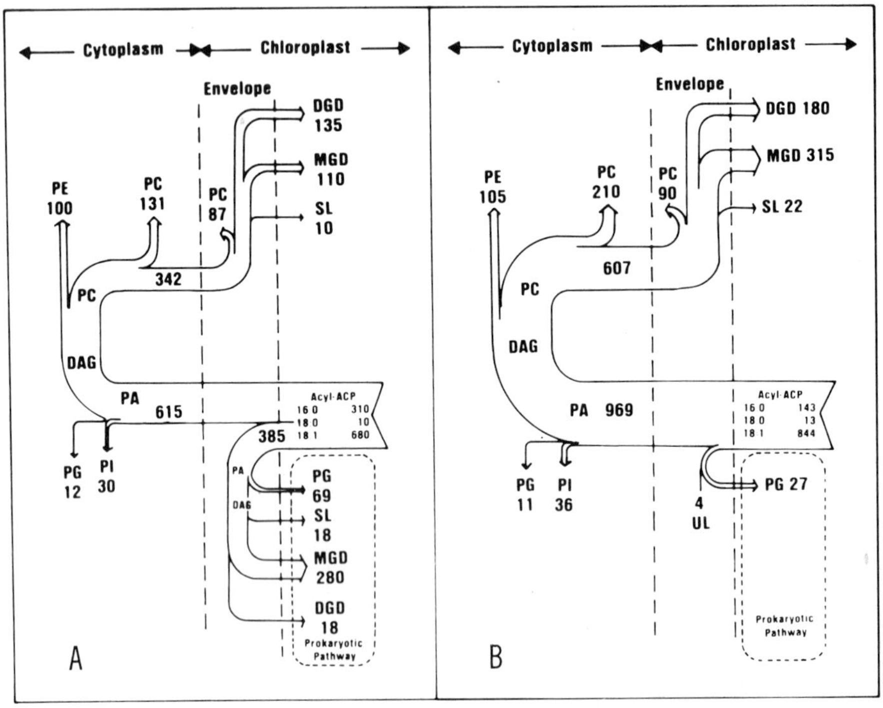


**Figure 2.** Flow diagrams showing the fluxes of fatty acids during lipid synthesis in *Arabidopsis* leaves of (A) the wild type and (B) the *act*1 mutant. Fluxes are expressed as mol/1000 mol.

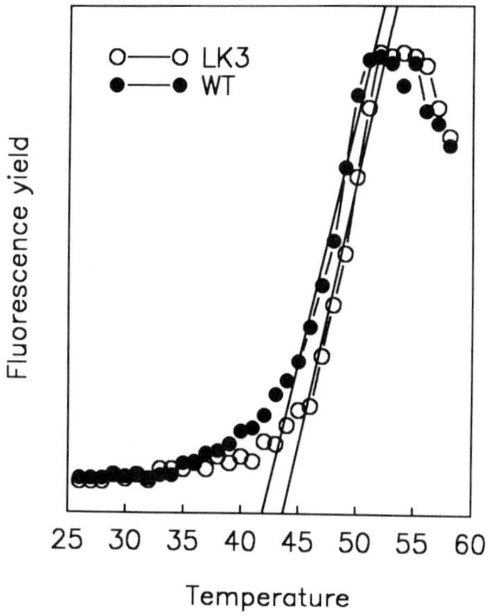


**Figure 3.** Temperature-induced fluorescence yield enhancement of leaves from wild type and *fad*C (LK3) mutant *Arabidopsis*. Data is from Hugly *et al.* (1989), but normalized to the same maximum fluorescence yield. Lines were fitted by least squares regression.

In all higher plants the prokaryotic pathway is responsible for the synthesis of chloroplast PG and in 16:3 plants PG synthesis involves the same pool of PA used for the synthesis of prokaryotic galactolipids and SL (Fig. 1). Nevertheless, the *act*1 mutants contain at least 75% as much PG as the wild type even though it exhibits less than 5% of the wild type level of chloroplast glycerol-P acyltransferase activity. It appears, therefore, that in the mutant, the small amount of PA formed is used preferentially for PG synthesis at the expense of DAG formation by the PA phosphatase (Kunst *et al.*, 1988).

## THE FUNCTIONAL SIGNIFICANCE OF LIPID COMPOSITION

The collection of *Arabidopsis* mutants with specific changes in membrane lipid composition provides a resource for the investigation of membrane structure and function. Detailed studies of several mutants (*fad*A, *fad*D, *act*1) indicate that photosynthetic electron transport is relatively insensitive to changes in thylakoid fatty acid composition (McCourt *et al.*, 1985, 1987; Kunst *et al.*, 1989). By contrast, several of the mutants exhibit significant changes in chloroplast ultrastructure. For example, the *fad*D mutant contains smaller, more numerous chloroplasts (McCourt *et al.*, 1987), while the *act*1 mutants show changes in the arrangement of thylakoid lamellae. These mutants indicate a role for lipid composition in the development and maintenance of membrane architecture.

More recently, our investigations of two other mutants, *fad*B and *fad*C, have revealed more direct effects of lipid composition on the thermal behavior of chloroplast membranes. In the *fad*C mutant, the substantial decrease in lipid unsaturation (Table 1) is accompanied by decreased chloroplast membrane synthesis and a resulting decrease in photosynthesis in plants grown and assayed at 22°C (Hugly *et al.*, 1989). However, based on two different criteria, the thylakoid membranes showed increased stability to thermal disruption. First, the temperature at which fluorescence yield enhancement occurred in the mutant was increased 2°C relative to wild type, from 42°C to 44°C (Fig. 3). Second, when thylakoid preparations were heat treated prior to the measurement of photosynthetic electron transport, membranes from the mutant consistently retained higher relative rates than membranes from wild type plants (Hugly *et al.*, 1989). The magnitude of the effects observed in the *fad*C mutant are comparable with observations on natural populations of adapted and non-adapted plants (Downton *et al.*, 1984) and consistent with the effects of catalytic hydrogenation on the thermal properties of thylakoid preparations (Thomas *et al.*, 1986).

The high temperature tolerance of photosynthesis in the *fad*C and *fad*B mutants is correlated with an increased sensitivity of these two lines to very low growth temperatures. The wild type as well as the mutant lines *fad*A, *fad*D and *act*1 are indistinguishable from one another in growth and appearance when plants are kept at 6°C. In contrast, the *fad*C and *fad*B mutants become chlorotic at this temperature and show a 30-40% reduction in growth rate relative to wild type (S. Hugly and C.R. Somerville, unpublished). The *fad*2 mutants which are deficient in the endoplasmic reticulum 18:1 desaturase are also severely affected by growth at low temperatures. Further investigations of all these mutants promises to provide insights into the structural requirements of the lipid components which help to maintain the integrity of plant cells and organelles.

## REFERENCES


Andrews, J., Ohlrogge, J.B., and Keegstra, K., 1985, Final step of phosphatidic acid synthesis in pea chloroplasts occurs in the inner envelope membrane, *Plant Physiol.*, 78:459-466.

Block, M.A., Dorne, A.J., Joyard, J., and Douce, R., 1983, The phosphatidic acid phosphatase of the chloroplast envelope is located on the inner envelope membrane, *FEBS Lett.*, 164:111-115.

Browse, J., Kunst, L., Anderson, S., Hugly, S., and Somerville, C.R., 1989, A mutant of *Arabidopsis* deficient in the chloroplast 16:1/18:1 desaturase, *Plant Physiol.*, 90:522-529.

Browse, J., McCourt, P., and Somerville, C.R., 1985a, A mutant of *Arabidopsis* lacking a chloroplast specific lipid, *Science*, 227:763-765.

Browse, J., McCourt, P., and Somerville, C.R., 1985b, Overall fatty acid composition of leaf lipids determined after combined digestion and fatty acid methyl ester formation from fresh tissue, *Anal. Biochem.*, 152:141-146.

Browse, J., McCourt, P., and Somerville, C.R., 1986a, A mutant of *Arabidopsis* deficient in $C_{18:3}$ and $C_{16:3}$ leaf lipids, *Plant Physiol.*, 81:859-864.

Browse, J., McCourt, P., and Somerville, C.R., 1986b, Fluxes through the prokaryotic and eukaryotic pathways of lipid synthesis in the 16:3 plant *Arabidopsis thaliana, Biochem. J.*, 235:25-31.

Browse, J.A., and Slack, C.R., 1981, Catalase stimulates linoleate desaturase activity in microsomes from developing linseed cotyledons, *FEBS Lett.*, 131:111-114.

Coves, J., Block, M.A., Joyard, J., and Douce, R., 1986, Solubilization and partial purification of UDP-galactose diacylglycerol galactosyl transferase activity from spinach chloroplast envelope, *FEBS Lett.*, 208:401-407.

Cronan, J.E., Jr., and Rock, C.O., 1987, Biosynthesis of membrane lipids, *in Escherichia coli* and *Salmonella typhimurium*, Vol. 1, pp. 474-497, (American Society of Microbiologists, Washington), J.L. Ingraham, K.B. Low, B. Magasanik, M. Schaechter and H.E. Umbarger, eds.

Downton, W.J.S., Berry, J.A., and Seemann, J.R.,1984, Tolerance of photosynthesis to high temperature in desert plants, *Plant Physiol.*, 74:786-790.
Frentzen, M., 1986, Biosynthesis and desaturation of the different diacylglycerol moieties in higher plants, *J. Plant Physiol.* 124:193-209.
Frentzen, M., Hares, W., and Schiburr, A., 1984, Properties of the microsomal glycerol-3-P and monoacylglycerol-3-P acyltransferases from leaves, *in* Structure, Function and Metabolism of Plant Lipids, P.A. Siegenthaler, and W. Eichenburger, eds., Elsevier, Amsterdam, pp. 105-110.
Frentzen, M., Heniz, E., McKeon, T.A., and Stumpf, P.K., 1983, Specificities and selectivities of glycerol-3-phosphate acyltransferase from pea and spinach chloroplasts, *Eur. J. Biochem.*, 129:629-636.
Galey, J., Franchke, B., and Bahl, J., 1980, Ultrastructure and lipid composition of etioplasts in developing dark-grown wheat leaves, *Planta*, 149:433-439.
Gounaris, K., and Barber, J., 1983, Monogalactosyldiacylglycerol: The most abundant polar lipid in nature, *Trends Biochem. Sci.*, 8:378-381.
Gounaris, K., Brain, A.P.R., Quinn, P.J., and Williams, W.P., 1984, Structural reorganisation of chloroplast thylakoid membranes in response to heat stress, *Biochim. Biophys. Acta*, 766:198-208.
Gounaris, K., Mannock, D.D., Sen, A., Brain, A.P.R., Williams, W.P., and Quinn, P.J., 1983, Polyunsaturated fatty acyl residues of galactolipids are involved in the control of bilayer/non-bilayer lipid transitions in higher plant chloroplasts, *Biochim. Biophys. Acta*, 732:229-242.
Heemskert, J.W.M., Bogemann, G., and Wintermans, J.F.G.M., 1985, Spinach chloroplasts: localization of enzymes involved in galactolipid metabolism, *Biochim. Biophys. Acta*, 835:212-220.
Heinz, E., and Roughan, P.G., 1983, Similarities and differences in lipid metabolism of chloroplasts isolated from 18:3 and 16:3 plants, *Plant Physiol.*, 72:273-279.
Howling, D., Morris, L.J., Gurr, M.I., and James, A.T., 1972, The specificity of fatty acid desaturases and hydroxylases. The dehydrogenation and hydroxylation of monoenoic acids, *Biochim. Biophys. Acta*, 260:10-19.
Hugly, S., Kunst, L., Browse, J., and Somerville, C., 1989, Enhanced thermal tolerance of photosynthesis and altered chloroplast ultrastructure in a mutant of *Arabidopsis* deficient in lipid desaturation, *Plant Physiol.*, 90:1134-1142.
Jamieson, G.R., and Reid, E.H., 1971, The occurrence of hexadeca-7,10,13-trienoic acid in the leaves of angiosperms, *Phytochemistry* 10:1837-1843.
Joyard, J., Blee, E., and Douce, R., 1986, Sulfolipid synthesis from $^{35}SO^{2-}{}_4$ and [1-$^{14}$C]acetate in isolated intact spinach chloroplasts, *Biochim. Biophys. Acta*, 879:78-87.
Kleppinger-Sparace, K., Mudd, J.M., and Bishop, D.G., 1985, Biosynthesis of sulfoquinovosyldiacylglycerol in higher plants, *Arch. Biochem. Biophys.*, 240:859-865.
Kunst, L., Browse, J., and Somerville, C., 1988, Altered regulation of lipid biosynthesis in a mutant of *Arabidopsis* deficient in chloroplast glycerol phosphate acyltransferase activity, *Proc. Natl. Acad. Sci.* USA, 85:4143-4147.
Kunst, L., Browse, J., and Somerville, C., 1989a, Altered chloroplast structure and function in a mutant of *Arabidopsis* deficient in plastid glycerol-3-phosphate acyltransferase activity, *Plant Physiol.*, 90:846-843.
Kunst, L., Browse, J., and Somerville, C., 1989b, A mutant of *Arabidopsis* deficient in desaturation of palmitic acid in leaf lipids, *Plant Physiol.*, 90:943-947.
Leech, R.M., Rumsby, M.G., and Thomson, W.W., 1973, Plastid differentiation, acyl lipid, and fatty acid changes in developing green maize leaves, *Plant Physiol.*, 52:240-245.

Leech, R.M., Walton, C.A., and Baker, N.R., 1985, Some effects of 4-chloro-5-dimethylamino-2-phenyl-3(2H)-pyridazinone (SAN9785) on the development of thylakoid membranes in *Hordeum vulgare* L., *Planta*, 165:277-283.

Lyons, J.M., Raison, J.K., and Steponkus, P.L., 1979, The plant membrane in response to low temperature, *in* Low Temperature Stress in Crop Plants: The Role of the Membrane, J.M. Lyons, D. Graham, and J.K. Raison, eds., Academic Press, NY, pp. 1-24.

McCourt, P., Browse, J., Watson, J., Arntzen, C.J., and Somerville, C.R., 1985, Analysis of photosynthetic antenna function in a mutant of *Arabidopsis thaliana* (L.) lacking *trans*-hexadecenoic acid, *Plant Physiol.* 78:853-858.

McCourt, P., Kunst, L., Browse, J., and Somerville, C.R., 1987, The effects of reduced amounts of lipid unsaturation on chloroplast ultrastructure and photosynthesis in a mutant of *Arabidopsis*, *Plant Physiol.*, 84:353-360.

McElhaney, R.N., 1989, The influence of membrane lipid composition and physical properties on membrane structure and function in *Acholephasma laidlawii*, *Crit. Rev. Microbiol.*, 17:1-32.

McKeon, T.A., and Stumpf, P.K., 1982, Purification and characterization of the stearoylacyl carrier protein desaturase and the acyl-acyl carrier protein thioesterase from maturing seeds of safflower, *J. Biol. Chem.*, 257:12141-12147.

Moore, T.S., 1982, Phospholipid biosynthesis, *Ann. Rev. Plant Physiol.*, 33:235-259.

Pang, P.P., and Meyerowitz, E.M., 1987, *Arabidopsis thaliana*: a model system for plant molecular biology, *Biotechnology* 5:1177-1181.

Quinn, P.J., and Williams, W.P., 1983, The structural role of lipids in photosynthetic membranes, *Biochim. Biophys. Acta* 737:223-266.

Rawyler, A., and Siegenthaler, P.A., 1981, Transmembrane distribution of phospholipids and their involvement in electron transport as revealed by phospholipase A2 treatment of spinach thylakoids, *Biochim. Biophys. Acta*, 635:348-368.

Roughan, P.G., Mudd, J.B., McManus, T.T., and Slack, C.R., 1979, Linoleate and a-linolenate synthesis by isolated chloroplasts, *Biochem. J.*, 184:571-574.

Roughan, P.G., and Slack, C.R., 1982, Cellular organization of glycerolipid metabolism, *Ann. Rev. Plant Physiol.*, 33:97-123.

Sen, A., Williams, W.P., and Quinn, P.J., 1981, The structure and thermotropic properties of pure 1,2-diacylgalactosylglycerols in aqueous systems, *Biochim. Biophys. Acta*, 663:380-389.

Slack, C.R., Roughan, P.G., and Terpstra, J., 1976, Some properties of a microsomal oleate desaturase from leaves, *Biochem. J.*, 155:71-80.

Sparace, S.P., and Mudd, J.B., 1982, Phosphatidylglycerol synthesis in spinach chloroplasts: characterization of the newly synthesized molecule, *Plant Physiol.*, 70:1260-1264.

Stymne, S., and Appelqvist, A., 1978, The biosynthesis of linoleate from oleoyl-CoA via oleoyl phosphatidylcholine in microsomes of developing safflower seeds, *Eur. J. Biochem.*, 90:223-229.

Thomas, P.G., Dominy, P.J., Vigh, L., Mansourian, A.R., Quinn, P.J., and Williams, W.P., 1986, Increased thermal stability of pigment-protein complexes of pea thylakoids following catalytic hydrogenation of membrane lipids, *Biochim. Biophys. Acta* 849:131-140.

Van Walraven, H.S., Koppenaal, E., Marvin, H.J.P., Hagendoorn, M.J.M., and Kraayenhot, R., 1984, Lipid specificity for the reconstitution of well coupled ATPase proteoliposomes and a new method for lipid isolation from photosynthetic membranes, *Eur. J. Biochem.*, 144:563-566.

Vigh, L., Joo, F., Droppa, M., Horvath, L.I., and Horvath, G., 1985, Modulation of chloroplast membrane lipids by homogeneous catalytic hydrogenation, *Eur. J. Biochem.*, 147:477-481.

# ALKALOIDS: A NEW TARGET FOR MOLECULAR BIOLOGY

Toni M. Kutchan

Lehrstuhl für Pharmazeutische Biologie
Universität München, Karlstrasse 29
D-8000 München 2, West Germany

## INTRODUCTION

We encounter many alkaloids in our everyday lives in the foods we eat and in the beverages we drink. Relatively little is known about the biosynthesis of these alkaloids at the enzyme level and virtually nothing is known of their regulation. More information has accumulated on the alkaloids of pharmaceutical interest concerning their biological mode of action, but again little is known about their biosynthesis in the plant. It is presently not possible to produce commercially by chemical syntheses many of the alkaloids currently of medicinal use due to their sophisticated structures containing multiple chiral centers. In the past, two main approaches have been taken for the production of alkaloids for industrial use: isolation from the field grown plant and production in cell culture. The limitations of both sources are well known: dependence on environmental and political conditions with field harvesting and the failure to produce many of the commercially interesting alkaloids in sufficient quantities in plant cell culture. New approaches are now feasible utilizing the methods of molecular genetics, these are the production of alkaloids in microorganisms and the alteration of the regulation of alkaloid biosynthesis in cell culture to accumulate higher concentrations of target alkaloids. Before the techniques of molecular biology can be applied to these plant systems, however, a detailed knowledge of the biochemical pathways leading to the alkaloids from simple primary metabolites must be gained. The complex structures of the alkaloids suggest that numerous enzymatic conversions (in many cases at least twenty or more) are involved. The key, limiting steps must be identified and the enzymes catalyzing these conversions must be characterized in order to approach these problems in a logical and specific manner. Here, select cases of historically and commercially significant alkaloids together with the information available concerning their biosynthesis will be presented. The successful example of the production of the key indole alkaloid intermediate, strictosidine, in *Escherichia coli* will also be given.

## CAFFEINE

Many of the pharmaceutically important plant principles in use today, caffeine included, were isloated in the period 1817-1820 by Pelletier and Caventou in Paris.[1] Caffeine is used widely and frequently as a stimulant by those of us who drink coffee and black tea. Caffeine, as well as the structurally related purine alkaloids, theophylline and theobromine, find clinical use as well in the treatment of migraines, asthma, and cardiac edema and angina pectoris, respectively. In view of its common occurrence

*Plant Molecular Biology* 2, Edited by R.G. Herrmann and B. Larkins, Plenum Press, New York, 1991

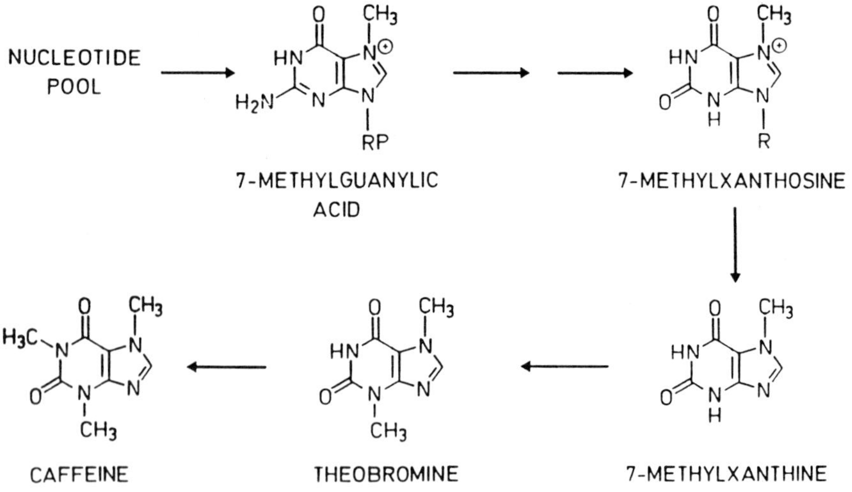


Figure 1. Biosynthesis of caffeine.

and use, it is surprising that the biosynthesis of caffeine has not yet been completely elucidated. The pathway as is currently accepted is given in Figure 1. The purine ring of caffeine is derived in coffee and tea leaves via purine nucleotide biosynthesis.[2] S-Adenosylmethionine is the donor of the N-methyl groups.[3-5] The precise sequence to 7-methylxanthosine is not known. The involvement of 7-methylxanthine and theobromine in the biosynthesis of caffeine has been demonstrated by pulse feeding experiments with *Camellia sinensis* shoot tips.[6] It is not clear whether adenine or guanine originating from nucleic acids or from the nucleotide pool serves as precursor to caffeine. Studies of specific enzymes involved in caffeine biosynthesis are lacking, thus leading to the confusion as to the correct sequence of methylation of the purine ring.

NICOTINE

One of the best studied alkaloids from a biosynthetic viewpoint is that of nicotine due largely to the efforts of E. Leete. Nicotine, isolated in 1809,[1] is one of the oldest known alkaloids. It is also one of the best known alkaloids today due to the large production of tobacco leaf (over five million tons annually). *Nicotiana tabacum* was introduced to Europe from Florida around 1560 and it is thought that smoking the tobacco leaf possibly started as a cure for ailments of the cardiovascular or respiratory system. Nicotine is now known to be toxic, 40 mg is fatal to humans. Due to this toxicity and the correlation of cigarette smoking with heart disease and lung cancer, tobacco plants low in nicotine content could be desirable. To this end, knowledge of the biosynthetic pathway which leads to nicotine is essential, not only for the production of higher quantities of alkaloid, but also for the knowledge of how to hem or inhibit completely production of the alkaloid in the plant. The biosynthetic pathway leading to nicotine in *Nicotiana* is represented in Figure 2. Based on the results of early tracer experiments with *Nicotiana*,[7-9] the route to the pyrrolidine ring of nicotine proceeds by two possible pathways. Ornithine is N-methylated to δ-N-methylornithine which is then decarboxylated to N-methylputrescine resulting in the assymetric incorporation of carbon atoms C-3 and C-4 of ornithine into the N-methyl-1-pyrrolinium salt. The alternate pathway to the N-methyl-1-pyrrolinium salt (decarboxylation of ornithine to putrescine followed by methylation to N-methylputrescine and symmetrical incorporation into nicotine) has now been substantiated[10] by an analysis of the enzymes present in alkaloid producing root cultures of *Hyoscyamus albus*. Activities for an ornithine-N-methyltransferase and an N-methylornithine decarboxylase could not be detected. Rather, activities for putrescine-N-methyltransferase and

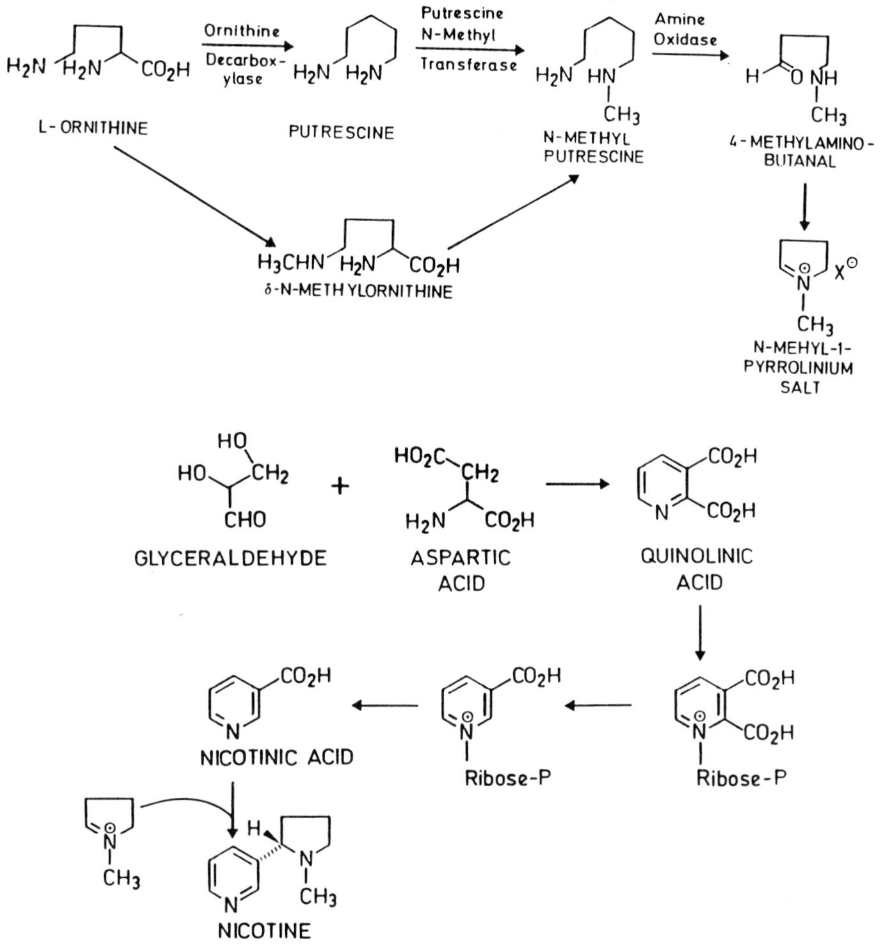


Figure 2. Biosynthesis of nicotine. A. Two proposed pathways leading to the N-methyl-1-pyrrolinium salt of nicotine. B. Formation of nicotinic acid in plants and condensation with the N-methyl-1-pyrrolinium salt.

ornithine decarboxylase were found. This pathway suggests a symmetrical incorporation of putrescine into the N-methyl-1-pyrrolinium salt. This discrepancy between the two methods of analysis in the two plant systems used has not yet been reconciled. Nicotinic acid originates from aspartic acid and glycerol in plants,[11] but the enzymes which catalyze this ring formation remain unknown. The key step in the formation of nicotine is the condensation of nicotinic acid and the N-methyl-1-pyrrolinium salt in a stereospecific manner. The enzyme which facilitates this condensation has been discovered (E. Leete, unpublished). Even as one of the oldest known alkaloids, the exact biosynthetic pathway leading to nicotine in plants remains somewhat ambiguous and it is only in very recent years that several of the enzymes involved in this pathway have been detected.

SCOPOLAMINE

The tropane alkaloids, *l*-hyoscyamine and atropine (the racemate of *l*-hyoscyamine), were isolated in 1833 from the extremely poisonous plants *Hyoscyamus niger* and *Atropa belladonna*, respectively.[1] It was not until 1881 that scopolamine was isolated from *Hyoscyamus muticus*. As with

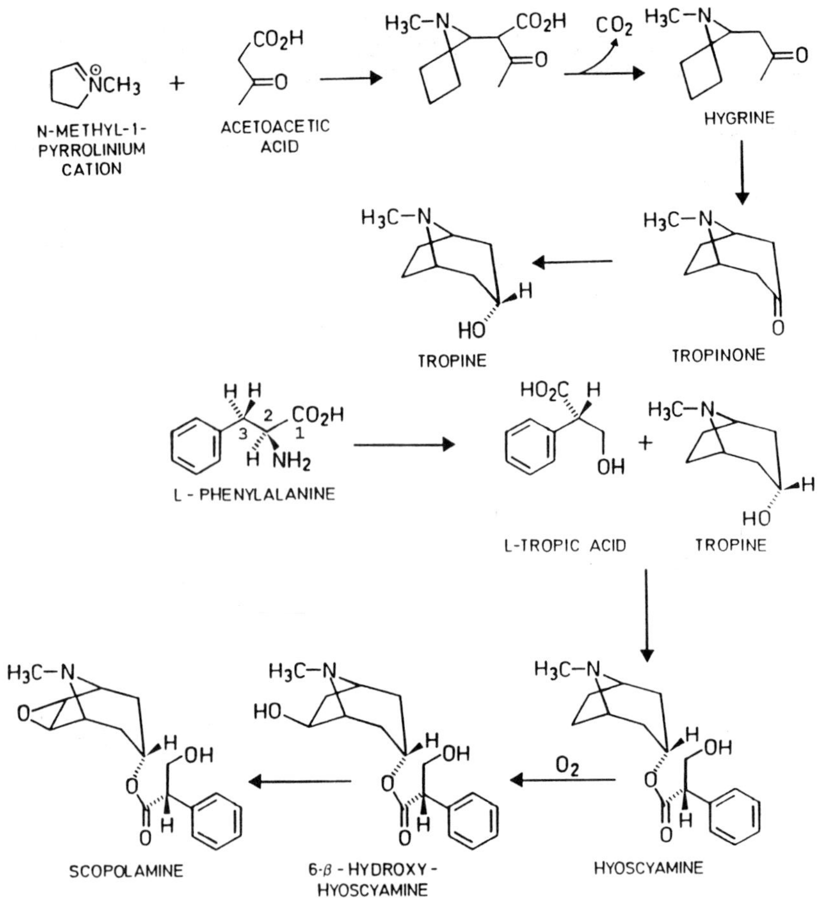


Figure 3. Biosynthesis of scopolamine. A. Proposed pathway leading to tropine. B. Formation of tropic acid and condensation with tropine.

hyoscyamine, it readily racemizes during isolation. *Duboisia leichardtii* serves as a commercial source of scopolamine. One of the most notorious tropane alkaloids today is cocaine, the use of which, in the form of coca chewing, can be traced back to precolumbian times in South America. Scopolamine is currently used in the prophylactic treatment of motion sickness, while cocaine has found legitimate use as a model in the design of improved anesthetics. Tropane alkaloid biosynthesis is also a relatively well studied system, due again in large part, to the efforts of E. Leete and Y. Yamada. The pathway (Figure 3) leading from ornithine and phenylalanine to scopolamine was first elucidated by tracer feeding experiments.[12-16] Several of the enzymes catalyzing individual steps in this pathway have been identified,[10] but the majority of the pathway remains to be verified in this mannner. The salient features are the following. The N-methyl-1-pyrrolinium salt arises as described for nicotine. This then condenses with acetoacetic acid to form hygrine. Hygrine then cylizes by an unknown mechanism to form tropinone, which is reduced to tropine. A key step in the pathway is the formation of tropic acid from phenylalanine. This proceeds via an intramolecular rearrangement which involves a 1,2-shift of the 3-*pro-S* proton accompanied by migration of the carboxyl group. In this process, the stereochemistry at C-3 is retained.[17-22] The enzyme which catalyzes this rearrangement has not yet been identified. The second key transformation is the esterification of tropic acid with tropine forming hyoscyamine. This enzyme is also unknown. The subsequent steps, hydroxylation of hyoscyamine and epoxide formation to scopolamine is catalyzed by a well characterized enzyme, hyoscyamine 6β-hydroxylase.[23-25] One enzyme was found to catalyze both the hydroxylation and ether bridge formation in extracts of *Hyoscyamus niger* root cultures.

QUININE

Quinine has been used in the treatment of malaria at least since 1633.[1] Extensive use of *Cinchona* bark depleated the South American supplies which fact lead to the development of plantations in the East Indies by the Dutch. When this main source of quinine from Java was interrupted during World War II, new synthetic antimalarials were developed. Although the use of quinine has deminished since that time, contemporary outbreaks of malaria have lead to the belief that malarial strains resistant to the synthetic drugs have developed. The historical importance of quinine goes back over 350 years in that the drug played a major role in both the exploration and colonization of the tropics. Even though quinine was a factor in the formation of world demographics as we know them today, surprisingly little is known of how *Cinchona* species produce this alkaloid. The little that is known is depicted in Figure 4. Although quinine contains a quinoline nucleus, it is in fact derived from the indole alkaloid, strictosidine.[26-29] The intermediates shown between corynantheal and quinine were derived from tracer feeding experiments,[30] although thorough proof of their intermediacy is still lacking. The naturally occurring diastereomer of quinine is quinidine. This change in stereochemistry results in an alkaloid which acts on the cardiac muscle more effectively than quinine and is therefore used in the treatment of cardiac arrythmias.[1] This point serves to illustrate the importance of stereochemical configuration with respect to mode and efficacy of drug action.

EMETINE

Emetine was one of the early alkaloids, isolated in 1817.[1] It is the active principle in the old, Brazilian medicinal, ipecacuanha (*Cephaelis ipecacuanha*). Brought to Europe in the late 1700s, it was used in the treatment of dysentery. Imports of ipecac today to the United States are up to 15 tons annually, the commercial source of the alkaloid still being extraction from the plant. Emetine shows antiviral as well as strong

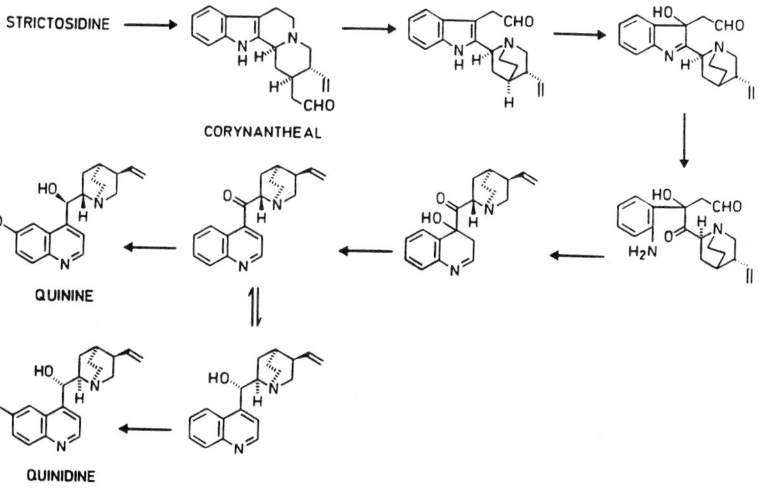


Figure 4. Biosynthesis of quinine.

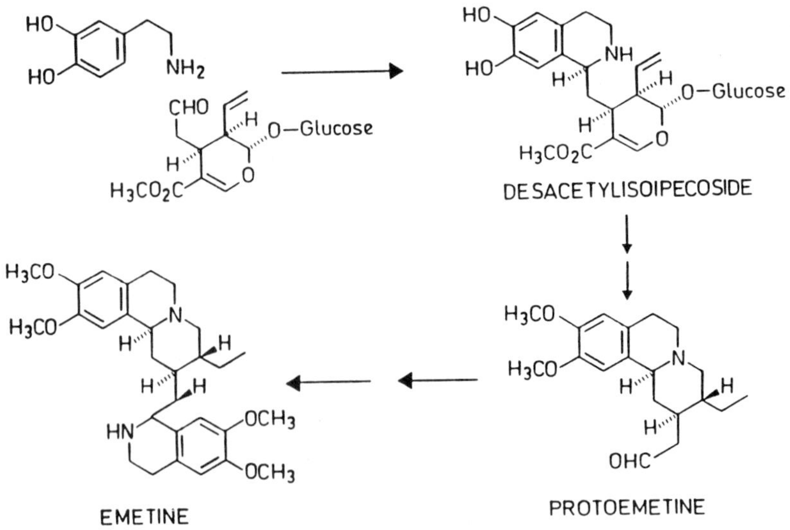


Figure 5. Biosynthesis of emetine.

antiamebic activity and is used in the treatment of amebic dysentery. Dihydroemetine has been effective in the treatment of certain types of malignancies. Interest in emetine and its derivatives stems from these anti -tumor and -viral activities. Preliminary studies on the biosynthesis of ipecac alkaloids was carried out by tracer feeding experiments with *C. ipecacuanha* and *Alangium lamarckii* plants and the condensation product of dopamine and secologanin, desacetylisoipecoside, was identified as the key intermediate leading to emetine[31,32] (Figure 5). The lower tetrahydroisoquinoline moiety supposedly originates from condensation with a second molecule of dopamine. Since the time of these early biosynthetic experiments, no further work has been done on this pathway. The key enzymes in this pathway would be those catalyzing the stereospecific condensations of dopamine and secologanin and of dopamine and protoemetine. Emetine is a pharmaceutically and commercially significant alkaloid which has recieved relatively little attention concerning its biosynthesis and development of alternate sources.

PRODUCTION OF STRICTOSIDINE IN *ESCHERICHIA COLI*

Indole alkaloids comprise a very large group (over 1800 individual structures) of chemically and pharmaceutically interesting compounds. The key intermediate leading to these alkaloids was discovered[26] as well as the enzyme catalyzing its formation.[33,34] This enzyme, strictosidine synthase, has been well characterized and purified to homogeneity,[35,36] cloned[37] and expressed in an enzymatically active form in *E. coli*.[38] Strictosidine synthase (Figure 6) was the first example of an enzyme of alkaloid biosynthesis in a higher plant to be cloned and expressed in an active form in a microbe. The heterologously expressed enzyme accumulates to 3% of the total protein in *E. coli*[39] and the alkaloidal intermediate, strictosidine, can be readily produced by addition of the precursors, tryptamine and secologanin, to intact, non-permeabilized bacteria.[38] The advantages of producing both enzyme and alkaloid in a bacterial system become apparent when the microbial system is compared to the plant cell suspension culture. Utilizing suspension cultures of *Rauvolfia serpentina* optimized for alkaloid production, strictosidine synthase accumulates to 0.1% of the soluble protein. As such, the *E. coli* system accumulates approximately thirty times as much enzyme as the *R. serpentina* cultures. When these same values are analyzed with respect to length of the culture periods of the

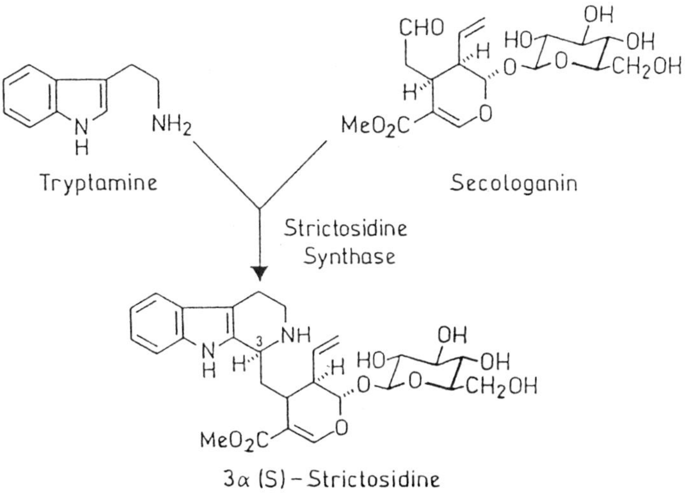


Figure 6. Enzymatic reaction catalyzed by strictosidine synthase.

bacteria and of the plant cells (Table 1), the bacteria produce approximately 1000 times more enzyme per hour than do the plant cell suspension cultures. The other advantage is that the bacteria do not modify strictosidine, whereas the plant cells contain enzymes which further metabolize the alkaloid. The quantity of strictosidine synthase in *R. serpentina* suspension cultures is the exceptional example. Typically the enzymes of alkaloid biosynthesis are not accumulating to such high levels. The more representative cases are shown in Table 1. When compared to these other cell culture systems, *E. coli* pJUB5 produces up to 24 x $10^3$ more enzyme.

Table 1. Comparison of Catalytic Activity of Cloned Strictosidine Synthase with That in Plant Cell Suspension Cultures

| Species | Activity (nkat/l) | %Total Protein | Growth Period | Productivity (nkat/l/h) |
|---|---|---|---|---|
| pJUB5 (*E. coli*) | 6600 | 3 | 7 hours | 939 |
| *Rauvolfia serpentina* | 300 | 0.11 | 12 days | 1 |
| *Catharanthus roseus* | 44 | 0.1 | 7 days | 0.26 |
| *Tabernanthe iboga* | 22 | - | 21 days | 0.04 |
| *Vinca herbacea* | 11 | - | 9 days | 0.05 |

There are many interesting alkaloid systems, interesting from a chemical/structural viewpoint, interesting from a medicinal viewpoint or interesting from the point of view of basic research. Even though many key alkaloids have been known for a century or more, even though these alkaloids have had a crucial role in the development of many classical organic reactions and in the development of many contemporary pharmaceuticals and even though they are involved in political, health and social issues worldwide today, relatively little is known of their biogenesis, *in vivo* function or regulation in plants. The example of strictosidine synthase illustrates the potential of the techniques of biochemistry and molecular genetics when applied to a study of alkaloid biosynthesis. It is now possible to produce alkaloids of higher plants in microorganisms through a thorough biochemical analysis of the enzymes and precusory metabolites involved in alkaloid biosynthesis followed by the application of cloning methodology.

ACKNOWLEDGEMENTS: The molecular biological investigations with strictosidine synthase were supported by a grant from the Bundesminister für Forschung und Technologie, Bonn, to Professor M.H. Zenk.

REFERENCES

1. For an excellent review of alkaloid history, biosynthesis and use, see: G. A. Cordell, "Introduction to Alkaloids: A Biogenetic Approach," John Wiley & Sons, New York (1981).
2. L. Anderson and M. Gibbs, The biosynthesis of caffeine in the coffee plant, J. Biol. Chem.237:1941 (1962).
3. E. Looser, T. W. Baumann and H. Wanner, The biosynthesis of caffeine in the coffee plant, Phytochemistry 13:2515 (1974).
4. T. Suzuki and E. Takahashi, Biosynthesis of caffeine be tea-leaf extracts, Biochem. J. 146:87 (1975).
5. T. Suzuki and E. Takahashi, Metabolism of methionine and biosynthesis of caffeine in the tea plant (*Camellia sinensis* L.), Biochem. J. 160:171 (1976).
6. T. Suzuki and E. Takahashi, Caffeine biosynthesis in *Camellia sinensis*, Phytochemistry 15:1235 (1976).
7. E. Leete and K. J. Siegfried, The biosynthesis of nicotine. III. Further observations on the incorporation of ornithine into the pyrrolidine ring, J. Am. Chem. Soc. 79:4529 (1957).
8. E. Leete, The biogenesis of nicotine. V. New precursors of the pyrrolidine ring, J. Am. Chem. Soc. 80:2162 (1958).
9. E. Leete and Ming-Li Yu, The incorporation of ornithine-[2,3-$^{13}C_2$] into nicotine and nornicotine established by NMR, Phytochemistry 19:1093 (1980) and references cited therein.
10. Y. Yamada and T. Hashimoto, Biosynthesis of tropane alkaloids, in: "Applications of plant cell and tissue culture," (Ciba Foudation Symposium 137) Wiley, Chichester (1988).
11. R. B. Herbert, "The Biosynthesis of Secondary Metabolites," Chapman and Hall, London (1989).
12. E. Leete, The stereospecific incorporation of ornithine into the tropine moiety of hyoscyamine, J. Am. Chem. Soc. 84:55 (1962).
13. F. E. Baralle and E. G. Gros, Biosynthesis of cuscohygrine and hyoscyamine in *Atropa belladonna* from DL-$\alpha$-*N*- methyl-[$^3$H]ornithine and DL-$\delta$-*N*-methyl-[$^3$H]ornithine, Chem. Commun. 721 (1969).
14. D. G. O´Donovan and M. F. Keogh, The role of hygrine in the biosynthesis of cuscohygrine and hyoscyamine, J. Chem. Soc. (C) 223 (1969).
15. E. Leete and D. H. Lucast, Loss of tritium during the biosynthesis of meteloidine and scopolamine from [N-methyl-$^{14}$C,6$\beta$,7$\beta$-$^3H_2$] tropine, Tetrahedron Lett. 3401 (1976).
16. B. A. McGaw and J. G. Woolley, Stereochemistry of tropane alkaloid formation in *Datura*, Phytochemistry 17:257 (1978).
17. E. Leete, The biogenesis of tropic acid and related studies on the alkaloids of *Datura stramonium*, J. Am. Chem. Soc. 82:612 (1960).

18. E. W. Underhill and H. W. Youngken, Jr., Biosynthesis of hyoscyamine and scopolamine in *Datura stramonium*, J. Pharm. Sci. 51:121 (1962).
19. M. L. Louden and E. Leete, The biosynthesis of tropic acid, J. Am. Chem. 84:1510 (1962).
20. E. Leete, N. Kowanko and R. A. Newmark, Use of carbon-13 nuclear magnetic resonance to establish that the biosynthesis of tropic acid involves an intramolecular rearrangement of phenylalanine, J. Am. Chem. Soc. 97:6826 (1975).
21. E. Leete, 1,2-Migration of hydrogen during the biosynthesis of tropic acid from phenylalanine, J. Am. Chem. Soc. 106:7271 (1984).
22. E. Leete, Stereochemistry of the 1,2-migration of the carboxyl group that occurs during the biosynthesis of tropic acid from phenylalanine, Can. J. Chem. 65:226 (1987).
23. T. Hashimoto and Y. Yamada, Hyoscyamine 6β-hydroxylase, a 2-oxoglutarate-dependent dioxygenase, in alkaloid-producing root cultures, Plant Physiol. 81:619 (1986).
24. T. Hashimoto and Y. Yamada, Purification and characterization of hyoscyamine 6β-hydroxylase from root cultures of *Hyoscyamus niger* L., Eur. J. Biochem. 164:277 (1987).
25. T. Hashimoto, J. Kohno and Y. Yamada, Epoxidation *in vivo* of hyoscyamine to scopolamine does not involve a dehydration step, Plant Physiol. 84:144 (1987).
26. G. N. Smith, Strictosidine: A key intermediate in the biogenesis of indole alkaloids, Chem. Commun. 912 (1968).
27. N. Kowanko and E. Leete, Biosynthesis of the cinchona alkaloids. I. The incorporation of tryptophan into quinine, J. Am. Chem. Soc. 84:4919 (1962).
28. E. Leete and J. N. Wemple, Biosynthesis of the *Cinchona* alkaloids. II. The incorporation of tryptophan-1-$^{15}N$,2-$^{14}C$ and geraniol-3-$^{14}C$ into quinine, J. Am. Chem. Soc. 91:2698 (1969).
29. A. R. Battersby and E. S. Hall, Biosynthesis of quinine from loganin, Chem. Commun. 194 (1970).
30. A. R. Battersby and R. J. Parry, Biosynthesis of the *Cinchona* alkaloids: Late stages of the pathway, Chem. Commun. 31 (1971).
31. N. Nagakura, G. Höfle and M. H. Zenk, Desacetylisoipecoside: The key intermediate in the biosynthesis of the alkaloids cephaeline and emetine, Chem. Commun. 896 (1978).
32. N. Nagakura, G. Höfle, D. Coggiola and M. H. Zenk, The biosynthesis of the ipecac alkaloids and of ipecoside and alangiside, Planta Med. 34:381 (1978).
33. J. Stöckigt and M. H. Zenk, Strictosidine (isovincoside): The key intermediate in the biosynthesis of monoterpenoid indole alakloids, Chem. Commun. 646 (1977).
34. J. Stöckigt and M. H. Zenk, Isovincoside (strictosidine), the key intermediate in the enzymatic formation of indole alkaloids, FEBS Lett. 79:233 (1977).
35. N. Hampp and M. H. Zenk, Homogeneous strictosidine synthase from cell suspension cultures of *Rauvolfia serpentina*, Phytochemistry 27:3811 (1988).
36. U. Pfitzner and M. H. Zenk, Homogeneous strictosidine synthase isozymes from cell suspension cultures of *Catharanthus roseus*, Planta Med. 55:525 (1989).
37. T. M. Kutchan, N. Hampp, F. Lottspeich, K. Beyreuther, and M. H. Zenk, The cDNA clone for stictosidine synthase from *Rauvolfia serpentina*: DNA sequence determination and expression in *Escherichia coli*, FEBS Lett. 237:40 (1988).
38. T. M. Kutchan, Expression of enzymatically active cloned strictosidine synthase from the higher plant *Rauvolfia serpentina* in *Escherichia coli*, FEBS Lett. 257:127 (1989).
39. T. M. Kutchan, Cloning the genes of alkaloid biosynthesis: Strictosidine synthase as prototype, in: "Norman R. Farnsworth Symposium on Natural Products Research," in press.

# THE SHIKIMATE PATHWAY'S FIRST ENZYME

Klaus M. Herrmann, José E.B.P. Pinto*, Lisa M. Weaver, and Jian-min Zhao

Department of Biochemistry
Purdue University, West Lafayette, IN 47907 U.S.A.
and *Department of Agriculture
E.S.A.L. Lavras, M.G. 37200 Brazil

## INTRODUCTION

The shikimate pathway, a major route of carbon metabolism, leads to the biosynthesis of the three aromatic amino acids phenylalanine, tyrosine, and tryptophan (Herrmann, 1983; Pittard, 1987). In some plants, more than 20% of the fixed carbon flows through this pathway, the bulk for biosynthesis of secondary metabolites such as lignins, phytoalexins, and alkaloids.

Another interesting feature of the shikimate pathway is the simultaneous requirement of two different glucose catabolic routes for the biosynthesis of aromatic amino acids. The carbon skeletons of all amino acids but histidine are derived from intermediates of glycolysis or the tricarboxylic acid cycle. In addition, four carbons of the benzene ring in aromatic amino acids originate from erythrose 4-phosphate, an intermediate of the hexose monophosphate shunt and the dark reactions of photosynthesis. Thus, high demands for aromatic compounds may influence the regulation of several catabolic and anabolic pathways that supply the necessary substrates.

3-Deoxy-D-*arabino*-heptulosonate 7-phosphate (DAHP) is the first intermediate of the shikimate pathway. This C7 compound was first described 35 years ago by David Sprinson (Srinivasan *et al.*, 1955). However, it was determined by NMR spectroscopy only six years ago that DAHP is a pure $\alpha$-anomer (Garner and Herrmann, 1984).

DAHP synthase catalyzes the condensation of phosphoenolpyruvate and erythrose 4-phosphate to DAHP. This reaction is the rate limiting step of the shikimate pathway in *E. coli* (Ogino *et al.*, 1982), and we assume this is also the case in higher plants. In bacteria, enzyme activity of DAHP synthase is modulated mainly by feedback inhibition (Ogino *et al.*, 1982), but transcriptional control by repression has also been demonstrated (Pittard, 1987; Garner and Herrmann, 1985).

Plant DAHP synthases are developmentally and environmentally regulated. The enzyme activity of DAHP synthase varies during growth of carrot and potato cells in suspension culture (Suzich *et al.*, 1984; Pinto *et al.*, 1988) and in whole potato plants with organ and age (Pinto, 1984). Wound-

*Plant Molecular Biology* 2, Edited by R.G. Herrmann and
B. Larkins, Plenum Press, New York, 1991

ing of potato tubers, leaves, and also of tomato fruit elevates DAHP synthase activity, amount, and mRNA (Dyer *et al.*, 1989). Similar results have been obtained for potato suspension culture cells treated with glyphosate (Pinto *et al.*, 1988). In this paper, we will address the expression of plant DAHP synthase in *E. coli* and the subcellular localization of the enzyme in various plant tissues.

## PROCESSING OF PRE-DAHP SYNTHASE

Isolated chloroplasts can synthesize aromatic amino acids (Schulze-Siebert and Schultz, 1989). *In vitro* synthesized EPSP synthase, the penultimate enzyme of the shikimate pathway, is transported into isolated chloroplasts (della-Cioppa *et al.*, 1986). Cell fractionation by density gradient centrifugation provided further evidence for a complete shikimate pathway in chloroplasts (Mousdale and Coggins, 1985). These results led us to assume that DAHP synthase is a plastidic enzyme.

A full length cDNA encoding DAHP synthase was obtained by screening a potato cell expression library with antibodies raised against pure DAHP synthase from potato tuber (Dyer *et al.*, 1990). The 5'-terminal 70-80 codons of the cDNA specify 26% hydroxylated amino acid residues and a net positive charge, both characteristics of previously described transit peptides (Keegstra *et al.*, 1989). These results substantiated our assumption that DAHP synthase is located in plastids.

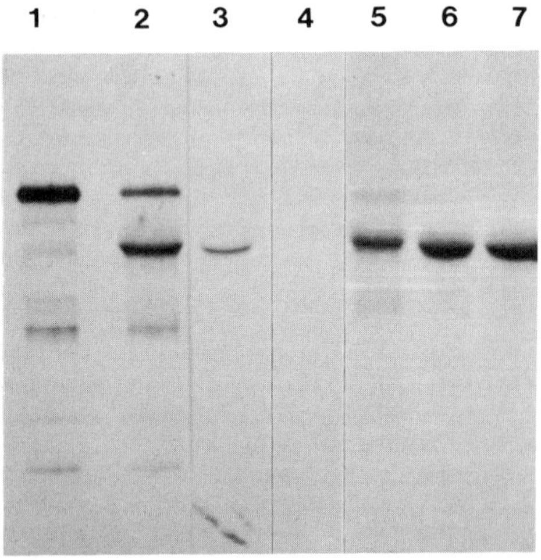


Fig. 1. Import of pre-DAHP synthase into isolated spinach chloroplasts. $^{35}S$-labelled pre-DAHP synthase was synthesized in an *in vitro* translation rabbit reticulocyte system. The pre-enzyme was incubated for 10 min with isolated spinach chloroplasts in the presence and absence of trypsin/hymorypsin. Translation products (lane 1); translation products after incubation with chloroplasts (lanes 2,5); soluble (lane 3) and membrane (lane 4) fractions of reisolated chloroplasts; total protein of untreated (lane 6) and protease treated (lane 7) reisolated chloroplasts.

The coding sequence for DAHP synthase was placed downstream of a T7 promoter. RNA was synthesized *in vitro* and used as a template for production of $^{35}S$-labelled protein. The $^{35}S$-labelled pre-DAHP synthase was incubated with chloroplasts isolated from fresh spinach leaves. The chloroplasts processed the 59 KD pre-DAHP synthase to a polypeptide of 53 KD and simultaneously protected the processed enzyme against degradation by added proteases (Fig. 1). Reisolation and fractionation of the chloroplasts indicated a stromal location for mature DAHP synthase. These experiments confirm that the cloned DAHP synthase is a plastidic protein.

## SYNTHESIS OF POTATO DAHP SYNTHASE IN *ESCHERICHIA COLI*

The coding region of the full length cDNA was placed downstream of the *E. coli* transcription/translation signal of plasmid pKK233-2 (Amann and Brosius, 1985). The resulting recombinant plasmid complements *E. coli* mutant strains devoid of DAHP synthase. Extracts of transformed cells contain potato DAHP synthase as revealed by enzyme activity measurements and immunoblots. The potato DAHP synthase made in *E. coli* was purified to electrophoretic homogeneity by conventional methods. The pure enzyme served as an antigen to raise polyclonal antibodies in rabbits.

In immunoblots we noticed one difference between the antibodies raised against the potato tuber enzyme and the antibodies raised against the cDNA product. While the latter recognized the cDNA product quite well, they were less powerful than the antibodies raised against the tuber enzyme in staining DAHP synthase from potato tuber. These blots suggested to us that the tuber enzyme and the enzyme expressed in cell suspension cultures are isoenzymes.

## PLANT DAHP SYNTHASE ISOENZYMES

The mature DAHP synthase from potato tubers has a blocked amino terminus. We digested the tuber enzyme (Pinto *et al.*, 1986) with cyanogen bromide and clostripain. Three peptides were sequenced by automated Edman degradation. The amino acid sequences confirmed the sequence deduced from the cDNA (Dyer *et al.*, 1990). However, when more proteolytic fragments of the tuber enzyme were subjected to amino acid sequence analysis, a few differences became apparent (Fig. 2). Although the study is incomplete, it represents clear evidence for isoenzymes of DAHP synthase.

```
Cells   MALSSTSTTN  SLLPNRSLVQ  NQPLLPSPLK 30

Cells   NAFFSNNSTK  TVRFVQPISA  VHSSDSNKIP 60

Cells   IVSDKPSKSS  PPAATATTAP  APAVTKTEWA 90

Cells   VDSWKSKKAL  QLPEYPNQEE  LRSVLKTIDE 120
Tuber        SLPAF  SLPEYPDKVK  LESVLDTLST

Cells   FPPIVFAGEA  RSLEERLGEA  AMGRAFLLQG 150
Tuber   YPPIVFAGEA  RSLEEKLGEA  ALGNAFLLQG
```

Fig. 2. Amino acid sequences of potato DAHP synthases. The sequence of the enzyme from cells grown in suspension culture is derived from a cDNA. The tuber enzyme sequence was determined by automated Edman degradation of suitable proteolytic fragments.

Previously, plant isoenzymes of DAHP synthase have been described as either $Mn^{2+}$ or $Co^{2+}$-dependent activities (Morris *et al.*, 1989). The two activities have been separated. However, all antibodies that we have raised so far against any DAHP synthase only recognize the $Mn^{2+}$ dependent enzyme. The $Co^{2+}$-dependent enzyme has not been characterized further.

IMMUNOCYTOCHEMISTRY

To confirm our *in vitro* import experiments, we attempted to localize DAHP synthase in chloroplasts by immunocytochemistry. Plant tissue sections were analyzed by light and electron microscopy. Thick sections from potato tuber, leaf, stem, stollen, and root were stained with primary antibodies raised in rabbits against the tuber DAHP synthase and with goat anti-rabbit gold conjugated secondary antibodies. Light microscopic examination of these sections showed a striking concentration of the enzyme in the vascular tissue of the plant (Fig. 3). Surprisingly, light microscopy suggested that chloroplasts were not stained. Electron microscopy fully confirmed these results (Fig. 4). Immunogold staining showed DAHP synthase mainly in the secondary cell walls of the xylem and phloem and at a lower concentration in the primary cell walls.

The location of the enzyme in the cell walls was unexpected, and we expended considerable effort to ensure the validity of this finding. We ran a number of controls with the following results: preimmune sera used

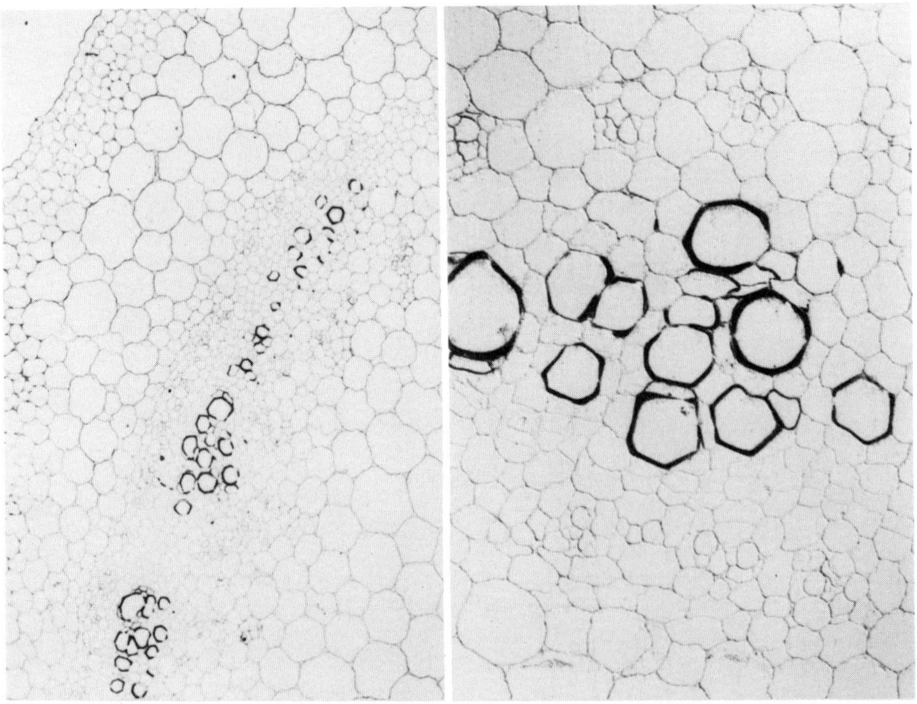

Fig. 3. Potato stem sections. Staining was with primary antibodies raised in rabbits against pure tuber DAHP synthase, with gold-conjugated goat anti-rabbit secondary antibodies, and enhanced by silver.

as primary antibodies did not stain the vascular tissue; as expected, antibodies raised against RuBisCO stained chloroplasts, not cell walls; preincubation of the primary antibodies with increasing amounts of pure cDNA encoded DAHP synthase titrated out the immunogold staining.

We repeated the immunocytochemical experiments with a high titer antibody raised against the cDNA encoded DAHP synthase. The results were essentially the same as those obtained with the antibody raised against the tuber enzyme. Collectively, these experiments demonstrate that there is a DAHP synthase in the cell walls and that the bulk of this isoenzyme is in the secondary walls of the vascular tissue.

## CONCLUSIONS AND OUTLOOK

The major DAHP synthase of potato cells grown in suspension culture is synthesized as a precursor. *In vitro* synthesized pre-DAHP synthase is processed by and transported into isolated chloroplasts. The enzyme from potato cell cultures complements *E. coli* mutants void of DAHP synthase.

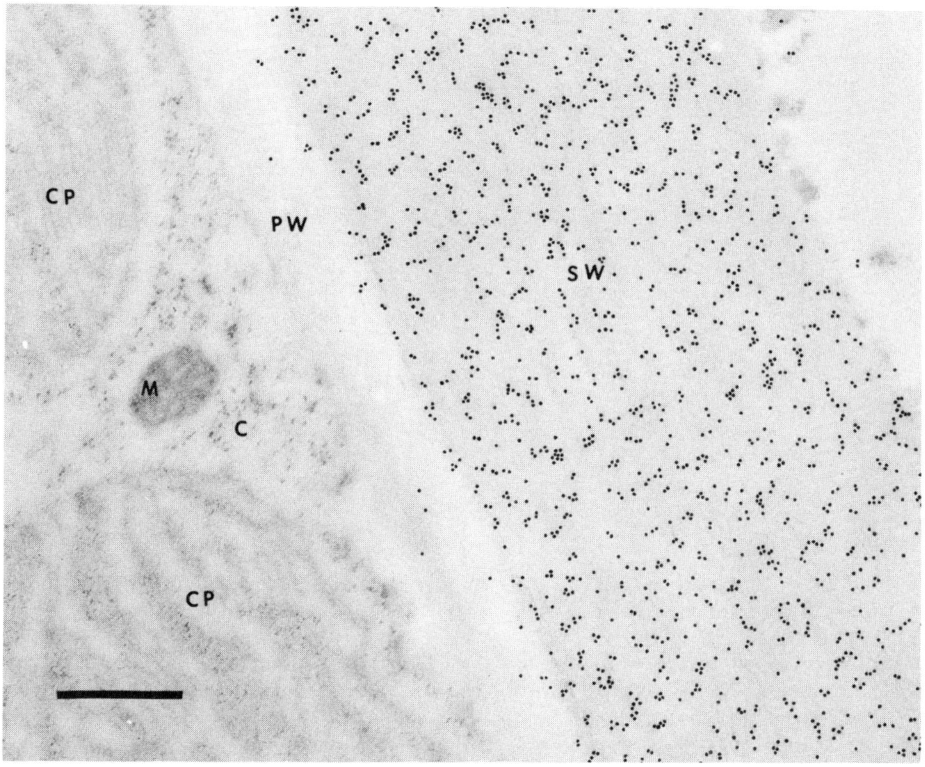


Fig. 4. Subcellular localization of a DAHP synthase in potato. A leaf section of 0.06-0.1 $\mu$m was stained with primary antibodies raised in rabbits against pure tuber DAHP synthase and with 10 nm gold-conjugated goat anti-rabbit secondary antibodies. Magnification 42,000 fold; bar = 0.5 $\mu$m; CP, chloroplasts; M, mitochondrion; C, cytoplasm; PW, primary cell wall; SW, secondary cell wall.

Immunogold staining of potato tissues demonstrates an isoenzyme of DAHP synthase in the secondary cell walls of the vascular tissue, with small amounts in the primary cell walls. Thus far, we have been unable to show by immunogold staining the presence of DAHP synthase in chloroplasts. Since potatoes have several DAHP synthase isoenzymes, the failure to localize the enzyme by immunocytochemical procedures in chloroplasts may be due to lack of antibody specificity or sufficient concentration of the enzyme. Current efforts are directed to sort out the isoenzymes, mainly through cloning and nucleotide sequence analysis of additional cDNA's. We plan to study the transport of the enzyme from the cytosol to the cell wall and the involvement of DAHP synthase in lignification, since xylem is the site of lignin biosynthesis (Rubery and Northcote, 1968).

**Acknowledgements:** We thank Dr. Gloria K. Muday and Mike Poling for preparation of $Mn^{2+}$-free $Co^{2+}$-dependent DAHP synthase, Prof. Charles E. Bracker and Dr. Craig R. Lending for much help with the microscopy, and Drs. George H. Lorimer and Tony Gatenby for antibodies against RuBisCO. This is Journal Paper No. 12487 from the Purdue University Agricultural Experiment Station.

## REFERENCES

Amann, E., and Brosius, J., 1985, ATG vectors for regulated high-level expression of cloned genes in *Escherichia coli*, *Gene* **40**:183-190.

della-Cioppa, G., Bauer, S.C., Klein, B.K., Shah, D.M., Fraley, R.T., and Kishore, G.M., 1986, Translocation of the precursor of 5-*enol*pyruvylshikimate-3-phosphate synthase into chloroplasts of higher plants *in vitro*, *Proc. Natl. Acad. Sci. U.S.A.* **83**:6873-6877.

Dyer, W.E., Henstrand, J.M., Handa, A.K., and Herrmann, K.M., 1989, Wounding induces the first enzyme of the shikimate pathway in *Solanaceae*, *Proc. Natl. Acad. Sci. U.S.A.* **86**:7370-7373.

Dyer, W.E., Weaver, L.M., Zhao, J., Kuhn, D.N., Weller, S.C., and Herrmann, K.M., 1990, A cDNA encoding 3-deoxy-D-*arabino*-heptulosonate 7-phosphate synthase from *Solanum tuberosum* L., *J. Biol. Chem.* **265**:1608-1614.

Garner, C.C and Herrmann, K.M., 1984, Structural analysis of 3-deoxy-D-*arabino*-heptulosonate 7-phosphate by $^{1}H$- and natural-abundance $^{13}C$-NMR spectroscopy, *Carbohydrate Res.* **132**:317-322.

Garner, C.C. and Herrmann, K.M., 1985, Operator mutations of the *Escherichia coli aroF* gene, *J. Biol. Chem.* **260**:3820-3825.

Herrmann, K.M., 1983, The common aromatic biosynthetic pathway, *in Amino Acids: Biosynthesis and Genetic Regulation* (Herrmann, K.M., and Somerville, R.L., eds.) pp. 301-322, Addison Wesley, Reading, MA.

Keegstra, K., Olsen, L.J., and Theg, S.M., 1989, Chloroplastic precursors and their transport across the envelope membranes, *Annu. Rev. Plant Physiol. Plant Mol. Biol.* **40**:471-501.

Morris, P.F., Doong, R.-L., and Jensen, R.A., 1989, Evidence from *Solanum tuberosum* in support of the dual pathway hypothesis of aromatic biosynthesis, *Plant Physiol.* **89**:10-14.

Mousdale, D.M. and Coggins, J.R., 1985, Subcellular localizaiton of the common shikimate pathway enzymes in *Pisum sativum* L., *Planta* **163**:241-249.

Ogino, T., Garner, C., Markley, J.L., and Herrmann, K.M., 1982, Biosynthesis of aromatic compounds: $^{13}C$ NMR spectroscopy of whole *Escherichia coli* cells, *Proc. Natl. Acad. Sci. U.S.A.* **79**:5828-5832.

Pinto, J.E.B.P., 1984, The first enzyme of the shikimate pathway from potato (*Solanum tuberosum* L. cv. superior) *Ph.D. Thesis*, Purdue University.

Pinto, J.E.B.P., Dyer, W.E., Weller, S.C., and Herrmann, K.M., 1988, Glyphosate induces 3-deoxy-D-*arabino*-heptulosonate 7-phosphate synthase in potato (*Solanum tuberosum* L.) cells grown in suspension culture, *Plant Physiol.* **87**:891-893.

Pinto, J.E.B.P., Suzich, J.A., and Herrmann, K.M., 1986, 3-Deoxy-D-*arabino*-heptulosonate 7-phosphate synthase from potato tuber (*Solanum tubersoum* L.), *Plant Physiol.* **82**:1040-1044.

Pittard, A.J., 1987, Biosynthesis of the aromatic amino acids, *in Escherichia coli and Salmonella typhimurium* (Neidhardt, F.C., ed.) pp 368-394, American Society of Microbiology, Washington, D.C.

Rubery, P.H., and Northcote, D.H., 1968, Site of phenylalanine ammonia-lyase activity and synthesis of lignin during xylem differentiation, *Nature* (London) **219**:1230-1234.

Schulze-Siebert, D. and Schultz, G., 1989, Formation of aromatic amino acids and valine from $^{14}CO_2$ or 3-[U-$^{14}$C] phosphoglycerate by isolated intact spinach chloroplasts. Evidence for a chloroplastic 3-phosphoglycerate → 2 phosphoglycerate → phosphoenolpyruvate → pyruvate pathway, *Plant Science* **59**:167-174.

Srinivasan, P.R., Katagiri, M., and Sprinson, D.B., 1955, The enzymatic synthesis of shikimic acid from D-erythose-4-phosphate and phosphoenolpyruvate, *J. Amer. Chem. Soc.* **77**:4943-4944.

Suzich, J.A., Ranjeva, R., Hasegawa, P.M., and Herrmann, K.M., 1984, Regulation of the shikimate pathway of carrot cells in suspension culture, *Plant Physiol.* **75**:369-371.

# ACC SYNTHASE GENES IN ZUCCHINI AND TOMATO[1]

Gary Peter, William Rottmann, Takahide Sato[2], Pung-Ling Huang, Paul Oeller, Julie Keller, Jean Murphy, and Athanasios Theologis

Plant Gene Expression Center, USDA-ARS; 800 Buchanan St., Albany, CA, 94710; and Dept. of Plant Biology, Univ. of Calif., Berkeley, 94720 USA

## Abstract

Ethylene is considered to be the plant hormone that regulates fruit ripening. The rate limiting step in its synthesis is the formation of the ethylene precursor 1-aminocyclopropane-1-carboxylic acid (ACC) from S-adenosylmethionine (SAM), catalyzed by ACC synthase. We have isolated a complementary DNA sequence (pACC1) encoding ACC synthase from zucchini fruits (*Cucurbita pepo*) by a novel experimental approach. The authenticity of the cDNA clone has been confirmed by expressing it and recovering ACC synthase activity in *E. coli*. Isolation of genomic sequences reveals the presence of two linked ACC synthase genes in the *Cucurbita* genome. Using the zucchini pACC1 as a probe, cDNAs and the corresponding gene were isolated from tomato (*Lycopersicon esculentum*). Its structure and expression have been characterized.

## Introduction

Ethylene is a plant hormone that influences many aspects of plant growth and development (Abeles, 1973). Ethylene production is augmented during several stages of plant growth, including fruit ripening, seed germination, leaf and flower senescence, and abscission (Abeles, 1973). It is also induced by a variety of external factors, including the application of auxins, wounding, anaerobiosis, viral infection, elicitor treatment, chilling injury, drought, $Cd^+$ and $Li^+$ ions (Yang and Hoffman, 1984).

The ethylene biosynthetic pathway was elucidated by the pioneering work of Yang and his associates (Yang and Hoffman, 1984). They established that methionine is the biological precursor of ethylene in all higher plants (Young and Hoffman, 1984) and is converted to ethylene via the following biosynthetic route: Methionine → S-adenosylmethionine → ACC → $C_2H_4$. The rate limiting step is the conversion of SAM to ACC, catalyzed by the enzyme ACC synthase. Induction of ethylene synthesis by a variety of agents is due to *de novo* synthesis of the enzyme (Bleecker *et. al.*, 1988). It has been partially purified from several sources but with wide discrepancies in its molecular weight reported (e.g., Bleecker *et al.*, 1986; Privelle and Graham, 1987; Mehta *et al.*, 1988; Nakajima *et al.*, 1988; Tsai *et al.*, 1988).

---

[1]This work was supported by grants to A.T. from NSF (DCB-8421167, 8819129, 8916286); NIH (GM-35447); and USDA (5835-23410-D002).

[2]Present address: Faculty of Horticulture, Chiba University, 648 Matsudo; MATSUDO 271, Japan.

Since the enzyme is unstable and present in low abundance, the enzyme has been difficult to purify to homogeneity (Bleecker et al., 1986). Here, we report the isolation of a complementary DNA sequence to ACC synthase mRNA from zucchini fruit (*Cucurbita pepo*) using a strategy that requires only partially purified enzyme. Using this cDNA, genomic sequences to ACC synthase from *Cucurbita* were isolated and structurally characterized. Furthermore, cDNA and genomic clones to a tomato ACC synthase have also been isolated and structurally characterized. These tomato sequences are induced by ethylene and during tomato fruit ripening

## Materials and Methods

### Plant Material

Slices 1 mm thick were prepared from zucchini fruits (*Cucurbita pepo*) and incubated for 24 hrs in an induction medium (Sato and Theologis, 1989). [$KPO_4$ buffer 50 mM, pH 6.8 – indoleacetic acid (IAA) 0.5 mM – benzyladenine (BA) 0.1 mM – LiCl 50 mM – amino-oxyacetic acid (AOA) 0.6 mM – chloramphenicol 50 $\mu$g/ml]. Tomato fruits (*Lycopersicon esculentum*) were obtained from a local supermarket or greenhouse grown plants.

### Purification of ACC Synthase

The purification of the enzyme is described in detail elsewhere (Sato and Theologis, 1989). The specific activity of the purified enzyme is 35,000 nmoles/hr/mg protein, representing a 6000-fold purification.

### Antibody Production and Purification

A New Zealand white rabbit was immunized at 3-week intervals with 5000 nmoles/hr ACC synthase activity (sp. act. 9000 nmoles ACC/hr/mgr protein). Crude antiserum was purified by incubation with Sepharose *4B* coupled with soluble proteins from intact zucchini fruit (Sato and Theologis, 1989).

### cDNA Library Construction and Immunoscreening

Poly(A)$^+$-RNA from induced zucchini tissue was used for constructing a cDNA library in $\lambda$gt11. The library was plated at a density of $3 \times 10^4$ pfu/90 mm petri dish and was screened with purified ACC synthase antiserum essentially as described by Huynh *et al.* (1985). In addition a tomato cDNA library was constructed in $\lambda$gt10 with poly(A)$^+$-RNA isolated from a ripening tomato fruit. The library was screened at low stringency with the zucchini cDNA pACC1 (Sato and Theologis, 1989).

### Expression of Zucchini ACC Synthase in *E. coli* – $\lambda$gt11 Lysogens

*E. coli* strain Y1089 was lysogenized with appropriate $\lambda$gt11 recombinant phages (Huynh *et al.*, 1985). Lysogens were grown in LB media containing 50 $\mu$gr/ml carbenicillin at 30°C for 2.5 hr ($Ab_{600}$ = 0.5); the temperature was shifted to 42°C for 20 minutes and IPTG was added (final concentration of 1 mM); and the culture grown for an additional 1.5 hr at 37°C. For SDS-PAGE analysis of the fusion proteins, three mls of the ± IPTG cultures were centrifuged at 14,000 × g for 1 min at 4°C. The bacterial pellet was resuspended in 120 $\mu$l SDS loading buffer [16% SDS – 10% sucrose – 2mM EDTA – 62.5 mM Trizma base – 0.9M $\beta$-mercaptoethanol – 0.0125% bromophenol blue] and then denatured at 100°C for 3 min. The proteins were separated on a 7.5% gel (Laemmli, 1970), 35 $\mu$l (~250 $\mu$g) were loaded per lane on the gel stained with Commassie blue and 4 $\mu$l (~30 $\mu$g) per lane were loaded on the gel used for immunoblotting (Burnette, 1981). Bacterial pellets of 100 mls of culture were frozen in liquid $N_2$ and kept at -80°C for 2 hrs. The cells were resuspended in

5 mls of Buffer A [Tris-HCl 100 mM, pH 8.0 – EDTA 20 mM – pyridoxal phosphate 10 $\mu$M – PMSF 0.5 mM – $\beta$-ME 20 mM]; sonicated 5 times (3 sec each); centrifuged at 16,000 × g for 25 min at 4°C; and the ACC synthase activity assayed (Lisada and Yang, 1979).

Expression of the Tomato ACC Synthase in *E. coli*

The open reading frame of the ptACC1 cDNA was fused in the correct orientation to the first ten amino acids of the alpha peptide of pUC19; the opposite orientation was cloned into pUC18. *E. coli* cells were grown to 0.1 $OD_{600}$ then either induced (+) or not (-) with 1.0 mM IPTG. At 0, 45, 90, 135, 180 minutes aliquots were taken, the cells pelleted and 15 $\mu$l of the supernatant was assayed for ACC (Lisada and Yang, 1979).

Isolation of Genomic Sequences

Genomic libraries were constructed with Sau3A partially digested zucchini or tomato DNA into the EMBL3 $\lambda$-cloning vector. Plaque filter hybridization with $P^{32}$-labeled pACC1 (zucchini) and ptACC1 (tomato) inserts allowed the isolation of recombinant clones containing ACC synthase genomic sequences (Davis *et al.*, 1980).

**Results**

Cloning Strategy

The cloning of ACC synthase mRNA has been difficult because traditional cloning approaches require that the protein be purified and at least partially sequenced. Unfortunately, ACC synthase has been difficult to purify to homogeneity because of its lability and low abundance [e.g., 0.0001% of the total protein in ripe tomatoes (Bleecker *et al.*, 1986)]. Our cloning strategy shown in Figure 1 does not demand that the protein be pure, only *inducible*.

The experimental approach consists of seven main steps: (1) identifying a highly inducible system; (2) partially purifying the ACC synthase from induced tissue by conventional protein purification procedures; (3) use of the partially purified enzyme to produce an antiserum that inhibits ACC synthase activity; (4) purifying the crude antiserum by affinity chromatography with total proteins from *uninduced* tissue; (5) immunoscreening $\lambda$gt11 cDNA libraries with the purified antiserum and isolating putative cDNA clones to ACC synthase; (6) identifying the cDNA clones by immunoblotting analysis of plant extracts with antibodies released from putative $\lambda$gt11 cDNA clones; and finally (7) by expressing and recovering ACC synthase activity in *E. coli* using putative ACC synthase cDNA clones.

The Inducible System

An extensive survey of several plants was conducted to identify one with highly inducible ACC synthase for partial purification of the enzyme. Thin slices of zucchini fruit tissue treated with indoleacetic acid (IAA) + benzyladenine (BA) + LiCl + aminooxyacetic acid (AOA) were found to develop higher levels of enzyme activity (15-20 nmoles/ACC/hr/gr fr wt). One of us (T. Sato) was able to devise a rapid and efficient protein purification scheme that led to 6000-fold purification of the enzyme (Sato and Theologis, 1990).

Antibody Production and Purification

Polyclonal antibodies were raised with 1500-fold purified ACC synthase preparation. The antiserum inhibits the enzyme activity and recognizes numerous polypeptides in highly impure enzyme preparation but only two polypeptides, molecular weight of 46 and 67 kd, in highly purified enzyme preparation (Sato and Theologis, 1989). The crude antiserum was purified by passing it through a Sepharose *4B* column containing total proteins from *uninduced* tissue to remove all antibodies to uninduced proteins present in the initial enzyme preparation. The

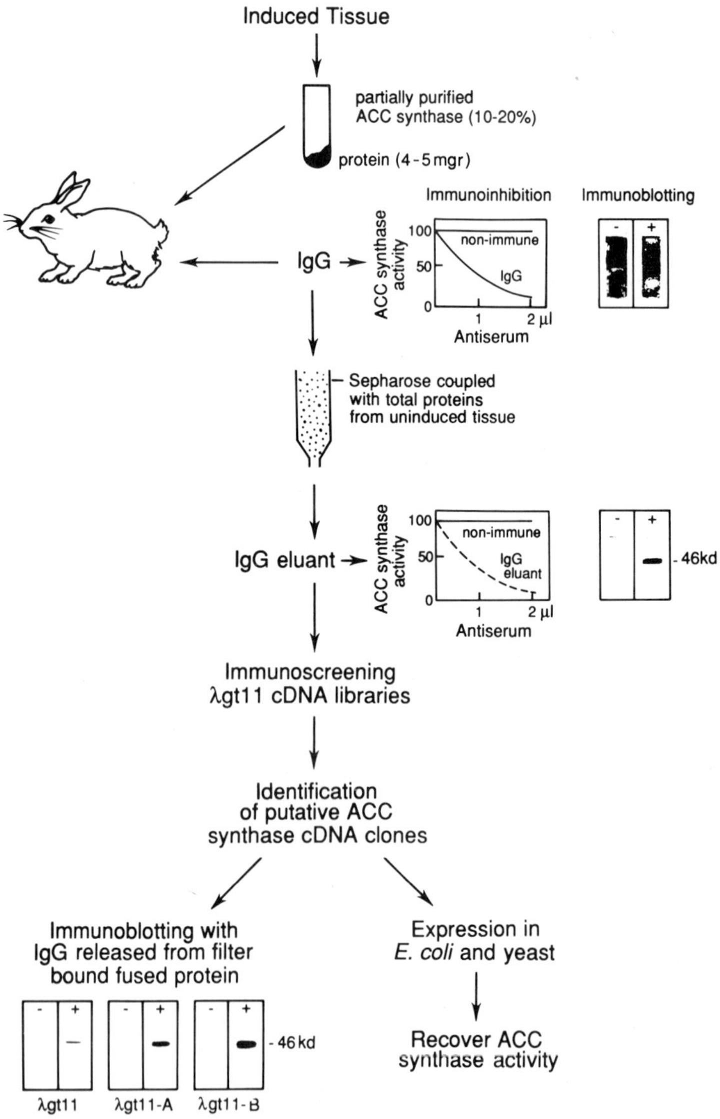


Fig. 1. Strategy for cloning the mRNA of zucchini ACC synthase.

column eluate is able to immunoinhibit ACC synthase activity with the same intensity as the non-purified antiserum (Sato and Theologis, 1989). The purified antiserum recognizes only two polypeptides, predominantly a 46 kd and weakly a 67 kd species in both crude and highly purified enzyme preparation.

The specificity of the antibody was tested by its ability to recognize an inducible polypeptide in (a) total protein extracts and (b) in *in vitro* translation products. Immunoblotting analysis reveals that the 46 kd is inducible whereas the 67 kd is constitutively expressed. The 46 kd polypeptide has been identified to be the ACC synthase subunit by disk gel electrophoresis followed by SDS-PAGE (Sato and Theologis, 1990). The molecular weight of the native enzyme is about 86 kd, suggesting that it is a dimer of two identical subunits, 46 kd each (Sato and Theologis, 1990). Furthermore, the purified antiserum immunoprecipitates an inducible *in vitro* translation product coding for a 53 kd polypeptide, suggesting that the ACC synthase is made as a larger precursor (Sato and Theologis, 1990).

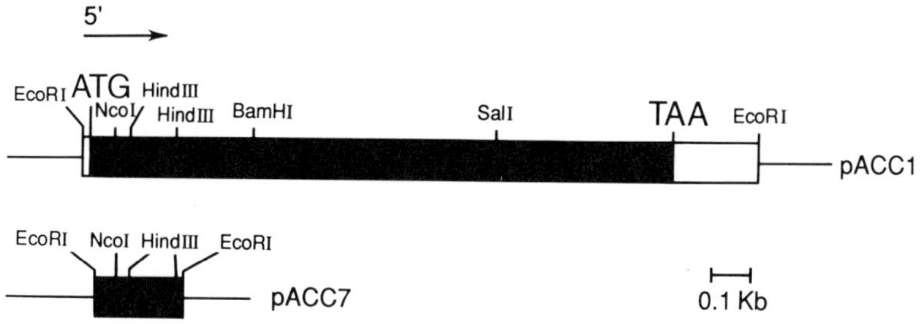


Fig. 2. Partial restriction analysis of two zucchini ACC synthase cDNA clones pACC1 and pACC7.

Immunoscreening λgt11 cDNA Libraries

Sixty-six immunoreactive clones were isolated by screening $1.4 \times 10^5$ λgt11 recombinant clones with the purified antiserum. Upon rescreening, only 30 were truly positive. Figure 2 shows a partial restriction map of two cDNA clones, pACC1 and pACC7. Clone pACC1 has an open reading frame coding for a 55.8 kd polypeptide. Clone pACC7 is a small cDNA clone (220 bp) that gave the strongest signal during immunoscreening and was used for antibody selection.

Expression of Zucchini ACC Synthase in *E. coli*

Expression experiments carried out in *E. coli* prove the authenticity of the ACC synthase cDNA clones. ACC synthase activity is present in *E. coli* lysogenized with λgt11-ACC1 (Table 1). The expression of enzyme activity depends on the presence of IPTG and on the correct orientation of the cDNA insert (Table 1). [The low levels of expression in the absence of IPTG are presumably due to insufficient amounts of *lac* repressor in strain Y1089 (Brosius, 1988)]. Formation of ACC is also detected in the lysates of λgt11-ACC1 lysogens (Table 1). Enzymatic activity was not detected in lysogens of the smaller cDNA clone λgt11-ACC7 or the non ACC synthase cDNA clone λgt11-W7.

Immunoblotting analysis of crude lysogen lysates with purified antiserum shows that λgt11-ACC1 lysogen accumulates a hybrid protein Mr 170,000 (Figure 3). Since low molecular weight degradation products were never detected, the enzymatic activity is most likely associated with the 170 kd chimeric protein, as has previously been observed for other fusion proteins (Kaufman *et al.*, 1986; Marullo *et al.*, 1988).

## Gene Organization of ACC Synthase in *Cucurbita*

Southern analysis of nuclear DNA with the cDNA clone pACC1 shows strong hybridization to a 14 kb BamHI fragment and to two EcoRI fragments, 9 and 7 kb in size,

TABLE 1

*Expression of the Zucchini ACC Synthase in E. coli*

| Lysogen | Size of cDNA | Orientation | Enzyme Activity -IPTG | Enzyme Activity +IPTG |
|---|---|---|---|---|
| | kb | | $nMoles \cdot mgr^{-1} \cdot hr^{-1}$ | |
| λgt11 | — | — | 0 | 0 |
| λgt11-ACC1 | 1.7 | correct | 6 | 84 |
| λgt11-ACC1 | 1.7 | opposite | 0 | 0 |
| λgt11-ACC7 | 0.2 | correct | 0 | 0 |
| λgt11-W7 | 1.7 | correct | 0 | 0 |

IPTG concentration: 1 mM

suggesting the presence of one or two gene copies (Sato and Theologis, 1989). This conclusion has been recently confirmed by the isolation of the genomic clone containing two ACC synthase genes separated by a 5.5 kb intergenic region (Figure 4). DNA sequence analysis reveals that a high degree of DNA homology (95%) exists between the two genes in their exon and introns; however, they differ in their 3' untranslated and 5' nontranscribed regions.

The pACC1 mRNA cDNA is the transcriptional product of the left gene (Huang and Theologis, unpublished data). It is not yet known when and how (type of inducer) the right gene is transcriptionally activated. The possibility exists that the left gene is induced by various hormones and environmental stimuli whereas the right gene is developmentally regulated.

## Isolation and Characterization of a Tomato ACC Synthase Gene

In order to study how ACC synthase gene expression is regulated during tomato fruit ripening, a cDNA library was constructed with mRNA from ripening tomato fruit. This library was screened at low stringency with the zucchini cDNA pACC1, and a number of clones were isolated. The longest cDNA, ptACC1 (1.8 kb), contains an open reading frame that encodes a 54 kDa polypeptide which is 64% homologous to the polypeptide encoded by the zucchini cDNA. The authenticity of the tomato cDNA was confirmed by expression in *E. coli* (Fig. 5). When the cDNA is expressed from the *lac* promoter of pUC in the correct orientation (sense), ACC is readily detected in the media; whereas no ACC is detected when it is expressed in the incorrect orientation (antisense). Structural characterization of genomic

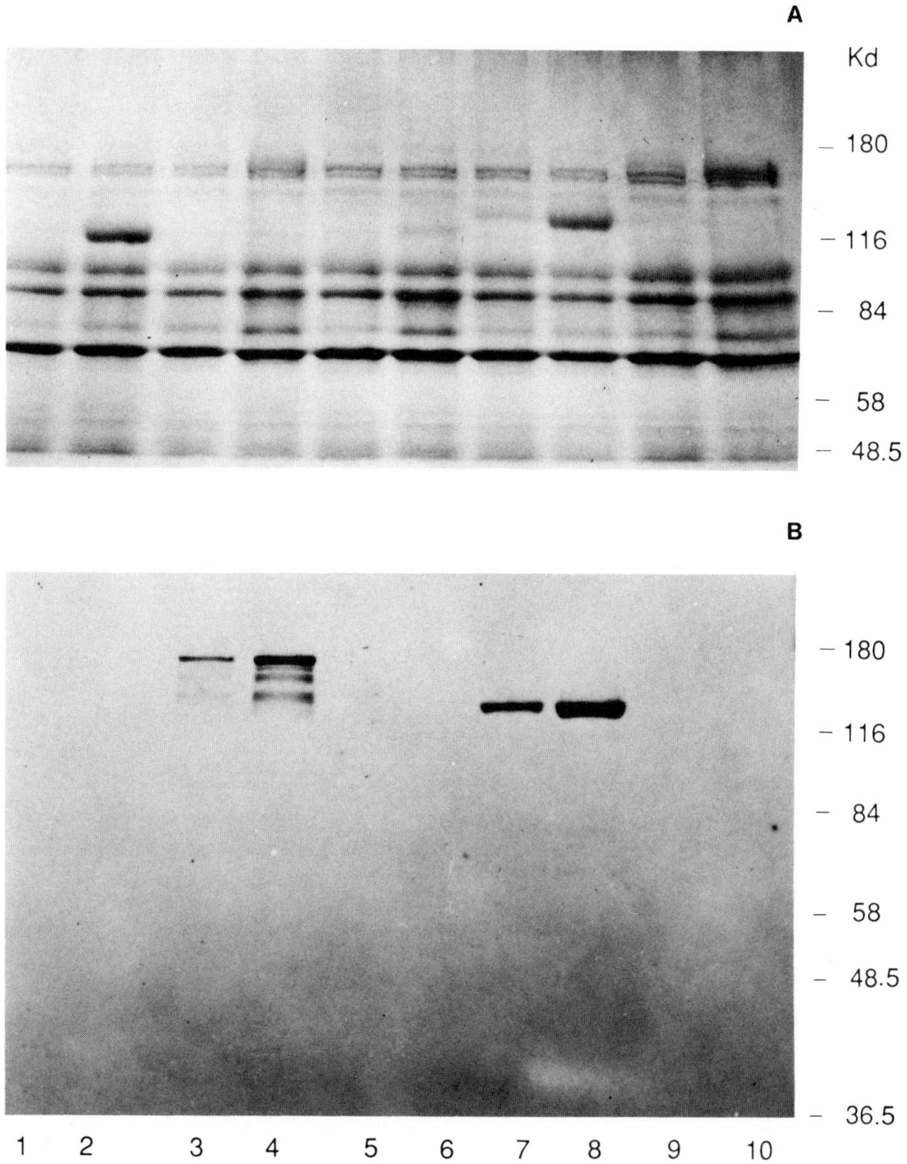


Fig. 3. Expression of zucchini ACC Synthase in *E. coli*. Expression of λgt11 recombinant lysogens. **A**: Commassie blue stained SDS-PAGE gel of crude protein extracts from induced recombinant λgt11 lysogens. **B**: Immunoblotting analysis of a replica gel as shown in **A**, with antibody selected by recombinant clone λgt11-ACC1. Lanes 1 and 2 are λgt11; 3 and 4, λgt11-ACC1 (correct orientation); 5 and 6, λgt11-ACC1 (opposite orientation); 7 and 8, λgt11-ACC7 (correct orientation); 9 and 10, λgt11-W7. The odd numbered lanes are -IPTG whereas the even numbered lanes are +IPTG.

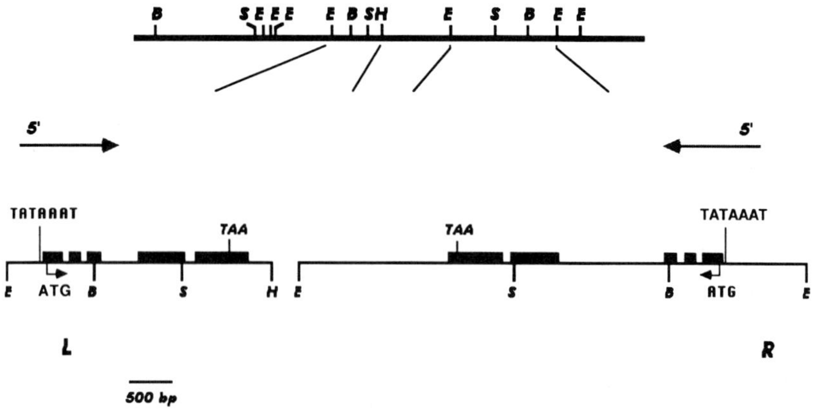


Fig. 4. Gene organization of ACC Synthase in *Cucurbita*. A partial restriction map is shown above. Below indicates the gene structures; the dark boxes are exons and the lines are introns, and the directions of transcription are indicated by the arrows.

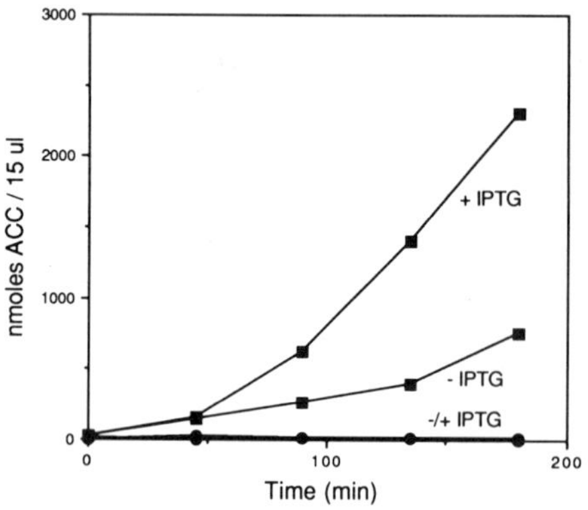


Fig. 5. Expression of the Tomato ACC synthase in *E. coli*. The orientation of cDNA is sense (■–■–■) and antisense (•–•–•).

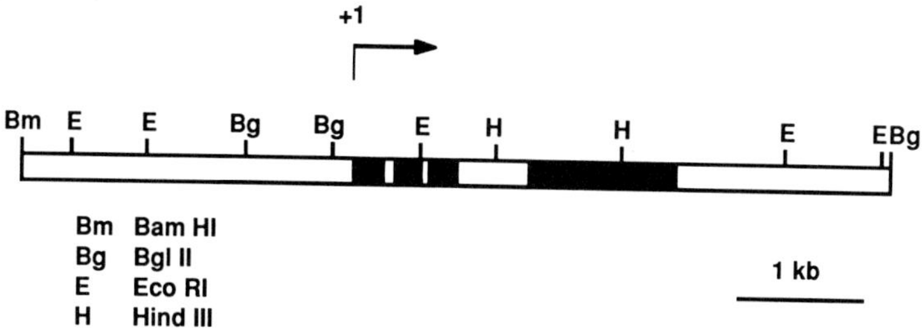


Fig. 6. Gene Organization of ACC Synthase in *Lycopersicon esculentum*. The exons are depicted in black and the arrow indicates the start site and direction of transcription.

clones show that the tomato gene LEACC1 contains 4 exons, 3 introns (Figure 6), and is present in a single copy. Studies of LEACC1 expression indicate that it gives rise to a 1900 nucleotide long mRNA that is developmentally regulated. The mRNA is undetectable during the early stages of floral organ development. It is first observed slightly before the onset and increases dramatically during senescence of anthers, petals and fruits. Studies during fruit ripening show that ptACC1 mRNA levels parallel the rate of ethylene produced throughout the ripening process and peak at the orange stage. Treatment of mature green fruits with 10 $\mu$l/l ethylene show a rapid increase in mRNA levels (Rottmann, Peter, and Theologis, unpublished data).

## Discussion

We have described the isolation of a cDNA encoding the ACC synthase in zucchini fruits. Isolation of the cDNA did not require the isolation and purification of ACC synthase to homogeneity which would have been difficult because of its low abundance.

The *Cucurbita* ACC synthase activity is associated with a 46 kd polypeptide similar in size to that isolated from tomato (Bleecker *et al.*, 1986; Privelle and Graham, 1987) and winter squash (Nakajima *et al.*, 1988). The protein constitutes 0.01% of the total protein in the induced state. The polypeptide is coded by a mRNA 1900 nucleotides long that is greatly induced by IAA + BA + LiCl + AOA (Sato and Theologis, 1990). The induction of ACC synthase mRNA by IAA, wounding, and $Li^+$ ions may be due to transcriptional activation since IAA and wounding have been shown to transcriptionally activate other plant genes (Theologis, 1986; Thornburg *et al.*, 1987). However, posttranscriptional induction is also possible. The *Cucurbita* ACC synthase is encoded by two highly homologous genes whose expression is regulated by different inducers. It is a task of the future to determine when and how the two genes are transcriptionally activated.

Using the zucchini cDNA as a probe we have isolated and characterized a tomato ACC synthase gene which is expressed during senescence of anthers, petals and fruits. Although the nucleic acid and amino acid sequences show somewhat low homology, the sizes of the transcripts and the sizes and positions of the introns are conserved. However, the tomato gene lacks the fourth intron that is present in the zucchini genes (compare Fig. 6 with Fig. 1). The cloning of the ACC synthase mRNA provides the potential to manipulate endogenous ethylene production and consequently to prevent senescence. This is of paramount importance in the agricultural industry where billions of dollars are lost worldwide each year due to overripening of fruits and vegetables during transportation from one locale to another. It may be possible to reduce ethylene production by reducing expression of the ACC synthase gene. This could potentially be accomplished by a variety of approaches including: gene disruption (Scherer and Davis, 1979); antisense RNA technology (Ecker and

Davis, 1986); dominant negative mutation (Herskowitz, 1987); expression of antibody genes (Carlson, 1988); or with anti-gene ribozyme (Haselhoff and Gerlach, 1988). All approaches will require the use of strong inducible and tissue specific promoters.

### References


Abeles, F. B., 1973, *Ethylene in Plant Biology*, Academic Press, New York.

Bleecker, A. B., Kenyon, W. H., Somerville, S. C. and Kende, H., 1986, Use of monoclonal antibodies in the purification and characterization of 1-aminocyclopropane-1-carboxylate synthase, an enzyme in ethylene biosynthesis. *Proc. Natl. Acad. Sci. USA*, **83**:7755.

Bleecker, A. B., Robinson, G. and Kende, H., 1988, Studies on the relation of 1-aminocyclopropane-1-carboxylate synthase in tomato using monoclonal antibodies. *Planta*, **173**:385.

Brosius, J., 1988, A survey of molecular cloning vectors and their uses. *in: Vectors*, R. L. Rodriguez and D. T. Denhardts, eds., p. 205, Butterworths, Stoneham MA.

Burnette, W. N., 1981, 'Western Blotting': Electrophoretic transfer of proteins from sodium dodecyl sulfate-polyacrylamide gels to unmodified nitrocellulose and radiographic detection with antibody and radioiodinated protein A. *Anal. Biochem.*, **112**:195.

Carlson, J. R., 1988, A new means of inducibly inactivating a cellular protein. *Mol. Cell. Biol.*, **8**:2638.

Davis, R. W., Botstein, D. and Roth, J. R., 1980, *Advanced Bacterial Genetics*. Cold Spring Harbor Laboratory, Cold Spring Harbor, NY.

Ecker, J. R. and Davis, R. W., 1986, Inhibition of gene expression in plant cells by expression of antisense RNA. *Proc. Natl. Acad. Sci. USA*, **83**:5372.

Haselhoff, J. and Gerlach, W. L., 1988, Simple RNA enzymes with new and highly specific endoribonuclease activities. *Nature*, **334**:585.

Herskowitz, I., 1987, Functional inactivation of genes by dominant negative mutations. *Nature*, **329**:219.

Huynh, T. V., Young, R. A. and Davis, R. W., 1985, Construction and screening cDNA libraries in λgt10 and λgt11, *in: DNA Cloning Techniques: A Practical Approach*, D. Glover, ed., p. 49, IRL Press, Oxford.

Kaufman, D. L., McGinnis, J. F., Krieger, N. R. and Tobin, A. L., 1986, Brain glutamate decarboxylase cloned in λgt11: fusion protein produces γ-aminobutyric acid. *Science*, **232**:1138.

Laemmli, U. K., 1970, Cleavage of structural proteins during the assembly of the head of bacteriophage T4. *Nature*, **227**:680.

Lisada, M. C. C. and Yang, S. F., 1979, A simple and sensitive assay for 1-aminocyclopropane-1-carboxylic acid. *Anal. Biochem.*, **100**:140.

Marullo, S., Klutchko, C. D., Eshdat, Y., Strosberg, D. A. and Emorine, L., 1988, Human β-adrenergic receptors expressed in *Escherichia coli* membranes retain their pharmacological properties. *Proc. Natl. Acad. Sci. USA*, **85**:7551.

Mehta, A. M., Jordan, R. L., Anderson, J. D. and Mattoo, A. K., 1988, Identification of a unique isoform of 1-aminocyclopropane-1-carboxylic acid synthase by monoclonal antibody. *Proc. Natl. Acad. Sci. USA*, **85**:8810.

Nakajima, N., Nakagawa, N. and Imaseki, H., 1988, Molecular size of wound-induced 1-aminocyclopropane-1-carboxylate synthase from *Cucurbita maxima* Dutch and change of translatable mRNA of the enzyme after wounding. *Plant Cell Physiol.*, **29**:989.

Privelle, L. S. and Graham, J. S., 1987, Radiolabeling of a wound-inducible pyridoxal phosphate utilizing enzyme: evidence for its identification as ACC synthase. *Arch. Biochem. Biophy.*, **253**:333.

Sato, T. and Theologis A., 1989, Cloning the mRNA encoding 1-aminocyclopropane-1-carboxylate synthase, the key enzyme for ethylene biosynthesis in plants. *Proc. Natl. Acad. Sci. USA* **86**:6621.

Sato, T. and Theologis A., 1990, The ACC synthase from *Cucurbita*. *J. Biol. Chem.*, submitted.

Scherer, S. and Davis, R. W., 1979, Replacement of chromosome segments with altered DNA sequences constructed *in vitro*. *Proc. Natl. Acad. Sci. USA*, **76**:4951.

Theologis, A., 1986, Rapid gene regulation by auxin. *Ann. Rev. Plant Physiol.*, **37**:407.

Thornburg, R. W., An, G., Cleveland, T. E., Johnson, R. and Ryan, C. A., 1987, Wound-inducible expression of a potato inhibitor II-chloramphenicol acetyltransferase gene fusion in transgenic tobacco plants. *Proc. Natl. Acad. Sci. USA*, **84**:744.

Tsai, D. S., Arteca, R. N., Bechman, J. M. and Phillips, A. T., 1988, Purification and characterization of 1-aminocyclopropane-1-carboxylate synthase from etiolated mung bean hypotyls. *Arch. Biochem. Biophys.*, **264**:632.

Yang, S. F. and Hoffman, N. E., 1984, Ethylene biosynthesis and its relation in higher plants. *Ann. Rev. Plant. Physiol.* **35**:155.

# MOLECULAR ANALYSIS OF ANTHOCYANIN GENES OF *ZEA MAYS*

Udo Wienand, Brian Scheffler[1], Philipp Franken, Andreas Schrell, Ursula Niesbach-Klösgen[2], Elvira Tapp, Javier Paz-Ares[3] and Heinz Saedler

Max-Planck-Institut für Züchtungsforschung, D-5000 Köln 30, FRG

[1] Inst. Plant Science Res. Cambridge, GB

[2] Max-Plank-Institut für Biochemie, D-8000 München, FRG

[3] Centro Invest. Biologicas, 28006, Madrid, Spain

## INTRODUCTION

The colorful pigmentation of plant tissues like flowers, fruits or seeds is due to the synthesis of anthocyanins, a subclass of flavonoids. The genetical and molecular analysis of anthocyanin biosynthesis in a number of plants, predominantly in *Zea mays* (maize) and *Antirrhinum majus*, showed that anthocyanins are sequentially synthesized from small precursor molecules. The synthesis of anthocyanins occurs in a well organized coordinated fashion. In maize, about a dozen regulatory and structural genes interact, which finally results in tissue-specific synthesis of pigments. Recently several anthocyanin genes of maize have been cloned and are under molecular investigation.

## ANALYSIS OF ANTHOCYANIN GENES

### 1. Genetical analysis of structural and regulatory anthocyanin genes in maize

Structural genes are directly involved in color formation by encoding enzymes which sequentially synthesize or modify the high molecular weight anthocyanins. Such genes in maize are *c2*, *a1*, *a2*, *bz1* and *bz2* (Styles and Ceska, 1977; for review: Coe et al., 1988). Regulatory genes have been identified to control the action of structural genes. Among those are genes which are tissue specifically expressed and thus distribute anthocyanin biosynthesis into various tissues of the plant. Two examples are *C1* and *Pl*. *C1* regulates the expression of anthocyanin only in kernels,

*Plant Molecular Biology 2*, Edited by R.G. Herrmann and B. Larkins, Plenum Press, New York, 1991

whereas the genetically homologous *Pl* locus acts in other tissues of maize. Further regulatory genes are *R1* and *B*. *R1* and *B* are rather complex loci and various alleles of both genes influence anthocyanin production in different plant tissues (for review: Coe et al., 1988). Beside these regulatory genes there are some, which generally intensify anthocyanin biosynthesis, like *In* (Reddy and Peterson, 1978) and *A3* (Styles and Coe, 1986) or which have a pleiotropic effect like *Vp1* (Dooner, 1985). *Vp1* for example is not only necessary for color formation in the kernel but also influences dormancy of the embryo as well as endosperm development of the maize kernel.

For pigmentation it needs the coordinate expression of several regulatory genes in the same tissue to initiate anthocyanin biosynthesis. In the case of kernel pigmentation *C1*, *R1* and *Vp1* must be functional. The regulatory nature of genes like *C1*, *R1* and *B* has been concluded from the fact, that the expression of (at least) three structural genes of the anthocyanin pathway is affected in the absence of the function of each of these regulatory genes. Transcripts of the genes *C2*, *A1* and *Bz1* are no longer detectable in a background where either of the regulatory genes is not expressed (Cone et al., 1986; Ludwig et al., 1989; Chandler et al., 1989).

## 2. Isolation of anthocyanin genes in maize

The molecular analysis of genes involved in this pathway is of particular interest, since it could provide further data on the structure of plant genes in general and more important on the interaction between regulatory and structural loci. Cloning of a number of these genes became possible using the transposon tagging strategy (for review: Wienand and Saedler, 1987; Döring, 1989). In such a procedure transposable element (TE) induced mutants of the particular gene of interest are used in the isolation procedure. This technique requires, that the element inducing the mutation has been cloned molecularly. Sequences of that element can then be used as specific probes to isolate fragments from a library constructed from the mutant of interest. One or several clones isolated from such library should, in addition to transposable element specific sequences, contain part of the gene of interest.

So far three maize transposable elements have been sucessfully used in the isolation of anthocyanin genes. This is the *Ac* element (or its *Ds* derivatives) (Fedoroff et al., 1983), the *En1/Spm* element (Pereira et al., 1985) and the *Mu1* element (Barker et al., 1984). These elements differ in size and sequence and do not cross hybridize. This is a very important feature, because cross hybridization would considerably complicate the molecular cloning of the genes. So far eight anthocyanin genes have been cloned using the transposon tagging strategy (see table 1) and four of those are regulatory genes, demonstrating the powerful potential of this procedure.

Table 1. Maize anthocyanin genes isolated by transposon tagging

| Gene | Function | TE | Reference |
|---|---|---|---|
| *C1* | regulatory | *En1* | Paz-Ares et al.,1986<br>Cone et al., 1986 |
| *R1* | regulatory | *Ac* | Dellaporta et al., 1987 |
| *Vp1* | regulatory | *Mu* | McCarty et al., 1988 |
| *P* | regulatory | *Ac* | Lechelt et al., 1986 |
| *C2* | structural | *Spm/En1* | Wienand et al., 1986 |
| *A1* | structural | *En1/Mu* | O'Reilly et al., 1985 |
| *Bz1* | structural | *Ac* | Fedoroff et al., 1984 |
| *Bz2* | structural | *Ds2* | Theres et al., 1987 |

## 3. Molecular analysis of the structural genes *A1*,*C2*, and *Whp*

The *A1* gene has been isolated using the transposable elements *Mu1* and *En1* in the cloning procedure (O'Reilly et al., 1985). The structure of the locus was determined by sequence analysis of the wildtype gene and the corresponding cDNA (Schwarz-Sommer et al., 1987). The *A1* gene contains four exons (219, 179, 194 and 834bp) and three short introns. Studies on the function of the gene indicated that it encodes the enzyme dihydroquercitin reductase (Rohde et al., 1986; Reddy et al., 1987).

The *C2* locus, which is active in an early step of anthocyanin biosynthesis was cloned from a variegated *Spm/En1* induced mutant (Wienand et al., 1986). The wildtype gene was isolated from a colored maize line and sequenced. The comparison of this sequence with a full size *C2* specific cDNA revealed two exons 427bp and 1210bp, separated by a 1.5kb intron. The sequence of the *C2* cDNA showed, that the locus encodes a 44kD protein (Niesbach-Klösgen et al., 1987). Data obtained by Dooner (Dooner, 1983) indicated, that the *C2* locus encodes chalcone synthase. Indeed, sequence comparison of the amino acid sequence derived from the *C2* cDNA clearly showed that the *C2* gene encodes chalcone synthase and that the protein sequence is highly conserved (Niesbach-Klösgen et al., 1987).

Recently a second maize chalcone synthase gene was isolated using *C2* specific probes. This gene, which has been identified as the white pollen gene (*Whp* ; Coe et al., 1981), is very similar to the *C2* gene and plays an important role in pollen fertility. The *Whp* locus and *Whp* specific cDNA fragments have been cloned and analyzed. This showed sequence conservation between the *C2* and the *Whp* gene concerning the exon structures and the gene products. However, the introns as well as promoter and 3' flanking sequences are considerably diverged (Franken et al., manuscript in preparation).

## 4. Molecular analysis of the regulatory gene *C1*

### Cloning and sequence analysis

Since the *C1* gene has a regulatory function, the isolation of this gene was of particular interest. The gene was cloned (Paz-Ares et al., 1986) from a variegated *En1* induced *C1* mutant containing an autonomous *En1* element at the locus (Peterson, 1978). Sequence analysis of the wildtype gene as well as of *C1* specific transcripts led to a definition of the structure of the locus (Paz-Ares et al., 1987). The *C1* gene consists of three exons 150bp, 129bp and at least 720bp in length separated by two small introns (88bp and 144 bp). The sequence of the putative protein encoded by the *C1* locus revealed a protein, 273 amino acids in length with a molecular weight of 29kD. The protein shows a basic domain (120 amino acids) at the aminoterminus and an acidic one (30 amino acids) at the carboxy teminus.

The basic domain of the putative *C1* encoded protein shows a 40% homology to *myb* proto-oncogenes from animals and human (Majello et al., 1986; Katzen et al., 1985). The *myb* proto-oncogenes have been shown to be DNA binding proteins and represent transcriptional activators ( Weston and Bishop, 1989; Klempnauer et al., 1989; Nishina et al., 1989). The DNA binding domain of these proteins is the sequence that shares homology to the *C1* encoded protein. The presence of two domains, a basic (DNA binding) and an acidic one, is very similar to the organisation of such domains in transcriptional activators like GAL4 and GCN4 in yeast (Hope and Struhl, 1986; Ma and Ptashne, 1987). Since the *C1* locus affects transcription of the structural genes involved in the anthocyanin pathway it is likely that this gene encodes a transcriptional activator as well. Further evidence that this might be the case comes from the analysis of various *C1* specific mutants.

### Functional analysis of the *C1* gene

The function and regulation of the *C1* gene has been analyzed by the isolation and characterization of *C1* specific mutants. Mutants that have been investigated so far are the dominant inhibitory allele *C1-I* (Paz-Ares et al., 1990), the light inducible allele *c1-p*, the anthocyanin overproducing allele *C1-S* as well as a series of TE induced *C1* alleles.

The dominant inhibitory allele *C1-I* is characterized by the loss of the acidic domain of the protein. Due to an eight base pair insertion, the reading frame at the carboxy-terminus of the protein is changed (Paz-Ares et al., 1990). Thus, instead of 273 amino acids specific for the wildtype protein, the putative *C1-I* encoded protein is 252 amino acids in length. Since the missing of acidic domains in transcriptional activators can result in loss of transcriptional activation it is assumed, that the same is true for the inhibitory function of the *C1-I* allele. Further evidence which supports this comes from the analysis of revertants of a TE induced *C1* mutant (Franken, unpublished). The position of the TE insertion in this mutant is close to the acidic domain of the protein. The imperfect excisions of that TE in germinal revertants of this mutant leads to pale, colorless (and colored) progeny. The pale revertants, as has been shown molecularly are frame shift mutations affecting length and charge distribution of the acidic domain of the *C1* protein. This further indicated, that the acidic domain of the protein is involved in the regulation of transcription.

The light inducible *C1* allele *c1-p* differs from the wildtype *C1* gene in various aspects. The phenotype of the mature kernels of *c1-p* is colorless. Upon germination in the light the *c1-p* allele becomes activated and as a consequence the germinating kernels produce anthocyanins. Major differences between the *c1-p* allele and the wildtype allele on the molecular level are large deletions at the 3' end of the gene. Minor differences concern the amino acid composition and the promoter sequence of this allele. The differences in the amino acid composition between the wildtype and the *c1-p* gene product do not change the length or charge profile of the *c1-p* encoded protein compared to the wildtype product. This has been deduced from sequence analysis of the genomic and PCR derived cDNA fragments (Scheffler, 1989; Wienand et al., 1990).

The anthocyanin overproducing *C1-S* allele is almost identical to the wildtype gene (Cone et al., 1986) and has only minor sequence differences in the 3' nontranslated part of the gene as well as in the promoter region (Tapp, unpublished).

Concerning the transcription rate of the different *C1* alleles it has been shown that in case of *C1-I* and *C1-S* more *C1* specific mRNA is synthesized compared to the wildtype level (Cone et al., 1986; Paz-Ares et al., 1990) whereas in case of *c1-p* no *C1* mRNA is produced upon maturation and less than wildtype level upon germination in the light. We think that these differences in levels of transcription can be correlated with small sequence differences in the promoters of the various *C1* alleles.

Interaction of the *C1* protein with cis regulatory sequences of the *A1* gene

The interaction of the *C1* encoded protein with structural genes under the control of *C1* was analyzed using in vitro synthesized *C1* specific protein and different fragments of the *A1* gene, which is a target gene of *C1*. Filter binding assays were performed to determine the interaction between in vitro synthesized *C1* protein and *A1* specific sequences (Wienand et al., 1990). These experiments showed interaction of the *C1* protein with promoter and 3' located sequences of the *A1* gene. In addition gel retardation assays were carried out using nuclear extracts and fragments derived from the *A1* promoter. These experiments also showed specific interaction of nuclear proteins with distinct promoter fragments. The minimal promoter length of the *A1* gene that is necessary to function in aleurone tissue was determined using the particle gun

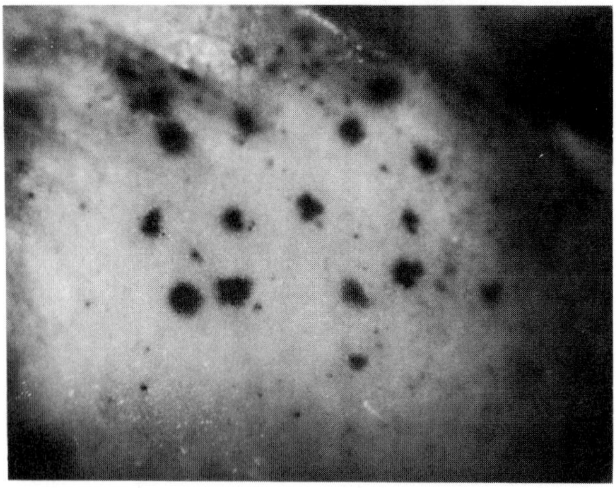

Fig. 1. Anthocyanin biosynthesis in aleurone cells after bombardment of kernels (homozygous recessive for the *A1* gene) with a genomic clone containing the *A1* gene. The expression of the gene leads to colored spots on a colorless background (shown here as dark spots on a colorless background).

approach ( Klein et al., 1989 ). In these experiments *A1* genomic clones containing promoter deletions were used. These constructs were delivered into colorless kernels not expressing the *A1* gene. Complementation of the mutation by expression of the delivered *A1* gene is visible by the synthesis of pigment in various cells (Fig.1). These experiments indicated, that the minimal promoter size which is still active in aleurone cells is approximately 300 base pairs in length (Schrell, unpublished).

Further analysis of cis regulatory sequences of the target genes under the control of the *C1* gene like *C2* should lead to a more detailed understanding of the interaction of regulatory and structural genes involved in this pathway.

REFERENCES

Barker,R.F., Thompson,D.V., Talbot,D.R., Swanson,J. and Bennetzen,J.L. (1984): Nucleotide sequence of the maize transposable element Mu1. Nucl. Acid. Res., 12, 5955-5967.

Chandler,V.L., Radicella,J.P., Robbins,T.P., Chen,J. and Turks,D. (1989): Two regulatory genes of the maize anthocyanin pathway are homologous: Isolation of B utilizing R genomic sequences. The Plant Cell, 1, 1175-1183.

Coe,E.H.,Jr., Neuffer,M.G. and Hoisington,D.A. (1988): The genetics of corn. in: Corn and Corn Improvement, eds. Sprague,G.F. and Dudley,J.W., pp.83-258.

Coe,E.H., McCormick,S.M. and Modena,S.A. (1981): White pollen in maize. Journ. of Heredity, 72, 318-320.

Cone,K.C., Burr,F.A. and Burr,B. (1986): Molecular analysis of the maize anthocyanin regulatory locus C1. Proc.Natl.Acad.Sci. USA, 83, 9631-9635.

Dellaporta,S.L., Greenblatt,I., Kermicle,J.L., Hicks,J.B., and Wessler,S.R. (1988): Molecular cloning of the maize R-nj allele by transposon tagging with Ac. in: Chromosome structure and Function, 18th Stadler Genetics Symposium. eds.:Gustafson, J.P. and Appels,R. (Plenum New York) pp.263-282.

Dooner, H.(1983): Coordinate genetic regulation of flavanoid biosynthetic enzymes in maize. Mol.Gen.Genet., 189, 136-141.

Dooner,H. (1985): Viviparous-1 mutation in maize conditions pleiotropic enzyme deficiencies in the aleurone. Plant Physiol., 77, 486-488.

Döring,H.P. (1989): Tagging genes with maize transposable elements. An overview. Maydica, 34, 73-88.

Fedoroff,N., Wessler,S., Shure,M. (1983): Isolation of the maize controlling elements Ac and Ds. Cell, 35, 235-242.

Fedoroff,,N., Furtek,D. and Nelson,O.E. (1984): Cloning of the bronze locus in maize by a simple and generalizable procedure using the transposable controlling element Ac. Proc.Natl.Acad.Sci. USA, 81, 3825-3829.

Hope,I.A. and Struhl,K. (1986): Functional dissection of a eucaryotic transcriptional activator protein, GCN4 of yeast. Cell, 46, 885-894.

Katzen,A.L., Kornberg,T.B. and Bishop,J.M., (1985): Isolation of the proto-oncogene c-myb from D. melanogaster. Cell, 41, 449-456.

Klein,T.M., Roth,B.A. and Fromm,M.E. (1989): Regulation of anthocyanin biosynthetic genes introduced into intact maize tissues by microprojectiles. Proc.Natl.Acad.Sci. USA, 86, 6681-6685.

Klempnauer,K.H., Arnold,H. and Biedenkapp,H. (1989): Activation of transcription by v-myb: evidence for two different mechanisms. in: Genes and Development, 3, Cold Spring Harbour Laboratory Press, pp. 1582-1589.

Lechelt,C., Laird,A. and Starlinger,P. (1986): Cloning of DNA from the P locus. Maize Gen.Coop.News.Lett., 60, 40.

Ludwig,S.R., Habera,L.F., Dellaporta,S.L. and Wessler,S.R. (1989): Lc, a member of the maize R gene family responsible for tissue-specific anthocyanin production, encodes a protein similar to transcriptional activators and contains the myc-homology region. Proc.Natl.Acad.Sci. USA, 86, 7092-7096.

Ma,J. and Ptashne,M. (1987): Deletion analysis of GAL4 defines two transcriptional activating segments. Cell, 48: 847-853.

Majello,B., Kenyon,L.C. and Dalla-Favera,R.D., (1986): Human c-myb protooncogene: nucleotide sequence of cDNA and organization of the genomic locus. Proc.Natl.Acad.Sci. USA, 83, 9636-9640.

McCarty,D.R., Carson,C.B., Stinard,P.S. and Robertson,D.S. (1989): Molecular analysis of viviparous-1:An abscisic acid-insensitive mutant of maize. The Plant Cell, 1, 523-532.

Niesbach-Klösgen,U., Barzen,E., Bernhardt,J., Rohde,W., Schwarz-Sommer,Zs., Reif,J., Wienand,U. and Saedler,H. (1987): Chalcone synthase genes in plants: a tool to study evolutionary relationships, J Mol. Evol., 26, 213-225.

Nishina,Y., Nagagoshi,H., Imamoto,F., Gonda,T.J. and Ishii,S.(1989): Trans-activation by the c-myb proto-oncogene. Nucl.Acid Res. 17, 107-117.
O'Reilly,C., Shepherd,N., Pereira,A., Schwarz-Sommer,Zs., Bertram,I., Robertson,D.S., Peterson,P.A. and Saedler,H. (1985): Molecular cloning of the a1 locus of Zea mays using the transposable elements En and Mu1. EMBO J., 4, 877-882.
Paz-Ares,J., Wienand,U., Peterson,P.A. and Saedler,H. (1986): Molecular cloning of the c locus of Zea mays: a locus regulating the anthocyanin pathway. EMBO J.,5, 829-833.
Paz-Ares,J., Ghosal,D., Wienand,U., Peterson, P.A. and Saedler,H. (1987): The regulatory c1 locus of Zea mays encodes a protein with homology to myb proto-oncogene products and with structural similarities to transcriptional activators. EMBO J.,6, 3553-3558.
Paz-Ares,J., Ghosal,D. and Saedler,H. (1990): Molecular analysis of the C1-I allele from Zea mays: a dominant mutant of the regulatory C1 locus. EMBO J., 9, 315-321.
Pereira,A., Schwarz-Sommer,Zs., Gierl.A., Bertram,I., Peterson,P.A. and Saedler,H. (1985): Genetic and molecular analysis of the Enhancer (En) transposable element system in Zea mays. EMBO J., 4, 17-23.
Peterson,P.A. (1978): Controlling elements: The induction of mutability at the A2 and C loci in maize. in: Maize Breedings and Genetics (Walden),pp.601-631.
Reddy.A.R. and Peterson,P.A. (1978): The action of the Intensifier (In) gene in flavonoid production in aleurone tissue of maize. Can.J.Genet.Cytol. 20, 337-347.
Reddy,A.R., Britsch,L., Salamini,F., Saedler,H. and Rohde,W. (1987): The A1 (anthocyanin-1) locus in Zea mays encodes Dihydroquercitin reductase. Plant Science, 52, 7-13.
Rhode,W., Barzen,E., Marocco,A., Schwarz-Sommer,Zs., Saedler,H. and Salamini,F. (1986): Isolation of genes that could serve as traps for transposable elements in Hordeum vulgare. Barley Genetics,V, 533.
Scheffler,B. (1989): The molecular analysis of c1-p and c1-m1, two recessive alleles of C1: A regulatory locus of the anthocyanin pathway in the aleurone tissue of Zea mays L., thesis, Iowa State University, Ames, Iowa
Schwarz-Sommer,Zs., Shepherd,N., Tacke,E., Gierl,A., Rohde,W., Leclerq,L., Mattes,M., Berndtgen,R., Peterson,P.A. and Saedler,H. (1987): Influence of transposable elements on the structure and function of the A1 gene of Zea mays. EMBO J., 6, 287-294.
Styles,D.E. and Ceska,O., (1977): The genetic control of flavonoid synthesis in maize. Can.J.Genet.Cytol., 19, 289-302.
Styles,E.D. and Coe,E.H.Jr.,(1986): Unstable expression of an R allele with a3 in maize. Journ. of Heredity, 77: 389-393.
Theres,N., Scheele,T. and Starlinger,P. (1987): Cloning of the bz2 locus of Zea mays using the transposable element Ds as a gene tag. Mol.Gen.,Genet., 209, 193-197.
Weston,K. and Bishop,J.M. (1989): Transcriptional activation by the v-myb oncogene and its cellular progenitor, c-myb. Cell, 58, 85-93.

Wienand,U., Weydemann,U., Niesbach-Klösgen,U., Peterson,P.A. and Saedler,H. (1986): Molecular cloning of the c2 locus of Zea mays, the gene coding for chalcone synthase. Mol. Gen.Genet., 203, 202-207.
Wienand,U. and Saedler,H. (1987): Plant transposable elements: unique structures for gene tagging and gene cloning. in: Plant DNA infectious agents; Plant Gene Research, eds.: Hohn,T. and Schell,J.(Springer Verlag Wien New York) pp.205-227.
Wienand,U., Paz-Ares,J. Scheffler,B. and Saedler,H. (1990): Molecular analysis of gene regulation in the anthocyanin pathway of Zea mays. in: Plant Gene Transfer, UCLA Symposia on Molecular and Cellular Biology, 129, eds.: Lamb,C.L. and Beachy,R.N.(Wiley-Liss),pp.111-124.

# AUTHOR INDEX

SUBJECT INDEX
(The first page of the respective article is listed)